STATISTICAL MECHANICS
OF CHAIN MOLECULES

STATISTICAL MECHANICS OF CHAIN MOLECULES

PAUL J. FLORY

J. G. Jackson–C. J. Wood Professor of Chemistry
Stanford University,
Stanford, California

INTERSCIENCE PUBLISHERS

a division of John Wiley & Sons, New York · London · Sydney · Toronto

LIBRARY OF CONGRESS CATALOG CARD NUMBER: 68-21490

SBN 470 264950

PRINTED IN THE UNITED STATES OF AMERICA

To Emily, my wife

If my hypothesis is not the truth, it is at least as naked;
for I have not with some of our learned moderns disguis'd my nonsense
in Greek, cloth'd it in algebra or adorned it with fluxions.

Benjamin Franklin, 1753

STATISTICAL MECHANICS OF CHAIN MOLECULES — Paul J. Flory

ERRATA

p. 5, line 2: Replace "l" by "$\{ l \}$"

p. 46: Remove "R. Chiang, J. Phys. Chem., **70**, 2348 (1966)" from footnote a, and append to footnote b.

p. 85, line 10: Replace "chain" with "chains"

p. 122: The first two lines should read:
"illustration in its application in Chapter VI to stereoirregular vinyl polymers, which may be treated as copolymers of units that are mirror images"

p. 157, line 3: Replace "kcal $mole^{-1}$" by "cal $mole^{-1}$"

p. 162: The first two lines should read:
"represent them. For the bond pair i, i + i centered about O in Fig. 15, and for bond pair i - 1, i centered about CH_2, we have respectively"

line 4: Replace "former" with "latter"

line 7: Replace "ω_a" with "ω_b"

line 12: Replace "ω_b" with "ω_a"

p. 193 third line following Eq. (34.1) replace "group 1" with "group 0"

p. 255, line 7: Replace "(Chap. I, p. 3)" with "(Chap. I, p. 13)"

p. 321 last line of Eq. (70) should read:

$$+(33/175)(r/nl)^4 + \ldots]\} \qquad (70)$$

p. 381: Replace "$\mathbf{E}_3$" with "$\mathbf{E}_9$" at each of the two places where $\mathbf{E}_3$ occurs in Eq. (128)

Preface

Comprehension of the configurational statistics of chain molecules is indispensible for a rational interpretation and understanding of their properties. The spatial configurations of linear macromolecules, natural or synthetic, are reflected in their average dimensions, either in a dilute solution or in the amorphous state in absence of a diluent. The facility with which cyclic structures are formed, by chemical reaction, from acyclic chains obviously is related to the statistical distribution of the two ends of the chain relative to one another, and hence depends on their configurational characteristics. Constitutive properties of a chain molecule which are dependent upon its configuration include the mean-square dipole moment, the optical anisotropy, and the spectral dichroism of the molecule. Mechanical properties of polymers, or macromolecules, also are intimately related to the configurational characteristics of their molecular chains, although the connections may not lend themselves so readily to explicit mathematical expression. Clearly, a firm grasp of the interrelationships between configuration and chemical structure is essential for the rational interpretation of the properties of chain molecules, including the special properties peculiar to macromolecules.

The spatial configurations of polymeric chains have long been treated in terms of hypothetical models, a practice which continues unabated at present. Foremost amongst models in this category is the freely jointed, or random flight, chain consisting of bonds of fixed length and uncorrelated directions. The late Werner Kuhn directed attention as early as 1934 to the correspondence between a real polymer chain of sufficient length and this mathematically tractable analog with origins extending back to the pioneering studies on radiation of random phase initiated by Lord Rayleigh a half century earlier. Formulation of theories of polymers in terms of this hypothetical model chain has been widespread throughout ensuing years, and this fact pays tribute to Kuhn's insight at a time when the prevalence of linear chain structures amongst polymeric materials, natural and synthetic, was only beginning to be perceived.

The correspondence between real chain molecules and this artificial model suffers definite limitations, especially for chains of comparatively short length (i.e., not exceeding several hundred bonds), or for "stiff" chains of low tortuosity. Other hypothetical models have appeared, notably the Porod-Kratky chain characterized by a curvature of fixed magnitude but of random direction. Irrespective of the particular model chosen, forfeiture of a connection with the actual chemical structure is the inevitable price of adoption of a hypothetical model for interpretation of the properties of the real chain. Thus, if the freely jointed chain is adopted, one is obliged to attribute properties of length, volume, electric moment, optical anisotropy, etc., to the bond, or segment, of this hypothetical model. The segment is a figment of the model. Properties ascribed to it in order to achieve agreement between experimental observations and relationships deduced for the model are, like the model itself, in the realm of artificiality, and they cannot be related explicitly to those of the constituent bonds and groups of the real chain. Thus, no matter how faithfully such a model may represent experimental observations, interpretations carried out in its terms are cast in a framework of unreality. The properties deduced for the hypothetical segment defy transcription to the real chain. Conversely, the inherent characteristics of the real chain, *viz.*, its structural geometry, rotational hindrances, group dipole moments, group polarizability tensors, and so forth, cannot be adequately represented within the limitations of the hypothetical segment. An alternative hypothetical model, more versatile than the freely jointed chain in its capacity to reproduce empirically the configuration-dependent features of a wider variety of real chains, would not necessarily fulfill to a greater degree the objective of rationally relating properties of chain molecules to their structures. A model meeting this need must conform closely, and in an identifiable manner, with the actual characteristics of the real chain, i.e., with its structural geometry and the potentials impeding rotations about its bonds

According to the precepts of statistical mechanics, properties of chain molecules should be evaluated by averaging over all configurations generated by varying all angles of rotation about the bonds of the structure in a continuous manner, and also by varying bond angles and bond lengths to the extents permitted by the structure. For a long chain the task would be formidable indeed. It can be greatly abbreviated, however, through adoption of the *rotational isomeric state scheme*, whereby only discrete values of each rotation angle are considered. The rotational states suitable for a given bond are often three in number. Integration over configuration variables, as required for evaluation of the configuration integral of statistical mechanics, for example, may then be replaced by summation over a discrete set of configurations of the chain molecule as a whole. This model, or scheme, has abundant precedent for the interpretation of conformations of small molecules, and of their conformation-dependent properties. Its use for this purpose by spectroscopists, organic chemists, and others concerned with small molecules subject to rotational isomerism is universal.

The basis and validity of the rotational isomeric state model were critically examined by Volkenstein in his monograph published a decade ago.* He and his collaborators pioneered in the adaptation of this model to polymeric chains. While the replacement of continuous ranges of rotation angle accessible to the real chain by discrete states constitutes an approximation, the error thus introduced unquestionably is small. In other respects the model admits of retention of the characteristics of the real chain, including in particular the geometrical parameters defining bond lengths and bond angles. Even the effects of the rotational potentials may be introduced in a realistic manner by assigning appropriate statistical weights to the discrete states.

Adaptation of the rotational isomeric state model to a polymeric chain is straightforward in concept and in principle. Its application is complicated, however, by the interdependence of rotations about neighboring bonds of the chain. The mathematical treatment of the Ising representation of a linear array of neighbor-dependent magnetic dipoles afforded a precedent for this situation. It was adapted independently in 1959 by Yu. Ya. Gotlib, T. M. Birshtein, O. B. Ptitsyn, and M. V. Volkenstein in the Soviet Union, by S. Lifson in Israel, and by K. Nagai in Japan to the considerably more complicated treatment of a sccond moment of the configuration of a simple polymer chain, e.g., to the treatment of the average square of its end-to-end length. The equivalent formulations derived by these authors are mathematically rigorous, and asymptotically exact for chains of sufficient length, subject only to whatever physical limitations may be imposed by the rotational isomeric state model. Newer developments during the past five years provide methods applicable to real chain molecules of any length, comprising repeating structural units of virtually any description; they can be applied also to copolymeric chains consisting of a variety of structural units occurring in irregular sequence. Other quantities that can be calculated, likewise without mathematical approximation, include the radius of gyration, the fourth moment of the vector connecting any pair of units, and the invariants of the optical polarizability tensor which are relevant to experimental quantities determined from optical measurements.

Thus, the seemingly insuperable complexities posed by the immense number of configurations admissible to a long polymer chain are solved, without recourse to limiting approximations, by the newer methods cited above. Since chains of any length are comprehended within the scope of

* M. V. Volkenstein, *Konfiguratsionnaya Statistika Polimernyikh Tsepei*, Izdatel'stvo Akad. Nauk SSR, Moscow, 1959 (published in translation as *Configurational Statistics of Polymeric Chains*, Interscience, New York, 1963).

the same model and mathematical formulation, the entire range of molecular species from small molecules to high polymeric homologs may be dealt with on a uniform basis. Equilibrium properties of long-chain molecules may thus be related to their structure and constitution without approximations beyond those inherent in the rotational isomeric state model, and hence with no greater compromise of reality than is incident upon the interpretation of corresponding properties of analogous low molecular species. Knowledge gained from study of long chain polymers can be directly applied to low molecular homologs, and *vice versa.*

The methods referred to are new and their applications have thus far been sparse in relation to the potentialities they offer. The present volume has been written with the primary object of making available a logical and coherent presentation of these methods for evaluating statistical mechanical averages of the moments of the chain vector distribution and of related invariants relevant to various physical measurements of equilibrium properties of chain molecules and macromolecules. A secondary objective, brought within reach of attainment by the availability of the moments of the distribution, is the critical analysis of characteristics of the distribution itself and of prevailing ideas pertaining thereto. It is hoped that the material herein presented will facilitate and encourage investigations affording deeper understanding and appreciation of the relationship between the properties of chain molecules and their chemical structures.

The more familiar model chains, of limited applicability to real chains, are treated in Chapter I, which also includes primary definitions, concepts, and some of the mathematical apparatus for later use. Principal experimental methods yielding average dimensions of polymeric chains and their temperature coefficients are summarized in Chapter II; tabulations of representative results are included. The main theory and methodology is set forth in Chapters III and IV. Applications to symmetric chains, with inclusion of examples illustrative of the principal kinds of chain molecules falling in this category, are presented in Chapter V. Vinyl chains having asymmetric centers at alternate atoms of the chain skeleton, and for which stereoregularity is a prominent feature, are treated in Chapter VI. Polypeptides and their analogs, including proteins, are the subjects of Chapter VII. The even moments of the end-to-end vector for various chain molecules, treated in Chapters IV–VII, provide information bearing on the statistical distribution of these chain vectors. Functions offered to represent this distribution are examined in Chapter VIII. The final chapter (IX) is concerned with optical properties: Rayleigh scattering, depolarization of scattered light, strain birefringence (and strain dichroism), and the Kerr effect. These topics serve to illustrate the versatility of the methods developed in earlier chapters. Supplementary material is presented in eight Appendices. These are followed by a glossary of mathematical conventions and of the more important symbols used in the text.

Principal references are given at the end of each chapter and appendix. The lists are not comprehensive bibliographies, however. Many significant contributions have been omitted, in keeping with the nature of the present work, which is not intended to be a treatise on the "state of the art" of each topic touched upon.

Every effort has been made to avoid confrontation of the reader with the familiar, frustrating phrase "It can be shown that ...." Derivations are given in full with few exceptions; those which are ancillary to the main discourse are located in the appendices. The utmost mathematical simplicity consistent with generality, rigor, and exactness has been attempted. If elegance and sophistication have suffered as a result of this predilection, our apologies can only be offered without regrets.

Matrix algebra is used extensively and some knowledge of matrix manipulations is assumed. The availability of excellent elementary texts on this subject seemed to obviate the need for including a resumé of matrix algebraic methods. The diagonalization of a matrix is illustrated in Appendix B, however. Direct matrix multiplication, a topic often omitted in elementary texts, is defined on p. 98, and rules relating thereto are explained adequately for the purposes of succeeding developments. Fourier transformations are used in Chapters VIII and IX. Additional higher mathematics is not necessary.

The title of the book notwithstanding, little knowledge of statistical mechanics is required of the reader beyond familiarity with the concept of a partition function, the configurational integral of classical statistical mechanics, and the elementary methods for deriving averages over an ensemble of identical systems. Abstract formalisms and niceties of the subject have been avoided without difficulty or sacrifice of substance. The chain molecule assumes the role of the system in the terminology of statistical mechanics. It is a small system in the sense of T. L. Hill.* The relevant canonical ensemble is the collection of an indefinitely large number of these systems, isolated from one another but subject to identical external conditions of temperature, force, electric field, etc.

Long chain molecules invariably occur in environments where they interact with other molecules, either of like kind (as in the undiluted amorphous polymer) or with a solvent (e.g., in a dilute solution). We shall not, however, be concerned with their mixtures with other molecules. Thus, macroscopic systems comprising mixtures of many chain molecules with molecules of a solvent (e.g., polymer solutions) are outside the purview of

* T. L. Hill, *Thermodynamics of Small Systems*, W. A. Benjamin, New York, 1964.

the present study. In focusing on the chain molecule as the statistical mechanical system, we regard the molecules making up the environment as a continuum whose interactions with the molecule designated as the system do not require examination in detail at the molecular level. This point of view is well supported, especially for the polymer chain molecule dispersed in a solvent providing an "ideal" environment, i.e., at the Θ point (see Chap. II). Even in the undiluted amorphous state, there is growing evidence that the average configurations of individual chain molecules are not much perturbed by the copious interactions with their neighbors. Hence, treatment of the chain molecule as an isolated entity is warranted, its association with other molecules in all physical environments of its occurrence notwithstanding.

The (small) systems here considered, being linear arrays of bonds or units, admit of exact treatment. They may readily serve to illustrate many of the features of three-dimensional statistical mechanical systems, which can be treated only in approximation. Because of the simplicity of the one-dimensional, or linear, systems which are the subjects of this volume, they are especially well suited as illustrative material for courses of study of statistical mechanics in its larger context. Some of the material in this volume should prove useful for that purpose.

The author is much in debt to his collaborators of recent years. They have contributed extensively to this work, as the lists of references and citations in the text testify. Especially to be mentioned are Drs. A. Abe, D. A. Brant, C. A. J. Hoeve, R. L. Jernigan, J. E. Mark, W. G. Miller, P. R. Schimmel, J. A. Semlyen, A. E. Tonelli, and A. D. Williams. The content of a field of science at a given stage in its development almost invariably rests on the contributions of many investigators, both predecessors and contemporaries. This field is no exception. The earlier work of Volkenstein, previously cited in this Preface, and the more recent book by Birshtein and Ptitsyn* have provided much useful material. Amongst recent research contributions, those of K. Nagai of the Government Industrial Research Institute, Osaka, Japan have been indispensible, and accordingly deserve special mention. Nagai has provided the basis for primary steps in the evaluation of higher moments of the chain vector. He introduced the Fourier transformation method for developing the expansion of the chain vector distribution in its even moments, and applied the same procedure to the treatment of the optical anisotropy for a system subjected to mechanical or electrical stress. The importance of Nagai's

* T. M. Birshtein and O. B. Ptitsyn, *Conformations of Macromolecules*, Translated from the 1964 Russian edition, by S. N. Timasheff and M. J. Timasheff, Interscience, New York, 1966.

contributions will be apparent from the frequent reference to his work in the text.

Professor J. E. Mark and Dr. R. L. Jernigan, who kindly agreed to read the manuscript critically, offered many helpful suggestions which were incorporated in the final version. The author is indebted to them for their generous assistance. Finally, it is a pleasure to acknowledge with deep gratitude the skillful assistance of Mrs. Heather Perry who typed initial drafts of much of the manuscript, and of Miss Ann M. Dudas who prepared the final copy. Daunted neither by mathematics nor by jargon, Miss Dudas willingly read proof with scrupulous attention to every detail. Errors may have escaped, but they are far fewer for her painstaking efforts.

PAUL J. FLORY

July, 1968

Contents

CHAPTER I

Analysis of the Spatial Configurations of Chain Molecules and Treatment of Simplified Model Chains

In a most primitive sense, a molecule may be regarded as a collection of certain atoms. The spatial arrangement of these atoms, designated by the term *configuration*, should accordingly be specified in terms of the coordinates of each. In a molecule comprising N atoms, this will in general require stipulation of $3N - 6$ internal coordinates; the absolute location of the molecule and its orientation in space determined by the remaining six coordinates may be ignored.

A molecule is more than a mere collection of atoms. A definite pattern of connections is implicit in the molecular formula and in the concepts it conveys. The existence of these connections, or chemical bonds, greatly restricts the configurations which the atoms of the molecule may assume. The coordinates of the various atoms cannot be assigned in a completely arbitrary manner; geometric requirements of the array of chemical bonds which provide the framework of the molecule must be observed. Specifically, lengths of the bonds are restricted within narrow limits, and likewise the angles between bonds meeting at a given atom, i.e., the bond angles, are effectively confined to small ranges. Atomic configurations in comparatively simple molecules may often be fully determined, within these limits, by their chemical structures. For molecules of greater complexity, and in particular for those having certain of their atoms bonded in concatenated sequence, angles of rotation about bonds of the molecular framework permit a diversity of molecular configurations, or *conformations*; the latter term has a more circumscribed connotation with specific reference to bond rotations (*cf. seq.*). The potentials affecting rotations about chemical bonds (specifically, single bonds) are much less restrictive than those operative with respect to bond lengths and bond angles. The nature of bond rotation potentials will be discussed in Chapter III. Here it suffices to point out that the bond rotation angles are the variables which assume the dominant role in the analysis of the spatial configurations of various chain molecules.

1. SPECIFICATION OF THE CONFIGURATION OF A CHAIN MOLECULE

For most chain molecules of interest, designation of the locations of the skeletal atoms which occur in the backbone of the molecule will adequately define a configuration. Consider, for example, the comparatively simple homologous series of linear chain molecules described by the chemical formula $CH_3—(CH_2)_{n-1}—CH_3$. If the number n of carbon-to-carbon bonds is less than about 100, the species may be referred to as a *normal alkane* (n-alkane), or as a straight-chain paraffin hydrocarbon. If n is of the order of thousands or greater and if the material comprises a distribution of species extending over a range of n, probably it will be designated as a *polymethylene* or a linear *polyethylene*, the latter term specifying the monomer, ethylene, from which the polymer may have been synthesized. Its structure is portrayed in condensed graphical form by

etc. $\diagdown$ CH_2 $\diagup$ CH_2 $\diagdown$ CH_2 $\diagup$ CH_2 $\diagdown$ CH_2 $\diagup$ CH_2 $\diagdown$ etc.

Designation of the relative positions of the carbon atoms will invariably suffice to specify a configuration of a molecule of this series of homologs. The locations of the pendant hydrogen atoms are fixed within narrow limits by the length of the C—H bond and by the angles $\angle$CCH and $\angle$HCH. Hence, specification of coordinates of atoms other than those of the chain skeleton is superfluous.

Similarly, the configuration of a polydimethylsiloxane chain

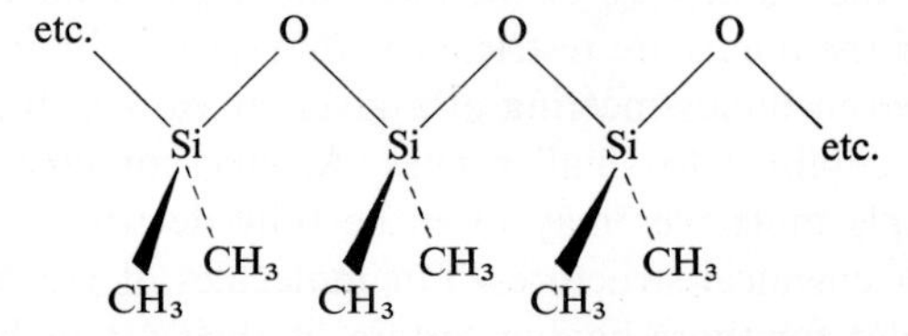

will be adequately defined for most purposes by giving the relative positions of its Si and O atoms; the angles of rotation of pendant methyl groups about the $Si—CH_3$ bonds usually may be ignored.*

The configuration of the skeletal atoms of a hypothetical chain consisting of n bonds is shown schematically in Fig. 1. Atoms A are numerated serially from zero to n, as denoted by subscripts. Bonds are represented by bond

* The molecular formulas shown above are not to be construed as literal representations of either structure or configuration. The bond angles at O and Si in the siloxane chain, for example, differ considerably, as will be pointed out in Chapter V. Stipulations regarding bond rotations likewise are not intended.

vectors $\mathbf{l}_i$ numbered from 1 to n. Thus, bond i leads from atom $i - 1$ to atom i, the chain being "read" from left to right.* The $n + 1$ atoms connected by n bonds may be identical or they may be of two or more kinds. In any event, the sequence of atoms of the several kinds is presumed to be uniquely prescribed by the molecular formula. The serial indexes on the A_i then serve to denote both the kind of atom (or group) and its location in the chain sequence.

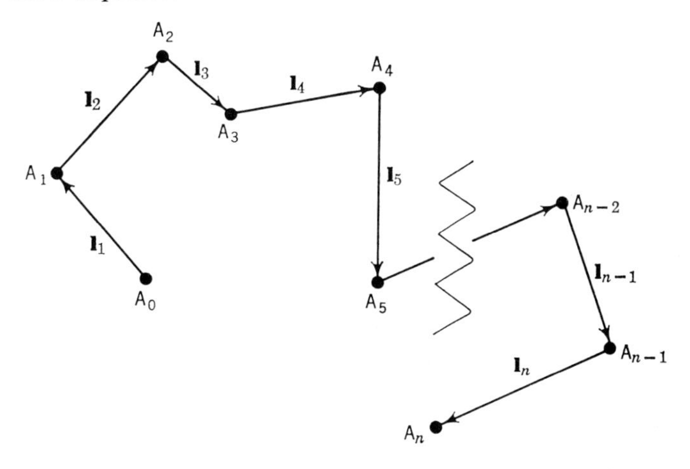

Fig. 1. Schematic representation of skeletal atoms of a chain.

In the interests of achieving the utmost generality, we shall for the present ignore restrictions on bond lengths and bond angles. Then, a sufficient specification of the configuration of the *chain skeleton* will consist in the designation of each skeletal bond vector $\mathbf{l}_i$. Let this set of vectors be represented symbolically by $\{\mathbf{l}\}$. The relative position of each of the skeletal atoms is defined by this set. The full set of n vectors $\mathbf{l}_i$ exceeds requirements inasmuch as the orientation of the configuration in space is immaterial. Thus, the directions of two of the vectors, $\mathbf{l}_1$ and $\mathbf{l}_2$, for example, may be omitted from the specification beyond stipulation of the angle between them; their magnitudes and the intervening angle will suffice, all other bond vectors being given in full.

The quantity of information included in $\{\mathbf{l}\}$ is excessive. Hence, the need

* This numeration of skeletal atoms unfortunately conflicts with the established chemical conventions for numbering sites of substitution in chain molecules, according to which the first atom in a chain is assigned the numeral 1. Adoption of the latter convention for our purposes would lead to intolerable complications from incongruities elsewhere in the scheme. For example, the angle at the atom (i) renumbered atom $i+1$ in Figs. 2 and 3 (*cf. seq.*) should retain the designation θ_i in order to comply with its usage in conjunction with bond rotation angle ϕ_i.

for a more concise measure of configuration becomes apparent. One that serves this purpose in some degree is the vector **r** which connects the ends of the chain. It is obvious that

$$\mathbf{r} = \sum_{i=1}^{n} \mathbf{l}_i \tag{1}$$

Ordinarily, only the scalar magnitude r of **r** will be of interest. Its square is given by the scalar product of **r** with itself. Thus

$$r^2 = \mathbf{r} \cdot \mathbf{r} = \sum_{i,j} \mathbf{l}_i \cdot \mathbf{l}_j \tag{2}$$

where the sums over i and j are from 1 to n. This expression may be written alternatively as follows

$$r^2 = \sum_i l_i^2 + 2 \sum_{0<i<j\leq n} \mathbf{l}_i \cdot \mathbf{l}_j \tag{3}$$

where the diagonal terms are relegated to the first summation; the remaining terms are summed over each pair ij, the redundancy arising from the equivalence of $\mathbf{l}_i \cdot \mathbf{l}_j$ and $\mathbf{l}_j \cdot \mathbf{l}_i$ having been removed by the stipulation $i < j$.

The scalar length r_{ij} of the vector connecting atoms i and j may be defined similarly. Thus

$$r_{ij}^2 = \sum_{i'=i+1}^{j} l_{i'}^2 + 2 \sum_{i<i'<j'\leq j} \mathbf{l}_{i'} \cdot \mathbf{l}_{j'} \tag{4}$$

It will be apparent that $r \equiv r_{0n}$.

A further quantity relevant to the characterization of a configuration is its radius of gyration s. It is defined as the root-mean-square distance of the collection of atoms, or groups, from their common center of gravity. Confining attention to the skeletal atoms (or groups), we have by this definition

$$s^2 = (n+1)^{-1} \sum_{0}^{n} s_i^2 \tag{5}$$

where s_i is the distance of atom i from the center of gravity of the chain in a specified configuration. The radius of gyration may be defined alternatively in terms of arbitrary chain segments or repeating units (e.g., —CH_2— or —$OSi(CH_3)_2$— in the examples above). On this basis

$$s^2 = x^{-1} \sum_{1}^{x} s_k^2 \tag{5'}$$

where x may be, for example, the number of repeating units in the chain.

Both r and s are defined for a given configuration $\{\mathbf{l}\}$ of the chain skeleton. They vary, in general, from one configuration to another. All of the distances r_{ij} are defined also by $\{\mathbf{l}\}$. A suitably chosen independent set of the r_{ij} would serve to define $\{\mathbf{l}\}$ apart from the orientation of the chain in space, and hence would define the internal configuration of the chain.

According to a theorem due to Lagrange (see Appendix A),

$$s^2 = (n+1)^{-2} \sum_{0 \le i < j \le n} r_{ij}^2 \tag{6}$$

That is, for a constellation of $n + 1$ particles the mean-square distance from their center of gravity is directly related to the sum of the squares of the distances between every pair of particles. A connection is thus established between the interatomic distances r_{ij} and the radius of gyration, which is a measure of the size of the spatial domain pervaded by the configuration.

2. SPATIAL DISTRIBUTIONS

The number of configurations which a long chain molecule may assume is very large. Consideration of each of them individually would be a task of scope beyond all possibility of realization. This brings us at once to the necessity of adopting a statistical point of view. Therefore, the deduction of appropriate averages over the total population of configurations for an ensemble of chain molecules will be our goal, rather than description in the detail implied by some of the formulation above.

Distribution functions are a concomitant of statistics and statistical mechanics. The expectation that a certain segment is situated at a specified location relative to another segment may be expressed by such a function of the vector reaching from the latter segment to the specified location. Functions of this kind assume special prominence in the elaboration of the subject at hand. For definiteness, consider the distribution $W_{ij}(\mathbf{r}_{ij})$ of the vector distance $\mathbf{r}_{ij}$ between skeletal atoms (or other chain elements, or units) i and j. Thus, $W_{ij}(\mathbf{r}_{ij})\, d\mathbf{r}_{ij}$ denotes the probability that the atom j is situated within an element of volume $d\mathbf{r}_{ij}$ located at $\mathbf{r}_{ij}$ relative to atom i.* This probability can be defined either as the time average incidence of $\mathbf{r}_{ij}$ within the specified range for a given molecule or as the average incidence for an ensemble of many identical molecules subject to identical conditions. If the functional form of W_{ij} is found for a given pair ij, presumably it can be determined for all pairs ij. It will suffice therefore to focus attention on the distribution for the end-to-end vector $\mathbf{r} \equiv \mathbf{r}_{0n}$ reaching from the zeroth to the nth atom of the chain.

For a very long chain unperturbed by self-interactions of long range

* If x, y, and z are the Cartesian components of $\mathbf{r}$, then $d\mathbf{r}$ stands for the volume element $dx\, dy\, dz$. This well-established notation for the volume element at a specified location will be used throughout.

(see Chap. II) and by external constraints, the end-to-end distribution, which we denote by $W(\mathbf{r})$, may be shown to be Gaussian, that is

$$W(\mathbf{r}) = (3/2\pi\langle r^2\rangle)^{3/2} \exp\left[-(3/2\langle r^2\rangle)r^2\right] \quad (7)$$

where $W(\mathbf{r})$ is expressed as the probability per unit volume as above, and $\langle r^2\rangle$ is the square of the magnitude of the end-to-end vector $\mathbf{r}$ averaged over all configurations. Here and elsewhere throughout following chapters *angle brackets are used to denote the statistical mechanical average* of the quantity therein, this average being taken over all configurations of the chain. The constant preceding the exponential in Eq. 7 normalizes the distribution function so that

$$\int_0^\infty W(\mathbf{r})4\pi r^2\, dr = 1$$

and of course

$$\int_0^\infty W(\mathbf{r})4\pi r^4\, dr = \langle r^2\rangle$$

Various arguments can be advanced in support of the foregoing assertion that the asymptotic form of the distribution $W(\mathbf{r})$ must be Gaussian for a sufficiently long chain molecule. A rigorous proof is given in Chapter VIII. Here we observe that the displacement $\mathbf{r}$ of one end of the chain from the other is governed by fluctuations. Intuitively, therefore, the distribution of $\mathbf{r}$ may be expected to conform to the mathematical representation of random errors. Further analogy is found in the velocity distribution among the molecules of a gas. In fact, Eq. 7 for $W(\mathbf{r})$ may be obtained by Maxwell's[1,2] original derivation of the velocity distribution which bears his name, with scarcely so much as an alteration of his notation.

Maxwell considered the components x, y, and z of the velocity $\mathbf{r}$ of a gas molecule. If $w(x)$ represents the distribution function for the x component, then isotropy of space dictates that the distributions for y and z must be represented by functions of the same form, i.e., by $w(y)$ and $w(z)$. The critical step in this derivation, a revelation of Maxwell's insight, is the assertion that the probability of an arbitrary value of velocity component x is independent of the value assigned to y or to z. The plausibility of this assertion rests on the magnitude of the very large number of molecules in a system of the kind considered. On this basis Maxwell concluded that the probability of velocity components x, y, z is the product of the independent probabilities, i.e., that it is given by

$$w(x)w(y)w(z)\, dx\, dy\, dz$$

But this probability can depend only on the magnitude of the velocity (or end-to-end distance) and not on its direction in space. Hence, Maxwell deduced that

$$w(x)w(y)w(z) = \phi(r^2)$$

where ϕ is a function of $r^2 = x^2 + y^2 + z^2$. The only functions $w(x)$ meeting this condition are of the form

$$w(x) = c \exp(ax^2)$$

where c and a are constants. It follows that

$$\phi(r^2) = c^3 \exp(ar^2)$$

which corresponds to our $W(\mathbf{r})$.

In a later paper[3] Maxwell conceded that the assumption of independence of the component distributions $w(x)$, etc. "... may appear precarious ...," and set about to produce another, more rigorous proof which appealed to the principle of energy conservation and other considerations having no obvious counterpart in our problem. Having realized the advantage of his insight to the extent that analogy permits, we abandon further pursuit of his classic investigations.

The independence of $w(x)$, $w(y)$, and $w(z)$ which underlies the essential step in Maxwell's derivation of the velocity distribution in a gas is plausible enough for a system of 10^{20} or more molecules. However, the range permitted for a component x of the vector $\mathbf{r}$ for a chain consisting of only some 10^4 bonds is much less extensive than the momentum accessible to a molecule. Hence, the assertion of independence of the distributions $w(x)$, $w(y)$, and $w(z)$ of the components of r for the long chain is more "precarious" than for the velocity of a gas molecule. If the chain is very long and the displacements x, y, z considered are much smaller than its contour length (r_{max}), then indeed the assertion is credible, as may be shown. Moreover, the intersegmental distribution function $W_{ij}(\mathbf{r}_{ij})$ for any pair of atoms ij far removed from one another in sequence must likewise be Gaussian.

In chains of typical tortuosity, departures from the Gaussian formula become appreciable for $|j - i|$ less than 20–50 bonds (see Chap. VIII). Distribution functions W_{ij} for short chains are not easily expressed in concise mathematical form. Ignoring this difficulty, we note that the set of intersegmental distribution functions W_{ij} for an arbitrary i and including all values of j from 0 to n must together determine the *spatial distribution of all segments* of the molecule with respect to one another. The latter distribution is conveniently pictured as a dispersion of skeletal atoms, or of chain units, about their common center of gravity. The density at a given point in the space occupied by the molecule comprises a contribution from each segment of the chain, each such contribution representing the frequency of occurrence of the segment in question within a volume element at the given point.

The spatial distribution of all segments collectively, unlike $W(\mathbf{r})$, does not reduce to a simple mathematical form without approximation, even in the limit $n \to \infty$.[4] We shall not pursue it beyond the observation that the distribution of segments in space is intimately related to $W(\mathbf{r})$, and is in fact determined by an appropriate set of intersegmental distributions W_{ij}, as noted above. Hence, the intersegmental distribution functions, typified by $W(\mathbf{r})$, also characterize the spatial configuration of the chain, albeit indirectly.

3. MOMENTS OF DISTRIBUTIONS

Distributions do not in general lend themselves to succinct representation, except, of course, in graphical form. Unless the distribution conforms to a simple mathematical expression—and this is the exception rather than the rule—numerical characterization may be cumbersome. Moments of distributions afford a general basis for characterizing and comparing distributions, and in principle any degree of refinement can be achieved by extending the number of moments to include those of successively higher powers of the variable (e.g., r).

A quantity which to a first order of approximation characterizes the spatial distribution, and hence the configuration, of a chain molecule is the second moment $\langle r^2 \rangle$ of its end-to-end distance, which has already appeared in Eq. 7. For a higher approximation, the fourth moment $\langle r^4 \rangle$ also would be required, etc. It is the even moments which are relevant for spherically symmetric distributions like those considered here (cf. Chap. VIII). If the distribution function is known, the moments may of course be evaluated from it. If not, moments of the distribution often can be evaluated independently without recourse to the distribution function, which usually is less readily susceptible to determination. Moreover, most of the configuration-dependent properties of chain molecules can be resolved in terms of functions of moments of distributions related to the configuration, without explicit formulation of the distribution function itself.

The second moment $\langle r^2 \rangle$ of the end-to-end distribution, which will be the key quantity we shall use to characterize the spatial configurations of chain molecules, can be expressed according to Eq. 3 as follows:

$$\langle r^2 \rangle = \sum_i \langle l_i^2 \rangle + 2 \sum_{i<j} \langle \mathbf{l}_i \cdot \mathbf{l}_j \rangle \tag{8}$$

where the second summation will be understood to be the double sum over i and j such that $0 < i < j \leq n$. The appearance of the l_i^2 as their statistical mechanical averages in the first sum in this equation is proper inasmuch as lengths of bonds have not been asserted to be fixed at this stage of the argument. If l^2 is defined as the average squared bond length for a given configuration of the chain, then $\sum_i l^2 = nl^2$ for that configuration and

$$\sum_i \langle l_i^2 \rangle = \langle \sum_i l_i^2 \rangle = n\langle l^2 \rangle$$

With this definition of l^2, Eq. 8 may be written

$$\langle r^2 \rangle = n\langle l^2 \rangle + 2\sum_{i<j} \langle \mathbf{l}_i \cdot \mathbf{l}_j \rangle \tag{9}$$

In the approximation that the bond lengths are fixed

$$\langle r^2 \rangle = nl^2 + 2\sum_{i<j} \langle \mathbf{l}_i \cdot \mathbf{l}_j \rangle \tag{10}$$

where l^2 will be understood to be the average squared length of the n skeletal bonds, the value for each being fixed, presumably at its mean value. A term of the sum appearing in Eqs. 9 and 10 may then be regarded as the product $l_i l_j$ of fixed scalar bond lengths and the cosine of the angle between their directions averaged over all configurations. Thus, $\langle \mathbf{l}_i \cdot \mathbf{l}_j \rangle$ is a measure of the correlation of the directions of the pair of bonds, i and j. Much of what is to follow will be concerned with the evaluation of these averaged scalar products, which obviously are essential for arriving at an estimate of $\langle r^2 \rangle$.

According to Eq. 6, the statistical mechanical average of the radius of gyration, i.e., its average taken over all configurations, is

$$\langle s^2 \rangle = (n+1)^{-2} \sum_{0 \le i < j \le n} \langle r_{ij}^2 \rangle \tag{11}$$

where $\langle r_{ij}^2 \rangle$ is given by an expression corresponding to Eq. 4.

The angular dissymmetry of light scattered by a dilute solution of macromolecules is directly related to their average squared radii of gyration (cf. Chap. IX). It will be apparent from Eq. 11 and the preceding discussion of the interatomic distributions that $\langle s^2 \rangle$ is closely related to $\langle r^2 \rangle$. For chains of finite length the relationship is not unique; it depends in general on features peculiar to the chain considered, as we shall show later.

4. THE UNPERTURBED STATE

The foregoing equations for $\langle r^2 \rangle$ and $\langle s^2 \rangle$ are applicable generally to a system of identical chain molecules subject to any conditions whatever. The molecules may be dispersed in solution, concentrated or dilute; they may be mutually intertwined in the amorphous state, or be the occupants of a crystal lattice. It is required merely that they be subject to uniform macroscopic constraints, and that the usual conditions for statistical mechanical equilibrium be satisfied.

As we proceed further, it will be essential to identify a particular state as the one of reference. Considerations presented later dictate that the choice be the state in which the molecule is free of constraints, such as those which might be imposed by an external force or by hydrodynamic interactions with the fluid in which the molecule is dispersed when the fluid is subject to shear; thermodynamic interactions with solvent must also be suppressed. In effect, the molecule is subject only to the local constraints relating to the geometrical features of the bond structure, and hindrances to rotation about bonds. This unperturbed state is discussed more fully in the following chapter. Inasmuch as the relationships which follow will be found on close inspection to refer specifically to this state, it is necessary to introduce the notational convention for its designation at this point. A subscript zero will serve this purpose. It will be appended to the angle brackets surrounding r^2 or s^2; thus, $\langle r^2 \rangle_0$ and $\langle s^2 \rangle_0$ will denote the values of the second moments averaged over a statistical mechanical ensemble of unperturbed molecules free of constraints, as stipulated above in qualitative terms. The requirements for fulfilling these conditions will be examined more critically in the following chapter.

5. THE FREELY JOINTED CHAIN

The hypothetical *random flight*, or *freely jointed*, chain consists of n bonds, each of fixed length and joined in linear succession, the angles at the bond junctions being free to assume all values with equal probability. Rotations about bonds are likewise free. The directions of neighboring bonds are thus completely uncorrelated in the sense that all directions for a given bond are equally probable, irrespective of the directions of its neighbors in the chain. In absence of correlations between bonds

$$\langle \mathbf{l}_i \cdot \mathbf{l}_j \rangle = 0, \qquad i \neq j \tag{12}$$

It follows from Eq. 10 that*

$$\langle r^2 \rangle_0 = nl^2 \tag{13}$$

This relationship holds for hypothetical *freely jointed chains of any length n.*

* The subscript zero is applied to $\langle r^2 \rangle$ to denote the unperturbed state in recognition of the disregard of contributions to $\langle r^2 \rangle$ from interactions of long range involving remote pairs ij, i.e., pairs for which $|i-j|$ is large; see Chapter II. On the other hand, it could be argued that absence of all interactions, including those of long range, is implicit in the adoption of the hypothetical freely jointed chain as a model. If this position is taken, then the subscript zero is superfluous.

If the various bonds of the chain differ in length, l^2 is understood to represent the mean-square bond length as noted above.

The *characteristic ratio* defined by

$$C_n = \langle r^2 \rangle_0 / nl^2 \tag{14}$$

is unity for all values of n for the freely jointed chain. In Chapter VIII we shall examine higher moments of r, i.e., $\langle r^4 \rangle_0$, $\langle r^6 \rangle_0$, etc., for this chain. Their analogous ratios $\langle r^4 \rangle_0 / n^2 l^4$, etc., unlike C_n, will be found to depend on n; these ratios, however, approach asymptotic values with increase in n.*

The mean-square radius of gyration for the freely jointed chain is readily derived from Eq. 11. Introduction of the relation $\langle r_{ij}^2 \rangle_0 = (j-i)l^2$ implied by Eq. 13, followed by replacement of $j - i$ by k gives

$$\begin{aligned}\langle s^2 \rangle_0 &= l^2(n+1)^{-2} \sum_{0 \le i < j \le n} (j-i) \\ &= l^2(n+1)^{-2} \sum_{j=1}^{n} \sum_{k=1}^{j} k \\ &= l^2(n+1)^{-2} \sum_{1}^{n} j(j+1)/2\end{aligned}$$

Evaluation of the summations over j^2 and j and rearrangement of the result yields

$$\langle s^2 \rangle_0 / nl^2 = \frac{1}{6}(n+2)/(n+1) \tag{15}$$

It will be observed that $\langle s^2 \rangle_0 / nl^2$, unlike $\langle r^2 \rangle_0 / nl^2$, depends on n. Specifically, the former ratio for the freely jointed chain *decreases* with n, reaching

$$(\langle s^2 \rangle_0 / nl^2)_\infty = \frac{1}{6} \tag{16}$$

in the limit $n = \infty$. Thus

$$\langle s^2 \rangle_0 = \langle r^2 \rangle_0 / 6, \qquad n \to \infty \tag{17}$$

a relationship due to Debye.[5]

* Adherence to Eq. 13 is frequently identified, erroneously, as the mark of a "Gaussian chain," with implied reference to the chain vector distribution $W(\mathbf{r})$. Freely jointed chains of finite length obey this equation, but the functions $W(\mathbf{r})$ describing them are not strictly Gaussian. The distribution $W(\mathbf{r})$ for a real chain may be approximately Gaussian, yet the relationship of $\langle r^2 \rangle_0$ to n for a series of its homologs may depart markedly from Eq. 13. The form of the distribution and the dependence of $\langle r^2 \rangle_0$ on the chain length are not universally related, and hence they must be separately characterized. These properties of chain molecules will be taken up in Chapter VIII.

Kuhn[6] invested this manifestly artificial model with a semblance of reality by pointing out that the correlation between bonds i and $i+k$ in any real chain endowed with a finite degree of flexibility must vanish as k increases. In random macromolecules of the usual large n, the correlations will vanish for $k \ll n$. The sum over j of terms $\langle \mathbf{l}_i \cdot \mathbf{l}_j \rangle$ for a given value of i converges, for unperturbed chains, and the value of this sum is independent of i, provided that i is well removed from both ends of the chain. The double sum in Eq. 10 may therefore be taken to be proportional to n for sufficiently large n. Asymptotic proportionality of $\langle r^2 \rangle_0$ to n is thus assured for any flexible, real chain. Unlike the freely jointed chain, however, the constant of proportionality will depart considerably from unity, in general. A real chain of sufficient length may be represented, therefore, by an *equivalent chain* comprising n' hypothetical bonds each of length l' connected by free joints as above. Arbitrariness in the choices of n' and l' is removed by requiring that

$$\begin{aligned} n'l' &= r_{\max} \\ n'(l')^2 &= \langle r^2 \rangle_0 \end{aligned} \tag{18}$$

where $r_{\max}$ is the fully extended length of the real chain and $\langle r^2 \rangle_0$ is its actual unperturbed mean square end-to-end length. Thus, for a polymethylene chain with n sufficiently large, $r_{\max} = 0.83nl$ and $\langle r^2 \rangle_0 = 6.7nl^2$ (see Chap. II, Table 1). It follows that $n/n' \cong 10$ real bonds per equivalent segment.*

The correspondence of the equivalent random flight chain to a real chain holds only for large n. Its use for short chains, even in the range where n is great enough to encompass several equivalent segments, may be quite misleading. The deficiencies of the equivalent random flight chain as a model for a real chain of finite n will become apparent in the course of the treatment of real chains presented in later chapters.

The asymptotic convergence of the real chain to its hypothetical freely jointed analog as $n \rightarrow \infty$ validates Eq. 17 for real chains in this limit.

6. CONFIGURATIONS OF CHAINS HAVING FIXED BOND LENGTHS AND BOND ANGLES

The lengths of bonds and the angles between contiguous bonds usually are restricted to fairly narrow ranges. Amplitudes of thermal oscillations at

* Other definitions of an "equivalent chain" are frequently used. For example, the bond length l' may be adjusted arbitrarily to satisfy the second of the relations in Eq. 18, while preserving the number of bonds $n' = n$. Then $l' = (\langle r^2 \rangle_0 / n)^{1/2}$, a quantity often designated by b.

ordinary temperature are typically on the order of 3% of the bond length, i.e., ca. ±0.05 Å. Bond angles are usually subject to comparable average fluctuations amounting to ±3 to ±5° at ordinary temperature. Among random configurations of free chains, fluctuations occur more or less symmetrically about the mean values of these parameters; hence, the effects of those of opposite signs tend to cancel one another. For most purposes it is permissible, therefore, to ascribe to each bond length and to each bond angle its mean value, as if each were fixed. The remaining configuration variables are the bond rotation angles.

Before proceeding further, conventions of notation for expressing bond angles and bond rotations need to be introduced. Consider again an n-alkane, or polymethylene, chain. One of its configurations, or conformations, amenable to representation on the printed page is shown in Fig. 2.

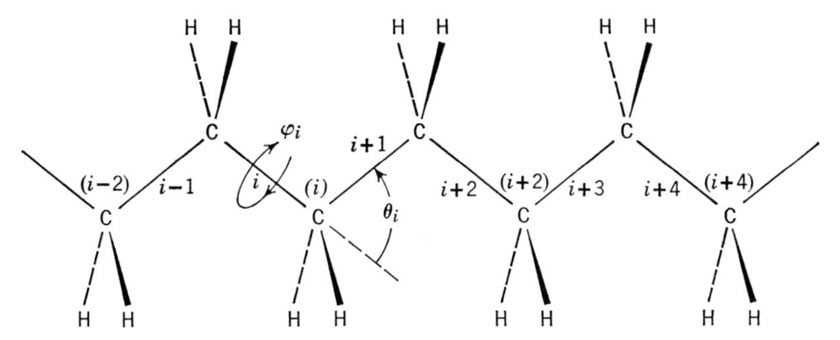

Fig. 2. Portion of a polymethylene chain in the *trans* form including bonds $i-1$ to $i+4$. Indexes for some of the skeletal atoms are shown in parentheses. Bond angle θ_i and bond rotation angle ϕ_i are indicated.

In this *planar trans* form, atoms of the *chain skeleton*, in this case, carbons, lie in a plane, chosen here to be the plane of the paper. Pendant atoms, hydrogens in this molecule, lie above and below the plane. Atoms above the plane are differentiated in Fig. 2 from those below it according to the convention whereby the bonds connecting the former to the skeletal carbons are shown in exaggerated perspective; bonds to hydrogens below the plane are shown as dashed lines. The angle θ_i is the supplement of the skeletal bond angle at chain atom i. The bond rotation angle ϕ_i measures the dihedral angle between the two planes defined, respectively, by bond pairs $i-1, i$ and $i, i+1$. The sign of the angle ϕ_i is taken as positive for a right-handed rotation.* It is measured relative to the *trans* conformation

* A right-handed rotation is simply defined as one which would increase the distance between the pair of groups if the bond joining them were a right-hand screw and the skeletal atoms were threaded to it.

shown in Fig. 2; i.e., in the configuration shown the rotation angles $\phi_i = 0$ for all i.*

In the generally acceptable approximation whereby bond lengths l_i and bond angle supplements θ_i are assigned fixed (mean) values, the various configurations of the chain are differentiated by the bond rotation angles. The skeletal configuration is then determined by the set $\phi_2, \phi_3, \ldots, \phi_{n-1}$, numbering $n-2$ in all, which will be designated symbolically by $\{\phi\}$. It is an alternative to the set $\{\mathbf{l}\}$ of vectors for specifying the configuration and, being a less extravagant set than the latter, will generally be preferred.†

Configurations of other simple chains can be similarly represented. Consider, for example, polyoxymethylene

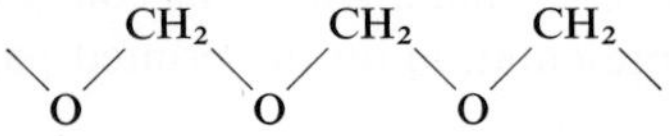

$$\diagdown O \diagup CH_2 \diagdown O \diagup CH_2 \diagdown O \diagup CH_2 \diagdown$$

and polyglycine

$$\diagdown CH_2 \diagup CO \diagdown NH \diagup CH_2 \diagdown CO \diagup NH \diagdown$$

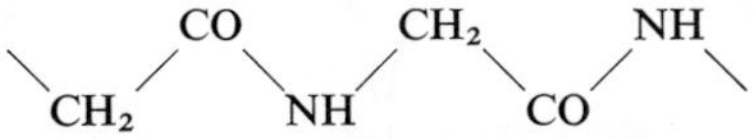

The former possesses a two-bond repeating sequence; all bonds are of the same length, but bond angles alternate between ∠OCO and ∠COC. Thus, there are two angles θ, which in this case, however, happen to be very nearly equal. The second example is characterized by a three-bond repeat unit. The three bonds differ in length, and there are three distinct bond angles.

The scheme illustrated in Fig. 2 for polymethylene is applicable to these

* The opposite convention by which $\phi = 0$ for the *cis* conformation is often used. In the field of spectroscopy it has a strong precedent for the description of the conformations of simple molecules containing at most only a few skeletal bonds. In chain molecules the assignment of $\phi_i = 0$ to the *trans* conformation is intuitively consistent with the parallelity of bonds $i-1$ and $i+1$ in this form for equal bond angles, $\theta_{i-1} = \theta_i$. The opposite convention would imply reversed directions for bond vectors $\mathbf{l}_{i-1}$ and $\mathbf{l}_{i+1}$ relative to the bond sequence in the chain, i.e., the vectors representing these two bonds would be directed towards opposite ends of the chain. The convention adopted here has the distinct advantage of assigning $\phi_i = 0$ to that conformation which is invariably chosen for representation on the printed page, and hence the one from which analysis of the conformation will most naturally proceed. Finally, under this convention, right-handed helices in simple chains are generated by introducing positive rotations and negative rotations yield left-handed helices. According to the alternative *cis* convention, left-handed helices are generated by positive rotations.

† The locations of the hydrogen atoms of the terminal CH_3 groups of an n-alkane chain $CH_3—(CH_2)_{n-1}—CH_3$ depend on rotations about bonds 1 and n, i.e., on ϕ_1 and ϕ_n, which we have excluded from the set $\{\phi\}$ on the grounds that they do not affect the configuration of the chain skeleton. If the precise positions of terminal hydrogen atoms should be of importance, these rotation angles may be restored to the set.

and other simple chains without elaboration. In general, it is necessary merely to introduce the appropriate values for the bond lengths l_i and angles θ_i according to the serial index i. The bond parameters l_i and θ_i may be assigned their mean values as revealed by structural studies on the chain molecules or, more usually, on their low molecular analogs. The set of bond rotation angles $\{\phi\}$ then completes the specification of a configuration.

If the chain bears substituents, the same formal description of the configuration often will be adequate. Vinyl polymers having the repeating unit —CH_2—CHR—, where R may be Cl, CH_3, C_2H_5, $COOCH_3$, or C_6H_5, are illustrative. The stereochemical configuration of the asymmetric carbon —CHR— complicates the full description of the chemical structure, but not the formal representation (Chap. VI) of the chain configuration. Specification of the skeletal configuration by the set of variables $\{\phi\}$, or by $\{\mathbf{l}\}$, defines the positions of atoms directly connected to the backbone, in the approximation of fixed bond lengths and angles. This will usually be sufficient. If circumstances should require a more detailed specification of the configuration of the molecule, the scheme may be elaborated.

The class of *configurations* which are generated by executing rotations about single bonds of a molecule often are referred to as *conformations*. These two terms, *configuration* and *conformation*, will be used somewhat interchangeably, in keeping with their dictionary definitions and broader usage, with only a minor difference in connotation.* We shall generally reserve the latter term for comparatively simple configurations of small molecules, and of small segments of long chains. We shall use it also in referring to the regular forms of macromolecules as they occur in the crystalline state, also including the stable helical forms of polypeptides and certain biological macromolecules. For the formless arrangements of the

* The term *conformation* connotes form and a symmetrical arrangement of parts. The alternative term, *configuration*, is perhaps the more general of the two in referring to the disposition of the parts of the object in question without regard for shape and symmetry. Our usage of the latter term may at times violate conventions of organic chemists, who presume to have preempted the term *configuration* for a more specific purpose, namely, to designate the stereochemical arrangement of atoms or groups about a structural element of optical asymmetry in the molecule. This term was widely used in the language of science well in advance of its appropriation for this specific purpose. Its use was well established in mechanics, and especially in statistical mechanics, the fields from which our usage of the term draws main precedent. We may, for example, conceive of a configuration of the chain as being specified by a point in configuration hyperspace, and set about to deduce average properties by averaging over an ensemble of systems (molecules) canonically distributed in that space.

Confusion with configuration about an asymmetric center of the kind associated with optical rotation can easily be avoided by use of an appropriate prefix such as *stereochemical.*

macromolecule which collectively are designated as the *random coil*, the term configuration is more appropriate.

7. THE FREELY ROTATING CHAIN

The hypothetical model chain here considered comprises n bonds of fixed length joined at fixed bond angles. For simplicity, all bond lengths and all bond angles will at the outset be taken to be equal. The condition of free rotation implies that the energy of each configuration $\{\phi\}$ of the molecule is equal to that of every other, and hence that each occurs with equal incidence in an ensemble of these systems. Unlike the freely jointed model, however, correlations between bond directions are imposed by fixing the angles θ.

The projection of bond $i+1$ on i is $l \cos\theta$; its projections in transverse directions average to zero according to the assumption of free rotation. It follows that

$$\langle \mathbf{l}_i \cdot \mathbf{l}_{i+k} \rangle = l^2 \alpha^k \tag{19}$$

where $\alpha = \cos\theta$. Thus, from Eq. 10

$$\langle r^2 \rangle_0 = nl^2 + 2l^2 \sum_{i<j} \alpha^{j-i}$$

where, as before, the range on the double summation is $0 < i < j \leq n$. Combination of terms for which $j - i$ has the same value k leads to

$$C_n \equiv \langle r^2 \rangle_0 / nl^2 = 1 + (2/n) \sum_{k=1}^{n-1} (n-k)\alpha^k$$

$$= 1 + 2 \sum_{k=1}^{n-1} \alpha^k - (2/n) \sum_{k=1}^{n-1} k\alpha^k$$

Evaluation* of these two sums gives[7,8]

$$C_n = 1 + 2(\alpha - \alpha^n)(1-\alpha)^{-1}$$

$$- (2/n)[\alpha(1-\alpha^n)(1-\alpha)^{-2} - n\alpha^n(1-\alpha)^{-1}]$$

$$= (1+\alpha)(1-\alpha)^{-1} - (2\alpha/n)(1-\alpha^n)(1-\alpha)^{-2} \tag{20}$$

---

* The second of these summations is easily obtained from the first by resort to the following general procedure. Let

$$S_0 = \sum_1^{n-1} \alpha^k; \quad S_1 = \sum_1^{n-1} k\alpha^k; \quad \text{etc.}$$

or in general

$$S_m = \sum_1^{n-1} k^m \alpha^k$$

According to Eq. 20 the characteristic ratio for the freely rotating chain varies with chain length approximately as $-1/n$. For $\theta<\pi/2$, which is the usual case, C_n increases with the chain length n. A similar dependence will be demonstrated in the following section for chains with hindered rotation treated in the approximation that neighboring bond rotations are independent. In real chains, for which bond rotations are strongly interdependent, the dependence of the characteristic ratio C_n on n is more complicated; it evades simple generalizations beyond the certainty that C_n is asymptotic with n, as will be shown in due course.†

In the limit of large n, Eq. 20 reduces to

$$C_\infty = (1+\alpha)/(1-\alpha) = (1+\cos\theta)/(1-\cos\theta) \tag{21}$$

as was first shown by Kuhn[9] following the work of Eyring.[7] For a tetrahedrally bonded chain, $\cos\theta = 1/3$ and $C_\infty = 2$.

The mean-square radius of gyration may be obtained from Eq. 11 through the use of Eq. 20 for $\langle r_{ij}^2\rangle_0$, with n therein replaced by $k = j - i$. Thus

$$\langle s^2\rangle_0 = \frac{l^2}{(n+1)^2}\sum_{j=1}^{n}\sum_{k=1}^{j}\left[\left(\frac{1+\alpha}{1-\alpha}\right)k - \frac{2\alpha(1-\alpha^k)}{(1-\alpha)^2}\right]$$

$$= \frac{l^2 n(n+2)(1+\alpha)}{6(n+1)(1-\alpha)} - \frac{2l^2\alpha}{(n+1)^2(1-\alpha)^2}\sum_{j=1}^{n}\left[j - \left(\frac{\alpha-\alpha^{j+1}}{1-\alpha}\right)\right]$$

Evaluation of the summation in the second term leads to the final result[10,11]

Then

$$S_1 = \alpha\, dS_0/d\alpha$$

or in general

$$S_{m+1} = dS_m/d\ln\alpha$$

Writing

$$S_0 = \sum_1^\infty \alpha^k - \alpha^{n-1}\sum_1^\infty \alpha^k$$

we have at once that

$$S_0 = (\alpha - \alpha^n)/(1-\alpha)$$

from which the higher sums may be evaluated by differentiation.

† G. Porod, *J. Polymer Sci.*, **10**, 157 (1953), has given a lengthy expression for the fourth moment $\langle r^4\rangle_0$ for a freely rotating chain as a function of the chain length n. We have not reproduced Porod's result inasmuch as a treatment of the fourth moment, which is generally applicable to chain molecules, and not subject to the assumption of free rotation, follows in Chapter IV.

$$\langle s^2\rangle_0/nl^2 = \frac{(n+2)(1+\alpha)}{6(n+1)(1-\alpha)} - \frac{\alpha}{(n+1)(1-\alpha)^2} + \frac{2\alpha^2}{(n+1)^2(1-\alpha)^3} - \frac{2\alpha^3(1-\alpha^n)}{n(n+1)^2(1-\alpha)^4} \tag{22}$$

For large n

$$(\langle s^2\rangle_0/nl^2)_\infty = \frac{1}{6}(1+\alpha)/(1-\alpha) = \frac{1}{6}(1+\cos\theta)/(1-\cos\theta) \tag{23}$$

which follows also from Eq. 21 and Eq. 17, the latter having been noted previously to be valid for random chains in general in the limit $n \to \infty$.

These methods are applicable also to chains having a two-bond repeat unit. Polydimethylsiloxane is such an example: its bonds are identical in length, but the skeletal bond angles at silicon and oxygen differ. The supplements of these bond angles are $\theta' \cong 70°$ and $\theta'' \cong 37°$, respectively. Then, assuming free rotation to prevail (which is by no means a valid assumption for this chain), we have

$$\langle \mathbf{l}_i \cdot \mathbf{l}_{i+k}\rangle = (\alpha'\alpha'')^{k/2} \qquad k \text{ even}$$

$$= \begin{cases} (\alpha'\alpha'')^{(k-1)/2}\alpha' \\ \quad\text{or} \\ \alpha''(\alpha'\alpha'')^{(k-1)/2} \end{cases} \qquad k \text{ odd}$$

where $\alpha' = \cos\theta'$ and $\alpha'' = \cos\theta''$. By executing summations similar to those used above and retaining terms only for the limit $n \to \infty$, one readily obtains[12]

$$C_\infty = (1+\alpha')(1+\alpha'')/(1-\alpha'\alpha'') \tag{24}$$

This chain will be discussed in more realistic terms in Chapter V. It is cited here for illustrative purposes only.

The foregoing procedure may be applied to more complicated chains, e.g., to polypeptides, to cellulose, and to chains with double bonded units (such as $—CH_2—CH{=}CH—CH_2—$), when treated in the approximation of free rotation about single bonds of the chain skeleton. It becomes excessively tedious, however, and treatment of such chains can be carried out in a more orderly fashion by the methods developed in later chapters. These methods do not require the assumption of free rotation, which seldom is admissible on physical grounds.

8. NEIGHBOR CORRELATIONS IN REAL CHAINS AND THE GEOMETRICAL INTERRELATION OF BOND VECTORS

In real chain molecules the bond vectors are subject to strong mutual correlations in the sense that the direction of a given bond is influenced by the directions of its neighbors in the chain skeleton. This correlation is brought about in the first instance by the restriction imposed by confinement of bond angles, $\pi - \theta_i$, to characteristic values within narrow limits. Even if free rotation prevailed, the restriction on θ_i would enforce a direct correlation between bonds which are first neighbors, i.e., between $\mathbf{l}_{i-1}$ and $\mathbf{l}_i$. Bonds i and $i+1$ are subject to a corresponding correlation dictated by θ_{i+1}, and so on.

Hindrances to free rotation, invariably operative in real chains, extend the range of correlation between bond directions. This is apparent from Fig. 2. The direction of bond $i+1$ not only depends on the direction of bond i in consequence of bond angle θ_i; it also depends on the direction of bond $i-1$ through the dihedral angle ϕ_i between the planes defined by the respective bond pairs $i-1, i$ and $i, i+1$. Owing to the hindrance potential affecting the rotation ϕ_i about bond i, this angle is not permitted to assume all values 0 to 2π with equal probability. The interdependence of bond directions holds equivalently for succeeding bonds as for those viewed as predecessors according to the arbitrary order of numeration adopted, for example, in Fig. 2. Thus, bond $i+1$ is subject also to correlations with bonds $i+2$ and $i+3$.

In most chain molecules the *rotation angle* about a given skeletal bond is correlated with rotations about the bonds which are its immediate neighbors on either side. This correlation is brought about by the dependence of the potential affecting the rotation about a given bond i on the rotation angles ϕ_{i-1} and ϕ_{i+1} for neighboring bonds. When such neighbor dependence of rotation angles prevails, as is usually the case (see Chap. III), the range of direct correlation between bond vectors extends to bonds which are third neighbors. This is apparent from models, or from diagrams like Fig. 2. Of course, the effective range of correlation may be much greater owing to influences transmitted from one tier of neighbors to the next, and so on.

Comprehensive treatment of these neighbor correlations, a task which will dominate much of the development in succeeding chapters, will not be undertaken here. The objectives of the present chapter do not go beyond the formulation of expressions for chains in which rotational potentials depend only on the parameters relating to the bond in question. It is expedient, however, to introduce at this juncture the general procedures for dealing with the geometrical relationships between neighboring bonds subject to correlations of one sort or another. These relationships are

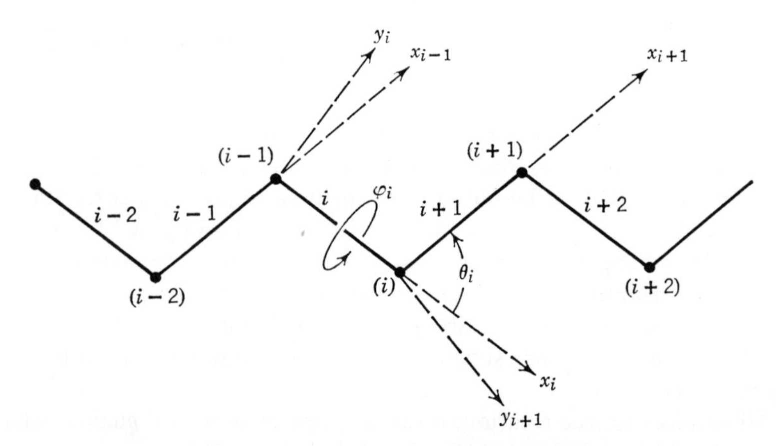

Fig. 3. Cartesian reference frames as arbitrarily defined for consecutive bonds of a chain. The z axes, not shown, are perpendicular to the plane, with directions alternating up and down from one frame to the next.

applicable to chain molecules of any description. They are required for the remainder of this chapter, and will find extensive use in those following.

In order to construct a scheme for formulating the required scalar products $\mathbf{l}_i \cdot \mathbf{l}_j$, it will be imperative to define a Cartesian coordinate system, or reference frame, for each bond of the chain.[7,13] To this end, let the axis x_i of the coordinate system affixed to bond i be taken in the direction of bond i, as shown in Fig. 3. The y_i axis is taken in the plane of bonds $i-1$ and i, with its positive direction chosen to render its projection on x_{i-1} positive. The direction of the z_i axis is that required to complete a right-handed Cartesian coordinate system. Thus, in the planar *trans* conformation with $\phi_i = 0$ for all i, all of the z axes are perpendicular to the plane of the diagram, with their directions alternating upwards and downwards.

Let $\mathbf{T}_i$ be the orthogonal matrix which by premultiplication transforms a vector $\mathbf{v}$ with components v_x, v_y, v_z expressed in reference frame $i+1$ to its representation $\mathbf{v}'$ in reference frame i. That is

$$\mathbf{v}' = \mathbf{T}_i \mathbf{v}$$

where $\mathbf{v}$ and $\mathbf{v}'$ are given in the form of column vectors. The matrix $\mathbf{T}_i$ is (see Appendix B)

$$\mathbf{T}_i = \begin{bmatrix} \cos\theta_i & \sin\theta_i & 0 \\ \sin\theta_i \cos\phi_i & -\cos\theta_i \cos\phi_i & \sin\phi_i \\ \sin\theta_i \sin\phi_i & -\cos\theta_i \sin\phi_i & -\cos\phi_i \end{bmatrix} \tag{25}$$

This matrix will be used extensively in later chapters.

The scalar product of bond vectors $\mathbf{l}_i$ and $\mathbf{l}_j$ appearing in Eq. 3 can now be formulated. We assume $j > i$. By transforming the representation of $\mathbf{l}_j$ successively from the reference frame for one of the intervening bonds to that of its predecessor, we obtain for the representation of this vector in the coordinate system of bond i

$$\mathbf{T}_i \cdots \mathbf{T}_{j-2}\mathbf{T}_{j-1}\mathbf{l}_j$$

where $\mathbf{l}_j$ is given in its own coordinate system by the column vector

$$\mathbf{l}_j = \begin{bmatrix} l_j \\ 0 \\ 0 \end{bmatrix} = l_j \begin{bmatrix} 1 \\ 0 \\ 0 \end{bmatrix} \tag{26}$$

Here and elsewhere l_j is the scalar magnitude of $\mathbf{l}_j$. A similar expression represents $\mathbf{l}_i$ in its coordinate system. Let $\mathbf{l}_i^T$ be the transpose of $\mathbf{l}_i$ (i.e., its row vector representation). The scalar product of $\mathbf{l}_i$ upon $\mathbf{l}_j$ is $\mathbf{l}_i^T\mathbf{l}_j$, the latter being expressed in the same coordinate system as the former.* Hence[7,13]

$$\mathbf{l}_i \cdot \mathbf{l}_j = \mathbf{l}_i^T(\mathbf{T}_i \cdots \mathbf{T}_{j-1})\mathbf{l}_j \tag{27}$$

$$= l_i l_j [1 \quad 0 \quad 0](\mathbf{T}_i \cdots \mathbf{T}_{j-1}) \begin{bmatrix} 1 \\ 0 \\ 0 \end{bmatrix}$$

$$= l_i l_j (\mathbf{T}_i \cdots \mathbf{T}_{j-1})_{11} \tag{28}$$

where the subscript denotes the 1,1 element of the matrix product.

The average projection on bond i of a unit vector along bond j is given by the 1,1 element of the matrix product averaged over all configurations. Thus[7,13]

$$\langle \mathbf{l}_i \cdot \mathbf{l}_j \rangle = l_i l_j \langle \mathbf{T}_i \cdots \mathbf{T}_{j-1} \rangle_{11} \tag{29}$$

where angle brackets again denote the statistical mechanical average over all configurations of the chain. The average of the product of orthogonal (axis transformation) matrices in Eq. 29 is expressed formally by

$$\langle \mathbf{T}_i \cdots \mathbf{T}_{j-1} \rangle = \frac{\int \cdots \int (\mathbf{T}_i \cdots \mathbf{T}_{j-1}) \exp[-E\{\mathbf{l}\}/RT]\, d\{\mathbf{l}\}}{\int \cdots \int \exp[-E\{\mathbf{l}\}/RT]\, d\{\mathbf{l}\}} \tag{30}$$

* The same symbol $\mathbf{l}$ serves to denote both the vector and its representation as a column matrix. It is used in the former sense on the left-hand side of Eq. 27, and in the latter sense on the right-hand side. Corresponding dual roles for the same symbol will recur in pages following. Confusion should not arise from this practice; which one of the alternative representations is intended will be obvious from the context.

where $E\{\mathbf{l}\}$ is the energy of the configuration $\{\mathbf{l}\}$, R is the gas constant, and T is the absolute temperature. The integrals extend over all rotation angles. The integrations in the numerator apply to each of the elements of the matrix product. In Chapters III and IV these integrals will be replaced by appropriate sums, which will be evaluated by methods that are rigorous and of broad applicability.

The foregoing expressions are applicable also, with only nominal alterations, to the scalar product of any pair of vectors associated with bonds i and j. Suppose $\mathbf{m}_i$ and $\mathbf{m}_j$ to be two such vectors which are uniquely defined in the respective reference frames of bonds i and j. That is, the representation of each in its reference frame is invariant to the chain configuration. The vectors $\mathbf{m}_i$ and $\mathbf{m}_j$, for example, might be dipole moments associated with the respective bonds and the substituents rigidly attached to them. Then

$$\langle \mathbf{m}_i \cdot \mathbf{m}_j \rangle = \mathbf{m}_i^T \langle \mathbf{T}_i \cdots \mathbf{T}_{j-1} \rangle \mathbf{m}_j \tag{31}$$

Equations 25–31 are quite general. No assumptions are implied concerning the interdependence, or independence, of rotational potentials. In fact, bond angles $\pi - \theta_i$ may be taken to be variable within the scope of these relations. Only the bond lengths l are assumed to be fixed, as is implicit in Eq. 29.

Although the transformation matrix given by Eq. 25 is universally applicable to any linear chain, subject only to the specification of coordinate systems as prescribed above, other schemes for defining coordinate systems may be preferred on occasion. It will prove advantageous, for example, to treat polypeptide chains (Chap. VII) in terms of "virtual bonds" each of which spans a semirigid peptide unit. The angle between two consecutive virtual bonds depends on the angles of rotation about two real bonds; consequently, an alternative scheme for relating one peptide unit (i.e., virtual bond) to its neighbors takes preference over that represented by Eq. 25. Without entering into details introduced later, we note merely that a matrix for transforming a vector from the reference frame of bond (or virtual bond) $i + 1$ to that of bond i can always be defined in such a way that this matrix $\mathbf{T}_i$ will be a function exclusively of geometrical parameters, such as θ_i and ϕ_i (or, symbolically, of $\mathbf{l}_i$), pertaining to the union of bond $i + 1$ with bond i. These preliminary remarks are offered at this point in order to give broader scope to the section which follows.

9. CHAINS WITH SEPARABLE CONFIGURATIONAL ENERGIES

Let the configurational energy $E\{\mathbf{l}\}$ for the chain under consideration be separable into a sum of energies E_i, one such energy contribution being associated with the conformation of each skeletal bond i in relation to its neighbors, terminal bonds $i = 1$ and n excepted. Inasmuch as l_i is taken to be fixed, the conformational energy for bond i is appropriately designated by $E_i(\theta_i, \phi_i)$, and according to the assertion (usually inadmissible; see Chap. III) that the energy is separable

$$E\{\mathbf{l}\} = \sum_{i=1}^{n-1} E_i(\theta_i, \phi_i) \tag{32}$$

with $E_1 = E_1(\theta_1)$, since ϕ_1 is undefined. If the bond angles also are fixed, as considered in the following section, then

$$E\{\mathbf{l}\} = \sum_{i=2}^{n-1} E_i(\phi_i) \tag{33}$$

where $E_i(\phi_i)$ represents the rotational potential for bond i. The probability that rotation angle ϕ_i occurs within the range ϕ_i to $\phi_i + d\phi_i$ is

$$p(\phi_i)\, d\phi_i = z_i^{-1} \exp\left[-E_i(\phi_i)/RT\right] d\phi_i \tag{34}$$

where E_i is expressed as the energy per mole, and $z_i = z_i(T)$ is the *bond rotational partition function* defined by

$$z_i = \int_0^{2\pi} \exp\left[-E_i(\phi_i)/RT\right] d\phi_i \tag{35}$$

In a stricter sense, $E_i(\phi_i)$ should be defined as a potential of mean force in a statistical mechanical system. Of main importance is the exlusive dependence of $p(\phi_i)$ on ϕ_i which follows from the postulated independence of the rotational potentials.

The separability of the configurational energy according to Eq. 32, or Eq. 33, in conjunction with the fact that the transformation $\mathbf{T}_h$ from the reference frame of bond $h + 1$ to that of bond h is a function only of angles such as θ_h and ϕ_h specifying the orientation of bond $h + 1$ relative to h, permits the integrals in both the numerator and denominator of Eq. 30 to be factored into separate integrals for each bond. Thus, subject to the condition stated, Eq. 30 may be replaced by a product of ratios

$$\langle \mathbf{T}_h \rangle = \frac{\int \mathbf{T}_h \exp\left[-E_h(\mathbf{l}_h)/RT\right] d\mathbf{l}_h}{\int \exp\left[-E_h(\mathbf{l}_h)/RT\right] d\mathbf{l}_h} \tag{36}$$

Inasmuch as the magnitude of l_h is fixed, $d\mathbf{l}_h$ may be replaced by $\sin\theta_h\, d\theta_h\, d\phi_h$ with limits 0 to π and 0 to 2π, respectively; or, if θ_h is also fixed, then $d\phi_h$ may represent $d\mathbf{l}_h$, and

$$\langle \mathbf{T}_h \rangle = \frac{\int_0^{2\pi} \mathbf{T}_h \exp\left[-E_h(\phi_h)/RT\right] d\phi_h}{\int_0^{2\pi} \exp\left[-E_h(\phi_h)/RT\right] d\phi_h} \tag{37}$$

In any event, the postulated independence of the rotational potentials (or the separability of the configurational energy) sanctions expression of the average product of **T**'s as the product of the averages of the individual **T**'s, i.e.,

$$\langle \mathbf{T}_i \cdots \mathbf{T}_{j-1} \rangle = \prod_{h=i}^{j-1} \langle \mathbf{T}_h \rangle \tag{38}$$

the averaged transformations for individual bonds being given by either Eq. 36 or Eq. 37.

Treatments of various properties of chain molecules are greatly simplified by the foregoing factorization based on the separability of the energy (or potential of mean force). Instances where such separation is justified are the exception rather than the rule, however.

Consider a *simple chain* comprising identical skeletal atoms joined by identical bonds. The subscript i may then be dropped, except as may be required to identify the serial order of a bond. Let a vector **m** be rigidly attached to each bond. The expression of this vector in the reference frame of the associated bond is taken to be invariant to the configuration of the chain. The vector **m** may or may not be the bond vector **l**. Further, let $\mathbf{M} = \sum_{i=1}^{n} \mathbf{m}_i$ be the resultant of all of the moments **m**. Under the stated conditions the average transformation matrices $\langle \mathbf{T}_i \rangle$ are the same for all i. Hence,

$$\begin{aligned} \langle M^2 \rangle / nm^2 &= 1 + (2/nm^2) \sum_{i<j} \mathbf{m}^T \langle \mathbf{T} \rangle^{j-i} \mathbf{m} \\ &= 1 + (2/nm^2) \mathbf{m}^T \left[\sum_{k=1}^{n-1} (n-k) \langle \mathbf{T} \rangle^k \right] \mathbf{m} \end{aligned} \tag{39}$$

The series in $\langle \mathbf{T} \rangle^k$ and that in $k\langle \mathbf{T} \rangle^k$ may be summed exactly like the corresponding series in powers of scalar quantities. For example

$$\sum_{1}^{n-1} \langle \mathbf{T} \rangle^k = (\langle \mathbf{T} \rangle - \langle \mathbf{T} \rangle^n)(\mathbf{E} - \langle \mathbf{T} \rangle)^{-1} \tag{40}$$

where **E** is the identity matrix. The result obtained by summing the series in Eq. 39 may be transcribed directly from Eq. 20. It is

$$\begin{aligned} \langle M^2 \rangle / n = \mathbf{m}^T [&(\mathbf{E} + \langle \mathbf{T} \rangle)(\mathbf{E} - \langle \mathbf{T} \rangle)^{-1} \\ &- (2\langle \mathbf{T} \rangle / n)(\mathbf{E} - \langle \mathbf{T} \rangle^n)(\mathbf{E} - \langle \mathbf{T} \rangle)^{-2}] \mathbf{m} \end{aligned} \tag{41}$$

For long chains it suffices to use

$$(\langle M^2 \rangle / n)_{n \to \infty} = \mathbf{m}^T (\mathbf{E} + \langle \mathbf{T} \rangle)(\mathbf{E} - \langle \mathbf{T} \rangle)^{-1} \mathbf{m} \tag{42}$$

If **m** is to be identified with the bond vector **l**, which is expressed in its own coordinate system by Eq. 26, then the preceding results take the form

$$C_n \equiv \langle r^2 \rangle_0 / nl^2 = [(\mathbf{E} + \langle \mathbf{T} \rangle)(\mathbf{E} - \langle \mathbf{T} \rangle)^{-1} - (2\langle \mathbf{T} \rangle / n)(\mathbf{E} - \langle \mathbf{T} \rangle^n)(\mathbf{E} - \langle \mathbf{T} \rangle)^{-2}]_{11} \tag{43}$$

and

$$C_\infty \equiv \lim_{n \to \infty} \left[\frac{\langle r^2 \rangle_0}{nl^2} \right] = [(\mathbf{E} + \langle \mathbf{T} \rangle)(\mathbf{E} - \langle \mathbf{T} \rangle)^{-1}]_{11} \tag{44}$$

where subscripts on the brackets denote the 1,1 element.

The corresponding ratio of the mean square radius of gyration, obtained by substitution of Eq. 43 in 11 (see the derivation of Eq. 22) is

$$\begin{aligned}\langle s^2 \rangle_0 / nl^2 = \frac{(n+2)}{6(n+1)} [(\mathbf{E} + \langle \mathbf{T} \rangle)(\mathbf{E} - \langle \mathbf{T} \rangle)^{-1}]_{11} \\ - \frac{[\langle \mathbf{T} \rangle (\mathbf{E} - \langle \mathbf{T} \rangle)^{-2}]_{11}}{n+1} + \frac{2[\langle \mathbf{T} \rangle^2 (\mathbf{E} - \langle \mathbf{T} \rangle)^{-3}]_{11}}{(n+1)^2} \\ - \frac{2[\langle \mathbf{T} \rangle^3 (\mathbf{E} - \langle \mathbf{T} \rangle^n)(\mathbf{E} - \langle \mathbf{T} \rangle)^{-4}]_{11}}{n(n+1)^2}\end{aligned} \tag{45}$$

which of course reduces to Eq. 22 if $\langle \mathbf{T} \rangle$ may be replaced by the scalar quantity $\alpha = \cos \theta$.

Equations 39–45 are applicable to any chain for which separate averaging of the **T** matrices is legitimate. They hold, irrespective of the nature of the geometrical constraints on the bond connections. Separability of energy according to Eq. 32 is the necessary and sufficient condition.

10. CHAINS WITH FIXED BOND ANGLES AND INDEPENDENT BOND ROTATIONAL POTENTIALS

We turn our attention now to the specific case of a chain in which angles between bonds are fixed and the potential affecting rotation about bond i depends only on ϕ_i. We shall assume further that this potential $E_i(\phi_i)$ is symmetric, i.e., that $E_i(-\phi_i) = E_i(\phi_i)$. A chain devoid of asymmetric centers, such as —CHR—, for example, will necessarily meet this condition, as is obvious from examination of simple models or by scrutiny of Fig. 2.

For a chain having a symmetric rotational potential, $\langle \sin \phi_i \rangle = 0$; hence, from Eq. 25

$$\langle \mathbf{T}_i \rangle = \begin{bmatrix} \cos \theta_i & \sin \theta_i & 0 \\ \sin \theta_i \langle \cos \phi_i \rangle & -\cos \theta_i \langle \cos \phi_i \rangle & 0 \\ 0 & 0 & -\langle \cos \phi_i \rangle \end{bmatrix} \tag{46}$$

Equation 37 is called upon to supply only the average value of $\cos \phi_i$. If the rotations are unhindered, i.e., if $E_i(\phi_i)$ = a constant, then $\langle \cos \phi_i \rangle = 0$ and Eq. 46 reduces to

$$\langle \mathbf{T}_i \rangle = \begin{bmatrix} \cos \theta_i & \sin \theta_i & 0 \\ 0 & 0 & 0 \\ 0 & 0 & 0 \end{bmatrix} \tag{47}$$

The following derivations are addressed to simple chains comprising $n + 1$ identical skeletal atoms joined by n bonds which are likewise identical apart from their sequential order.* Hence, the subscript i may be dropped. Substitution of Eq. 47 into Eqs. 43 and 45 then yields Eqs. 20 and 22, respectively, for freely rotating chains. This reduction is easily established by observing that according to Eq. 47 the 1,1 element of $\langle \mathbf{T} \rangle^k$ is $(\cos \theta)^k = \alpha^k$.

Substitution of Eq. 46 for $\langle \mathbf{T} \rangle$ into Eq. 44 leads directly to the important result first derived by Oka[13] (see also refs. 14 and 15)

$$\begin{aligned} C_\infty &= \left(\frac{1 + \cos \theta}{1 - \cos \theta}\right)\left(\frac{1 + \langle \cos \phi \rangle}{1 - \langle \cos \phi \rangle}\right) \\ &= \left(\frac{1 + \alpha}{1 - \alpha}\right)\left(\frac{1 + \eta}{1 - \eta}\right) \end{aligned} \tag{48}$$

where $\alpha = \cos \theta$ as before, and $\eta = \langle \cos \phi \rangle$. This is the general expression for the characteristic ratio for a symmetric chain of large n and comprising identical bonds subject to independent rotational potentials. For free rotation, $\eta = 0$ and Eq. 48 reduces to Eq. 21.

Resolution of the complete expression, Eq. 43, for chains of finite length in terms of the scalar parameters is somewhat more involved.[11,15] The task may be carried out by first diagonalizing $\langle \mathbf{T} \rangle$ as given by Eq. 46 (see Appendix B). To this end let $\mathbf{B}\langle \mathbf{T} \rangle \mathbf{A} = \mathbf{\Lambda}$, where $\mathbf{\Lambda} = \mathrm{diag}(\lambda_1, \lambda_2, \lambda_3)$ is the diagonal matrix of the eigenvalues λ_k of $\langle \mathbf{T} \rangle$, and $\mathbf{A} = \mathbf{B}^{-1}$ is the matrix of the eigenvectors of $\langle \mathbf{T} \rangle$. The eigenvalues (see Appendix B) obtained from the secular equation of $\langle \mathbf{T} \rangle$ are [13,16]

$$\begin{aligned} \lambda_{1,2} &= \tfrac{1}{2}[\alpha(1 - \eta) \pm \sqrt{\alpha^2(1 - \eta)^2 + 4\eta}] \\ \lambda_3 &= -\eta \end{aligned} \tag{49}$$

* The resultant of bond $i + 1$ averaged over all angles of rotation consists not only of its projection on bond i, as in the case of free rotation, but also includes a component perpendicular to bond i which lies in the plane of bonds $i - 1$ and i. If, as here assumed, the rotational potential is symmetric, then the average component perpendicular to this plane will be null. The correlation (see Sect. 8) imposed by $E_i(\phi_i)$ on the direction of bond $i + 1$ with bond $i - 1$ may be understood in these terms.

Substitution of $\langle \mathbf{T} \rangle = \mathbf{A\Lambda B}$ in Eq. 43 and the usual cancellations from $\mathbf{AB} = \mathbf{E}$ lead to

$$C_n = \{\mathbf{A}[(\mathbf{E} + \mathbf{\Lambda})(\mathbf{E} - \mathbf{\Lambda})^{-1} - (2/n)\mathbf{\Lambda}(\mathbf{E} - \mathbf{\Lambda}^n)(\mathbf{E} - \mathbf{\Lambda})^{-2}]\mathbf{B}\}_{11} \quad (50)$$

The 1,1 element may be abstracted by premultiplying with [1 0 0] and postmultiplying with the corresponding column. Execution of these operations yields

$$C_n = \sum_{k=1,2,3} A_{1k} B_{k1} \left[\frac{1 + \lambda_k}{1 - \lambda_k} - \frac{2\lambda_k(1 - \lambda_k^n)}{n(1 - \lambda_k)^2} \right] \quad (51)$$

where A_{1k} and B_{k1} are elements of the matrices **A** and **B**.

From **A** and its inverse **B** (see Appendix B) we find

$$\begin{aligned} A_{11} B_{11} &= (\alpha - \lambda_2)/(\lambda_1 - \lambda_2) = (\alpha\eta + \lambda_1)/(\lambda_1 - \lambda_2) \\ A_{12} B_{21} &= (\alpha - \lambda_1)/(\lambda_2 - \lambda_1) = (\alpha\eta + \lambda_2)/(\lambda_2 - \lambda_1) \\ A_{13} B_{31} &= 0 \end{aligned} \quad (52)$$

It will be apparent that the sum over the first term in the brackets in Eq. 51 should represent the limiting result for $n = \infty$. Substitution of Eq. 52 in this part of the sum, followed by suitable rearrangements, yields Eq. 48 as required. The final result of substitution of Eq. 52 into Eq. 51 may therefore be written

$$C_n = \left(\frac{1 + \alpha}{1 - \alpha}\right)\left(\frac{1 + \eta}{1 - \eta}\right) - \left(\frac{\alpha\eta + \lambda_1}{\lambda_1 - \lambda_2}\right)P_1 + \left(\frac{\alpha\eta + \lambda_2}{\lambda_1 - \lambda_2}\right)P_2 \quad (53)$$

where

$$P_k = (2\lambda_k/n)(1 - \lambda_k^n)/(1 - \lambda_k)^2, \qquad k = 1, 2 \quad (54)$$

Equations equivalent to Eq. 53 were derived by Benoit[11,15] and also by Volkenstein and Ptitsyn.[16,17] The resemblance of these equations to Eq. 20 is apparent; P_k duplicates the second term of Eq. 20, with α replaced by λ_k. For free rotation, $\eta = 0$, $\lambda_1 = \alpha$, $\lambda_2 = 0$, $P_2 = 0$, and Eq. 53 reduces to Eq. 20.

Similar treatment of the radius of gyration leads to[11]

$$\langle s^2 \rangle_0 / nl^2 = \left(\frac{1}{6}\right)\left(\frac{n + 2}{n + 1}\right)\left(\frac{1 + \alpha}{1 - \alpha}\right)\left(\frac{1 + \eta}{1 - \eta}\right) - \left(\frac{\alpha\eta + \lambda_1}{\lambda_1 - \lambda_2}\right)Q_1 + \left(\frac{\alpha\eta + \lambda_2}{\lambda_1 - \lambda_2}\right)Q_2 \quad (55)$$

where

$$Q_k = \frac{\lambda_k}{(n + 1)(1 - \lambda_k)^2} - \frac{2\lambda_k^2}{(n + 1)^2(1 - \lambda_k)^3} + \frac{2\lambda_k^3(1 - \lambda_k^n)}{n(n + 1)^2(1 - \lambda_k)^4} \quad (56)$$

A correspondence to Eq. 22 for the case of free rotation is again apparent, with λ_k in Q_k assuming the role of α in the last three terms of Eq. 22. For large n

$$(\langle s^2\rangle_0/nl^2)_\infty = \left(\frac{1}{6}\right)\left(\frac{1+\alpha}{1-\alpha}\right)\left(\frac{1+\eta}{1-\eta}\right) \tag{57}$$

which reaffirms the general validity of Eq. 17.

Equations 49 and 53–56 permit exact calculations of $\langle r^2\rangle_0$ and $\langle s^2\rangle_0$, for chains of any length comprising identical bonds subject to mutually independent rotational potentials which are symmetrical with respect to $\phi = 0$. The behavior of $\langle r^2\rangle_0/nl^2$ and $\langle s^2\rangle_0/nl^2$ with n thus calculated will be examined in Chapter VIII in conjunction with analogous relationships for chains having interdependent rotational potentials.

If the chain skeleton is constructed from two or more kinds of atoms or groups, then bond angles and bond rotational potentials may be expected to differ correspondingly. These differences will manifest themselves in the transformation matrices $\langle \mathbf{T}_i\rangle$ for the various pairs of bonds occurring in the chain. The summation procedures used above for chains of bonds represented by identical averaged transformation matrices $\langle \mathbf{T}\rangle$ are not readily adaptable in general to chains comprising a plurality of bond types. If, however, the chain is described by a periodically repeating unit, as, for example, in

$$(\text{—CH}_2\text{—CH}_2\text{—O—})_x$$

or

$$[\text{—(CH}_2)_5\text{CO—NH—}]_x$$

then the foregoing methods are applicable in principle, although at the expense of considerable elaboration in practice. In the former example, three $\langle \mathbf{T}\rangle$ matrices are required, and they occur in cyclic order in the transformations required to bring $\mathbf{l}_j$ into the coordinate system of $\mathbf{l}_i$. Thus, a term $\langle \mathbf{l}_i \cdot \mathbf{l}_j\rangle$ involves the product of the repeating triad of matrices $\langle \mathbf{T}\rangle$ raised to a power determined by the integral number of intervening units; additional factors $\langle \mathbf{T}\rangle$ are required as determined by the additional bonds which complete the sequence i to j. The summations of terms $\langle \mathbf{l}_i \cdot \mathbf{l}_j\rangle$ must then be carried out separately for every class of bond pairs from the several types present. Difficulties mount enormously if the chain possesses no regularly repeating sequence of bonds of the two or more kinds present. The present procedures are then ineffectual.

Methods developed later for the express purpose of taking account of the neighbor dependence of rotational potentials turn out to be easily adapted to the treatment of chains comprising a plurality of bond types.

Even chains devoid of cyclically repeating order fall within the scope of these methods. We therefore postpone further analysis of more complicated chain structures pending their treatment in better approximation.

REFERENCES

1. J. C. Maxwell, *Phil. Mag.*, **[4]19**, 19 (1860).
2. See E. A. Guggenheim, *Elements of the Kinetic Theory of Gases*, Pergamon Press, New York, 1960.
3. J. C. Maxwell, *Phil. Trans. Roy. Soc. (London)*, **157**, 49 (1867).
4. A. Isihara, *J. Phys. Soc. Japan*, **5**, 201 (1950); *J. Polymer Sci.*, **8**, 573 (1952); P. Debye and F. Bueche, *J. Chem. Phys.*, **20**, 1337 (1952).
5. P. Debye, *J. Chem. Phys.*, **14**, 636 (1946).
6. W. Kuhn, *Kolloid-Z.*, **76**, 258 (1936); **87**, 3 (1939).
7. H. Eyring, *Phys. Rev.*, **39**, 746 (1932).
8. F. T. Wall, *J. Chem. Phys.*, **11**, 67 (1943).
9. W. Kuhn, *Kolloid-Z.*, **68**, 2 (1934).
10. R. A. Sack, *Nature*, **171**, 310 (1953).
11. H. Benoit and P. M. Doty, *J. Phys. Chem.*, **57**, 958 (1953).
12. P. J. Flory, L. Mandelkern, J. B. Kinsinger, and W. B. Shultz, *J. Am. Chem. Soc.*, **74**, 3364 (1952).
13. S. Oka, *Proc. Phys. Math. Soc. Japan*, **24**, 657 (1942).
14. W. J. Taylor, *J. Chem. Phys.*, **15**, 412 (1947); **16**, 257 (1948); H. Kuhn, *J. Chem. Phys.*, **15**, 843 (1947).
15. H. Benoit, *J. Chim. Phys.*, **44**, 18 (1947).
16. M. Volkenstein, *Configurational Statistics of Polymeric Chains* (translated from the Russian ed., S. N. Timasheff and M. J. Timasheff), Interscience, New York, 1963, pp. 184 *et seq.*
17. M. V. Volkenstein and O. B. Ptitsyn, *Dokl. Akad. Nauk SSSR*, **78**, 657 (1951); *Zh. Fiz. Khim.*, **26**, 1061 (1952).

CHAPTER II

Random-Coil Configurations and Their Experimental Characterization

1. THE PREFERRED CONFORMATION AND THE RANDOM COIL

Out of the multitude of configurations realizable through execution of rotations about the various bonds of the skeleton of a long-chain molecule, one might wish to ascertain which one of them would be the most stable. This would entail a mapping of the energy as a function of the manifold of variables required to specify the configuration, and then a search for the combination of these variables which minimizes the energy. What at first sight may appear to be an overwhelming task will, on further reflection, yield to simplification owing to the circumstance that the preferred configuration for a chain consisting of identical repeating units will, with rare exceptions, be one in which all repeating units adopt the same conformation. The resulting configuration for the chain as a whole will consequently conform to a simple geometric pattern, e.g., the planar zigzag or one or another of various helical forms. Thus, the preferred conformation will be a symmetrical one by virtual necessity. The task of exploring all configurations of the chain as a whole therefore ordinarily reduces to the analysis of the conformations of a single unit, with due account being taken of its relationship with neighboring units in the chain when all of them are required to adopt the same conformation.

A priori deduction of the most stable form would require adequate knowledge of the intramolecular energy of a unit in a long chain as a function of the conformation. The needed bond rotational potentials, being strongly affected by interactions between atoms and groups of neighboring units, elude reliable, quantitative formulation (see Chap. V). Inspection of models, preferably supplemented by semiempirical calculations, may afford certain inferences concerning allowed conformations, but unambiguous deduction of the most stable one is seldom possible from existing knowledge of intramolecular energies associated with bond rotations.

The procedure may be reversed to advantage if the conformation of the chain is known in its stable crystalline form. If the chains occur in parallel array in the crystal, as is almost invariably the case (globular proteins excepted), the configuration of the chain is necessarily a symmetrical one, with equivalent units in equivalent conformations. Moreover, this conformation will usually approximate the one of greatest stability for the free molecule when it is dispersed in a solvent and consequently freed of interactions with neighbors in the crystal. Thus, knowledge of the conformation in the crystal affords important clues concerning the most stable form and the intramolecular potentials affecting bond rotations.

At the opposite extreme we have the random coil configuration, or rather the array of configurations referred to collectively by this term. Differences of energy between alternative rotational conformations about a given bond usually are of the order of RT. Hence, in absence of the constraints imposed by neighboring molecules in an ordered crystalline array, departures from the preferred conformation will be abundant. The configuration of the molecule when dispersed in a dilute solution will consequently be devoid of discernible geometric form and, superficially at least, will bear scant resemblance to the preferred conformation.* Likewise, the configurations of a system of chain molecules at high density, as in a concentrated solution or in the amorphous (noncrystalline) state, will include a great diversity of irregular forms.

These spatial configurations, which have no recurrent pattern of replication of bond conformations along the chain, constitute the random-coil configurations. They comprise the overwhelming majority of all possible configurations of the long-chain molecule. In fact, from a statistical viewpoint the random-coil category may appropriately be considered to include *all* configurations, even including the select few which can be classified as regular conformations, one of which is the preferred form. Each of the vast number of configurations of the free molecule usually has a very small probability of occurrence relative to all others combined. Hence, in the absence of special circumstances (such as the existence of crystalline organization, or strong neighbor correlations between successive units of the chain which confer exceptional stability on a preferred helical form; see Chap. VII), the preferred configuration contributes negligibly to the total.

We shall mainly be concerned with chain molecules under conditions in which they are free of external constraints that would affect their configurations. Hence, the random coil is the focus of our attention throughout most of the chapters which follow. In particular, we shall treat the average properties of free chain molecules. These averages represent the

* Exceptions occur among certain biological macromolecules and stereoregular synthetic polypeptides, as we shall note later.

complete array of configurations which collectively comprise the random coil.*

2. INTERACTIONS OF LONG RANGE: THE EFFECT OF EXCLUDED VOLUME

The analogy between the configuration of a chain molecule and a random walk, or flight, consisting of steps of fixed length was pointed out in Section 5 of the preceding chapter. Inasmuch as the direction of each bond is subject to partial correlations with the directions of its neighbors in the chain sequence, a *semirandom* flight would perhaps be a more apt designation for the analog of a molecular chain. The correlations imposed by the bond angles and the potentials hindering rotations about bonds are, in any event, of limited range (see p. 12). As the treatment of various model chains considered in Chapter I makes clear, the correlation between the directions of two bonds depends on the number of intervening bonds. Thus, for any chain having a finite degree of conformational flexibility, $\langle \mathbf{l}_i \cdot \mathbf{l}_j \rangle$ diminishes with $|j-i|$, approaching zero as $|j-i|$ increases without limit. In fact, if the correlations between a pair of bonds were solely determined by the factors considered thus far, namely, bond angles and bond rotational potentials, then $\langle \mathbf{l}_i \cdot \mathbf{l}_j \rangle$ would vanish exponentially with $|j-i|$ for large values of this difference.

Polymer chain configurations depart in a most important respect from their random flight analogs (as defined, for example, by Eq. I-18). Whereas a random flight may cross its own path, a chain molecule is obviously forbidden from doing so. To be sure, such intersections in the trajectory of the random flight analog to a polymer chain molecule are of rare occurrence in the sense that the likelihood for a given step in the random flight to intersect another is very small. Yet, the total number of such events in a flight of very many steps may exceed unity considerably, on the average. Out of the total number of random flight configurations, therefore, only a very small fraction will be altogether free of self-intersections and hence acceptable configurations for a real chain molecule. As a consequence of retention of the latter configurations to the exclusion of all others, the average spatial configuration of the real molecule is perturbed relative to its random flight analog. Average dimensions of the chain (e.g., $\langle r^2 \rangle$ and $\langle s^2 \rangle$) are increased. It is possible to show that the contribution to the sum

* The term *random coil* is sometimes taken to signify compliance with a particular model, e.g., with the freely jointed chain, or with the freely rotating chain (see Chap. I). The term as here defined agrees with its wider usage. The concept it is intended to convey is relevant to real chains, and free of the connotations of hypothetical models.

in Eq. I-9 from terms $\langle \mathbf{l}_i \cdot \mathbf{l}_j \rangle$ for large $|j-i|$ no longer vanishes.* That is to say, correlations of long range exist. In fact, it is the interactions between pairs of units which are far apart in sequence that are mainly responsible for the perturbation of the chain dimensions.

The problem of the spatial configuration of a long chain molecule separates quite naturally into two parts, each fairly distinct from the other. One of them has to do with the bond structure and local interactions between atoms and groups which are near neighbors in sequence along the chain. We call these the *short-range* interactions. The other part of the problem pertains to the interactions of *long range*, i.e, to interactions involving pairs of units which are remote in the chain sequence, though, of course, near to one another in space when involved in mutual interaction. In order to give substance to these concepts, we let $\langle r^2 \rangle_0$ (introduced provisionally in Chap. I) denote the mean square of the end-to-end distance which would obtain in the absence of perturbation of the distribution by the long-range effects. By definition $\langle r^2 \rangle_0$ is determined solely by interactions of short range. Let α be the factor by which a linear dimension of the average configuration is altered as a result of the long-range effects; i.e., α represents the linear perturbation attributable to the so-called excluded volume effect. If $\langle r^2 \rangle$ is the actual value of the second moment of r, then

$$\langle r^2 \rangle = \alpha^2 \langle r^2 \rangle_0 \tag{1}$$

according to these definitions. The separation of the two effects in this manner is, at this stage, only a formal one. Conceptually, at least, the distinction is substantially free of ambiguity. The short-range effects are determined overwhelmingly by interactions between groups separated by only a few bonds. The long-range "volume effects" are dominated by interactions between pairs which are separated by many bonds. It turns out that long-range interactions between pairs separated by numbers of bonds commensurate with the total chain length n assume dominant importance.

The unperturbed average dimension of the chain as expressed by $\langle r^2 \rangle_0$ is susceptible to calculation from structural data in conjunction with estimates of the potentials affecting bond rotations—a task which occupies much of what follows. But the quantity determined experimentally is $\langle r^2 \rangle$ (or the associated quantity $\langle s^2 \rangle$). The foundation of a

* The magnitudes of individual terms $\langle \mathbf{l}_i \cdot \mathbf{l}_j \rangle$ vanish with increase in $|j-i|$, but their combined contribution to the sum in Eq. I-9 does not do so when the chain is perturbed by excluded volume interactions. The decrease of the term value with $|j-i|$ is slowed by the excluded volume effect. Instead of being asymptotically exponential in $-|j-i|$, these terms decrease *approximately* as $(|j-i|)^{-1+\varepsilon}$ for very large differences $|j-i|$, where $0 < \varepsilon < \frac{1}{5}$.

common basis for comparing theoretical calculations with experimental observations is our immediate concern at this point.

The long-range effect (or volume effect) depends not only on the actual volume of the chain unit but also on its interaction with the solvent. Thus, as was emphasized when the significance of the excluded volume effect was first comprehended,[1] it is the effective covolume for a pair of units immersed in the given solvent, and not the actual volume of the chain unit, which determines the perturbation of the configuration. Consequently, α^2 depends markedly on the solvent, and, incidentally, also on the temperature. This circumstance might appear to complicate matters further. It certainly eliminates all prospect for evaluating α by calculations carried out on the basis of the sizes of the structural units comprising the chain.

The situation is by no means as unfavorable as the foregoing remarks may suggest. Just as the covolume for the chain unit can be enhanced by use of a good solvent for the polymer, it may also be diminished by choice of a poor one barely capable of dissolving the polymer. Through judicious selection of solvent and temperature, the finite volume of the unit can be compensated exactly by the mutual attractions between chain units when immersed in the poor solvent. The covolume is then zero and the excluded volume effect vanishes[1]; α equals unity, and $\langle r^2 \rangle = \langle r^2 \rangle_0$.

This device finds an exact parallel in the Boyle point for a real gas. At the Boyle temperature the repulsion (volume exclusion) between a pair of molecules is exactly compensated by their mutual attraction. The excluded volume integral for the pair is then zero, and the real gas obeys the ideal gas isotherm up to fairly high pressures. The corresponding "Theta (Θ) point" for a polymer solution, in further analogy to the Boyle point for a gas, can be identified as the temperature at which the second coefficient vanishes in the virial expansion of the osmotic pressure; i.e., at the Theta point the osmotic pressure of the solution obeys the law of van't Hoff up to concentrations of several per cent. Other experimental means are available also for accurately locating the Θ point.[2,3]

The vanishing of the covolume (i.e., the excluded volume) for a pair of chain units at the Θ point assures that the perturbation of the configuration must likewise vanish. Theory is unequivocal on this score.[3] The complications which would otherwise enter as a consequence of interactions between chain units of "long range" in sequence along the chain thus may be avoided altogether by suitable choice of the conditions under which experiments to determine the chain dimensions are conducted. The validity of this procedure has received abundant confirmation since its inception.

Although direct determination of chain dimensions at the Θ point is generally preferable, it is sometimes impractical to conduct the necessary experiments under these conditions. Measurements carried out in good solvents, where the covolume is large and the configuration is perturbed by long-range interactions, may be corrected to "ideal," or Θ-point, conditions by use of proper procedures prescribed by valid theory. Experimental results determined at the Θ point, or appropriately corrected thereto, will be designated by a subscript zero as in $\langle r^2 \rangle_0$ or $\langle s^2 \rangle_0$ in accord with the conventions introduced above.

The expansion of the configuration of a large, linear macromolecule owing to long-range interactions is peculiar to the environment afforded by a dilute solution. When the concentration is increased to the range where the domains of the random coils overlap copiously, the expansion should be largely suppressed. Thus, in the bulk amorphous state perturbation of the configuration may be predicted to vanish. These assertions follow from considerations of a theoretical nature which are at once simple and virtually incontrovertible.[1,3] They are not readily susceptible to experimental verification, however, owing to difficulties attending determination of molecular dimensions at high concentrations.[4–6]

Whereas the excluded volume perturbation of the chain configuration is indicated by theory to be ineffectual over a wide range of concentration, it becomes maximal at high dilutions in good solvents. The latter are the conditions used for most of the conventional physical chemical measurements applicable to polymers. The excluded volume effect has assumed importance on this account in the investigation of macromolecules.

In summary, the interpretation of the configuration of a linear polymeric chain molecule dispersed in a dilute solution can be resolved into short-range and long-range parts. The former is determined by the geometrical parameters l and θ introduced in Chapter I, together with the potentials affecting rotations about bonds, including the effects of steric interactions between atoms and groups which are near neighbors in the chain sequence. The latter is dominated by interactions involving units which are remote from one another in the chain sequence. The long-range effect can be eliminated on the experimental side of the ledger by the procedures cited above. The unperturbed dimensions thus obtained may then be interpreted in terms of the short-range features of the chain molecule in question. This is the strategy which we shall pursue.

With these few general remarks we leave the subject of the excluded volume effect, a detailed treatment being outside the scope of the present volume.

3. EXPERIMENTAL DETERMINATION OF THE UNPERTURBED DIMENSIONS OF CHAIN MOLECULES

Two principal methods are currently in use for determining the dimensions of macromolecules. Both are operative in very dilute solutions only.

One depends upon determination of the angular dependence of light scattered by the dispersed macromolecules. The other involves the determination of the intrinsic viscosity $[\eta]$ defined by

$$[\eta] = \lim_{c \to 0} [(\eta_{\mathrm{rel}} - 1)/c] \tag{2}$$

Here η_{rel} is the relative viscosity, or ratio of the viscosity of the solution to that of the pure solvent, and c is the concentration expressed in weight per unit volume; we shall use g deciliter^{-1}.

In the light scattering dissymmetry method the intensity of scattered radiation is measured as a function of the concentration and of the angle between the incident beam and the direction of observation. The scattering intensity for the pure solvent is deducted from that for the solution. Other corrections depend on the geometry and optics of the apparatus, and on the polarization of the scattered light. The ratio of the concentration c to the corrected intensity $I(\vartheta)$ of light scattered at angle ϑ is customarily extrapolated to infinite dilution to obtain $[c/I(\vartheta)]_{c=0}$. These limiting ratios for various angles are then plotted against $\sin^2(\vartheta/2)$. The slope of this plot at $\vartheta = 0$ is directly proportional to the mean-square radius of gyration $\langle s^2 \rangle$ (see Chap. IX). The mean-square end-to-end length $\langle r^2 \rangle$ usually can be calculated from $\langle s^2 \rangle$. For very long chains in the unperturbed state, for example, we have $\langle r^2 \rangle_0 = 6\langle s^2 \rangle_0$ according to Eq. I-17.

If the sample comprises species covering a wide range of molecular weights, which is often the case, determination of the limiting slope of $c/I(\vartheta)$ with $\sin^2(\vartheta/2)$ may be complicated by the molecular heterogeneity. Even if the extrapolation is successful, care must be exercised in treating the resulting value of $\langle s^2 \rangle$ as the appropriate average over the distribution of contributing molecular species.

The light scattering method is fraught with a number of sources of error: optical artifacts, molecular heterogeneity, and the hazards implicit in the dual extrapolations required with respect to concentration and angle, respectively. If the molecule is optically anisotropic (i.e., if its refractive indices in different directions differ appreciably on the average), the polarization and the intensity of the scattered radiation may be affected to extents which elude simple correction for scattering angles other than $\vartheta = 90°$. Large errors may in some cases occur on this account.[7] With exercise of great pains, however, reliable results can be obtained by the light scattering dissymmetry method.

The viscosity method depends upon the relationship

$$[\eta] = \Phi \langle r^2 \rangle^{3/2} / M \tag{3}$$

suggested a number of years ago[1,3] for random coiled polymers of high molecular weight and since confirmed in a number of instances. Here M

is the molecular weight and Φ is a constant reasoned to be the same for all systems. If the measurements are carried out at the Θ point, or if they are corrected thereto, this expression is appropriately written

$$[\eta]_\Theta = \Phi(\langle r^2 \rangle_0/M)^{3/2} M^{1/2} \tag{4}$$

For chain molecule homologs of a given kind, and of sufficient length ($n >$ ca. 400 bonds; *cf. seq.*), the unperturbed second moment $\langle r^2 \rangle_0$ is proportional to the chain length, and therefore to M. Hence, Eq. 4 may be written

$$[\eta]_\Theta = KM^{1/2} \tag{5}$$

where K is a constant for the given series of homologs. Thus,

$$(\langle r^2 \rangle_0/M)_\infty = (K/\Phi)^{2/3} \tag{6}$$

The subscript ∞ serves as a reminder that the ratio has been assigned its limiting value.

The proportionality of $[\eta]_\Theta$ to $M^{1/2}$ over several decades of M has been confirmed by numerous investigations. To the extent that $(\langle r^2 \rangle_0/M)_\infty$ may depend on the solvent, this dependence will be reflected in the value of K. The effect of the solvent generally is small, however. It is usually legitimate to regard K as being characteristic of the polymer homologous series (e.g., of polymethylene); the solvent which was used is of secondary importance at most (see below).

According to some of the most painstaking light scattering experiments, critically interpreted,[7,8] the value of Φ applicable at or near the Θ point* is ca. $2.5(\pm 0.1) \times 10^{21}$, with r in centimeters and $[\eta]$ in deciliters per gram. Recent theory[9] yields $\Phi = 2.66 \times 10^{21}$. We shall adopt an intermediate value of 2.6×10^{21}, which agrees with the empirical result within limits of experimental error and departs from the theoretical value by an amount which probably does not exceed its residual uncertainty.

The characteristic ratio $C_n = \langle r^2 \rangle_0/nl^2$ introduced in Chapter I is generally preferred over $\langle r^2 \rangle_0/M$ as a basis for comparing the average dimensions of various random-coil chains. It may be calculated from $\langle r^2 \rangle_0/M$, of course. Thus, in the limit for long chains

$$C_\infty = (\langle r^2 \rangle_0/M)_\infty (M_b/l^2) \tag{7}$$

* As hydrodynamic considerations show, the value of Φ must depend on the *form* of the spatial distribution of the units of the chain molecule. The volume effect perturbs not only the average size of the molecule; to a lesser degree it also may distort the form of the distribution. Hence, retention of Eq. 3 for systems removed from the Θ point appears to require a small downward adjustment of Φ owing to perturbation of the form of the spatial distribution. Present considerations are limited to the value of Φ applicable to the unperturbed spatial distribution.

where M_b is the mean molecular weight per skeletal bond. Or

$$C_\infty = (K/\Phi)^{2/3} M_b / l^2 \tag{8}$$

As will be recalled (see Eq. I-13), nl^2 represents the value which $\langle r^2 \rangle_0$ would assume if correlations between bond directions did not exist; i.e., nl^2 is the mean-square end-to-end length for a random flight chain of n bonds, each being of length l. Thus, C_n, and likewise C_∞, express ratios of the actual unperturbed mean square of r to the value it would assume if all correlations between bond directions were abolished, bond lengths being preserved.*

Other hydrodynamic methods are available for determining $\langle r^2 \rangle$ or $\langle s^2 \rangle$. Measurement of either the sedimentation velocity or the diffusion constant for macromolecules in dilute solution yields the translational frictional constant, from which both of the foregoing quantities can be deduced for particles of sufficiently large n. These procedures are more elaborate than viscosity determinations and, for the purpose here in mind, they do not offer compensating advantages.

Experimental values of the characteristic ratio C_∞ for representative polymers are presented in Table 1. They have been deduced from intrinsic viscosities in every case with the exception of poly(isopropyl acrylate), for which results of the light scattering dissymmetry method are relied upon. For all other polymers the value of $K = [\eta]_\Theta / M^{1/2}$ is given in the penultimate column. Most of these results were obtained by direct measurements in Θ solvents; the solvent and temperature (Θ) are specified in the third and fourth columns. Those instances in which viscosities were measured in good solvents and then corrected for perturbation (α^3) by intramolecular interactions of long range are indicated by an asterisk appended to the solvent in the third column.

The specific effect of the solvent on the dimensions of the unperturbed chain is remarkably small. For polymethylene the range for C_∞ is only

* Another quantity sometimes used[10] to characterize the spatial configurations of chain molecules is the ratio $\langle r^2 \rangle_0 / \langle r^2 \rangle_{0,f}$ where $\langle r^2 \rangle_{0,f}$ is the unperturbed average r^2 which would obtain for free rotation. This ratio reflects the effects of hindrances to rotations exclusive of the effects of the fixed bond angles. The characteristic ratio $\langle r^2 \rangle_0 / nl^2$ is subject to less ambiguity and is more readily calculable from suitable experimental results giving $\langle r^2 \rangle_0$. Whereas bond lengths usually are known within narrow limits, bond angles $\pi - \theta$ are often subject to uncertainties of several degrees—enough to affect the calculated value of $\langle r^2 \rangle_{0,f}$ by 10% or more. Thus, according to Eq. I-21, $(\langle r^2 \rangle_{0,f} / nl^2)_\infty = 2.00$ for a simple tetrahedral chain, with $\theta = 70.53°$; if $\theta = 68°$, then $(\langle r^2 \rangle_{0,f} / nl^2)_\infty = 2.20$. Moreover, the calculation of $\langle r^2 \rangle_{0,f}$ is often difficult for chains of complicated structural geometry. Even for simple chains the calculation is a tedius one in the range of finite n (see Eq. I-20), and expression of experimental results as the ratio $\langle r^2 \rangle_0 / \langle r^2 \rangle_{0,f}$ would obscure the dependence of $\langle r^2 \rangle_0$ on chain length.

3% and the data quoted are supported in this respect by results obtained for a number of other Θ solvents for this polymer. (See refs. a, b, and c at the foot of Table 1.) The range is similar for polystyrene, for poly(vinyl acetate), and for isotactic poly(methyl methacrylate). The greater variation for isotactic polypropylene may be due in part to the effect of temperature on the unperturbed dimensions. The lower value of C_∞ for this polymer in diphenyl ether, if confirmed, would indicate a substantial solvent effect. Polydimethylsiloxane appears to manifest appreciably larger unperturbed dimensions in the fluorinated solvent mixture of exceptionally low cohesive energy density.

Inasmuch as specific effects of solvents ordinarily are quite small, it is usually legitimate to regard the value of K and the unperturbed dimension it denotes as being intrinsic characteristics of the polymer molecule at the given temperature. The interactions which determine the conformational energy, and hence the average dimensions of the chain, evidently are fairly insensitive to the solvent medium. There are, to be sure, instances of departures from these generalizations. Cellulose derivatives, not included in Table 1, are notable in this respect. Their average unperturbed dimensions in solution depend markedly on the solvent.[11] Some reported cases of large effects of solvents on the unperturbed sizes of other polymers in solution have been found, upon reexamination, to be spurious.

The characteristic ratios C_∞ for polymers usually are in the range of 4–10. They considerably exceed the values which would obtain for free rotation. Among stereoregular vinyl polymers, C_∞ for the isotactic chain structure is greater than that for its atactic isomer, but by a margin which is surprisingly small. To the extent that syndiotactic chains have been investigated, their dimensions appear to approximate those for their atactic analogs. Rationalization of these results must be postponed pending presentation of theoretical methods suitable for treating chain molecules in a realistic manner in the chapters to follow.

4. TEMPERATURE COEFFICIENTS OF UNPERTURBED DIMENSIONS OF MACROMOLECULES

If reliably determined, the temperature coefficient of the mean-square unperturbed length, i.e., $d \ln \langle r^2 \rangle_0 / dT$, can contribute important information on the chain configuration, and especially on the energetics of the bond conformations. In principle, it could be obtained directly by determining $\langle r^2 \rangle_0$ at several temperatures, using one of the methods cited above. The temperature coefficient seldom exceeds about 1×10^{-3} deg^{-1} in magnitude. If solvents are chosen offering Θ points over a suitable range of temperatures, the effect of temperature on the molecular dimensions is

Table 1

Characteristic Ratios for Representative Polymer Chains at the Limit of High Chain Length

Polymer	Structural unit	Solvent	Temp, °C	$K \times 10^4 = ([\eta]_\Theta/M^{1/2}) \times 10^4$ deciliters g^{-1} (g mol wt)$^{-1/2}$	$C_\infty = (\langle r^2 \rangle_0/nl^2)_\infty$
Polymethylene	$—CH_2—$	Dodecanol-1	138	30.7[a], 31.6[b]	6.7
		Diphenylmethane	142	31.5[a], 32.3[c]	6.8
		α-Cl-Naphthalene*	140	30.5[d]	6.6
Polystyrene, atactic	$—CH_2—CH(C_6H_5)—$	Cyclohexane	34.8	8.2[e], 8.4[f], 8.7[g]	10.2
		$Cl—(CH_2)_{11}—H$	32.8	7.8_6[g]	10.0
		Diethyl malonate	35.9	7.7[g]	9.9
Polypropylene, atactic	$—CH_2—CH(CH_3)—$	α-Cl–Naphthalene	74	18.2[h]	7.0
		Cyclohexane	92	17.2[h]	6.8
		Diphenyl ether	153	12.0[h]	5.3
Polypropylene, isotactic	$—CH_2—CH(CH_3)—$	Diphenyl ether	145	13.2[h]	5.7
Poly(*n*-pentene-1), isotactic	$—CH_2—CH(C_3H_7)—$	Isoamyl acetate	31.5	14.1[i]	10.0
		2-Pentanol	62.4	12.1[j]	9.0
Polyisobutylene	$—CH_2—C(CH_3)_2—$	Benzene	24	10.7[k]	6.6

Polymer	Repeat unit	Solvent	Temp. (°C)		
Poly(vinyl acetate), atactic	$-CH_2-CH(OCOCH_3)-$	Ethyl *n*-butyl ketone	29	9.3[l]	9.2
		i-Pentanone–hexane	25	8.8[l]	8.9
		i-Pentanone–heptane	25	8.8[m]	8.9
Poly(isopropyl acrylate), atactic	$-CH_2-CH(COOC_3H_7)$	Benzene*	25		9.7 ± 0.8[n]†
isotactic	$-CH_2-CH(COOC_3H_7)$	Bromobenzene*	60		7.1 ± 0.6[n]†
syndiotactic	$-CH_2-CH(COOC_3H_7)$	Bromobenzene*	60		7.2 ± 1.0[n]†
Poly(methyl methacrylate), atactic	$-CH_2-C(CH_3)(COOCH_3)-$	Various solvents	4–70	4.8 ± 0.5[o,p]	6.9 ± 0.5
isotactic	$-CH_2-C(CH_3)(COOCH_3)-$	Acetonitrile	27.6	7.5[q]	9.3
		n-Butyl chloride	26.5	7.7[r]	9.5
		Butanone–isopropanol	25	7.2[r]	9.1
syndiotactic	$-CH_2-C(CH_3)(COOCH_3)-$	*n*-Butyl chloride	35	5.1[r]	7.2
		Butanone–isopropanol	8	4.4[r]	6.5

(*continued*)

Table 1 (*continued*)

Polymer	Structural unit	Solvent	Temp, °C	$K \times 10^4 = ([\eta]_\Theta/M^{1/2}) \times 10^4$ deciliters g^{-1} (g mol wt)$^{-1/2}$	$C_\infty = (\langle r^2\rangle_0/nl^2)_\infty$
Polyoxyethylene	—O—CH_2—CH_2—	Aqueous K_2SO_4	35	11.5 ± 1.5[s]	4.0
		Aqueous $MgSO_4$	45	11.5 ± 1.5[s]	
Poly-dimethylsiloxane	—O—Si(CH_3)(CH_3)—	Butanone	20	7.8[t]	6.2
		$C_8F_{18} + C_2Cl_4F_2$	22.5	10.6[t]	7.6
		Various solvents	2–90	7.0–7.8[u]	5.7–6.2
Polyhexamethylene adipamide	—NH$(CH_2)_6$NH— —CO$(CH_2)_4$CO—	HCOOH, KCl, H_2O	25	19.2[v]	5.9
L-Polypeptides‡	—NHCH(CH_2R')—CO—	Various solvents*	25–100		8.5–9.5[w]‡
Polyphosphate	—O—$\bar{P}O_2$—	Aq LiBr	25	7.5[x] (Li salt)	7.1
		Aq NaBr	25	5.2[y] (Na salt)	6.6

* These results of measurements carried out in the good solvents indicated have been corrected to the Θ point on the basis of the second virial coefficient found from either light scattering or osmotic pressure experiments.

† These results for poly(isopropyl acrylate) were obtained from light scattering determinations of $\langle r^2\rangle$ in benzene, with correction to the Θ point.

‡ Stereoregular polypeptides for which R′ = —COOBz, —CH_2COOBz, —CH_2COO^-, and —$(CH_2)_3NH_3^+$. Intrinsic viscosities of the latter two were determined in aqueous solutions; the former in *m*-cresol and in dichloroacetic acid. Results were corrected to

the Θ point. The characteristic ratio is expressed as $(\langle r^2\rangle_0/xl_u^2)_\infty$, where x is the number of peptide units and l_u is the distance (3.80 Å) between α carbons of successive units in planar *trans* form (see Chap. VII).

[a] R. Chiang, *J. Phys. Chem.*, **70**, 2348 (1966). Results for several other Θ solvents are presented and compared in this reference.

[b] C. J. Stacy and R. L. Arnett, *J. Phys. Chem.*, **69**, 3109 (1965).

[c] A. Nakajima, F. Hamada, and S. Hayashi, *J. Polymer Sci. C*, **15**, 285 (1966).

[d] R. Chiang, *J. Phys. Chem.*, **69**, 1645 (1965); P. J. Flory, A. Ciferri, and R. Chiang, *J. Am. Chem. Soc.*, **83**, 1023 (1961).

[e] W. R. Krigbaum and P. J. Flory, *J. Polymer Sci.*, **11**, 37 (1953).

T. Altares, D. P. Wyman, and V. R. Allen, *J. Polymer Sci. A*, **2**, 4533 (1964).

[g] T. A. Orofino and J. W. Mickey, Jr., *J. Chem. Phys.* **38**, 2513 (1963).

[h] J. B. Kinsinger and R. E. Hughes, *J. Chem. Phys.*, **67**, 1922 (1963). Somewhat lower values have been found by H. Inagaki, T. Miyamoto and S. Ohta, *ibid.*, **70**, 3420 (1966).

[i] G. Moraglio and J. Brzezinski, *J. Polymer Sci. B*, **2**, 1105 (1964).

[j] J. E. Mark and P. J. Flory, *J. Am. Chem Soc.*, **87**, 1423 (1965).

[k] T. G Fox, Jr., and P. J. Flory, *J. Am. Chem. Soc.*, **73**, 1909 (1951); see also ref. e.

[l] M. Matsumoto and Y. Ohyanagi, *J. Polymer Sci.*, **50**, S1 (1961).

[m] A. R. Shultz, *J. Am . Chem. Soc.*, **76**, 3422 (1954).

[n] J. E. Mark, R. A. Wessling, and R. E. Hughes, *J. Phys. Chem.*, **70**, 1895, 1903, 1909 (1966).

[o] T. G Fox, *Polymer*, **3**, 111 (1962).

[p] G. V. Schulz and R. Kirste, *Z. Physik. Chem.* (*Frankfurt*), **30**, 171 (1961).

[q] S. Krause and E. Cohn-Ginsberg, *J. Phys. Chem.*, **67**, 1479 (1963).

[r] G. V. Schulz, W. Wunderlich, and R. Kirste, *Makromol. Chem.*, **75**, 22 (1964).

[s] F. E. Bailey, Jr., and R. W. Callard, *J. Appl. Polymer Sci.*, **1**, 56 (1959); see also J. E. Mark and P. J. Flory, *J. Am. Chem. Soc.*, **87**, 1415 (1965).

[t] V. Crescenzi and P. J. Flory, *J. Am. Chem. Soc.*, **86**, 141 (1964).

[u] G. V. Schulz and A. Haug, *Z. Physik. Chem.* (*Frankfurt*), **34**, 328 (1962).

[v] P. R. Saunders, *J. Polymer Sci. A*, **2**, 3765 (1964).

[w] D. A. Brant and P. J. Flory, *J. Am. Chem. Soc.*, **87**, 2788 (1965).

[x] U. P. Strauss and P. Ander, *J. Phys. Chem.*, **66**, 2235 (1962).

[y] U. P. Strauss and P. L. Wineman, *J. Am. Chem. Soc.*, **80**, 2366 (1958).

likely to be masked by specific effects of the solvents, even though these may be quite small, as pointed out above. Errors incident on the light scattering method are intolerably large for this purpose. Reliable values of $d \ln \langle r^2 \rangle_0 / dT$ may be obtained however by determining the temperature coefficient of the intrinsic viscosity in an athermal solvent (not at a Θ point) and applying appropriate corrections for molecular expansion by the effect of excluded volume.[12] Alternatively, values of $[\eta]_\Theta$ may be compared for a series of chemically similar solvents in which specific solvent effects may be presumed to be the same, and which offer a suitable range of Θ temperatures.[13]

A method which has proved particularly effective for the determination of $d \ln \langle r^2 \rangle_0 / dT$ involves measurement of the stress–temperature coefficient for the amorphous polymer converted to a network by cross-linking and maintained at fixed strain. The most convenient type of strain for the purpose is simple elongation. This method rests on the connection between $\langle r^2 \rangle_0^{-1}$ and the average force exerted by the ends of a polymer chain held at fixed length. The dependence of the force on $\langle r^2 \rangle_0$ was intimated in the early work of Bresler and Frenkel,[14] and later elaborated by Volkenstein and Ptitsyn.[15] On this basis, it is readily shown[15–17] that

$$d \ln \langle r^2 \rangle_0 / dT = -[\partial \ln (f/T)/\partial T]_{L,V} \tag{9}$$

where f is the tensile force exerted by a sample maintained at length L.

Determination of the tension–temperature coefficient at constant volume presents obvious experimental difficulties. Experiments conducted at constant pressure can be used by resorting to the following relationship derived[16–18] from the theory of rubber elasticity for a network of Gaussian chains

$$d \ln \langle r^2 \rangle_0 / dT = -[\partial \ln (f/T)/\partial T]_{p,L} - \beta_L(\alpha^3 - 1)^{-1} \tag{10}$$

where α is the relative extension (related to, but to be distinguished from, the α used earlier in this chapter), and $\beta_L = (\partial \ln V/\partial T)_p$ is the coefficient of thermal expansion of the sample.*

Experimental values of $d \ln \langle r^2 \rangle_0 / dT$ are given in Table 2 for various polymers investigated with due regard for the sources of error attending the measurements (see also ref. 20). Dilution by swelling has no discernible effect on the results of the stress–temperature method, indicated by "df/dT" in column two. The dilute-solution-viscosity method yields results in close agreement with those obtained from stress–temperature measure-

* Failure to take account of the controversial C_2 term in the expression of the stored energy used to derive Eq. 10 has been held by Krigbaum and Roe[19] to invalidate this equation. A proper analysis, however, will show that this deliberate disregard of C_2 can scarcely have an effect which is significant.

Table 2

Temperature Coefficients of Unperturbed Chain Dimensions

Polymer	Method	Temp range, °C	$10^3\, d \ln \langle r^2 \rangle_0 / dT$, deg^{-1}
Polymethylene	df/dT	140–190	−1.0 (±0.1)[a]
	df/dT, swollen with $n\text{-}C_{32}H_{66}$	120–170	-1.1_5 (±0.1)[a]
	df/dT, swollen with DEHA*	130–180	−1.0 (±0.2)[a]
	$d[\eta]/dT$ in $n\text{-}C_{16}H_{34}$ and $n\text{-}C_{28}H_{58}$	110–170	−1.2 (±0.2)[b]
Polystyrene,			
atactic	df/dT	120–170	0.3_7[c]
	$[\eta]_\Theta$ in Cl—$(CH_2)_m$—H, with $m = 10, 11, 12$	6.6–58.6	0.4_4[c]
Polyisobutylene	df/dT	20–95	−0.08 (±0.06)[a]
	df/dT, swollen with $n\text{-}C_{16}H_{34}$	20–60	−0.09 (±0.07)[a]
	df/dT	18	−0.27 (±0.12)[d]
	$d[\eta]/dT$ in $n\text{-}C_{16}H_{34}$	30–130	−0.28 (±0.10)[e]
Poly(n-butene-1)			
atactic	df/dT	140–200	0.50 (±0.04)[f]
isotactic	df/dT	140–200	0.09 (±0.07)[f]

(*continued*)

Table 2 (*continued*)

Polymer	Method	Temp range, °C	$10^3\ d \ln \langle r^2 \rangle_0/dT$, deg^{-1}
Poly(*n*-pentene-1),			
atactic	df/dT	40–140	0.53 (±0.05)[f]
isotactic	df/dT	80–140	0.34 (±0.04)[f]
	$d[\eta]/dT$ in $n\text{-}C_{16}H_{34}$	35–90	0.52 (±0.2)[f]
Polydimethylsiloxane	df/dT	40–100	0.78 (±0.06)[g]
	$d[\eta]/dT$ in silicone fluid	30–105	0.71 (±0.13)[g]
Polyoxyethylene	df/dT	30–90	0.23 (±0.02)[h]
Natural rubber	df/dT	−20 to +25	0.41 (±0.04)[i,a]

* DEHA = di(2-ethylhexyl) azelate.

[a] A Ciferri, C. A. J. Hoeve, and P. J. Flory, *J. Am. Chem. Soc.*, **83**, 1015 (1961).

b R. Chiang, *J. Phys. Chem.*, **70**, 2348 (1966).

[b] P. J. Flory, A. Ciferri, and R. Chiang, *J. Am. Chem. Soc.*, **83**, 1023 (1961).

[c] T. A. Orofino and A. Ciferri, *J. Phys. Chem.*, **68**, 3136 (1964).

[d] G. Allen, G. Gee, M. C. Kirkham, C. Price, and J. Padget, Intern. Symp. Macromol. Chem., Tokyo, Kyoto, 1966 (*J. Polymer Sci. C*, **23**, 201 (1968)).

[e] J. E. Mark and G. B. Thomas, *J. Phys. Chem.*, **70**, 3588 (1966).

[f] J. E. Mark and P. J. Flory, *J. Am. Chem. Soc.*, **87**, 1423 (1965).

[g] J. E. Mark and P. J. Flory, *J. Am. Chem. Soc.*, **86**, 138 (1964).

[h] J. E. Mark and P. J. Flory, *J. Am. Chem. Soc.*, **87**, 1415 (1965).

[i] L. A. Wood and F. L. Roth, *J. Appl. Phys.*, **15**, 781 (1944).

ments on cross-linked networks prepared from the same polymer. The temperature coefficients in the last column may be either positive or negative, with positive values being more prevalent. A correlation with chemical structure is not obvious. We shall find later, however, that these temperature coefficients are closely related to the bond structure and rotational potentials associated therewith.

5. OTHER METHODS FOR CHARACTERIZING RANDOM-CHAIN CONFIGURATIONS

The experimental methods briefly indicated above are applicable only to chain molecules of great length. The dimensions of chains of intermediate length, i.e., with $n = 10\text{–}10^2$, are not subject to determination by the standard methods available at present. Low-angle x-ray scattering offers some promise.[21] Use of chain molecules labeled with heavy atoms at both ends renders the method applicable to the bulk polymer.[6,22] If the chains are terminated with suitably reactive functional groups, the study of cyclization equilibria may afford information on the macromolecular configuration, in bulk as well as in solution (see Appendix D).

The dearth of methods for measuring chain dimensions over the entire range of molecular lengths invites consideration of other properties which are configuration sensitive and hence deserve inspection pursuant to the acquisition of an understanding of the statistical mechanics of chain molecules. The dipole moment is such a property.* It is treated in due course and illustrated by applications to appropriate data where available. Optical anisotropy and electric birefringence are additional properties which may contribute to further information bearing significantly on the configurations of chain molecules. They are treated in Chapter IX.

REFERENCES

1. P. J. Flory, *J. Chem. Phys.*, **17**, 303 (1949); T. G Fox, Jr., and P. J. Flory, *J. Phys. Chem.*, **53**, 197 (1949).
2. P. J. Flory and T. G Fox, Jr., *J. Am. Chem. Soc.*, **73**, 1904 (1951); T. G Fox, Jr., and P. J. Flory, *ibid.*, **73**, 1909, 1915 (1951).
3. P. J. Flory, *Principles of Polymer Chemistry*, Cornell University Press, Ithaca, 1953, pp. 546–554, 600–602.
4. P. J. Flory, *Lectures in Materials Science*, P. Leurgans, Ed., Benjamin, New York, 1963, pp. 36–42.
5. C. A. J. Hoeve and M. K. O'Brien, *J. Polymer Sci. A*, **1**, 1947 (1963).

* For a summary of results on dipole moments of macromolecules, see Birshtein and Ptitsyn, reference 20, Chapter I.

6. W. R. Krigbaum and R. W. Godwin, *J. Chem. Phys.*, **43**, 4523 (1965).
7. D. McIntyre, A. Wims, L. C. Williams, and L. Mandelkern, *J. Phys. Chem.*, **66,** 1932 (1962).
8. G. C. Berry, *J. Chem. Phys.*, **46,** 1338 (1967).
9. C. W. Pyun and M. Fixman, *J. Chem. Phys.*, **44,** 2107 (1966).
10. M. Kurata and W. H. Stockmayer, *Fortschr. Hochpolymer Forsch.*, **3,** 196 (1963).
11. P. J. Flory, O. K. Spurr, Jr., and D. K. Carpenter, *J. Polymer Sci.*, **27**, 231 (1958); P. J. Flory, *Makromol. Chem.*, **98**, 128 (1966).
12. P. J. Flory, A. Ciferri, and R. Chiang, *J. Am. Chem. Soc.*, **83**, 1023 (1961).
13. T. A. Orofino and A. Ciferri, *J. Phys. Chem.*, **68**, 3136 (1964).
14. S. E. Bresler and Ya. I. Frenkel, *Zh. Eksperim. i Teor. Fiz.*, **9**, 1094 (1939).
15. M. V. Volkenstein and O. B. Ptitsyn, *Zh. Tekhn. Fiz.*, **25**, 662 (1955); see also M. V. Volkenstein, *Configurational Statistics of Polymeric Chains* (translated from the Russian ed., S. N. Timasheff and M. J. Timasheff), Interscience, New York, 1963, pp. 501–507.
16. P. J. Flory, C. A. J. Hoeve, and A. Ciferri, *J. Polymer Sci.*, **34**, 337 (1959); **45**, 235 (1960).
17. P. J. Flory, *Trans. Faraday Soc.*, **57**, 829 (1961).
18. T. N. Khasanovich, *J. Appl. Phys.*, **30**, 948 (1959).
19. W. R. Krigbaum, and R.-J. Roe, *Rubber Chem. Technol.* **38**, 1039 (1965); R.-J. Roe, *Trans. Faraday Soc.*, **62**, 312 (1966).
20. T. M. Birshtein and O. B. Ptitsyn, *Conformations of Macromolecules* (translated from the 1964 Russian ed., S. N. Timasheff and M. J. Timasheff), Interscience, New York, 1966, Chap. 8.
21. O. Kratky, *Progr. Biophys.*, **13**, 105 (1963); *Kolloid-Z.*, **182**, 7 (1962).
22. G. W. Brady, E. Wasserman, and J. Wellendorf, *J. Chem. Phys.*, **47**, 855 (1967).

CHAPTER III

Configurational Statistics of Chain Molecules with Interdependent Rotational Potentials

General relationships were derived in the final section of Chapter I for chain molecules in which the potentials affecting bond rotations are (*1*) mutually independent and (*2*) symmetric about the planar conformation ($\phi = 0$). Compliance with the latter condition is determined by the structure of the chain; absence of structural asymmetry in the stereochemical sense is a sufficient guarantee of its fulfillment. If the bond rotational potentials are independent (*1*), then the conformational energy of the molecule as a whole may be separated into a sum of energies E_i, one for each bond, after the manner of Eq. I-33, with E_i for bond i a function of rotation angle ϕ_i only. Compliance with both conditions (*1*) and (*2*) assures that $\langle \sin \phi_i \rangle = 0$ for all i. The entire effect of each bond potential is then embodied in $\langle \cos \phi_i \rangle$. Inquiry into the nature of the rotational potential, apart from its symmetry, could therefore be disregarded within the restricted scope of Chapter I; it was sufficient to formulate a bond partition function z_i in the general terms of Eq. I-35.

The primary assumption of independence of bond rotations is seldom acceptable, as we have emphasized earlier. Interdependence of bond rotations is manifested in a dependence, often marked, of E_i on ϕ_{i-1} and ϕ_{i+1}, as well as on ϕ_i. The dependence might conceivably extend to second neighbors and possibly beyond; however, such effects usually can be ignored with confidence (*cf. seq.*). The interdependence of bond rotations, even though limited to first neighbors, precludes separation of the conformational energy of the molecule into a sum of terms as above, and, correspondingly, the molecular partition function cannot be factored into a set of bond partition functions.

The obvious complications of interdependence necessitate compensating simplifications. This will be achieved by the device of artificially replacing the continuous range of ϕ by several discrete values, judiciously chosen. In effect, the integrals occurring in Eq. I-30 will be replaced by sums over discrete values of the arguments. Physical reality suffers little compromise by reliance on this device, which enables adoption of extremely versatile mathematical methods. An appropriate set of rotational "states" must be chosen, and to this end examination of the character of the potentials hindering rotation about skeletal bonds, or their analogs in simple compounds, becomes necessary. We therefore consider bond rotation potentials at the outset, before proceeding with the main task of the present chapter, which is to formulate methods for treating one-dimensional, statistical mechanical systems of neighbor-dependent elements. Linear macromolecules are unique representatives of such systems.

1. BOND ROTATIONAL POTENTIALS FOR SIMPLE MOLECULES

Information bearing on the nature of bond rotational potentials has been made available largely by spectroscopic studies. Early investigations relied principally on Raman and infrared spectroscopy,[1] and on comparisons of thermodynamic properties of molecules such as ethane and its higher homologs with results calculated from structural data and from normal vibration frequencies, including torsional modes.[2,3] More recently, microwave spectroscopy applied to simple compounds in the gaseous state has produced the most definitive information.[4,5] Electron diffraction[6,7] has provided valuable supplementary evidence bearing on rotational potentials in the lower *n*-alkanes.

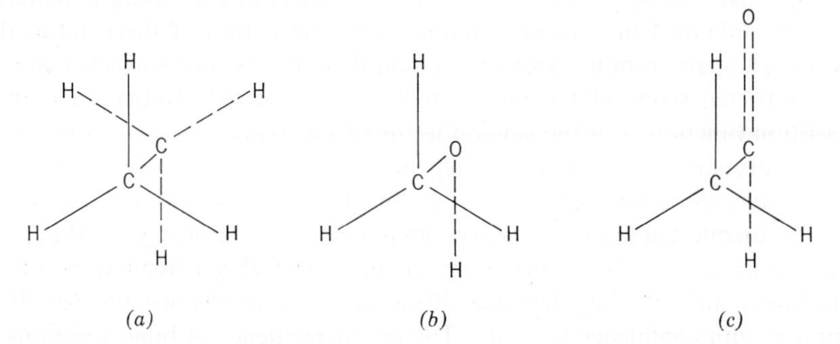

Fig. 1. Conformations of minimum energy for (*a*) ethane, (*b*) methanol, and (*c*) acetaldehyde.

Rotation potentials for simple molecules like H_3C—CH_3, H_3C—OH, H_3C—NH_2, and H_3C—SH are necessarily threefold and symmetric. This follows from the presence of a group, CH_3, having C_{3v} symmetry joined to another group which is either C_{3v} also or C_{1v}. The energy minima occur when the substituents, hydrogens in these examples, of the respective groups are in the staggered conformations shown in Fig. 1*a* and *b*; maxima occur at the eclipsed conformations. This rule appears to hold universally for single bonds joining groups of this kind. For a single bond connecting a tetravalent carbon atom with a doubly bonded one, as in the structure

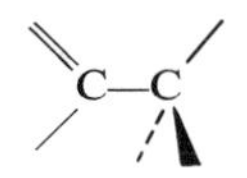

the conformation of lowest energy is that in which one of the pendant bonds of the latter eclipses the double bond. The stable conformation of acetaldehyde shown in Fig. 1*c* is illustrative. These properties of rotational potentials for single bonds adjoining double bonds can be reconciled with the bond staggering rule, as exemplified in Fig. 1*a* and *b*, by regarding the double bond as two single bonds drawn together from their normal (sp^3) positions.

In the examples cited above the rotational potential may be approximated by

$$E(\phi) = (E^\circ/2)(1 - \cos 3\phi) \tag{1}$$

where E° is the height of the rotational barrier and ϕ, as defined in Chapter I, is the angle of rotation measured from the staggered form.

For molecules such as $H_3C—NO_2$ in which a C_{3v} and a C_{2v} (planar with twofold axis) group are joined, the rotational potential is sixfold,[4,5] as must be expected from the symmetry about the C—N bond. The height of the barrier is only of the order of 10 cal mole^{-1}, however. This suggests that the next higher term (sixfold) in the Fourier expansion of $E(\phi)$ for predominantly threefold potentials likewise may be very small, and hence that Eq. 1 may be a good approximation to the form of the torsional potential.[8]

Heights of rotational barriers, i.e., energy differences between maxima and minima in the potentials, are summarized in Tables 1 and 2 for several classes of simple compounds. Results given to the second decimal in kilocalories per mole are derived from microwave frequency measurements. Those quoted to one decimal only have been deduced from thermodynamic studies and/or from measurements of infrared or Raman intensities as functions of temperature. Potentials for the molecules listed are threefold symmetric in every case except *n*-butane (Table 1). Here, as expected, the energy maximum for the *cis* conformation separating the two *gauche* forms at $\phi \cong 2\pi/3$ and $-2\pi/3$, respectively, is much higher than the energy for the two maxima separating the *trans* from each of the *gauche* forms. The figure quoted refers to the latter barrier.

The suggestion has frequently been made that the hindrance potential can be interpreted in terms of van der Waals repulsions between pendant atoms. Table 2 offers comparisons of barrier heights with distances $d_{X...Y}$ of closest approach of the pair of atoms X and Y, identified in column two, when the bond assumes the conformation

X Y
C—C

Table 1

Barrier Heights for Representative Bond Rotational Potentials (principally from Herschbach's[5] tabulations)

Compound and bond	Barrier height, kcal $mole^{-1}$
CH_3—CH_3	2.9
CH_3—CH=CH_2	1.98
CH_3—CH=O	1.17
CH_3CH_2—CH_2CH_3	3.5[a]
CH_3—OH	1.07
CH_3—OCH_3	2.72
CH_3—SH	1.27
CH_3—PH_2	1.96

[a] The value given for *n*-butane refers to the two lower barriers for this molecule. For further information on rotational potentials in *n*-butane and related hydrocarbons, see Chapter V, Table 1.

in which this atom pair is eclipsed. These distances are given in the third column. They may be compared with the sums of the van der Waals radii of the atom pair, listed in the fourth column. The differences between entries in the fourth and third columns, as tabulated in the fifth column, afford a measure of the degree of overlap of these atoms at the maximum in the potential. These differences cannot be accorded absolute significance in view of uncertainties in the assignment of van der Waals radii.[9] They are derived, however, from a consistent set of these radii, and hence should be indicative of the relative degrees of overlap. The staggered form is the stable one in every case, including CH_3SiH_3 for which the H $\cdots$ H distances considerably exceed the sum of the van der Waals radii.

Comparison of the last two columns of Table 2 reveals the futility of seeking an explanation for the rotational potentials in terms of repulsions between atoms estimated on the basis of van der Waals radii. The data in Table 1 corroborate this conclusion. The barrier height does indeed exhibit a partial correlation with the size of the atom, or group, estimated on this basis. However, the main source of the potential obviously cannot be thus explained.[4] Substitution of a halogen for hydrogen in ethane raises the barrier (Table 2), but to a much lesser degree than differences in van der Waals radii would suggest. On the other hand, in more highly substituted chains such as

$$\begin{array}{c} \quad CH_3 \qquad CH_3 \\ \quad | \qquad\quad | \\ -CH_2-C-CH_2-C-\text{etc.} \\ \quad | \qquad\quad | \\ \quad CH_3 \qquad CH_3 \end{array}$$

Table 2

Rotational Barrier Heights in Comparison with Interatomic Distances for Juxtaposed Atom Pairs in the Eclipsed Conformation (from tabulations of Herschbach[5] except as noted)

Molecule	Juxtaposed atom pair X, Y	$d_{X\ldots Y}$ eclipsed form, Å	Sum of van der Waals radii,[a] Å	Difference (overlap), Å	Barrier height,[5] kcal $mole^{-1}$
CH_3—CH_3	H,H	2.26	2.40	+0.14	2.9
CH_3—SiH_3	H,H	2.75	2.40	−0.35	1.66
CH_3—CH_2F	H,F	2.37	2.55	+0.18	3.31
CH_3—CHF_2	H,F	2.37	2.55	+0.18	3.18
CH_2F—CF_3	F,F	2.45	2.70	+0.25	4.2
CF_3—CF_3	F,F	2.45	2.70	+0.25	ca. 4.0[b]
CH_3—CH_2Cl	H,Cl	2.57	3.00	+0.43	3.69
CH_3—CH_2Br	H,Br	2.67	3.15	+0.48	3.57

[a] From Pauling, ref. 9.

[b] Estimated from thermodynamic data and from infrared torsional frequency. See E. L. Pace and J. G. Aston, *J. Am. Chem. Soc.*, **70**, 566 (1948); and D. E. Mann and E. K. Plyler, *J. Chem. Phys.*, **21**, 1116 (1953).

the threefold rotational potential usually associated with the skeletal C—C bond may be obliterated altogether by much stronger steric repulsions.

The origin of torsional potentials affecting rotations about single bonds has been one of the most baffling problems confronting physical chemistry for a number of years.[2,4] Hypotheses and theories advanced to account for the phenomenon have been numerous and diverse. A remarkably simple electrostatic model suggested by Karplus and Parr[10] correctly predicts the staggered form to be the most stable one, and leads to barriers of about the right magnitude for molecules such as CH_3—OH, CH_3—SH, CH_3—SiH_3, and CH_3—NH_2, as well as CH_3—CH_3, in which all of the substituent atoms are hydrogens. Electrostatic repulsions between nuclei (protons in these examples) of the pendant atoms were implicated as the major source of the potential, attractions between each nucleus and the electrons of pendant atoms of the opposite group contributing a term of smaller magnitude which favors the eclipsed form. More elaborate quantum mechanical calculations[11] show this interpretation to be oversimplified. Electron–electron repulsions also contribute to the barrier, and rotation-dependent polarization effects are appreciable. Thus, according to Pedersen and Morokuma,[11] "No single factor emerges ... as the source of the barrier."

Quantum mechanical studies of rotational barriers, thus far confined to bonds situated between atoms bearing hydrogen atoms as the only substituents, offer little basis for the interpretation of rotational barriers in more complicated molecules in which the bond in question is adjoined by substitutents of other kinds (e.g., halogens, —CH_3, —CH_2—, etc.). It may be inferred, however, that any bond joining tetrahedrally bonded (sp^3) atoms will be subject to an "inherent" threefold torsional potential (see Eq. 1) arising from interactions between the orbitals and substituents (or atomic cores) associated with one of the atoms and those of the other. Superimposed thereon are the interactions involving larger substituents attached to the respective atoms joined by the bond. These "nonbonded interactions" may render the resultant potential unsymmetrical, the three minima being of unequal energy. Beyond these conjectures, we are obliged to rely mainly on empirical evidence for representation of rotational potentials about skeletal bonds of chain molecules.

It will be observed that all of the barrier heights recorded in Tables 1 and 2 exceed RT at ordinary temperatures. The distribution of rotation angles at equilibrium will by no means be uniform over the 2π range. In molecules possessing C—C bonds, values of the rotation angle in the neighborhood of the potential minima are strongly favored over those near the maxima. On the other hand, none of the barriers cited is of sufficient height to prevent rapid interconversion amongst the several forms.

Torsion potential barriers E° for bonds such as Si—O,[12] P—O, and P—N are almost certainly lower than any of those quoted above, and they probably are less than RT. In molecules constructed of these bonds, nonbonded interactions and dipolar interactions must assume overriding importance in determining conformational preferences.

2. THE ROTATIONAL ISOMERIC STATE APPROXIMATION

The effects of nonbonded interactions of substituent groups on the torsional potentials of bonds in larger molecules should be manifested rudimentarily in *n*-butane. The rotational potential for the central bond in this molecule may be considered to consist of the combination of a threefold symmetric, ethane-like potential and the superimposed effects (mainly repulsive) of interactions involving the terminal methyl groups. The resultant rotational potential about this bond is portrayed conjecturally by the curve in Fig. 2 for the range 0 to π. The function over the range

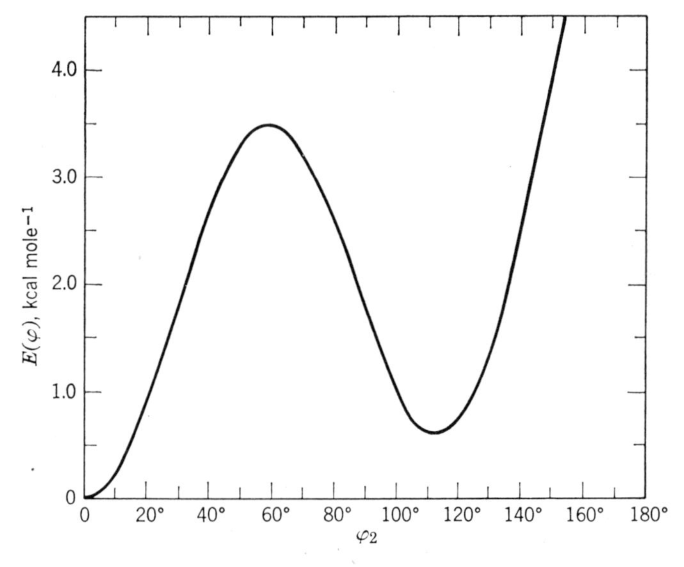

Fig. 2. The conformational energy of *n*-butane as a function of the rotation angle ϕ_2 about the central C—C bond. The terminal methyl groups are presumed to be fixed in staggered conformations.

$-\pi$ to 0 is obtained by reflection through the ordinate axis. The height of the barrier occurring at ca. 60° has been set at the value given in Table 1. The barrier at 180° (*cis*) is unknown but is believed to be large. The curve shown is intended to be illustrative only.

The coordinates of the three minima, one for the *trans* (0°) and two for *gauche* conformations (ca. ±120°), are the features of most importance for our purposes. Measurements of temperature coefficients of Raman intensities of liquid *n*-butane indicate the energy of the latter minima to be 800 cal mole^{-1} higher than the former one.[13] More extensive results for higher *liquid n*-alkanes confirm a value of 500 ± 100 cal mole^{-1} for the energy of the *gauche* minima relative to the *trans* minimum for a given bond.[1,3,14,15] The *gauche* minimum for *n*-butane in Fig. 2 has been placed 600 cal mole^{-1} above the *trans*. Calculations[16,17] and experiments[6,7] indicate a displacement of the *gauche* minima of perhaps 5–10° from their symmetrical locations at ±120°, owing to the incipience of the large repulsions occurring in the *cis* conformation. In Fig. 2, the minimum has been situated at 112°.

The potential wells are sufficiently steep to confine the preponderance of the molecules of *n*-butane at ordinary temperatures to states of torsional oscillation near one or another of the minima in $E(\phi)$. The proportion of them having sufficient energy to surmount the lower barriers (ca. 3.5 kcal mole^{-1}) is small. In classical mechanical language, most of the molecules will at any instant occur in the ranges of configuration within about ±20° of one of the minima. Molecules confined to the regions of the three potential minima are well differentiated, and in this sense three conformations of *n*-butane are clearly distinguishable. They are designated *trans* (t), *gauche*$^+$ (g^+), and *gauche*$^-$ (g^-), respectively. The latter two are related to the *trans* form by right- and left-handed rotations, respectively, of *approximately* 120°. It is important to observe that g^+ and g^- *n*-butane are nonsuperimposable rotational isomers; they are right- and left-handed species related by a mirror reflection. Of course, the *trans* and *gauche* forms cannot be separately isolated, owing to the extreme rapidity of their interconversion. Their separate identification is possible only by spectroscopic methods operating in the frequency range above 10^{10} sec^{-1}.

On the same basis, we may expect six distinguishable rotational isomers, or conformers, in *n*-pentane: tt, tg^+, tg^-, g^+g^+, g^-g^-, and g^+g^-. The conformers g^+t, g^-t, and g^-g^+ are identical with tg^+, tg^-, and g^+g^-, respectively. Two pairs of mirror images, namely, tg^+, tg^- and g^+g^+, g^-g^-, are included. The last conformation, g^+g^- (alias g^-g^+), is largely suppressed owing to steric repulsions, as will be pointed out below.

Description of the conformations of these and other simple molecules in terms of distinct rotational isomeric forms is both convenient and well justified by physical circumstances. In long chain molecules similar potentials affect the rotations of the more numerous skeletal bonds. Concomitant with the increased complexity of larger molecules is the proliferation of intermolecular interactions with neighboring molecules of the condensed phase, liquid or crystalline. These perturb the intramolecular potentials and obscure the representation of bond rotations and of the associated internal modes by a set of discrete quantum mechanical states. Conceptualization in classical mechanical terms gains validity and utility, however. In a crystalline phase, the intermolecular interactions may conspire to systematically displace the (average) rotation angle from its minimum. That is to say, the requirements of efficient packing with neighbors and optimization of the intermolecular energy may conceivably compromise the intramolecular conformation from its state of lowest energy. This appears to be the exception rather than the rule, however. Ordinarily, the conformation found to occur in the crystalline state seems to correspond very nearly to the form having minimum intramolecular energy. Minor departures from this minimum may, however, be imposed by the exigencies of efficient intermolecular packing, and this must be borne in mind.

In the liquid state, where disorder prevails, it can be reasoned that the intramolecular potential faithfully represents the average state of affairs confronting a given molecule or one of its bonds. The instantaneous local structure in the immediate neighborhood of the molecule, or that part of it associated with a given bond, may, to be sure, perturb the rotational potential. But such perturbations will be random. The *average* potential should therefore correspond closely to its unperturbed form, and the system of molecules may be pictured as populating the configuration space, i.e., the ϕ space of the various rotation angles (and possible other internal coordinates) of the molecule, according to a Boltzmann distribution over the intramolecular energy, intermolecular effects being ignored. With reference to Fig. 2, which represents a one-dimensional configuration space, the configurations of most of the molecules will gravitate to the regions of the three potential minima estimated for the free molecule, without a discernible bias due to intermolecular interactions.

In the *rotational isomeric state* approximation[18] each molecule, or bond, is treated as occurring in one or another of several *discrete* rotational states. These states ordinarily are chosen to coincide with potential minima. Fluctuations about the minima are ignored. Their occurrence is not denied, however. Rather, the foregoing assumption derives its justification from recognition that these fluctuations, being of random sign, will be mutually compensatory, or nearly so, in their effect on the average properties of the chain molecule. The rotational isomeric state approximation, whereby the continuous distribution over ϕ is replaced by a distribution over several discrete states (usually three in consequence of the prevalence of the threefold character of rotational potentials), is well founded for those bonds having distinct rotational minima separated by barriers substantially greater than RT. For bonds not subject to rotational energy barriers exceeding RT, the rotational isomeric state approximation is deprived of its primary *physical* basis. It may, however, be adopted as a suitable *mathematical* device. Its use in such cases finds justification in the fact that the integral of a continuous function, e.g., the integral in Eq. I-35, may be approximated by a summation. Thus, the rotational isomeric scheme may be used legitimately, even in those cases where rotational barriers are low, or negligible, in height. A judicious choice of rotation states is obviously necessary.

The rotational partition function for *n*-butane expressed in compliance with the rotational isomeric state approximation is

$$z = (1 + \sigma + \sigma) \tag{2}$$

where

$$\sigma = \exp(-E_g/RT) \tag{3}$$

E_g being the energy of a *gauche* state relative to *trans*.* Thus, Eq. 2 replaces the previous integral, Eq. I-35, over the 2π range of ϕ. In general, for a bond whose conformations are represented by v rotational states

$$z = u_1 + u_2 + \cdots + u_v \tag{4}$$

Here u_1, u_2, etc. are the *statistical weights* applicable to the respective isomeric states. It will be convenient in practice to assign the value unity to one of these and to express the other statistical weights relative to the one so chosen. If $u_1 = 1$, then

$$u_\eta = \exp[-(E_\eta - E_1)/RT] \tag{5}$$

where E_η is the energy (or free energy) for state η. Equations 2 and 3 are expressed on this basis.

For a symmetrical chain, selection of rotational isomeric states so that they occur symmetrically about $\phi = 0$ will be obligatory. If for such a chain we find the statistical weights to be equal, then $u_\eta = 1$ for all $\eta = 1$ to v, and $z = v$. If, further, the rotational states are spaced at equal intervals of angle (e.g., at 0° and ±120°), then $\langle \cos \phi \rangle = 0$, and the average dimensions reduce to those for the freely rotating chain (compare Eqs. I-50 and I-20). These are the conditions under which the properties of the rotational isomeric state model for a symmetric chain reduce to those for free rotation.

* Since the shapes of the potential minima are very nearly the same (see Fig. 2), the factor which should otherwise multiply the exponential is effectively unity and may be ignored. In other words, the configuration integrals (of classical statistical mechanics) over the respective potential wells are approximately equal.

If the rotational potentials for the various bonds of a chain molecule are treated in the approximation of mutual independence as in Sections 9 and 10 of Chapter I, then adoption of the rotational isomeric state representation allows the configuration partition function of the chain as a whole to be written

$$Z = \sum_{\{n_\eta\}} \frac{(n-2)!}{n_1! \cdots n_\nu!} u_1^{n_1} \cdots u_\nu^{n_\nu}$$

where n_η is the number of bonds in rotational state η and $\{n_\eta\} = n_1, n_2 \cdots n_\nu$. The sum includes all sets $\{n_\eta\}$ such that $\sum n_\eta = n - 2$, the number of internal bonds in the chain skeleton. According to the multinomial theorem, the required sum is just z, given by Eq. 4, raised to the power $n - 2$, i.e.,

$$Z = z^{n-2} \tag{6}$$

The fraction of internal bonds in state η will be (see Eq. I-34)

$$p(\phi_\eta) = p_\eta = n_\eta/(n-2) = u_\eta/z \tag{7}$$

and the average value of any function $f(\phi)$ of ϕ is

$$\langle f \rangle = z^{-1} \sum_\eta u_\eta f(\phi_\eta) \tag{8}$$

Let us explore the interpretation of a polymethylene chain in this approximation, which is tantamount to ascribing to each internal bond a rotational potential like that for the central bond of *n*-butane. For simplicity, let the three rotation states be taken at 0°, 120°, −120°. Then from Eqs. 2, 3, 5, and 8

$$\begin{aligned} \langle \cos \phi \rangle &= z^{-1} \sum u_\eta \cos \phi_\eta \\ &= [1 - 2(1/2)\sigma]/(1 + 2\sigma) \\ &= (1 - \sigma)/(1 + 2\sigma) \end{aligned} \tag{9}$$

If we take $E_g = 500$ cal mole^{-1}, then at 140°C, the approximate temperature of relevant experiments, $\sigma = 0.54$ and $\langle \cos \phi \rangle = 0.22$. The characteristic ratio calculated from Eq. I-48 with $\theta = 68°$ is 3.4, which is much smaller than the value 6.7 found experimentally (Table II-1). To amend the discrepancy, it is necessary to take $E_g = 1100$ cal mole^{-1}. Not only is this figure inconsistent with a considerable body of evidence on liquid *n*-alkanes; additionally, it yields a temperature coefficient of $\langle r^2 \rangle_0$ which is more than twice that observed (see Table II-2). Alteration of rotation angles, e.g., placement of the *gauche* states at $\phi_g = \pm 110°$, does not diminish appreciably the serious disagreement with experiment. Thus, treatment of the configuration of the polymethylene chain on the foregoing basis fails utterly to achieve agreement with experiments.

The cause of this failure becomes readily apparent upon examination of the conformations of higher *n*-alkane homologs of butane. Consider *n*-pentane, which has two internal bonds. Three of its rotational isomeric conformations are shown in Fig. 3. The conformation in Fig. 3*b* is generated

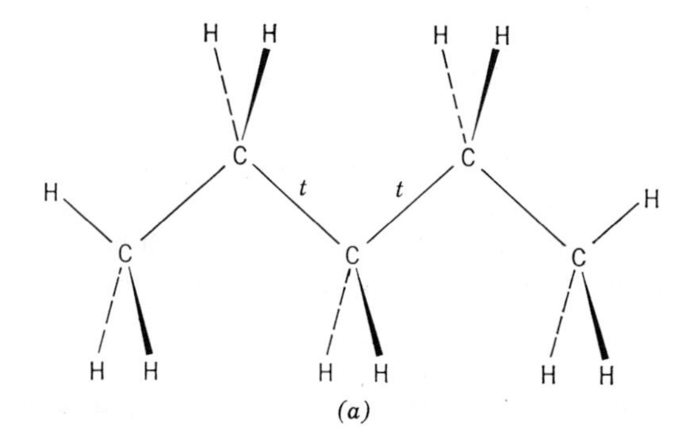

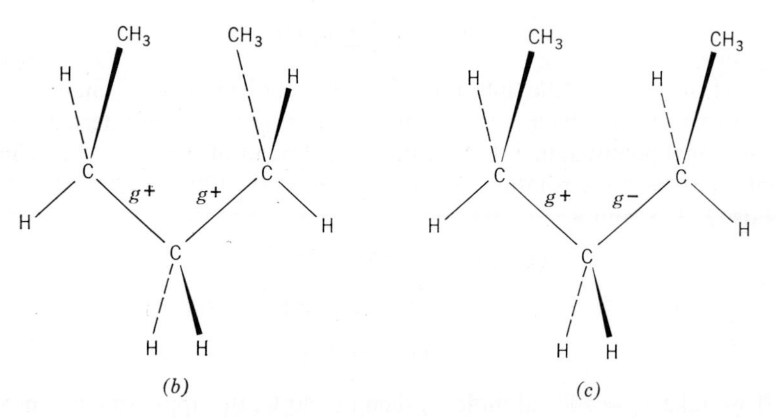

Fig. 3. Conformations generated by rotations about the internal C—C bonds of *n*-pentane: (*a*) *trans*, *trans* with $\phi_2 = \phi_3 = 0°$; (*b*) *gauche*$^+$, *gauche*$^+$ with $\phi_2 = \phi_3 \cong 120°$; and (*c*) *gauche*$^+$, *gauche*$^-$ with $\phi_2 = -\phi_3 \cong 120°$.

from that in Fig. 3*a* by executing right-hand (positive) rotations about bonds 2 and 3; the one in Fig. 3*c* results from rotations of opposite sign about the two bonds, i.e., a right-hand rotation about bond 2 and a left-hand rotation about bond 3. In diagrams such as these, it is convenient to retain the pair of bonds subject to rotation in the plane of the drawing. Adjoining bonds take up positions above or below the plane as indicated. If both bonds sustain rotations of the same sign, then the groups subtended by the adjoining pair of bonds are placed by these operations on opposite sides of this plane of the figure, as in Fig. 3*b*, for example; if of opposite sign as in Fig. 3*c*, the groups occur on the same side of the plane.

Examination of models shows the terminal methyl groups in the g^+g^+ conformation, and in its mirror image g^-g^-, to be at such a distance of separation, ca. 3.6 Å, as to suggest that their mutual interaction is approximately neutral, i.e., neither attractive nor repulsive to an appreciable degree. *Gauche* rotations of opposite sign, shown in Fig. 3*c*, subject the pendant methyl groups to severe steric conflict; their distance of separation is only about 2.5 Å. The g^+g^- conformation, and likewise g^-g^+, consequently are of rare occurrence. Suppression of conformations of this kind by steric overlaps has sometimes been referred to as the "pentane effect."

Equivalent considerations apply to higher homologs, polymethylene included. If a given bond assumes a *gauche* rotational state, the occurrence of either of its immediate neighbors in a *gauche* state of opposite sign is strongly discouraged. The rotational potential for a given bond thus acquires dependence on the rotational states of its neighbors. The virtual suppression of *gauche* rotations of opposite sign for pairs of bonds which are first neighbors is a feature manifested in most chain molecules. It supersedes all other characteristics of bond rotational potentials in its effect on the configurations of chain molecules. Interdependence of rotational potentials is the rule; exceptions are comparatively few.

3. STATISTICAL WEIGHT MATRICES FOR INTERDEPENDENT BONDS

A configuration $\{\phi\}$ of an n-bond chain molecule may be specified in the terms of the rotational isomeric state scheme by $n-2$ numerals from a ν-digital system, ν being the number of rotational isomeric states chosen to characterize the rotational potential of a bond. Thus, if $\nu = 2$, a configuration may be specified by the series of digits

$$2\ 1\ 1\ 2\ 1\ 2\ 2 \text{ etc.}$$

Let us assume that the rotational potential affecting any given bond i depends exclusively on ϕ_{i-1}, ϕ_i, and ϕ_{i+1}, interactions of longer range being unimportant. It will be apparent that the total configurational energy may then be expressed as a sum of energies for first-neighbor pairs. That is, for the binary state configuration specified above

$$E\{\phi\} = E_2 + E_{21} + E_{11} + E_{12} + \text{etc.}$$

The first term E_2 carries a single index owing to the absence of a predecessor in the sequence. Each succeeding term carries as its first index the latter index of its predecessor in the polynomial. In general

$$E\{\phi\} = \sum_{i=2}^{n-1} E_i(\phi_{i-1}, \phi_i) = \sum_{i=2}^{n-1} E_{\zeta\eta;\,i} \tag{10}$$

where ζ denotes the state of bond $i - 1$ and η that of bond i, it being understood that the first term in the sum is a function of one angle only, and hence is indexed by η alone. The energy $E_{\zeta\eta;\,i} = E_i(\phi_{i-1}, \phi_i)$ is appropriately regarded as *the contribution to* $E\{\phi\}$ *associated with the assignment of bond* i *to state* η, *bond* $i - 1$ *being in state* ζ. In adopting this viewpoint we do not overlook the dependence of the rotational potential for bond i on ϕ_{i+1}. Account of the dependence of the energy on ϕ_{i+1} is merely reserved for the next term in the sum. The total energy may thus be reckoned systematically as a sum of terms each dependent upon a pair of consecutive rotation angles.

Specifically, $E_{\zeta\eta;\,i}$ will include the intrinsic torsional potential of bond i, this potential being a function solely of ϕ_i as expressed, for example, by Eq. 1. Nonbonded interactions which depend exclusively on ϕ_i will be assigned to this term. Those which depend jointly on ϕ_{i-1} and ϕ_i may logically be included in it also. If interactions dependent upon three consecutive angles were to contribute appreciably (which is seldom the case), those determined by ϕ_{i-2}, ϕ_{i-1}, and ϕ_i should be included in $E_{\zeta\eta;\,i}$, and so forth for interactions of even longer range. The significance of subscripts ζ and η would then require revision (*cf. seq.*).

For chains devoid of large substitutents which engender copious interactions with one another and with atoms and groups comprising the chain skeleton, a somewhat different convention may be preferred with regard to the allocation of those contributions to the conformational energy which depend jointly on two adjacent rotation angles. For such chains the planar *trans* conformation is a convenient reference state, and it may be chosen to serve in this capacity, even though another conformation may be of lower energy. Commencing with the chain in this reference state, we assign bonds in successive order, proceeding from left to right, to their respective rotational states. To energy $E_{\zeta\eta;\,i}$ are allotted all contributions arising from alteration of ϕ_i from the *trans* state to state η, preceding bonds being in their previously assigned states and all succeeding bonds $j > i$ being retained tentatively in *trans* states. Thus, interactions precipitated, or alleviated, by rotating bond i from $\phi = 0$ to ϕ_η, bonds $i - 1$ and $i + 1$ being in states ϕ_ζ and $\phi_t = 0$, respectively, are assigned to $E_{\zeta\eta;\,i}$. The effect of the artifice of considering bond $i + 1$ to be *trans* at this stage in the construction of the energy is removed when this bond is subsequently assigned to its actual state, in conjunction with evaluation of E_{i+1}. Adoption of this convention renders $E_{\zeta\eta;\,i} = 0$ if η is *trans*, irrespective of ζ.

The energies for various skeletal bond pairs in an n-alkane, or polymethylene, reckoned on the foregoing basis assume values as follows:

$$\begin{aligned} E_{\zeta t} &= 0 \qquad \text{for} \qquad \zeta = t, g^+, g^- \\ E_{tg^\pm} &= E_{g^\pm g^\pm} = E_g \cong 500 \text{ cal mole}^{-1} \\ E_{g^\pm g^\mp} &\approx 3 \text{ kcal mole}^{-1} \end{aligned} \tag{11}$$

Assignment of the same energy to $E_{g^\pm g^\pm}$ as to $E_{tg^\pm}$ rests on the observation that a pair of *gauche* rotations of the same sign does not promote significant interaction between the pendant groups (see Fig. 3*b*) of the n-alkane chain. Conformation pairs $g^\pm g^\pm$ are considered to entail a total energy of $2E_g$. This energy is divided between successive terms of the sum; the portion assigned to bond i is E_g.*

It will be apparent that various conventions could be adopted for the assignment of the energies for neighboring bond pairs. The particular choice affects only the method of accounting and not the resultant $E\{\phi\}$, which must comprise all contributions for any configuration specified by $\{\phi\}$.

Statistical weights $u_{\zeta\eta}$ corresponding to the energies $E_{\zeta\eta}$ may be defined by invoking the relationship

$$u_{\zeta\eta;\, i} = \exp(-E_{\zeta\eta;\, i}/RT) \tag{12}$$

These are conveniently expressed in the form of a *statistical weight matrix*

$$\mathbf{U}_i = [u_{\zeta\eta}]_i \tag{13}$$

with states (ζ) for bond $i - 1$ indexing the rows and those (η) for bond i the columns. The subscript i may be omitted if all bonds of the chain are of identical character.

The statistical weight of a configuration of the chain as a whole is given by

$$\Omega_{\{\phi\}} = \prod_{i=2}^{n-1} u_{\zeta\eta;\, i} \tag{14}$$

which follows at once from Eqs. 10 and 12. The first term in the product carries a single index for the reason noted earlier. The statistical weights $u_{\zeta\eta}$, rather than the energies $E_{\zeta\eta}$, could have been chosen as the primary quantities for characterizing a configuration. Equation 12 would then have served to define the energies in *terms of the statistical weights*,† instead of

* These energies will be treated in greater detail in Chapter V. They are introduced here mainly for illustration.

† It will be apparent that the energies should in fact be regarded as free energies. The distinction will be unimportant, however. See the footnote on p. 58.

the reverse. The statistical weights must, of course, appropriately take account of neighbor dependence, and they must yield the correct statistical weight for any configuration of the molecule as a whole when they are multiplied in the combination prescribed by the configuration. Conventions corresponding to those introduced for the energy are applicable. Thus, $u_{\zeta\eta;\,i}$ denotes the factor entering into $U\{\phi\}$ as a consequence of assigning bond i to state η, bond $i-1$ being in state ζ.

On the basis of the set of energies given in first approximation by Eq. 11 for rotational state pairs for n-alkane homologs, the statistical weight matrix takes the form

$$\begin{array}{c} \qquad\qquad (t) \quad (g^+) \quad (g^-) \\ \mathbf{U} = \begin{array}{c} (t) \\ (g^+) \\ (g^-) \end{array} \begin{bmatrix} 1 & \sigma & \sigma \\ 1 & \sigma & 0 \\ 1 & 0 & \sigma \end{bmatrix} \end{array} \tag{15}$$

where σ is defined by Eq. 3 and $\exp(-E_{g^+g^-}/RT)$ is approximated by zero. States of the preceding bond are shown to the left of each row; those for the bond considered are shown above the columns. Elements in the first column equate to unity according to the convention introduced above for reckoning energies relative to the *trans* state for bond i.

Generalization of **U** as given by Eq. 15 to render it applicable to bonds of *any* symmetrical chain (i.e., any chain devoid of asymmetric centers) characterized by threefold rotation potentials requires the designation of two additional statistical weight parameters, one to represent g^+g^+ and g^-g^- pairs and the other for g^+g^- and g^-g^+. Let $\sigma\psi$ denote the statistical weight associated with assignment of bond i to g^+ or g^- when bond $i-1$ is in the *gauche* state of the *same* sign; $\sigma\omega$ will denote the corresponding statistical weight when bond $i-1$ is in the *gauche* state of opposite sign. Reasons for this choice of parameters will be brought forth in Chapter V. Then the generalized form of the statistical weight matrix for this case is

$$\mathbf{U} = \begin{bmatrix} 1 & \sigma & \sigma \\ 1 & \sigma\psi & \sigma\omega \\ 1 & \sigma\omega & \sigma\psi \end{bmatrix} \tag{16}$$

Symmetry requires that

$$u_{12} = u_{13}$$

$$u_{21} = u_{31}$$

$$u_{22} = u_{33}$$

$$u_{23} = u_{32}$$

The convention adopted above, whereby statistical weights (and the associated energies) are assigned at each step on the premise that all succeeding bonds are *trans*, renders

$$u_{11} = u_{21} = u_{31} = 1$$

Equation 16 meets these conditions.* It contains three independent parameters, the minimum number required for a general representation of bonds of the kind considered.

It has been assumed above that the bond rotational interdependence does not extend beyond first neighbors. Correlations of longer range would imply interactions contingent upon a set of rotational states for three or more consecutive skeletal bonds. Conformations yielding such interactions within a set of five to eight skeletal bonds, and depending therefore on three to six rotation angles, involve at least one g^+g^- or g^-g^+ pair of neighbors, as may be verified by critical study of molecular models. The conformation $g^+g^-g^+$, for example, entails a severe five-bond repulsion, i.e., a repulsion between groups separated by five skeletal bonds. Inasmuch as two *gauche* pairs of opposite sign are included in this conformation, it is relegated to rare occurrence by the associated four-bond interactions, and further suppression in acknowledgement of the five-bond interaction is superfluous. The conformation $g^+g^+g^-g^-$ engenders a steric overlap between groups separated by six skeletal bonds. If tetrahedral valence angles are preserved and rotations $|\phi| = 120°$ are enforced, then the resulting repulsion is comparable to the four-bond repulsion for a g^+g^- pair. The six-bond repulsion may be readily relieved, however, by compromising the

* Without the convention cited above the form of **U** would be, in general

$$\mathbf{U}' = \begin{bmatrix} 1 & u'_{12} & u'_{12} \\ u'_{21} & u'_{22} & u'_{23} \\ u'_{21} & u'_{23} & u'_{22} \end{bmatrix}$$

all weights being normalized to $u'_{11} = 1$. This matrix may be converted to the form of Eq. 16 by the similarity transformation

$$\begin{bmatrix} 1 & & \\ & u_{21}'^{-1} & \\ & & u_{21}'^{-1} \end{bmatrix} \mathbf{U}' \begin{bmatrix} 1 & & \\ & u'_{21} & \\ & & u'_{21} \end{bmatrix} = \begin{bmatrix} 1 & u'_{12}u'_{21} & u'_{12}u'_{21} \\ 1 & u'_{22} & u'_{23} \\ 1 & u'_{23} & u'_{22} \end{bmatrix} = \mathbf{U}$$

The matrices **U**′ and **U** being thus related, either could be used in the development presented in Section 4. This follows from the cancellation of the matrices surrounding **U**′when multiplied sequentially as in Eq. 23; only the terminal matrix factors remain. Hence, use of **U**′ rather than **U** merely requires an appropriate revision in the account of "end effects" at terminal bonds of the chain sequence. These effects are more easily handled by use of **U** rather than **U**′.

It will be apparent that the variants of **U** obtained by similarity transformations differ according to the manner in which the energy (and corresponding statistical weight) associated with a given bond rotation is allocated between neighboring pairs.

four intervening rotation angles, and hence it is of lesser consequence than the steric overlap associated with the g^+g^- pair. It is thus apparent that significant steric repulsions between skeletal groups separated by five to eight bonds in most chains of interest are reduced to rare occurrence by interactions of lower range. They may safely be ignored except in unusual circumstances. Intramolecular encounters of much longer range are statistically improbable and, moreover, are nullified by the reduction of experimental measurements to ideal conditions such that the covolume for an encounter of two uncorrelated (sequentially remote) chain elements is zero, as explained in Chapter II.

If interactions beyond those dependent upon rotation angles which are first neighbors should be important, they can be included in the present scheme by a well-known device. Suppose that triads of bond rotations need to be considered. The "state" of bond $i-2$ becomes relevant to the rotational energy accountable in the assignment of bond i. Accordingly, the "state" of bond $i-1$ is redefined to include the rotational state of bond $i-2$ as well as bond $i-1$. There are ν^2 of these "states" for each bond. The order of the statistical weight matrices is thereby increased to $\nu^2 \times \nu^2$ instead of $\nu \times \nu$. Obviously, a state thus defined for bond i is eligible to succeed one of them for bond $i-1$ only if bond $i-1$ is accorded the same rotation in both states. Only those elements of the enlarged statistical weight matrix are nonzero which meet this condition, i.e., those representing "transitions" $(\phi_{i-2}, \phi_{i-1}) \rightarrow (\phi'_{i-1}, \phi'_i)$ for which $\phi'_{i-1} = \phi_{i-1}$. The formal scheme is adaptable in this manner to chains which are subject to interactions of longer, but finite, range. The order of the required statistical weight matrices increases geometrically with the range of the interdependence.

4. THE CONFIGURATION PARTITION FUNCTION: GENERAL FORMULATIONS

The configuration partition function is given formally (see Eq. 14) by

$$Z = \sum_{\{\phi\}} \Omega_{\{\phi\}} = \sum_{\{\phi\}} \prod_{i=2}^{n-1} u_{\zeta\eta;\, i} \tag{17}$$

where the summations are taken over all configurations. The task of evaluating the partition function by forming the product of statistical weights $u_{\zeta\eta}$ for each configuration and then summing all of them as directed by this equation would be prohibitive for a chain having more than about 20 bonds, even with the aid of high-speed computers. Other methods for evaluating this sum of products fortunately are available. They are applicable to chains of any length and require only moderate computational effort.

It was noted in connection with Eq. 14 that terms of the sum in Eq. 17 consist of products $\cdots u_{\beta\alpha} u_{\alpha\gamma} u_{\gamma\alpha} \cdots$ in which the first index of each factor u is required to be identical with the second index of the factor preceding it in the product. This is implicit in the scheme for translating a configuration $\{\phi\}$ to the corresponding set of statistical weights. Products of this kind are produced by matrix multiplication. In fact, the entire sum of products can be generated by matrix multiplication, as we proceed to show. Matrix methods for treating the statistical mechanical properties of one-dimensional systems were introduced by Kramers and Wannier[19] in 1941. These authors were concerned with the Ising[20] ferromagnet, which consists of a hypothetical one-dimensional system of magnetic dipoles subject to neighbor interactions and quantum restrictions, a problem bearing close similarities to ours. The great power and versatility of matrix methods[21] for evaluating Eq. 17 will be made apparent in due course.

Consider a simple chain comprising $n + 1$ identical skeletal atoms or groups joined by n identical skeletal bonds. Each bond, terminal bonds excepted, is permitted to occur in one or the other of two rotational states (i.e., $\nu = 2$). Rotational states for terminal bonds being undefined, the statistical weights u_α and u_β for the second bond of the chain carry a single index. They may, if desired, be presented as the elements of a diagonal matrix, i.e.,

$$\mathbf{U}_2 = \begin{bmatrix} u_\alpha & 0 \\ 0 & u_\beta \end{bmatrix} \tag{18}$$

Those for succeeding bonds may be embodied in the 2×2 matrix

$$\mathbf{U} = \begin{bmatrix} u_{\alpha\alpha} & u_{\alpha\beta} \\ u_{\beta\alpha} & u_{\beta\beta} \end{bmatrix} \tag{19}$$

Greek letter subscripts are used to denote rotational states, Roman letters and numerical subscripts being reserved in general for the ordinal number of the bond.

The configuration partition function Z for a chain of three bonds (e.g., n-butane) is given by the sum of elements in $\mathbf{U}_2$. For $n = 4$, Z is the sum of elements in

$$\mathbf{U}_2\,\mathbf{U} = \begin{bmatrix} u_\alpha u_{\alpha\alpha} & u_\alpha u_{\alpha\beta} \\ u_\beta u_{\beta\alpha} & u_\beta u_{\beta\beta} \end{bmatrix}$$

which correspond to the states $\alpha\alpha$, $\alpha\beta$, $\beta\alpha$, and $\beta\beta$. Similarly, the statistical weights for the eight states of a five-bond chain are reproduced by the elements of the matrix product

$$\mathbf{U}_2\,\mathbf{U}^2 = \begin{bmatrix} u_\alpha u_{\alpha\alpha}^2 + u_\alpha u_{\alpha\beta} u_{\beta\alpha} & u_\alpha u_{\alpha\alpha} u_{\alpha\beta} + u_\alpha u_{\alpha\beta} u_{\beta\beta} \\ u_\beta u_{\beta\alpha} u_{\alpha\alpha} + u_\beta u_{\beta\beta} u_{\beta\alpha} & u_\beta u_{\beta\alpha} u_{\alpha\beta} + u_\beta u_{\beta\beta}^2 \end{bmatrix}$$

By induction, Z for an n-bond chain equals the sum of the elements of $\mathbf{U}_2\mathbf{U}^{n-3}$. This sum may be extracted by pre-multiplication with [1 1] and postmultiplication with the corresponding column. That is

$$Z = [1 \quad 1]\mathbf{U}_2\,\mathbf{U}^{n-3}\begin{bmatrix}1\\1\end{bmatrix} \tag{20}$$

The generalization of this result to a chain comprising bonds for which $\nu > 2$ is obvious.

The same result is obtained by rewriting $\mathbf{U}_2$ as follows:

$$\mathbf{U}_2 = \begin{bmatrix}u_\alpha & u_\beta\\0 & 0\end{bmatrix} \tag{21}$$

and replacing Eq. 20 by

$$Z = [1 \quad 0]\mathbf{U}_2\,\mathbf{U}^{n-3}\begin{bmatrix}1\\1\end{bmatrix} \tag{22}$$

This formulation offers a distinct advantage arising from the correspondence, in general, of the first row of elements in $\mathbf{U}$ with those required for $\mathbf{U}_2$. In the usual case of $\nu = 3$, the elements in the first row of $\mathbf{U}$ represent statistical weights for bond i when bond $i - 1$ is *trans*. In this conformation interactions with preceding groups are often negligible. The statistical weights in the first row of $\mathbf{U}$ will then take on values virtually equal to those for bond $i = 2$, that is, for the first internal bond of the chain. If this is the case, then $\mathbf{U}$ may replace $\mathbf{U}_2$, inasmuch as elements from rows other than the first will be rejected through premultiplication by [1 0 0], or by [1 0] if $\nu = 2$. When the foregoing condition is met, and this will usually be the case, Eq. 22 simplifies to

$$Z = [1 \quad 0]\mathbf{U}^{n-2}\begin{bmatrix}1\\1\end{bmatrix} \tag{23}$$

wherein a matrix $\mathbf{U}_2$ differing from $\mathbf{U}$ is not required.

This result admits of immediate generalization to chains of bonds occurring in any variety and characterized by any number ν of rotational states. Thus, in general

$$Z = \mathbf{J}^*\left[\prod_{i=2}^{n-1}\mathbf{U}_i\right]\mathbf{J} \tag{24}$$

where $\mathbf{J}^*$ and $\mathbf{J}$ are the row and column vectors

$$\mathbf{J}^* = [1 \quad 0 \quad \cdots \quad 0]; \qquad \mathbf{J} = \begin{bmatrix}1\\1\\ \vdots \\ 1\end{bmatrix} \tag{25}$$

of order $1 \times \nu$ and $\nu \times 1$, respectively. Successive bond pairs may or may not be the same. Differences will be reflected in the various $\mathbf{U}_i$, the subscript being retained on this account. There will be as many *different* $\mathbf{U}_i$ as there are different kinds of bond pairs, and the respective $\mathbf{U}_i$ must be multiplied in the serial order dictated by the structure of the chain. If the numbers ν of rotational states appropriate for the various kinds of bonds present should differ, it is required merely to construct appropriate rectangular $\mathbf{U}_i$ matrices in which the number of rows equals the number of rotational states for bond $i - 1$ and the number of columns equals the number of states for bond i.

If the chain comprises a sequence of identical structural units such as —NH—CH_2—CO— for example, and these units repeat in regular succession throughout the chain, then the $\mathbf{U}$ matrices for the several types of bond pairs will repeat in cyclic order in the product occurring in Eq. 24. Let the repeating units be ξ bonds in length, and let the product of the corresponding statistical weight matrices, taken in the required order, be represented by $\mathbf{U}^{(\xi)}$. Then the product occurring in Eq. 24 can be expressed in terms of a power of the matrix thus defined. That is,

$$Z = \mathbf{J}^* \mathbf{U}^{(\xi')} [\mathbf{U}^{(\xi)}]^{x-2} \mathbf{U}^{(\xi'')} \mathbf{J} \tag{26}$$

where $\mathbf{U}^{(\xi')}$ and $\mathbf{U}^{(\xi'')}$ are products which take account of bonds preceding and following the regular succession of $x - 2$ *internal* units within the chain, and $n - 2 = \xi' + \xi'' + (x - 2)\xi$. The evaluation of Z may be simplified considerably through use of the matrix $\mathbf{U}^{(\xi)}$ for the repeating unit, which, according to Eq. 26, is raised to the power $x - 2$. This factor may be generated by successively squaring $\mathbf{U}^{(\xi)}$. The number of matrix multiplications prescribed by Eq. 24 is thereby greatly reduced.

For a chain in which all bonds and all bond angles are identical, i.e., for a simple chain, Eqs. 24 and 26 reduce to

$$Z = \mathbf{J}^* \mathbf{U}^{n-2} \mathbf{J} \tag{27}$$

which is also the generalization of Eq. 23 for $\nu \geq 2$. The partition function for such a chain may be reduced to a scalar algebraic expression. To this end let $\mathbf{U}$ be diagonalized by the similarity transformation

$$\mathbf{A}^{-1} \mathbf{U} \mathbf{A} = \mathbf{\Lambda} \tag{28}$$

(see Appendix B for illustration of the method). Here $\mathbf{\Lambda}$ is the diagonal matrix comprising the eigenvalues λ_η of $\mathbf{U}$ as elements, the eigenvalues being the solutions of the secular equation

$$|\mathbf{U} - \lambda \mathbf{E}| = 0 \tag{29}$$

(see the following section) where **E** is the identity of order ν. For later convenience, we rewrite Eq. 28 as follows:

$$\mathbf{BUA} = \mathbf{\Lambda}$$

or

$$\mathbf{U} = \mathbf{A\Lambda B} \tag{28'}$$

where $\mathbf{B} = \mathbf{A}^{-1}$. Substitution of this equation in Eq. 27 yields

$$Z = \mathbf{J}^*\mathbf{A\Lambda}^{n-2}\mathbf{BJ} \tag{30}$$

Recalling the definitions of $\mathbf{J}^*$ and $\mathbf{J}$ given in Eq. 25, we find at once that

$$Z = [A_{11}\lambda_1^{n-2}, A_{12}\lambda_2^{n-2}, \ldots, A_{1\nu}\lambda_\nu^{n-2}]\begin{bmatrix} \sum_{\eta=1}^{\nu} B_{1\eta} \\ \vdots \\ \sum_{\eta=1}^{\nu} B_{\nu\eta} \end{bmatrix}$$

where $A_{\zeta\eta}$ and $B_{\zeta\eta}$ are elements of **A** and **B**. It follows that

$$Z = \sum_{\zeta=1}^{\nu} \Gamma_\zeta \lambda_\zeta^{n-2} \tag{31}$$

where

$$\Gamma_\zeta = A_{1\zeta} \sum_{\eta=1}^{\nu} B_{\zeta\eta} \tag{32}$$

This result is exact within the scope of validity of the rotational isomeric state model. It may be obtained by the alternative method given in Appendix C, which adheres more closely to the conventional methods of Ising lattice statistics.[19,21]

The partition function for a simple chain of n bonds may be expressed according to Eq. 31 as a polynomial in the eigenvalues λ_ζ raised to the power $n-2$. The coefficients Γ_ζ are determined by the corresponding eigenvectors and eigenrows of **U**. They may be expressed in terms of the eigenvalues, as is shown in the following section. For a long chain, the dominant term in Eq. 31 is the one for the largest eigenvalue, which we take to be λ_1, and if the chain is sufficiently long, this term dwarfs all others. The partition function is then given in satisfactory approximation by

$$Z \cong \Gamma_1 \lambda_1^{n-2}, \qquad \text{large } n \tag{33}$$

In most applications it is the logarithm of Z which is required. Hence, for sufficiently large n it will be permissible to ignore the coefficient Γ_1 and let

$$Z \cong \lambda_1^{n-2}, \qquad n \to \infty \tag{34}$$

The largest eigenvalue λ_1 of $\mathbf{U}$ assumes a role analogous to that of the bond partition function z for a chain in which rotations are mutually independent (see Eqs. 2, 6, and I-35). The two are not interchangeable, however, as this would imply equivalence of relationships for the two kinds of chains, namely, those for which the bond rotations are neighbor dependent and those for which they are mutually independent. The astonishingly simple result expressed by Eq. 34 could have been anticipated by inspection of Eq. 27, since the dominant element of a matrix raised to the nth power approaches the nth power of its largest eigenvalue as n increases without limit. We have chosen to develop the intervening equations in order to have at our disposal the scalar algebraic expressions, Eqs. 31 and 32, which are applicable, without approximation, to chains of any length.

The partition function, Eq. 26, for a chain of regularly repeating units, each unit comprising ξ bonds, can be treated similarly in terms of eigenvalues $\lambda_\zeta^{(\xi)}$ of the matrix $\mathbf{U}^{(\xi)}$ defined above. The partition function Z can be expressed as a polynomial in powers of these eigenvalues, i.e., in $(\lambda_\zeta^{(\xi)})^{x-2}$, after the manner of Eq. 31. Evaluation of the coefficients is complicated by the factors $\mathbf{U}^{(\xi')}$ and $\mathbf{U}^{(\xi'')}$ for terminal bond sequences. For long chains, however, the partition function is again approximated by

$$Z \cong (\lambda_1^{(\xi)})^x \tag{35}$$

the analog of Eq. 34.

In summary, the insuperable mathematical difficulties confronting direct evaluation of the configuration partition function according to Eq. 17 for large n are resolved by the methods set forth above. The configuration partition function may be calculated for any chain having a specified structural sequence of bonds through the use of Eq. 24. If, as in a copolymer, several types t of structural units are present, a corresponding number of matrices $\mathbf{U}_t^{(\xi)}$ will be required; each $\mathbf{U}_t^{(\xi)}$ represents the product of matrices for the ξ bonds comprising unit t. In order to obtain Z, the $\mathbf{U}_t^{(\xi)}$ must be multiplied in the order dictated by the structure of the particular molecule considered. They cannot, in general, be averaged in such a way as to allow representation of a collection of copolymer molecules, statistically distributed in structure, by an averaged matrix $\mathbf{U}^{(\xi)}$ raised to the power x of the number of units in the chain (see Chap. VI). Performance of the matrix multiplications of the $\mathbf{U}_t^{(\xi)}$ stipulated by Eq. 24 is easily carried out for chains comprising up to several hundreds of units by use of a digital computer.

For regular or homopolymeric chains made up of structural units of only one kind, a single product of matrices $\mathbf{U}^{(\xi)}$ is required to describe the body of the chain, and Eq. 26 may be applied. The computing labor may be greatly reduced by matrix squaring. That is, the product $[\mathbf{U}^{(\xi)}]^{x-2}$ required in Eq. 26 may be generated by successively squaring products of $\mathbf{U}^{(\xi)}$. The number of operations is of the order of $\ln x$ for large x, and hence does not become exorbitant for chains of any conceivable length. The same applies of course to simple chains ($\xi = 1$), for which Eq. 27 is applicable.

Calculation of the partition function by scalar methods will usually be preferred where applicable, and this is feasible for chains of uniform, or regularly repeating structure by use of Eq. 31. This equation is exact, within the limitations set by use of the rotational isomeric state approximation, and it is applicable to chain molecules of any length. The simplified expressions for Z provided by Eqs. 33 and 34 for very long chains of regular structure will prove especially useful. Their application to finite chains would entail neglect of the influence of the chain terminals, where the continuity of neighbor dependence is interrupted.

5. THE CONFIGURATION PARTITION FUNCTION FOR CHAINS WITH INTERDEPENDENT THREEFOLD POTENTIALS

The statistical weight matrix for a symmetrical chain subject to threefold potentials, with interdependence limited to first neighbors, is given with complete generality by Eq. 16. The characteristic equation obtained by substitution of Eq. 16 in Eq. 29 can be put in the form*

$$[\lambda - \sigma(\psi - \omega)][\lambda^2 - (1 + \sigma\psi + \sigma\omega)\lambda + \sigma(\psi + \omega - 2)] = 0 \quad (36)$$

The solutions of Eq. 36 are

$$\left.\begin{aligned} \lambda_{1,2} &= (\tfrac{1}{2})\left[1 + \sigma(\psi + \omega) \pm \sqrt{[1 - \sigma(\psi + \omega)]^2 + 8\sigma}\right] \\ \lambda_3 &= \sigma(\psi - \omega) \end{aligned}\right\} \quad (37)$$

* The factorability of this equation into a linear and a quadratic expression in λ indicates that the matrix $\mathbf{U}$ could have been simplified to 2×2 order. This form, obtained by combining the two *gauche* states, is

$$\mathbf{U} = \begin{bmatrix} 1 & 2\sigma \\ 1 & \sigma(\psi + \omega) \end{bmatrix}$$

These two states are equivalent by symmetry, but not identical, and when the bonds are considered singly, and not as successive pairs or larger sequences, no sacrifice of content would result from use of this expression for $\mathbf{U}$. In following chapters, where we shall be concerned with developing averages for products of transformation matrices $\mathbf{T}$, it will be imperative to maintain the distinction between g^+ and g^- states. We therefore retain $\mathbf{U}$ in the 3×3 form of Eq. 16.

The eigenvector and eigenrow matrices **A** and $\mathbf{B} = \mathbf{A}^{-1}$, readily found by usual procedures as illustrated in Appendix B, are

$$\mathbf{A} = \begin{bmatrix} 1-\lambda_2 & -(\lambda_1-1) & 0 \\ 1 & 1 & -1 \\ 1 & 1 & 1 \end{bmatrix} \tag{38}$$

$$\mathbf{B} = (\lambda_1-\lambda_2)^{-1}\begin{bmatrix} 1 & (\lambda_1-1)/2 & (\lambda_1-1)/2 \\ -1 & (1-\lambda_2)/2 & (1-\lambda_2)/2 \\ 0 & -(\lambda_1-\lambda_2)/2 & (\lambda_1-\lambda_2)/2 \end{bmatrix} \tag{39}$$

From Eqs. 31, 32, 38, and 39

$$Z = [(1-\lambda_2)/(\lambda_1-\lambda_2)]\lambda_1^{n-1} + [(\lambda_1-1)/(\lambda_1-\lambda_2)]\lambda_2^{n-1} \tag{40}$$

Γ_3 being equal to zero. Inasmuch as $\lambda_1 \geq 1 \geq \lambda_2$, for any combination of σ, ψ, ω with each ≥ 0, the coefficients of both λ_1^{n-1} and λ_2^{n-1} are non-negative. However, λ_2 may be negative, in which case the sign of the second term in Eq. 40 will be alternatingly positive and negative with n. Convergence ordinarily will be rapid.

If the bond rotational potentials are independent, then $\psi = \omega = 1$, giving $\lambda_1 = 1 + 2\sigma$ and $\lambda_2 = \lambda_3 = 0$. In this event, λ_1 reduces to z of Eq. 2, and Z is given by Eq. 6.

6. AVERAGE BOND CONFORMATIONS

The *statistical weight* $\Omega_{\{\phi\}}$ of a configuration $\{\phi\}$ represents the relative probability of its incidence, or its frequency of occurrence in a statistical mechanical ensemble of molecules at equilibrium. The *probability* that a given molecule occurs in this configuration equals the statistical weight divided by the sum of statistical weights for all possible configurations. This sum is just the configuration partition function Z (see Eq. 17). Hence the probability of a given configuration is

$$p_{\{\phi\}} = Z^{-1}\prod_{i=2}^{n-1} u_{\zeta\eta;\, i} \tag{41}$$

The distinction between statistical weights and probabilities is important, as is demonstrated by examples in the following sections.

Let us consider the probability $p_{\eta;\, i}$ that bond i is in state η. Thus defined, $p_{\eta;\, i}$ is an *a priori* probability in contrast to a *conditional* probability which would be subject to other contingencies such as the occurrence of certain nearby bonds in specified states. It will equal the quotient of the sum of the statistical weights for all configurations for which bond i is in state η, divided by Z. Algebraically expressed

$$p_{\eta;\,i} = Z^{-1}\mathbf{J}^*\left[\prod_{h=2}^{i-1}\mathbf{U}_h\right]\mathbf{U}'_{\eta;\,i}\left[\prod_{j=i+1}^{n-1}\mathbf{U}_j\right]\mathbf{J} \tag{42}$$

where $\mathbf{U}'_{\eta;\,i}$ is the matrix obtained from $\mathbf{U}_i$ by striking all elements except those of column η, those stricken being replaced by zeros. It will be apparent that by this device we retain precisely those terms meeting the condition $\phi_i = \phi_{\eta;\,i}$ in the sum of statistical weights $\Omega_{\{\phi\}}$ over all configurations $\{\phi\}$ of the chain.

Alternatively, $\mathbf{U}'_{\eta;\,i}$ may be defined as the partial derivative

$$\mathbf{U}'_{\eta;\,i} = \partial\mathbf{U}_i/\partial \ln \mathbf{u}_{\eta;\,i} \tag{43}$$

$\mathbf{U}_i$ being treated as a function of its column vectors $\mathbf{u}_{\eta;\,i}$.

The probability that bonds $i-1$ and i occur simultaneously in states ζ and η, respectively, is, similarly

$$p_{\zeta\eta;\,i} = Z^{-1}\mathbf{J}^*\left[\prod_{h=2}^{i-1}\mathbf{U}_h\right]\mathbf{U}'_{\zeta\eta;\,i}\left[\prod_{j=i+1}^{n-1}\mathbf{U}_j\right]\mathbf{J} \tag{44}$$

where $2 < i < n$, and

$$\mathbf{U}'_{\zeta\eta;\,i} = \partial\mathbf{U}_i/\partial \ln u_{\zeta\eta;\,i}$$

i.e., $\mathbf{U}'_{\zeta\eta;\,i}$ is the matrix obtained by striking all elements of $\mathbf{U}_i$ with the exception of $\mathrm{u}_{\zeta\eta;\,i}$. It will be observed that

$$p_{\eta;\,i} = \sum_{\zeta=1}^{\nu} p_{\zeta\eta;\,i} = \sum_{\zeta=1}^{\nu} p_{\eta\zeta;\,i+1} \tag{45}$$

Let the number of all bonds $i = 2$ to $n-1$ in rotational state η averaged over all configurations at equilibrium be

$$\langle n_\eta \rangle = (n-2)p_\eta \tag{46}$$

As defined by this equation, p_η is the *a priori* probability of state η averaged over all internal bonds $i = 2$ to $n-1$ of the chain. That is

$$p_\eta = (n-2)^{-1}\sum_{i=2}^{n-1} p_{\eta;\,i} \tag{47}$$

The sum resulting from substitution of Eq. 42 in 47 involves a set of terms which are matrix products. Its evaluation provides a simple illustration of a method which will find extensive use later for summing more complicated series of this general character.

This series representing $(n-2)p_\eta$ comprises each of the terms obtained from the product $\mathbf{U}_2 \cdots \mathbf{U}_{n-1}$ by replacing one of its factors by the $\mathbf{U}'$ corresponding to that factor. The only property of these matrices that needs to be heeded is their noncommutativity; in other respects it would suffice for the present to regard them as scalar quantities. Let the required terms be generated by starting with $\mathbf{U}_2$ and $\mathbf{U}_2'$, then appending factors $\mathbf{U}_3$ and $\mathbf{U}_3'$ to the former and $\mathbf{U}_3$ to the latter, thereby obtaining $\mathbf{U}_2\mathbf{U}_3$, $\mathbf{U}_2\mathbf{U}_3'$, and $\mathbf{U}_2'\mathbf{U}_3$. The next set of terms is obtained by appending $\mathbf{U}_4$ and $\mathbf{U}_4'$ to the first of the preceding three terms, and $\mathbf{U}_4$ to both of the latter two. Consider the set of distinct terms comprising $i-1$ factors generated in this manner. Let $\mathbf{S}_i$ denote the term $\mathbf{U}_2 \cdots \mathbf{U}_i$ having no primed factor and let $\mathbf{S}_i'$ denote the sum of all terms that have acquired one primed factor. Then let the subsets $\mathbf{S}_i$ and $\mathbf{S}_i'$ so defined be written as the elements of a row matrix and multiplied as follows:

$$[\mathbf{S}_i \quad \mathbf{S}_i'] \begin{bmatrix} \mathbf{U}_{i+1} & \mathbf{U}_{i+1}' \\ \mathbf{0} & \mathbf{U}_{i+1} \end{bmatrix}$$

where $\mathbf{0}$ is the null matrix of $v \times v$ order. The result of this multiplication is the set of terms for a sequence of $i+1$ bonds, i.e.,

$$[\mathbf{S}_{i+1} \quad \mathbf{S}_{i+1}']$$

these terms being separated into two subsets on the same basis as before. Repetition of the multiplication by the corresponding matrix for bond $i+2$ yields terms of the next higher degree in $\mathbf{U}$, etc. At each stage of the multiplication the term $\mathbf{S}_i$ begets a term $\mathbf{S}_{i+1}$ devoid of a primed factor, and also a term $\mathbf{S}_i\mathbf{U}_{i+1}'$ which belongs to the subset $\mathbf{S}_{i+1}'$. The former term emerges from the first column, and at the following stage of the generation process it will duplicate the performance of its predecessor; the latter term emerges from the second column and can therefore only interact with the second row of the succeeding square matrix. By the rule of matrix multiplication, terms emerging from the respective *columns* are multiplied by factors in the corresponding *rows* of the square matrix used in the following step. The square matrix is so arranged as to assure that once a term acquires a $\mathbf{U}'$, this term and its progeny will acquire only $\mathbf{U}$ matrices as succeeding factors.

It will be apparent that the required sum is generated identically by

$$[\mathbf{E} \quad \mathbf{0}] \left[\prod_{i=2}^{n-1} \hat{\mathbf{U}}_{\eta;\, i} \right] \begin{bmatrix} \mathbf{0} \\ \mathbf{E} \end{bmatrix} \tag{48}$$

where $\hat{\mathbf{U}}_{\eta;\, i}$ is the "super matrix"

$$\hat{\mathbf{U}}_{\eta;\, i} = \begin{bmatrix} \mathbf{U} & \mathbf{U}_\eta' \\ \mathbf{0} & \mathbf{U} \end{bmatrix}_i, \qquad 1 < i < n \tag{49}$$

the elements of which are matrices. The ordinal index i appended to the bracket is understood to apply to the quantities within. The desired mean probability is given therefore by

$$p_\eta = (n-2)^{-1} Z^{-1} \mathscr{J}^* \left[\prod_{i=2}^{n-1} \hat{\mathbf{U}}_{\eta;\,i} \right] \mathscr{J} \tag{50}$$

where $\mathscr{J}^*$ is the row consisting of a first element of unity and $2\nu - 1$ succeeding elements equal to zero; $\mathscr{J}$ is the column comprising ν zeros followed by ν elements equal to unity. That is

$$\begin{aligned} \mathscr{J}^* &= [\mathbf{J}^* \quad 0 \quad \cdots \quad 0] \\ &= [1 \quad 0 \quad \cdots \quad 0] \end{aligned} \tag{51}$$

and

$$\mathscr{J} = \begin{bmatrix} 0 \\ \vdots \\ 0 \\ \mathbf{J} \end{bmatrix} = \begin{bmatrix} 0 \\ \vdots \\ 0 \\ 1 \\ \vdots \\ 1 \end{bmatrix} \tag{52}$$

These definitions of $\mathscr{J}^*$ and $\mathscr{J}$ will be generalized for future use to represent a row and a column, respectively, *of whatever orders may be required to conform* with the matrices with which they are associated in a given instance. The first element of the former will invariably be unity and all remaining elements will be zeros; the latter will consist of the required number of zeros followed by $\mathbf{J}$, i.e., by ν final elements of unity.

The average incidence of pairs $\zeta\eta$ is given similarly by

$$p_{\zeta\eta} = \langle n_{\zeta\eta} \rangle / (n-3) = (n-3)^{-1} Z^{-1} \mathscr{J}^* \left[\prod_{i=2}^{n-1} \hat{\mathbf{U}}_{\zeta\eta;\,i} \right] \mathscr{J} \tag{53}$$

where

$$\hat{\mathbf{U}}_{\zeta\eta;\,i} = \begin{bmatrix} \mathbf{U} & \mathbf{U}'_{\zeta\eta} \\ \mathbf{0} & \mathbf{U} \end{bmatrix}_i, \qquad 2 < i < n \tag{54}$$

For $i = 2$ we let $\mathbf{U}'_{\zeta\eta} = \mathbf{0}$ in recognition of the absence of a preceding bond assigned to a rotational state.

This procedure is readily applicable to determination of temperature coefficients.[22] For example

$$dZ/dT = \mathbf{J}^* \left\{ \mathbf{U}'_{T;\,2} \prod_{i=3}^{n-1} \mathbf{U}_i + \mathbf{U}_2 \mathbf{U}'_{T;\,3} \prod_{i=4}^{n-1} \mathbf{U}_i + \cdots + \left[\prod_{i=2}^{n-2} \mathbf{U}_i \right] \mathbf{U}'_{T;\,n-1} \right\} \mathbf{J}$$

where $\mathbf{U}'_{T;i} = d\mathbf{U}_i/dT$. The lengthy expression for dZ/dT can be condensed to

$$dZ/dT = \mathscr{J}^* \left[\prod_{i=2}^{n-1} \hat{\mathbf{U}}_{T;i} \right] \mathscr{J} \tag{55}$$

where

$$\hat{\mathbf{U}}_{T;i} = \begin{bmatrix} \mathbf{U} & \mathbf{U}'_T \\ \mathbf{0} & \mathbf{U} \end{bmatrix}_i \tag{56}$$

The temperature coefficient of the numerator in Eq. 50 takes the similar form

$$d[Z(n-2)p_\eta]/dT = \mathscr{J}^* \left[\prod_{i=2}^{n-1} \hat{\hat{\mathbf{U}}}_{\eta,T;i} \right] \mathscr{J} \tag{57}$$

where

$$\hat{\hat{\mathbf{U}}}_{\eta,T;i} = \begin{bmatrix} \hat{\mathbf{U}}_\eta & \hat{\mathbf{U}}'_{\eta,T} \\ \mathbf{0} & \hat{\mathbf{U}}_\eta \end{bmatrix}_i \tag{58}$$

and

$$\hat{\mathbf{U}}'_{\eta,T;i} = d\hat{\mathbf{U}}_{\eta;i}/dT$$

The temperature coefficient dp_η/dT may be calculated by use of these general relationships applicable to chains of any length. The temperature coefficient of $p_{\zeta\eta}$ can be treated similarly.

For a homopolymer chain consisting of regularly repeating units, the matrix products can be expressed as powers of a factor corresponding to the product of matrices for the succession of bonds in a single unit. The methods discussed above are then applicable.

As prototypes of homopolymeric chains made up of repeating units of a single type, we consider simple chains having a repeating unit consisting of one bond only. Ordinal indexes may then be deleted in Eqs. 49, 54, 56, and 58 and the various products of matrices reduce to the powers of their arguments. Thus, for example, Eq. 50 for p_η may be replaced by

$$p_\eta = (n-2)^{-1} Z^{-1} \mathscr{J}^* \hat{\mathbf{U}}_\eta^{n-2} \mathscr{J} \tag{59}$$

The matrix $\hat{\mathbf{U}}$ is irreducible, i.e., it cannot be diagonalized by a similarity transformation. Hence, p_η cannot be expressed as a polynomial in powers of the eigenvalues of $\hat{\mathbf{U}}_\eta$. Simple expressions for the mean *a priori* probabilities $p_{\zeta\eta}$ and p_η in the limit of large n (or large x), however, may be obtained by resort to a more conventional procedure, albeit a less general one, as set forth below.

To this end, we consider Eq. 17 for Z. The average number of bond pairs in states $\zeta\eta$ is

$$\langle n_{\zeta\eta} \rangle = Z^{-1} \sum_{\{\varphi\}} n_{\zeta\eta} \Omega_{\{\phi\}}$$

where $n_{\zeta\eta}$ is the number of $\zeta\eta$ pairs in configuration $\{\phi\}$. The factor $u_{\zeta\eta}$

will occur raised to the power $n_{\zeta\eta}$ in $\Omega_{\{\phi\}}$. It follows from Eq. 17, and the use of a manipulation which is commonplace in statistical mechanics, that

$$\langle n_{\zeta\eta} \rangle = u_{\zeta\eta} Z^{-1}\, \partial Z/\partial u_{\zeta\eta} = \partial \ln Z/\partial \ln u_{\zeta\eta} \tag{60}$$

The average fraction of the $n-3$ relevant bond pairs in paired states $\zeta\eta$ is

$$p_{\zeta\eta} = \langle n_{\zeta\eta} \rangle/(n-3) = (n-3)^{-1}(\partial \ln Z/\partial \ln u_{\zeta\eta}) \tag{61}$$

Partial differentiation of Eq. 31 with respect to the element $u_{\zeta\eta}$ of $\mathbf{U}$, as required by Eqs. 60 and 61, leads to an inordinately complicated expression. We therefore confine our attention to the limiting result for large n where use of Eq. 34 for Z is legitimate. Then

$$p_{\zeta\eta} = \langle n_{\zeta\eta} \rangle/n = \partial \ln \lambda_1/\partial \ln u_{\zeta\eta}, \qquad n \to \infty \tag{62}$$

This equation may be used directly to calculate the various *a priori* pair probabilities. Another rendition of this result may sometimes be preferred. It is obtained through use of the relationship

$$\partial \lambda_\kappa/\partial \ln u_{\zeta\eta} = u_{\zeta\eta} A_{\eta\kappa} B_{\kappa\zeta} \tag{63}$$

where $A_{\eta\kappa}$ and $B_{\kappa\zeta}$ are elements of the eigenvector and eigenrow matrices $\mathbf{A}$ and $\mathbf{B}$. It follows that

$$p_{\zeta\eta} = u_{\zeta\eta} A_{\eta 1} B_{1\zeta}/\lambda_1, \qquad n \to \infty \tag{64}$$

A proof of Eq. 63 given by Birshtein and Ptitsyn[23] follows. According to Eq. 28, with $\mathbf{B}$ replacing $\mathbf{A}^{-1}$

$$\lambda_\kappa = \sum_{\alpha,\beta} B_{\kappa\alpha} u_{\alpha\beta} A_{\beta\kappa} \tag{65}$$

Hence

$$\partial \lambda_\kappa/\partial u_{\zeta\eta} = B_{\kappa\zeta} A_{\eta\kappa} + \sum_{\alpha,\beta} [u_{\alpha\beta} A_{\beta\kappa}\, \partial B_{\kappa\alpha}/\partial u_{\zeta\eta} + B_{\kappa\alpha} u_{\alpha\beta}\, \partial A_{\beta\kappa}/\partial u_{\zeta\eta}] \tag{66}$$

But from $\mathbf{UA} = \mathbf{A\Lambda}$ and $\mathbf{BU} = \mathbf{\Lambda B}$, respectively, we have

$$\left.\begin{aligned} &\sum_\beta u_{\alpha\beta} A_{\beta\kappa} = A_{\alpha\kappa} \lambda_\kappa \\ \text{and} \\ &\sum_\alpha B_{\kappa\alpha} u_{\alpha\beta} = \lambda_\kappa B_{\kappa\beta} \end{aligned}\right\} \tag{67}$$

Substitution of Eqs. 67 in the summation in Eq. 66 yields

$$\lambda_\kappa \Big[\sum_\alpha A_{\alpha\kappa}\, \partial B_{\kappa\alpha}/\partial u_{\zeta\eta} + \sum_\beta B_{\kappa\beta}\, \partial A_{\beta\kappa}/\partial u_{\zeta\eta}\Big] = \lambda_\kappa \sum_\alpha \partial(B_{\kappa\alpha} A_{\alpha\kappa})/\partial u_{\zeta\eta} = 0$$

since $\mathbf{BA} = \mathbf{E}$. With elimination of this term, Eq. 66 yields Eq. 63.

The fraction of the bonds of a long chain which occur in state η is

$$p_\eta = \sum_\zeta p_{\zeta\eta} = \sum_\zeta \partial \ln \lambda_1 / \partial \ln u_{\zeta\eta}, \qquad n \to \infty \tag{68}$$

Use of Eq. 63 gives

$$p_\eta = \lambda_1^{-1} A_{\eta 1} \sum_\zeta u_{\zeta\eta} B_{1\zeta}$$

which by Eq. 67 reduces to

$$p_\eta = A_{\eta 1} B_{1\eta}, \qquad n \to \infty \tag{69}$$

The probability that bond 2 of a simple chain of *any length n* occurs in state η is easily reduced to algebraic form. According to Eq. 42 with $i = 2$

$$\begin{aligned} p_{\eta;\,2} &= Z^{-1}\mathbf{J}^*\mathbf{U}'_{\eta;\,2}\,\mathbf{U}^{n-3}\mathbf{J} \\ &= Z^{-1}\mathbf{J}^*\mathbf{U}'_{\eta;\,2}\,\mathbf{A}\boldsymbol{\Lambda}^{n-3}\mathbf{B}\mathbf{J} \end{aligned} \tag{70}$$

Since $\mathbf{U}_2$ is to be expressed in the manner of Eq. 21, it follows that $\mathbf{J}^*\mathbf{U}'_{\eta;\,2}$ is the row matrix with one nonzero element, namely $u_{\eta;\,2}$, this being the $1,\eta$ element of $\mathbf{U}_2$. (For reasons given earlier, we expect $u_{\eta;\,2}$ to equal element $u_{1\eta}$ of $\mathbf{U}$.) Hence, the *a priori* probability for bond 2 is

$$\begin{aligned} p_{\eta;\,2} &= Z^{-1} u_{\eta;\,2}[A_{\eta 1} A_{\eta 2} \cdots A_{\eta\nu}]\boldsymbol{\Lambda}^{n-3} \begin{bmatrix} \sum B_{1\zeta} \\ \vdots \\ \sum B_{\nu\zeta} \end{bmatrix} \\ &= u_{\eta;\,2} \sum_{\kappa,\,\zeta} A_{\eta\kappa} B_{\kappa\zeta} \lambda_\kappa^{n-3} \Big/ \sum_\kappa \Gamma_\kappa \lambda_\kappa^{n-2} \end{aligned} \tag{71}$$

where the Γ_κ are given by Eq. 32. The same expression must obviously apply to the penultimate bond at the opposite end of a chain of the kind considered here, i.e., to bond $n - 1$. For chains of such a length that it is legitimate to retain only terms in the largest eigenvalue λ_1, we have the simple expression

$$p_{\eta;\,2} = p_{\eta;\,n-1} = u_{\eta;\,2} A_{\eta 1} / \lambda_1 A_{11}, \qquad n \to \infty \tag{72}$$

which follows from Eqs. 32 and 71.

The *a priori* probabilities $p_{\eta;\,i}$ and $p_{\zeta\eta;\,i}$ for $2 < i < n - 1$ do not lend themselves to similar reduction to simple form. They may be computed readily for chains of any length by use of Eqs. 42 and 44 however.

The average lengths of sequences of bonds in a given rotational state are sometimes of interest. They are obtainable directly from quantities derived above. For example, the average length of sequences of *trans* bonds in a long chain subject to threefold potentials is

$$\begin{aligned} \langle y_t \rangle &= \langle n_t \rangle / (\langle n_{tg^+} \rangle + \langle n_{tg^-} \rangle) \\ &= \langle n_t \rangle / 2\langle n_{tg^+} \rangle = p_t / 2p_{tg^+} \end{aligned} \tag{73}$$

The average length of *gauche* sequences of the same sign is

$$\langle y_{g^+}\rangle = \langle y_{g^-}\rangle = \langle n_{g^+}\rangle/(\langle n_{g+t}\rangle + \langle n_{g+g^-}\rangle)$$
$$= p_{g^+}(p_{g+t} + p_{g+g^-})^{-1} \tag{74}$$

7. SOME ILLUSTRATIVE APPLICATIONS

Consider a hypothetical chain with threefold potentials for which the statistical weights (see Eq. 16) are specified by $\sigma = \psi = 1$ and $\omega = 0$. Then

$$\mathbf{U} = \begin{bmatrix} 1 & 1 & 1 \\ 1 & 1 & 0 \\ 1 & 0 & 1 \end{bmatrix} \tag{75}$$

This chain departs from the rotational isomeric state analog of free rotation, for which all states have equal weights, only through the replacement of the 2,3 and 3,2 elements in **U** by zero. The incidences of various bond conformations differ markedly from the free rotation analog, however, as will be apparent from the numerical results which follow.

From Eqs. 37, 38, and 39

$$\lambda_{1,2} = 1 \pm \sqrt{2}, \qquad \lambda_3 = 1$$

$$\mathbf{A} = \begin{bmatrix} \sqrt{2} & -\sqrt{2} & 0 \\ 1 & 1 & -1 \\ 1 & 1 & 1 \end{bmatrix}; \qquad \mathbf{B} = \left(\frac{1}{4}\right)\begin{bmatrix} \sqrt{2} & 1 & 1 \\ -\sqrt{2} & 1 & 1 \\ 0 & -2 & 2 \end{bmatrix}$$

The second-order *a priori* probabilities for large n, calculated according to Eq. 64 and expressed for convenience in the form of a matrix, are

$$\mathbf{P} = [p_{\zeta\eta}] = \begin{bmatrix} (\sqrt{2}-1)/2 & (2-\sqrt{2})/4 & (2-\sqrt{2})/4 \\ (2-\sqrt{2})/4 & (\sqrt{2}-1)/4 & 0 \\ (2-\sqrt{2})/4 & 0 & (\sqrt{2}-1)/4 \end{bmatrix} \tag{76}$$

$$= \begin{bmatrix} 0.207 & 0.1465 & 0.1465 \\ 0.1465 & 0.1035 & 0 \\ 0.1465 & 0 & 0.1035 \end{bmatrix}$$

The first-order probabilities calculated from Eq. 69 are

$$p_t = 1/2; \qquad p_{g^+} = p_{g^-} = 1/4$$

The fraction of penultimate bonds which are *trans* is

$$p_{t;\,2} = \sqrt{2} - 1 = 0.414$$

according to Eq. 72.

The necessity for distinguishing probabilities from statistical weights will be evident from comparison of Eq. 76 with Eq. 75. Whereas the statistical weight for an allowed *gauche* state equals that for *trans*, its frequency of occurrence in a long chain is only half as great. Penultimate bonds occur in *gauche* states more frequently than do internal bonds, the former being subject to effects of only one neighbor. Note also the departure of the second-order *a priori* probabilities from products of the first-order probabilities. Thus,

$$\begin{aligned} p_{tt} &= 0.207 \quad < p_t^2 &&= 0.250 \\ p_{tg^+} &= 0.1465 > p_t\, p_{g^+} &&= 0.125 \\ p_{g^+g^+} &= 0.1035 > p_{g^+}^2 &&= 0.0625 \\ p_{g^+g^-} &= 0 \qquad < p_{g^+}\, p_{g^-} &&= 0.0625 \end{aligned}$$

These inequalities emphasize the fallacy of considering the bonds individually, without regard for their couplings with neighbors.

For a second example, let $\sigma = \frac{1}{2}$, $\psi = 1$, and $\omega = 0$; i.e.,

$$\mathbf{U} = \begin{bmatrix} 1 & \frac{1}{2} & \frac{1}{2} \\ 1 & \frac{1}{2} & 0 \\ 1 & 0 & \frac{1}{2} \end{bmatrix} \tag{77}$$

These weights approximate those for the polymethylene chain. Then

$$\lambda_{1,2} = (3 \pm \sqrt{17})/4 = 1.781, \ -0.281$$

and the required eigenvector and eigenrow are

$$\mathbf{A}_1 = \begin{bmatrix} 1.281 \\ 1 \\ 1 \end{bmatrix}; \qquad \mathbf{B}_1^* = [0.485, 0.1894, 0.1894]$$

giving values of $p_{\zeta\eta}$ as follows:

$$\mathbf{P} = \begin{bmatrix} 0.349 & 0.136 & 0.136 \\ 0.136 & 0.053 & 0 \\ 0.136 & 0 & 0.053 \end{bmatrix}$$

Also, $p_t = 0.621$, $p_{g^+} = p_{g^-} = 0.189$, and $p_{t;\,2} = 0.562$. This example again demonstrates the disparity between statistical weights and probabilities, and further illustrates the inequalities $p_{\zeta\eta} \neq p_\zeta p_\eta$.

The foregoing simple illustrative calculations are restricted to chains of infinite length. Corresponding quantities may be calculated for chains of any length through use of Eqs. 42 and 44. Rotational state populations for *n*-alkanes, calculated by Jernigan[22] using these equations, are shown in

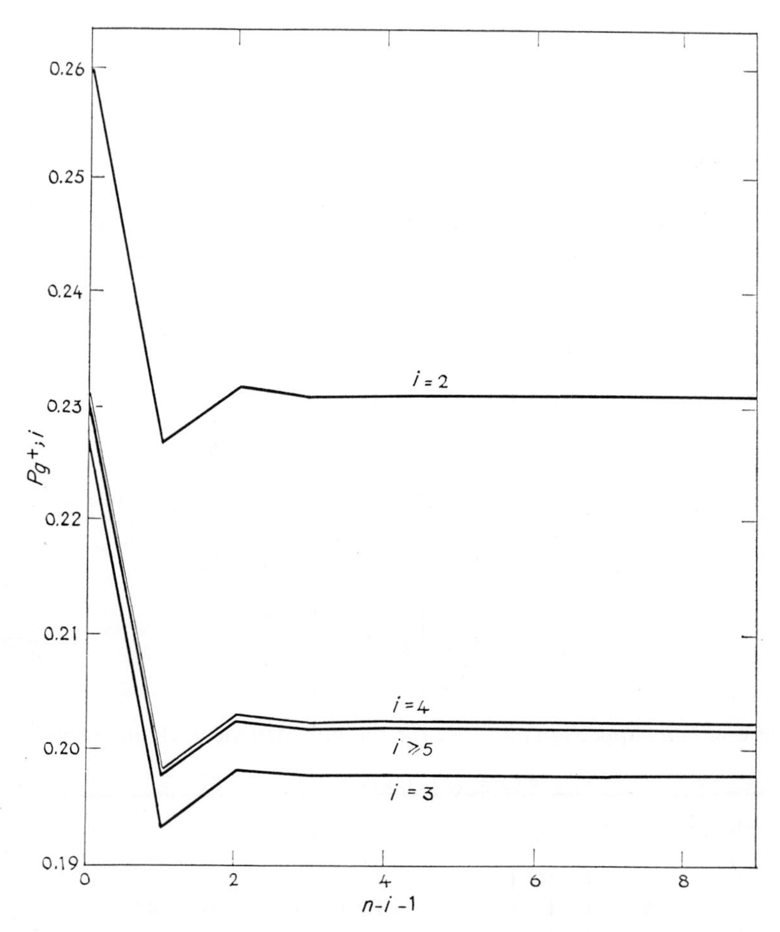

Fig. 4. Fraction of bonds $i = 2, 3, 4$, and ≥ 5 of a polymethylene chain which are in one of the *gauche* states ($p_{g+;i} = p_{g-;i}$) plotted as a function of the total chain length expressed as the number of bonds in excess of the minimum number $i + 1$ for an internal bond i. Calculations carried out[22] according to Eq. 42 with $\sigma = 0.54$, $\psi = 1.00$ and $\omega = 0.088$; see Chapter V.

Figs. 4, 5, and 6. Values of the parameters chosen were $\sigma = 0.54$, $\psi = 1.00$, $\omega = 0.088$, these being appropriate for the polymethylene chain at temperatures of 400–450°K (see Chap. V). The first order *a priori* probability $p_{g^+;i}$ [which equals $p_{g^-;i} = (1 - p_{t;i})/2$] is plotted in Fig. 4 against the chain length, expressed as $n - i - 1$, for the several values of i indicated. The quantity $n - i - 1$ expresses the number of bonds in excess of the minimum number required for occurrence of an ith internal bond. The points for various values of n are connected with straight lines. Redundancies

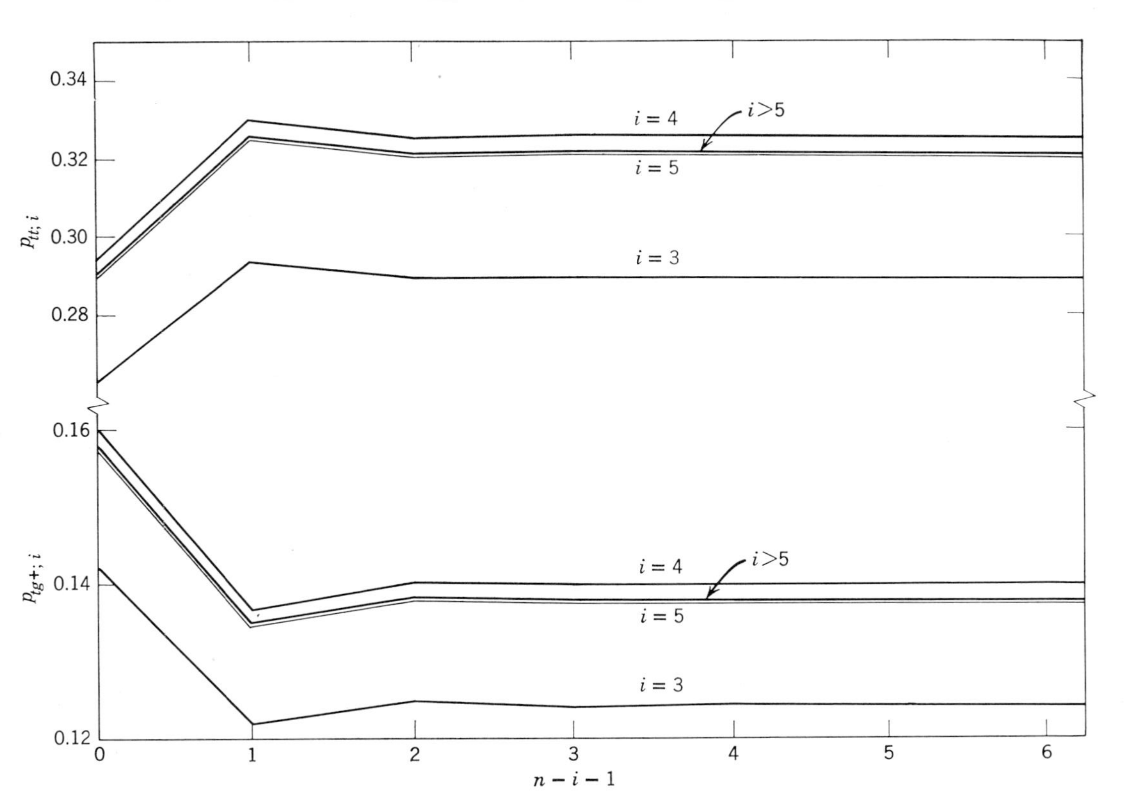

Fig. 5. Second-order probabilities $p_{tt;i}$ and $p_{tg+;i}$ for bond pair $i-1$, i of a PM chain, calculated for the values of i indicated, with each line plotted against $n-i-1$. Calculations were carried out according to Eq. 44 using parameters given in the legend for Fig. 4.[22] The points for tg^+ pairs $i-1$, i are also representative of the g^+t for pair $i'-1$, i' with $i' = n-i+2$.

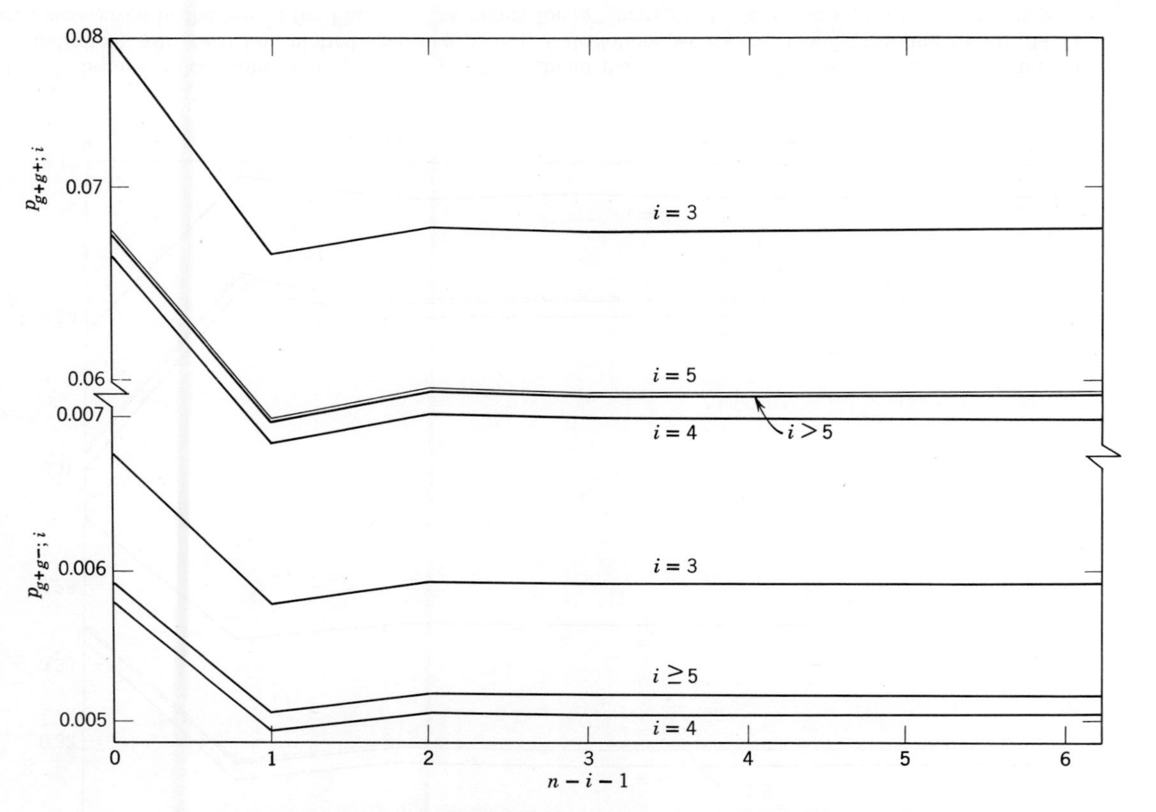

Fig. 6. Second-order probabilities $p_{g+g+;i}$ and $p_{g+g-;i}$ plotted as in Fig. 5.

occur for different values of i. Thus, $p_{g^+;i}$ for $i = 2$, $n = 4$ ($n - i - 1 = 1$) equals $p_{g^+;i}$ for $i = 3$, $n = 4$ ($n - i - 1 = 0$); likewise, $p_{g^+;i}$ is the same for $n = 5$, $i = 2$, and for $n = 5$, $i = 4$, etc. The values of $p_{g^+;i}$ display a pronounced oscillation both with i and with n. For given i, they rapidly approach their asymptotes with increase in n. The dependence of these asymptotic values on i can be ascertained from the intercepts of the lines at the right-hand margin of Fig. 4. Bonds beyond the fourth differ imperceptibly from those more removed from the end of the chain.

Second-order *a priori* probabilities for various states of bond pairs $i - 1$, i in polymethylene chains are plotted similarly in Figs. 5 and 6. Calculations were carried out through the use of Eq. 44.[22] A pronounced dependence on proximity to the ends of the chain is again apparent. Asymptotic values ($p_{tt} = 0.321$, $p_{g^+t} = 0.138$, $p_{g^+g^+} = 0.059$, $p_{g^+g^-} = 0.0052$) for internal bond pairs in a very long chain are rapidly approached with increase in i. The proportion of g^+g^- pairs is smaller than might be inferred from the numerical value of $\sigma^2\omega$. The effect of their occurrence on the average dimensions of the chain will be found to be disproportionately large, however (see Chap. V).

A further example of interest is the polyoxymethylene (POM) chain:

$$\diagup \overset{O}{\wedge} \diagdown CH_2 \diagup \overset{O}{\wedge} \diagdown CH_2 \diagup \overset{O}{\wedge} \diagdown CH_2 \diagup$$

It represents the simplest case of a chain having a two-bond repeat unit. Two statistical weight matrices are required for a description of its configurations, one for the pair of bonds C—O and O—C flanking an O atom and the other for the pair O—C and C—O flanking a CH_2 group. Expressed in the notation of Eq. 16 and in the form generally appropriate for any symmetric chain having a two-bond repeat unit, these matrices are, respectively

$$\mathbf{U}_a = \begin{bmatrix} 1 & \sigma & \sigma \\ 1 & \sigma\psi_a & \sigma\omega_a \\ 1 & \sigma\omega_a & \sigma\psi_a \end{bmatrix}; \qquad \mathbf{U}_b = \begin{bmatrix} 1 & \sigma & \sigma \\ 1 & \sigma\psi_b & \sigma\omega_b \\ 1 & \sigma\omega_b & \sigma\psi_b \end{bmatrix} \tag{78}$$

The partition function for the molecule terminated with methyl groups and conforming therefore to the formula $CH_3(—O—CH_2)_x—H$ where $x = n/2$, can be written

$$Z = \mathbf{J}^*(\mathbf{U}^{(2)})^{x-1}\mathbf{J} \tag{79}$$

according to Eq. 26, with $\xi = 2$. Thus

$$\mathbf{U}^{(2)} = \mathbf{U}_a \mathbf{U}_b \tag{80}$$

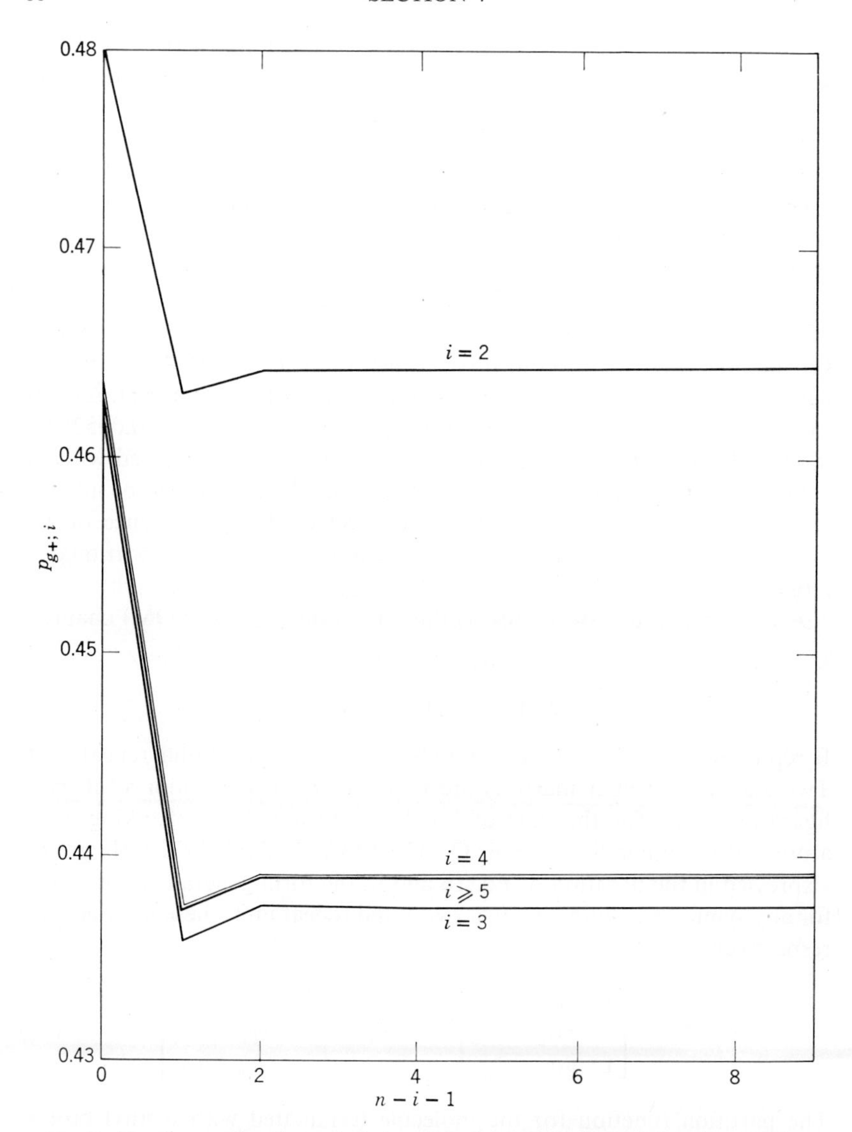

Fig. 7. Fractions of bonds $i = 2, 3, 4$, and 5 of the POM chain CH_3—$(O—CH_2)_{n/2-1}$—O—CH_3 which occur in *gauche* states ($p_{g+;i} = p_{g-;i}$) plotted against $n - i - 1$ as in Fig. 4. Calculations carried out[22] according to Eqs. 42 and 78, with $\sigma = 12$ and $\omega_a = 0.05$.

As shown in Chapter V, Section 7, the structural geometry of the POM chain assures that $\psi_a \cong \psi_b \cong 1$ and that $\omega_b \cong 0$. Plausible choices for the remaining parameters are $\sigma = 12$ and $\omega_a = 0.05$. These assignments are introduced here for illustrative purposes; the basis for their selection will be discussed in Chapter V, Section 7. Polyoxymethylene is illustrative of a chain in which *gauche* rotational states are preferred over *trans*; the preference is pronounced. The contrast to the PM chain is most striking in this respect.

The rotational state *a priori* probabilities for a given bond i or bond pair $i-1, i$ may be deduced by the methods of Eqs. 42 and 44. Similarly, the mean values of these probabilities for all bonds, or bond pairs, are calculable by use of Eqs. 50 and 53. Fractions $p_{g^+;i} = p_{g^-;i}$ of bonds $i = 2,3,4,$ and 5 in *gauche* states thus computed[22] for POM are plotted in Fig. 7 against $n - i - 1$ after the manner of Fig. 4 for PM. Convergence is rapid, due in part to the overriding preference for the *gauche* conformation. The representation of *a priori* probabilities for the two kinds of bond pairs comprising POM requires a much more extensive array of data, which is not reproduced here.

For long chains it suffices to let

$$Z \cong (\lambda_1^{(2)})^x = (\lambda_1^{(2)})^{n/2} \tag{81}$$

in accordance with Eq. 35, $\lambda_1^{(2)}$ being the largest eigenvalue of $\mathbf{U}^{(2)}$. The *a priori* probabilities of various conformational states within the chain are then conveniently obtained by differentiation of this eigenvalue. (See Eqs. 62 and 68.)

The required eigenvalue $\lambda_1^{(2)}$ of $\mathbf{U}^{(2)}$ is most readily obtained by first simplifying $\mathbf{U}_a$ and $\mathbf{U}_b$ through abolition of the distinction between g^+ and g^- states. This can be accomplished by adding the third column to the second and striking the third row and third column in both $\mathbf{U}_a$ and $\mathbf{U}_b$. (See the footnote on p. 72.) We thus reduce $\mathbf{U}^{(2)}$ to 2×2 order as follows:

$$\mathbf{U}^{(2)} = \begin{bmatrix} 1 + 2\sigma & 2(\sigma + \sigma^2\beta) \\ 1 + \sigma\alpha & 2\sigma + \sigma^2\alpha\beta \end{bmatrix} \tag{82}$$

where

$$\begin{aligned} \alpha &= \psi_a + \omega_a \\ \beta &= \psi_b + \omega_b \end{aligned} \tag{83}$$

The eigenvalues* are

$$\lambda_{1,2}^{(2)} = 2^{-1}[(1 + 4\sigma + \sigma^2\alpha\beta) \pm \sqrt{(1 - \sigma^2\alpha\beta)^2 + 8\sigma(1 + \sigma\alpha)(1 + \sigma\beta)}] \tag{84}$$

The average number of bonds in *gauche* states, inclusive of both g^+ and g^-, is

$$\langle n_g \rangle = \partial \ln Z / \partial \ln \sigma$$

* The third eigenvalue, eliminated in the reduction of order of the statistical weight matrices, is readily found to be $\lambda_3^{(2)} = \sigma^2(\psi_a - \omega_a)(\psi_b - \omega_b)$.

The *a priori* probability of *gauche* states

$$p_g = p_{g^+} + p_{g^-} = \langle n_g \rangle / n$$

is therefore

$$p_g = (\tfrac{1}{2})(\partial \ln \lambda_1^{(2)} / \partial \ln \sigma) \tag{85}$$

for a very long chain.

Second-order *a priori* probabilities are similarly obtained as follows:

$$\begin{aligned} p_{g^+g^+;a} + p_{g^-g^-;a} &= \partial \ln \lambda_1^{(2)} / \partial \ln \psi_a \\ &= (\psi_a / \lambda_1^{(2)})(\partial \lambda_1^{(2)} / \partial \alpha) \end{aligned} \tag{86}$$

Likewise

$$p_{g^-g^+;a} + p_{g^+g^-;a} = (\omega_a / \lambda_1^{(2)})(\partial \lambda_1^{(2)} / \partial \alpha) \tag{87}$$

The additional relations

$$\begin{aligned} p_{g^+t;a} &= p_{g^+} - p_{g^+g^+;a} - p_{g^+g^-;a} \\ p_{g^+t;a} &= p_{g^-t;a} = p_{tg^+;a} = p_{tg^-;a} \\ p_{tt;a} &= p_t - p_{tg^+;a} - p_{tg^-;a} \end{aligned} \tag{88}$$

together with $p_{g^+} = (\tfrac{1}{2})p_g$, complete the set of second-order *a priori* probabilities for bond pair *a* in a chain for which $n \gg 1$. The analogous relations for bond pair *b* are given by corresponding equations with *a* and *b* interchanged. Thus, all of the *a priori* probabilities can be deduced from the three partial derivatives $\partial\lambda_1^{(2)}/\partial\sigma$, $\partial\lambda_1^{(2)}/\partial\alpha$, and $\partial\lambda_1^{(2)}/\partial\beta$.

Results of numerical calculations carried out for the POM chain in the limit $n = \infty$ for values of the statistical weight parameters quoted above are presented in Table 3.

Table 3

Bond Conformational Probabilities Calculated for an Infinite POM Chain[22,a]

	Bond pair *b* flanking CH_2	Bond pair *a* flanking O
$p_g = 1 - p_t$	0.8777	
$p_{g^+g^+} = p_{g^-g^-}$	0.3820	0.3639
$p_{g^+g^-} = p_{g^-g^+}$	0	0.0182
$p_{tg^+} = p_{g^+t}$	0.0568	0.0567
p_{tt}	0.0087	0.0089

[a] Calculated according to Eqs. 85–88 using the statistical weight matrices 78 and the parameters $\sigma = 12$, $\psi_a = \psi_b = 1$, $\omega_a = 0.05$, $\omega_b = 0$.

The value of p_t, namely 0.1223, is larger than might be inferred from the ratio $1/\sigma = 1/12$ of the statistical weights for *trans* and *gauche* states. The greater incidence of *trans* bonds can be understood by observing that this state offers the adjoining bond a choice of two *gauche* states, each with a large weight σ.

The average length of an uninterrupted sequence of *gauche* states of the same sign can be calculated according to Eq. 74, adapted to a chain having a two-bond repeat unit. Thus, the average number of bonds in such a helical sequence is

$$\langle y_{g^+} \rangle = \langle y_{g^-} \rangle = 2p_{g^+}/(p_{g^+t;\,a} + p_{g^+g^-;\,a} + p_{g^+t;\,b} + p_{g^+g^-;\,b}) \tag{89}$$

From the results above we find $\langle y_{g^+} \rangle = 6.7$ bonds. The POM chain may thus be regarded as a random succession of helical segments of this average length. As the tabulation above shows, these sequences are terminated most frequently by incidence of a *trans* bond (rarely by a succession of two or more *trans* bonds), and, somewhat less frequently, by a *gauche* bond of opposite sign (the g^+g^- or g^-g^+ pair being of kind a, i.e., flanking an oxygen atom). Two helical segments joined by a g^+g^- or g^-g^+ pair will be of opposite chirality (hand); junction through a *trans* bond permits the adjoining segments to be either of the same or opposite chirality.

The parameters used for the foregoing calculations are arbitrary to some extent. Experimental evidence relating to the POM chain does not permit an accurate specification of them (see Chap. V, Sect. 7).

8. CONDITIONAL PROBABILITIES AND THE NON-MARKOFFIAN CHARACTER OF THE CONFIGURATIONAL STATISTICS

Let $q_{\zeta\eta;\,i}$ be the *conditional probability* that bond i is in state η, given that bond $i-1$ is in state ζ. That is, $p_{\zeta\eta;\,i} = p_{\zeta;\,i-1}q_{\zeta\eta;\,i}$; or

$$q_{\zeta\eta;\,i} = p_{\zeta\eta;\,i}/p_{\zeta;\,i-1} \tag{90}$$

which may be looked upon as a general definition of the conditional probability. For simple chains of large n, the indexes i and $i-1$ may be discarded and Eqs. 64 and 69 give

$$q_{\zeta\eta} = u_{\zeta\eta}A_{\eta 1}/A_{\zeta 1}\lambda_1, \qquad n \to \infty \tag{91}$$

For the first of the examples introduced in Section 7 above, we have according to Eq. 91

$$\mathbf{Q} \equiv [q_{\zeta\eta}] = \begin{bmatrix} \sqrt{2}-1 & 1-1/\sqrt{2} & 1-1/\sqrt{2} \\ 2-\sqrt{2} & \sqrt{2}-1 & 0 \\ 2-\sqrt{2} & 0 & \sqrt{2}-1 \end{bmatrix} \tag{92}$$

Equations 75, 76, and 92 illustrate the differences among statistical weights, *a priori* probabilities, and conditional probabilities.

In the case of finite chains the conditional probability depends on the length of the chain as well as on the location of the bond in relation to the chain ends. It is necessary, therefore, to retain the ordinal indexes in Eq. 90 and to use Eqs. 42 and 44 for the *a priori* probabilities, even if all **U** matrices are the same.

Probabilities of higher order are readily obtained from the results above. Thus, the *a priori* probability that bonds $i-2$, $i-1$, and i are in states $\phi_{i-2} = \phi_\varepsilon$, $\phi_{i-1} = \phi_\zeta$, and $\phi_i = \phi_\eta$ is

$$p_{\varepsilon\zeta\eta;\,i} = p_{\varepsilon\zeta;\,i-1}q_{\zeta\eta;\,i} = p_{\varepsilon\zeta;\,i-1}p_{\zeta\eta;\,i}/p_{\zeta;\,i-1} \tag{93}$$

These relationships, as presented, rest on the assumption that the interdependence of the bond rotations does not extend beyond first neighbors (see p. 65). If interactions of longer range should be important, these can be taken into account without revising the preceding development. It is necessary merely to redefine the states represented by Greek letter subscripts in the manner previously described (see pp. 65 and 66), a "state" being specified in terms of the rotation angles for a sequence of two or more bonds.

According to the definition of the conditional probability $q_{\zeta\eta}$, the sum of them for given ζ must equal unity. That is

$$\sum_\eta q_{\zeta\eta} = 1, \qquad \zeta = 1,2 \cdots \nu \tag{94}$$

That the $q_{\zeta\eta}$ meet this requisite may be confirmed by summing Eq. 91 with the aid of Eq. 67. They therefore form a stochastic matrix, each row of which sums to unity, a property conveniently expressed by

$$\mathbf{QJ} = \mathbf{J} \tag{95}$$

where **J** is the column vector defined in Eq. 25. The *a priori* probabilities sum to unity only after summation over both indexes, i.e.,

$$\sum_{\zeta,\eta} p_{\zeta\eta} = 1 \tag{96}$$

The statistical weights $u_{\zeta\eta}$, on the other hand, do not in general sum to unity in either manner. In particular, the elements of **U** cannot be reconciled with conditions like those set forth in Eqs. 94 or 95, even after scalar normalization of **U**. The statistical weight matrix is nonstochastic, and it is this feature which sets it apart fundamentally from the conditional probability matrix **Q**. Equations 75, 76, and 92 serve to illustrate these diverse characteristics of **U**, **P**, and **Q**.

Proceeding from Eq. 91, we find that

$$\mathbf{Q} = [u_{\zeta\eta} A_{\eta 1}/A_{\zeta 1}\lambda_1] = \lambda_1^{-1}\mathbf{D}_1^{-1}\mathbf{U}\mathbf{D}_1 \tag{97}$$

where $\mathbf{D}_1$ is the diagonal matrix comprising the elements of the eigenvector $\mathbf{A}_1$ of **U** corresponding to its largest eigenvalue λ_1, i.e.,

$$\mathbf{D}_1 = \begin{bmatrix} A_{11} & & \\ & \ddots & \\ & & A_{\nu 1} \end{bmatrix} \tag{98}$$

Thus, the matrix $\mathbf{D}_1$ provides the similarity transformation which converts **U** to stochastic form, after normalization by the largest eigenvalue λ_1. A stochastic matrix immediately suggests a Markoff process, and the fact that the statistical weight matrix is convertible to this form may seem to imply a resemblance between the statistics of the configuration of a chain molecule and a Markoff chain of events. Differences between the two overshadow their superficial similarities, however, as we proceed to show.

Each step, or element, in a Markoffian sequence of events is subject to *conditional probabilities* dependent upon the options selected for a finite number of its predecessors. Preceding events cast their spell; those to follow are not relevant to the one at hand. In this sense past and future differ fundamentally and are not interchangeable. Chemical synthesis of a copolymeric chain by successive addition of monomer units offers an example of a Markoff process.[24] The choice of chemical unit, and (or) the structure (*d* or *l*, *cis* or *trans*) of the one chosen, may be influenced by the preceding unit in the chain. Once a unit is incorporated in the growing chain, its selection is irrevocable and its chemical structure is permanently fixed, or at least we take this to be the case for purposes of this illustrative example.

In contrast, the conformational state of a bond is not assigned permanently. The configuration of a chain molecule is transitory, being subject to rapid changes brought about by thermal motions.* Moreover, numeration

* There is a finite chance, although it is unbelievably small, that if we could wait long enough and could follow the movements of all of the polymer molecules in our sample, we might get a glimpse of some specified configuration. But the event would be fleeting as well as rare, as follows from the rapidity of thermal motions on the one hand and from the incomprehensibly large number of configurations on the other.

of bonds in serial order from one end of the molecule rather than from the other has been introduced as an arbitrary convention. If reversal of the direction of numbering the bonds should lead to different results, then the analysis must certainly be fallacious. The methods devised for generating statistical weights are directional, to be sure; the analysis proceeds consecutively from bond 2 to bond $n - 1$, and depending on conventions adopted, the statistical weight matrix $\mathbf{U}_{n-1}$ may differ from $\mathbf{U}_2$. Also, the statistical weights have been formulated in such a way as to take into account the state of the preceding bond but not the succeeding one. These are matters of mere procedural convenience, however. They have been adopted in interests of developing a scheme for generating statistical weights in a systematic fashion. They do not in any way compromise the objective of generating statistical weights valid for the molecule *as a whole*, and they do not discriminate in favor of one end of the chain over the other. We have taken pains on pp. 62 and 63 to emphasize that the quantities required are the *total* energy of a configuration of the molecule, and the associated statistical weight for that configuration. Inasmuch as the range of interactions giving rise to neighbor dependence is short, the total configuration energy can be written as a sum of terms, each spanning the range of neighbor dependence (usually two bond rotations). The component terms in the total energy determine the corresponding statistical weights for neighboring bonds. These statistical weights have been construed to denote the weights appropriate for the given bond when the state of its predecessor has been specified. Their meaning in this respect should be confined, however, to the context of their usage for the purpose of generating the total statistical weight of a configuration for the molecule as a whole. Conformations adopted by bonds $j > i$ are just as important as those of bonds $j < i$ in affecting the choices open to bond i, and the *probabilities* relevant to bond i must take cognizance of the "future" as well as the "past" in the arbitrary left-to-right sequence.

The configurational statistics of a chain molecule do not, therefore, meet the terms of a Markoff process. Yet, we have seen how the basic statistical weight matrix $\mathbf{U}$ can be converted to the stochastic form appropriate for a Markoff chain. According to Eq. 97, the matrix $\mathbf{U}$ could be replaced by $\lambda_1 \mathbf{D}_1 \mathbf{Q} \mathbf{D}_1^{-1}$. For long chains the essential part would be $(\lambda_1 \mathbf{Q})^n$, and since the largest eigenvalue of $\mathbf{Q}$ is unity, this simplifies to λ_1^n. The important constitutent is λ_1, and this is an eigenvalue of $\mathbf{U}$ rather than of $\mathbf{Q}$. The statistical weight matrix $\mathbf{U}$ defines the stochastic matrix $\mathbf{Q}$, but the latter is not a substitute for the former in an operational sense.

Although the generation of statistical weights for chain molecules bears a resemblance to a Markoff process, the similarity is confined to superficialities of procedure. The Markoff *chain* and the polymer *chain con-*

figuration do not fit the same description and they require different interpretations.

9. CONCLUDING REMARKS

Systems of elements coupled to one another in a linear, or one-dimensional, sequence could have been the subjects of treatment of the foregoing development in the broadest sense of its application. The mathematical account given for such systems at equilibrium is comprehensive and capable of dealing with almost any detail of interest. With respect to its application to chain molecules, the treatment rests squarely on the rotational isomeric state model. Depending on physical features of the chain considered, one may regard this model either as a physical idealization or as a mathematical approximation of integrals by sums over small numbers of terms. No additional approximations are invoked beyond adoption of the rotational isomeric state model.[1,18]

Inaccuracies resulting from this model appear to be minor. They are undeniably outweighed by the advantage it offers in providing a tractable means for taking account of neighbor dependent effects, without which a realistic treatment of representative chain molecules cannot be achieved. Proof of the efficacy of the present procedure must await extension of the theory to chain dimensions and other properties of chain molecules, as treated in subsequent chapters. There we show that agreement with experiment is generally satisfactory and gratifyingly confirmatory of model and method.

One aspect of the rotational isomeric state model should be borne in mind. We have assumed, implicitly, that the set of rotational angles chosen to represent the states accessible to a given bond is valid for that bond, irrespective of the states of its neighbors. Rotations of neighbors are conceded to affect the energies of the states of the given bond, but *not the locations* of these states denoted by the ϕ_η. This assumption is of little relevance to the formulation of the configuration partition function and the probabilities for various local conformations considered above. It is a matter of greater concern in connection with treatments of average dimensions and other properties of chain molecules. Effects of rotations of neighboring bonds on the positions of the minima have turned out to be of marginal importance, however, in those instances where such effects have been explored. Should this not be true in a given case, incorporation of additional rotational states into the scheme may be required in order to bring into account the effects of perturbation of the locations of rotational states by neighboring rotations.

REFERENCES

1. S. Mizushima, *Structure of Molecules and Internal Rotation*, Academic Press, New York, 1954.
2. K. S. Pitzer, *Discussions Faraday Soc.*, **10**, 66 (1951).
3. W. B. Person and G. C. Pimentel, *J. Am. Chem. Soc.*, **75**, 532 (1953).
4. E. B. Wilson, Jr., *Advan. Chem. Phys.*, **2**, 367 (1959); *Pure Appl. Chem.*, **4**, 1 (1962).
5. D. R. Herschbach, *Intern. Symp. Mol. Struct. Spectry., Tokyo, 1962*, Butterworths, London (1963).
6. R. A. Bonham and L. S. Bartell, *J. Am. Chem. Soc.*, **81**, 3491 (1959); L. S. Bartell and D. A. Kohl, *J. Chem. Phys.*, **39**, 3097 (1963).
7. K. Kuchitsu, *J. Chem. Soc. Japan*, **32**, 748 (1959).
8. T. M. Birshtein and O. B. Ptitsyn, *Conformations of Macromolecules* (translated from the 1964 Russian ed., S. N. Timasheff and M. J. Timasheff), Interscience, New York, 1966.
9. L. Pauling, *The Nature of the Chemical Bond*, 3rd ed., Cornell University Press, Ithaca, 1960.
10. M. Karplus and R. G. Parr, *J. Chem. Phys.*, **38**, 1547 (1963); R. E. Wyatt and R. G. Parr, *ibid.*, **44**, 1529 (1966); J. P. Lowe and R. G. Parr, *ibid.*, **44**, 3001 (1966).
11. R. M. Pitzer and W. N. Lipscomb, *J. Chem. Phys.*, **39**, 1995 (1963); E. Clementi and D. R. Davis, *ibid.*, **45**, 2593 (1966); L. Pedersen and K. Morokuma, *ibid.*, **46**, 3941 (1967) ; R. M. Pitzer, *Ibid.*, **47**, 965 (1967).
12. D. W. Scott et al., *J. Phys. Chem.*, **65**, 1320 (1961).
13. G. J. Szasz, N. Sheppard, and D. H. Rank, *J. Chem. Phys.*, **16**, 704 (1948).
14. N. Sheppard and G. J. Szasz, *J. Chem. Phys.*, **17**, 86 (1949).
15. S. Mizushima and H. Okazaki, *J. Am. Chem. Soc.*, **71**, 3411 (1949).
16. N. P. Borisova and M. V. Volkenstein, *J. Struct. Chem. USSR* (*English Transl.*), **2**, 346, 437 (1961); N. P.Borisova, *Vysokomolekul. Soedin.*, **6**, 135 (1964).
17. A. Abe, R. L. Jernigan, and P. J. Flory, *J. Am. Chem. Soc.*, **88**, 631 (1966); R. A. Scott and H. A. Scheraga, *J. Chem. Phys.*, **44**, 3054 (1966).
18. M. V. Volkenstein, *Configurational Statistics of Polymeric Chains* (translated from the Russian ed., Serge N. Timasheff and M. J. Timasheff), Interscience, New York, 1963.
19. H. A. Kramers and G. H. Wannier, *Phys. Rev.*, **60**, 252 (1941).
20. E. Ising, *Z. Physik*, **31**, 253 (1925).
21. G. F. Newell and E. W. Montroll, *Rev. Mod. Phys.*, **25**, 353 (1953).
22. R. L. Jernigan and P. J. Flory, unpublished.
23. See ref. 8, Chapter IV.
24. L. Peller, *J. Chem. Phys.*, **43**, 2355 (1965).

CHAPTER IV

Moments of Chain Molecules

Most of the configuration-dependent properties of chain molecules of interest are combinations of vectorial quantities associated with the various elements of the chain. Thus, the dipole moment may be taken as the sum of contributions from individual bonds or groups, the end-to-end distance **r** as the sum of skeletal bond vectors, etc. Usually it is the square of the scalar magnitude of such a quantity which is of foremost interest. Stated more precisely, the quantity sought is the statistical mechanical average of this square over all configurations of the chain as a whole, each configuration being accorded its proper statistical weight. Mean-square dimensions (either $\langle r^2 \rangle_0$ or $\langle s^2 \rangle_0$), for example, may be constructed from sums of scalar products of bond vectors $\mathbf{l}_h \cdot \mathbf{l}_j$, which then must be averaged in the stated manner. Required for the purpose are the averaged products $\langle \mathbf{T}_h \cdots \mathbf{T}_{j-1} \rangle$ of transformation matrices for the intervening bonds (see Eq. I-29). Each transformation $\mathbf{T}_i$ is a function of the supplement θ_i of the bond angle and of the bond rotation angle ϕ_i. Since the former is taken to be fixed, $\mathbf{T}_i$ may be considered to be a function of ϕ_i; thus, $\mathbf{T}_i = \mathbf{T}_i(\phi_i)$. For a chain in which bond rotations are independent, the average of this product may be factored into the product of averages of the individual **T**'s (see Eq. I-38). For chains of interdependent bonds such a separation is not possible. The unfactored product must be averaged over all configurations of the chain. This is the problem to which we now turn our attention.

1. AVERAGES OF PRODUCTS OF THE TRANSFORMATION MATRICES

Consider a function $f_i(\phi_i)$ of the angle of rotation about bond i. Its average $\langle f_i \rangle$ over all configurations of the chain will be obtained by multiplying the statistical weight $\Omega_{\{\phi\}} = \prod_i u_{\zeta\eta;i}$ for each configuration by the value of $f_i(\phi_i)$ for the state of bond i in that configuration, and summing over all states of the chain. Through adoption of the following device, these operations can be incorporated into the well-known scheme presented in Chapter III for generating all of the statistical weights by multiplication of matrices **U**. Let $\mathbf{F}_i$ be the diagonal matrix

$$\mathbf{F}_i = \begin{bmatrix} f(\phi_\alpha) & & & \\ & f(\phi_\beta) & & \\ & & \ddots & \\ & & & f(\phi_\nu) \end{bmatrix}_i \equiv \begin{bmatrix} f_\alpha & & & \\ & f_\beta & & \\ & & \ddots & \\ & & & f_\nu \end{bmatrix}_i \tag{1}$$

of the values of the function $f_i(\phi_i)$ for the several rotational states. Insertion of $\mathbf{F}_i$ after $\mathbf{U}_i$ in the series of multiplications prescribed by Eq. III-24 for generating the partition function Z will accomplish the desired result. The product

$$\mathbf{U}_i\mathbf{F}_i = \begin{bmatrix} u_{\alpha\alpha} f_\alpha & \cdots & u_{\alpha\nu} f_\nu \\ \vdots & & \vdots \\ u_{\nu\alpha} f_\alpha & \cdots & u_{\nu\nu} f_\nu \end{bmatrix}$$

joins each $u_{\zeta\eta;i}$ with the value of f_i for state η, as required. It follows that†

$$\langle f_i \rangle = Z^{-1}\mathbf{J}^*\left(\prod_{h=2}^{i-1}\mathbf{U}_h\right)\mathbf{U}_i\mathbf{F}_i\left(\prod_{j=i+1}^{n-1}\mathbf{U}_j\right)\mathbf{J} \tag{2}$$

where $\mathbf{J}^*$ and $\mathbf{J}$ are the row and column defined by Eq. III-25.

The average of the product of functions $f(\phi_i)$ and $f(\phi_{i+1})$ of two consecutive bond angles, respectively, is obtained similarly by inserting $\mathbf{F}_i$ after $\mathbf{U}_i$ and $\mathbf{F}_{i+1}$ after $\mathbf{U}_{i+1}$. The average of the product of k such functions of consecutive bond angles is obtained by interdigitating the **F**'s with the appropriate **U**'s. Thus,

$$\langle f_i \cdots f_{j-1} \rangle = Z^{-1}\mathbf{J}^*[(\mathbf{U}_2 \cdots \mathbf{U}_{i-1})(\mathbf{U}_i\mathbf{F}_i\,\mathbf{U}_{i+1}\mathbf{F}_{i+1} \cdots \mathbf{U}_{j-1}\mathbf{F}_{j-1}) \times (\mathbf{U}_j \cdots \mathbf{U}_{n-1})]\mathbf{J} \tag{3}$$

where $j - i = k$. Expressions of this kind are conveniently written

$$\langle f_i^{(k)} \rangle = Z^{-1}\mathbf{J}^*\mathbf{U}_2^{(i-2)}(\mathbf{UF})_i^{(j-i)}\mathbf{U}_j^{(n-j)}\mathbf{J} \tag{4}$$

† The average value $\langle f_{i-1,i} \rangle$ of a function $f(\phi_{i-1}, \phi_i)$ of two consecutive rotation angles can be formulated by use of the matrix $[u_{\varsigma\eta} f_{\varsigma\eta}]_i$ in place of $\mathbf{U}_i\mathbf{F}_i$ in Eq. 2. The *a priori* probabilities p_η and $p_{\varsigma\eta}$ of the previous chapter are special cases of these averages; i.e., p_η is obtained from Eq. 2 by setting $f_\eta = 1$ and $f_\alpha = 0$ for $\alpha \neq \eta$, and $p_{\varsigma\eta}$ is obtained by taking $f_{\varsigma\eta} = 1$ and $f_{\alpha\beta} = 0$ for all other values of α and β.

where, for example, $\mathbf{U}_h^{(\mu)}$ means the *serial product of μ factors, the first of which is* $\mathbf{U}_h$. If all of the **U** matrices are identical and all of the **F** matrices are the same, these subscript indices may be omitted, and we have

$$\langle f_i^{(j-i)} \rangle = Z^{-1}\mathbf{J}^*\mathbf{U}^{i-2}(\mathbf{UF})^{j-i}\mathbf{U}^{n-j}\mathbf{J} \tag{5}$$

where the superscripts, written without parentheses, are exponents. The subscript index of f_i is retained in recognition of terminal effects. That is, the average of the product $f_i f_{i+1} \cdots f_{i+k-1}$ depends in general on its location within the bond sequence 2 to $n-1$; hence, the index i is significant.

Suppose the function $f_i(\phi_i)$ is one of the elements, e.g., $T_{xy;i}$, of $\mathbf{T}_i$. Equation 2 may be used to find its average value. Application of the same equation to each of the nine elements of $\mathbf{T}_i$ would yield the averaged matrix $\langle \mathbf{T}_i \rangle$. The labor could be condensed if a scheme were devised for combining the operations involved in taking these various averages separately. To this end we need to multiply each product $\Omega_{\{\phi\}}$ of statistical weights by the transformation matrix $\mathbf{T}_i(\phi_\eta)$ corresponding to the rotational state of bond i as represented by the second index of the contributing factor $u_{\zeta\eta;i}$. Summation of all such terms and division by Z would yield the desired average, namely, $\langle \mathbf{T}_i \rangle$. However, Eq. 2 as it stands is unsuitable for generating these terms by matrix multiplication. The difficulty arises from the fact that $\mathbf{T}_i$ is a 3×3 matrix and not a scalar like f. Thus, the analog of the diagonal matrix **F** of order ν defined by Eq. 1, is the pseudodiagonal matrix of order $3\nu \times 3\nu$

$$\begin{bmatrix} \mathbf{T}(\phi_1) & & \\ & \ddots & \\ & & \mathbf{T}(\phi_\nu) \end{bmatrix}_i \equiv \|\mathbf{T}_i\| \tag{6}$$

obtained by placing the **T** matrices for the various rotational states of bond i in diagonal array. It will be designated by the symbol on the right-hand side of Eq. 6. Being of a different order, this matrix cannot be interdigitated with the statistical weight matrices **U**.

Methods for circumventing this operational difficulty have been devised by Gotlib, Birshtein and Ptitsyn,[1,2] by Hoeve,[3] by Lifson,[4] and by Nagai.[5] Following these authors, we expand the order of **U** through replacement of each of its elements $u_{\zeta\eta}$ by a scalar matrix

$$u_{\zeta\eta;\,i}\mathbf{E}_3 = \begin{bmatrix} u_{\zeta\eta} & & \\ & u_{\zeta\eta} & \\ & & u_{\zeta\eta} \end{bmatrix}_i$$

where $\mathbf{E}_3$ is the identity of order three. We thus replace $\mathbf{U}_i$ by its *direct product*† with $\mathbf{E}_3$, i.e., by

$$\mathbf{U}_i \otimes \mathbf{E}_3 = \begin{bmatrix} u_{\alpha\alpha}\mathbf{E}_3 & \cdots & u_{\alpha\nu}\mathbf{E}_3 \\ \vdots & & \vdots \\ u_{\nu\alpha}\mathbf{E}_3 & \cdots & u_{\nu\nu}\mathbf{E}_3 \end{bmatrix}_i = \begin{bmatrix} u_{\alpha\alpha} & & & \cdots & u_{\alpha\nu} & & \\ & u_{\alpha\alpha} & & & & u_{\alpha\nu} & \\ & & u_{\alpha\alpha} & & & & u_{\alpha\nu} \\ \vdots & & & \ddots & & & \vdots \\ u_{\nu\alpha} & & & & u_{\nu\nu} & & \\ & u_{\nu\alpha} & & & & u_{\nu\nu} & \\ & & u_{\nu\alpha} & \cdots & & & u_{\nu\nu} \end{bmatrix}_i \tag{7}$$

This matrix is conformable with $\|\mathbf{T}\|$, and the product of $(\mathbf{U}_i \otimes \mathbf{E}_3)$ and $\|\mathbf{T}_i\|$ is just

$$(\mathbf{U}_i \otimes \mathbf{E}_3)\|\mathbf{T}_i\| = \begin{bmatrix} u_{\alpha\alpha}\mathbf{T}(\alpha) & \cdots & u_{\alpha\nu}\mathbf{T}(\nu) \\ \vdots & & \vdots \\ u_{\nu\alpha}\mathbf{T}(\alpha) & \cdots & u_{\nu\nu}\mathbf{T}(\nu) \end{bmatrix}_i$$

which combines the statistical weights $u_{\zeta\eta}$ and the transformations $\mathbf{T}$ in precisely the required manner. Statistical weight factors $u_{\zeta\eta}$ for units preceding and succeeding unit i may be introduced through premultiplication by $\mathbf{U}_{i-1} \otimes \mathbf{E}_3$ and postmultiplication by $\mathbf{U}_{i+1} \otimes \mathbf{E}_3$. By extending this scheme in the same manner used to arrive at Eq. 2 for the average of a scalar function, we have at our disposal a means for generating all of the statistical weight terms entering the molecular partition function Z, each term carrying the appropriate $\mathbf{T}(\phi_\eta)$ for bond i. These terms are included in the elements of

$$(\mathbf{U} \otimes \mathbf{E}_3)_2^{(i-2)}[(\mathbf{U}_i \otimes \mathbf{E}_3)\|\mathbf{T}\|_i](\mathbf{U} \otimes \mathbf{E}_3)_{i+1}^{(n-i-1)}$$

Owing to the absence of a skeletal rotation angle preceding bond 2, only those terms initiated by first row elements of $\mathbf{U}_2$ are of interest (see Chap. III). These comprise the first pseudorow, i.e., the first row of "elements" of order 3×3, and they may be selected to the exclusion of all others through premultiplication by

$$\mathbf{J}^* \otimes \mathbf{E}_3 = [\mathbf{E}_3 \mathbf{0} \cdots \mathbf{0}]$$

† The direct product, or Kronecker product, $\mathbf{C} \otimes \mathbf{D}$ of matrices $\mathbf{C}$ and $\mathbf{D}$ of orders $\mu \times \mu'$ and $\nu \times \nu'$, respectively, is the matrix of order $(\mu\nu) \times (\mu'\nu')$ obtained by replacing C_{st} by $C_{st}\mathbf{D}$ for each s, t. See Eq. 7 for illustration.

where the $\mathbf{0}$ denote null matrices of order three. All of the resulting terms are collected as a sum by postmultiplication with the $3v \times v$ "column"

$$\mathbf{J} \otimes \mathbf{E}_3 = \begin{bmatrix} \mathbf{E}_3 \\ \vdots \\ \mathbf{E}_3 \end{bmatrix}$$

The result thus obtained comprises the sum of every product of $\Omega_{\{\phi\}}$ with the $\mathbf{T}_i$ matrix corresponding to the state of bond i; this is the sum stated above to be required. The average of $\mathbf{T}_i$ is given therefore by

$$\langle \mathbf{T}_i \rangle = Z^{-1}(\mathbf{J}^* \otimes \mathbf{E}_3)(\mathbf{U} \otimes \mathbf{E}_3)_2^{(i-2)}[(\mathbf{U} \otimes \mathbf{E}_3)\|\mathbf{T}\|]_i(\mathbf{U} \otimes \mathbf{E}_3)_{i+1}^{(n-i-1)} \times (\mathbf{J} \otimes \mathbf{E}_3) \quad (8)$$

where $1 < i < n$. The subscript i appended to the bracketed expression will be understood to apply to both $\mathbf{U}$ and $\mathbf{T}$ therein. This expression is easily verified by trial, with the order of the $\mathbf{U}$ matrices taken to be 2×2 for simplicity.

Equation 8 may be condensed to

$$\langle \mathbf{T}_i \rangle = Z^{-1}[(\mathbf{J}^* \mathbf{U}_2^{(i-2)}) \otimes \mathbf{E}_3][(\mathbf{U} \otimes \mathbf{E}_3)\|\mathbf{T}\|]_i[(\mathbf{U}_{i+1}^{(n-i-1)}\mathbf{J}) \otimes \mathbf{E}_3] \quad (9)$$

by use of the *theorem on direct products*. According to this theorem, which is easily verified, $(\mathbf{A} \otimes \mathbf{B})(\mathbf{C} \otimes \mathbf{D}) = (\mathbf{AC}) \otimes (\mathbf{BD})$, where $\mathbf{A}$, $\mathbf{B}$, $\mathbf{C}$, and $\mathbf{D}$ are appropriately conformable matrices.†

By arguments corresponding exactly to those leading from Eqs. 2 to 4, the statistical mechanical average of a serial product of the transformation matrices $\mathbf{T}_i \cdots \mathbf{T}_{i+j-1}$ is found to be

$$\langle \mathbf{T}_i^{(j-i)} \rangle = Z^{-1}[(\mathbf{J}^* \mathbf{U}_2^{(i-2)}) \otimes \mathbf{E}_3][(\mathbf{U} \otimes \mathbf{E}_3)\|\mathbf{T}\|]_i^{(j-i)}[(\mathbf{U}_j^{(n-j)}\mathbf{J}) \otimes \mathbf{E}_3] \quad (10)$$

where $1 < i < j \leq n$. The case $j - i = 2$ which yields $\langle \mathbf{T}_i^{(2)} \rangle = \langle \mathbf{T}_i \mathbf{T}_{i+1} \rangle$ will be found to be illustrative of the operations prescribed by this equation.

In Eqs. 8, 9, and 10, and in others following as well, further use is made of the convention introduced above for expressing serial products of factors, whereby the ordinal number of the first factor is denoted by a subscript and the total number of factors in the product by a superscript in

† Multiple direct products, like ordinary matrix products, are associative; i.e., $(\mathbf{A} \otimes \mathbf{B}) \otimes \mathbf{C} = \mathbf{A} \otimes (\mathbf{B} \otimes \mathbf{C})$. The same does not hold ordinarily for mixed products like $(\mathbf{AB}) \otimes \mathbf{C}$. Parentheses are therefore required in Eq. 9 and elsewhere to specify the order in which the two operations are performed.

parentheses. Subscripts external to brackets or parentheses apply to all quantities such as **U** and **T** enclosed therein.

For a simple chain of identical bonds, ordinal subscripts on the right-hand side of Eqs. 8–10 (but not on the left-hand side; see p. 97) may be omitted and the parentheses may be removed from superscripts, which then become powers. In this case, in analogy to Eq. 5, we have for the average of the product of transformation matrices commencing with the ith

$$\langle \mathbf{T}_i^{(j-i)} \rangle = Z^{-1}[(\mathbf{J}^*\mathbf{U}^{i-2}) \otimes \mathbf{E}_3][(\mathbf{U} \otimes \mathbf{E}_3)\|\mathbf{T}\|]^{j-i}[(\mathbf{U}^{n-j}\mathbf{J}) \otimes \mathbf{E}_3] \tag{11}$$

Equations 8–11 are generalizations of expressions for infinite chains derived originally by Gotlib,[1] by Birshtein and Ptitsyn,[1,2] and also by Hoeve.[3] A somewhat different formalism was used by Lifson[4] and by Nagai.[5] Their results, also for infinite chains, may be shown to be equivalent to those of Gotlib, of Birshtein and Ptitsyn, and of Hoeve.†

2. MEAN-SQUARE MOMENTS[1–7]

Let $\mathbf{m}_i$ and $\mathbf{m}_j$ be two vectors associated with bonds i and j, respectively. As in Chapter I, Section 8, these may be the bond vectors $\mathbf{l}_i$ and $\mathbf{l}_j$; they may be dipole moments, or other properties expressible as vectors and associated with specific skeletal bonds. In any case, we take $\mathbf{m}_i$ and $\mathbf{m}_j$ to be uniquely defined in their respective reference frames. That is, the components of $\mathbf{m}_i$ expressed in the coordinate system for bond i are asserted to be invariant to the configuration $\{\phi\}$ of the chain as a whole.

The scalar product of these two vectors for a given configuration of the chain is

$$\mathbf{m}_i \cdot \mathbf{m}_j = \mathbf{m}_i^T \mathbf{T}_i^{(j-i)} \mathbf{m}_j \tag{12}$$

† In the scheme of Lifson[3] and Nagai,[4] Eq. 11 would take the form

$$\langle \mathbf{T}_i^{(j-i)} \rangle = \mathbf{Z}^{-1} \, [\mathbf{E}_3 \otimes (\mathbf{J}^* \mathbf{U}^{i-2})][(\mathbf{E}_3 \otimes \mathbf{U})\mathscr{T}]^{j-i} \, [\mathbf{E}_3 \otimes (\mathbf{U}^{n-j}\mathbf{J})]$$

where $\mathscr{T}$ is the square matrix of order 3ν comprising pseudoelements

$$\mathscr{T}_{xy} = \begin{bmatrix} \mathrm{T}_{xy}(\phi_1) & & & \\ & \cdot & & \\ & & \cdot & \\ & & & \cdot \\ & & & & \mathrm{T}_{xy}(\phi_\nu) \end{bmatrix}$$

in 3×3 array; x, y are representative Cartesian coordinates. Although this expression for $\langle \mathbf{T}_i^{(j-i)} \rangle$ differs from Eq. 11 in form, the two may be shown to be related through transformations involving elementary matrices.

where $\mathbf{m}_i^T$ and $\mathbf{m}_j$ on the right-hand side are the representations of these vectors by their components expressed in row and column form in their respective reference frames, i and j. The statistical mechanical average of this scalar product, i.e., the average over all configurations of the chain as a whole, is given by Eq. I-31, which in present notation may be written

$$\langle \mathbf{m}_i \cdot \mathbf{m}_j \rangle = \mathbf{m}_i^T \langle \mathbf{T}_i^{(j-i)} \rangle \mathbf{m}_j \tag{13}$$

The result obtained by substitution of Eq. 10 into Eq. 13 is

$$\langle \mathbf{m}_i \cdot \mathbf{m}_j \rangle = Z^{-1} \mathbf{m}_i^T [(\mathbf{J}^* \mathbf{U}_1^{(i-1)}) \otimes \mathbf{E}_3][(\mathbf{U} \otimes \mathbf{E}_3) \| \mathbf{T} \|]_i^{(j-i)} \times [(\mathbf{U}_j^{(n-j)} \mathbf{J}) \otimes \mathbf{E}_3] \mathbf{m}_j \tag{14}$$

An additional factor $\mathbf{U}_1$ has been incorporated into Eq. 14 in order to allow the first bond of the chain to be included, i.e., so that the ranges of i and j become $1 \leq i < j \leq n$. Preservation of the statistical weights requires setting $\mathbf{U}_1 = \mathbf{E}_\nu$. The additional factor in Eq. 14 is therefore of no consequence whatever if $i > 1$. If $i = 1$, then the meaning ascribed to $\mathbf{m}_1$ and $\mathbf{T}_1$ bears examination. Since definition of a coordinate system (see Fig. I-3) for bond 1 would require a preceding bond in the chain, neither the transformation matrix $\mathbf{T}_1$ nor the vector $\mathbf{m}_1$ can be specified in full. In fact, $\mathbf{m}_1$ can only have a component along bond 1, i.e., $\mathbf{m}_1^T = [m_1 \quad 0 \quad 0]$. To specify other components would necessarily imply the presence of a bond preceding bond 1, and hence an additional bond would need to be taken into account for specification of the chain configuration. Renumbering of the skeletal bonds would then be required. Without sacrifice of generality, therefore, we may assume a vector $\mathbf{m}_1$ associated with the first bond to be represented by a single component, as above. It follows that only the first row of $\mathbf{T}_1$ is called into service. The elements of this row (see Eq. I-25) depend only on θ_1, which is defined, and not on ϕ_1, which is not. All ambiguities are avoided therefore by taking $\mathbf{U}_1 = \mathbf{E}_\nu$ and acknowledging $\mathbf{m}_1$ to consist (at most) of a single component along the bond direction. If $\mathbf{m}_1 = \mathbf{l}_1$, fulfillment of this condition is automatic.

In the interests of simplifying subsequent operations, it is desirable to juxtapose $\mathbf{m}_j$ with the $\mathbf{U}_j$ in Eq. 14, and likewise to bring $\mathbf{m}_i^T$ into juxtaposition with $\mathbf{U}_i$. All quantities representing a given bond of the chain may thereby be introduced in the logical order of the bond sequence. This objective may be achieved through use of the identities

$$\begin{aligned} [(\mathbf{U}_j^{(n-j)} \mathbf{J}) \otimes \mathbf{E}_3] \mathbf{m}_j &= (\mathbf{E}_\nu \otimes \mathbf{m}_j) \mathbf{U}_j^{(n-j)} \mathbf{J} \\ \mathbf{m}_i^T [(\mathbf{J}^* \mathbf{U}_1^{(i-1)}) \otimes \mathbf{E}_3] &= \mathbf{J}^* \mathbf{U}_1^{(i-1)} (\mathbf{E}_\nu \otimes \mathbf{m}_i^T) \end{aligned} \tag{15}$$

which may be verified by trial. Alternatively, these equations may be derived algebraically by an appeal to the theorem on direct products, supplemented by the observation that the direct product of a matrix with unity is the matrix itself. Thus,

$$[(\mathbf{U}_j^{(n-j)}\mathbf{J}) \otimes \mathbf{E}_3]\mathbf{m}_j = [(\mathbf{U}_j^{(n-j)}\mathbf{J}) \otimes \mathbf{E}_3](1 \otimes \mathbf{m}_j)$$

Since a column is conformable with unity, the right-hand member condenses according to the direct product theorem to

$$(\mathbf{U}_j^{(n-j)}\mathbf{J}) \otimes \mathbf{m}_j$$

which may then be separated to

$$(\mathbf{E}_\nu \otimes \mathbf{m}_j)[(\mathbf{U}_j^{(n-j)}\mathbf{J}) \otimes 1] = (\mathbf{E}_\nu \otimes \mathbf{m}_j)\mathbf{U}_j^{(n-j)}\mathbf{J}$$

The second of Eqs. 15 can be proved similarly. Rearrangements of this type will be used frequently in the following pages.

Substitution of Eq. 15 into Eq. 14 gives[7]

$$\langle \mathbf{m}_i \cdot \mathbf{m}_j \rangle = Z^{-1}\mathbf{J}^*\mathbf{U}_1^{(i-1)}(\mathbf{E}_\nu \otimes \mathbf{m}_i^T)[(\mathbf{U} \otimes \mathbf{E}_3)\|\mathbf{T}\|]_i^{(j-i)}(\mathbf{E}_\nu \otimes \mathbf{m}_j)\mathbf{U}_j^{(n-j)}\mathbf{J} \tag{16}$$

Not only do all factors for each bond occur in Eq. 16 in the order of the bond sequence in the chain; we have also reduced the order of the matrices preceding and following the central factor in brackets in this equation. The orders of matrices required for following operations are greatly reduced by this rearrangement of Eq. 14 to Eq. 16.

Suppose that we wish to find the square of the magnitude of the vector

$$\mathbf{M} = \sum_{i=1}^{n} \mathbf{m}_i \tag{17}$$

averaged over all configurations of the chain. The square is given by

$$M^2 = \mathbf{M} \cdot \mathbf{M} = \sum_{i,j} \mathbf{m}_i \cdot \mathbf{m}_j$$

with the summations running from 1 to n. Its average is

$$\begin{aligned}\langle M^2 \rangle &= \sum_{i,j} \langle \mathbf{m}_i \cdot \mathbf{m}_j \rangle \\ &= \sum_1^n m_i^2 + 2 \sum_{0<i<j\leq n} \langle \mathbf{m}_i \cdot \mathbf{m}_j \rangle \end{aligned} \tag{18}$$

By substitution of Eq. 16 into Eq. 18

$$\begin{aligned}\langle M^2 \rangle = \sum_1^n m_i^2 + 2Z^{-1}\mathbf{J}^* \sum_{0<i<j\leq n} \mathbf{U}_1^{(i-1)}(\mathbf{E}_\nu \otimes \mathbf{m}_i^T) \\ \times [(\mathbf{U} \otimes \mathbf{E}_3)\|\mathbf{T}\|]_i^{(j-i)}(\mathbf{E}_\nu \otimes \mathbf{m}_j)\mathbf{U}_j^{(n-j)}\mathbf{J}\end{aligned} \tag{19}$$

The double sum in this equation is of the form

$$\sum_{0<i<j\leq n} U_1^{(i-1)}P_i S_i^{(j-i)} Q_j U_j^{(n-j)}$$

where U, P, S, and Q correspond to quantities in Eq. 19 which are identifiable by comparison. These quantities are here represented as scalars; their noncommutativity is respected, however. Each term of this double sum is characterized by a block of factors S starting at i, and running consecutively for a total of $j - i$ such factors. Bonds preceding and following the block are represented by powers of U. The procedure which follows for evaluation of a sum of this general character requires separation of factors of the same ordinal index. Hence, the foregoing sum is preferably written

$$\sum_{0<i<j\leq n} U_1^{(i-1)}(P_i S_i) S_{i+1}^{(j-i-1)}(Q_j U_j) U_{j+1}^{(n-j-1)} \tag{20}$$

Such a summation can be evaluated exactly[6,7] by extension of the method introduced in the preceding chapter (see Eqs. III-49 and III-50), and applied there to summation over a single index. Consider the operations embodied in

$$[1 \quad 0 \quad 0] \begin{bmatrix} U & PS & 0 \\ 0 & S & QU \\ 0 & 0 & U \end{bmatrix}_1^{(n-1)} \begin{bmatrix} 0 \\ Q \\ 1 \end{bmatrix}_n \tag{21}$$

Multiplication of the row in expression 21 into the first of the $n - 1$ square matrices yields $[U_1, P_1S_1, 0]$. The next multiplication yields the row

$$[U_1U_2, U_1P_2S_2 + P_1S_1S_2, P_1S_1Q_2U_2]$$

Terms in a given element of this row multiply elements in the corresponding row of the square matrix at the next step. Thus, at each step a new sequence of S's is started. The term $U_1^{(i-1)}P_iS_i$, which initiates the new sequence, is delivered to the second element of the product row. This element includes all terms of the character $U_1^{(i-h-1)}P_{i-h}S_{i-h}^{(h+1)}$, with $0 \leq h < i$. Each of these terms gives rise in the next step to two terms: $U_1^{(i-h-1)}P_{i-h}S_{i-h}^{(h+2)}$ and $U_1^{(i-h-1)}P_{i-h}S_{i-h}^{(h+1)}Q_{i+1}U_{i+1}$. The former term remains in the second element; the latter is relegated to the third element, where its fate is sealed, for it is destined to acquire none but U factors throughout all remaining multiplications.

After the $n - 1$ multiplications by the square matrix have been completed, the resulting row is multiplied by the final column in expression 21. The first element of the row consists of the single term $U_1^{(n-1)}$. It is of no interest and will be rejected by the column. Most of the required terms occur in the third element of the row; they represent completed sequences of S's, terminated by Q and followed by one or more U's. They are collected without alteration by the third element of the final column. Terms of the second element of the row comprise sequences of S's which have not acquired a factor Q, but are due to be concluded with such a factor at the final bond of the chain. The second element of the column performs this function.

Thus, the operations stipulated in expression 21 reproduce identically the double sum represented in expression 20, this being the prototype of the double sum in Eq. 19.

Pursuant to the transcription of these results to the problem of present interest, we identify S in expressions 20 and 21 with

$$(\mathbf{U} \otimes \mathbf{E}_3)\|\mathbf{T}\|$$

of Eq. 19. Similarly, we identify PS with

$$(\mathbf{E}_\nu \otimes \mathbf{m}^T)[(\mathbf{U} \otimes \mathbf{E}_3)\|\mathbf{T}\|] = (\mathbf{U} \otimes \mathbf{m}^T)\|\mathbf{T}\|$$

and QU with

$$(\mathbf{E}_\nu \otimes \mathbf{m})\mathbf{U} = \mathbf{U} \otimes \mathbf{m}$$

The required square matrix is therefore[6,7]

$$\mathscr{F}_i = \begin{bmatrix} \mathbf{U} & (\mathbf{U} \otimes \mathbf{m}^T)\|\mathbf{T}\| & \mathbf{0} \\ \mathbf{0} & (\mathbf{U} \otimes \mathbf{E}_3)\|\mathbf{T}\| & \mathbf{U} \otimes \mathbf{m} \\ \mathbf{0} & \mathbf{0} & \mathbf{U} \end{bmatrix}_i \tag{22}$$

where the $\mathbf{0}$ represent null matrices of the orders (not necessarily square; see p. 105) required to conform. For $i = 1$, the statistical weight matrix is to be equated to $\mathbf{E}_\nu$ as stipulated above. Evaluating the double sum in Eq. 19 according to expression 21, we have therefore

$$\langle M^2 \rangle = \sum_i m_i^2 + 2Z^{-1}\mathbf{J}^*[\mathbf{E}_\nu \mathbf{0} \cdots \mathbf{0}]\mathscr{F}_1^{(n-1)} \begin{bmatrix} \mathbf{0} \\ \mathbf{E}_\nu \otimes \mathbf{m}_n \\ \mathbf{E}_\nu \end{bmatrix} \mathbf{J} \tag{23}$$

The first term in Eq. 23 may be incorporated into the main term representing contributions to $\langle M^2 \rangle$ from $\langle \mathbf{m}_i \cdot \mathbf{m}_j \rangle$ for $i \neq j$ by replacing the generator matrix with

$$\mathscr{G}_i = \begin{bmatrix} \mathbf{U} & (\mathbf{U} \otimes \mathbf{m}^T)\|\mathbf{T}\| & (m^2/2)\mathbf{U} \\ \mathbf{0} & (\mathbf{U} \otimes \mathbf{E}_3)\|\mathbf{T}\| & \mathbf{U} \otimes \mathbf{m} \\ \mathbf{0} & \mathbf{0} & \mathbf{U} \end{bmatrix} \tag{24}$$

Then

$$\langle M^2 \rangle = 2Z^{-1}\mathbf{J}^*[\mathbf{E}_\nu \mathbf{0} \cdots \mathbf{0}]\mathscr{G}_1^{(n-1)} \begin{bmatrix} (m_n^2/2)\mathbf{E}_\nu \\ \mathbf{E}_\nu \otimes \mathbf{m}_n \\ \mathbf{E}_\nu \end{bmatrix} \mathbf{J} \tag{25}$$

This result may be simplified by introduction of the following extensions of previous definitions. Let

$$\begin{bmatrix} (m_n^2/2)\mathbf{E}_\nu \\ \mathbf{E}_\nu \otimes \mathbf{m}_n \\ \mathbf{E}_\nu \end{bmatrix} \mathbf{J} = \mathscr{G}_n \mathscr{J} \tag{26}$$

where $\mathscr{J}$ is the column defined in Chapter III (see Eq. III-52) and comprising (in this instance) 4ν zeros followed by ν elements equal to unity. If we take $\mathbf{U}_n = \mathbf{E}_\nu$, then $\mathscr{G}_n$ may be included within the definition given by Eq. 24. The last pseudocolumn only of $\mathscr{G}_n$ is required by Eq. 26; hence, the fact that $\mathbf{T}$ is undefined for the nth bond is of no consequence. We observe also that

$$\mathbf{J}^*[\mathbf{E}_\nu \mathbf{0} \cdots \mathbf{0}] = \mathscr{J}^* \tag{27}$$

where $\mathscr{J}^*$ is the row defined by Eq. III-51; it consists of an initial element of unity, with all other elements zero. Then Eq. 25 reduces to the succinct expression

$$\langle M^2 \rangle = 2Z^{-1}\mathscr{J}^*\mathscr{G}_1^{(n)}\mathscr{J} \tag{28}$$

Equation 28 is exact in the sense that no mathematical approximations have been introduced in its derivation.[6,7] It is applicable to chains of any length consisting of any variety of skeletal bonds in any specified order, provided only that their rotational potentials admit of approximation by a set of discrete rotational states. In view of the scope of this result, the nature of its constituent factors, and especially of the generator matrix $\mathscr{G}_i$, merit detailed examination.

The orders of the submatrices comprising this supermatrix are as follows:

$$\begin{bmatrix} \nu \times \nu & \nu \times 3\nu & \nu \times \nu \\ 3\nu \times \nu & 3\nu \times 3\nu & 3\nu \times \nu \\ \nu \times \nu & \nu \times 3\nu & \nu \times \nu \end{bmatrix}$$

The order of $\mathscr{G}$ as a whole is $5\nu \times 5\nu$. The premultiplying row $\mathscr{J}^*$ in Eq. 28 is of order $1 \times 5\nu$, and the postmultiplying column $\mathscr{J}$ is $5\nu \times 1$. If ν differs for successive bonds, the $\mathbf{U}$ matrices for such bond pairs will be rectangular instead of square (see Chap. III, p. 69); the corresponding $\mathscr{G}$ matrices will be rectangular also.

The generator matrix $\mathscr{G}_i$ contains all required information relating to bond i. It combines geometrical parameters (θ_i and the $\phi_{\eta;i}$) from the $\mathbf{T}_i$ for the various rotational states accorded to bond i, the statistical weights from $\mathbf{U}_i$, and the bond moment $\mathbf{m}_i$ associated with bond i. One such matrix is required for each type of bond pair occurring in the chain. The terminal $\mathscr{G}$ matrices are constructed according to Eq. 24, with $\mathbf{U}_1 = \mathbf{U}_n = \mathbf{E}_\nu$. Inasmuch as only the first row of $\mathscr{G}_1$ and the final pseudocolumn (ν actual columns) of $\mathscr{G}_n$ are called upon by Eq. 28, the undefined transformation $\mathbf{T}_n$ and the undefined rows of $\mathbf{T}_1$ (see p. 101) are not required. The $\mathscr{G}$'s corresponding to the several kinds of bond pairs present are multiplied in Eq. 28 precisely in the order of occurrence of these bond pairs in the chain. Thus, the mathematical operations required for computation of $\langle M^2 \rangle$ bear a correspondence to the structure of the chain, which is both direct and intuitive.

The matrix $\mathscr{G}$ cannot be diagonalized, i.e., it is irreducible. The foregoing relationships do not therefore offer the possibility of expression of the results they convey in terms of eigenvalues of the matrices. The customary calculation of numerical results from eigenvalues is therefore unfeasible. Necessary computations are easily carried out by computer, however.† The temperature coefficient of $\langle M^2 \rangle$, or of $\langle r^2 \rangle_0$, may be put in a form suitable for direct computation by application of the methods introduced in Chapter III (see Eqs. III-55 and III-57).

If all bond pairs are identical, the ordinal indexes on the $\mathscr{G}_i$ for $1 < i < n$ may be deleted, and Eq. 28 may be written

$$\langle M^2 \rangle = 2Z^{-1}\mathscr{J}^*\mathscr{G}_1\mathscr{G}^{n-2}\mathscr{G}_n\mathscr{J} \tag{29}$$

where $\mathscr{G}^{n-2}$ is the matrix $\mathscr{G}$ raised to the power $n-2$. The central factor in Eq. 29 may be generated by successive squaring (see Chap. III), and second moments for chains of any length of possible interest can be computed readily, only a small number of multiplications being required.[6,7]

A similar reduction is possible if the chain consists of regularly repeating sequences of bonds, examples of which are cited in Chapter I, p. 14. By analogy with the treatment of the partition function for such chains given in Chapter III (see Eq. III-26), we define a product of $\mathscr{G}$ matrices corresponding to the repeating unit as follows:

$$\mathscr{G}^{(\xi)} = \mathscr{G}_{i+1}\mathscr{G}_{i+2} \cdots \mathscr{G}_{i+\xi} \tag{30}$$

† In order to avoid excessively large or excessively small numbers in computations carried out for very long chains, it is expedient to normalize the matrices $\mathbf{U}_i$ through division of each by its largest eigenvalue λ_i. The factor used for normalization of course may depart somewhat from λ_i; it is essential merely that the same factors be used in generating the numerator in Eq. 28 as are used to compute Z. Normalization in this manner is usually required for computations of other quantities according to relationships introduced in the remainder of this chapter and in those to follow.

where $i + 1$ indexes the first bond of a unit of the chain comprising x repeating units; $i + \xi$ indexes the last bond of the same unit consisting of ξ skeletal bonds. Then, in further analogy to Eq. III-26, we write

$$\langle M^2 \rangle = 2Z^{-1} \mathscr{J}^* \mathscr{G}_1^{(\xi_1)} (\mathscr{G}^{(\xi)})^{x-2} \mathscr{G}_x^{(\xi_x)} \mathscr{J} \tag{31}$$

where ξ_1 and ξ_x are the numbers of bonds in the terminal units. Matrix squaring obviously is applicable here, as well as in Eq. 29.

If $\mathbf{m}_i = \mathbf{l}_i$, then $\langle M^2 \rangle$ becomes $\langle r^2 \rangle_0$. Nominal simplifications occur within the $\mathscr{G}_i$ owing to the zero elements comprising the vector

$$\mathbf{l}_i = l_i \begin{bmatrix} 1 \\ 0 \\ 0 \end{bmatrix}$$

The characteristic ratio C_n (see Eq. I-14) may be obtained from $\langle r^2 \rangle_0$ calculated according to Eqs. 28, 29, or 31. If the bonds are of variable length, this ratio is appropriately defined by

$$C_n = \langle r^2 \rangle_0 / n\overline{l^2}$$

where $\overline{l^2}$ is the mean-square bond length.

Returning to the general expression Eq. 28 for $\langle M^2 \rangle$, we observe that the mean-square moment calculated by this equation need not include the moments $\mathbf{m}_i$ for all bonds $i = 1$ to n. In fact, any set of these moments could be selected to form $\mathbf{M}$, all other group moments $\mathbf{m}_i$ being equated to zero in the construction of the $\mathscr{G}_i$ according to Eq. 24.

A case of particular interest is that in which the moments $\mathbf{m}_i$ chosen to be included represent a set of consecutive bonds. Let this set be those of indexes $h + 1$ to k, inclusive. Then, the mean-square moment considered is

$$\begin{aligned} \langle M_{hk}^2 \rangle &= \sum_{i=h+1}^{k} \sum_{j=h+1}^{k} \langle \mathbf{m}_i \cdot \mathbf{m}_j \rangle \\ &= \sum_{h<i\le k} m_i^2 + 2 \sum_{h<i<j\le k} \langle \mathbf{m}_i \cdot \mathbf{m}_j \rangle \end{aligned} \tag{32}$$

which comprises the contributions from bonds between skeletal atoms h and k. The result obtained by substitution of Eq. 16 into Eq. 32 can be expressed as

$$\begin{aligned} \langle M_{hk}^2 \rangle = \sum_{h<i\le k} m_i^2 + 2Z^{-1} \mathbf{J}^* \mathbf{U}_1^{(h)} \Big\{ \sum_{h<i<j\le k} \mathbf{U}_{h+1}^{(i-h-1)} (\mathbf{E}_\nu \otimes \mathbf{m}_i^T) \\ \times [(\mathbf{U} \otimes \mathbf{E}_3) \| \mathbf{T} \|]_i^{(j-i)} (\mathbf{E}_\nu \otimes \mathbf{m}_j) \mathbf{U}_j^{(k-j)} \Big\} \mathbf{U}_k^{(n-k)} \mathbf{J} \end{aligned} \tag{33}$$

The double sum within braces in this equation is identical with that in Eq. 19, except for alterations of indexes. The previous results therefore are directly applicable, and transcription of Eq. 25 to the present indexes leads to

$$\langle M_{hk}^2 \rangle = 2Z^{-1}\mathbf{J}^*\mathbf{U}_1^{(h)}[\mathbf{E}_\nu \mathbf{0} \cdots \mathbf{0}]\mathscr{G}_{h+1}^{(k-h-1)} \begin{bmatrix} (m_k^2/2)\mathbf{E}_\nu \\ \mathbf{E}_\nu \otimes \mathbf{m}_k \\ \mathbf{E}_\nu \end{bmatrix} \mathbf{U}_k^{(n-k)}\mathbf{J} \tag{34}$$

This result reduces to Eq. 25 for $h = 0$ and $k = n$. Equation 34 may be expressed alternatively as follows:

$$\langle M_{hk}^2 \rangle = 2Z^{-1}\mathbf{J}^*\mathbf{U}_1^{(h)}[\mathbf{E}_\nu \mathbf{0} \cdots \mathbf{0}]\mathscr{G}_{h+1}^{(k-h)} \begin{bmatrix} \mathbf{0} \\ \vdots \\ \mathbf{0} \\ \mathbf{E}_\nu \end{bmatrix} \mathbf{U}_{k+1}^{(n-k)}\mathbf{J} \tag{35}$$

where a factor $\mathbf{U}_n = \mathbf{E}_\nu$ has been appended as above.

These equations are especially important for the calculation of the unperturbed mean-square distance $\langle r_{hk}^2 \rangle_0$ between two elements h and k in a chain of finite length n. They will be called upon for this purpose in the treatment of the radius of gyration presented in Section 5.

3. SECOND MOMENTS FOR SIMPLE CHAINS OF INFINITE LENGTH

Consider again the average value of a product of functions $f_i(\phi_i)$ of rotation angles for the succession of bonds, i to $j-1$ inclusive, given by Eq. 5 for a simple chain of bonds which are alike. Adopting the procedure set forth in Appendix C, we may equate Z, a scalar quantity, to its own trace, thereby obtaining

$$Z = \text{trace}\,(\mathbf{J}^*\mathbf{U}^{n-2}\mathbf{J}) \tag{36}$$

Substitution of $\mathbf{U} = \mathbf{A\Lambda B}$ (see Eq. III-28′) and cyclic permutation of the argument of the trace gives

$$Z = \text{trace}\,(\mathbf{\Lambda}^{n-2}\mathbf{BJJ^*A})$$

The quantity representing $\langle f_i^{(j-i)} \rangle Z$ on the right-hand side of Eq. 5 may similarly be equated to its trace. After substituting for $\mathbf{U}$ as above and permuting the argument, we obtain

$$\langle f_i^{(j-i)} \rangle = \frac{\text{trace}\,[\mathbf{\Lambda}^{i-2}\mathbf{B}(\mathbf{UF})^{j-i}\mathbf{A\Lambda}^{n-j}\mathbf{BJJ^*A}]}{\text{trace}\,(\mathbf{\Lambda}^{n-2}\mathbf{BJJ^*A})} \tag{37}$$

For very long chains it will suffice to retain only terms in the largest eigenvalue λ_1 of $\mathbf{U}$. Thus, in the limit $n \to \infty$, where also $1 \ll i < j \ll n$, Eq. 37 reduces to

$$\langle f^{j-i} \rangle = \lambda_1^{i-2} \mathbf{B}_1^* (\mathbf{UF})^{j-i} \mathbf{A}_1 \lambda_1^{n-j} (\mathbf{BJJ^*A})_{11} / \lambda_1^{n-2} (\mathbf{BJJ^*A})_{11}$$

$\mathbf{B}_1^*$ and $\mathbf{A}_1$ being the (mutually normalized) eigenrow and eigencolumn of $\mathbf{U}$ corresponding to λ_1. Equation 37 therefore simplifies to

$$\langle f^{j-i} \rangle = \mathbf{B}_1^* (\mathbf{UF}/\lambda_1)^{j-i} \mathbf{A}_1 \tag{38}$$

for very long chains.

If the function f is identified with one of the elements of $\mathbf{T}$, then obviously $\mathbf{B}_1^*(\mathbf{UF}/\lambda_1)\mathbf{A}_1$ gives the average value of that element. Proceeding as before, we readily find

$$\langle \mathbf{T} \rangle = (\mathbf{B}_1^* \otimes \mathbf{E}_3) \mathbf{S} (\mathbf{A}_1 \otimes \mathbf{E}_3) \tag{39}$$

where

$$\mathbf{S} = (\mathbf{U} \otimes \mathbf{E}_3) \| \mathbf{T} \| / \lambda_1 \tag{40}$$

The average of the product $\mathbf{T}^k$ for a sequence of transformation matrices is

$$\langle \mathbf{T}^k \rangle = (\mathbf{B}_1^* \otimes \mathbf{E}_3) \mathbf{S}^k (\mathbf{A}_1 \otimes \mathbf{E}_3) \tag{41}$$

where $k = j - i$. Subscript indexes are omitted on the supposition that $1 \ll i < i + k \ll n$. The validity of Eqs. 39 and 41 may, if desired, be verified by expansion of the matrices.

It follows from Eqs. 13, 18, and 41 that

$$\begin{aligned} \langle M^2 \rangle &= nm^2 + 2\mathbf{m}^T \sum_{i<j} (\mathbf{B}_1^* \otimes \mathbf{E}_3) \mathbf{S}^{j-i} (\mathbf{A}_1 \otimes \mathbf{E}_3) \mathbf{m} \\ &= nm^2 + 2\mathbf{m}^T (\mathbf{B}_1^* \otimes \mathbf{E}_3) \left[\sum_{k=1}^{n} (n-k) \mathbf{S}^k \right] (\mathbf{A}_1 \otimes \mathbf{E}_3) \mathbf{m} \end{aligned} \tag{42}$$

again in the limit $n \to \infty$. But in this limit the summation may be reduced further to give

$$\begin{aligned} (\langle M^2 \rangle / nm^2)_\infty &= 1 + (2/m^2) \mathbf{m}^T (\mathbf{B}_1^* \otimes \mathbf{E}_3) \mathbf{S} (\mathbf{E}_{3\nu} - \mathbf{S})^{-1} (\mathbf{A}_1 \otimes \mathbf{E}_3) \mathbf{m} \\ &= (\mathbf{m}^T/m) (\mathbf{B}_1^* \otimes \mathbf{E}_3) (\mathbf{E}_{3\nu} + \mathbf{S}) (\mathbf{E}_{3\nu} - \mathbf{S})^{-1} \\ &\quad \times (\mathbf{A}_1 \otimes \mathbf{E}_3) (\mathbf{m}/m) \end{aligned} \tag{43}$$

If the vector $\mathbf{m}$ is the bond itself, then $\langle M^2 \rangle = \langle r^2 \rangle_0$ and the foregoing result reduces to

$$(\langle r^2 \rangle_0 / nl^2)_\infty = [(\mathbf{B}_1^* \otimes \mathbf{E}_3)(\mathbf{E}_{3\nu} + \mathbf{S})(\mathbf{E}_{3\nu} - \mathbf{S})^{-1} (\mathbf{A}_1 \otimes \mathbf{E}_3)]_{11} \tag{44}$$

where the 1,1 element only is required as indicated by the subscript on the brackets.

These are the results obtained by Birshtein and Ptitsyn,[1,2] and also by Hoeve.[3] Equivalent expressions differing in form from those above have been derived by Lifson[4] and by Nagai.[5] These results, in contrast to Eq. 28, are restricted of course to uniform chains of large n. The presence of a regularly repeating structure is essential for generation of chain moments and the partition function Z by raising a matrix such as $\mathbf{U}$ to a power, n or x, equal to the number of repeating units. The eigenvalue method is otherwise of no use. Additionally, the chain must be sufficiently long to justify the omission of terms in powers of eigenvalues other than λ_1, and the evaluation of the summation in Eq. 42 by extending its limit to infinity as is required to obtain Eq. 43.

The foregoing development is applicable also to derivation of the mean-square distance $\langle r_{hk}^2 \rangle_0$ between two skeletal atoms, h and k, separated by a *finite* number $k - h$ of bonds within a simple chain of *infinite* length, i.e., for $0 \ll h < k \ll n$. Thus, from Eq. 42 with $\mathbf{m}$ replaced by $\mathbf{l}$, and with n therein replaced by $k - h$, we have

$$\langle r_{hk}^2 \rangle_0 = (k-h)l^2 + 2\mathbf{l}^T(\mathbf{B}_1^* \otimes \mathbf{E}_3)\left[\sum_{g=1}^{k-h}(k-h-g)\mathbf{S}^g\right](\mathbf{A}_1 \otimes \mathbf{E}_3)\mathbf{l}$$

which may be summed exactly as Eq. I-39 to yield (compare Eq. I-41)

$$[\langle r_{hk}^2 \rangle_0/(k-h)l^2]_\infty = l^{-2}\mathbf{l}^T(\mathbf{B}_1^* \otimes \mathbf{E}_3)[(\mathbf{E}_{3\nu} + \mathbf{S})(\mathbf{E}_{3\nu} - \mathbf{S})^{-1}$$
$$- 2(k-h)^{-1}\mathbf{S}(\mathbf{E}_{3\nu} - \mathbf{S}^{k-h})(\mathbf{E}_{3\nu} - \mathbf{S})^{-2}](\mathbf{A}_1 \otimes \mathbf{E}_3)\mathbf{l} \quad (45)$$

A relationship equivalent to this one, but formulated in the Lifson[4]-Nagai[5] scheme, has been given by Nagai.[8]

Equation 45 is the counterpart, for an infinite chain, of Eq. 35 for the second moment $\langle M_{hk}^2 \rangle$, or $\langle r_{hk}^2 \rangle_0$, for a finite chain. Comparison of calculations according to Eq. 35 for a finite chain with those given by Eq. 45 should reveal explicitly the end effects imposed by chain terminals. The correspondence of Eq. 45, with $h = 0$ and $k = n$, to Eq. I-20 for the freely rotating chain and to Eq. I-41 for the chain in which hindrance potentials are independent will be apparent. These equations in Chapter I refer specifically to the second moment for an entire chain of n bonds, but they are applicable without alteration to a finite sequence of n (or $j - i$) bonds in a chain of any length. This equivalence follows immediately from the independence of the rotation potentials there considered. Numerical comparisons of these relationships are presented in the following chapter.

4. THE PERSISTENCE LENGTH

A related quantity of interest is the *persistence length*,[9] defined as the average sum of the projections of all bonds $j \geq i$ on bond i in an indefinitely long chain. The bond i is taken to be remote from either end of the chain, i.e., $1 \ll i \ll n$. The scalar projection of bond $i + k$ on i is

$$\langle \mathbf{T}^k \rangle_{11} l = [(\mathbf{B}_1^* \otimes \mathbf{E}_3)\mathbf{S}^k(\mathbf{A}_1 \otimes \mathbf{E}_3)]_{11} l$$

according to Eq. 41. The persistence length a is the infinite sum of these expressions for $k \geq 0$. That is

$$a = l[(\mathbf{B}_1^* \otimes \mathbf{E}_3)(\mathbf{E}_{3\nu} - \mathbf{S})^{-1}(\mathbf{A}_1 \otimes \mathbf{E}_3)]_{11} \tag{46}$$

Whereas the mean-square end-to-end distance $\langle r^2 \rangle_0$ comprises the sum of the averaged scalar products of each bond with every bond in the chain, the persistence length is the sum of averaged scalar products involving *one* bond with itself and its *successors only* in the infinite chain. Hence, in the limit $n \to \infty$

$$\langle r^2 \rangle_0 = 2nla - nl^2 \tag{47}$$

or

$$(\langle r^2 \rangle_0 / nl^2)_\infty = 2a/l - 1 \tag{48}$$

which serves to define the persistence length in terms of the limiting characteristic ratio for the infinite chain. This relationship is applicable to any type of chain of simple constitution. It may be deduced alternatively from Eqs. 44 and 46.

5. THE MEAN-SQUARE RADIUS OF GYRATION[10]

According to the relationship attributed in Chapter I to Lagrange (see Eq. I-11), the unperturbed mean-square radius of gyration is

$$\begin{aligned}\langle s^2 \rangle_0 &= (n+1)^{-2} \sum_{0 \leq h < k \leq n} \langle r_{hk}^2 \rangle_0 \\ &= (n+1)^{-2} \sum_{0 \leq h < k \leq n} \sum_{i=h+1}^{k} \sum_{j=h+1}^{k} \langle \mathbf{l}_i \cdot \mathbf{l}_j \rangle \end{aligned} \tag{49}$$

The indexes occurring in this expression are correlated with atoms and bonds of the chain molecule in Fig. 1. The double summation over i and j was performed in Section 3 to obtain Eq. 35. If $\mathbf{m}$ is replaced by $\mathbf{l}$ in

Eq. 24, which defines $\mathscr{G}_i$, then $\langle M_{hk}^2\rangle$ of Eq. 35 represents $\langle r_{hk}^2\rangle_0$, and by substitution of Eq. 35 into the first of Eqs. 49 we obtain

$$\langle s^2\rangle_0 = 2(n+1)^{-2}Z^{-1}\mathbf{J}^* \sum_{0\le h<k\le n} \mathbf{U}_1^{(h)}[\mathbf{E}_\nu \mathbf{0}\cdots \mathbf{0}]\mathscr{G}_{h+1}^{(k-h)}\begin{bmatrix}\mathbf{0}\\ \vdots\\ \mathbf{0}\\ \mathbf{E}_\nu\end{bmatrix}\mathbf{U}_{k+1}^{(n-k)}\mathbf{J} \quad (50)$$

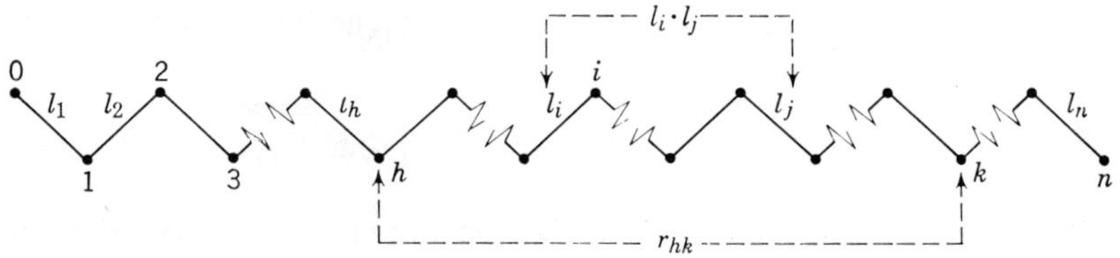

Figure 1.

The prototype of the double sum in Eq. 50 is

$$\sum_{0\le h<k\le n} U_1^{(h)} P S_{h+1}^{(k-h)} Q U_{k+1}^{(n-k)} \quad (51)$$

where P and Q represent $[\mathbf{E}_\nu \mathbf{0}\cdots \mathbf{0}]$ and $\begin{bmatrix}\mathbf{0}\\ \vdots\\ \mathbf{0}\\ \mathbf{E}_\nu\end{bmatrix}$, respectively. This expression bears a close resemblance to the previous expression 20, which represents the sum occurring in Eq. 19. The indexes and their ranges differ, but these differences are found to be superficial on close examination. In fact, expression 51 could be summed by expression 21. A more important distinction resides in the absence of indexes on the P and Q in the present sum. This feature permits expression 51 to be summed by

$$[1 \quad 0 \quad 0]\begin{bmatrix} U & PS & PSQ\\ 0 & S & SQ\\ 0 & 0 & U\end{bmatrix}_1^{(n)}\begin{bmatrix}0\\0\\1\end{bmatrix} \quad (52)$$

Expression 52 is not applicable to the preceding sum, expression 20, owing to the fact that the quantities P and Q must be introduced into expression 20 with the indexes specified therein.

Applying this result to Eq. 50, we find

$$\langle s^2\rangle_0 = 2(n+1)^{-2}Z^{-1}\mathbf{J}^*[\mathbf{E}_\nu\mathbf{0}\cdots\mathbf{0}]\mathscr{S}_1^{(n)}\begin{bmatrix}\mathbf{0}\\ \vdots\\ \mathbf{0}\\ \mathbf{E}_\nu\end{bmatrix}\mathbf{J} \tag{53}$$

$$= 2(n+1)^{-2}Z^{-1}\mathscr{J}^*\mathscr{S}_1^{(n)}\mathscr{J} \tag{54}$$

where $\mathscr{J}^*$ and $\mathscr{J}$ are defined by Eqs. III-51 and III-52, and

$$\mathscr{S}_i = \begin{bmatrix}\mathbf{U} & [\mathbf{E}_\nu\mathbf{0}]\mathscr{G} & [\mathbf{E}_\nu\mathbf{0}]\mathscr{G}\begin{bmatrix}\mathbf{0}\\ \mathbf{E}_\nu\end{bmatrix}\\ \mathbf{0} & \mathscr{G} & \mathscr{G}\begin{bmatrix}\mathbf{0}\\ \mathbf{E}_\nu\end{bmatrix}\\ \mathbf{0} & \mathbf{0} & \mathbf{U}\end{bmatrix}_i \tag{55}$$

where matrices such as $[\mathbf{E}_\nu\mathbf{0}\cdots\mathbf{0}]$ have been condensed in interests of concision. From Eq. 24 for $\mathscr{G}_i$, with $\mathbf{m}_i = \mathbf{l}_i$,

$$\mathscr{S}_i = \begin{bmatrix}\mathbf{U} & \mathbf{U} & (\mathbf{U}\otimes\mathbf{l}^T)\|\mathbf{T}\| & (l^2/2)\mathbf{U} & (l^2/2)\mathbf{U}\\ \mathbf{0} & \mathbf{U} & (\mathbf{U}\otimes\mathbf{l}^T)\|\mathbf{T}\| & (l^2/2)\mathbf{U} & (l^2/2)\mathbf{U}\\ \mathbf{0} & \mathbf{0} & (\mathbf{U}\otimes\mathbf{E}_3)\|\mathbf{T}\| & \mathbf{U}\otimes\mathbf{l} & \mathbf{U}\otimes\mathbf{l}\\ \mathbf{0} & \mathbf{0} & \mathbf{0} & \mathbf{U} & \mathbf{U}\\ \mathbf{0} & \mathbf{0} & \mathbf{0} & \mathbf{0} & \mathbf{U}\end{bmatrix}_i \tag{56}$$

where $0 < i \le n$. For $i = 1$ or n, the statistical weight matrix $\mathbf{U}$ is represented by the identity $\mathbf{E}_\nu$, as previously stated. The order of $\mathscr{S}_i$ is $7\nu \times 7\nu$. Like $\mathscr{G}_i$, it contains all information relating to bond i, and it is irreducible.

The performance of this matrix in generating $\langle s^2\rangle_0$ when used in Eq. 54 may readily be rationalized by reference to Eq. 49 and to Fig. 1. Thus, the first column (i.e., pseudocolumn) in Eq. 56 takes account of bonds preceding atom h in Fig. 1; the second column accounts for bonds between atoms h and i; the central column represents bonds within the span of the term $\mathbf{l}_i \cdot \mathbf{l}_j$; termination of the sequence i to j occurs in the fourth column, which thereafter perpetuates bonds between j and k; and the final column gathers factors for bonds beyond k. Equations 54 and 56 could have been deduced intuitively from Eq. 49 and Fig. 1.

Equations 53–56 are general in scope; they are applicable to chains of any length consisting of any specified sequence of bonds. The temperature coefficient of $\langle s^2\rangle_0$ may be computed by the methods introduced in Chapter III (see Eqs. III-53 to III-58).

For very long chains the dominant terms $\langle r_{hk}^2 \rangle_0$ entering the sum giving $\langle s^2 \rangle_0$ (see Eq. I-11 or Eq. 49) will be those for which $k \gg h$. For those terms $\langle r_{hk}^2 \rangle_0$ can be taken to be proportional to $k - h$. For sufficiently large n it will be permissible therefore to let

$$\langle s^2 \rangle_0 \cong (n+1)^{-2}(\langle r^2 \rangle_0/nl^2)_\infty l^2 \sum_{0 \le h < k \le n} (k-h)$$

$$(\langle s^2 \rangle_0/nl^2)_\infty = (1/6)(\langle r^2 \rangle_0/nl^2)_\infty \tag{57}$$

This result was derived in Chapter I (see Eqs. I-16 and I-17) for the freely jointed chain, and found to hold for random coiled chains in general (see pp. 12 and 28). Separate treatment of the radius of gyration in this limit is not therefore required.

6. CHAINS WITH INDEPENDENT ROTATION POTENTIALS

The discussion in Chapter I of chains of independent bonds was devoted primarily to simple chains. The treatment there presented becomes excessively complicated for chains comprising structural units spanning more than two bonds, even if these units repeat regularly throughout the chain. If the chain is devoid of regularly repeating sequences of bonds, the methods of Chapter I, Section 8, are utterly inapplicable. It is of more than passing interest therefore to examine the forms assumed by the preceding general relationships when simplified appropriately for chains in which the rotational potentials may be regarded as independent.

For such chains we have according to Eqs. I-31 and I-38 transcribed to present notation

$$\langle \mathbf{m}_i \cdot \mathbf{m}_j \rangle = \mathbf{m}_i^T \langle \mathbf{T} \rangle_i^{(j-i)} \mathbf{m}_j \tag{58}$$

which replaces Eq. 16. By substituting into Eq. 18 and treating the double sum as above, we obtain

$$\langle M^2 \rangle = 2[1 \quad 0 \quad \cdots \quad 0] \mathfrak{G}_1^{(n)} \begin{bmatrix} 0 \\ \vdots \\ \cdot \\ 0 \\ 1 \end{bmatrix} \tag{59}$$

where

$$\mathfrak{G}_i = \begin{bmatrix} 1 & \mathbf{m}^T \langle \mathbf{T} \rangle & m^2/2 \\ \mathbf{0} & \langle \mathbf{T} \rangle & \mathbf{m} \\ 0 & \mathbf{0} & 1 \end{bmatrix}_i \tag{60}$$

$\mathfrak{G}_1^{(n)}$ is the serial product of n factors defined by Eq. 60; null submatrices of orders exceeding 1×1 are shown in boldface. Equations 59 and 60 are obvious analogs of Eqs. 28 and 24, respectively. If $\mathbf{m} = \mathbf{l}$, they yield $\langle r^2 \rangle_0$ for a chain of independent bonds of any variety and bond sequence whatever. The characteristic ratio $\langle r^2 \rangle_0 / n\overline{l^2}$ follows at once.

Similarly, the squared radius of gyration is

$$\langle s^2 \rangle_0 = 2(n+1)^{-2} [1 \quad 0 \quad \cdots \quad 0] \mathfrak{S}_1^{(n)} \begin{bmatrix} 0 \\ \vdots \\ 0 \\ 1 \end{bmatrix} \tag{61}$$

where

$$\mathfrak{S}_i = \begin{bmatrix} 1 & 1 & \mathbf{l}^T\langle \mathbf{T} \rangle & l^2/2 & l^2/2 \\ 0 & 1 & \mathbf{l}^T\langle \mathbf{T} \rangle & l^2/2 & l^2/2 \\ \mathbf{0} & \mathbf{0} & \langle \mathbf{T} \rangle & \mathbf{l} & \mathbf{l} \\ 0 & 0 & \mathbf{0} & 1 & 1 \\ 0 & 0 & \mathbf{0} & 0 & 1 \end{bmatrix}_i \tag{62}$$

These equations, like their analogs, Eqs. 54 and 56, accomplish the fourfold summations required by Eq. 49. They may be rationalized with aid of diagrammatic representations of the terms required to be summed, as in Fig. 1. The order of $\mathfrak{S}_i$ is 7×7. It contains all information pertinent to bond i. Equation 61, like Eq. 59, is applicable to any chain of specified constitution in which rotations about its bonds are independent.

Chains in which bond rotations within a given structural unit are interdependent, but in which interdependence does not extend from bonds of one unit to those of the next, are of interest in the present connection. Examples are *cis*- and *trans*-1,4-polybutadiene,[11] polyamides,[12] and poly(ethylene terephthalate),[13] which will be discussed in the following chapter. Specifically, we consider a chain consisting of x structural units, the kth of which spans ξ_k skeletal bonds; $n = \sum_{k=1}^{x} \xi_k$. For simplicity the index k will be deleted from ξ in the following equations. Rotational potentials for bond pairs (1,2), (2,3), ..., $(\xi - 1, \xi)$ within the given unit may be interdependent, but pairs comprising bonds from different units—specifically, a pair of terminal bonds of adjoining units—are independent. When this condition holds, the serial product of the matrices in Eq. 28 can be separated into factors, one for each unit of the chain, after the manner of Eqs. 30 and 31. Such a factor for the kth unit is*

$$\mathscr{G}_k^{(\xi)} = \mathscr{G}_{i+1} \mathscr{G}_{i+2} \cdots \mathscr{G}_{i+\xi} \tag{63}$$

* The notation for expressing $\mathscr{G}_k^{(\xi)}$ and $\mathbf{U}_k^{(\xi)}$ departs in one respect from earlier usage. Here k denotes the *structural unit* comprising the ξ bonds represented in these serial products, and not the first bond thereof as in previous notation.

where the $\mathscr{G}$'s are defined by Eq. 24, and $i + 1$ indexes the first bond of the kth unit comprising ξ bonds in all. This equation differs from Eq. 30 only through designation of the particular unit by the subscript k. The serial product of $\mathbf{U}$ matrices in Eq. III-24 for the partition function can be similarly separated into products such as

$$\mathbf{U}_k^{(\xi)} = \mathbf{U}_{i+1}\mathbf{U}_{i+2} \cdots \mathbf{U}_{i+\xi} \tag{64}$$

The first factor, $\mathbf{U}_{i+1}$, will consist of v identical rows, owing to the independence of the statistical weights for this bond on states of its predecessors. The product, $\mathbf{U}_k^{(\xi)}$, likewise will consist of v identical rows. Letting $\mathbf{U}_k^{(\xi)*}$ represent a single row of this latter matrix, obtained from Eq. 64, for example, by expressing its leading factor as a row, $\mathbf{U}_{i+1}^*$, we have

$$\mathbf{U}_k^{(\xi)} = \mathbf{J}\mathbf{U}^{(\xi)*} \tag{65}$$

With aid of this equation, Eq. III-24 may be simplified as follows[12]:

$$Z = \mathbf{J}^* \left[\prod_{k=1}^{x} (\mathbf{J}\mathbf{U}_k^{(\xi)*}) \right] \mathbf{J} \tag{66}$$

and since $\mathbf{J}^*\mathbf{J} = 1$,

$$Z = \prod_{1}^{x} (\mathbf{U}_k^{(\xi)*}\mathbf{J}) = \prod_{1}^{x} z_k^{(\xi)} \tag{67}$$

where

$$z_k^{(\xi)} = \mathbf{U}_k^{(\xi)*}\mathbf{J} \tag{68}$$

is the partition function for the kth independent unit. If there are terminal bonds extending beyond the integral set of x units in the chain, these may of course be included as members of terminal units consisting of ξ' and ξ'' bonds, respectively, after the manner of Eq. III-26.

If $\mathbf{U}_{i+1}$ occurring in the first $\mathscr{G}$ factor on the right-hand side of Eq. 63 is expressed in $v \times v$ order, the product $\mathscr{G}_k^{(\xi)}$ will consist of five rows of $5v$ elements, each repeated identically v times. If, however, the row $\mathbf{U}_{i+1}^*$ is used instead, the redundance of rows is suppressed in $\mathscr{G}_{i+1}$ and also in the resulting product which we designate $\mathscr{G}_k^{(\xi)*}$. The order of the matrix thus constructed is $5 \times 5v$. It is related as follows to $\mathscr{G}_k^{(\xi)}$, which contains redundant rows and is of order $5v \times 5v$, by

$$\mathscr{G}_k^{(\xi)} = \begin{bmatrix} \mathbf{J} & \mathbf{0} & \mathbf{0} \\ \mathbf{0} & \mathbf{J} \otimes \mathbf{E}_3 & \mathbf{0} \\ \mathbf{0} & \mathbf{0} & \mathbf{J} \end{bmatrix} \mathscr{G}_k^{(\xi)*} \tag{69}$$

Substitution of Eqs. 67 and 69 into Eq. 28 leads to[12]

$$\langle M^2 \rangle = 2[1 \quad 0 \quad 0 \quad 0 \quad 0](\mathfrak{G}^{(\xi)})_1^{(x)} \begin{bmatrix} 0 \\ 0 \\ 0 \\ 0 \\ 1 \end{bmatrix} \tag{70}$$

where

$$\mathfrak{G}_k^{(\xi)} = \mathscr{G}_k^{(\xi)*} \begin{bmatrix} \mathbf{J} & \mathbf{0} & \mathbf{0} \\ \mathbf{0} & \mathbf{J} \otimes \mathbf{E}_3 & \mathbf{0} \\ \mathbf{0} & \mathbf{0} & \mathbf{J} \end{bmatrix} (z_k^{(\xi)})^{-1} \tag{71}$$

and $(\mathfrak{G}^{(\xi)})_1^{(x)}$ is the serial product of such factors commencing with $k = 1$. Equation 70 is of the same form as Eq. 59 for a chain with all bonds independent. The matrix $\mathfrak{G}_k^{(\xi)}$ may be looked upon as the simplified and normalized representation of the product of matrices $\mathscr{G}_k^{(\xi)}$ for the kth independent unit. According to Eq. 71, this matrix for unit k is obtained by taking the serial product of the $\mathscr{G}$ matrices for the individual bonds, as in Eq. 63, but with $\mathbf{U}$ for the first bond of the unit expressed as a row of elements. The resulting product $\mathscr{G}_k^{(\xi)*}$ is to be postmultiplied by the matrix displayed in Eq. 71, and divided by the partition function factor $z_k^{(\xi)}$ for unit k. Its order is thus reduced to 5×5, the same as the order of $\mathfrak{G}_i$ given by Eq. 60 for one bond in a chain, all bonds of which are independent. The matrix $\mathfrak{G}_k^{(\xi)}$ reduces to the latter for $\xi = 1$.

Equation 70 is generally applicable to any chain meeting the terms of rotational independence between units stipulated above.[12,13] The units may differ in length ξ, and in other characteristics as well. Their sequential order must of course be specified. For a homopolymer, the $\mathfrak{G}_k^{(\xi)}$ are the same for all k (terminal units excepted; see p. 116), the subscript can be dropped, and the superscript (x) in Eq. 70 may be treated as an exponent. Copolymers in which the sequence of the units is statistically variable are considered in the following section.

The representation of the mean-square radius of gyration of a chain of independent bonds given by Eq. 61 above can be generalized in a like manner to include chains in which interdependence of bond rotations is confined within structural units.

7. RANDOM AND MARKOFFIAN COPOLYMERS

Let the copolymer consist of structural units of kinds $a, b, \ldots,$ comprising $\xi_a, \xi_b, \ldots$ bonds, respectively. For the present we assume that rotational independence obtains between units, as more fully set forth in the preceding section. If the sequence of units is uniquely specified, as in many proteins, then the second moment $\langle M^2 \rangle$ may be generated according to Eq. 70, using factors $\mathfrak{G}_q^{(\xi)}$ for each kind q of unit, where $q = a, b, \ldots$. If the copolymer consists of an array of molecular species differing in the numbers and sequences of the several kinds of units, then $\langle M^2 \rangle$ should be averaged over the population, or distribution, of molecules of specified length n or degree of polymerization x. In such circumstances, this further operation will be indicated by an overbar, as in $\langle \overline{M^2} \rangle$ and corresponding quantities. Specifically, the overbar will denote averaging over all copolymeric species of given chain length x. Angle brackets will be retained to denote the average over the configurations of a given species. The variation of x, or n, among different species can be separately dealt with and need not be treated explicitly. It will suffice to consider molecules of a given degree of polymerization x.

Consider first the case of a *random copolymer* in which the units occur in random succession without correlation one to the next. Then the expectation that the kth unit is of kind q will be w_q, irrespective of k and of the kinds of units preceding unit k in the chain. Moreover, we assume w_q to be the same for all chains. Then w_q can be identified with the fraction of units of kind q in the system. Of course,

$$\sum_{q=a,b\cdots} w_q = 1 \tag{72}$$

The probability of occurrence of a molecule in which the sequence of units is $bca \cdots$ is given by $w_{bcca\cdots} = w_a^{x_a} w_b^{x_b} \cdots$, where x_a, x_b, etc., are the numbers of units of kinds a, b, etc., in the given molecule. This species is to be represented through Eq. 70 by the product

$$\mathfrak{G}_b^{(\xi)} \mathfrak{G}_c^{(\xi)} \mathfrak{G}_c^{(\xi)} \mathfrak{G}_a^{(\xi)} \cdots$$

taken in the serial order of occurrence of the units, and multiplied by $w_{bcca\cdots}$. (For simplicity, subscripts b, c, etc., on the ξ are again omitted.) The sum of serial products for all molecules containing x units chosen from the set a, b, etc., with each such product weighted according to its composition, is given by the xth power of the sum

$$\overline{\mathfrak{G}} = w_a \mathfrak{G}_a^{(\xi)} + w_b \mathfrak{G}_b^{(\xi)} + \cdots \tag{73}$$

Hence, the value of the second moment $\langle M^2 \rangle$ averaged over all species of the random copolymer can be determined through use of an equation of the form of Eqs. 59 and 70 without alteration except for substitution of $\overline{\mathfrak{G}}$ for each factor in the serial product, i.e.,

$$\langle \overline{M^2} \rangle = 2[1 \quad 0 \quad 0 \quad 0 \quad 0]\overline{\mathfrak{G}}^x \begin{bmatrix} 0 \\ 0 \\ 0 \\ 0 \\ 1 \end{bmatrix} \tag{74}$$

If the sequence of units in the copolymer chain is determined by kinetic factors operative in the growth of the polymer chain during its synthesis, the expectation that unit k is of kind q may depend on its predecessor in the chain. Barring subsequent reorganization of the sequence of units, this expectation does not depend on the kinds of units succeeding it in the chain.[14] The generation of sequences then fits the pattern of a Markoff process. If the dependence on preceding units does not extend beyond the immediate predecessor (and this will generally hold true), then the matrix of sequence of probabilities takes the form

$$\mathbf{w} = [w_{ab}] \tag{75}$$

where w_{ab} is the conditional probability that a unit of kind b succeeds one of kind a. For simplicity, we shall assume a binary copolymer consisting of units a and b; generalization to a greater variety of units will be obvious. For the first unit of the chain $\mathbf{w}$ is expressed by

$$\mathbf{w}_1 = [w_{a;1}, w_{b;1}] \tag{76}$$

The index 1 is appended to $w_{a;1}$ and $w_{b;1} = 1 - w_{a;1}$ to acknowledge that $w_{a;1}$ may differ from the average composition w_a. The matrix to be used for generating $\langle \overline{M^2} \rangle$ is in this case

$$\overline{\mathfrak{G}} = \begin{bmatrix} w_{aa}\,\mathfrak{G}_a^{(\xi)} & w_{ab}\,\mathfrak{G}_b^{(\xi)} \\ w_{ba}\,\mathfrak{G}_a^{(\xi)} & w_{bb}\,\mathfrak{G}_b^{(\xi)} \end{bmatrix} \tag{77}$$

for units beyond the first; or in general

$$\overline{\mathfrak{G}} = (\mathbf{w} \otimes \mathbf{E}_5) \begin{bmatrix} \mathfrak{G}_a^{(\xi)} & \mathbf{0} \\ \mathbf{0} & \mathfrak{G}_b^{(\xi)} \end{bmatrix} \tag{78}$$

which applies to all units provided that $\mathbf{w}_1$ given by Eq. 76 is used for the first. The product of such matrices taken to the power x as in Eq. 74 must be postmultiplied by $\begin{bmatrix} \mathbf{E}_5 \\ \mathbf{E}_5 \end{bmatrix}$. As a further variant of Eq. 59, adapted in this instance to Markoffian copolymers, we have therefore*

$$\langle M^2 \rangle = 2[1 \quad 0 \quad 0 \quad 0 \quad 0]\bar{\mathfrak{G}}_1 \bar{\mathfrak{G}}^{x-1} \begin{bmatrix} 1 \\ 1 \end{bmatrix} \otimes \begin{bmatrix} 0 \\ 0 \\ 0 \\ 0 \\ 1 \end{bmatrix} \tag{79}$$

Adaptation of these equations to Markoffian copolymers comprising any number of kinds of units is readily achieved by revision of **w** to the order required to include the variety of units present, by corresponding expansion of the pseudodiagonal matrix displayed in Eq. 78, and by lengthening the postmultiplying column $\begin{bmatrix} 1 \\ 1 \end{bmatrix}$ in Eq. 79 as required to match the order of $\bar{\mathfrak{G}}$.

Equations 73 and 74 for random copolymers, and Eqs. 76, 78, and 79, modified to accommodate any number of kinds of units, are quite general, and exact as well,† for copolymers meeting the condition of interunit configurational independence asserted above to be required. They are readily applied in practice.‡ It is of the utmost importance to observe, however, that separation of the general expressions Eqs. 28 or 31 for the second moment $\langle M^2 \rangle$ into factors for each successive unit of a copolymeric chain, after the manner of Eqs. 74 and 79, is permissible only if the rotations within each unit are independent of those of neighboring units. The solidity of this assertion can be demonstrated by expressing $\langle M^2 \rangle$ in a form suitable for a copolymer molecule of specified constitution in which the configurations of neighboring units are interdependent. The demonstration follows.

* The result of Miller, Brant, and Flory[15] for Markoffian copolymers in which each unit is spanned by one bond of fixed length (as in polypeptides) differs drastically in form from Eq. 79. The two results may be shown to be equivalent, however.

† It is gratifying to observe that the equations above comprehend every eligible copolymeric species, each being averaged over all of its configurations. For a copolymeric chain of a hundred or more units the number of possible species vastly exceeds the number of molecules in any conceivable specimen. The equations therefore are more comprehensive than the material specimen they represent; the latter is but a relatively small, random collection from the larger set. This circumstance is not without parallel elsewhere in statistics and statistical mechanics.

‡ Polypeptides consisting of two or more kinds of amino acid residues are a special class of copolymeric chain molecules whose skeletal configurations can be represented by a succession of *virtual bonds*, one for each peptide unit. The mutual orientation of each neighbor pair of virtual bonds depends explicitly on the intervening amino acid residue. The equations set forth above for copolymers admit of simplification in this special case, which is treated in Chapter VII.

Let statistical weight matrices $\mathbf{U}_{q'q}^{(\xi)}$ be specified for every pair of units of kinds q and q'. Here ξ denotes the number of bonds in a unit of kind q, and q' identifies the preceding unit in the chain. The corresponding $\mathscr{G}_{q'q}^{(\xi)}$ matrices, defined as in Eq. 63, are assumed to be given as well. Then, for a copolymer *of specified constitution* the second moment $\langle M^2 \rangle$ may be computed through use of the relationship

$$\langle M^2 \rangle = 2Z^{-1} \mathscr{J}^* (\mathscr{G}^{(\xi)})_1^{(x)} \mathscr{J} \tag{80}$$

where each $\mathscr{G}_k^{(\xi)}$ is assigned from the set of $\mathscr{G}_{q'q}^{(\xi)}$ in accordance with the kinds of units occurring at positions $k - 1$ and k in the chain. The partition function Z is generated in like manner, and for the identical sequence of units, from the $\mathbf{U}_{q'q}$. Equation 80, which differs from Eq. 28 only in the grouping of factors $\mathscr{G}$, may be used to calculate $\langle M^2 \rangle$ for a copolymer having any sequence of units, provided that the sequence is given.

The serial product $(\mathscr{G}^{(\xi)})_1^{(x)}$ occurring in the numerator of Eq. 80 may, to be sure, be averaged over all species of a Markoffian copolymer by adaptation of the foregoing methods to a chain of bonds with interdependence. The partition function Z may be averaged in a like manner. But the result obtained by dividing one average by the other does *not* yield $\overline{\langle M^2 \rangle}$. For a chain with neighbor-dependent rotations which persist without interruption from one bond to the next, the quantities required are the ratios $\mathscr{G}_1^{(n)}/Z$ *for each species* of the copolymer. These ratios must be weighted according to the incidence of the various species and the average taken over all species of the copolymer. The ratio of averaged products will differ in general from the average of the ratios. Use of the former instead of the latter may commit a serious error, as is apparent if the matter is carefully considered. Avoidance of the temptation to take the ratio of the averaged quantities is essential for proper analysis of stereoirregular vinyl polymers (Chap. VI) and helix–coil transitions in polypeptides and polynucleotides (Chap. VII), as well as configurational properties of conventional copolymers.

A copolymer in which the configurations of neighbor units are interdependent may nevertheless be treated in an approximate way as follows, the sequence of units being either random or Markoffian. Given the elements of $\mathbf{w}$ (Eq. 75 for a Markoffian copolymer), a representative sequence of x units may be generated using random number methods. Both Z and the serial product $(\mathscr{G}^{(\xi)})_1^{(x)}$ are computed exactly for this sequence through use of the appropriate factors, $\mathbf{U}_{q'q}$ and $\mathscr{G}_{q'q}^{(\xi)}$, for each successive pair of units. The ratio of these two quantities substituted in Eq. 80 gives $\langle M^2 \rangle$. The computation is repeated for another similarly generated chain, etc., until the mean of results for these "Monte Carlo" trials converges within limits which are acceptable. This method finds illustration in its application in to stereoirregular vinyl polymers in Chapter VI, which may be treated as copolymers of units that are mirror images of one another.

8. THE FOURTH MOMENT OF A VECTOR M

The statistical mechanical average of the fourth moment of a vector **M**, defined according to Eq. 17 as the sum of vectors $\mathbf{m}_i$ associated with each bond, may be separated into summations as follows[16]:

$$\begin{aligned}\langle M^4 \rangle &= \sum_{i_1,i_2,j_1,j_2} \langle (\mathbf{m}_{i_1} \cdot \mathbf{m}_{j_1})(\mathbf{m}_{i_2} \cdot \mathbf{m}_{j_2}) \rangle \\ &= \sum_i^{\mathrm{I}} m_i^4 + 2 \sum_{i<j}^{\mathrm{II}} m_i^2 m_j^2 \\ &\quad + 4 \sum_{i_1;\, i_2<j_2}^{\mathrm{III}} m_{i_1}^2 \langle \mathbf{m}_{i_2} \cdot \mathbf{m}_{j_2} \rangle + 8 \sum_{i_1<j_1 \le i_2<j_2}^{\mathrm{IV}} \langle (\mathbf{m}_{i_1} \cdot \mathbf{m}_{j_1})(\mathbf{m}_{i_2} \cdot \mathbf{m}_{j_2}) \rangle \\ &\quad + 4 \sum_{i<j}^{\mathrm{V}} \langle (\mathbf{m}_i \cdot \mathbf{m}_j)(\mathbf{m}_i \cdot \mathbf{m}_j) \rangle \\ &\quad + 8 \sum_{i_1<i_2<j_2 \le j_1}^{\mathrm{VI}} \langle (\mathbf{m}_{i_1} \cdot \mathbf{m}_{j_1})(\mathbf{m}_{i_2} \cdot \mathbf{m}_{j_2}) \rangle \\ &\quad + 8 \sum_{i_1 \le i_2<j_1<j_2}^{\mathrm{VII}} \langle (\mathbf{m}_{i_1} \cdot \mathbf{m}_{j_1})(\mathbf{m}_{i_2} \cdot \mathbf{m}_{j_2}) \rangle \end{aligned} \tag{81}$$

The Roman numeral superscripts are appended for later identification of the several summations, which may be diagrammed, respectively, as follows:

$$\begin{aligned}\langle M^4 \rangle &= (|)^{\mathrm{I}} + 2(|\,|)^{\mathrm{II}} + 4(|\sqcap)^{\mathrm{III}} + 8(\sqcap\sqcup)^{\mathrm{IV}} \\ &\quad + 4(\sqsubset\!\sqsupset)^{\mathrm{V}} + 8(\overline{\sqcup})^{\mathrm{VI}} + 8(\overline{\quad\sqcup\quad})^{\mathrm{VII}}\end{aligned} \tag{82}$$

where $\sqcap$ and $\sqcup$ represent pairs i_1, j_1 and i_2, j_2, respectively; coincident pairs $i_1 = j_1$ and/or $i_2 = j_2$ are represented by vertical lines $|$.

The first and second terms do not entail averaging over the configurations of the chain. The averages of the quantities required for the third and fourth terms are obtainable at once by application of Eq. 16. Thus,

$$m_{i_1}^2 \langle\ \rangle^{\mathrm{III}} = m_{i_1}^2 Z^{-1} \mathbf{J}^* \mathbf{U}_1^{(i_2-1)} [(\mathbf{U} \otimes \mathbf{m}^T) \|\mathbf{T}\|]_{i_2} [(\mathbf{U} \otimes \mathbf{E}_3) \|\mathbf{T}\|]_{i_2+1}^{(j_2-i_2-1)} \times (\mathbf{U} \otimes \mathbf{m})_{j_2} \mathbf{U}_{j_2+1}^{(n-j_2-1)} \mathbf{J} \tag{83}$$

and

$$\begin{aligned}\langle\ \rangle^{\mathrm{IV}} &= Z^{-1} \mathbf{J}^* \mathbf{U}_1^{(i_1-1)} [(\mathbf{U} \otimes \mathbf{m}^T) \|\mathbf{T}\|]_{i_1} [(\mathbf{U} \otimes \mathbf{E}_3) \|\mathbf{T}\|]_{i_1+1}^{(j_1-i_1-1)} (\mathbf{U} \otimes \mathbf{m})_{j_1} \\ &\quad \times \mathbf{U}_{j_1+1}^{(i_2-j_1-1)} [(\mathbf{U} \otimes \mathbf{m}^T) \|\mathbf{T}\|]_{i_2} [(\mathbf{U} \otimes \mathbf{E}_3) \|\mathbf{T}\|]_{i_2+1}^{(j_2-i_2-1)} \\ &\quad \times (\mathbf{U} \otimes \mathbf{m})_{j_2} \mathbf{U}_{j_2+1}^{(n-j_2-1)} \mathbf{J}\end{aligned} \tag{84}$$

for $j_1 < i_2$. For the term with $j_1 = i_2$ required in Eq. 81, the product

$$(\mathbf{U} \otimes \mathbf{m})_{j_1} \mathbf{U}_{j_1+1}^{(i_2-j_1-1)} [(\mathbf{U} \otimes \mathbf{m}^T) \| \mathbf{T} \|]_{i_2} = (\mathbf{E}_\nu \otimes \mathbf{m}_{j_1}) \mathbf{U}_{j_1}^{(i_2-j_1)} \times [(\mathbf{U} \otimes \mathbf{m}^T) \| \mathbf{T} \|]_{i_2}$$

is replaced by

$$[(\mathbf{U} \otimes \mathbf{m}\mathbf{m}^T) \| \mathbf{T} \|]_{i_2}$$

For the treatment of remaining sums, we adopt the device[17] of expressing the product of scalars in Eq. 81 as their direct matrix product,[7] i.e., we let

$$\langle \ \ \rangle = \langle (\mathbf{m}_{i_1} \cdot \mathbf{m}_{j_1})(\mathbf{m}_{i_2} \cdot \mathbf{m}_{j_2}) \rangle = \langle (\mathbf{m}_{i_1}^T \mathbf{T}_{i_1} \cdots \mathbf{T}_{j_1-1} \mathbf{m}_{j_1}) \otimes (\mathbf{m}_{i_2}^T \mathbf{T}_{i_2} \cdots \mathbf{T}_{j_2-1} \mathbf{m}_{j_2}) \rangle \quad (85)$$

The purpose of this artifice is to enable the factors to be permuted in the desired manner, as will be apparent at once. Applying the conditions $i_1 = i_2$ and $j_1 = j_2$ for terms of $\sum^{\mathrm{V}}$, we have by the theorem on direct products

$$\langle \ \ \rangle^{\mathrm{V}} = \langle (\mathbf{m}_i^T \otimes \mathbf{m}_i^T)(\mathbf{T}_i \otimes \mathbf{T}_i) \cdots (\mathbf{T}_{j-1} \otimes \mathbf{T}_{j-1})(\mathbf{m}_j \otimes \mathbf{m}_j) \rangle \quad (86)$$

Straightforward adaptation of Eq. 16 to vectors of ninth order (i.e., to $\mathbf{m}_i^T \otimes \mathbf{m}_i^T$ and $\mathbf{m}_j \otimes \mathbf{m}_j$), with $\mathbf{T} \otimes \mathbf{T}$ replacing $\mathbf{T}$, yields

$$\langle \ \ \rangle^{\mathrm{V}} = Z^{-1} \mathbf{J}^* \mathbf{U}_1^{(i-1)} [(\mathbf{U} \otimes \mathbf{m}^T \otimes \mathbf{m}^T) \| \mathbf{T} \otimes \mathbf{T} \|]_i \times [(\mathbf{U} \otimes \mathbf{E}_9) \| \mathbf{T} \otimes \mathbf{T} \|]_{i+1}^{(j-i-1)} \times (\mathbf{U} \otimes \mathbf{m} \otimes \mathbf{m})_j \mathbf{U}_{j+1}^{(n-j-1)} \mathbf{J} \quad (87)$$

where $\| \mathbf{T} \otimes \mathbf{T} \|_i$ is defined by analogy to Eq. 6 as the pseudodiagonal matrix whose "elements" are the matrices $\mathbf{T}_i \otimes \mathbf{T}_i$ for the various rotational states of bond i.

The summand for the sixth term of Eq. 81 may be put in a form in which transformations of the same index are adjacent by application of the theorem on direct products to the expression on the right-hand side of Eq. 85. The result is

$$(\mathbf{m}_{i_1}^T \otimes \mathbf{m}_{i_2}^T)(\mathbf{T}_{i_1} \otimes \mathbf{E}_3) \cdots (\mathbf{T}_{i_2-1} \otimes \mathbf{E}_3)(\mathbf{T}_{i_2} \otimes \mathbf{T}_{i_2}) \cdots (\mathbf{T}_{j_2-1} \otimes \mathbf{T}_{j_2-1}) \times (\mathbf{T}_{j_2} \otimes \mathbf{E}_3) \cdots (\mathbf{T}_{j_1-1} \otimes \mathbf{E}_3)(\mathbf{m}_{j_1} \otimes \mathbf{m}_{j_2})$$

The product of transformation matrices in this expression can be averaged by the same method used to obtain Eqs. 10 and 14. The matrices $\mathbf{T} \otimes \mathbf{E}_3$ and $\mathbf{T} \otimes \mathbf{T}$ must be expanded to the pseudodiagonal arrays

$$\| \mathbf{T} \otimes \mathbf{E}_3 \| = \| \mathbf{T} \| \otimes \mathbf{E}_3$$

and

$$\| \mathbf{T} \otimes \mathbf{T} \|$$

These matrices are then combined by premultiplication with the expanded statistical weight matrix $\mathbf{U} \otimes \mathbf{E}_9$. The result (compare Eq. 14) for the quantity to be averaged in this case is

$$\langle\ \rangle^{\text{VI}} = Z^{-1}(\mathbf{m}_{i_1}^T \otimes \mathbf{m}_{i_2}^T)[(\mathbf{J}^*\mathbf{U}_1^{(i_1-1)}) \otimes \mathbf{E}_9][(\mathbf{U} \otimes \mathbf{E}_9)(\|\mathbf{T}\| \otimes \mathbf{E}_3)]_{i_1}^{(i_2-i_1)} \times [(\mathbf{U} \otimes \mathbf{E}_9)\|\mathbf{T} \otimes \mathbf{T}\|]_{i_2}^{(j_2-i_2)}[(\mathbf{U} \otimes \mathbf{E}_9)(\|\mathbf{T}\| \otimes \mathbf{E}_3)]_{j_2}^{(j_1-j_2)} \times [(\mathbf{U}_{j_1}^{(n-j_1)}\mathbf{J}) \otimes \mathbf{E}_9](\mathbf{m}_{j_1} \otimes \mathbf{m}_{j_2}) \quad (88)$$

Pursuant to rearrangement of the factors for bond j_2 and succeeding bonds, we observe that since $\mathbf{U} \otimes \mathbf{E}_9 = \mathbf{U} \otimes \mathbf{E}_3 \otimes \mathbf{E}_3$

$$(\mathbf{U} \otimes \mathbf{E}_9)(\|\mathbf{T}\| \otimes \mathbf{E}_3) = [(\mathbf{U} \otimes \mathbf{E}_3)\|\mathbf{T}\|] \otimes \mathbf{E}_3$$

By a rearrangement corresponding to the first of Eqs. 15

$$(\mathbf{U}_{j_1}^{(n-j_1)}\mathbf{J} \otimes \mathbf{E}_9)(\mathbf{m}_{j_1} \otimes \mathbf{m}_{j_2}) = (\mathbf{E}_\nu \otimes \mathbf{m}_{j_1} \otimes \mathbf{m}_{j_2})\mathbf{U}_{j_1}^{(n-j_1)}\mathbf{J}$$

Through use of these identities and further rearrangements we obtain

$$\begin{aligned}
&[(\mathbf{U} \otimes \mathbf{E}_9)(\|\mathbf{T}\| \otimes \mathbf{E}_3)]_{j_2}^{(j_1-j_2)}[(\mathbf{U}_{j_1}^{(n-j_1)}\mathbf{J}) \otimes \mathbf{E}_9](\mathbf{m}_{j_1} \otimes \mathbf{m}_{j_2}) \\
&\quad = \{[(\mathbf{U} \otimes \mathbf{E}_3)\|\mathbf{T}\|]_{j_2}^{(j_1-j_2)} \otimes \mathbf{E}_3\}(\mathbf{E}_\nu \otimes \mathbf{m}_{j_1} \otimes \mathbf{m}_{j_2})\mathbf{U}_{j_1}^{(n-j_1)}\mathbf{J} \\
&\quad = (\mathbf{E}_\nu \otimes \mathbf{E}_3 \otimes \mathbf{m}_{j_2})[(\mathbf{U} \otimes \mathbf{E}_3)\|\mathbf{T}\|]_{j_2}^{(j_1-j_2)}(\mathbf{E}_\nu \otimes \mathbf{m}_{j_1})\mathbf{U}_{j_1}^{(n-j_1)}\mathbf{J} \\
&\quad = [(\mathbf{U} \otimes \mathbf{E}_3 \otimes \mathbf{m})\|\mathbf{T}\|]_{j_2}[(\mathbf{U} \otimes \mathbf{E}_3)\|\mathbf{T}\|]_{j_2+1}^{(j_1-j_2-1)}(\mathbf{U} \otimes \mathbf{m})_{j_1}\mathbf{U}_{j_1+1}^{(n-j_1-1)}\mathbf{J}
\end{aligned} \quad (89)$$

The product of factors preceding the central one in Eq. 88 yields similarly to rearrangement, the result being

$$\begin{aligned}
&(\mathbf{m}_{i_1}^T \otimes \mathbf{m}_{i_2}^T)[(\mathbf{J}^*\mathbf{U}_1^{(i_1-1)}) \otimes \mathbf{E}_9][(\mathbf{U} \otimes \mathbf{E}_9)(\|\mathbf{T}\| \otimes \mathbf{E}_3)]_{i_1}^{(i_2-i_1)} \\
&\quad = \mathbf{J}^*\mathbf{U}_1^{(i_1-1)}[(\mathbf{U} \otimes \mathbf{m}^T)\|\mathbf{T}\|]_{i_1}[(\mathbf{U} \otimes \mathbf{E}_3)\|\mathbf{T}\|]_{i_1+1}^{(i_2-i_1-1)}(\mathbf{E}_\nu \otimes \mathbf{E}_3 \otimes \mathbf{m}_{i_2}^T)
\end{aligned} \quad (90)$$

Substitution of Eqs. 89 and 90 into Eq. 88 gives[7] for $j_2 < j_1$

$$\begin{aligned}
\langle\ \rangle^{\text{VI}} = {} & Z^{-1}\mathbf{J}^*\mathbf{U}_1^{(i_1-1)}[(\mathbf{U} \otimes \mathbf{m}^T)\|\mathbf{T}\|]_{i_1}[(\mathbf{U} \otimes \mathbf{E}_3)\|\mathbf{T}\|]_{i_1+1}^{(i_2-i_1-1)} \\
& \times [(\mathbf{U} \otimes \mathbf{E}_3 \otimes \mathbf{m}^T)\|\mathbf{T} \otimes \mathbf{T}\|]_{i_2}[(\mathbf{U} \otimes \mathbf{E}_9)\|\mathbf{T} \otimes \mathbf{T}\|]_{i_2+1}^{(j_2-i_2-1)} \\
& \times [(\mathbf{U} \otimes \mathbf{E}_3 \otimes \mathbf{m})\|\mathbf{T}\|]_{j_2} \\
& \times [(\mathbf{U} \otimes \mathbf{E}_3)\|\mathbf{T}\|]_{j_2+1}^{(j_1-j_2-1)}(\mathbf{U} \otimes \mathbf{m})_{j_1}\mathbf{U}_{j_1+1}^{(n-j_1-1)}\mathbf{J}
\end{aligned} \quad (91)$$

For $j_2 = j_1$, the factor with this index is $[(\mathbf{U} \otimes \mathbf{m} \otimes \mathbf{m})]_{j_1}$, as may readily be verified.

This lengthy result may be deduced semi-intuitively as follows: Rearrange the summand as given by Eq. 85 to

$$\langle\ \rangle^{\mathrm{VI}} = \langle \mathbf{m}_{i_1}^T \mathbf{T}_{i_1} \cdots \mathbf{T}_{i_2-1}(\mathbf{E}_3 \otimes \mathbf{m}_{i_2}^T)(\mathbf{T}_{i_2} \otimes \mathbf{T}_{i_2}) \cdots (\mathbf{T}_{j_2-1} \otimes \mathbf{T}_{j_2-1}) \times (\mathbf{E}_3 \otimes \mathbf{m}_{j_2})\mathbf{T}_{j_2} \cdots \mathbf{T}_{j_1-1}\mathbf{m}_{j_1}\rangle \tag{92}$$

Then, as implied by Eq. 16, introduce statistical weight factors $\mathbf{J}^*\mathbf{U}_1^{(i_1-1)}$ and $\mathbf{U}_{j_1}^{(n-j_1)}\mathbf{J}$, respectively, to precede i_1 and to follow j_1. Replace each factor $\mathbf{p}$ such as $\mathbf{m}^T$, $\mathbf{E}_3 \otimes \mathbf{m}$, etc., by $\mathbf{E}_\nu \otimes \mathbf{p}$, exactly as in Eq. 16. Replace each $\mathbf{T}$ by $(\mathbf{U} \otimes \mathbf{E}_3)\|\mathbf{T}\|$ as in Eq. 16; in like manner replace each $\mathbf{T} \otimes \mathbf{T}$ by $(\mathbf{U} \otimes \mathbf{E}_9)\|\mathbf{T} \otimes \mathbf{T}\|$. Division by the partition function Z and condensation of consecutive factors bearing the same serial index [e.g., condensation of $(\mathbf{E}_\nu \otimes \mathbf{m}_{i_1}^T)(\mathbf{U} \otimes \mathbf{E}_3)\|\mathbf{T}_{i_1}\|$ to $(\mathbf{U}_{i_1} \otimes \mathbf{m}_{i_1}^T)\|\mathbf{T}_{i_1}\|$] reproduces Eq. 91.

The final term in Eq. 81 may be formulated by first rearranging the summand in Eq. 85 to

$$\langle\ \rangle^{\mathrm{VII}} = \langle \mathbf{m}_{i_1}^T \mathbf{T}_{i_1} \cdots \mathbf{T}_{i_2-1}(\mathbf{E}_3 \otimes \mathbf{m}_{i_2}^T)(\mathbf{T}_{i_2} \otimes \mathbf{T}_{i_2}) \cdots (\mathbf{T}_{j_1-1} \otimes \mathbf{T}_{j_1-1}) \times (\mathbf{m}_{j_1} \otimes \mathbf{E}_3)\mathbf{T}_{j_1} \cdots \mathbf{T}_{j_2-1}\mathbf{m}_{j_2}\rangle \tag{93}$$

Proceeding according to the rules enunciated above, we arrive at the result

$$\begin{aligned}
\langle\ \rangle^{\mathrm{VII}} &= Z^{-1}\mathbf{J}^*\mathbf{U}_1^{(i_1-1)}(\mathbf{E}_\nu \otimes \mathbf{m}_{i_1}^T)[(\mathbf{U} \otimes \mathbf{E}_3)\|\mathbf{T}\|]_{i_1}^{(i_2-i_1)} \\
&\quad \times (\mathbf{E}_\nu \otimes \mathbf{E}_3 \otimes \mathbf{m}_{i_2}^T)[(\mathbf{U} \otimes \mathbf{E}_9)\|\mathbf{T} \otimes \mathbf{T}\|]_{i_2}^{(j_1-i_2)}(\mathbf{E}_\nu \otimes \mathbf{m}_{j_1} \otimes \mathbf{E}_3) \\
&\quad \times [(\mathbf{U} \otimes \mathbf{E}_3)\|\mathbf{T}\|]_{j_1}^{(j_2-j_1)}(\mathbf{E}_\nu \otimes \mathbf{m}_{j_2})\mathbf{U}_{j_2}^{(n-j_2)}\mathbf{J} \\
&= Z^{-1}\mathbf{J}^*\mathbf{U}_1^{(i_1-1)}[(\mathbf{U} \otimes \mathbf{m}^T)\|\mathbf{T}\|]_{i_1}[(\mathbf{U} \otimes \mathbf{E}_3)\|\mathbf{T}\|]_{i_1+1}^{(i_2-i_1-1)} \\
&\quad \times [(\mathbf{U} \otimes \mathbf{E}_3 \otimes \mathbf{m}^T)\|\mathbf{T} \otimes \mathbf{T}\|]_{i_2}[(\mathbf{U} \otimes \mathbf{E}_9)\|\mathbf{T} \otimes \mathbf{T}\|]_{i_2+1}^{(j_1-i_2-1)} \\
&\quad \times [(\mathbf{U} \otimes \mathbf{m} \otimes \mathbf{E}_3)\|\mathbf{T}\|]_{j_1}[(\mathbf{U} \otimes \mathbf{E}_3)\|\mathbf{T}\|]_{j_1+1}^{(j_2-j_1-1)}(\mathbf{U} \otimes \mathbf{m})_{j_2} \\
&\quad \times \mathbf{U}_{j_2+1}^{(n-j_2-1)}\mathbf{J}
\end{aligned} \tag{94}$$

for $i_1 < i_2$. The modification required for $i_1 = i_2$ will be obvious. A rigorous proof of this result is given in Ref. 7.

After substitution of Eqs. 83, 84, 87, 91, and 94 into Eq. 81, it remains to execute the summations over the various terms of Eq. 81. All of these operations may be carried out simultaneously through use of Matrix 95.

$$
\mathscr{K}_i = \begin{bmatrix}
\mathbf{U} & \mathrm{m}^2\mathbf{U} & 2(\mathbf{U}\otimes\mathbf{m}^T)\|\mathbf{T}\| & \mathrm{m}^2(\mathbf{U}\otimes\mathbf{m}^T)\|\mathbf{T}\| & \mathbf{0} & (\mathbf{U}\otimes\mathbf{m}^T\otimes\mathbf{m}^T)\|\mathbf{T}\otimes\mathbf{T}\| & \mathbf{0} & \mathbf{0} & \left(\frac{m^4}{4}\right)\mathbf{U} \\
\mathbf{0} & \mathbf{U} & \mathbf{0} & (\mathbf{U}\otimes\mathbf{m}^T)\|\mathbf{T}\| & \mathbf{0} & \mathbf{0} & \mathbf{0} & \mathbf{0} & \left(\frac{m^2}{2}\right)\mathbf{U} \\
\mathbf{0} & \mathbf{0} & (\mathbf{U}\otimes\mathbf{E}_3)\|\mathbf{T}\| & \left(\frac{\mathrm{m}^2}{2}\right)(\mathbf{U}\otimes\mathbf{E}_3)\|\mathbf{T}\| & \mathbf{U}\otimes\mathbf{m} & (\mathbf{U}\otimes\mathbf{E}_3\otimes\mathbf{m}^T)\|\mathbf{T}\otimes\mathbf{T}\| & [\mathbf{U}\otimes(\mathbf{m}\mathbf{m}^T)]\|\mathbf{T}\| & \mathbf{0} & \left(\frac{m^2}{2}\right)(\mathbf{U}\otimes\mathbf{m}) \\
\mathbf{0} & \mathbf{0} & \mathbf{0} & (\mathbf{U}\otimes\mathbf{E}_3)\|\mathbf{T}\| & \mathbf{0} & \mathbf{0} & \mathbf{0} & \mathbf{0} & \mathbf{U}\otimes\mathbf{m} \\
\mathbf{0} & \mathbf{0} & \mathbf{0} & \mathbf{0} & \mathbf{U} & \mathbf{0} & (\mathbf{U}\otimes\mathbf{m}^T)\|\mathbf{T}\| & \mathbf{0} & \left(\frac{m^2}{2}\right)\mathbf{U} \\
\mathbf{0} & \mathbf{0} & \mathbf{0} & \mathbf{0} & \mathbf{0} & (\mathbf{U}\otimes\mathbf{E}_9)\|\mathbf{T}\otimes\mathbf{T}\| & (\mathbf{U}\otimes\mathbf{E}_3\otimes\mathbf{m})\|\mathbf{T}\| & (\mathbf{U}\otimes\mathbf{m}\otimes\mathbf{E}_3)\|\mathbf{T}\| & \mathbf{U}\otimes\mathbf{m}\otimes\mathbf{m} \\
\mathbf{0} & \mathbf{0} & \mathbf{0} & \mathbf{0} & \mathbf{0} & \mathbf{0} & (\mathbf{U}\otimes\mathbf{E}_3)\|\mathbf{T}\| & \mathbf{0} & \mathbf{U}\otimes\mathbf{m} \\
\mathbf{0} & \mathbf{0} & \mathbf{0} & \mathbf{0} & \mathbf{0} & \mathbf{0} & \mathbf{0} & (\mathbf{U}\otimes\mathbf{E}_3)\|\mathbf{T}\| & \mathbf{U}\otimes\mathbf{m} \\
\mathbf{0} & \mathbf{0} & \mathbf{0} & \mathbf{0} & \mathbf{0} & \mathbf{0} & \mathbf{0} & \mathbf{0} & \mathbf{U}
\end{bmatrix}_i \quad (95)
$$

The fourth moment is given by

$$\langle M^4 \rangle = 4Z^{-1} \mathscr{J}^* \mathscr{K}_1^{(n)} \mathscr{J} \tag{96}$$

where $\mathscr{J}^*$ and $\mathscr{J}$ are the row and column introduced in Chapter III and adapted to operations in conjunction with the $\mathscr{K}$ matrices according to previous usage in other connections. As in the corresponding expression for the second moment (see Eqs. 28 and 54), the initial and final matrices $\mathscr{K}_1$ and $\mathscr{K}_n$ of the serial product are to be expressed with $\mathbf{U}_1 = \mathbf{U}_n = \mathbf{E}_\nu$; only the first row of $\mathscr{K}_1$ and the final pseudocolumn of $\mathscr{K}_n$ are required. The order of $\mathscr{K}_i$ is $25\nu \times 25\nu$. Computations are readily performed by computer.[7]† These equations for evaluation of $\langle M^4 \rangle$ are equivalent in generality to Eqs. 28 and 54 for $\langle M^2 \rangle$ and $\langle s^2 \rangle_0$, respectively.

If the bond rotations are independent, expression of the fourth moment takes the simpler form

$$\langle M^4 \rangle = 4[1 \quad 0 \quad \cdots \quad 0] \mathfrak{K}_1^{(n)} \begin{bmatrix} 0 \\ \vdots \\ 0 \\ 1 \end{bmatrix} \tag{97}$$

where $\mathfrak{K}$ is Matrix 98 (see p. 128).

Equations 97 and 98 are analogous to Eqs. 59 and 60 for the second moment of a chain with independent rotations. They may be derived directly by rearranging the summands in the various terms of Eq. 81, through use of Eq. 85 and the direct-product theorem, so that factors of the same serial index are juxtaposed in each case, as in Eqs. 86, 92, and 93. The assumption of rotational independence permits the average of the product to be factored into the product of the individual averages $\langle \mathbf{T}_i \rangle$ and $\langle \mathbf{T}_i \otimes \mathbf{T}_i \rangle$. The result expressed by Eqs. 97 and 98 follows at once. The correspondence of the matrix $\mathfrak{K}_i$ of Eq. 98 to $\mathscr{K}_i$ of Eq. 95 is apparent.

The sixth moment $\langle M^6 \rangle$ has been treated by Jernigan.[18] The matrices required, of orders up to 231×231, exceed the limits readily handled by digital computers.

† The matrix squaring method is applicable to Eq. 96 for chains of regular repeating structure, and may be used to advantage for very long chains. Squaring must of course be carried out in advance of premultiplication by $\mathscr{J}^*$ whereby the product is converted to a row. Alternatively, multiplication by $\mathscr{J}^*$ may be performed at the outset with the advantage that the premultiplying factor for subsequent multiplications is a row instead of a square matrix. The multiplications must then be carried out sequentially, however, with forfeiture of the advantage of successive squaring. R. L. Jernigan points out that for large matrices like $\mathscr{K}_i$ sequential multiplication beginning with $\mathscr{J}^*$ requires less numerical labor for chains up to about 1000 bonds or repeating units.

$$\mathfrak{R}_i = \begin{bmatrix} 1 & \mathrm{m}^2 & 2\mathbf{m}^T\langle\mathbf{T}\rangle & m^2\mathbf{m}^T\langle\mathbf{T}\rangle & 0 & (\mathbf{m}^T\otimes\mathbf{m}^T)\langle\mathbf{T}\otimes\mathbf{T}\rangle & \mathbf{0} & \mathbf{0} & \frac{m^4}{4} \\ 0 & 1 & \mathbf{0} & \mathbf{m}^T\langle\mathbf{T}\rangle & 0 & \mathbf{0} & \mathbf{0} & \mathbf{0} & \frac{m^2}{2} \\ \mathbf{0} & \mathbf{0} & \langle\mathbf{T}\rangle & \left(\frac{m^2}{2}\right)\langle\mathbf{T}\rangle & \mathbf{m} & (\mathbf{E}_3\otimes\mathbf{m}^T)\langle\mathbf{T}\otimes\mathbf{T}\rangle & \mathbf{m}\mathbf{m}^T\langle\mathbf{T}\rangle & \mathbf{0} & \left(\frac{m^2}{2}\right)\mathbf{m} \\ \mathbf{0} & \mathbf{0} & \mathbf{0} & \langle\mathbf{T}\rangle & \mathbf{0} & \mathbf{0} & \mathbf{0} & \mathbf{0} & \mathbf{m} \\ 0 & 0 & \mathbf{0} & \mathbf{0} & 1 & \mathbf{0} & \mathbf{m}^T\langle\mathbf{T}\rangle & \mathbf{0} & \frac{m^2}{2} \\ \mathbf{0} & \mathbf{0} & \mathbf{0} & \mathbf{0} & \mathbf{0} & \langle\mathbf{T}\otimes\mathbf{T}\rangle & (\mathbf{E}_3\otimes\mathbf{m})\langle\mathbf{T}\rangle & (\mathbf{m}\otimes\mathbf{E}_3)\langle\mathbf{T}\rangle & \mathbf{m}\otimes\mathbf{m} \\ \mathbf{0} & \mathbf{0} & \mathbf{0} & \mathbf{0} & \mathbf{0} & \mathbf{0} & \langle\mathbf{T}\rangle & \mathbf{0} & \mathbf{m} \\ \mathbf{0} & \mathbf{0} & \mathbf{0} & \mathbf{0} & \mathbf{0} & \mathbf{0} & \mathbf{0} & \langle\mathbf{T}\rangle & \mathbf{m} \\ 0 & 0 & \mathbf{0} & \mathbf{0} & 0 & \mathbf{0} & \mathbf{0} & \mathbf{0} & 1 \end{bmatrix}_i \quad (98)$$

REFERENCES

1. Yu. Ya. Gotlib, *Zh. Tekhn. Fiz.*, **29**, 523 (1959); T. M. Birshtein and O. B. Ptitsyn, *ibid.*, **29**, 1048 (1959); T. M. Birshtein, *Vyosokomolekul. Soedin.*, **1**, 798 (1959).
2. T. M. Birshtein and O. B. Ptitsyn, *Conformations of Macromolecules* (translated from the 1964 Russian ed., S. N. Timasheff and M. J. Timasheff), Interscience, New York, 1966, Chap. 5.
3. C. A. J. Hoeve, *J. Chem. Phys.*, **32**, 888 (1960).
4. S. Lifson, *J. Chem. Phys.*, **30**, 964 (1959).
5. K. Nagai, *J. Chem. Phys.*, **31**, 1169 (1959).
6. P. J. Flory, *Proc. Natl. Acad. Sci. U.S.*, **51**, 1060 (1964).
7. P. J. Flory and R. L. Jernigan, *J. Chem. Phys.*, **42**, 3509 (1965).
8. K. Nagai, *J. Chem. Phys.*, **45**, 838 (1966).
9. G. Porod, *Monatsh. Chem.* **80**, 251 (1949); O. Kratky and G. Porod, *Rec. Trav. Chim.*, **68**, 1106 (1949).
10. P. J. Flory and R. L. Jernigan, unpublished.
11. J. E. Mark, *J. Am. Chem. Soc.*, **88**, 4354 (1966); **89**, 6829 (1967).
12. P. J. Flory and A. D. Williams, *J. Polymer Sci. A-2*, **5**, 399 (1967).
13. A. D. Williams and P. J. Flory, *J. Polymer Sci. A-2*, **5**, 417 (1967).
14. L. Peller, *J. Chem. Phys.*, **43**, 2355 (1965).
15. W. G. Miller, D. A. Brant, and P. J. Flory, *J. Mol. Biol.*, **23**, 67 (1967).
16. K. Nagai, *J. Chem. Phys.*, **38**, 924 (1963).
17. K. Nagai, *J. Chem. Phys.*, **40**, 2818 (1964).
18. R. L. Jernigan, Ph.D. dissertation, Stanford University, 1967.

CHAPTER V

Symmetric Chains

The preceding chapters are chiefly concerned with the elaboration of a rationale for comprehending the configurational characteristics of chain molecules and with the construction of a framework of theory for the quantitative treatment of their properties. Here we undertake to explore the application of the theory set forth in those chapters to representative chain molecules whose structural units do not possess centers of asymmetry. Molecules in the category thus excluded are treated in the following two chapters.

The projected analysis is conceived in terms of the structure of the molecule. Hence, structural data consisting of bond lengths and bond angles are prerequisites for its application to a given series of chain homologs. Data of this nature usually are at hand, with ample accuracy for the purpose. They are not, however, sufficient. Unambiguous deduction of the properties of a chain molecule further requires an adequate understanding of the intramolecular energy as a function of the conformation. The conformational energy usually is described in terms of potentials associated with the various bond rotations, or combinations of such rotations. Enough must be known concerning these potentials to permit a judicious selection of rotational isomeric states at or near potential minima, and also to allow the statistical weights for the states so chosen to be evaluated. Only in the case of the *n*-alkanes does the information independently available approach fulfillment of this requirement. The energy difference between *gauche* and *trans* states, known to an accuracy of ± 100 cal mole^{-1}, determines the approximate value of the essential statistical weight parameter for molecules of this homologous series. In general, however, the best that can be done is to outline the qualitative character of the conformational energy on the basis of the structure, aided by inferences from analogous molecules and by approximate estimates of the principal contributions to the energy. These contributions include the inherent torsional potentials (see p. 54) of the various bonds, steric interactions, etc. Appropriate sets of rotational isomeric states can usually be selected. Their statistical weights ordinarily cannot be evaluated quantitatively, however, beyond estimation of their orders of magnitude. Thus, instead of attempting *a priori* prescription of statistical weights, we shall be obliged to introduce them as parameters whose values are to be determined experimentally from $\langle r^2 \rangle_0$, dipole moments, etc., and from the temperature coefficients of these quantities.

The number of statistical weight parameters can be reduced by symmetry considerations. Ranges of values which they may reasonably assume usually can be narrowed by inspection of scale models and by resort to numerical estimates of nonbonded interactions and other contributions to the molecular energy. Similarities between different chain molecules can be perceived, and these serve to relate statistical weights (and the associated conformational energies) for one type of chain to those for another. The pattern of rotational preferences amongst various related structures which thus emerges goes beyond mere empirical description.

We first consider procedures for estimating the energy of a molecule as a function of its conformation, with particular attention to the interactions between nonbonded atoms or groups. Although the methods available are crude, they yield numerical results affording a measure of comprehension of the character of these *nonbonded interactions* which is more refined than can be gleaned from inspection of models alone. Normal alkane chains will be treated in some detail. They serve to illustrate the kinds of interactions encountered in most other chain molecules of interest.

1. METHODS FOR ESTIMATING CONFORMATIONAL ENERGIES[1]

The character of the potential affecting rotation about a single bond in a polyatomic molecule was discussed briefly in Chapter III (pp. 50–57). It was pointed out that an *inherent* threefold potential may be ascribed in general to an sp^3 bond. This potential may be presumed to be adequately represented by a simple cosine function with the form of Eq. III-1. If the bond bears polyatomic substituents such as —CH_3 or —CH_2—, then interactions involving them must be taken into account as well. These interactions involve *atoms* separated one from the other by four or more intervening bonds. Hence, they should resemble *intermolecular* interactions between corresponding atoms and groups when they are members of different molecules. The rotational potential may be markedly affected by substituents of the size of those cited. However, the total potential associated with rotation about a single bond cannot in general be attributed entirely to nonbonded interactions between the substituents attached to the respective bonded atoms. As was pointed out on p. 52, the dependences of the rotational potentials in simple molecules on the sizes of atomic substituents and on their distance of separation provides compelling evidence against any interpretation founded on this basis. Rather, the total rotational potential is more appropriately considered to consist of the combination of an inherent bond potential and the energy of interaction between nonbonded substituents. The separation of one contribution from the other is unquestionably naive and oversimplified. The so-called inherent potential must obviously depend on the atoms directly connected to the bond. If, however, most of its substituents are hydrogen atoms throughout a series of compounds in which the bond occurs, the postulation of the same inherent torsional potential for all is not an unreasonable approximation. Support for this scheme is found in its applications, which also serve to illustrate its limitations. It is noteworthy that in ethane the preponderance of the total rotational potential is attributable to the inherent potential of the C—C bond (*cf. seq.*).

The repulsions between nonbonded atom pairs are customarily represented[1–3] either by an expression of the form a'_{kl}/r_{kl}^{m} or by $a_{kl} \exp(-br_{kl})$ where k, l index the atom pair, r_{kl} is the distance between them, a'_{kl} and a_{kl} are constants characteristic of the atom pair, and m and b are parameters which, to a usually satisfactory approximation, may be taken to be the same for all pairs. Thus, a frequent choice is $m = 12$ in the first expression. For atoms such as H, C, O, and N, a value of b of 4×10^{-8} to 5×10^{-8} cm^{-1} is usually acceptable in the latter (*cf. seq.*).[4] Either expression then requires assignment of only one remaining parameter to characterize the given atom pair. The exponential function of r_{kl} affords a somewhat better representation of the repulsive energy for intermolecular pairs of atoms.[2]

The contributions to the intramolecular energy from London dispersion interactions (attractions) between nonbonded atoms are comparatively small and their sum is rendered less dependent on conformation by the longer range of these forces. The usual inverse sixth-power law for these interactions yields terms $-c_{kl}/r_{kl}^{6}$ where c_{kl} is a constant for the specified atom pair.

Combining these contributions, we have for the intramolecular energy associated with the configuration specified by the set of rotation angles ϕ[1,4–6]

$$E_{\{\phi\}} = \sum_{i} (E_i^{\circ}/2)(1 - \cos 3\phi_i) + \sum_{k,l} [a_{kl} \exp(-br_{kl}) - c_{kl}/r_{kl}^{6}] \qquad (1)$$

The first sum includes all bonds subject to rotation; the second includes all atom pairs k, l whose distance of separation r_{kl} depends on one or more of the ϕ_i. Terms of the latter sum, as well as those of the former, are determined by the set of rotation angles $\{\phi\}$. The bond lengths and bond angles having been fixed, these rotation angles comprise the independent variables.

The parameters c are least important for the present purpose. They may be calculated from atomic polarizabilities, as compiled by Ketelaar,[7] using the Slater-Kirkwood equation.[8] Only the a parameters remain to be assigned. They may be so chosen as to minimize the interaction energy for the atom pair k, l as expressed by the pair potential

$$E_{kl} = a_{kl} \exp(-br_{kl}) - c/r_{kl}^6 \tag{2}$$

at the distance r_{kl}^* corresponding to the sum of their van der Waals radii. There is of course some latitude in the choice of these radii. Alternatively, the choice of the a parameters may be guided by known conformational energies of molecules. Thus, values may be deliberately selected to reproduce the known barriers to rotation for simple molecules or the differences in energy for the various conformations, e.g., the energy for the *gauche* as compared with the *trans* form, if this difference in energy is known.

The functions cited are inexact and the procedure for evaluating the necessary parameters is mainly empirical. It follows that calculations carried out in this manner can be meaningful only within the class of structures inclusive of the compounds on which the empirical information was derived. Such calculations may nevertheless be useful in translating experimental results on a few relatively simple compounds to their chain molecular analogs.†

2. CONFORMATIONAL ENERGIES OF *n*-ALKANES

Calculations for the normal paraffin hydrocarbons are illustrative of application of the methods outlined above.

It is first necessary to consider the geometric parameters for these molecules. The C—C and C—H bond lengths are 1.53 and 1.10 Å, respectively.[9] The carbon–carbon bond angle, i.e., ∠CCC, is 112° according to evidence provided by analysis of the x-ray diffraction of crystalline *n*-alkanes,[9,10] by electron diffraction of gaseous *n*-alkanes from propane to heptane[11] and by the microwave spectrum of propane.[12] Other bond angles, namely, ∠CCH and ∠HCH, may for simplicity, and with negligible error in result, be treated as being equal to one another in the normal alkanes. This implies a value of 109.0° for these angles.

A number of calculations of the conformational energies of *n*-alkane chains have been carried out by combining the several contributions after the manner of Eq. 1.[6,13–15] Principal differences occur in the magnitude chosen for the height of the intrinsic bond torsional barrier E°, and in the form adopted for the repulsive terms. Abe et al.[14] chose parameters which would reproduce the observed barrier to rotation in ethane (ca. 3000 cal mole^{-1}) and the energy of the *gauche* conformation relative to *trans* (ca. 500 ± 100 cal mole^{-1}; see p. 56) in the liquid *n*-alkanes. They were

† For a more extensive discussion of methods for estimating conformational energies see ref. 1.

thus led to adopt the comparatively large value of 2800 cal mole^{-1} for E°, leaving only ca. 200 cal mole^{-1} of the total barrier to be charged to nonbonded interactions between the hydrogens of one methyl group and those of the other. This small contribution represents, of course, the difference between the resultant of the repulsions and attractions involving these atoms in the staggered and in the eclipsed conformations.

The parameter b governing the steepness of the mutual repulsion of nonbonded atom pairs has been correlated with the atomic number by Scott and Scheraga.[4] For atoms such as hydrogen and carbon, values in the range of 4–5 Å^{-1} are usual. The precise value chosen is of no importance[14] if the procedure outlined in the preceding section is followed. The effect of a change of b within this range will be compensated at the next step in the procedure wherein the a_{kl} are so chosen as to reproduce the observed energy difference between *gauche* and *trans* conformations, while at the same time placing the minimum of the interaction energy for each nonbonded atom pair at a distance in approximate accord with accepted van der Waals radii.

Taking $b \cong 4.55$ Å^{-1} for the several atom pairs, Abe et al.[14] found $a_{kl} = 10.0 \times 10^3$, 86×10^3, and 909×10^3 kcal mole^{-1} for H···H, C···H, and C···C, respectively. With these choices for the a parameters, the minima occur at the interatomic distances 2.6, 3.1, and 3.6 Å, corresponding to van der Waals radii of 1.3 and 1.8 Å for the hydrogen and carbon atoms, respectively. Energy barriers to rotation were calculated by taking the difference between the conformational energies $E_{\{\phi\}}$ for $\phi = 60°$ and for $\phi = 0°$ for the bond in question. (Rotations of other bonds do not affect this energy difference significantly in the molecules considered.) Values thus obtained are given in the second column of Table 1 where they are compared with results deduced from experiment and given in the last column. The observed trends with structure are well reproduced by the calculations. The energy difference $E_g - E_t$ between *gauche* and *trans* conformations of *n*-butane, calculated using Eq. 1 and the same parameters, is 530 cal mole^{-1} (*cf. seq.*).

The van der Waals radii adopted for these calculations (see above) are ca. 0.1 Å greater than the usual values, such as those deduced by Bondi[16] from equilibrium intermolecular distances in crystals, for example. The reason for the discrepancy has been clarified by Brant et al.[17] The present calculations pertain to a pair of atoms hypothetically disembodied from their respective molecules, and hence not subject to the effects of interactions between other atom pairs. Attractive interactions extend over a greater range than repulsions. Those operative between various atoms of two molecules in contact in a crystal must act to diminish slightly the equilibrium distance between the pair of atoms which are nearest neighbors on the periphery of the two molecules. Consequently, the distance observed for the latter pair is less than would obtain in the absence of interactions involving the other atoms of the two polyatomic molecules in contact. This difference between the van der Waals radii as determined by conventional methods and those presumed to represent a pair of atoms exclusive of the effects of all other interactions has generally been overlooked. It is this difference which necessitates the use of somewhat greater radii for the purpose at hand.

Table 1

Potential Barriers to Internal Rotation for Simple Alkanes[14]
(barrier heights in kcal $mole^{-1}$)

Hydrocarbon	Calculated	Experimental
CH_3—CH_3	3.0	2.9[e], 3.0[f]
CH_3—CH_2CH_3	3.4[a]	3.4[e]
CH_3—$CH(CH_3)_2$	3.7[b]	3.6[e], 3.9[g]
CH_3—$C(CH_3)_3$	4.3[c]	4.3[e]
CH_3CH_2—CH_2CH_3[d]	3.6[a]	3.3[h], 3.6[i]

[a] $\angle CCC = 112°$.
[b] $\angle CCC = 111°$ (see D. R. Lide, Jr., *J. Chem. Phys.*, **33**, 1519 (1960).
[c] $\angle CCC = 109.5°$.
[d] Values given refer to the lower barriers located symmetrically with respect to *trans* (see Figs. 1 and III-2).
[e] K. S. Pitzer, *Discussions Faraday Soc.*, **10**, 66 (1951); estimated from thermodynamic data.
[f] D. R. Lide, Jr., *J. Chem. Phys.*, **29**, 1426 (1958); from infrared spectroscopy.
[g] D. R. Lide, Jr., and D. E. Mann, *J. Chem. Phys.*, **29**, 914 (1958); from microwave spectroscopy.
[h] K. S. Pitzer, *Ind. Eng. Chem.*, **36**, 829 (1944); W. B. Person and G. C. Pimental, *J. Am. Chem. Soc.*, **75**, 539 (1953); estimated from thermodynamic data.
[i] K. S. Pitzer, *J. Am. Chem. Soc.*, **63**, 2413 (1941); B. P. Dailey and W. A. Felsing, *ibid.*, **65**, 44 (1943); from thermodynamic data.

The conformation of *n*-butane, the simplest *n*-alkane of interest, is described by three rotation angles; see Fig. 1. Conformation energies computed[14] for this molecule in the neighborhood of the *trans* and *gauche* minima for bond 2 are shown in Fig. 2. Calculations were carried out according to Eq. 1, the threefold, inherent torsional terms being included. Inasmuch as the terminal bonds 1 and 3 are equivalent, the dimensionality of the energy surface was simplified without sacrifice of significant information by setting $\phi_1 = \phi_3$.

The strong repulsions encountered in the *cis* conformation (see Figs. 1 and III-2) are incipient at the *gauche* conformation located approximately

60° from *cis*. The *gauche* minima may consequently be displaced a few degrees to an angle $|\phi_g|$ somewhat less than 120°, as was pointed out by Borisova and Volkenstein.[5] According to the calculations shown in Fig. 2, $\phi_g = \pm 112.5°$. Precision cannot be claimed for this result, which can only be taken as indicative of a displacement in the range of 5–10°. This conclusion finds partial confirmation in the electron diffraction of gaseous *n*-butane.[18,19] Calculations carried out by Scott and Scheraga[15] using Lennard-Jones "6–12" potentials for nonbonded interactions yield similar results.

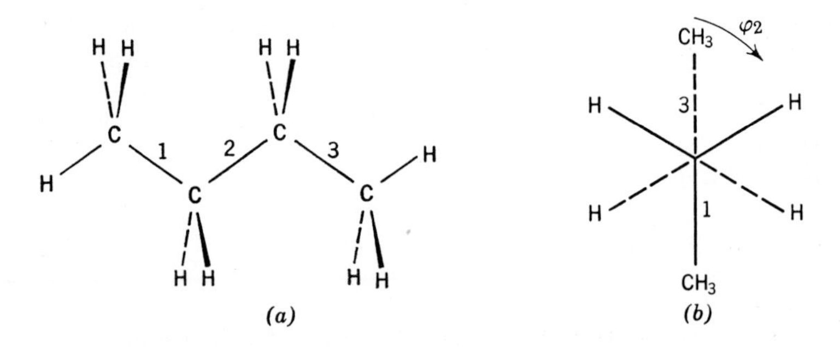

Fig. 1. *Trans* conformation of *n*-butane: (*a*) view with carbons in the plane of the figure; (*b*) view along C—C bond 2, with positive rotation about that bond indicated by ϕ_2.

A second feature of importance is the similarity in breadths of the potential wells for the *trans* and *gauche* minima (see Fig. 2). The domain of the latter is displaced a few degrees from $\phi_1 = \phi_3 = 0$, and its direction of major breadth is tilted with respect to the ϕ_2 axis. The curvature of the energy well with ϕ_2 is almost identical for the two minima; the *gauche* well is somewhat more confined in the direction of the $\phi_1 = \phi_3$ axis, however. This minor difference is sufficiently small to be ignored. It is thus warranted to equate the relative statistical weight to a Boltzmann factor of the energy difference as in Eq. III-3, the preexponential factor being, by implication, unity.

The indicated small displacement of $\phi_1 = \phi_3$ from 0 to about 4° when ϕ_2 is *gauche** would suggest, if taken literally, that the precise location of the *trans* state in a longer chain is perturbed if its neighbor is *gauche*. The effect is small and may reasonably be disregarded.

Energy contours for *n*-pentane drawn at 1 kcal $mole^{-1}$ intervals up to

* Scott and Scheraga[15] calculated a larger displacement of 11–12°.

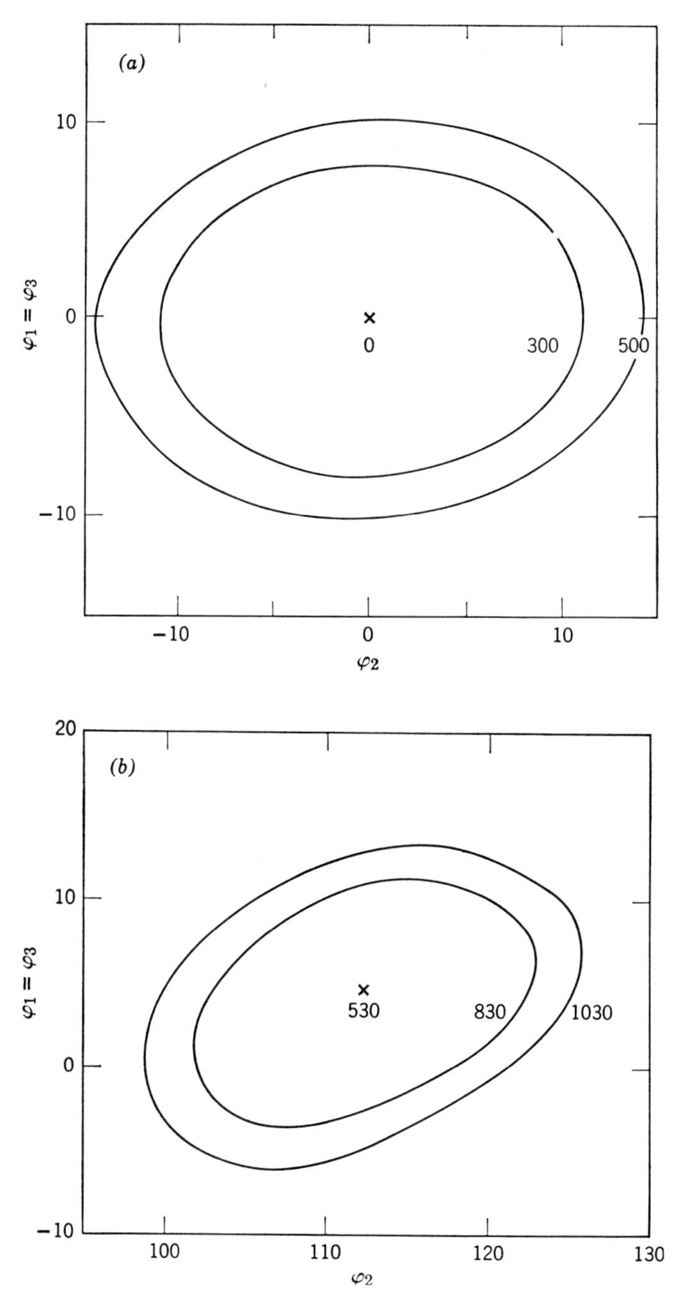

Fig. 2. Conformational energy for *n*-butane in the vicinity of its (*a*) *trans* and (*b*) *gauche* minima. Energy contours were calculated by Abe, Jernigan, and Flory[14] using Eq. 1 as described in the text. Locations of minima are indicated by ×. Energies are expressed in calories per mole relative to the *trans* minimum.

5 kcal mole^{-1} above the *tt* conformation ($\phi_2 = \phi_3 = 0$) are shown in Fig. 3. Calculations were carried out[14] on the basis of Eq. 1, using the same parameters as were chosen for *n*-butane in the calculations shown in Fig. 2. For the purposes of this diagram, the terminal rotations were fixed at $\phi_1 = \phi_4 = 0$, i.e., the terminal methyl groups were fixed in their staggered conformations. The portion of the energy surface for $\phi_2 < 0$ is produced by inversion through the origin in the ϕ_2, ϕ_3 plane.

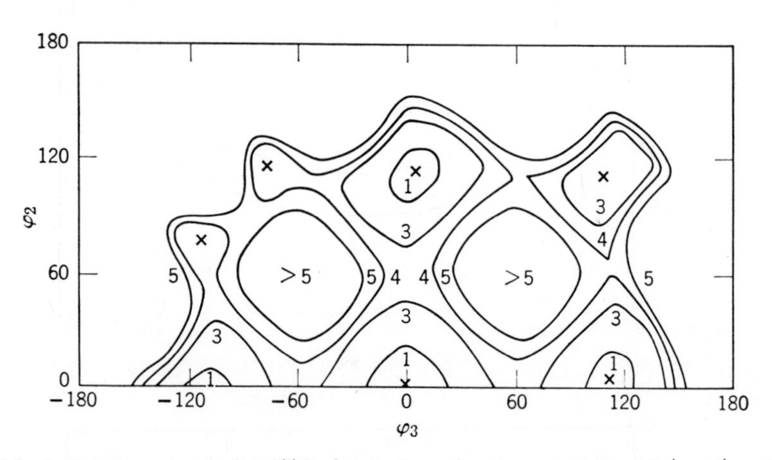

Fig. 3. Energy map calculated[14] for internal rotations in *n*-pentane, with $\phi_1 = \phi_4 = 0$; contours are shown at intervals of 1 kcal mole^{-1}. Minima are indicated by ×.

The $tg^{\pm}$ and $g^{\pm}t$ minima are equivalent to those for *n*-butane. The g^+g^+ and g^-g^- (not shown) minima occur in the vicinity of $\phi_2 = \phi_3 = \pm 110°$. Thus, the *gauche* minima for two adjoining bonds in *gauche* conformations of the same sign are mutually displaced a few degrees from the values (ca. 112.5°) which would be assumed by each if both of its neighbors were *trans*. The *trans* minima for neighboring bonds 1 and 4 are also perturbed a few degrees. These effects, arising from subtle interactions between pairs of H atoms on third neighbor carbons (see Abe et al.[14]), are small. More to the point is the fact that the calculated energy, 1180 cal mole^{-1}, for the g^+g^+ pair is close to twice that for one *gauche* bond alone. Hence, the energy for neighboring bonds in *gauche* states of the same sign may be treated in good approximation as additive. Also, the breadth of the *gauche* well is little affected by its neighbor.

Instead of a single g^+g^- minimum, Fig. 3 reveals two shallow minima located on either side of the anticipated g^+g^- state. Although of fairly high energy, they are much below the undistorted g^+g^- conformation at 120, −120°. These minima are shown in greater detail in Fig. 4. They are located at $\phi_2 = 115°$, $\phi_3 = -77°$ and at $\phi_2 = 77°$, $\phi_3 = -115°$; a corresponding pair, related to g^-g^+, occur in the half of the diagram not included in Fig. 3. Similar minima were found by Scott and Scheraga.[15] Inspection of models confirms their plausibility. In short, the severe steric overlap for $g^{\pm}g^{\mp}$ pairs may obviously be alleviated by diminishing $|\phi_{i-1}|$ and/or $|\phi_i|$.

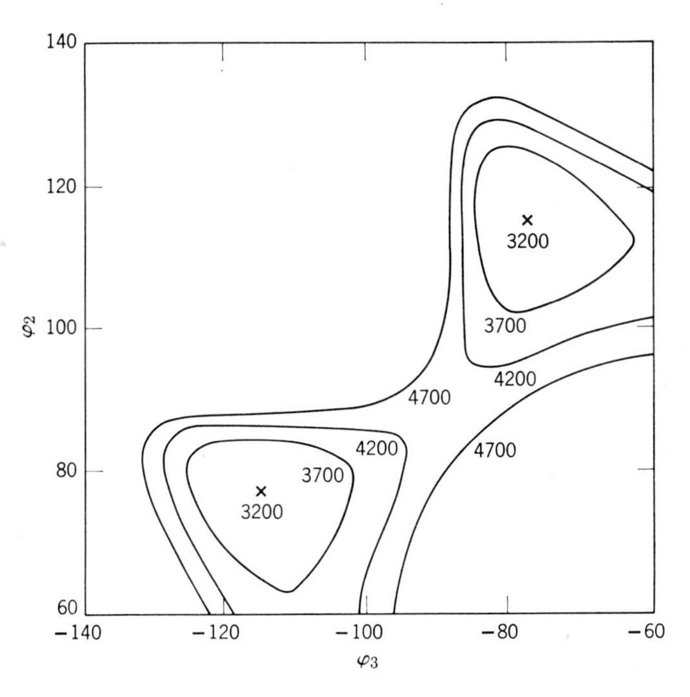

Fig. 4. The energy contour map calculated[14] for *n*-pentane in the neighborhood of its g^+g^- conformation shown in enlarged detail; see Fig. 3. Energies of the contours and minima ($\times$) are in calories per mole relative to the *tt* conformation.

A decrease in one of the angles is a more efficient means to this end, however, than simultaneous decreases in both angles. Contributions from the inherent torsional potentials for the bonds thus rotated must, of course, be taken into account when molecular models are used for rationalization of the conformational energy.

These minima occur at a calculated energy of 3200 cal mole^{-1} relative to the *tt* state ($\phi_2 = \phi_3 = 0$), or about 2000 cal mole^{-1} above g^+g^+. An energy of this magnitude is sufficient to reduce the population of $g^{\pm}g^{\mp}$ bond pairs to a low level, but not to the point where they may be disregarded. This becomes apparent in the following sections. The calculated energy is only a rather crude estimate at best. The reduction of the repulsive energy which relaxation of the condition of constancy of bond angles (θ)

would allow has, for example, been ignored. In conformations of large steric overlap, acceptance of this condition is no longer legitimate.

Although the values of the energies calculated for the various minima revealed in Fig. 3 are approximate owing to the limitations of the potential functions used, the general features of the energy surface are doubtless correctly portrayed.

3. STATISTICAL WEIGHTS FOR POLYMETHYLENE CHAINS

As was pointed out in Chapter III, the interdependence of bond rotations is limited to first neighbors, in good approximation, for real chains of main interest. It follows that the conformational energies for *n*-pentane, having two internal C—C bonds, can be transcribed to *n*-alkane molecules of any length. The results surveyed above provide the basis for assigning statistical weights appropriate for polymethylene (PM) chains.

The energy surface depicted in Fig. 3 suggests a fivefold scheme ($\nu = 5$) with rotational states at angles $\phi_i \cong 0, 77, 115, -115, -77°$. These could be labeled t, g^{*+}, g^+, g^-, g^{*-}, with combinations $g^{\pm}g^{\mp}$ forbidden by their large energies. Computations[14] of the characteristic ratio $(\langle r^2 \rangle / nl^2)_\infty$ on this basis yield results which differ very little from those obtained using the three-state scheme from which the g^{*+} and g^{*-} states are deleted and $g^{\pm}g^{\mp}$ is accorded an energy similar to that computed for $g^{*\pm}g^{\mp}$ and $g^{\pm}g^{*\mp}$. The reasons for this approximate equivalence of the two procedures will be apparent. The four minima of high energy in the fivefold scheme are replaced by only two in the threefold representation. On the other hand, in each of the former states one bond ($g^{*\pm}$) departs less drastically from the *trans* conformation, and hence is less effective in reducing the chain dimensions. Thus, the smaller number of minima in the threefold scheme is compensated by the greater effect of each on the configuration. Most important is the fact that these high-energy states occur so infrequently as to permit their inclusion in crude approximation. The following development is confined to the threefold representation, which finds extensive application to other chains.

Having considered the detailed treatment of the conformational energy of *n*-pentane wherein account was taken of the interaction between every pair of atoms in the molecule (apart from simplifications of trivial effect; see Abe et al.[14]), we may turn at this stage to the simplified representation of the chain shown in Fig. 5. The scheme here adopted proves especially useful in application to the molecules dealt with later. The chain is represented as a linearly connected sequence of groups, the identities of individual atoms being ignored. The groups, in this case methylene CH_2, rather than their constituent atoms will, for descriptive purposes, be regarded as the entities which engage in mutual interactions.

The planar form of the chain is shown by solid lines in Fig. 5. Bonds turned out of the plane by *gauche* rotations about intervening skeletal bonds are shown as dashed lines; the longer lines represent those above

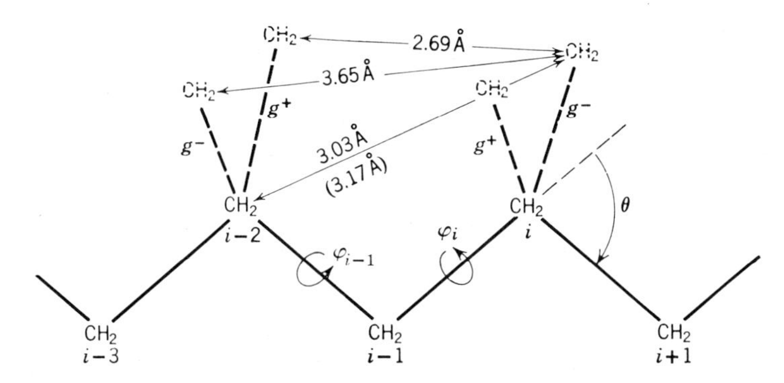

Fig. 5. Portion of a polymethylene chain shown in its planar *trans* conformation by solid lines. Positions of C—C bonds $i-2$ and $i+1$ for *gauche* states about bonds $i-1$ and i, respectively, are shown by dashed lines. Intergroup distances indicated have been calculated for $l_{C-C} = 1.53$ Å, $\theta = 68°$, and $\phi_g = \pm 120°$; the one given in parentheses refers to $\phi_g = 112.5°$.

the plane and the shorter ones those below it. The pair of skeletal bonds subject to rotations is depicted as remaining in the plane of the figure. Intergroup distances expressed in Angstrom units and measured from carbon to carbon, are shown for some of the CH_2 pairs for the nonplanar conformations of the chain backbone. These distances have been calculated taking $\theta = 68°$ and $\phi_g = \pm 120°$, except that the value in parentheses is for $\phi_g = \pm 112.5°$. The larger rotation angle has been given preference in this diagram in order to facilitate comparisons with other chains considered in the sections that follow.

The distance between two CH_2 groups separated by three skeletal bonds depends on one rotation angle ϕ. Interactions between a pair of groups so situated in sequence will be called *three-bond* interactions (see p. 65), or, following Birshtein and Ptitsyn,[1] interactions of *first order*. Distances between groups separated by four skeletal bonds depend on two bond rotations. Interactions between such pairs are termed *four-bond*, or *second order*; they include the $g^{\pm}g^{\mp}$ pairs, the repression of which plays a major role in various chain molecules.

Consider the first-order interaction associated with rotation about bond i, illustrated in Fig. 5. In the *gauche* conformation, CH_2 group $i+1$ is placed within range of repulsion by CH_2 group $i-2$. We assign this conformation a statistical weight

$$\sigma = \exp\left(-E_\sigma/RT\right) \tag{3}$$

relative to a weight of unity for the *trans* form (compare Eq. III-3). Interactions of higher order are ignored at this juncture in the analysis; they will in due course be combined with the first-order effects under consideration at present. Thus, a factor σ is to be included in each element of the second column (g^+) of the statistical weight matrix $\mathbf{U}$ for bond i. Since the groups comprising the polymethylene chain are symmetric, identical statistical weight factors σ must hold for g^- states, and hence will occur likewise in elements of the third (g^-) column of $\mathbf{U}$. The statistical weights representing first-order interactions are therefore $1, \sigma, \sigma$. They are conveniently colligated as the elements of a diagonal matrix, which may be written simply as

$$\mathbf{D} = \operatorname{diag}(1, \sigma, \sigma) \tag{4}$$

Second-order interactions, involving groups separated by four skeletal bonds and dependent upon two bond rotation angles, may be represented by

$$\mathbf{V} = \begin{bmatrix} 1 & 1 & 1 \\ 1 & \psi & \omega \\ 1 & \omega & \psi \end{bmatrix} \tag{5}$$

Elements of unity in the first row and first column follow automatically from the convention introduced in Chapter III, whereby statistical weights are reckoned relative to the *trans* conformation. This expression is generally applicable to symmetric chains consisting of bonds characterized by $\nu = 3$.

The statistical weight matrix $\mathbf{U}$, inclusive of both first and second-order interactions, is obtained as the product

$$\mathbf{U} = \mathbf{VD} = \begin{bmatrix} 1 & \sigma & \sigma \\ 1 & \sigma\psi & \sigma\omega \\ 1 & \sigma\omega & \sigma\psi \end{bmatrix} \tag{6}$$

This general expression for $\mathbf{U}$ for a symmetric chain with $\nu = 3$ was introduced in Chapter III, Eq. III-16. The foregoing method of derivation will find further use in the treatment of other chains, including especially asymmetric chains treated in the following chapter.

According to the calculations shown in Fig. 3, second-order interactions in $g^\pm g^\pm$ pairs of the same sign are negligible in the PM chain, i.e., the calculated energy for such a pair approximates the sum $2E_\sigma$ of energies for the two first-order interactions involved. It follows that $\psi \cong 1$. As we have seen, however, $0 < \omega \ll 1$, i.e.,

$$\omega = \exp(-E_\omega/RT) \tag{7}$$

where E_ω is the energy for a $g^\pm g^\mp$ pair in excess of the energy $2E_\sigma$. The computations of Abe et al.,[14] yield 3200 cal mole^{-1} for the approximate energy of a $g^\pm g^\mp$ bond pair. Hence

$$\begin{aligned} E_\omega &\cong 3200 - 2E_\sigma \\ &\cong 2200 \text{ cal mole}^{-1} \end{aligned}$$

At 400°K, for example, $\omega \approx 0.06$ according to these calculations.

On the basis of critical examination of models supplemented by the foregoing calculations, we let

$$\mathbf{V} = \begin{bmatrix} 1 & 1 & 1 \\ 1 & 1 & \omega \\ 1 & \omega & 1 \end{bmatrix} \tag{8}$$

and

$$\mathbf{U} = \begin{bmatrix} 1 & \sigma & \sigma \\ 1 & \sigma & \sigma\omega \\ 1 & \sigma\omega & \sigma \end{bmatrix} \tag{9}$$

for the PM chain. The approximation $\omega = 0$ reduces this result to Eq. III-15.

Abandonment of detailed accounting of interactions between pairs of atoms and consideration instead of interactions between groups as set forth above need not entail additional approximations. To be sure, the interactions between two groups must depend to some extent on the disposition of their constituent atoms. The interactions of group $i + 1$ with $i - 2$ in Fig. 5, for example, depend not only on ϕ_i, but also on ϕ_{i-1} and ϕ_{i+1} since these latter rotations determine the positions of the H atoms of these groups. This effect happens to be very small according to the computations cited above. In any event, however, the convention introduced in Chapter III, whereby all bonds succeeding the one (bond i) under consideration are tentatively maintained in *trans* states, offers the means for taking full account of whatever effect rotation about bond $i + 1$ may have on the interactions of group $i + 1$ with its predecessors in the chain. If ϕ_{i+1} is assigned a state other than *trans*, the resulting alteration of the first-order interaction between groups $i - 2$ and $i + 1$, separated by three skeletal bonds, can be incorporated in the statistical weight for bond $i + 1$. Thus, a simplified analysis in terms of group interactions does not predicate forfeiture of refinement.

4. RESULTS FOR POLYMETHYLENE CHAINS

Characteristic ratios $C_\infty = (\langle r^2 \rangle_0 / nl^2)_\infty$ calculated for infinite PM chains by Abe, Jernigan, and the author[14] are shown in Figs. 6 and 7. The curves illustrate the dependence of C_∞ on the parameters, ϕ_g, E_σ (or σ), and E_ω (or ω). Some of the computations[14] were carried out according to

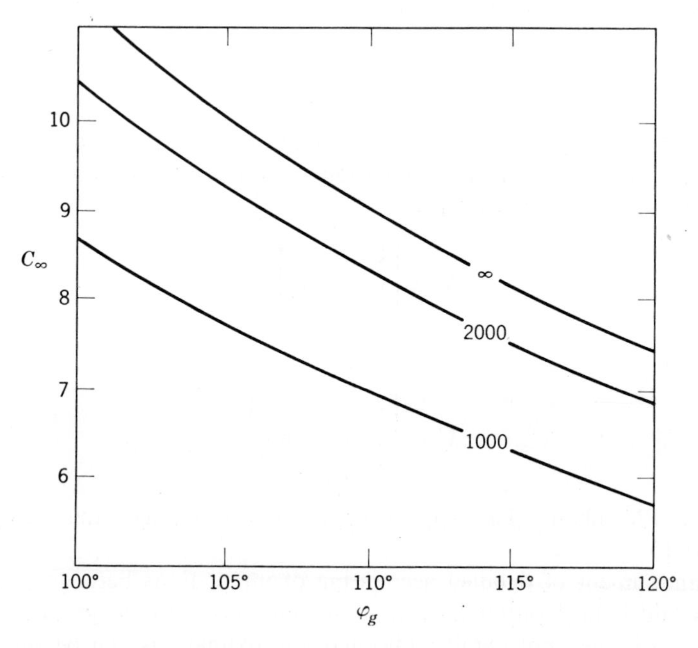

Fig. 6. Dependence of the characteristic ratio C_∞ for polymethylene on the angle $\pm\phi_g$ chosen for *gauche* rotational states. All calculations are for 140°C, with $E_\sigma = 500$ cal mole^{-1} ($\sigma = 0.54$), $E_\psi = 0$ ($\psi = 1.00$), and with E_ω taking on the values indicated in in calories per mole for each curve. (From Abe, Jernigan, and Flory.[14])

Eq. IV-28, matrix squaring being used to reach values of n sufficiently great ($>10^4$) to assure convergence of the characteristic ratio. Others were performed through the use of Eq. IV-44, which gives the asymptotic value of the characteristic ratio directly. All calculations refer to a temperature of 140°C. The numerical values of the temperature coefficients shown in Fig. 8 were determined by varying the statistical weights σ and ω over small intervals about their values for 140°C.

A fairly strong dependence of the characteristic ratio on ϕ_g is depicted by the curves in Fig. 6. These curves were calculated for $E_\sigma = -RT \ln \sigma = 500$ cal mole^{-1}, and for the three values of $E_\omega = -RT \ln \omega$ indicated with the curves. The experimental value of 6.7 ± 0.3 for the ratio at 140°C (see

Table II-1) requires $E_\omega < 2000$ cal mole^{-1} for any value of the angle $\phi_g < 120°$. Thus, a value of $\omega > 0$ appears to be required for good agreement with experimental results, as was first pointed out by Hoeve.[20] The characteristic ratio and its temperature coefficient are plotted against

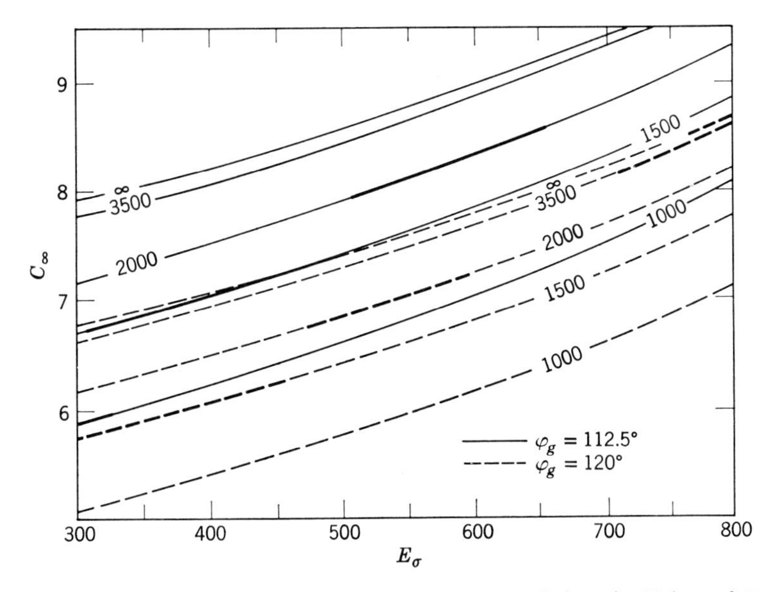

Fig. 7. Influence of energies E_σ and E_ω on the characteristic ratio. Values of E_ω are indicated with each curve. Calculations are for 140°C. Heavy-line portions of the curves represent ranges of E_σ within which $d \ln \langle r^2 \rangle_0 / dT = -1.1 (\pm 0.1) \times 10^{-3}$ in agreement with experiment (Table II-2). (From Abe, Jernigan, and Flory.[14])

E_σ in Figs. 7 and 8. Dashed lines are for $\phi_g = \pm 120°$ and solid lines for the more probable value of $\pm 112.5°$. The several curves of each character are for the values of E_ω indicated. The heavy-line portions of the curves in Fig. 7 represent ranges of E_σ within which $(d \ln \langle r^2 \rangle_0 / dT) \times 10^3 = -1.1 \pm 0.1$, in agreement with experiment (see Table II-2). The heavy-line sections of curves in Fig. 8 similarly denote ranges over which the characteristic ratio matches the experimental value. The temperature coefficient of $\langle r^2 \rangle_0$ is sensitive to both E_σ and E_ω, but is affected relatively little by ϕ_g.

Ranges of the parameters E_σ and E_ω affording agreement with experiments on the characteristic ratio and its temperature coefficient are as follows:

for $\phi_g = 120°$:

$$E_\sigma = 430\text{–}590 \text{ cal mole}^{-1}$$

$$E_\omega = 1700\text{–}2000 \text{ cal mole}^{-1}$$

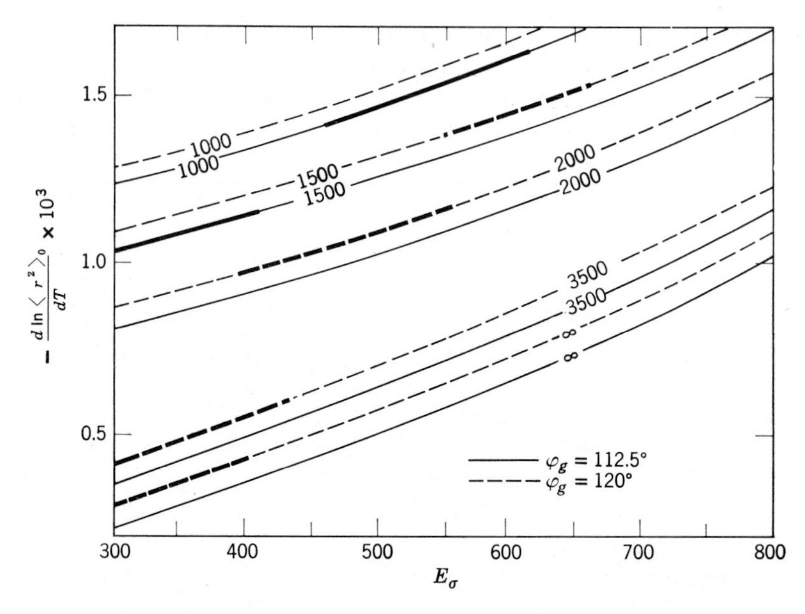

Fig. 8. Influence of E_σ and of E_ω on the temperature coefficient of $\langle r^2 \rangle_0$. Values of E_ω are indicated with each curve. Heavy-line portions represent ranges within which $C_\infty = 6.7 \pm 0.3$ in approximate agreement with experiment (see Table II-1). (From Abe, Jernigan, and Flory.[14])

for $\phi_g = 112.5°$:

$$E_\sigma = 260\text{–}450 \text{ cal mole}^{-1}$$

$$E_\omega = 1300\text{–}1600 \text{ cal mole}^{-1}$$

The more realistic value of ϕ_g suggests $E_\sigma \cong 400$ cal mole^{-1}, which is less than the figure of 500 cal mole^{-1} deduced from the Raman spectral intensities of *n*-pentane and *n*-hexane[21] and corroborated by analyses of the entropies of *n*-hexane, *n*-heptane, and *n*-octane.[22] Each of these values for E_σ is subject to an uncertainty approximating the difference between them. The intimation of disagreement is therefore of no moment. The energy E_ω required to fit the experimental results on the PM chain is about 1500 cal mole^{-1}, which agrees very well with the value of 2000 cal mole^{-1} indicated by the conformational energy calculations for *n*-pentane.

It is apparent that the rotational isomeric state scheme offers an eminently satisfactory account of experimental observations on the average dimensions, and on the temperature coefficients thereof, for polymethylene chains of high molecular weight. The parameters required stand in good accord with the considerable body of evidence from other sources bearing on the conformational characteristics of *n*-alkane chains.

The fractions of various C—C bonds of *n*-alkanes, or of polymethylene chains, which occur in the several rotational states, are treated in Chapter III. Calculated fractions of bonds in various conformations are shown in Fig. III-4, and the fractions of neighbor pairs in Figs. III-5 and III-6. The calculations were carried out using $\sigma = 0.54$ and $\omega = 0.088$, which correspond, respectively, to $E_\sigma = 500$ cal mole^{-1} and $E_\omega = 2000$ cal mole^{-1} at 140°C. These rounded values of the energies are somewhat higher than are indicated by the analysis of chain dimensions presented here. They afford approximate agreement with experimental results on the average dimensions if used in conjunction with $|\phi_g| = 120°$, the effect of this larger value of the rotation angle being compensated by adoption of slightly larger energies. This choice of parameters is satisfactory for the following illustrative calculations of configuration-dependent properties.

The solid curve in Fig. 9 portrays the rise of the characteristic ratio C_n, with n for the PM chain; values for short chains are shown in the inset

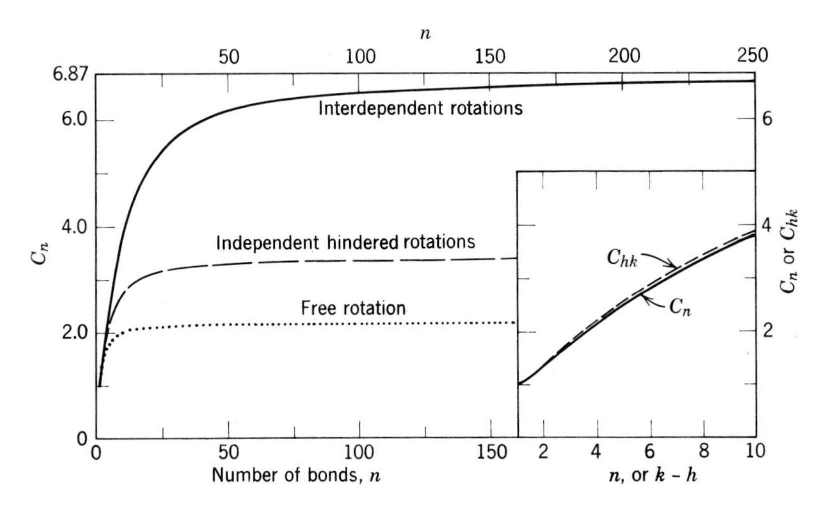

Fig. 9. The characteristic ratio calculated for a polymethylene chain at ca. 140°C as a function of chain length n. The solid curve has been calculated using Eq. IV-28, which takes full account of interdependence of rotations ($\theta = 68°$, $\phi_g = \pm 120°$, $\sigma = 0.54$, $\psi = 1.00$, $\omega = 0.088$). The upper boundary of the figure is its asymptote. The dotted curve represents the freely rotating chain with $\theta = 68°$, and the long dashed curve the chain subject to independent hindrance potentials, with σ assuming the same value, 0.54, as for the solid curve. The inset shows C_n (solid curve) in the range of small n, and (short dashed curve) the corresponding characteristic ratios C_{hk} for a finite sequence in an infinite chain (see text). Ratios C_n and C_{hk} for integral numbers of bonds are connected by continuous curves for clarity of representation. (From Jernigan and Flory.[23,24])

(solid line). These solid curves have been calculated[23,24] according to Eq. IV-28. The dotted curve in Fig. 9 represents the freely rotating chain having the same bond angle $\pi - \theta$; the dashed curve is for the chain subject to independent rotation potentials. The dotted curve has been calculated from Eq. I-20, and the dashed curve from Eq. I-53, with $\alpha = \cos\theta$ and $\eta = \langle\cos\phi\rangle = (1-\sigma)/(1+2\sigma)$, in accordance with Eq. III-9. The parameters θ and σ were assigned the same values as above. Rotational hindrance increases the asymptote and slows the rate of attainment of it. Interdependence further increases C_∞ and postpones the approach to this asymptote to even larger n. The curve for the independent bond model could, however, be adjusted to approximate the solid curve (interdependent bonds) by empirical alteration of the parameter σ, i.e., by arbitrarily increasing the energy E_σ to about 1100 cal mole^{-1} and thus decreasing σ. The temperature coefficient of C_∞ would be seriously exaggerated by this adjustment, however, as previously noted (see p. 59).

According to the solid curve in Fig. 9, which can be accepted as reliably representing the mean dimensions of normal alkane molecules, a chain length n of at least 200 bonds is required before the asymptote, located at the upper boundary of the figure, is substantially attained. For smaller n, the ratio depends on n. Hence, the random flight, or freely jointed equivalent chain (see Chap. I, pp. 10–12) is not an acceptable model analog in this range.

The dashed curve in the inset in Fig. 9 represents the ratio $C_{hk} = \langle r_{hk}^2\rangle_0/(k-h)l^2$ for a sequence of $k-h$ bonds within an infinite chain. This curve has been calculated according to Eq. IV-45. The differences between $C_n = \langle r_n^2\rangle_0/nl^2$, shown by the solid curve, and C_{hk}, with $k-h=n$, represent end effects. These differences are extremely small, even for the lowest values of n. The dependence of C_n on n must therefore be attributed to correlations of the directions of near-neighbor bonds and *not to end effects*.* This observation is of foremost importance in reference to the configurational statistics of long-chain molecules. It permits $\langle r_{hk}\rangle_0$ for a finite sequence to be equated to $\langle r_n^2\rangle_0$ in good approximation, for any value of $k-h=n$; the lengths of the adjoining sequences of h and of $n-k$ bonds may be ignored.

The mean-square radius of gyration for PM, calculated[23] according to Eq. IV-54 using the same parameters as above, is shown as a function of n in Fig. 10. The approach of the ratio $\langle s^2\rangle_0/nl^2$ to its asymptote, again located at the upper boundary of the figure, is more protracted than is that of $\langle r^2\rangle_0/nl^2$. The curve for the latter quantity, shown by the dashed line for comparison, has been taken from Fig. 9. It will be recalled that the radius of gyration depends on the sum of quantities $\langle r_{hk}^2\rangle$ where h, k lie

* The convergence of C_{hk} to C_n for much larger values of $k-h=n$ has been pointed out by K. Nagai, *J. Chem. Phys.*, **45**, 838 (1966).

within the interval 0—n. Terms for sequences $|h-k|$ considerably less than n are dominant in this sum, and this fact furnishes the reason for the slower approach of the radius of gyration to its asymptote with increase in n.

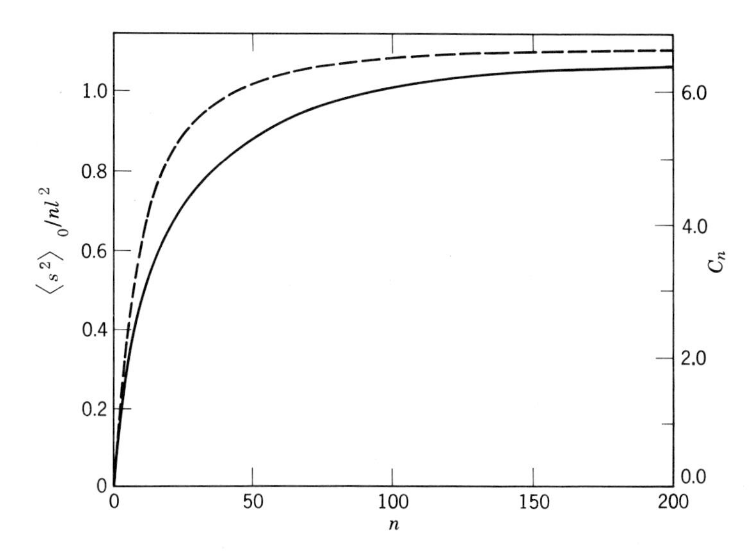

Fig. 10. The characteristic ratio of the unperturbed mean-square radius of gyration $\langle s^2 \rangle_0$ for PM chains calculated from Eq. IV-54 using the parameters given in the legend for Fig. 9. The dashed curve represents $C_n = \langle r^2 \rangle_0/nl^2$, as taken from the solid curve in Fig. 9. The asymptotes of the curves are coincident with the upper boundary of the figure. (From Jernigan and Flory.[23])

The fourth moment of r, expressed as its ratio to $\langle r^2 \rangle_0^2$, is plotted against n in Fig. 11. These calculations[23,24] were carried out according to Eqs. IV-95 and IV-96. The ratio $\langle r^4 \rangle_0/\langle r^2 \rangle_0^2$ must approach $\frac{5}{3}$ as $n \rightarrow \infty$. This follows from the fact that the distribution function $W(\mathbf{r})$ must be asymptotically Gaussian in the limit of large n, and the ratio of the fourth moment to the square of the second is $\frac{5}{3}$ for a Gaussian distribution. It is evident that the ratio $\langle r^4 \rangle_0/n^2l^4$ approaches its asymptote more slowly than does $\langle r^2 \rangle_0/nl^2$.

These characteristics of various moments related to the distribution $W(\mathbf{r})$ are discussed further in Chapter VIII.

Additional evidence bearing on the configurational statistics of polymethylene chains may be secured from the dipole moments of n-alkanes bearing dipolar substituents at their termini. The dibromo-n-alkanes, $Br—(CH_2)_{n-1}—Br$, whose dipole moments have been determined by

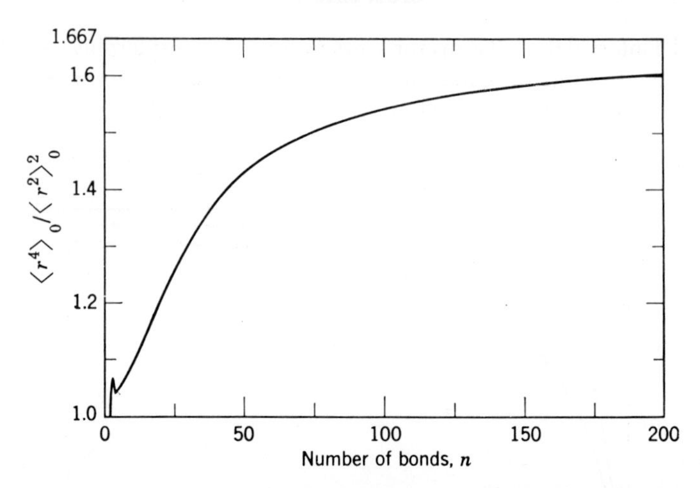

Fig. 11. The fourth moment $\langle r^4\rangle_0$ for PM chains, expressed as its ratio to the square of the second moment, plotted against n. Calculations were carried out according to Eqs. IV-95 and IV-96 using the parameters given in the legend to Fig. 9. The upper boundary is again the asymptote to the curve. (From Jernigan and Flory.[23,24])

Hayman and Eliezer,[25] are illustrative. Bonds of the chain are conveniently numbered as follows:

The molecular dipole moment $\boldsymbol{\mu}$ may be taken as the vector sum of moments $\mathbf{m}_1$ and $\mathbf{m}_n$ ascribed to the terminal C—X bonds. Polarization effects, formally ignored, may be included in good approximation by assigning to each terminal bond moment the magnitude of the dipole moment observed for the analogous monofunctional compounds, i.e., for the higher n-alkyl bromides in the case of the examples considered for which X = Br. This value of the bond moment m includes a polarization contribution which should approximate that occurring in the dibromoalkanes.

The mean-square molecular dipole moment $\langle \mu^2\rangle$ may be averaged over all configurations of the chain, treated in the rotational isomeric state approximation, by the methods of Chapter IV as adapted for this purpose by Leonard, Jernigan, and the author.[26] Thus

$$\boldsymbol{\mu} = \mathbf{m}_1 + \mathbf{T}_1^{(n-1)}\mathbf{m}_n \tag{10}$$

where

$$\mathbf{m}_1 = -\mathbf{m}_n = m\begin{bmatrix}1\\0\\0\end{bmatrix} \tag{11}$$

with $m = |\mathbf{m}_1| = |\mathbf{m}_n|$. Hence†

$$\langle \mu^2 \rangle = 2m^2 - 2\mathbf{m}_1 \langle \mathbf{T}_1^{(n-1)} \rangle \mathbf{m}_1 \tag{12}$$

Substitution of Eq. IV-14 for the product of **T** matrices averaged over all configurations yields

$$\langle \mu^2 \rangle / m^2 = 2 - 2Z^{-1}\{(\mathbf{J}^* \otimes \mathbf{E}_3)[(\mathbf{U} \otimes \mathbf{E}_3)\|\mathbf{T}\|]_1^{(n-1)}(\mathbf{J} \otimes \mathbf{E}_3)\}_{11} \tag{13}$$

where the postsubscript denotes the 1,1 element, and $\mathbf{U}_1 = \mathbf{E}_v$, as stipulated in Chapter IV.

The statistical weight matrices $\mathbf{U}_i$ for $3 < i < n - 1$ are given by Eq. 9. For bond 2 it suffices to let

$$\mathbf{U}_2 = \begin{bmatrix} 1 & \sigma' & \sigma' \\ 0 & 0 & 0 \\ 0 & 0 & 0 \end{bmatrix} \tag{14}$$

since only the first row is required; $\sigma' = \exp(-E_{\sigma'}/RT)$ is the statistical weight for the first-order interaction of X with CH_2 in the *gauche* conformation of bond 2. For the third skeletal bond

$$\mathbf{U}_3 = \begin{bmatrix} 1 & \sigma & \sigma \\ 1 & \sigma & \sigma\omega' \\ 1 & \sigma\omega' & \sigma \end{bmatrix} \tag{15}$$

where ω' is the statistical weight factor for the second-order interaction in which X is involved. Consistency dictates that

$$\mathbf{U}_{n-1} = \begin{bmatrix} 1 & \sigma' & \sigma' \\ 1 & \sigma' & \sigma'\omega' \\ 1 & \sigma'\omega' & \sigma' \end{bmatrix} \tag{16}$$

for bond $n - 1$. In these expressions σ is the parameter defined in Eq. 9 for internal bonds of a polymethylene chain.

Leonard et al.[26] treated the dipole moments of the α,ω-dibromo-*n*-alkanes in the approximation $\omega = \omega' = 0$. The greater length of the C—Br bond (1.94 Å) compared with C—C suggests that any repulsion in the *gauche* conformation for penultimate bonds should be quite small. Analysis of temperature coefficients of infrared spectral frequencies associated with *trans* and *gauche* conformers of normal alkyl bromides, specifically H—$(CH_2)_n$—Br with $n = 3$ and 4, indicates that $E_{\sigma'}$ is zero[27] or negative,[28]

† See p. 101 with regard to the specifications of $\mathbf{T}_1$, only the first row of which is required.

possibly by as much as -300 cal mole^{-1}. Taking $E_{\sigma'} = -100$ cal mole^{-1}, $E_{\sigma} = 500$ cal mole^{-1} as above, and $m = 1.956$ D as found for alkyl bromides,[25] Leonard et al.[26] calculated dipole moments which are in fair agreement with the experimental measurements of Hayman and Eliezer[25] for the alkylene dibromides with $n - 1 = 6$ to 10.* The root-mean-square molecular moment $\langle\mu^2\rangle^{1/2}$ calculated for $n - 1 = 10$ is 2.64 D compared with 2.57 observed. The calculated moment increases with n toward its asymptotic value $\sqrt{2}m = 2.77$ D. The calculated temperature coefficient, $(\frac{1}{2})\, d \ln \langle\mu^2\rangle/dT$, is small[26] ($0$ to 5×10^{-4} deg^{-1}) in agreement with observation.[25]

5. POLYTETRAFLUOROETHYLENE AND ITS OLIGOMERS

The perfluoro-n-alkanes are close analogs of the corresponding hydrocarbons. The greater van der Waals radius of fluorine, 1.35 Å vs. 1.20 Å for hydrogen, and the polarity of the C—F bond[29–31] pose the principal differences. Geometrical parameters,[32] namely, $l_{\text{C—C}} = 1.53$ Å, $l_{\text{C—F}} = 1.36$ Å, and $\theta = 64°$, are similar apart from the somewhat greater length of the C—F bond compared with C—H. Flourines on adjacent carbons are subjected in the eclipsed conformation to a steric overlap of 0.25 Å, which exceeds that for hydrogens in n-alkanes by about 0.10 Å; see Table III-2. The rotational barrier for perflouroethane, estimated at 4 kcal mole^{-1} (see Table III-2), is only marginally greater than that for ethane. Fluorines attached to second-neighbor carbon atoms engage in a steric overlap of about 0.10 Å in the staggered conformation ($\phi = 0$) shown below:

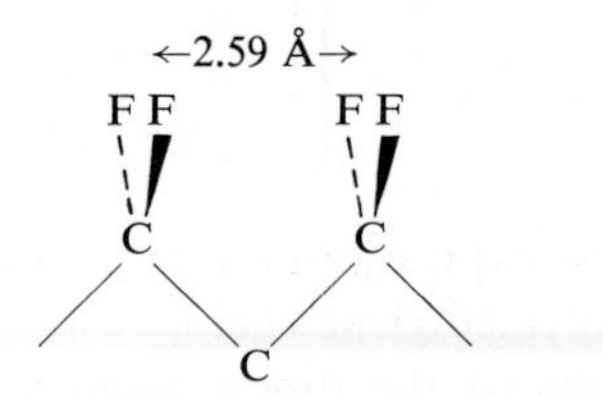

whereas in the n-alkane chain the corresponding H···H distance, 2.54 Å, *exceeds* the sum of the van der Waals radii by about 0.15 Å. The indicated overlaps appear to have little effect on the rotation of CF_3 groups in perfluoropropane, which are subject to a hindrance potential of only 3.35 kcal mole^{-1}, as estimated by Pace and Plaush[33] from the experimental entropy of the fluorocarbon. This barrier is about the same as for propane

* For $n - 1 > 5$, the effect on $\langle\mu^2\rangle$ of perturbation of the distribution of configurations by mutual interactions between terminal dipoles was shown[26] to be negligible.

(see Table 1, p. 135). On the other hand, the polytetrafluoroethylene (PTFE) chain, in contrast to PM, assumes the form of a helix in the crystalline state.[32] The helix is of low pitch, being generated by a rotation ϕ of about 17°, or $-17°$, around every skeletal bond. The indicated preference for a conformation departing from the planar zigzag offers compelling evidence[32] that the pairs of fluorine atoms cited above are subject to mutual repulsions in the planar conformation in which bonds are staggered. The repulsions between fluorine atoms are enhanced by coulombic repulsions of the C—F dipoles.[30]

The high melting point (327–340°C) of PTFE, its low solubility, and the high viscosity of its melt are generally construed to be indicative of extraordinary stiffness of the PTFE chain. The information presented above on rotational barriers and steric effects in low molecular perfluorocarbons, on the other hand, would seem to suggest close resemblance to the conformational energy of hydrocarbon chains.

The interactions in perfluoroalkanes and the conformational characteristics of these chains have been brought into sharper focus by recent studies of Bates and Stockmayer.[30,31] They have calculated intramolecular energies of *n*-perfluorobutane, CF_3—CF_2—CF_2—CF_3, in the vicinity of its *trans* and *gauche* conformations by the methods outlined in Section 1. In addition to an inherent torsional potential characterized by a barrier $E°$ (see Eq. 1) and nonbonded interactions between fluorine atoms estimated according to Eq. 2, Bates[30] took account of the interactions of the electric dipoles associated with each C—F bond ($m = 1.2$ D). The effect of variation of $E°$ was explored by so adjusting a_{FF} in Eqs. 1 and 2 as to reproduce a rotational barrier of 4.0 kcal mole^{-1} for C_2F_6 for each value of $E°$. The energy contours shown in Fig. 12 were calculated in this way (compare Fig. 2 for *n*-C_4H_{10}) for two values of the inherent barrier, namely, $E° = 0$ and 2.8 kcal mole^{-1}. Contours are expressed in kilocalories per mole relative to the *trans* conformation $\phi_1 = \phi_2 = \phi_3 = 0$. For $E° = 2.8$ kcal mole^{-1}, the minima resemble those for *n*-butane shown in Fig. 2; the small steric overlaps experienced by the fluorine atoms in the *trans* ($\phi = 0$) state have little effect on the *trans* minimum. The energy of the *gauche* minimum is 1.1 kcal mole^{-1} above the *trans*, however, and this greater separation of the two states when H is replaced by F is indicative of a greater steric repulsion in the *gauche* state. Reduction of $E°$ to zero causes the *trans* minimum to spread into two very shallow minima which occur near $\phi_1 = \phi_2 = \phi_3 = \pm 10°$. This is of course a manifestation of the repulsions between fluorine atoms attached to second-neighbor carbons. The location of the *gauche* minimum is displaced somewhat, as shown in Fig. 12, by taking $E° = 0$; of greater importance is the increase of its energy to 2.3 kcal mole^{-1} relative to the *trans* state.

Bates[30] also has calculated intramolecular energies for helical conformations of PTFE generated by assigning the same values of ϕ to every bond of the chain skeleton. Dipolar interactions of longer range are included in the energy thus calculated, and chiefly on this account the calculations differ from those for *n*-perfluorobutane in Fig. 12. These interactions are

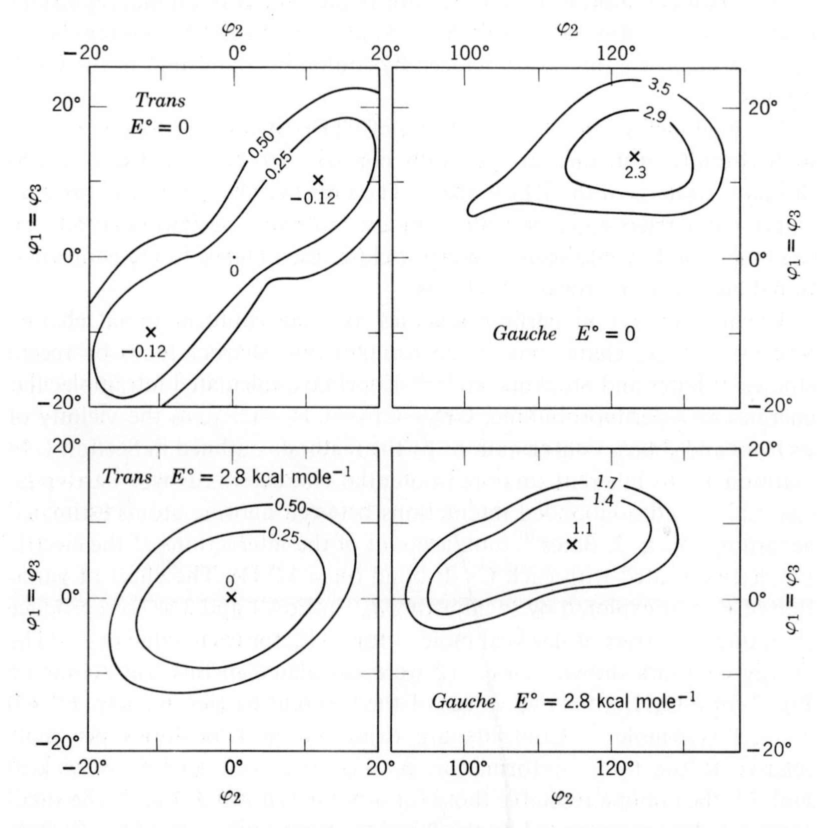

Fig. 12. Conformational energy of *n*-perfluorobutane as calculated by Bates[30] in the vicinity of its *trans* and *gauche* minima for two values of the height of the inherent torsional potential $E°$. Contours are labeled in kilocalories per mole relative to the *trans* conformation $\phi_1 = \phi_2 = \phi_3 = 0$.

predominantly repulsive, and they are diminished by displacing ϕ from 0°. For a value of $E°$ less than 1.5 kcal mole^{-1}, they cause the minimum at 0° to be replaced by a pair of minima equally displaced from 0° and occurring in the range $11° < |\phi| < 17°$, the same sign being perpetuated throughout consecutive bonds. If, however, $E°$ exceeds 1.5 kcal mole^{-1}, then the repulsions between fluorine atoms are dominated by the inherent torsional potential, and the minimum remains at $\phi = 0°$. Even for $E° = 0$, the energy at the displaced *trans* minima is only ca. 0.5 kcal mole^{-1} below the energy for $\phi = 0$. Bunn and Howells'[32] deduction of $\phi = 17°$ for the crystalline conformation would thus appear to lend credence to a low value for $E°$. However, the estimated energy depressions for the displaced *trans* states are in any case small, and present methods for estimating conformational energies are not sufficiently reliable to justify conclusions resting on small differences. It must be acknowledged that the energy computations are indecisive as to the separation of the *trans* state into two minima.

If the *trans* state is indeed so separated, as the crystalline conformation seems to suggest, and if this separation prevails also for the chain when free of the constraints of packing in the crystal which dictate a regular conformation, then a four-state model may be indicated for analysis of the random coil. Bates and Stockmayer[31] have treated the chain on this basis, taking $\phi_{t^\pm} = \pm 15°$ and $\phi_{g^\pm} = \pm 120°$. Conformations $g^\pm g^\mp$, $t^\pm g^\mp$, and $g^\pm t^\mp$ are disallowed by large steric overlaps. *Gauche* states $g^\pm$ are given statistical weights $\sigma < 1$ in consideration of the first-order interactions between CF_2 groups; pairs $g^\pm t^\pm$ and $g^\pm g^\pm$ do not entail second-order interactions and are permitted without further weighting. Bates and Stockmayer[31] point out that $t^\pm t^\mp$ conformations, denoting reversal of helix sense, subject the pair of fluorine atoms attached to second-neighbor carbons to steric repulsions which therefore discourage such conformations. They were accordingly weighted with a factor less than unity, which we call χ.

The barrier separating the assumed t^+ and t^- states must in any case be low and fluctuations about these minima will be large. In view of the small displacement between them, the adoption of a single state at $\phi = 0$ to represent both would seem an acceptable compromise for representation of the properties of the free chain. This minimum should perhaps be broadened to account for the range of variation of ϕ_t. The only characteristic of the chain which this approximation ignores is its propensity to perpetuate departures $\Delta\phi$ from $\phi = 0$, which are of the same sign,[31] for the reasons stated at the close of the preceding paragraph. The three-state model corresponds formally to the model for polymethylene. In view of the severity of $g^\pm g^\mp$ overlaps, ω may assuredly be equated to zero. The statistical weight σ is indicated to be substantially less than for PM, both by the higher energy (see Fig. 12) calculated for the *gauche* conformation and by the effective breadth of the *trans* state. This probably represents the most important difference in comparison with the PM chain.

The insolubility of PTFE at temperatures below about 300°C renders experimental determination of its dimensions difficult. Bates and Stockmayer[31] have determined instead the dipole moments of α,ω-dihydroperfluoroalkanes, H—$(CF_2)_{n-1}$—H, with $n-1 = 4, 6, 7, 8$, and 10. Their measurements are represented by the points in Fig. 13. The molecular dipole moment μ may be attributed to the terminal C—H bonds, inasmuch as the dipole moment of the corresponding fluorocarbon is essentially zero. The treatment of dipole moments of compounds X—$(CH_2)_{n-1}$—X presented in the preceding section is therefore immediately applicable. Because H is smaller than F (whereas X is larger than H in the examples treated in the preceding section), we expect $\sigma' > 1$ and $\omega' > 1$ in the three-state representation (see Eqs. 14–16).

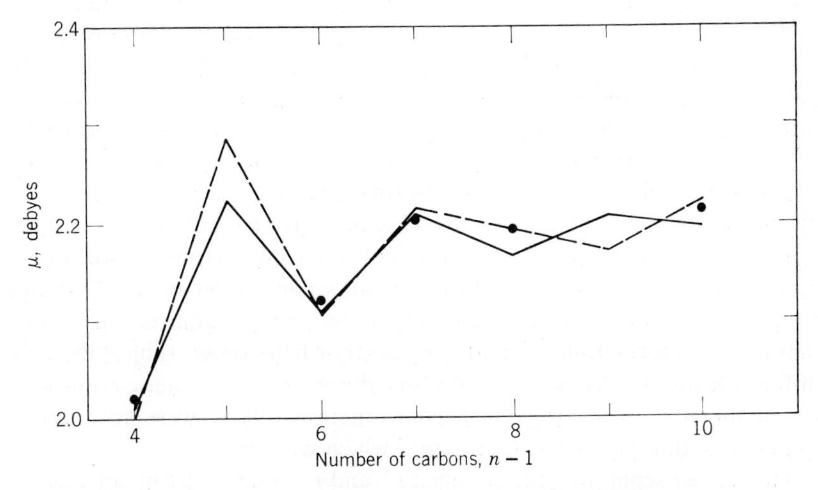

Fig. 13. Dipole moments of α,ω-dihydroperfluorocarbons. Filled circles are experimental values determined[31] in benzene at 25°C. The solid and dashed lines represent calculations carried out for three-state and four-state models, respectively. Parameters are given in the text. (From Bates and Stockmayer.[31])

Calculations carried out by Bates and Stockmayer[31] are represented by the lines in Fig. 13. The solid one is for the three-state model, with $\theta = 64°$, $\phi_g = \pm 115°$, $\sigma = 0.30$, $\sigma' = 2.0$, $\omega = 0$, and $\omega' = 0.33$; the dashed curve was computed using the four-state model, with $\theta = 64°$, $\phi_t = \pm 15°$, $\phi_g = \pm 120°$, $\sigma = 0.2$, $\sigma' = 2.0$, $\omega' = 0.50$, and $\chi = 0.05$ (see above). The group dipole moment m was set equal to the value, 1.60 D, observed[31] for H—$(CF_2)_7$—F. The experimental results are fairly well represented by calculations on either basis. The values chosen for ω' are at variance with predictions from van der Waals radii, however; see the paragraph above.

The three-state model yields a characteristic ratio of ca. 11; the four-state model commends a higher value of ca. 30, which is believed to be more nearly in accord with such meager experimental evidence as is available.[34] The ranges of parameters which permit reasonable agreement with the dipole moments are fairly large.[31] A value of σ between 0.1 and 0.3 is required by either model, however; the indicated (free) energy E_σ is 700–1400 cal mole^{-1}. This result agrees satisfactorily with the conformational energy calculations (Fig. 12) carried out assuming $E^\circ = 2.8$ kcal mole^{-1}.

In conclusion, the rotational potentials for the fluorocarbons differ only quantitatively from those for the hydrocarbon chain. This in conjunction with the similarity of the potential barriers in simple fluorocarbons and in hydrocarbons appears at first sight to be incompatible with the marked contrast in the physical properties of PTFE and PM. *Gauche* conformations which engender tortuous configurations of the chain are appreciably diminished in PTFE as compared with PM, and although the difference is not very great, it may yet be responsible for the striking differences in their physical properties cited above. It was emphasized by the author[35] some years ago that a chain molecule must be endowed with a minimum tortuosity as a requisite for packing to high density in the random-coiled form. Rough estimates suggest that this minimum tortuosity may be only marginally greater than is prevalent in representative macromolecules. The lower weighting of the *gauche* states of the PTFE chain may suffice to place it beyond this limit. The very high melting point of this polymer (in which intermolecular forces are lower than in any other polymer), and the pecularities of its melt may be explicable on this basis.

6. POLYMERIC SULFUR AND SELENIUM[36]

These final examples of simple chains (i.e., chains having a one-bond repeat) are of the ultimate simplicity from a structural point of view; they are devoid of substituent atoms. Experimental results relating to the configurational statistics are meager, but the potentials affecting rotation can be fairly well defined and the crystal structures of their linear polymeric (i.e., catenated) forms are known. Physical properties and structural information bearing on the rotational potentials have been summarized by Semlyen[36] and Abrahams.[37]

The rotational potential is in marked contrast to those encountered in other chains. Mutual repulsions between lone-pair $p\pi$ electrons on adjacent atoms give the rotational potential a twofold character.[38] The stable forms are those for which $|\phi| = 90^\circ \pm$ ca. 10° for both polycatenasulfur and polycatenaselenium.[36,37] Judging from information on various disulfides, HS_2H, $CH_3S_2CH_3$, and ClS_2Cl, for example, the barriers at $\phi = 0$ and 180° are large—probably on the order of 10 kcal mole^{-1}. The Se—Se bond is subject to a similar barrier.[37–39] The two rotational states are appropriately designated + and −, corresponding to the right- and left-handed rotations by which they are reached from the planar *trans* form, i.e., the conformation of reference, which in these cases is one of high energy. The statistical weight matrix takes the form[36]

$$U = \begin{bmatrix} 1 & \omega \\ \omega & 1 \end{bmatrix} \tag{17}$$

where ω is the statistical weight for $+-$ or $-+$ bond pairs relative to $--$ or $++$ combinations.

Geometrical parameters[36,37] for sulfur are: $l_{S-S} = 2.06$ (± 0.02) Å, $\theta = 74°$ ($\pm 2°$); for selenium,[36,37] $l_{Se-Se} = 2.34$ Å and $\theta = 76°$. Their van der Walls radii are ca. 1.80 and 1.90 Å, respectively.[16] Distances between atoms, even in the $+-$ conformations, exceed the van der Waals diameters. The conformation of polycatenasulfur[40] and of polycatenaselenium[37] in their fibrous crystalline forms are helices generated by rotations $\phi = 96$ and $\phi = 78°$, respectively. In consideration of the sizes of the atoms, the London dispersion forces should be comparatively large, and may be expected to favor the more compact conformations $\pm\mp$ for the random coil. The effect of these attractions may be diminished, of course, by competing interactions with neighboring molecules, e.g., with a solvent. The more compact $\pm\mp$ conformations may nevertheless be preferred. This prediction finds partial support in the dipole moments of dialkyl tri- and tetrasulfides. The limited experimental evidence in conjunction with the expectation of enhanced van der Waals attractions suggests a value of ω in the range of 1–2.

Characteristic ratios C_∞ for polycatenasulfur calculated by Semlyen[36] are plotted in Fig. 14. They decrease monotonically with ω as expected, ultimately falling below unity, the value for a freely jointed chain. With $\phi = 90°$, this occurs for $\omega > 2$. Characteristic ratios less than unity are exceptional for polymeric chains. For $\omega = 1$ (rotations independent) and $\phi = 90°$, the characteristic ratio is given by Eq. I-21, the equation for free rotation.

Dimensions of the random-coil forms of these polymers are unknown, but evidence affording crude comparisons of theory with experiment is forthcoming from the accurately determined standard entropies of cyclization. For the cyclization of polymeric sulfur to cyclooctasulfur[41–44]

$$—S_x— \rightleftarrows —S_{x-8}— + \lfloor S_8 \rfloor$$

$\Delta S° = -4.63$ cal deg^{-1} mole^{-1}. For the similar cyclization of polycatenaselenium,[45] $\Delta S° = -5.47$ cal deg^{-1} mole^{-1}. Approximate entropies for these processes can be calculated from $\langle r^2 \rangle_0$ for a chain of eight atoms according to the theory of Jacobson and Stockmayer[46] (see Appendix D). In the case of sulfur satisfactory agreement obtains for $\omega = 1.0$ to 1.5.[36] A

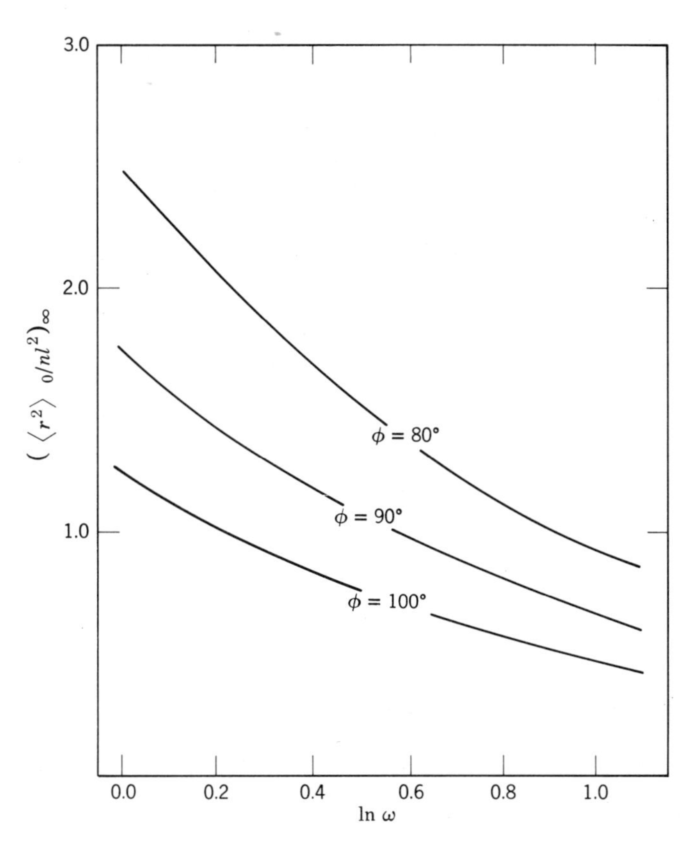

Fig. 14. The characteristic ratio for polycatenasulfur plotted against ln ω. Calculations were carried out using the statistical weight matrix given by Eq. 17, with $\theta = 74°$ and $|\phi|$ assigned the several values indicated on the curves. (From Semlyen.[36])

larger value of ω is indicated for selenium, and this may be caused by stronger attractions between selenium atoms.[36]

7. POLYOXYMETHYLENE

The polyoxymethylene (POM) chain $(—CH_2—O—)_x$ is shown in Fig. 15 following the diagrammatic scheme used in Fig. 5. The configurational statistics of this chain were discussed in Chapter III, Section 7. Its repeat unit embraces two bonds, which of course are identical, apart from their alternating directions with respect to the arbitrary order of numeration of bonds. The length of the C—O bond,[9] 1.43 Å, is 0.10 Å shorter than the C—C bond of the PM chain. The bond angles $\pi - \theta_a$ and $\pi - \theta_b$ at methylene and oxygen happen to be very nearly identical and approximately tetrahedral, i.e., $\theta_a \cong \theta_b \cong 70°$ within an uncertainty of $\pm 2°$.[9] The corresponding transformation matrices $\mathbf{T}_a$ and $\mathbf{T}_b$ may therefore be expressed identically as functions of the respective pairs of rotation angles. A further consequence of the near equality of θ_a and θ_b is the virtual retention of rectilinear form for the all-*trans* chain. The significance of this circumstance becomes apparent in the treatment, presented in Section 10, of the polydimethylsiloxane chain for which the alternating bond angles differ considerably.

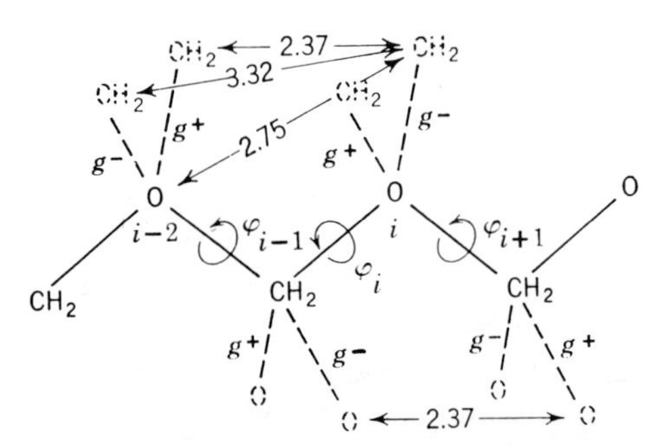

Fig. 15. The polyoxymethylene (POM) chain shown in the diagrammatic manner of Fig. 5. Intergroup distances, expressed in Angstrom units, have been calculated[47] assuming equal bond angle supplements of 70° and $l_{C-O} = 1.43$ Å.

The occurrence of a well-defined threefold rotational potential for bonds of this chain is demonstrated by analysis of the microwave spectrum of dimethyl ether (see Chap. III, Table 1). The rotation of a single skeletal bond from *trans* to *gauche* ($\pm 120°$) reduces the distance between the $CH_2 \cdots O$ pairs separated by three bonds from 3.60 to 2.75 Å,[47] as shown in Fig. 15. These distances are approximately 0.3 Å less than the corresponding ones in the PM chain, owing principally to the shorter length of the C—O bond, and in part also to the 2° larger value for θ. The van der Waals radius of oxygen, ca. 1.4 Å, is less than that for CH_2 by about 0.6 Å. From the intergroup distances in comparison with their van der Waals radii, therefore, the repulsion for the first-order *gauche* interaction should be much less than the 400–500 cal mole^{-1} for PM. Intergroup distances are somewhat misleading in this instance as inspection of models for POM and PM will show; distances between individual atoms need to be considered. Calculations[48] carried out after the manner of those applied to the PM chain in the preceding section, with full account of *all* interactions between pairs of H, C, and O atoms, indicate only a small decrease in the repulsive energy of the *gauche* state relative to PM. The contribution from van der Waals interactions to the energy difference between *gauche* and *trans* states, as estimated in this way,[48] is positive and at least 300 cal mole^{-1}. Coulombic interactions, estimated from the bond dipole moments, or from a corresponding array of equivalent charges, may lower the energy of the *gauche* state by as much as 500 cal mole^{-1}—enough to place the energy of the *gauche* state slightly below that of the *trans*. These are the predictions from analysis of the model.

That the *gauche* conformation in POM chains is in fact markedly preferred over *trans* is shown by two independent, but supplementary, pieces of experimental evidence. First, analysis of the dipole moments of $CH_3OCH_2OCH_3$ and $CH_3OCH_2OCH_2OCH_3$ by Uchida, Kurita, and Kubo[49] led them to the conclusion that *gauche* conformations are more stable than *trans*. From the temperature coefficient of the dipole moment, they estimated the energy of the former to be -1.7 kcal mole^{-1} relative to the latter. Second, investigations[50–52] on the crystalline forms of POM reveal an unerring preference for conformations in which each bond sustains a rotation of the same sign, this rotation being in the range of 100–120°. The preferred conformations consist therefore of equivalent right- and left-handed helices represented by $\cdots g^+g^+g^+\cdots$ and $\cdots g^-g^-g^-\cdots$. In the more stable crystalline form, which is hexagonal, the chain conformation according to Huggins[50] and to Tadokoro and co-workers[51] is a 9_5 helix (i.e., 9 units in 5 turns) in which each bond rotation ϕ is $\pm 102.5°$. The conformation in the alternative, orthorhombic form is the 2_1 helix generated by *gauche* rotations of ca. 117°, according to the analysis of the crystal structure carried out by Carazzolo and Mammi.[52] The conformation adduced from x-ray diffraction of the stable hexagonal form departs appreciably from the expected minimum near $\phi = 120°$. In absence of any other explanation, this deviation may be presumed to be imposed by interactions peculiar to the crystal arrangement. Occurrence of the *gauche* minima in the neighborhood of $\pm 120°$ seems most likely for the free molecule, and we proceed on the assumption that they are so situated.

The empirical evidence cited above clearly indicates that the parameter σ to be associated with *gauche* states in the POM chain exceeds unity; according to the analysis of the dipole moments[49] of the lower homologs at ordinary temperature, it should be at least as great as 10. The explanation for this pronounced preference for the *gauche* conformation, which greatly surpasses predictions from estimates of nonbonded interactions, is not evident. Irrespective of the origin of the large value of σ, we may again express the statistical weight factors for first-order interactions by

$$\mathbf{D} = \mathrm{diag}\,(1, \sigma, \sigma)$$

Inasmuch as second-order interactions differ for the two pairs of bonds, we are obliged to compose two separate statistical weight matrices to represent them. For the bond pair $i-1, i$ centered about CH_2 in Fig. 15, and for bond pair $i, i+1$ centered about O, we have, respectively

$$\mathbf{V}_a = \begin{bmatrix} 1 & 1 & 1 \\ 1 & \psi_a & \omega_a \\ 1 & \omega_a & \psi_a \end{bmatrix}$$

$$\mathbf{V}_b = \begin{bmatrix} 1 & 1 & 1 \\ 1 & \psi_b & \omega_b \\ 1 & \omega_b & \psi_b \end{bmatrix} \tag{18}$$

as the most general expressions for these matrices. The $g^{\pm}g^{\mp}$ states for the former bond pair place the two adjoining CH_2 groups at the intolerably close distance of 2.37 Å. This overlap is more severe than for the corresponding conformation of the PM chain (compare Figs. 5 and 15). We may confidently take $\omega_a = 0$, therefore. The $g^{\pm}g^{\mp}$ states for the bond pair $i, i+1$ place oxygen atoms likewise at a distance of 2.37 Å, which is not much less than the sum of their van der Waals radii (ca. 2.8 Å). While this conformation doubtless is repulsive (and the repulsion may be enhanced by coulombic interaction of the electronegative oxygen atoms; see ref. 47) it should not be suppressed altogether. We therefore let $\omega_b = \omega$, anticipating that $0 < \omega < 1$.

The $g^{\pm}g^{\pm}$ states of the same sign for bond pair $i-1, i$ appear to be free of overlaps and probably are not much affected by attractive interactions. The four-bond interactions for the corresponding conformation about the oxygen atom must be quite small. Computations[47] show the dimensions to be comparatively insensitive to the ψ parameters. It is permissible therefore to take $\psi_a = \psi_b = 1$. Consequently, the statistical weight matrices for the POM chain take the forms[47]

$$\mathbf{U}_a = \mathbf{V}_a\mathbf{D} = \begin{bmatrix} 1 & \sigma & \sigma \\ 1 & \sigma & \sigma\omega \\ 1 & \sigma\omega & \sigma \end{bmatrix}$$

$$\mathbf{U}_b = \mathbf{V}_b\mathbf{D} = \begin{bmatrix} 1 & \sigma & \sigma \\ 1 & \sigma & 0 \\ 1 & 0 & \sigma \end{bmatrix} \tag{19}$$

Characteristic ratios $C_\infty = (\langle r^2 \rangle_0 / nl^2)_\infty$, calculated using the geometric data ($\theta_a = \theta_b = 70°$; $\phi_g = \pm 120°$) quoted above and the statistical weight matrices given in Eq. 19, are plotted against $-\ln \sigma = E_\sigma/RT$ in Fig. 16 for the several values of ω indicated. The dashed curve in Fig. 16 for $\omega = 1$

represents the dependence of the characteristic ratio on σ which would obtain if $g^{\pm}g^{\mp}$ states about oxygen (i.e., for bond pairs like $i, i+1$ in Fig. 15) were freely permitted, the corresponding states for pairs about CH_2 being suppressed (see $\mathbf{U}_b$). This curve decreases monotonically with increase in σ, and for large values of σ it falls below the value (ca. 2.0) of the

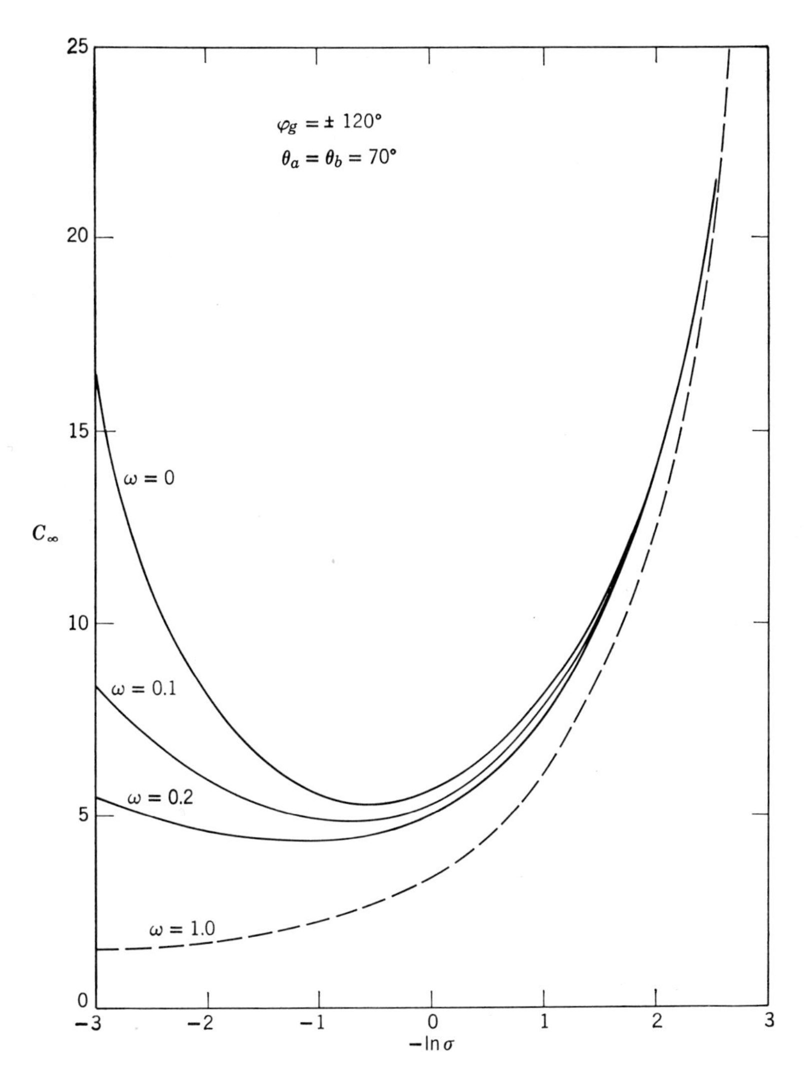

Fig. 16. Characteristic ratios C_∞ for the POM chain calculated[23,47] according to Eq. IV-28 using the statistical weight matrices in Eq. 19, with ω assigned the values indicated with each curve, and with $\phi_g = \pm 120°$.

characteristic ratio for free rotation about all bonds. Only by repression of $g^{\pm}g^{\mp}$ conformations for *both* kinds of bond pairs are large values obtained for the computed characteristic ratio when $\sigma > 1$.

The preference for *gauche* conformations is pronounced in the POM chain. If exercised indiscriminately as to sign, without regard for neighbor bonds, this preference could lead to $C_\infty < 1$, as will be apparent from Eqs. I-48 and III-9. The exigencies of interdependence of neighboring bonds, whereby *gauche* pairs of opposite sign are largely suppressed, intervene to preclude low values of the characteristic ratio.[47] The profound effect of suppression of these neighbor pairs is well illustrated in the POM chain.

It will be observed from Fig. 16 that even for $\sigma = 1$ the characteristic ratio given by the full lines for several values of $\omega \ll 1$ is well above the value, about 2, which would hold for free rotation (i.e., for $\sigma = 1$ and $\omega_a = \omega_b = 1$). With increase in σ to values $\gg 1$, *trans* states occur only infrequently, and the preponderant conformation consists of a succession of right- and left-handed helical sequences, $g^+g^+\cdots g^+$ and $g^-g^-\cdots g^-$, respectively, separated from one another by one or more *trans* bonds, or by a $g^{\pm}g^{\mp}$ pair to the extent that these are permitted. If σ is large, the sequence of *trans* bonds will seldom exceed one. The handedness of successive helical *gauche* sequences will be uncorrelated, and therefore random, if they are separated by one or more t units. The effect of an increase in σ is to raise the average lengths of the helical sequences. It is this increase which is mainly responsible for the rise of the characteristic ratio on the left side of Fig. 16.

Experimental determinations of the molecular weight and average dimensions of POM are difficult to carry out owing to the high melting point (ca. 200°C) and consequent limited solubility of this polymer at practicable temperatures. The results of Kokle and Billmeyer[53] yield ca. 7.5 for the characteristic ratio at 90°C. Stockmayer and Chan[54] found 10.5 ± 1.5 at 25°C, but their result is subject to uncertainty arising from the method of extrapolation used to eliminate the effect of long-range interactions, which are large for this polymer in the solvents required for it. If C_∞ is taken to be in the range 8–10, then according to Fig. 16 a value of $\sigma > 8$ is required, and for any reasonable value of σ the parameter ω cannot exceed 0.15. The combination $\sigma = 12$ and $\omega = 0.05$ is consistent with both the approximate value of C_∞ and with the temperature coefficients of the dipole moments for the lower homologs (Kubo et al.[49]; see above), both observed at ordinary temperatures. The corresponding energy for the *gauche* state relative to *trans* is $E_\sigma = -RT \ln \sigma \cong -1.5$ kcal mole^{-1}.

Rotational state populations for the POM chain calculated on this basis were presented in Chapter III, Section 7. As pointed out there, the chain may be viewed as a succession of helical segments comprising sequences of *gauche* bonds of the same sign and averaging about seven bonds in length.*

Experimental evidence from the diverse sources to which reference has been made establish beyond question that the *gauche* conformation of POM is strongly preferred over *trans*. The POM chain consequently presents a striking contrast to PM. An increase in the proportion of *gauche* bonds brought about by an increase in σ increases the dimensions of POM; it has the opposite effect on PM. The mean extension of the POM chain correlates with the length of *gauche* sequences; for the PM chain it depends on the proportion of *trans* bonds. Hence, the dependence of the mean extension on the fraction p_g of *gauche* bonds, and therefore on σ, is reversed for the two cases. As the temperature is increased, σ must approach unity. A fairly large negative temperature coefficient is predicted[47] for POM, and in this respect the POM chain should correspond to PM.

8. POLYOXYETHYLENE

The polyoxyethylene (POE) chain $(\text{—}CH_2CH_2O\text{—})_x$ presents a relatively simple structure having a repeat unit of three bonds. The comparative wealth of experimental information pertaining to POE homologs enhances interest in the analysis of the configurations of these chains. Not only have the characteristic ratio and its temperature coefficient been determined; dipole moments and their temperature coefficients also are available for homologs within the range $x = 1\text{–}120$.

There are two kinds of bonds, C—O and C—C, in the POE chain. Their lengths are 1.43 and 1.53 Å, respectively.[9] Supplements of the two bond angles $\angle$OCC and $\angle$COC are nearly the same and may be approximated by $\theta = 70°$, as in the case of the POM chain. Each of the three **T** matrices formally required is therefore the same function of the rotation angle ϕ for the relevant bond.

The principal conformations of bond pairs in the POE chain are shown in Fig. 17. The distances given are calculated[55] for *gauche* states at $\pm 120°$ for all bonds. Rotation of a C—O, or O—C, bond to a *gauche* state places the adjoining pair of CH_2 groups separated by three skeletal bonds at a distance of 2.81 Å. This is 0.22 Å less than the corresponding distance in

* The characteristic ratio of the chain cannot, however, be estimated merely from the lengths of these segments assuming their axes to be joined end to end, without regard for the displacements between the ends of the axes of consecutive segments and for the correlations of their directions. In other words, the spatial configuration cannot be correctly represented as an assemblage of freely jointed "bonds" identified with the axes of the helical segments. The geometry of the POM chain does not lend itself to this simplification, as may readily be shown.

polymethylene. The statistical weight σ associated with such conformations may therefore be expected to be appreciably less than the corresponding parameter for PM (ca. 0.5) and certainly much below unity. Repulsions between electric dipoles of the C—O bonds in this conformation may further reduce the statistical weight, but the effect should be small.

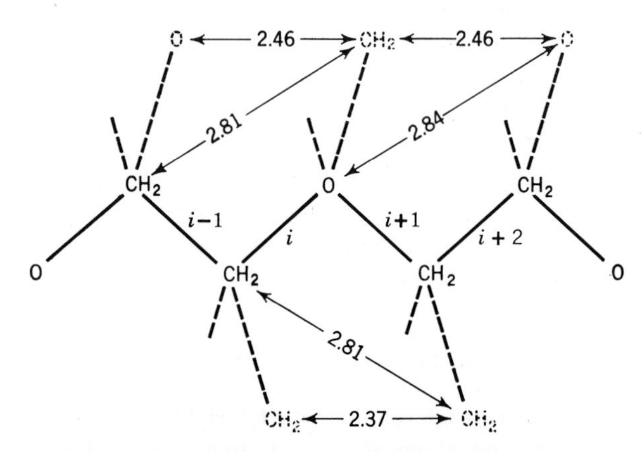

Fig. 17. The polyoxyethylene (POE) chain. Distances calculated for $l_{C-C} = 1.53$ Å, $l_{C-O} = 1.43$ Å, $\theta_a = \theta_b = \theta_c = 70°$, and $\phi_g = \pm 120°$. Compare Figs. 5 and 15.

The corresponding rotation about a C—C bond places the adjoining pair of oxygen atoms at the comfortable range of 2.84 Å, which is almost exactly the sum of their van der Waals radii. At this distance the interaction between these atoms should be attractive. However, the conformation energy estimated[48] according to Eq. 2, applied to *all atom pairs*, is only 0 to -100 cal mole^{-1} for the *gauche* state of this bond relative to *trans*, its neighbors being *trans*. Coulombic repulsions calculated from the partial charges assignable to the oxygen atoms contribute an energy disfavoring the *gauche* conformation by some 500 cal mole^{-1}. The resultant energy of the *gauche* conformation about C—C skeletal bonds is thus predicted to be positive by several hundred calories per mole, and the associated statistical weight, which we denote by σ', to be less than unity. This prediction is not confirmed by the analysis[55] of experimental results, presented below.

Irrespective of the magnitudes of the parameters σ and σ', the statistical weights for first-order interactions dependent upon rotations about bonds i, $i + 1$, and $i + 2$ may be introduced by the respective diagonal matrices $\mathbf{D}_a$, $\mathbf{D}_b$, and $\mathbf{D}_c$ defined by

$$\begin{aligned} \mathbf{D}_a &= \mathbf{D}_b = \text{diag}\,(1, \sigma, \sigma) \\ \mathbf{D}_c &= \text{diag}\,(1, \sigma', \sigma') \end{aligned} \tag{20}$$

The exact locations of the *gauche* minima are of course unknown. Inasmuch as we shall find $\sigma < 1$ and $\sigma' > 1$, the *gauche* minima for C—C bonds may be predicted to occur at angles somewhat less than 120°, while those for C—C bonds may be situated at angles a few degrees in excess of this figure. We shall take $\phi_g = \pm 120°$ for all bonds. Drawing analogy from the calculations for PM, the small error this assignment may entail can be absorbed without difficulty into the values deduced for the statistical weight parameters, the slight vitiation of which will prove unimportant.

Second-order (i.e., four-bond) interactions precipitated by rotations about bond pairs $i-1, i$ and $i+1, i+2$, denoted a and c, respectively, are identical (see Fig. 17). Using unprimed statistical weight factors ψ and ω for these bond pairs, and primed factors ψ' and ω' for bond pairs like $i, i+1$ (denoted b), we have for the most general expression of the statistical weights for a chain of this type

$$\mathbf{U}_a = \begin{bmatrix} 1 & \sigma & \sigma \\ 1 & \sigma\psi & \sigma\omega \\ 1 & \sigma\omega & \sigma\psi \end{bmatrix}$$

$$\mathbf{U}_b = \begin{bmatrix} 1 & \sigma & \sigma \\ 1 & \sigma\psi' & \sigma\omega' \\ 1 & \sigma\omega' & \sigma\psi' \end{bmatrix} \qquad (21)$$

$$\mathbf{U}_c = \begin{bmatrix} 1 & \sigma' & \sigma' \\ 1 & \sigma'\psi & \sigma'\omega \\ 1 & \sigma'\omega & \sigma'\psi \end{bmatrix}$$

There are six parameters in all.

States $g^{\pm}g^{\pm}$ of like sign, not shown in Fig. 17, involve CH_2 and O at a distance of 3.38 Å for bond pair $i-1$, i (a) and for bond pair $i+1$, $i+2$ (c); the corresponding states for pair, i, $i+1$ (b) place the adjoining CH_2 groups at a distance of 3.34 Å. The former conformations may be subject to some attraction; the latter may be slightly repulsive. Neither effect is likely to be large, and computations[55] show the numerical results for $\langle r^2 \rangle_0$ to be relatively insensitive to ψ and ψ'.* Hence, these parameters may be set equal to unity.

The $g^{\pm}g^{\mp}$ pairs (a) and (c) place CH_2 and O at a distance of 2.46 Å (Fig. 17). The comparatively small steric overlaps of C with O and of H with O at this distance can be diminished considerably by compromising one or the other of the two rotation angles just as in the case of PM (see

* The dipole moment is appreciably dependent upon ψ, though less dependent on ψ'.[56] It would be unwarranted to assume that all configuration-dependent properties are insensitive to the ψ parameters.

Figs. 3 and 4). The value of ω must therefore exceed zero, although it may be appreciably smaller than unity. We accordingly retain ω with the expectation that $0 < \omega < 1$. The $g^{\pm}g^{\mp}$ conformations for the i, $i+1$ pair (b) would place two CH_2 groups at the intolerable distance of 2.37 Å, which is well below the distance (2.69 Å) for the corresponding conformations of PM. Hence, it is justified to let $\omega' = 0$.

These considerations warrant reduction of the general expressions Eqs. 21 to[55]

$$\mathbf{U}_a = \begin{bmatrix} 1 & \sigma & \sigma \\ 1 & \sigma & \sigma\omega \\ 1 & \sigma\omega & \sigma \end{bmatrix}$$

$$\mathbf{U}_b = \begin{bmatrix} 1 & \sigma & \sigma \\ 1 & \sigma & 0 \\ 1 & 0 & \sigma \end{bmatrix} \tag{22}$$

$$\mathbf{U}_c = \begin{bmatrix} 1 & \sigma' & \sigma' \\ 1 & \sigma' & \sigma'\omega \\ 1 & \sigma'\omega & \sigma' \end{bmatrix}$$

which involve three parameters. Mean-square chain dimensions[55] and dipole moments[57] have been calculated according to this scheme using the geometrical parameters given above and a value of 0.99 D for the C—O bond dipole moment m. The dimensions[55] turn out to be sensitive to the value of σ' but comparatively insensitive to σ. The dipole moment[57] is nearly 20 times more sensitive than $\langle r^2 \rangle_0$ to σ. Simultaneous agreement with the temperature coefficient of $\langle r^2 \rangle_0$ and with its characteristic ratio requires $\omega > 0$.

Values of the parameters selected to afford optimum agreement with the experimental characteristic ratio (Table I-1), with the mean-square dipole moment ratio $\langle \mu^2 \rangle / nm^2$, and with the temperature coefficients (see Table II-2) of these quantities, are given in Table 2. Results calculated from them

Table 2

Parameters for POE[55,57]

Energies, cal mole^{-1}	Statistical weights at 60°C
$E_\sigma = 900\ (\pm 70)$	$\sigma = 0.26\ (\pm 0.03)$
$E_{\sigma'} = -430\ (\pm 70)$	$\sigma' = 1.90\ (\pm 0.2)$
$E_\omega = 350\ (\pm 200)$	$\omega = 0.60\ (\pm 0.2)$

are compared with experiment in Table 3. The agreement is very good. The discrepancy between the calculated temperature coefficient of $\langle r^2 \rangle_0$ and the value observed, a difference of ca. 0.1×10^{-3} deg^{-1}, is not significant, and could be diminished by refinement of the parameters.[57] The comparatively large positive temperature coefficient calculated for the mean-square dipole moment appears to be well corroborated by experimental measurements.[59,60]

Table 3

Comparison of Calculated and Observed Results for POE Chains

	Temp, °C	Calc[55,57]	Obs
$\left(\frac{\langle r^2 \rangle_0}{nl^2}\right)_\infty$	40	4.0	4.0 (±0.4)[55]
$\frac{d \ln \langle r^2 \rangle_0}{dT} \times 10^3$, deg^{-1}	60	0.12	0.23 (±0.02)[55]
$\left(\frac{\langle \mu^2 \rangle}{nm^2}\right)_\infty$	25	0.58	0.58[58]; 0.62(20°)[59]
$\left(\frac{d \ln \langle \mu^2 \rangle}{dT}\right)_6 \times 10^3$, deg^{-1}	25	1.7	2.0[60]
$\left(\frac{d \ln \langle \mu^2 \rangle}{dT}\right)_\infty \times 10^3$, deg^{-1}	25	2.5	2.6[59]

The characteristic ratio for $HO(—CH_2CH_2O—)_xH$, calculated using the parameters given in Table 2, is plotted against $n = 3x$ in Fig. 18. Shown for comparison by the dashed line is the corresponding curve for polymethylene. The characteristic ratio for POE converges more rapidly with n than does the corresponding curve for PM.

The mean-square dipole moment ratio $\langle \mu^2 \rangle / nm^2$, similarly calculated for the series of POE homologs having hydroxyl terminal groups, is shown as a function of the degree of polymerization x by the solid curve in Fig. 19. Bond dipole moments $m_{C—O} = 0.99$ D and $m_{H—O} = 1.7$ D were used (with due regard for the dependence of the sign on the direction of the bond in the chain) in these calculations on finite chains whose terminal moments differ from internal ones. The quantity nm^2 occurring in the denominator of the dipole moment ratio represents the sum of the squares of the individual dipole moments in the molecule. The maximum exhibited by the

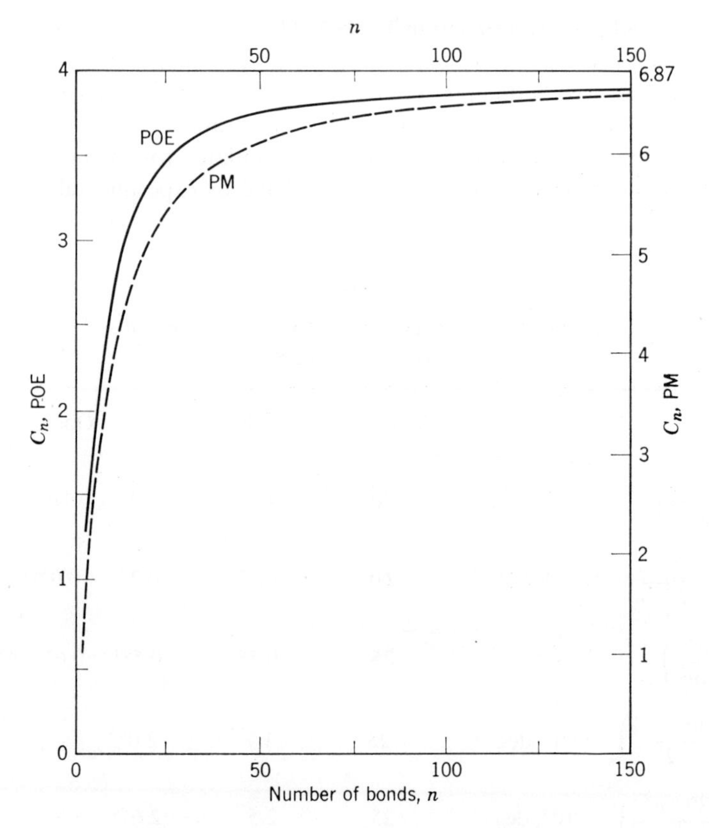

Fig. 18. The characteristic ratio C_n calculated as a function of chain length n; $HO(—CH_2CH_2O—)_{n/3}—H$, solid line; polymethylene, dashed line.[57] The upper margin is the asymptote for each curve.

experimental results in the vicinity of $x = 4$ is reproduced by the computation. Also shown is the curve (dashed) for homologs of the series

$$C_2H_5O(—CH_2CH_2O—)_xC_2H_5$$

bearing terminal ethyl groups. Agreement with results for lower members of the series is again satisfactory.

The two sets of data, one for chain dimensions and the other for the dipole moments, are well correlated by the rotational isomeric state treatment. The statistical weight parameters required for this purpose are consistent with other evidence of a less direct nature. In the first place, they are in qualitative agreement with the conformation found for crystalline POE. According to Tadokoro[62] this conformation is a 7_2 helix described by $(ttg)_x$, i.e., C—O and O—C bonds approximate *trans* ($\phi = -8°$) and

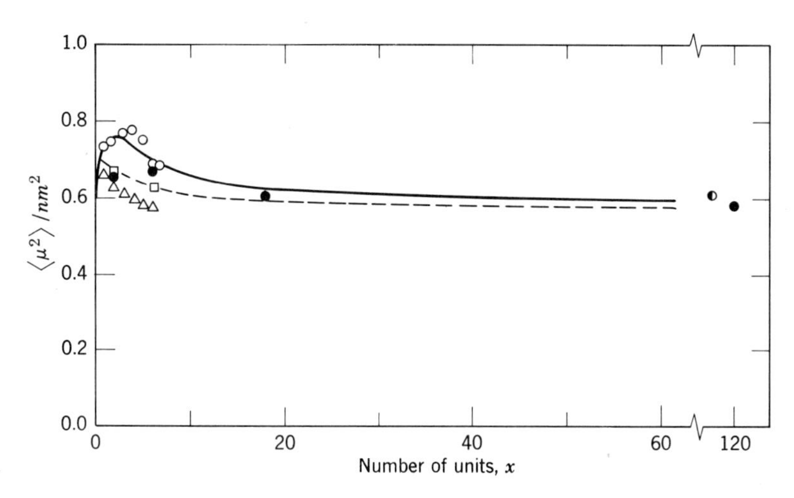

Fig. 19. Dipole moment ratios at 25° calculated[57] as functions of the degree of polymerization x, using the parameters given in Table 2: HO(—CH_2CH_2O—)$_x$ H, solid line; C_2H_5O(—CH_2CH_2O—)$_x$ C_2H_5, dashed line. Experimental results for the former series are shown by: ○ (Uchida et al.[61]; in dioxane at 20°C); ● (Marchal and Benoit[58]; in benzene at 25°C); ◐ (Bak, Elefante, and Mark[59]; in benzene at 20°C). Those for the latter series are shown by: □ (Marchal and Benoit[58]; in benzene at 20°); △ (Kotera et al.[60]; in benzene at 25°).

C—C bonds approximate the *gauche* ($\phi = 115°$) states. The occurrence of this bond conformation in the crystal is in accord with $\sigma < 1$ and $\sigma' > 1$ as found by analysis of properties of the random coil. The energy E_σ for the three-bond interactions of two CH_2 groups is about twice its value for PM where the distance is 0.22 Å greater, the comparison being made with $\phi_g = \pm 120°$ for both chains (see Figs. 5 and 17). The three-bond conformation responsible for σ' places the pair of oxygen atoms at a distance of 2.84 Å which is slightly greater than the sum of their van der Waals radii, and 0.09 Å greater than the corresponding $CH_2 \cdots O$ distance in the *gauche* conformation of POM (Fig. 15). As was pointed out above, coulombic repulsion would have been expected to render $E_{\sigma'} > 0$, and correspondingly $\sigma' < 1$. Experimental observations are emphatic in asserting the opposite, i.e., σ' is undeniably greater than unity; neither the characteristic ratio of the chain dimensions nor the molecular dipole moment would be accountable otherwise.

These findings on the O···O interaction (three-bond) in POE, which favors the *gauche* form despite coulombic repulsion, call to mind the unexpectedly large negative energy for the O···CH_2 interaction in POM. The rather striking departures from conformational energies estimated by current methods as applied to these two chain molecules may be related. These deductions in conjunction with evidence presented in the following section suggest that a pendant oxygen atom, attached by a single bond, acts to lower the energy of the *gauche* conformation so that is it preferred over *trans*. The conclusion reached[63] from the microwave spectrum of *n*-propyl fluoride that its *gauche* conformation is lower than the *trans* by 500 ± 300 cal mole^{-1} indicates that a fluorine substituent may act similarly. The basis for these observations may be presumed to reside in aspects of the torsional potentials as yet poorly understood. The pattern of behavior is consistent, however, throughout the series of related molecules. Analysis of configuration-dependent properties of chain molecules as a means to evaluate some of the characteristics of bond rotational potentials is thus validated.

9. HIGHER POLYOXYALKANES

J. E. Mark[59,64,65] has investigated the next higher analogs of POM and POE, namely, poly(trimethylene oxide), $[-(CH_2)_3-O-]_x$, and poly-(tetramethylene oxide), $[-(CH_2)_4-O-]_x$, which we designate POM_3 and POM_4, respectively. Their chain structures and the parameters adopted to describe them are shown in Figs. 20 and 21. First-order parameters (σ) are

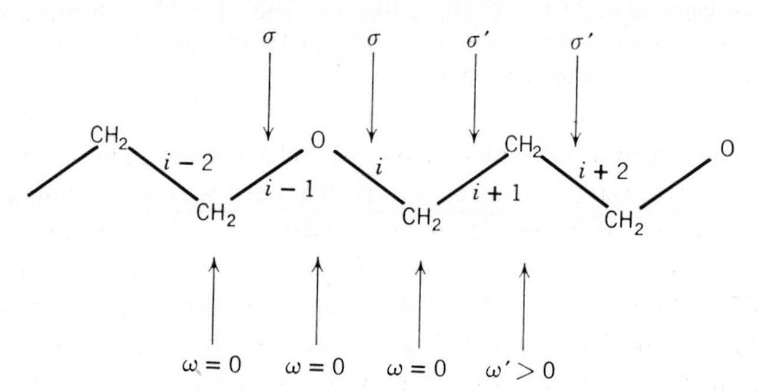

Fig. 20. The poly(trimethylene oxide) chain.

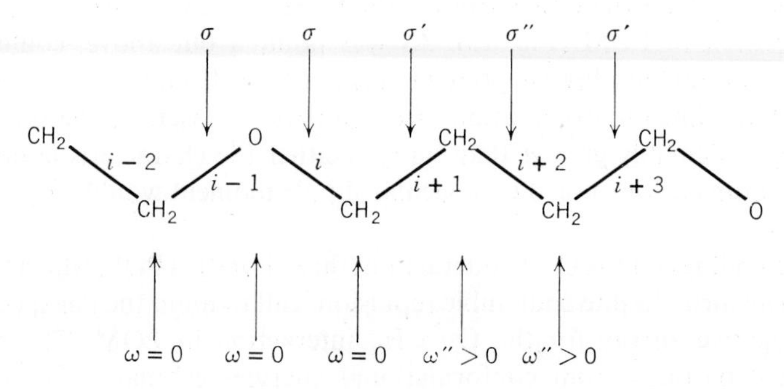

Fig. 21. The poly(tetramethylene oxide) chain.

indicated above the plan of the diagrams, the second-order parameters (ω) below. The latter are understood to apply to the bond pair flanking the skeletal atom or group indicated. Parameters of the same designation in the two diagrams should be approximately equivalent and will be so considered. Assignments of zero to the various ω's, other than ω' and ω'', is justified by the steric overlaps in $g^{\pm}g^{\mp}$ conformations for the bond pairs indicated. The parameters shown furnish the elements for all of the required statistical weight matrices (with all ψ's equated to unity). The formulation of these matrices will be obvious; hence they are not displayed.

Conformational energies and associated statistical weights chosen by Mark[64,65] through examination of interatomic distances, allowance for polar interactions of partial charges, and comparison with corresponding parameters for POM and POE are given in Table 4. The parameter σ

Table 4

Parameters for Poly(trimethylene oxide) and Poly(tetramethylene oxide)[64,65]

Energies, cal mole^{-1}	Statistical weights (20–25°C)
$E_{\sigma} = 900$ (as in POE)	$\sigma = 0.21$
$E_{\sigma'} = -200\ (\pm 200)$	$\sigma' = 1.4\ (\pm 0.5)$
$E_{\sigma''} = 500$	$\sigma'' = 0.43$
$E_{\omega'} = 250\ (\pm 250)$	$\omega' = 0.65\ (\pm 0.3)$
$E_{\omega''} = 340\ (\pm 250)$	$\omega'' = 0.55\ (\pm 0.3)$

corresponds to σ for POE and E_{σ} is accordingly assigned the same value. Mark[64] concluded that $E_{\sigma'}$, like $E_{\sigma'}$ for POM, should be negative, but that its magnitude should be smaller because of the greater distance between O and CH_2 in the *gauche* form and because of the smaller partial charge on CH_2. The argument is admittedly tenuous; justification for the choice of σ' in Table 4 must rest on comparison with experimental results. The parameter σ'' should simulate σ for PM, although it may be reduced somewhat by small partial charge repulsions in POM_4; it has been chosen on this basis. The $g^{\pm}g^{\mp}$ states of bonds $i+1$, $i+2$ of POM_3 juxtapose the two oxygen atoms, but, owing to the small van der Waals radius of oxygen, they do not impose a steric overlap. Electrostatic repulsion may render the energy positive, though certainly small. The $g^{\pm}g^{\mp}$ states for bond pairs $i+1$, $i+2$ or $i+2$, $i+3$ of POM_4 produce small overlaps of CH_2 with O, which, however, may be partially alleviated by coulombic interactions that should be attractive in this instance.

The results of calculations, and their comparison with experimental results are summarized in Table 5. Ranges in the calculated values correspond to ranges of the parameters in Table 4. Within the bounds of experimental data available, the agreement is satisfactory.

Table 5

Comparison of Calculated and Observed Results for POM_3 and POM_4[a]

	POM_3 (25°C)		POM_4 (20°C)	
	Calcd[65]	Obs	Calcd[64]	Obs
$\left(\frac{\langle r^2\rangle_0}{nl^2}\right)_\infty$	3.4 to 4.3	ca. 4.2[66]	4.6 to 5.3	ca. 4.8[64,67]
$\frac{d\ln\langle r^2\rangle_0}{dT}\times 10^3$, deg^{-1}	0.1 to 0.9		−1.3 to −1.2	−1.33[59]
$\left(\frac{\langle\mu^2\rangle}{nm^2}\right)_\infty$	0.47		0.5 to 0.6	0.59[59]
$\frac{d\ln\langle\mu^2\rangle}{dT}\times 10^3$, deg^{-1}	1.4		1.0 to 1.5	2.7[59]

[a] All calculations were carried out using the geometrical parameters given in Section 7 and the statistical weights from Table 4.

The preference of a C—C bond bearing a pendant oxygen for the *gauche* and of the O—C and C—O bonds for the *trans* conformation is confirmed by the occurrence of bonds $i-1$ to $i+2$ of the POM_3 chain (see Fig. 20) in the helical conformation $ttg^\pm g^\pm$ in its stable (orthorhombic) crystalline form.[68] Two other crystal forms of POM_3 occur. These have the respective conformations $tttg^\pm$ and *tttt*.[68] According to Mark's analysis, their intramolecular energies should be only slightly greater than for $ttg^\pm g^\pm$.

The conformations of POM_4 predicted to be lowest in energy are $ttg^\pm tg^\pm$ and $ttg^\pm tg^\mp$. They should differ little from the more symmetrical form *ttttt*, however. The latter prevails in the stable crystalline modification,[68] perhaps due to its more efficient packing in the lattice.

10. POLYDIMETHYLSILOXANE

This chain, $[\text{—Si(CH}_3)_2\text{—O—}]_x$, depicted in Fig. 22, departs from others considered thus far in having alternating bond angles which differ considerably. Within an uncertainty of two or three degrees $\theta_a = 37°$ at the oxygen atom and $\theta_b = 70°$ at the silicon atom.[9,69] As a result, the planar *trans* conformation does not generate a rectilinear axis. Instead, it closes upon itself in the course of about eleven units. This will be evident by continuation of the diagrams in Fig. 22. Of course, the siloxane chain is

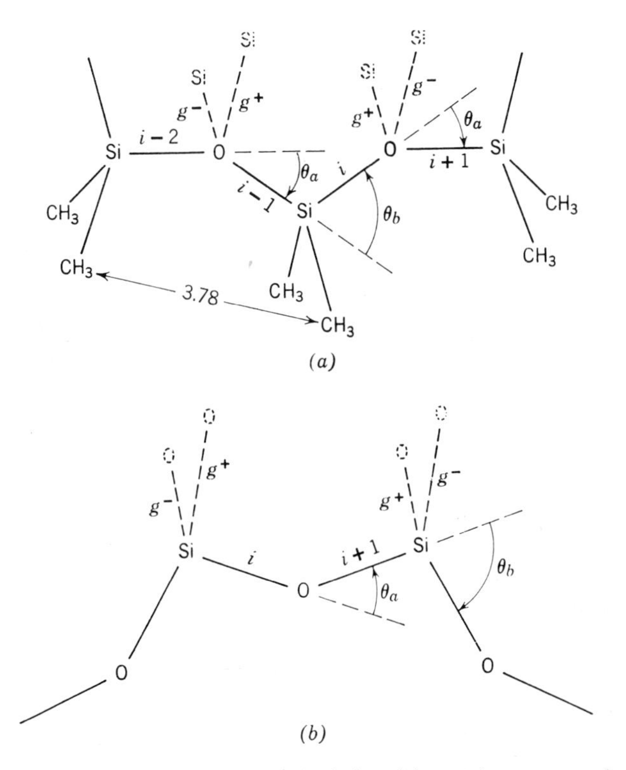

Fig. 22. Conformations of the PDMS chain: (*a*) rotations centered about Si; (*b*) rotations centered about O. Intergroup distances for the various conformations are indicated.

flexible in the sense that many configurations are accessible to it, and adherence to planarity over the span of such a number of units should be exceedingly rare in any case. The cyclic character of the planar conformation of polydimethylsiloxane (PDMS) might therefore be dismissed as an irrelevancy. We shall find, however, that this characteristic provides the clue to rationalization of its configuration. Birshtein, Ptitsyn, and Sokolova[70] were the first to point out the importance of the inequality of θ_a and θ_b in this connection.

Planar and *gauche* conformations are shown in Fig. 22. The comparatively long Si—O and Si—C bonds, 1.64 and 1.90 Å, respectively,[9] tend to relieve the congestion which would otherwise occur. The inherent rotational barrier about the Si—O bond probably is of the order of RT at ordinary temperature.[71] Rotational isomeric states therefore are not well defined, and adoption of the discrete state model must rest mainly on its use as a device for approximation of the configuration integral (see Chap. III). The choice of rotational states symmetrically situated with respect to *trans*, and located at equal intervals (120°), is indicated under such circumstances.

In the *trans* conformation the neighbor pairs of methyl groups on the same side of the plane are at a favorable distance of 3.78 Å. Their interaction should be attractive, though small if estimated by calculation according to the methods described in Section 1. Pairs on opposite sides of the plane are 4.88 Å apart, a distance beyond the range of significant interaction. A *gauche* rotation destroys one $CH_3\cdots CH_3$ pair, replacing it by $CH_3\cdots O$ at a distance at the outer limit of the range of attraction for such a pair. The partial ionic character of the Si—O bond is estimated[72] to be 0.37. Electrostatic interactions may therefore contribute appreciably to the energy. The effect of the attraction between the partial charges assignable to the silicon and oxygen atom pair separated by three skeletal bonds appears to be largely compensated, however, by repulsions between other atom pairs affected by the rotation. As above, we let σ denote the statistical weight associated with rotation of one skeletal bond, its neighbors being *trans*.

The $g^{\pm}g^{\mp}$ conformation for the pair of bonds on either side of silicon (bonds $i-1$ and i in Fig. 22*a*) places the neighboring pair of silicon atoms at a distance of 4.29 Å, with the adjoining methyl groups overlapping excessively. Hence, this conformation may be assumed to be suppressed. In the $g^{\pm}g^{\pm}$ conformation the Si···Si distance is 4.62 Å and the small steric overlaps of the adjoining methyl groups are readily alleviated by compromising bond angles, as will be apparent from inspection of models. Inasmuch as precise weighting of these $g^{\pm}g^{\pm}$ states is unimportant, it suffices to let[69]

$$\mathbf{U}_a = \begin{bmatrix} 1 & \sigma & \sigma \\ 1 & \sigma & 0 \\ 1 & 0 & \sigma \end{bmatrix} \tag{23}$$

The distance (3.69 Å) between the out-of-plane pair of oxygen atoms shown in Fig. 22*b* for the g^+g^- conformation of bond pair $i, i+1$ is well above the sum of their van der Waals radii. Coulombic repulsions between their partial charges could contribute as much as 1000 cal mole^{-1}, however.[69] On the other hand, one proximate $CH_3\cdots CH_3$ pair is retained in this conformation. The net effect of the two *gauche* rotations is the elimination of only one such pair. This should enhance ω. Obviously this parameter cannot be legitimately suppressed. Hence, we write

$$\mathbf{U}_b = \begin{bmatrix} 1 & \sigma & \sigma \\ 1 & \sigma & \sigma\omega \\ 1 & \sigma\omega & \sigma \end{bmatrix} \tag{24}$$

Exploratory calculations[69] illustrating the effect of the inequality of bond angles on C_∞ are shown in Fig. 23. Here ω has been set equal to zero, which renders $\mathbf{U}_a$ and $\mathbf{U}_b$ equivalent. For angles $\theta_a \neq 70°$, the characteristic ratio

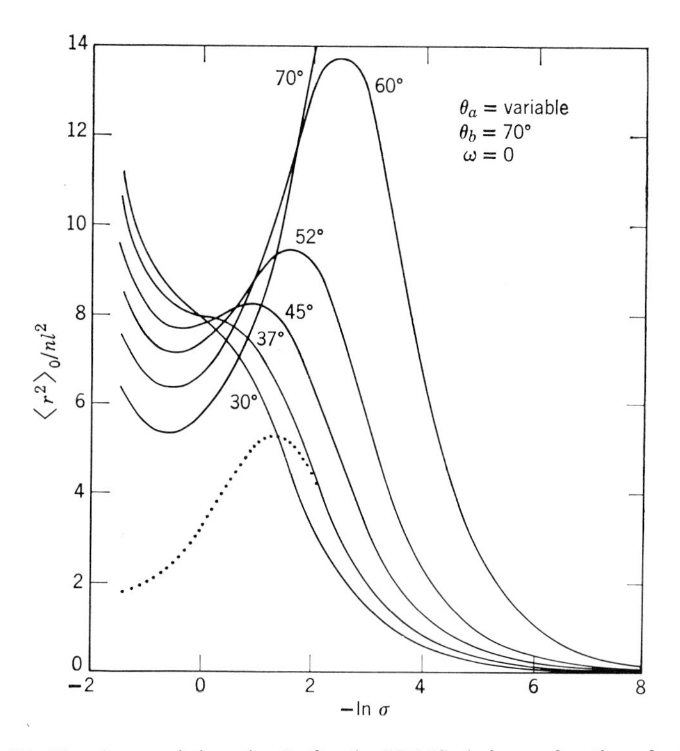

Fig. 23. The characteristic ratio C_∞ for the PDMS chain as a function of $-\ln \sigma$ for various angles θ_a; other parameters are indicated. The dotted curve has been calculated for independent bond rotations, with $\theta_a = 37°$ and $\theta_b = 70°$. (From Flory, Crescenzi, and Mark.[69])

goes to zero as σ vanishes, i.e., as the chain is forced into the planar cyclic form. For angles $40° < \theta_a < 70°$, the curve passes through a maximum beyond which it decreases monotonically with further decrease in σ. For $\theta_a < 40°$, the maximum is absent and the characteristic ratio decreases with $-\ln \sigma$ over the entire range of σ. Since σ must be expected to change in the direction of unity with increase of temperature, the decrease of the characteristic ratio with decrease in σ implies a positive temperature coefficient, which is therefore directly traceable to the inequality $\theta_a \neq \theta_b$.

The slope of the curve for $\theta_a = 37°$ at the point where $\langle r^2 \rangle_0 / nl^2 = 6.2$, the experimental value at ca. 70°C (see Table II-1), yields a temperature coefficient too large by a factor of about 2. The calculated temperature coefficient is decreased by taking $\omega > 0$. Both the experimental temperature coefficient, 0.75×10^{-3} deg^{-1}, and the characteristic ratio are reproduced by letting $\sigma = 0.29$ and $\omega = 0.20$ at 70°C. These values correspond to $E_\sigma = 850$ cal mole^{-1} and $E_\omega = 1100$ cal mole^{-1}.[69]

The value of E_σ is unexpectedly large. A marginal preference for the planar *trans* conformation is rendered plausible by inspection of the structure, but so large a difference would not have been anticipated from conventional estimates of nonbonded intramolecular interactions. Analysis of the crystal structure of crystalline PDMS (stretched at −90°C) by Damaschun[73] reveals a flat helical conformation, thought to be a 6_1 helix, the projection of the Si—O bond on the helix axis being only 0.69 Å. A helix of this description may be generated by introducing a rotation ϕ of ca. 35–40° about each skeletal bond, as calculated by the methods of Shimanouchi and Mizushima.[74] Thus, the conformation adopted by PDMS in the crystalline state would appear to depart from the impossible planar form only to the extent required to meet the spatial requirements of its constituent groups. On the assumption that the conformation in the crystal is indicative of the form of lowest energy, the x-ray crystallographic results support the conclusion that the *trans* conformation is of low energy.

The characteristic ratio C_n for PDMS, calculated[76] on the basis of parameters given above through use of Eq. IV-28, is plotted against the degree of polymerization $x = n/2$ in Fig. 24. The two curves are referred to the abscissa scales shown on the upper and lower margins, respectively. They connect points for homologs having even numbers of bonds, the chain sequence being terminated at either end by silicon atoms. Points for odd species terminated by Si at one end and by O at the other describe a slightly lower curve (not shown).[76] The upper boundary of the graph marks the limiting value $C_\infty = 6.43$. Convergence of C_n with n is somewhat more rapid than for PM (compare Fig. 9, solid curve).

The unperturbed dimensions of the PDMS chain appear to be affected by the solvent medium to a much greater degree than is true for PM and other nonpolar polymers. In a fluorinated solvent of low cohesive energy and low dielectric constant, C_∞ was found to be 7.6 (Table II-1). The polarity of the Si—O bond, and perhaps also of Si—C, may render this chain more sensitive to influence by the medium.[69] The possibility of such effects should be borne in mind, especially when the chain is highly polar, or carries polar substituents.

Polydimethylsiloxane may contain cyclic polymers

$$\boxed{-[\,Si(CH_3)_2-O\,]_x-}$$

as well as linear species. Equilibrium amongst various species, linear and cyclic, can be established under the influence of acidic or basic catalysts for the interchange of Si—O bonds. Applying chromatographic methods,

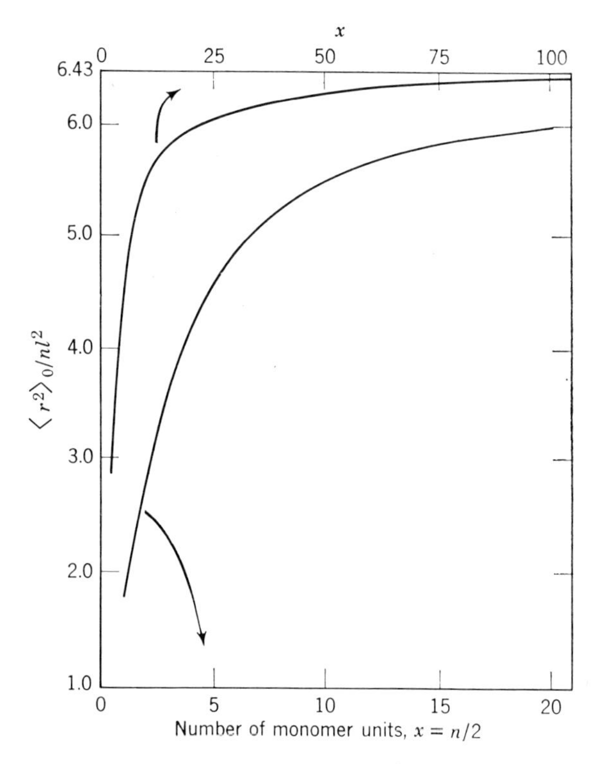

Fig. 24. The characteristic ratio $C_n = \langle r_2 \rangle_0/nl^2$ for PDMS chains CH_3—[Si $(CH_3)_2$—O—$]_{n/2}$—$Si(CH_3)_3$, with r measured from terminal Si atoms. The curves have been calculated using the parameters given in the text. Upper and lower abscissa scales are for the respective curves as indicted. (From ref. 76.)

Brown and Slusarczuk[75] have determined the concentration of the cyclic siloxanes from $x = 4$ to about 200 at equilibrium. Cyclization depends of course on the occurrence of atoms separated by x units in suitable juxtaposition for bond formation, and hence for closure of the ring. The probability of their occurrence depends in first approximation on $\langle r_x^2 \rangle_0$ (see Appendix D).[46,76] The concentrations of various cyclic species at equilibrium are thus related to the geometry and configurational statistics of the chain. Inasmuch as cyclization equilibrium can be established in concentrated solutions or in the bulk polymer, analysis of the equilibrium mixture affords information bearing on the configurational character of the chain in concentrated media, and hence in circumstances that differ from the dilute solution environment prevailing in more conventional measurements on macromolecular dimensions.

The theory of macrocyclization, originated by Jacobson and Stockmayer,[46] is presented in Appendix D, and it is there applied to PDMS.[76] Calculations carried out on the basis of the experimental characteristic ratio determined in dilute solution are in good agreement with observed cyclization equilibria. It is thus confirmed that the configuration of the PDMS chain in concentrated solutions and in the bulk polymers, in absence of diluent, approximates that of the unperturbed chain in dilute solution.[76]

11. THE POLYPHOSPHATE CHAIN[77]

The linear polyphosphate (PP) chain shown in the diagram in Fig. 25 resembles the chain of PDMS in several respects. The repeating unit comprises two bonds, alternate skeletal atoms bear two identical substituents,

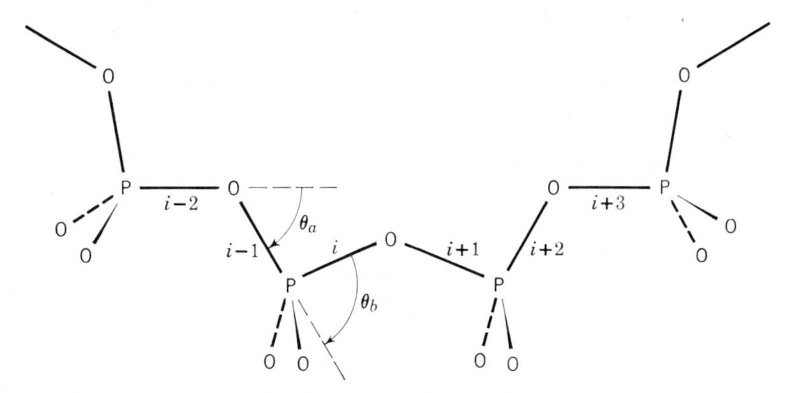

Fig. 25. The polyphosphate chain.

the intervening skeletal atoms are oxygens, and valence angles are unequal, i.e., $\theta_a = 50°$ (at oxygen) and $\theta_b = 78°$ (at phosphorus). On the other hand, geometrical parameters differ in detail ($l_{P—O} = 1.62$ Å for skeletal bonds and 1.48 Å for pendant bonds[77]). Of foremost importance, each unit of the PP chain carries an electronic charge. Coulombic interactions therefore are large. Their range is limited, however, by the shielding provided by the salt present at fairly high ionic strength in the aqueous media chosen for experimentation at or near the Θ point.[78] The range of the strong coulombic interactions is therefore restricted to distances of several Angstrom units. Like the bonds in PDMS, the skeletal P—O bonds probably are free of large inherent potentials. Steric and electrostatic interactions must therefore be dominant. In adopting a three-state scheme with $\phi = 0°, \pm 120°$, we do so as a means of approximating a continuous range of accessible rotational angles.

Inspection of distances between nonbonded atoms[77] shows steric overlaps to be absent throughout the range $\phi_i = 0$ to 2π of rotation about *one bond only*, its several neighbors on either side being maintained in the *trans* ($\phi = 0$) conformation. Moreover, estimates of the electrostatic energy, though large, reveal it to depend remarkably little on ϕ_i.[77] Simultaneous rotations of opposite sign about the pair of bonds $i - 1, i$ in Fig. 25 lead to severe steric overlaps; those of the same sign (e.g., $g^{\pm}g^{\pm}$), though permitted sterically, are opposed by strong coulombic repulsions. The statistical weight matrix for this bond pair is therefore expressed by

$$\mathbf{U}_a = \begin{bmatrix} 1 & \sigma & \sigma \\ 1 & 0 & 0 \\ 1 & 0 & 0 \end{bmatrix} \qquad (25)$$

with the expectation that σ will prove to be of the order of unity.

Rotations about the bond pair $i, i + 1$ flanking oxygen are less subject to steric obstruction. Rotations of opposite sign are opposed by formidable electrostatic repulsions, however. Rotations of the same sign ($g^{\pm}g^{\pm}$) engender smaller repulsions, which however are sufficient to reduce ψ to the order of magnitude of 10^{-2}. We accordingly write[77]

$$\mathbf{U}_b = \begin{bmatrix} 1 & \sigma & \sigma \\ 1 & \sigma\psi & 0 \\ 1 & 0 & \sigma\psi \end{bmatrix} \qquad (26)$$

In satisfactory approximation, we may let $\psi = 0$, whereupon $\mathbf{U}_b = \mathbf{U}_a$.

Calculations[77] of the characteristic ratio on this basis ($\psi = 0$) are shown in Fig. 26. Introduction of a separate matrix $\mathbf{U}_b$ with $\psi \leq 0.10$ has only a trivial effect on the curve.[77] Comparison with the experimental results of Strauss and co-workers[78] (see Table II-1) confirms that $\sigma \cong 1$ as expected.*

The PP chain appears to be one in which substantial rotations (e.g., *gauche*

* Treatment of the chain in terms of two states, *trans* ($\phi = 0$) and *cis* ($\phi = 180°$), is equally satisfactory.[77] Then

$$\mathbf{U}_a = \mathbf{U}_b = \begin{bmatrix} 1 & \sigma \\ 1 & 0 \end{bmatrix}$$

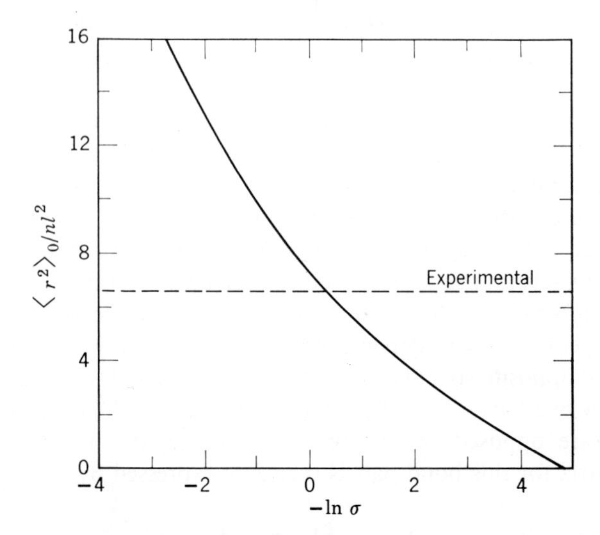

Fig. 26. The characteristic ratio for the polyphosphate chain calculated from the statistical weight matrices $\mathbf{U}_a$ and $\mathbf{U}_b$ given by Eqs. 25 and 26, with $\psi = 0$.[77]

or even *cis*) are permitted for an isolated bond, but simultaneous rotations of two consecutive bonds, irrespective of sign, are strongly forbidden.

12. POLYAMIDES AND POLYESTERS

We next consider polyamides and polyesters of the types

$[—NH(CH_2)_\eta CO—]_x$	A-I
$[—NH(CH_2)_\eta NH—CO(CH_2)_{\eta'}CO—]_x$	A-II
$[—O(CH_2)_\eta CO—]_x$	E-I
$[—O(CH_2)_\eta O—CO(CH_2)_{\eta'}CO—]_x$	E-II

consisting of polymethylene sequences connected by amide or ester groups.

These polymers provide examples of chains wherein neighbor dependence of rotations is confined to bonds within a given repeating unit. The rigidity of the amide and ester groups assures independence of rotations of the bonds adjoining on either side. Hence, the methods of Chapter IV, Section 6, are applicable. Examples of principal interest here are those for which the numbers η and η' of methylene groups are at least two. The polypeptides comprising α-amino acid residues (type A-I) with $\eta = 1$ are reserved for later consideration (see Chap. VII).

The conformations of both the amide and the ester group are planar and preponderantly *trans*, as shown in Fig. 27. Their planarity is a consequence of the partial double bond character of the CO—NH and CO—O bonds. This is manifested in a shortening of these bonds from 1.46 to 1.33 Å in the amide and from 1.44 (or 1.43) to 1.33 Å in the ester.[9] The degree of double-bond character is thus indicated to be large in both groups, although somewhat less in the ester. The planarity of these groups is demonstrated by a variety of evidence—crystallographic, electron diffraction of gases, and spectroscopic—principally on low molecular analogs. Most definitive is the evidence

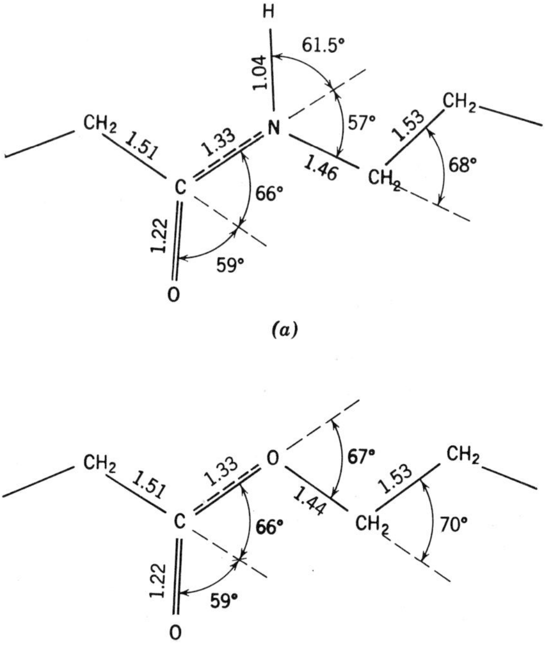

Fig. 27. Geometrical parameters for the amide (*a*) and the ester (*b*) group. Bond lengths are given in Angstrom units along each bond. These values have been used in the calculations described.

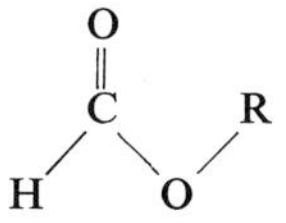

from microwave spectroscopy which, however, is limited to the simplest analogs. Formamide has been shown[79] to be planar, and methyl formate[80] and ethyl formate[81] to be preponderantly in the conformation in which R (i.e., methyl or ethyl) and the carbonyl oxygen occur on the same side of the ester bond. This conformation we designate as *trans* in reference to the disposition of the chain skeleton of higher homologs depicted as in Fig. 27. Nuclear magnetic resonance has furnished the most decisive evidence that the amides and esters are *trans*. LaPlanche and Rogers[82] found no detectable resonance in *N*-alkylacetamides which could be identified with a *cis* conformation. The *cis* conformer, if present, does not exceed 5%. An even greater preference for the *trans* form is indicated for the esters.[83] When incorporated in a polymer chain, the *cis* form of the amide or ester group should be further suppressed as a consequence of the fewer configurations available to skeletal bonds neighboring a *cis* as compared with a *trans* linkage in a chain molecule.

Bond lengths and angles given in Fig. 27 are best values for typical aliphatic amide and ester groups gleaned from various sources and used in the calculations to follow. The difference of 2° between the assigned values of ∠NCC and ∠OCC is not significant. The dipole moment of typical amides is 3.7 D[84]; for esters it is 1.8 D.[85] Both values have been obtained from measurements of dielectric constants of solutions. The direction of the dipole moment in the *N*-alkyl amide group shown in Fig. 27*a*, as deduced by addition of bond dipole moments,[85] is approximately parallel to the C═O and to the N—H bonds. This deduction is substantiated by the direction of the moment in formamide as determined by analysis of the microwave Stark spectrum of this molecule.[79,86]* Interactions of the amide group moment may be very large. They have an important effect on the dimensions of polypeptides (see Chap. VII).[86] For polyamides with $\eta > 3$ the effect of dipole–dipole interactions on the chain configuration in solvents of high dielectric constant may be ignored.

The dipole moment in methyl formate is inclined at an angle of 39° to C═O according to Curl's[80] interpretation of the microwave Stark spectrum of this ester. The orientation of the moment may differ in esters of other carboxylic acids bearing an α-carbon atom, which in formic acid is replaced by hydrogen. By analogy with the corresponding amides discussed above, the angle of inclination of the dipole moment in relevant aliphatic esters should be less than 39°, and probably is in the range of 20–30°. Dipolar interactions between ester groups in most polyesters of interest should not affect their configurations appreciably.

Potential barriers to rotation about CH_2—CO and NH—CH_2 bonds adjoining the amide bond are believed to be low. Symmetry dictates that

* The dipole moment of formamide makes an angle of 22° with the C═O bond according to Kurland and Wilson.[79] Allowance for the much smaller bond moment of the N—C (alkyl) bond as compared to N—H brings the resultant approximately parallel to the C═O bond in the *N*-alkyl amides.[86]

they be threefold. The *trans* conformation of the former bond certainly represents one of its minima, and the same probably holds for the latter bond as well.[86] The remaining minima may therefore be designated as *gauche*, and they may be represented approximately by states placed at $\phi = \pm 120°$. A similar disposition of minima doubtless applies to the ester group. That the O—CH_2 bond of the ester group possesses a minimum in the *trans* conformation, rather than at $\pm 60°$ as might have been assumed, has been demonstrated in the case of ethyl formate by analysis of its microwave spectrum.[81] The barrier height is ca. 1100 cal mole^{-1}, and the energy of the $g^{\pm}$ states relative to the *trans* is only ca. 200 cal mole^{-1}. The *gauche* angle is $\pm 95°$.[81]

As an illustrative example, we consider the polyamide, poly-ε-amino-caproamide, having the repeating unit represented in Fig. 28. The skeletal atoms are numbered 0 to 7; in this case, $\eta = 5$ and $\xi = \eta + 2 = 7$ bonds per unit. In keeping with custom, we designate this polymer as polyamide-6, the numeral referring to the number of carbon atoms in the unit. The first-order interaction parameter σ is indicated above each bond; the second-order parameter ω applicable to each bond pair is shown below. The σ's and ω's without subscripts are equivalent to those for polymethylene.

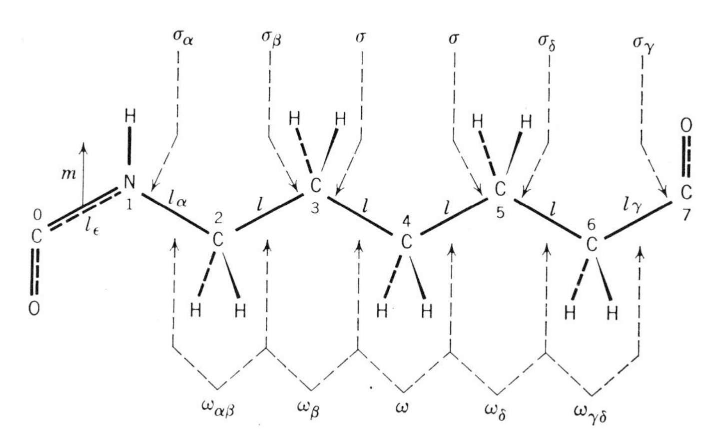

Fig. 28. The polyamide-6 chain. Statistical weight parameters σ and ω, and the dipole moment m of the amide group are indicated.[87] Numerals adjacent to skeletal atoms designate serial order within the unit.

Since bond 1 of the repeating unit occurs exclusively in the *trans* conformation, it is permissible to let $\mathbf{U}_1 = 1$ and to express $\mathbf{U}_2$ as the row[87]

$$\mathbf{U}_2 = [1 \quad \sigma_\alpha \quad \sigma_\alpha] \tag{27.2}$$

Succeeding matrices are[87]

$$\mathbf{U}_3 = \begin{bmatrix} 1 & \sigma_\beta & \sigma_\beta \\ 1 & \sigma_\beta & \sigma_\beta \omega_{\alpha\beta} \\ 1 & \sigma_\beta \omega_{\alpha\beta} & \sigma_\beta \end{bmatrix} \tag{27.3}$$

$$\mathbf{U}_4 = \begin{bmatrix} 1 & \sigma & \sigma \\ 1 & \sigma & \sigma\omega_\beta \\ 1 & \sigma\omega_\beta & \sigma \end{bmatrix} \tag{27.4}$$

$$\mathbf{U}_5 = \begin{bmatrix} 1 & \sigma & \sigma \\ 1 & \sigma & \sigma\omega \\ 1 & \sigma\omega & \sigma \end{bmatrix} \tag{27.5}$$

$$\mathbf{U}_6 = \begin{bmatrix} 1 & \sigma_\delta & \sigma_\delta \\ 1 & \sigma_\delta & \sigma_\delta \omega_\delta \\ 1 & \sigma_\delta \omega_\delta & \sigma_\delta \end{bmatrix} \tag{27.6}$$

$$\mathbf{U}_7 = \begin{bmatrix} 1 & \sigma_\gamma & \sigma_\gamma \\ 1 & \sigma_\gamma & \sigma_\gamma \omega_{\gamma\delta} \\ 1 & \sigma_\gamma \omega_{\gamma\delta} & \sigma_\gamma \end{bmatrix} \tag{27.7}$$

as will be obvious from Fig. 28. The partition function $z^{(\xi)}$ for the repeating unit is therefore

$$\begin{aligned} z^{(\xi)} &= \mathbf{U}_1\mathbf{U}_2 \cdots \mathbf{U}_7\mathbf{J} \\ &= \mathbf{U}_2 \cdots \mathbf{U}_7\mathbf{J} \end{aligned} \tag{28}$$

in accordance with Eq. IV-68.

Transformation matrices $\mathbf{T}$ may be formulated from the geometrical data in Fig. 27, *gauche* states being taken at $\phi_g = \pm 120°$ for all bonds, bond 1 excepted of course. The $\mathscr{G}_i$ matrix corresponding to each of the $\mathbf{U}_i$'s specified above may then be constructed according to Eq. IV-24, the bond vector $\mathbf{l}_i$ being substituted for $\mathbf{m}_i$ in each. Then

$$\mathscr{G}^{(\xi)*} = \mathscr{G}_1\mathscr{G}_2 \cdots \mathscr{G}_7 \tag{29}$$

With $\mathbf{U}_1$ and $\mathbf{U}_2$ expressed as stipulated above, the orders of $\mathscr{G}_1$ and $\mathscr{G}_2$ are 5×5 and 5×15, respectively. The remaining $\mathscr{G}$ matrices are 15×15. The order of $\mathscr{G}^{(\xi)*}$ is therefore 5×15; the asterisk is appended (see Chap. IV, Sect. 6) for the purpose of distinguishing it from the corresponding square matrix of order 15×15 containing redundant rows.

The 5×5 matrix $\mathfrak{G}^{(\xi)}$ representing the repeating unit of $\xi = 7$ bonds is obtained from $z^{(\xi)}$ and $\mathscr{G}^{(\xi)}$ by resorting to Eq. IV-71. The calculation of $\langle r^2 \rangle_0$ proceeds as prescribed by Eq. IV-70.

Distances between groups engaged in first- and second-order interactions in the polyamide-6 chain are presented in Fig. 29 for the $g^{\pm}$ and $g^{\pm}g^{\mp}$ states, respectively. As in Figs. 5, 15, and 17, the *trans* form is shown by heavy solid lines; the various *gauche* conformations are represented by heavy dashed lines. The bonds about which the given rotations take place remain in the plane of the figure; adjoining bonds, shown dashed, are displaced out of the plane. Associated statistical weight parameters are

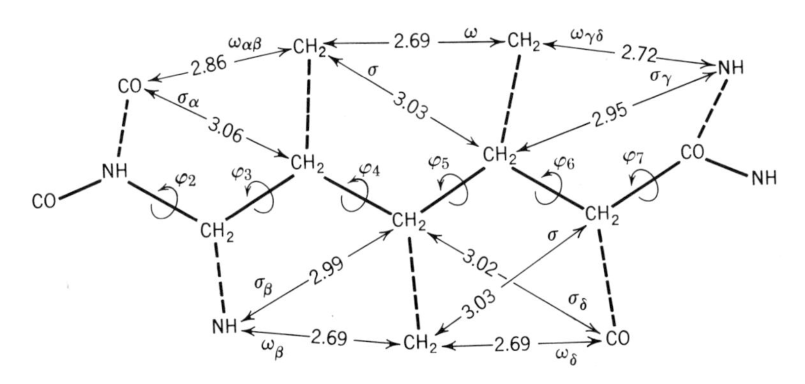

Fig. 29. Distances between groups separated by three and by four bonds, respectively, in the $g^{\pm}$ and in $g^{\pm}g^{\mp}$ conformations, indicated by dashed lines, of the polyamide-6 chain. Bonds rotated are to be construed as remaining in the plane of the figure; pendant bonds, shown dashed, are out of the plane. Related statistical weights are shown as well (compare Fig. 28).

included in the diagram. Identifying σ and ω with the corresponding parameters for PM (chosen for use in conjunction with $\phi_g = \pm 120°$; see p. 147) and expressing them for a temperature of 25°C, we take $\sigma = 0.43$ and $\omega = 0.034$. The other first-order distances are about the same (ca. 3.0 Å) or slightly smaller than that for the polymethylene distance. One of the participating groups, CO or NH, is smaller, however. Moreover, if NH acts like O in the polyethers considered in Sections 7–9, it may favor the *gauche* conformation. These considerations, supplemented by more detailed examination of distances between interacting atom pairs,[87] lead to the conclusion that the various subscripted σ's should exceed the value for PM, but that they should not be much greater than unity. For a tentative calculation, therefore, we take $\sigma_\alpha = \sigma_\beta = \sigma_\gamma = \sigma_\delta = 1.00$.

Second-order distances for $g^{\pm}g^{\mp}$ states equal or exceed slightly the corresponding distance for PM, but one of the participating groups of each pair is smaller; see Fig. 29. Hence, we expect the subscripted ω's to exceed ω for PM. Values in the range 0.05–0.20 are plausible. We therefore take $\omega_{\alpha\beta} = \omega_\beta = \omega_\delta = \omega_{\gamma\delta} = 0.10$.

Calculations[87] for polyamide-6 carried out using the σ and ω parameters specified above give $C_\infty = (\langle r^2 \rangle_0 / nl^2)_\infty = 6.08$, where l^2 is the mean of the squared lengths of the seven bonds comprising the repeating unit. The similarly calculated dipole moment ratio is $(\langle \mu^2 \rangle / xm^2)_\infty = 0.98$, which is very near the value (unity) which would obtain for uncorrelated dipole moment vectors $\mathbf{m}$.

The analysis of analogous polyamides, including those of type A-II, can be carried out similarly. If the number of methylene groups η in the unit exceeds three, the same set of parameters will serve. Polyamide-66, the type A-II polyamide with $\eta = 6$ and $\eta' = 4$, is illustrative. Its repeating unit is sketched in Fig. 30. Retention of the same geometrical parameters for corresponding bonds predicates interactions identical to those occurring in polyamide-6, as comparison with Fig. 28 will verify. Hence, the same statistical weight parameters should apply. They occur in different order, however, and the various statistical weight matrices will differ accordingly. The construction of the matrices from Fig. 30 is straightforward.[87]

The repeating unit of this polymer consists of two mutually independent parts separated by an amide bond. The partition function for the unit may be expressed accordingly as the product of two scalar factors as follows:

$$z^{(\xi)} = \mathbf{U}_2\mathbf{U}_3 \cdots \mathbf{U}_8\mathbf{J}\mathbf{U}_{10} \cdots \mathbf{U}_{14}\mathbf{J} \tag{30}$$

where $\mathbf{U}_2$ and $\mathbf{U}_{10}$ are expressed as rows after the manner of Eq. 27.2. The factors $\mathbf{U}_1$ and $\mathbf{U}_9$ are omitted inasmuch as they equate to unity in this expression. Alternatively, the partition function for the repeating unit may be written in terms of a product of one consecutive set of factors $\mathbf{U}_1$–$\mathbf{U}_{14}$, inclusive, i.e.,

$$z^{(\xi)} = \mathbf{U}_1\mathbf{U}_2 \cdots \mathbf{U}_{14}\mathbf{J} \tag{31}$$

Here the identifications $\mathbf{U}_1 = 1$ and $\mathbf{U}_9 = \mathbf{J}$ are required. Construction of each of the matrices $\mathscr{G}, \mathscr{G}_2, \ldots, \mathscr{G}_{14}$ is carried out as prescribed by Eq. IV-24 using the $\mathbf{U}_i$ matrices specified in Eq. 31. The normalized matrix $\mathfrak{G}^{(\xi)}$ for the repeating unit consisting of 14 bonds follows according to Eq. IV-71.

Calculations[87] thus carried out for polyamide-66 using the same set of parameters as above give $C_\infty = 6.10$ and $(\langle \mu^2 \rangle / xm^2)_\infty = 0.97$. The calculated characteristic ratio is in good agreement with the experimental results of Saunders[88] on the average dimensions of polyamide-66 at the Θ point at 25°C, which lead to $C_\infty = 5.95$.[87]

By way of comparison, the characteristic ratio calculated for PM at the same temperature using $\sigma = 0.43$ and $\omega = 0.034$, the values applicable at 25°C, is $C_\infty = 8.0$. Adherence of the amide group to the *trans* conformation might have been expected to raise C_∞ for the polyamides above its value

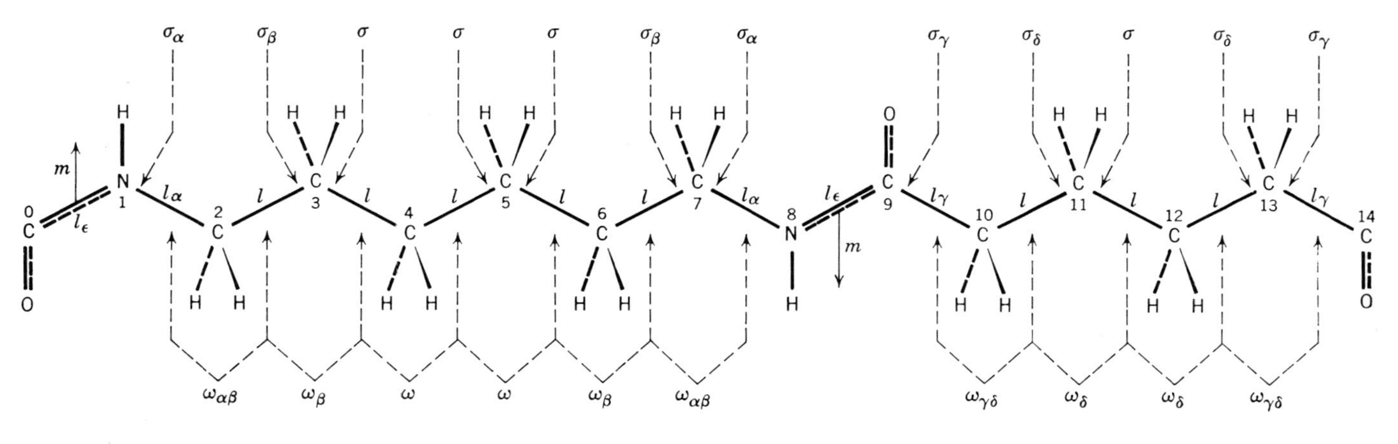

Fig. 30. Polyamide-66 repeating unit.[87]

for PM. The rigid, extended conformation of the amide group does indeed have the effect of increasing C_∞, as calculations show.[87] However, this conformation of the amide group also diminishes intergroup interactions associated with rotations about nearby bonds, even including those which are third neighbors of the amide bond. This is reflected in the **U** matrices and the statistical weights occurring therein. The effect of the reduction of the hindrances to rotation turns out to be overriding, and it is for this reason that the C_∞ for the foregoing polyamides turn out to be less than C_∞ for PM.

The corresponding polyesters of types E-I and E-II admit of similar treatment. The absence of the amide hydrogen should enhance σ_β and σ_γ somewhat. On the other hand, the greater angle $\theta = 67°$ at oxygen compared with $\theta = 57°$ at NH (see Fig. 27) diminishes the distance between $(CO)_0$ and $(CH_2)_3$ in the *gauche* conformation of bond 2 (see Fig. 30) and thus must be expected to reduce σ_α. These effects are partially compensatory. Average dimensions calculated[87] for aliphatic polyesters consequently are close to those for the corresponding polyamides. Suitable experimental results with which to compare these predictions are unavailable.

A polyester of a different kind and one presenting geometrical features of particular interest is poly(ethylene terephthalate). Its repeating unit is diagrammed in Fig. 31. In addition to two planar, *trans* ester groups, the repeating unit includes a segment, or "virtual bond," connecting carbons

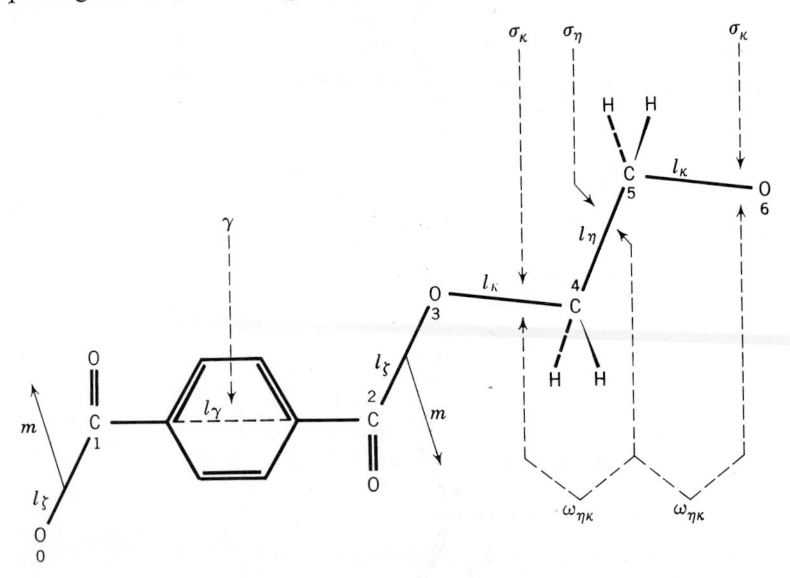

Fig. 31. The poly(ethylene terephthalate) unit.[89] Serial numbering of atoms within the repeating unit and statistical weights are indicated.

1 and 2 of length $l_2 \equiv l_\gamma = 5.74$ Å. Other data given in Fig. 27*b* complete the geometrical description of the chain unit.

The terephthaloyl residue is restricted to *trans* and *cis* conformations. Dipole moments of diesters suggest that the two forms are about equally populated,[89] as would be expected. For generality, we assign a weight γ to the *cis* conformer relative to *trans*. The statistical weight matrices are appropriately expressed as follows:

$$\mathbf{U}_1 = 1 \tag{32.1}$$

$$\mathbf{U}_2 = [1 \quad \gamma] \tag{32.2}$$

$$\mathbf{U}_3 = \begin{bmatrix} 1 \\ 1 \end{bmatrix} \tag{32.3}$$

$$\mathbf{U}_4 = [1 \quad \sigma_\kappa \quad \sigma_\kappa] \tag{32.4}$$

$$\mathbf{U}_5 = \begin{bmatrix} 1 & \sigma_\eta & \sigma_\eta \\ 1 & \sigma_\eta & \sigma_\eta \omega_{\eta\kappa} \\ 1 & \sigma_\eta \omega_{\eta\kappa} & \sigma_\eta \end{bmatrix} \tag{32.5}$$

$$\mathbf{U}_6 = \begin{bmatrix} 1 & \sigma_\kappa & \sigma_\kappa \\ 1 & \sigma_\kappa & \sigma_\kappa \omega_{\eta\kappa} \\ 1 & \sigma_\kappa \omega_{\eta\kappa} & \sigma_\kappa \end{bmatrix} \tag{32.6}$$

Formulation of $\mathfrak{G}^{(\xi)}$ is carried out as above, and $\langle r^2 \rangle_0$ may again be calculated according to Eq. IV-70.[89]

Rotation of ϕ_4 or ϕ_6 to $\pm 120°$ places CH_2 and CO at a distance of separation of 2.83 Å. This distance is less than for the corresponding first-order interaction between the same groups in the polyamide chain (see Fig. 29), owing principally to the difference in the angle θ at oxygen in the ester compared to that at nitrogen in the amide group. It is also less than the distance (3.03 Å) between the CH_2 groups adjoining a *gauche* bond in PM. The steric overlap between the CH_2 and CO groups of the polyester is more readily relieved by adjustment of rotation angles, however. A reasonable estimate is therefore $\sigma_\kappa = 0.5$. The parameter σ_η should resemble σ' for POE. Hence, a value of 1.5 to 2.0 is indicated for this parameter.

Calculations[89] carried out with $\sigma_\kappa = 0.5$, $\sigma_\eta = 1.5$, $\omega_{\eta\kappa} = 0.10$, and $\gamma = 1$ are shown in Fig. 32, where the ratio $\langle r^2 \rangle_0 / M$ is plotted against the degree of polymerization x. Convergence is somewhat more rapid than for PM, the comparison being made at the same number of bonds. At the asymptote $(\langle r^2 \rangle_0 / M)_\infty = 0.93$ Å^2 (g mol wt)$^{-1}$, compared with experimental values of

1.05[89,90] and of 0.95[91] in the same units. Expressed on the basis of the six bonds enumerated in Fig. 31, the calculated characteristic ratio $(\langle r^2\rangle_0/nl^2)_\infty$ is 4.15, a comparatively low value. A larger value might have been expected in consequence of the presence of two *trans* planar ester groups. Their direct effect is superseded by the enhanced freedom of rotation they confer on adjoining bonds.

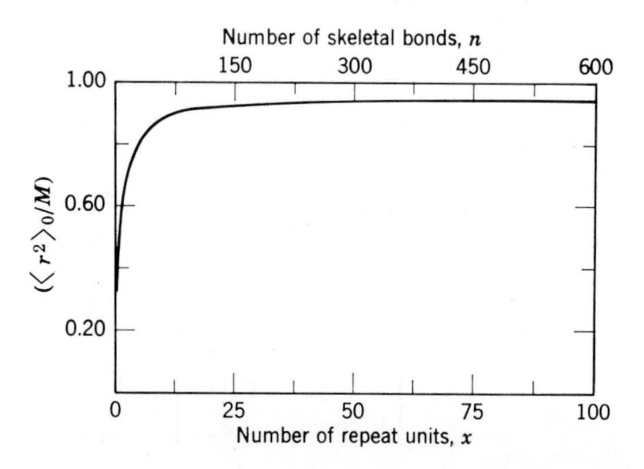

Fig. 32. The ratio $\langle r^2\rangle_0/M$ for poly(ethylene terephthalate) calculated from parameters given in the text and plotted against the degree of polymerization x (lower abscissa scale) and number of skeletal bonds $n = 6x$ (upper scale).[89]

13. 1,4 POLYMERS OF BUTADIENE AND ISOPRENE[92,93]

The polymers of butadiene and of isoprene in which the configuration about the residual double bond is either *trans* or *cis* furnish further examples of chain molecules in which rotations about bonds in separate units are mutually independent. The structure of the repeating unit of the *trans*-1,4-diene chain is presented in Fig. 33. All single bonds are shown in their *trans* conformations, thereby rendering the entire chain skeleton planar. The unit is that of *trans*-1,4-polybutadiene (PBD). Replacement of the methine hydrogens at the equivalent positions 0 and 4 by the methyl groups shown in parentheses in Fig. 33 yields the *trans*-1,4-polyisoprene (PIP) structure. We first consider *trans*-polybutadiene (PBD).

It must be observed at the outset that the planar conformation shown in Fig. 33 is not one of minimum energy for the 1,4-diene chain. Minima in the rotational potential for a single bond adjoining a double bond, e.g., the single bond in $>C{=}CH{-}CH_2R$, occur when one of the pendant atoms of the tetrahedral carbon eclipses the double bond and the other two pendant atoms are staggered with respect to the nearest methine hydrogen.[94,95] The inherent torsional potential for bonds 2 and 4 in Fig. 33 is therefore of the form

$$E_{\text{torsion}} = (E^\circ/2)[1 - \cos(3\phi + \pi)] \tag{33}$$

The torsional barrier in propylene[8,94,95] is 1.98 kcal mole^{-1}. Hence, E° may be assumed to be ca. 2.0 kcal mole^{-1} for these bonds, the minima occurring at $\phi = \pm 60$ and 180°.

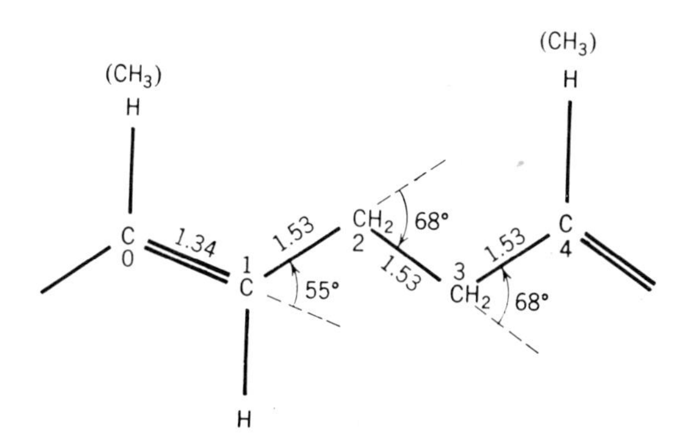

Fig. 33. The *trans*-polybutadiene (PBD) unit with geometric parameters and ordinal numeration of skeletal atoms indicated. Substitution of CH_3 groups, shown in parentheses, for the H atoms at the positions 0 and 4 yields the *trans*-polyisoprene (PIP) unit.

The following analysis of the configurational statistics of the *trans*-PBD chain is based principally on the work of J. E. Mark.[92] Starting with the *trans* double bond 1 connecting atoms 0 and 1 in Fig. 33, we let

$$\mathbf{U}_1 = 1 \tag{34.1}$$

in recognition of the fixed configuration of this bond. The ±60° conformations of bond 2 are quite free of interactions; in the *cis* (180°) conformation, methylene group 3 is involved with methine (CH) group 1, but overlaps are small. Following Mark,[92] we assign weights of unity to the former states and α to the latter, with the expectation that α will be less than, but near, unity. On this basis $\mathbf{U}_2$ may be expressed as the row

$$\begin{matrix} & 60^\circ & 180^\circ & -60^\circ \\ \mathbf{U}_2 = [& 1 & \alpha & 1 \;] \end{matrix} \tag{34.2}$$

Rotational states for bond 3 are appropriately taken at 0 and ±120°. The first-order interaction in the *gauche* states involves the two methine groups 1 and 4. The interaction is smaller[92,93] than for the *gauche* conformation in PM. We accordingly assign it a weight σ expected to be near unity. If, however, bond 2 is *cis*, thereby replacing the H attached to carbon 1 in Fig. 33 by $(CH)_0$, the *gauche* states for bond 3 are disfavored, though not disallowed. Assigning β to represent this second-order interaction, we have

$$\mathbf{U}_3 = \begin{array}{c} \\ 60^\circ \\ 180^\circ \\ -60^\circ \end{array} \begin{array}{c} \begin{array}{ccc} 0^\circ & 120^\circ & -120^\circ \end{array} \\ \begin{bmatrix} 1 & \sigma & \sigma \\ 1 & \sigma\beta & \sigma\beta \\ 1 & \sigma & \sigma \end{bmatrix} \end{array} \tag{34.3}$$

Finally, consistency requires that*

$$\mathbf{U}_4 = \begin{array}{c} \\ 0^\circ \\ 120^\circ \\ -120^\circ \end{array} \begin{array}{c} \begin{array}{ccc} 60^\circ & 180^\circ & -60^\circ \end{array} \\ \begin{bmatrix} 1 & \alpha & 1 \\ 1 & \alpha\beta & 1 \\ 1 & \alpha\beta & 1 \end{bmatrix} \end{array} \tag{34.4}$$

The *trans*-PIP chain is susceptible to similar treatment.[92] In order to amend Fig. 33 so that it represents the PIP chain, let the hydrogen atoms attached to carbons 0 and 4 be replaced by CH_3 (these carbons being corresponding members of consecutive units). The *cis* conformation about bond 2 is rendered untenable by the overlap of $(CH_3)_0$ with $(CH_2)_3$. Hence, in this case

$$\mathbf{U}_2 = [1 \quad 0 \quad 1] \tag{35.2}$$

The pendant methyl groups may be expected to affect the *gauche* states for bond 3 unfavorably, but detailed calculations of distances between atoms

* Interactions associated with *gauche* conformations of bond 3 obviously must depend to some extent on the signs of rotation ϕ_2 and ϕ_4 relative to the sign of $\phi_3 = \pm 120^\circ$, if either or both of the former are $\pm 60^\circ$. For example, conformational energies for $\phi_2 = -60^\circ$, $\phi_3 = 120^\circ$ and for $\phi_2 = 60^\circ$, $\phi_3 = 120^\circ$ must differ. Interactions in some of the conformations of the *trans*-PBD unit depend, moreover, upon all three angles, and hence upon their signs in relation to one another. The *trans*-diene polymers furnish examples of the uncommon occurrence of third-order interdependence. Molecular models indicate the effects of the signs of ϕ_2 and ϕ_4 relative to that of ϕ_3 to be small in *trans*-PBD, however. Disregard of these effects in the scheme of statistical weights incorporated in Eqs. 34.3 and 34.4 may therefore be acceptable as a first approximation. A somewhat greater (and more complex) interdependence on the signs of rotations may be anticipated in *trans*-PIP.

It is to be noted further that simultaneous occurrence of bonds 2 and 4 in *cis* conformations is represented by Eqs. 34.3 and 34.4 as if each contributes multiplicatively to the statistical weight, i.e., the energy for the conformation is tacitly assumed to be additive in contributions associated with the assignments $\phi_2 = \phi_4 = 180^\circ$. When $\phi_3 = \pm 120^\circ$, this assumption is invalid, but its effect is mitigated by the rarity of occurrence of these conformations for the bond triad.

of the interacting groups for various permitted conformations of neighboring bonds indicate the effect to be small.[92,93] A different parameter σ' is nevertheless introduced to replace σ in the expectation that σ' may be somewhat less than σ. Then

$$\mathbf{U}_3 = \begin{bmatrix} 1 & \sigma' & \sigma' \\ 0 & 0 & 0 \\ 1 & \sigma' & \sigma' \end{bmatrix} \tag{35.3}$$

The second row of $\mathbf{U}_3$ is represented as null inasmuch as its elements are made superfluous by U_2 as expressed in Eq. 35.2. The matrix $\mathbf{U}_4$ is unaffected by the substituent methyl group, and the same parameters α and β should apply therein for *trans*-PIP as for *trans*-PBD.[92]

The partition function for the repeating unit spanning the four bonds identified above is

$$z^{(\xi)} = \mathbf{U}_1 \cdots \mathbf{U}_4 \mathbf{J} \tag{36}$$

The corresponding $\mathscr{G}$ matrices may be formulated according to Eq. IV-24, and from them the normalized matrix $\mathfrak{G}^{(\xi)}$ defined by Eq. IV-71 may be obtained in the manner amply illustrated by examples in the preceding section. Values of $\langle r^2 \rangle_0$ may then be computed by resort to Eq. IV-70.

Experimental results for PBD and PIP polymers gathered by Mark[92,93] are listed in Table 6. Some of the results are tentative; others are subject to

Table 6

Experimental Results on Unperturbed Dimensions of 1,4-Diene Polymers[92,93]

	trans-PBD	*trans*-PIP[a]	*cis*-PBD	*cis*-PIP[b]
$\dfrac{\langle r^2 \rangle_0}{nl^2}$	5.8	7.35	4.9	4.7
$\dfrac{10^3 d \ln \langle r^2 \rangle_0}{dT}$, deg^{-1}	−0.6	−0.3	0.4	0.4

[a] Gutta-percha.
[b] Natural rubber.

rather large uncertainties. Comparison with calculated values therefore must be undertaken with reservations. Such comparisons are nevertheless of interest. They lend support to a set of parameters in reasonable accord with expectations on the basis of the analysis of the structure and its interactions.

The combination of parameters $\sigma = 1$, $\alpha = 0.96$, and $\beta = 0.24$ reproduces the results for *trans*-PBD in Table 6. Choice of $\sigma' = 0.54$ in conjunction with the same values of α and β reproduces the characteristic ratio for *trans*-PIP, but yields $d \ln \langle r^2 \rangle_0 / dT = -1.4 \times 10^{-3}$ deg^{-1}, which differs considerably from the tentative value -0.3×10^{-3} observed.[92] The set of parameters is within the bounds of predictions from the structure.

Analysis of the *cis* chains permits extension of the scope of correlations with experiment; only one additional parameter not related to those introduced above is required. The *cis*-PBD chain is shown in Fig. 34. Geometrical

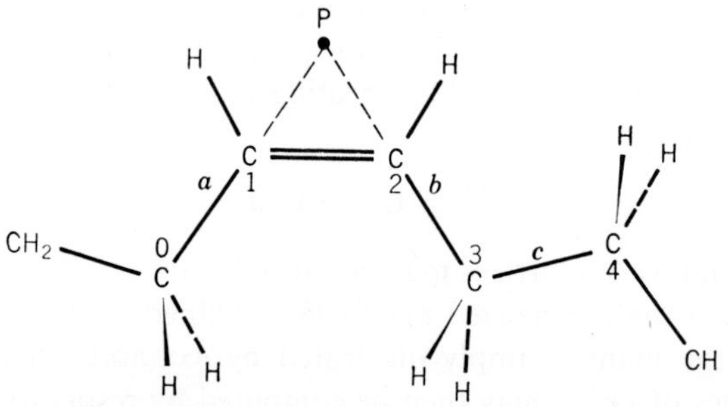

Fig. 34. The *cis*-polybutadiene (PBD) unit. Bond lengths and angles θ given in Fig. 33 are applicable. The dashed lines represent extensions of bonds 1 and 3 to form virtual bonds a and b meeting at point P, each of length 2.70 Å.

parameters are omitted from the figure inasmuch as they equate to those shown in Fig. 33. The numeration scheme defining the arbitrary sequence comprising the repeating unit has been altered, for reasons which will be apparent, from the sequence adopted in the preceding figure. Rotations about bonds 1 and 3, i.e., the pair flanking the double bond, are interdependent. *Cis* conformations about these bonds are forbidden by major steric overlaps. In this situation, rotation about bond 4, the CH_2—CH_2 bond should be essentially independent of its neighbors. Bonds 1 and 3 being coplanar, it is expedient to introduce hypothetical *virtual* bonds[96] formed by extending bonds 1 and 3 to their intersection at the point denoted by P in Fig. 34. The resulting pair of virtual bonds of length 2.70 Å, which are designated a and b, replaces bonds 1, 2, and 3. To complete the revision to a unit consisting of three bonds, two of them artificial, we designate bond 4 as bond c (see Fig. 34).

As noted above, *cis* conformations about bonds a and b are precluded. In any combination, $\pm 60°$ with $\pm 60°$, of the remaining conformations, a pair of methylenic hydrogens from groups $(CH_2)_0$ and $(CH_2)_3$ is coplanar with carbons 0, 1, 2, and 3; their distance apart is only 1.83 Å compared with 2.45 Å when situated in the *trans* conformation shown in Fig. 34. The repulsion at $\pm 60°$, $\pm 60°$ may be largely relieved if one of the bonds a or b assumes a rotation angle $\phi = 0$. Mark[93] estimates the difference in the repulsions in states $\pm 60°$, $\pm 60°$ compared with $0°$, $\pm 60°$, or $\pm 60°$, $0°$ to be about 1.2 kcal mole^{-1}.* Of course, the inherent torsional energy for the latter states is $E° \cong 2$ kcal mole^{-1}. Mark suggests that $0°$, $\pm 60°$ and $\pm 60°$, $0°$ states may nevertheless need to be considered for the *cis* chain. Assigning them a weight of ζ, we may represent bonds a and b in these terms by defining statistical weight matrices as follows†:

$$\begin{matrix} & 0° & 60° & -60° \\ \mathbf{U}_a = & [1 & 1 & 1] \end{matrix} \tag{37a}$$

$$\begin{matrix} & 0° \quad 60° \quad -60° \\ \mathbf{U}_b = & \begin{bmatrix} 0 & \zeta & \zeta \\ \zeta & 1 & 1 \\ \zeta & 1 & 1 \end{bmatrix} \end{matrix} \tag{37b}$$

Rotation about bond c being independent of the rotations of neighboring bonds, $\mathbf{U}_c$ could be expressed as a row. In order to render $\mathbf{U}_c$ conformable with $\mathbf{U}_b$, however, we present it in 3×3 form as follows:

$$\mathbf{U}_c = \begin{bmatrix} 1 & \sigma & \sigma \\ 1 & \sigma & \sigma \\ 1 & \sigma & \sigma \end{bmatrix} \tag{37c}$$

where σ is the parameter introduced in treating the *trans*-PBD chain. The same formulation should hold for *cis*-PIP, apart from replacement of σ by σ' in Eq. 37c; the same value of ζ should apply.

For the three-bond unit of either *cis* chain

$$z^{(\xi)} = \mathbf{U}_a \mathbf{U}_b \mathbf{U}_c \mathbf{J} \tag{38}$$

The remainder of the treatment proceeds along familiar lines.

Comparison of calculations carried out by Mark[93] over wide ranges of σ and ζ with the experimental results in Table 6 for *cis*-PBD suggests $\zeta = 0.05$ to 0.10 and $\sigma \cong 1.4$, which is consistent with the prediction that σ

* The importance of this repulsion is attested by the low rotational barriers observed for *cis* disubstituted ethylenes such as CH_3—CH=CH—CH_3.[94,95] In this case the resultant barrier is ca. 0.8 kcal mole^{-1}. The reduction below the value found for *trans* isomers, ca. 2.0 kcal mole^{-1}, has been attributed[94,95] to the steric repulsion cited in the text.

† Our formulation differs somewhat from Mark's,[93] but is equivalent to it. In particular, our ζ is to be identified numerically with the reciprocal of his γ.

should approximate unity. Both the characteristic ratio of *cis*-PBD and its temperature coefficient are well reproduced by this combination of parameters.[93]

For the treatment of *cis*-PIP, the parameter σ is appropriately replaced by $\sigma' = 0.54$, the value used for *trans*-PIP. The further choice of $\zeta = 0.10$ gives $\langle r^2 \rangle_0 / nl^2 = 5.2$ and $d \ln \langle r \rangle^2{}_0 / dT = 0.21 \times 10^{-3}$ deg^{-1}, in reasonable accord with the experimental values for natural rubber in Table 6.[92,93]

The greater value of C_∞ for *trans* as compared with *cis*-PBD is of course due to the more extended form of the *trans* configuration about the double bond. That the difference is only 18% is traceable to the inaccessibility of compact *cis* conformations to the single bonds adjoining the double bond in the *cis* diene polymers, as Mark[92] has pointed out. Whereas the rotational states about the CH_2—CH_2 bond are nearly equally weighted in *trans*-PBD, the *gauche* states are disfavored in *trans*-PIP (i.e., $\sigma' < \sigma$). It is principally on this account that the characteristic ratio for *trans*-PIP exceeds that of *trans*-PBD. The *trans*-PIP and *cis*-PIP chains accordingly present the most striking contrast. This may be viewed as a consequence of the greater prevalence of *trans* states amongst CH_2—CH_2 bonds of *trans*-PIP, since $\sigma' < 1$. Bonds in *trans* conformations conduce extended spatial configurations. The preponderance of the bonds in *trans*-PIP being in $\phi = 0$ or $\phi = \pm 60°$ conformations, the effect of "isomerizing" the double bonds from $\phi = 0$ to $\phi = 180°$ (*cis*) is particularly marked.[92] Hence, the dimensions of PIP chains are more sensitive than PBD to the configuration of the double bond.

14. POLYISOBUTYLENE AND ITS ANALOGS

The presence of a substituent in the repeating unit increases the interactions within the chain molecule, and if the substituent is much larger than hydrogen the number of configurations available to the chain may be greatly reduced as a result thereof. The vinyl polymers

$$(\text{—}CH_2\underset{}{\overset{\overset{\displaystyle R}{|}}{C}}H\text{—})_x$$

are illustrative. Inasmuch as their units are asymmetric, the treatment of these chains is reserved for the following chapter. If alternate skeletal atoms bear two substituents which are substantially larger than hydrogen, steric repulsions are prevalent in all conformations of the chain units, and these may obliterate all other contributions to the rotational potentials. Polyisobutylene, $[\text{—}CH_2\text{—}C(CH_3)_2\text{—}]_x$, and poly(vinylidene chloride), $(\text{—}CH_2\text{—}CCl_2\text{—})_x$, are illustrative of highly substituted chains of this character.

The structure of polymers of this class is shown in Fig. 35. The chains are symmetrical provided that both substituents on a given skeletal atom are the same, as is the case in the examples cited above. The considerations which follow are qualitatively applicable also to chains bearing dissimilar substituents, e.g., to poly(methyl methacrylate)

$$\left[-CH_2-\underset{\displaystyle COOCH_3}{\overset{\displaystyle CH_3}{\underset{|}{\overset{|}{C}}}}- \right]_x$$

Chains of this character are, of course, asymmetric.

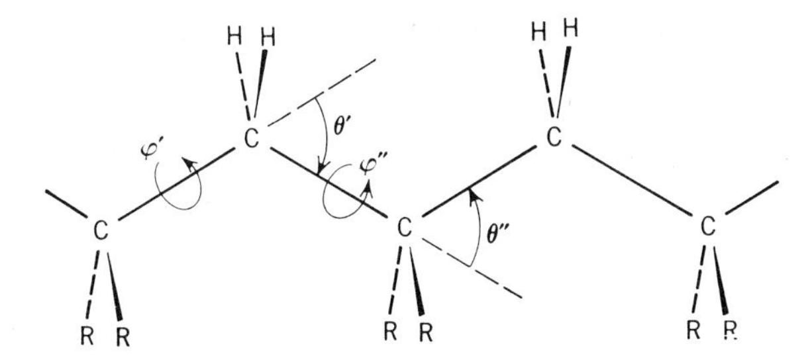

Fig. 35. Planar conformation of a disubstituted vinyl polymer chain.

The van der Waals radius of Cl is about 1.8 Å; hence, the overlap of nearest-neighbor pairs in the planar conformation of poly(vinylidene chloride) is about 1.0 Å. The repulsions can be relieved by rotations of the skeletal bonds, but they cannot be eliminated altogether. The crystalline conformation of this polymer is nonplanar. Miyazawa and Ideguchi[97] have suggested a conformation, based on the far infrared spectrum and x-ray crystallographic data, having the bond rotation sequence tg^+tg^- with *gauche* angles of 100–110°. In this conformation only one pair of chlorine atoms between neighboring CCl_2 groups is subject to a large overlap, the Cl···Cl distance for this pair being 2.35 Å. Although the repulsion at this distance is large, a more favorable conformation may not be available to the chain.

Steric repulsions are much more severe in polyisobutylene (PIB) owing to the greater size of the CH_3 group compared with Cl. They can be diminished somewhat by bond rotations, but the relief thus gained is only partial. The severity of these interactions is reflected in the thermochemical energy of formation of the PIB chain which is increased about 9 kcal per mole of C_4H_8 on their account.[98]

X-ray crystallographic studies of oriented crystalline PIB[99,100] reveal a helical conformation having eight repeating units in five turns (8_5 helix).

An average skeletal bond angle $\pi - \theta = 114°$ and uniform rotations $\phi = 82°$ would comply with this observation deduced from the layer-line repeat distance by Liquori.[99] Rotations of negative sign generate the corresponding left-handed helix.

In absence of the intermolecular constraints imposed by the uniform array of the crystal, perpetuation of bond rotations of the same sense from one unit to the next is not required. The regular helix of the crystal can be converted to a tortuous random coil by reversing signs of rotation at frequent intervals along the chain without necessarily altering the magnitudes of the rotations. Ptitsyn and Sharanov[101] pointed out that a satisfactory interpretation of the configuration of the PIB chain can be realized by choosing $\phi = \pm 82°$ for every bond, subject to the following conditions: (*1*) both skeletal bonds of the pair between substituted chain atoms must adopt the same sign, and (*2*) signs of neighboring pairs must be essentially uncorrelated, i.e., statistical weights for $(++)\,(--)$ and $(++)\,(++)$ sequences are about equal, and hence occur with nearly equal frequency.[102] Examination of models lends strong support for the validity of both of these conditions, provided, however, that the limitation to rotations of a single magnitude, 82°, can be justified. Neighbor dependence may thus be confined to distinct bond pairs, each such pair being independent of its neighbors. It is easily shown that[101]

$$\left(\frac{\langle r^2\rangle_0}{nl^2}\right)_\infty = 2\left(\frac{1+\cos\theta}{1-\cos\theta}\right)\left(\frac{1+\cos\phi}{1-\cos\phi}\right) \tag{39}$$

under these circumstances. With $\theta = 66°$ and $\phi = 82°$, we have $(\langle r^2\rangle_0/nl^2)_\infty = 6.3$, in satisfactory agreement with the experimental value of about 6.6. Condition (2) above connotes a temperature coefficient of zero, and this is also in approximate accord with experiments showing $d \ln \langle r^2\rangle_0/dT$ to be small (see Table II-2).

The foregoing account of the PIB chain appears on closer examination to be oversimplified. DeSantis et al.[100] calculated the conformational energy associated with rotations ϕ' and ϕ'' about a pair of skeletal bonds between substituted atoms. In addition to energy minima near the two states used above, they found four others of similar energy at (25°, 135°), at (135°, 25°), at (−25°, −135°), and at (−135°, −25°). Although such calculations are subject to uncertainty, they justifiably call into question the premise that the conformations occurring in the crystalline state are exclusively representative of the low energy forms. In a chain wherein steric interactions are severe this may not be the case.

A further parameter of concern in such chains is the bond angle $\pi - \theta$, assumed constant above. The steric repulsion energy (ca. 9 kcal mole^{-1} of units in PIB) is easily large enough to distort θ by 10° or more. Bunn and

Holmes[32] have concluded from the x-ray diffraction pattern of stretched, crystalline PIB that the bond angles are in fact distorted. They suggest $\pi - \theta' = 126°$ and $\pi - \theta'' = 107°$ at $—CH_2—$ and at $—C(CH_3)_2—$, respectively, with rotation angles of $\phi' = 51°$ and $\phi'' = 102.5°$. These parameters reproduce an 8_5 helix in conformity with the observed fiber repeat distance. Birshtein, Ptitsyn, and Sokolova[70] calculated the characteristic ratio using these parameters, based on the interpretation of the x-ray fiber diagram by Bunn and Holmes,[32] and assuming as above that the signs of the neighboring bond pairs are uncorrelated, i.e., that each bond pair adopts one of the four eligible rotational states without bias by the state of its neighbor. They found $\langle r^2 \rangle_0/nl^2 = 12$, which is much larger than the value observed by experiment. Rotations about neighboring pairs of bonds are interdependent for this set of conformations, and the assumption of their independence may readily lead to a large error.

The interpretation of the configurations of these highly substituted chains must await further inquiry. It is apparent that such chains depart in two important respects from others thus far considered: (*1*) the intrinsic bond torsional potentials are obliterated by the copious repulsions between nonbonded atoms prevalent in all conformations, and (*2*) the assumption of constancy of the angles θ is not a legitimate approximation. Despite the large steric interactions, the energy minima probably are not sharply defined, and there may be competing conformations of similar energy. Both of these factors impart to the chain a degree of flexibility which might not otherwise be expected in view of the large steric hindrances. The configurations of such chains may therefore be tortuous and irregular, the severity of steric interactions notwithstanding.

REFERENCES

1. T. M. Birshtein and O. B. Ptitsyn, *Conformations of Macromolecules* (translated from the 1964 Russian ed., S. N. Timasheff and M. J. Timasheff), Interscience, New York, 1966, Chap. 2.
2. E. A. Mason and M. M. Kreevoy, *J. Am. Chem. Soc.*, **77**, 5808 (1955); E. A. Mason, *J. Chem. Phys.*, **23**, 49 (1955).
3. T. L. Hill, *J. Chem. Phys.*, **16**, 399, 938 (1948); L. S. Bartell, *ibid.*, **32**, 827 (1960); A. I. Kitaigorodskii, *Dokl. Akad. Nauk SSSR*, **137**, 116 (1961); N. P. Borisova and M. V. Volkenstein, *Zh. Strukt. Khim.*, **2**, 346 (1961).
4. R. A. Scott and H. A. Scheraga, *J. Chem. Phys.*, **42**, 2209 (1965).
5. N. P. Borisova and M. V. Volkenstein, *Zh. Strukt. Khim.*, **2**, 469 (1961).
6. J. B. Hendrickson, *J. Am. Chem. Soc.*, **83**, 4537 (1961).
7. J. Ketelaar, *Chemical Constitution*, Elsevier, New York, 1958, p. 91.
8. K. Pitzer, *Advan. Chem. Phys.*, **2**, 49 (1959).
9. H. J. M. Bowen and L. E. Sutton, *Tables of Interatomic Distances and Configurations in Molecules and Ions*, The Chemical Society, London, 1958; *Supplement*, 1965.

10. H. M. M. Shearer and V. Vand, *Acta Cryst.*, **9**, 379 (1956).
11. R. A. Bonham, L. S. Bartell, and D. A. Kohl, *J. Am. Chem. Soc.*, **81**, 4765 (1959).
12. D. R. Lide, Jr., *J. Chem. Phys.*, **33**, 1514 (1960).
13. N. P. Borisova, *Vysokomolekul. Soedin.*, **6**, 135 (1964).
14. A. Abe, R. L. Jernigan, and P. J. Flory, *J. Am. Chem. Soc.*, **88**, 631 (1966).
15. R. A. Scott and H. A. Scheraga, *J. Chem. Phys.*, **44**, 3054 (1966).
16. A. Bondi, *J. Phys. Chem.*, **68**, 441 (1964).
17. D. A. Brant, W. G. Miller, and P. J. Flory, *J. Mol. Biol.*, **23**, 47 (1967).
18. K. Kuchitsu, *J. Chem. Soc. Japan*, **32**, 748 (1959).
19. R. A. Bonham and L. S. Bartell, *J. Am. Chem. Soc.*, **81**, 3491 (1959).
20. C. A. J. Hoeve, *J. Chem. Phys.*, **35**, 1266 (1961).
21. S. Mizushima and H. Okazaki, *J. Am. Chem. Soc.*, **71**, 3411 (1949); N. Sheppard and G. J. Szasz, *J. Chem. Phys.*, **17**, 86 (1949).
22. W. B. Person and G. C. Pimentel, *J. Am. Chem. Soc.*, **75**, 532 (1953).
23. R. L. Jernigan and P. J. Flory, unpublished.
24. P. J. Flory and R. L. Jernigan, *J. Chem. Phys.*, **42**, 3509 (1965).
25. H. J. G. Hayman and I. Eliezer, *J. Chem. Phys.*, **35**, 644 (1961); **28**, 890 (1958).
26. W. J. Leonard, R. L. Jernigan, and P. J. Flory, *J. Chem. Phys.*, **43**, 2256 (1965).
27. T. Yoshino and H. J. Bernstein, *Can. J. Chem.*, **35**, 339 (1957).
28. C. Komaki, I. Ichishima, K. Kuratani, T. Miyazawa, T. Shimanouchi, and M. Mizushima, *Bull. Chem. Soc. Japan.* **28**, 330 (1955); Yu. A. Pentin and V. M. Tatevskii, *Vestn. Mosk. Univ. Ser. Fiz. Mat. i Estestven. Nauk No. 2*, **3**, 63 (1955); *Dokl. Akad. Nauk SSSR*, **108**, 290 (1956); *Zh. Fiz. Khim.*, **31**, 1830 (1957).
29. M. Iwasaki, *J. Polymer Sci. A.*, **1**, 1099 (1963).
30. T. W. Bates, *Trans. Faraday Soc.*, **63**, 1825 (1967).
31. T. W. Bates and W. H. Stockmayer, *J. Chem. Phys.*, **45**, 2321 (1966); *Macromolecules*, **1**, 12 (1968).
32. C. W. Bunn and E. R. Howells, *Nature*, **174**, 549 (1954); C. W. Bunn and D. R. Holmes, *Discussions Faraday Soc.*, **25**, 95 (1958).
33. E. L. Pace and A. C. Plaush, *J. Chem. Phys.*, **47**, 38 (1967).
34. W. H. Stockmayer and T. W. Bates, *Macromolecules*, **1**, 17 (1968).
35. P. J. Flory, *Proc. Roy. Soc.* (*London*), **A234**, 60 (1956).
36. J. A. Semlyen, *Trans. Faraday Soc.*, **63**, 743 (1967); **64**, 1396 (1968).
37. S. C. Abrahams, *Quart. Rev.* **10**, 407 (1956).
38. L. Pauling, *Proc. Natl. Acad. Sci.*, **35**, 495 (1949); L. Pauling, *The Nature of the Chemical Bond*, 3rd. ed., Cornell Univ. Press, Ithaca, N.Y., 1960.
39. G. Bergson, *Arkiv Kemi*, **13**, 11 (1959).
40. F. Tuinstra, *Acta Cryst.*, **20**, 341 (1966).
41. G. Gee, *Trans. Faraday Soc.*, **48**, 515 (1952); G. Gee, in *Inorganic Polymers*, The Chemical Society, London, 1961.
42. F. Fairbrother, G. Gee, and G. T. Merrall, *J. Polymer Sci.*, **16**, 459 (1955).
43. A. V. Tobolsky and A. Eisenberg, *J. Am. Chem. Soc.*, **81**, 780 (1959).
44. T. Doi, *Rev. Phys. Chem. Japan*, **35**, 18 (1965).
45. A. Eisenberg and A. V. Tobolsky, *J. Polymer Sci.*, **46**, 19 (1960).
46. H. Jacobson and W. H. Stockmayer, *J. Chem. Phys.*, **18**, 1600 (1950).
47. P. J. Flory and J. E. Mark, *Makromol. Chem.*, **75**, 11 (1964).
48. J. A. Semlyen and P. J. Flory, unpublished calculations.
49. T. Uchida, Y. Kurita, and M. Kubo, *J. Polymer Sci.*, **19**, 365 (1956); S. Mizushima, Y. Morino, and M. Kubo, *Physik. Z.*, **38**, 459 (1937).
50. M. L. Huggins, *J. Chem. Phys.*, **13**, 37 (1945).

51. T. Uchida and H. Tadokoro, *J. Polymer Sci. A-2*, **5**, 63 (1967); H. Tadokoro, T. Yasumoto, S. Murahashi, and I. Nitta, *J. Polymer Sci.*, **44**, 266 (1960); see also G. Carazzolo, *J. Polymer Sci. A*, **1**, 1573 (1963).
52. G. Carazzolo and M. Mammi, *J. Polymer Sci. A*, **1**, 965 (1963).
53. V. Kokle and F. W. Billmeyer, *J. Polymer Sci. B*, **3**, 47 (1965).
54. W. H. Stockmayer and L. L. Chan, *J. Polymer Sci. A-2*, **4**, 437 (1966).
55. J. E. Mark and P. J. Flory, *J. Am. Chem. Soc.*, **87**, 1415 (1965).
56. J. E. Mark, private communication.
57. J. E. Mark and P. J. Flory, *J. Am. Chem. Soc.*, **88**, 3702 (1966).
58. J. Marchal and H. Benoit, *J. Chim. Phys.*, **52**, 818 (1955); *J. Polymer Sci.*, **23**, 223 (1957).
59. K. Bak, G. Elefante, and J. E. Mark, *J. Phys. Chem.*, **71**, 4007 (1967).
60. A. Kotera, K. Suzuki, K. Matsumura, T. Nakano, T. Oyama, and U. Kambayashi, *Bull. Chem. Soc. Japan*, **35**, 797 (1962).
61. T. Uchida, Y. Kurita, N. Koizumi, and M. Kubo, *J. Polymer Sci.*, **21**, 313 (1956).
62. H. Tadokoro, Y. Chatani, T. Yoshihara, S. Tahara, and S. Murahashi, *Makromol. Chem.*, **73**, 109 (1964).
63. E. Hirota, *J. Chem. Phys.*, **37**, 283 (1962).
64. J. E. Mark, *J. Am Chem. Soc.*, **88**, 3708 (1966).
65. J. E. Mark, *J. Polymer Sci. B*, **4**, 825 (1966).
66. K. Yamamoto, A. Teramoto, and H. Fujita, *Polymer*, **7**, 267 (1966).
67. M. Kurata, H. Utiyama, and K. Kamada, *Makromol. Chem.*, **88**, 281 (1965).
68. H. Tadokoro, Y. Takahashi, and Y. Chatani, IUPAC International Symposium, Japan, 1966.
69. P. J. Flory, V. Crescenzi, and J. E. Mark, *J. Am. Chem. Soc.*, **86**, 146 (1964).
70. T. M. Birshtein, O. B. Ptitsyn, and E. A. Sokolova, *Vyosokomolekul. Soedin.*, **1**, 852 (1959).
71. D. W. Scott et al., *J. Phys. Chem.*, **65**, 1320 (1961).
72. F. Liebau, *Acta Cryst.*, **14**, 1103 (1961).
73. G. Damaschun, *Kolloid-Z.*, **180**, 65 (1962).
74. T. Shimanouchi and S. Mizushima, *J. Chem. Phys.*, **23**, 707 (1955).
75. J. F. Brown, Jr., and G. M. J. Slusarczuk, *J. Am. Chem. Soc.*, **87**, 931 (1965).
76. P. J. Flory and J. A. Semlyen, *J. Am. Chem. Soc.*, **88**, 3209 (1966).
77. J. A. Semlyen and P. J. Flory, *Trans. Faraday Soc.*, **62**, 2622 (1966).
78. U. P. Strauss and P. Ander, *J. Phys. Chem.*, **66**, 2235 (1962); U. P. Strauss and P. L. Wineman, *J. Am. Chem. Soc.*, **80**, 2366 (1958).
79. R. J. Kurland and E. B. Wilson, Jr., *J. Chem. Phys.*, **27**, 585 (1957).
80. R. F. Curl, Jr., *J. Chem. Phys.*, **30**, 1529 (1959).
81. J. M. Riveros and E. B. Wilson, Jr., *J. Chem. Phys.*, **46**, 4605 (1967).
82. L. A. LaPlanche and M. T. Rogers, *J. Am. Chem. Soc.*, **86**, 337 (1964).
83. A. D. Williams, unpublished results.
84. R. M. Meighan and R. H. Cole, *J. Phys. Chem.*, **68**, 503 (1964).
85. C. P. Smyth, *Dielectric Behavior and Structure*, McGraw-Hill, New York, 1955.
86. D. A. Brant and P. J. Flory, *J. Am. Chem. Soc.*, **87**, 2791 (1965).
87. P. J. Flory and A. D. Williams, *J. Polymer Sci. A-2*, **5**, 399 (1967).
88. P. R. Saunders, *J. Polymer Sci. A*, **2**, 3765 (1964).
89. A. D. Williams and P. J. Flory, *J. Polymer Sci. A-2*, **5**, 417 (1967).
90. W. A. Lanka, unpublished results, quoted by W. R. Krigbaum, *J. Polymer Sci.*, **28**, 213 (1958).
91. M. L. Wallach, *Makromol. Chem.*, **103**, 19 (1967).

92. J. E. Mark, *J. Am. Chem. Soc.*, **89**, 6829 (1967).
93. J. E. Mark, *J. Am. Chem. Soc.*, **88**, 4354 (1966).
94. J. E. Kilpatrick and K. S. Pitzer, J. *Res. Natl. Bur. Std.*, **37**, 163 (1946).
95. D. R. Lide, Jr., and D. E. Mann, *J. Chem. Phys.*, **27**, 868 (1957); D. R. Lide, Jr., *Ann. Rev. Phys. Chem.*, **15**, 225 (1964).
96. H. Benoit, *J. Polymer Sci.*, **3**, 376 (1948).
97. T. Miyazawa and Y. Ideguchi, *J. Polymer Sci. B*, **3**, 541 (1965).
98. P. J. Flory, *Principles of Polymer Chemistry*, Cornell Univ. Press, Ithaca, N.Y., 1953, p. 252.
99. A. M. Liquori, *Acta Cryst.*, **8**, 345 (1955).
100. P. DeSantis, E. Giglio, A. M. Liquori, and A. Ripamonti, *J. Polymer Sci. A*, **1**, 1383 (1963).
101. O. B. Ptitsyn and I. A. Sharanov, *Zh. Tekhn. Fiz.*, **27**, 2744, 2762 (1957).
102. C. A. J. Hoeve, *J. Chem. Phys.*, **32**, 888 (1960).

CHAPTER VI

Asymmetric Vinyl Chains

1. STEREOCHEMICAL CONFIGURATIONS

The presence of an asymmetric center in a chain molecule introduces a distinction, not otherwise present, between right- and left-handed rotations. The interactions precipitated by rotations ϕ and $-\phi$ about a skeletal bond near the asymmetric center will differ, in general, and the two states thus related will not occur with equal frequency, as is the case for symmetric chains.

Of the various types of chain molecules comprising repeat units having one or more asymmetric centers in the stereochemical sense of this term, we shall confine our attention to two classes, namely, the vinyl polymers

$$H(\text{—}CH_2\text{—}CHR)_{n/2}\text{—}CH_3 \tag{A}$$

considered here and the polypeptides to be discussed in the following chapter. Amongst typical vinyl polymers, R may be Cl, CH_3, C_2H_5, C_6H_5, $COOCH_3$, etc.

A portion of a vinyl chain is shown in its planar *trans*, or fully extended form in Fig. 1. The chain shown is an atactic one in which successive R groups occur more or less indiscriminately on the same and on opposite sides of the plane of the figure. The position of a given R group in this respect is determined (for the chosen orientation of the fully extended chain; see below) by the stereochemical configuration of the asymmetric carbon (CHR) to which it is attached. The stereochemical configuration of the asymmetric atom is a feature of the structure which is invariant to conformational changes. For later convenience we have arbitrarily designated those carbon atoms bearing R groups which occur in front of the plane in Fig. 1 as "*d*" centers, and those with R behind it as "*l*."

It is important to observe that the dispositions of the C—R and C—H bonds of each CHR group are reversed with respect to the plane of the figure by a rotation of 180° about a vertical axis in that plane; R groups occurring in front of the plane are thereby placed behind it, and vice versa. The ends of the chain are reversed in position, and the sequence of units and bonds, taken in left-to-right order, is likewise reversed by this operation. The asymmetric *d* centers are converted to *l* and the *l* to *d*. These designations are therefore arbitrary.* They take on absolute significance only if the ends of the chain are differentiable, a distinction which we assume to be absent or trivial if indeed present. The relationship of the stereochemical configuration of a given asymmetric center to its neighbors in the chain *is* significant, however. Thus, pairs of like designation, i.e., *dd* and *ll*, though interconvertible by the axis rotation cited above, are not convertible to an unlike pair, *dl* or *ld*, by any symmetry operation. We call the former isotactic pairs (*ll* and *dd*) and the latter syndiotactic pairs (*dl* and *ld*), following the terminology introduced by Natta and his co-workers.[1] In particular, we shall designate a chain characterized by

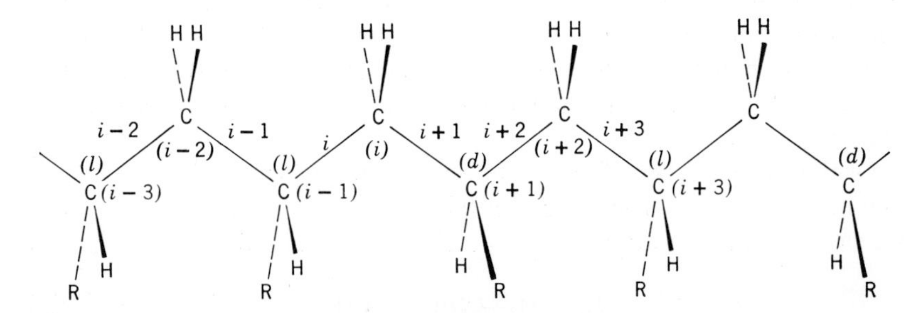

Fig. 1. Portion of an atactic vinyl chain. Asymmetric centers are given the designations *d* and *l* according to the arbitrary convention adopted in the text. Serial indexes referring to chain atoms are shown in parentheses.

indefinitely long sequences of units sharing the same stereochemical configuration of their asymmetric centers as an *isotactic* chain; one in which the configurations of these centers alternate with perfect regularity (or nearly so) will be called a *syndiotactic* chain.[1] Such chains are represented in Figs. 2 and 3, respectively. It must be borne in mind that an isotactic *d* chain, or sequence, becomes an isotactic *l* chain, or sequence, under the rotation operation identified above. A *dl* pair is invariant to rotation about a vertical axis in Fig. 3; by such rotation the *d* center is changed to *l* and the *l* to *d*, but simultaneously their order in the sequence is reversed, thus restoring the *dl* character of the pair. The pairs *dl* and *ld* differ intrinsically, being interconvertible only by an imaginary operation like mirror reflection.

The artificiality of the distinction between *d* and *l* centers in vinyl chains could be avoided, insofar as neighbor pairs (dyads) are concerned, through adoption of well-established stereochemical terminology of organic

* Asymmetric centers differing in an absolute sense will be designated D and L. See Chapter VII.

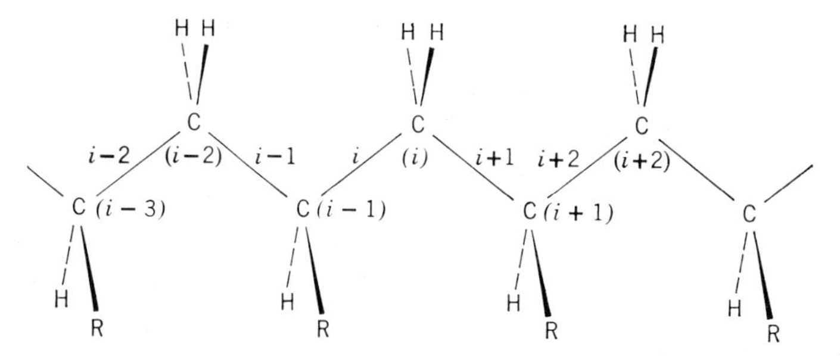

Fig. 2. An isotactic chain shown in the planar *trans* conformation, and oriented to represent the asymmetric centers in the (arbitrary) *d* configuration.

chemistry according to which a *dd* or *ll* pair is called *meso* (physically identical pair) and a *dl* or *ld* pair racemic (mirror image pair). We shall in due course show that vinyl chains of any specified stereosequence can be treated in terms of matrices for *meso* and racemic dyads without reference to *d* or *l* configurations of the asymmetric centers. At the outset, however, it is expedient to adopt the artifice of arbitrarily selecting one of the terminal skeletal atoms of the chain molecule as atom zero whereupon arbitrariness in the distinction between *d* and *l* is removed. The analysis must, of course, meet the condition of equivalence of result irrespective of the end of the chain chosen for the initial end.

The presence of the substituent R on alternate atoms of the chain skeleton proliferates steric interactions. Although increased in number, the interactions in the vinyl chains considered here have their counterparts in the *n*-alkanes treated in Sections 2 and 3 of the preceding chapter. Results presented there can consequently be drawn upon for purposes of the following analysis.

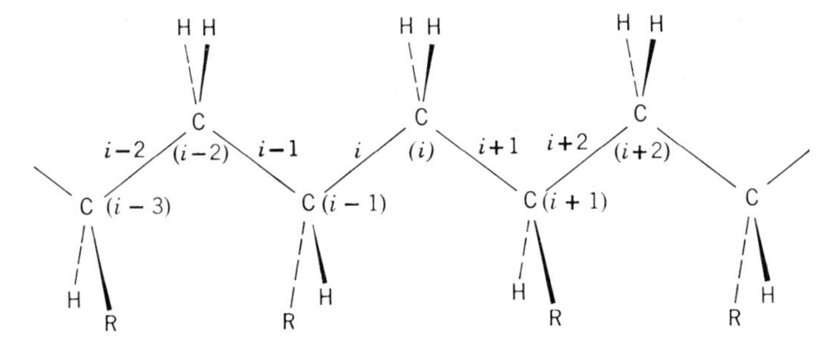

Fig. 3. A syndiotactic vinyl chain.

2. INTERACTIONS OF FIRST ORDER*

Having arbitrarily selected one terminus of the chain for designation as atom zero, we consider a pair of bonds such as i and $i + 1$ in Fig. 1 or 2 between two substituted carbons, namely, atoms $i - 1$ and $i + 1$. Let both of these centers be *d*, as in Fig. 2. Atoms and groups involved in first-order (i.e., three-bond) interactions dependent upon one rotation angle only are shown in Figs. 4*a* and 4*b*; the former is a view in the direction of bond i, the latter in the direction of bond $i + 1$. In the approximation that CH_2 and CH may be regarded as structureless domains, the first-order interactions associated with ϕ_i and ϕ_{i+1}, respectively, do not depend on

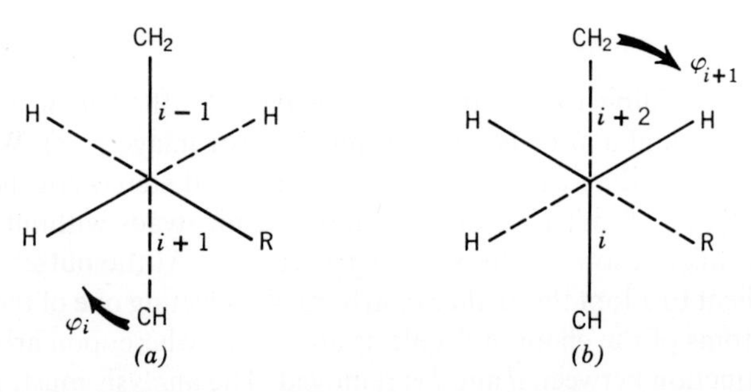

Fig. 4. Diagrams illustrating nonbonded interactions dependent on a single rotation angle: (*a*) ϕ_i, and (*b*) ϕ_{i+1}. The bonds in the background, i.e., those attached to the skeletal atom of higher index (i in Fig. 4*a* and $i + 1$ in Fig. 4*b*), are shown as dashed lines. Both asymmetric centers are represented in the *d* configuration.

rotations about adjoining bonds. In the *trans* conformation shown in Fig. 4*a* the CH group attached to bond $i + 1$ is close to the substituent R; in the g^+ state resulting from clockwise rotation of CH and the two H atoms subtended by dashed lines, the CH group is moved away from R and into corresponding proximity with CH_2; and in the g^- state CH is situated between CH_2 and R, where it is within range of interaction with both of them.

A statistical weight σ may be chosen to represent the contribution from three-bond interaction in the g^+ conformation, in keeping with the scheme applied to the *n*-alkanes and to polymethylene. To the extent that CH may be considered the equivalent of CH_2, the interaction in this conformation corresponds to that for the g^+ or the g^- state in the *n*-alkanes treated in the preceding chapter. The analogous interaction between CH and R in the

* The treatment of vinyl polymer chains given in this section and the ones following has been taken in large part from reference 2.

trans conformation will be given the weight $\sigma\eta$. If R is equivalent to CH_2, then $\eta = 1$; otherwise, η represents the ratio of the statistical weight for the three-bond interaction involving R to that involving CH_2. Finally, we let $\sigma\tau$ denote the statistical weight of the g^- conformation in which CH is involved in interactions with *both* CH_2 and R. Dividing each of these statistical weights by σ we have η, 1, and τ for the set representing first-order interactions associated with t, g^+, and g^- states, respectively, about bond i. These may be incorporated in a diagonal matrix

$$\mathbf{D}'_d = \text{diag}\,(\eta, 1, \tau) \tag{1}$$

following the procedure introduced in the preceding chapter. The subscript d designates the asymmetry of CHR group $i - 1$.

From analysis of thermodynamic functions and of Raman and infrared spectra of 2-methylbutane, Scott and co-workers[3] concluded that the conformation of C_s symmetry is at least several kilocalories per mole less stable than the C_1 form (see Fig. 5). Dipole moments for the analogous

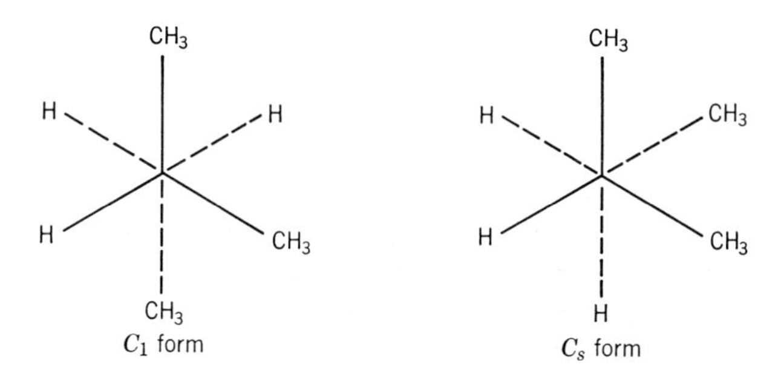

Fig. 5. Conformers of 2-methylbutane.

chlorinated ethane, 1,1,2-trichloroethane, determined over a range of temperatures by Thomas and Gwinn,[4] suggested an energy difference of about 4 kcal mole^{-1} between the corresponding conformers. Coulombic repulsions between C—Cl dipoles may contribute appreciably to the preference for the C_1 form, but if indeed the energy difference between them is of this magnitude then steric repulsions between Cl substituents must be invoked as the major factor. Clearly, steric repulsions must be responsible for the similar preference of 2-methylbutane for the C_1 conformation. The three-bond repulsion in the C_1 form, corresponding to the t and the g^+ forms in Fig. 4*a*, can be relieved by rotation to an angle ϕ differing somewhat, perhaps by 10° (see below), from the symmetrical conformation shown. Such relief is not possible in the C_s form (Fig. 5), corresponding to g^- in Fig. 4*a*, where a terminal methyl group suffers repulsion from each of two groups acting in opposing directions. Consistent with this explanation is the observation that the energy difference between the two conformers of 2,3-dimethylbutane[3] shown in Fig. 6 does not exceed 0.1 kcal mole^{-1}. Here the steric interactions between substituents cannot be relieved by rotational adjustment in either form. The energy difference between the corresponding conformers of 1,1,2,2-tetrachloroethane is likewise small.[4]

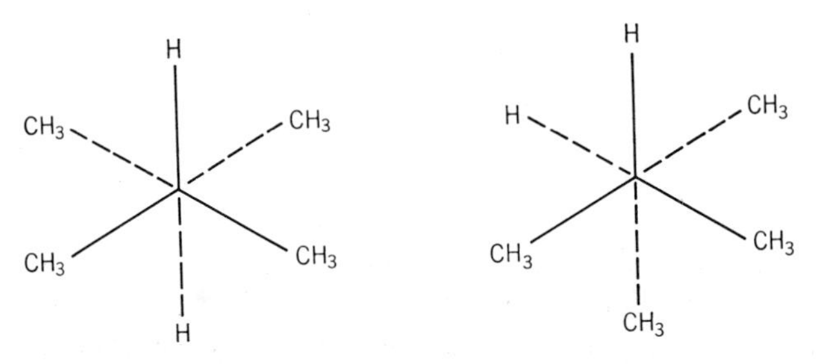

Fig. 6. Conformers of 2,3-dimethylbutane.

Other evidence suggests that the estimates cited above for the energy difference between the C_s and C_1 forms of 2-methylbutane may be too large. The Raman spectrum of the liquid and crystalline hydrocarbon, investigated by Brown and Sheppard,[5] favors the presence of two coexisting conformers in the liquid even at its freezing point. The conformational energies of the C_s and C_1 conformers estimated[2,6] by the methods described in the preceding chapter point to a difference of about 1 kcal mole^{-1}. In light of these somewhat conflicting observations, we can only conclude that the statistical weight τ, representing the effect of repulsions in the g^- state for bond i shown in Fig. 4*a* relative to the repulsions obtaining in the g^+ state, should be smaller than σ. A value in the range $0.1 < \tau < 0.4$ at ordinary temperature seems indicated, rather than $\tau \leq 0.01$ as implied by the analysis of thermodynamic data on 2-methylbutane.[3]

A corresponding analysis of the rotations about bond $i+1$ in Fig. 4*b* leads to

$$\mathbf{D}_d'' = \text{diag}\,(\eta, \tau, 1) \tag{2}$$

Here the subscript d denotes the configuration of CHR group $i+1$. Thus, the subscript applicable to $\mathbf{D}'$ in Eq. 1 refers to the CHR group at the left of the dyad and the subscript applied to $\mathbf{D}''$ in Eq. 2 to the CHR group on the right. For l units it will be apparent that

$$\mathbf{D}_l' = \text{diag}\,(\eta, \tau, 1) \tag{3}$$

$$\mathbf{D}_l'' = \text{diag}\,(\eta, 1, \tau) \tag{4}$$

The relations

$$\mathbf{D}'_d = \mathbf{D}''_l$$
$$\mathbf{D}'_l = \mathbf{D}''_d \qquad (5)$$

are consistent with invariance to end-for-end rotation of the chain.

Repulsions between the sterically interacting groups may be expected to alter the positions of the rotational minima after the manner adduced for polymethylene chains in the preceding chapter. As is apparent from Fig. 4*a*, the t and g^+ states about bond i may be shifted on this account to $\phi' = \Delta\phi$ and $2\pi/3 - \Delta\phi$, respectively, $\Delta\phi$ being taken positive, i.e., the angles ϕ' for these two minima are brought somewhat closer to one another. In the approximation of steric equivalence of CH_2 and R, the position of g^- should remain unchanged at $-2\pi/3$ for the reasons given above. In the same approximation, the t and g^- rotational states about bond $i + 1$ (Fig. 4*b*) should be shifted to $\phi''_t = -\Delta\phi$, and to $\phi''_{g-} = -(2\pi/3 - \Delta\phi)$. Alterations of the positions of the various rotational minima by three-bond interactions may thus be specified by a single parameter $\Delta\phi$ in the approximation of steric equivalence of R and CH_2. These positions are represented by*

$$\begin{aligned} \phi'_t &= \Delta\phi \\ \phi'_{g+} &= 2\pi/3 - \Delta\phi \\ \phi'_{g-} &= -2\pi/3 \end{aligned} \qquad (6)$$

and

$$\begin{aligned} \phi''_t &= -\Delta\phi \\ \phi''_{g+} &= 2\pi/3 \\ \phi''_{g-} &= -2\pi/3 + \Delta\phi \end{aligned} \qquad (7)$$

* The assumption that the positions of the unperturbed *trans* and *gauche* states occur at equal intervals of 120° is strictly valid only if all bond angles are tetrahedral. The CCC angles in isobutane[7] are 111°, and the *skeletal* bonds in isotactic poly-α-olefines have been estimated[8] to be 114.5°. If *all* CCC angles are taken to be 112° in polypropylene, then geometrical requirements about secondary and tertiary carbons of the skeleton are met by taking[2]

$$\angle CCH = 106.8° \text{ at tertiary carbons}$$

$$\angle CCH = \angle HCH = 110° \text{ at secondary carbons.}$$

If symmetrical staggering of bonds dictates the location of the *gauche* rotational minima, they will then occur at $\phi = \pm 126.8°$ instead of at $\pm 120°$. In interests of simplicity, we proceed in the approximation that the *gauche* states would occur at $\pm 120°$ if unperturbed by three- and four-bond interactions. That is, the positions of the various rotational states are treated under the simplifying assumption of tetrahedral symmetry. This assumption will not preclude adoption of a skeletal bond angle $\pi - \theta$ differing from the symmetrical angle in numerical calculations of the characteristic ratio.

for d configurations of the asymmetric centers. For l centers, the sets of values for ϕ' and ϕ'' are to be interchanged in correspondence with the symmetry rules expressed in Eq. 5. The magnitude of $\Delta\phi$ eludes reliable estimation.[2] Analysis of first-order interactions suggests a value less than 10°. Consideration of second-order interactions points to a somewhat larger value, however (see Sect. 3).

3. INTERACTIONS OF SECOND ORDER

Interactions between groups separated by four bonds play a role of the foremost importance in vinyl chains. We first consider those associated with rotations about a bond pair such as $i-1$, i in Figs. 1, 2, and 3, i.e., a bond pair flanking a CHR member of the chain. The relevant four-bond interaction dependent upon ϕ_{i-1} and ϕ_i involves the pair of CH groups $i-3$ and $i+1$. The associated statistical weight factor will be represented by ω with the implication that this interaction should resemble that of CH_2 with CH_2 in the g^+g^- conformation of PM, for which the same symbol is used in Chapter V. The dependence of the steric overlap of the two CH groups on the rotation angles is independent of the stereochemical configuration of the CHR member, and of its neighbors as well. Hence these interactions may be expressed in further analogy to the treatment of the PM chain (compare Eq. V-8) as

$$\mathbf{V}'_d = \mathbf{V}'_l = \mathbf{V}' = \begin{bmatrix} 1 & 1 & 1 \\ 1 & 1 & \omega \\ 1 & \omega & 1 \end{bmatrix} \tag{8}$$

If R is an articulated substituent of the type CH_2—Y, e.g., CH_2—CH_3, CH_2—CH_2—Y′, or CH_2—O—Y′, in which the atom or group separated from the chain backbone by two bonds is comparable to CH_2 in size, then $\mathbf{V}'$ requires revision.[2] The group Y may be accommodated without difficulty provided that one, or both, of the skeletal bonds adjoining the substituted carbon atom is in a *gauche* conformation, in the manner shown in Fig. 7*a*. Thus, if either bond $i-1$ or bond i is $g^\pm$, an intervening R group of this kind will have available to it a conformation unencumbered by steric repulsions warranting a factor comparable to τ or ω. If, however, both of these skeletal bonds are *trans*, then the situation depicted in Fig. 7*b* arises. Here the α carbon of the substituent has been placed (arbitrarily) in front of the plane occupied by the sequence of the five skeletal carbon atoms shown. Of the three rotational states for the C—CH_2Y bond, only the one shown avoids a four-bond steric overlap between Y and one or the other of the skeletal methylene groups (shown as C only) at the extremities of the sequence shown, and in this conformation the group Y is involved in three-bond interactions with *both* of the neighboring skeletal carbon atoms. The conformation in Fig. 7*b* therefore requires a statistical weight factor commensurate with τ, but not necessarily equal to it in view of differences between the groups involved. Introduction of a factor τ^*, with $\tau^* < 1$, for the skeletal conformation shown in Fig. 7*b* is therefore required in acknowledgment of the difficulty of accommodating the group Y in any of the rotational states of the pendant bond. This factor is to be applied for each

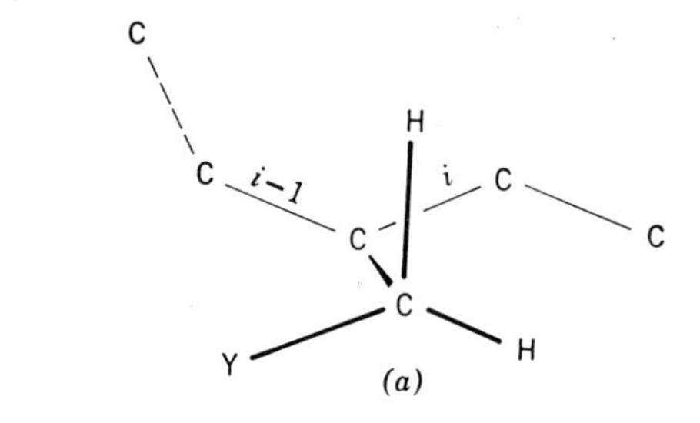

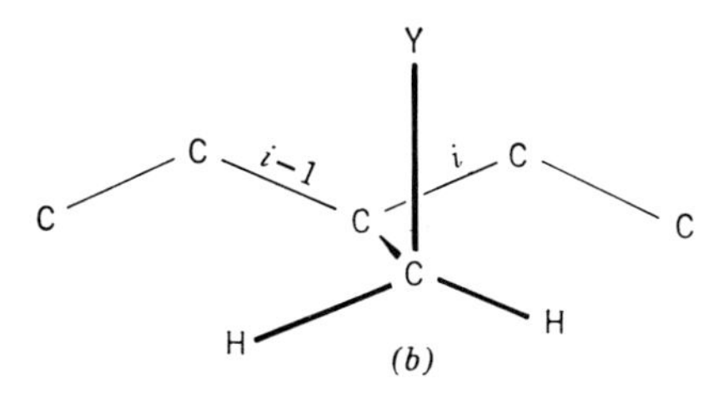

Fig. 7. Conformations for an articulated side chain situated between two skeletal bonds (*a*) in $g^+;t$ and (*b*) in $t;t$ conformations.[2]

occurrence of a $t;t$ pair of bonds. The semicolon is here included in order to indicate that the pair of bonds whose states are specified are situated on either side of a substituted chain atom. A pair of *trans* bonds between two consecutive CHR groups, designated *tt*, does not require a factor τ^*. It will be apparent also that the first unit of the chain having the formula A on p. 205 is not subject to this factor.

Introduction of τ^* in this manner can be accomplished by rewriting $\mathbf{V}'$ as follows[2]:

$$\mathbf{V}' = \begin{bmatrix} \tau^* & 1 & 1 \\ 1 & 1 & \omega \\ 1 & \omega & 1 \end{bmatrix} \tag{9}$$

for each unit except the first in the chain. This revision is of little significance for perfect isotactic chains, where *trans* pairs are of rare occurrence. In syndiotactic and in atactic chains, $t;t$ pairs of the kind specified are not otherwise suppressed; hence, acknowledgment of the factor $\tau^* < 1$ assumes importance. If, however, R is an unarticulated group like CH_3, then τ^* may be replaced by unity and Eq. 8 is restored.

Four-bond interactions associated with a bond pair such as i, $i+1$ between two CHR members are more numerous. They involve groups connected directly to the skeletal carbon atoms $i-1$ and $i+1$. These groups are CH_2 and R. Interactions in which R is involved depend primarily on that part of R which is bonded directly to the chain. Thus, if R is of the type CH_2—Y, then the CH_2 group is primarily involved in the relevant interactions of R. The nature of Y ordinarily will be of little concern in the present connection. If R is a group of the kind O—Y, steric overlaps for the second-order interactions involving R will be diminished compared with those for CH_2—Y, owing to the smaller van der Waals radius of O. If, however, the group directly attached to the backbone of the chain is larger than CH_2, e.g., if it is C_6H_5 or C_6H_4—Y, the repulsion opposing second-order overlaps will be increased.

The following analysis is addressed specifically to chains wherein the substituent R is of the kind CH_2—Y. The results should be adaptable, however, to chains in which the member of R directly attached to the chain exceeds CH_2 in size.

For simplicity, we shall assume that the four-bond interactions of CH_2 with CH_2, of CH_2 with R, and of R with R need not be differentiated, that the same statistical weight factor can be used for each, and further that this factor can be identified with ω introduced above for the four-bond interaction of CH with CH. The consequences of these rather sweeping approximations are mitigated by the fact that, for the substituents R of main interest, each of the interactions which is represented by ω is so severe as to make ω approach a vanishing magnitude for all of them. Hence, contributions of configurations affected by the approximations are very small.

The combinations of rotational states ϕ_i and ϕ_{i+1} which precipitate the various four-bond interactions cited above depend on the stereochemical configurations of *both* asymmetric centers of the dyad joined by this bond pair. Let us first consider a *dd* dyad shown in Fig. 2. Inspection of the various conformations leads at once to[2]

$$\mathbf{V}''_{dd} = \begin{bmatrix} \omega & \omega & 1 \\ 1 & \omega & \omega \\ \omega & \omega^2 & \omega \end{bmatrix}$$

The interactions represented by elements ω are easily identified from Fig. 2

or by study of molecular models. The factor ω^2 records the fact that two such interactions occur for the g^-g^+ rotational pair; one of these involves a pair of CH_2 groups, the other a pair of R's. The statistical weight for this paired state may be presumed to be effectively zero, and will be so regarded henceforth. The statistical weight matrix $\mathbf{V}''_{dd}$ may therefore be represented by

$$\mathbf{V}''_{dd} = \begin{bmatrix} \omega & \omega & 1 \\ 1 & \omega & \omega \\ \omega & 0 & \omega \end{bmatrix} \tag{10}$$

Analysis of other dyads yields

$$\mathbf{V}''_{dl} = \begin{bmatrix} 1 & \omega & \omega \\ \omega & 1 & \omega \\ \omega & \omega & 0 \end{bmatrix} \tag{11}$$

which may be obtained by a cyclic permutation of the columns of $\mathbf{V}''_{dd}$,

$$\mathbf{V}''_{ld} = \begin{bmatrix} 1 & \omega & \omega \\ \omega & 0 & \omega \\ \omega & \omega & 1 \end{bmatrix} \tag{12}$$

obtained by a cyclic permutation of the rows of $\mathbf{V}''_{dd}$, and

$$\mathbf{V}''_{ll} = \begin{bmatrix} \omega & 1 & \omega \\ \omega & \omega & 0 \\ 1 & \omega & \omega \end{bmatrix} \tag{13}$$

obtainable from $\mathbf{V}''_{dd}$ by applying both operations, or by interchanging second and third rows and columns. In each instance, a state involving two second-order interactions is given a weight of zero.

Although the matrices $\mathbf{V}''$ of Eqs. 10–13 should be suitable for most vinyl polymers, certain exceptions may be noted. If R is Cl, second-order interactions in which this comparatively small substituent is involved appear to be permitted to an appreciable extent according to studies on model compounds.[9-11] It may be necessary in this instance to distinguish the several second-order interactions, different weights being assigned depending on the pair of groups involved.[12] Analogs of poly(vinyl alcohol), in which R is OH, show a preference for conformations in which neighboring pairs of OH groups are juxtaposed for hydrogen bonding.[9,13] In this case, the four-bond interaction of OH with OH is favorable ($\omega > 1$), and the ω's for the several second-order interactions must be distinguished. The elaboration of the present analysis required for treatment of these examples will not be pursued here.

4. STATISTICAL WEIGHT MATRICES AND PREFERRED CONFORMATIONS

Interactions dependent upon more than two consecutive skeletal bonds may at most assume only a minor role, as has been emphasized previously. Interactions of higher order (in this sense) may therefore be ignored, and statistical weights can be generated from the 3 × 3 matrices $\mathbf{U}' = \mathbf{V}'\mathbf{D}'$ and $\mathbf{U}'' = \mathbf{V}''\mathbf{D}''$, respectively. Thus, from Eqs. 1, 2, 9, and 10 we have for the two bonds of a *dd* dyad[2]

$$\mathbf{U}_d' = \begin{bmatrix} \eta\tau^* & 1 & \tau \\ \eta & 1 & \tau\omega \\ \eta & \omega & \tau \end{bmatrix}; \qquad \mathbf{U}_{dd}'' = \begin{bmatrix} \eta\omega & \tau\omega & 1 \\ \eta & \tau\omega & \omega \\ \eta\omega & 0 & \omega \end{bmatrix} \tag{14}$$

The former statistical weight matrix is applicable to bonds which, like i, join CHR to CH_2; the matrix $\mathbf{U}''$ is applicable to bonds which, like $i + 1$, connect CH_2 with CHR. Each element of $\mathbf{U}'$ represents the factor entering into the total statistical weight when bond i is assigned to state ϕ_i, bond $i - 1$ having been assigned to state ϕ_{i-1}. Elements of $\mathbf{U}''$ perform the analogous role with respect to bond $i + 1$.

The corresponding statistical weight matrices for an *ll* dyad are given according to Eqs. 3, 4, 9, and 13 by

$$\mathbf{U}_l' = \begin{bmatrix} \eta\tau^* & \tau & 1 \\ \eta & \tau & \omega \\ \eta & \tau\omega & 1 \end{bmatrix}; \qquad \mathbf{U}_{ll}'' = \begin{bmatrix} \eta\omega & 1 & \tau\omega \\ \eta\omega & \omega & 0 \\ \eta & \omega & \tau\omega \end{bmatrix} \tag{15}$$

For *dl* and *ld* placements (syndiotactic) of the pair of asymmetric carbon atoms joined by bonds i and $i + 1$ (see Eqs. 2, 4, 11, and 12)

$$\mathbf{U}_{dl}'' = \begin{bmatrix} \eta & \omega & \tau\omega \\ \eta\omega & 1 & \tau\omega \\ \eta\omega & \omega & 0 \end{bmatrix} \tag{16}$$

and

$$\mathbf{U}_{ld}'' = \begin{bmatrix} \eta & \tau\omega & \omega \\ \eta\omega & 0 & \omega \\ \eta\omega & \tau\omega & 1 \end{bmatrix} \tag{17}$$

The $\mathbf{U}'$ matrices to be used in conjunction with $\mathbf{U}_{dl}''$ and $\mathbf{U}_{ld}''$ are those given in Eqs. 14 and 15, respectively. As noted above, τ^* should be replaced by unity in the $\mathbf{U}'$ matrix for the first unit of the chain. This adjustment is of concern only in Section 9.

The matrices $\mathbf{U}_l'$, $\mathbf{U}_{ll}''$, and $\mathbf{U}_{ld}''$ are obtained from $\mathbf{U}_d'$, $\mathbf{U}_{dd}''$, and $\mathbf{U}_{dl}''$, respectively, by interchanging second and third rows and columns, as follows directly from symmetry relations. We shall not at this juncture introduce the simplifications permitted by these connections, but will continue to treat d and l centers as if the distinction between them were absolute.

Let us consider a chain for which $\omega < 1$, $\tau\omega < 1$ and $\eta \cong 1$. These conditions will hold quite generally if the substituent R is CH_3, CH_2—Y, or a larger substituent. According to the elements of the $\mathbf{U}''$ matrices, the preferred conformations for bond pair i, $i+1$ within a dyad of specified stereochemical asymmetry are those given in the second column of Table 1.

Table 1

Preferred Conformations for Vinyl Chains

Stereochemical configuration of dyad i, $i+1$	Preferred dyad conformations	Corresponding stereoregular chain	Preferred conformations for stereoregular chain	Statistical weight per unit (2 bonds)
dd	g^+t	Isotactic, d	$(g^+t)(g^+t)$ etc.	η
	tg^-		$(tg^-)(tg^-)$ etc.	η
ll	g^-t	Isotactic, l	$(g^-t)(g^-t)$ etc.	η
	tg^+		$(tg^+)(tg^+)$ etc.	η
dl	g^+g^+	Syndiotactic, $dldl$, etc.	$(g^+g^+)(tt)(g^+g^+)(tt)$ etc.	η
	tt		$(tt)(g^-g^-)(tt)(g^-g^-)$ etc.	η
ld	g^-g^-		$(tt)(tt)(tt)(tt)$ etc.	$\eta^2\tau$*
	tt			

The acceptability of these conformations for the dyad in question depends also on neighbor bond pairs represented by the $\mathbf{U}'$ matrices. The appropriate elements to be selected from each of them depends in turn on the conformations of bonds outside the dyad considered, and hence on the stereochemical configurations of neighbor units. To proceed further it is necessary therefore to stipulate the stereochemical configuration of adjoining asymmetric centers.

For a chain which is uniformly isotactic the conformations given in the second column of Table 1 can be perpetuated through $\mathbf{U}_d'$ and $\mathbf{U}_{dd}''$ in alternating order without engaging those statistical weight factors, τ and ω, presumed to be small. The resulting conformations given in the fourth column are in fact the only ones meeting this requirement. This result is implicit in the statistical weight matrices. Examination of the structural formula shown in Fig. 2, or of scale models, likewise shows them to be the only ones free of major steric overlaps. These conformations are the right- and left-handed Natta-Corradini[14,15] 3_1 helices found to occur in the stable crystalline forms for isotactic chains[15] bearing substituents $R = CH_3$, C_2H_5, or C_6H_5.[16]

Succession of the right-handed g^+t helix by its left-handed equivalent, tg^-, incurs an additional factor τ^* for the $t;t$ bond pair as discussed on p. 213. For an unarticulated substituent such as CH_3 with $\tau^* \cong 1$, this conformation may readily occur. The reverse "transition" $(tg^-) \rightarrow (g^+t)$ would require at least one four-bond overlap and hence at least one factor ω. Again, this result follows from the statistical weight matrices, or it may be deduced by examination of models for the isotactic chain. Thus, a right-handed helix can be succeeded by a left-handed one subject only to the penalty of a single factor τ^*, but the reverse transition is strongly disfavored. The preferred conformations for the isotactic molecule as a whole consist therefore of two helical sections, one right-handed and the other left. They can be represented as follows:

$$(g^+t)^y; (tg^-)^{x-y-1}$$

where $0 \leq y \leq x - 1$.

The foregoing deductions are for an isotactic-*d* chain. Corresponding ones for an *l* chain are given in Table 1. The two pairs of conformations comply with the equivalence of *d* and *l* chains, the one being converted to the other by end-for-end rotation, as will be apparent.

A greater diversity of conformations is available to the syndiotactic chain. The preferred ones, selected on the same basis as above, are indicated in the fourth column of Table 1. These comprise an equivalent pair of right- and left-handed helices having a statistical weight η per structural unit, and the all-*trans* form for which the weight is $\eta^2\tau^*$. Preference for these conformations is confirmed by crystallographic studies.[17] The former conformation is more prevalent amongst crystalline syndiotactic polymers than the latter. The preference for one over the other may be expected to depend on both η and τ^*. These parameters play a more prominent role amongst syndiotactic chains than for isotactic ones. To be observed also is the facility with which the right- and left-handed helical forms and the *trans* form can be interspersed within the same syndiotactic chain; transitions from one form to the other are accommodated without difficulty.

If the interactions of R equate to those of CH_2 and if $\tau^* = 1$, conditions which should hold approximately in the case $R = CH_3$, then all of the preferred conformations listed in Table 1 for the various kinds of dyads become equivalent. Taking isotactic polypropylene* as the prototype, they are exemplified by

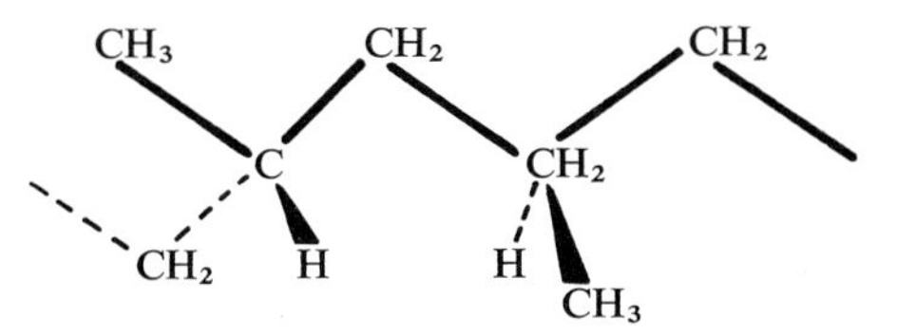

Interchanges of CH_3 and terminal CH_2 groups of the sequence shown reproduce the preferred conformations of other dyads.

Results of conformation energy calculations carried out for isotactic polypropylene by Borisova and Birshtein[18] for bond pair i, $i+1$, in the vicinity of the preferred conformation are shown in Fig. 8. Similar calculations for the low molecular analog, 2,4-dimethylpentane, by Flory, Mark, and Abe[2] are shown in Fig. 9. The same functions and parameters used in calculations on n-pentane described in the previous chapter were employed in computing Fig. 9. The conformation energy surfaces represented in

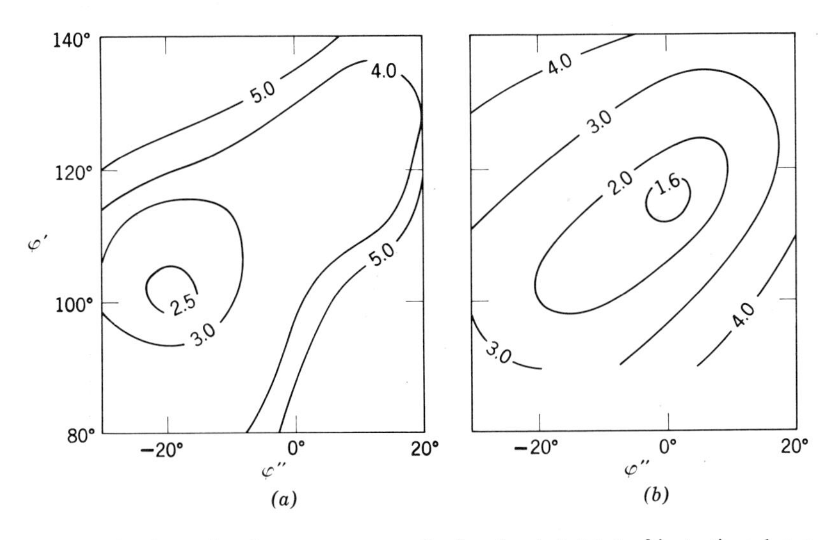

Fig. 8. Conformational energy contours for bond pair i, $i+1$ of isotactic polypropylene, expressed in kilocalories per mole. Calculations by Borisova and Birshtein[18] for (*a*) tetrahedral bond angles, and (*b*) for bond angles $\angle\,CCC = \pi - \theta = 114°$.

* A more detailed analysis taking account of the intramolecular interactions between hydrogen atoms attached to carbons separated by two bonds in polypropylene suggests that η for this polymer may be somewhat less than unity. Support for this deduction is found in the occurrence of the $\cdots ttgg \cdots$ conformation in the stable crystalline form of syndiotactic polypropylene.[17]

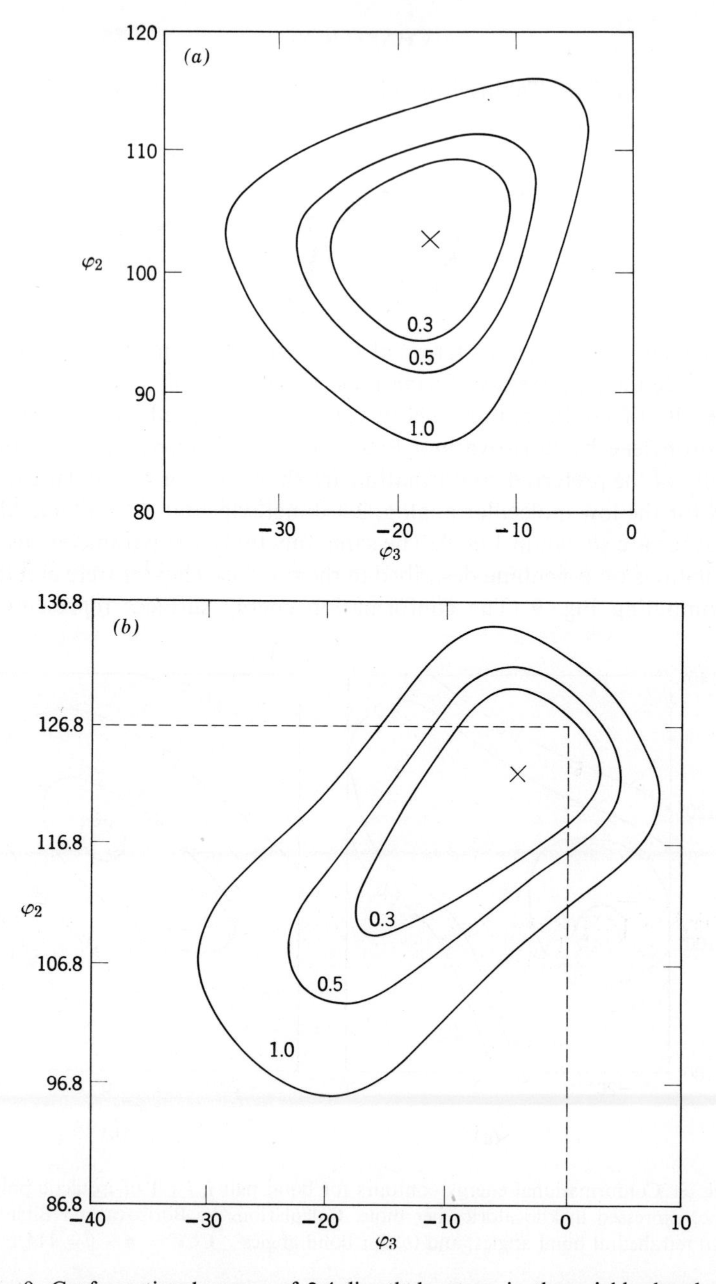

Fig. 9. Conformational energy of 2,4-dimethylpentane in the neighborhood of its preferred conformation, according to calculations of Abe[2]: (*a*) all bond angles tetrahedral, and (*b*) $\angle CCC = 112°$, $\angle CCH = 106.8°$ at tertiary carbons, and $\angle HCH = 110°$ (see footnote on p. 211). Energies of contours are given in kilocalories per mole relative to an energy of zero for the minimum located at $\times$.

Figs. 8 and 9 are similar despite the use of different functions for representing atom pair interactions. A displacement of both angles by a $\Delta\phi$ of about 15–20° from their symmetrical locations, 120 and 0°, respectively, is indicated by the calculations carried out assuming tetrahedral angles (Figs. 8*a* and 9*a*). Adoption of somewhat larger skeletal angles, 114° in Fig. 8*b*, and 112° for *all* $\angle$CCC in Fig. 9*b*, relocates the minimum near the symmetrical g^+t position. However, a prominent "valley" appears along a line $\Delta\phi_2 = \Delta\phi_3 = \Delta\phi$, with $\Delta\phi$ measured relative to the symmetrical position located by the intersection of the dashed lines in Fig. 9*b*. The computed energy is less than *RT* over a range of about 20° along this line. A considerable latitude in the conformation is thus indicated in the vicinity of each of the preferred conformations g^+t and tg^-.

Bond angles and other information required for calculation of the energies are not known with sufficient accuracy to justify a choice between the pairs of diagrams shown in Figs. 8 and 9. These calculations suggest, however, that (1) the optimum location for the preferred g^+t (or tg^-) conformation may differ by a $\Delta\phi$ of as much as 15 or 20°, affecting both angles in the way prescribed by Eqs. 6 and 7, and (*2*) permissible fluctuations about the minimum may be fairly large.[2,18]

Conformations other than the preferred one cannot in general be assessed in terms of a single diagram for angles ϕ_i and ϕ_{i+1} as in Fig. 8 or 9. Where four-bond interactions are involved, two such diagrams are of course required, one for bond pair $i-1, i$ and another for $i, i+1$.* The four-bond interaction associated with rotations about bond pair $i-1, i$ involve one skeletal CH with another, and hence duplicate approximately, those for *n*-alkanes. Each of the more numerous four-bond interactions for bond pair $i, i+1$ being of similar kind, analysis of these conformations individually in detail is unnecessary. Some of these steric overlaps are alleviated by shifts $\Delta\phi$, introduced in accordance with Eqs. 6 and 7. The matrix elements (ω's) for states thus affected may take on somewhat larger values. Interactions in other states are not diminished significantly by minor alterations $\Delta\phi$. Inspection of models indicates that none

* A single contour diagram suffices to express the energies of the various helical conformations generated by assignment of the same rotational angles to every dyad pair of a stereoregular vinyl chain. The axes of the diagram calculated subject to this condition represent values of rotation angles ϕ' and ϕ'' for *every unit* of the chain, and not merely for the single dyad, to which attention is here confined. Contour diagrams purporting to represent conformations perpetuated throughout the chain must obviously include contributions from interactions of one unit with *all* of its neighbors inclusive of those which, at the next turn of the helix, are near to it. Conformation energies for regular conformations have been investigated by Liquori and co-workers.[19] Diagrams of this nature should not be confused with those representing the localized interactions determined by two rotation angles only, e.g., by ϕ_i and ϕ_{i+1}.

of the states subject to second-order interactions is likely to have its energy markedly lowered in this manner. It is possible to show[2] that the effects of such revision of the locations of the rotational states by amounts $\Delta\phi$ as specified by Eqs. 6 and 7 can be taken into account in good approximation merely by altering the values of the parameters τ and ω without modifying the form of either $\mathbf{U}'$ or $\mathbf{U}''$. Hence, statistical weight matrices expressed as in Eqs. 14–17 can be used irrespective of the value of $\Delta\phi$.

For most vinyl chains we expect $\tau\omega \ll 1$. When it is justified to take $\tau\omega = 0$, the $\mathbf{U}$ matrices may be reduced to 2×2 order. This follows from the manner in which these matrices are multiplied one with another to generate the partition function. For example, the matrix $\mathbf{U}''_{dd}$ must be preceded and followed by $\mathbf{U}'_d$ in the required product. Inasmuch as the second column of $\mathbf{U}''_d$ is null when $\tau\omega = 0$, the second row of $\mathbf{U}'_d$ may be deleted. Combination of the third column of $\mathbf{U}'_d$ with the third row of $\mathbf{U}''_{dd}$ generates factors $\tau\omega$. Hence, this column and row may be deleted also. Inspection of the various combinations of the statistical weight matrices in this manner shows that they may be reduced when $\tau\omega = 0$ to [2]

$$\mathbf{U}'_d = \begin{matrix} t \\ g^- \end{matrix}\overset{\begin{matrix} t & g^+ \end{matrix}}{\begin{bmatrix} \eta\tau^* & 1 \\ \eta & \omega \end{bmatrix}}; \qquad \mathbf{U}'_l = \begin{matrix} t \\ g^+ \end{matrix}\overset{\begin{matrix} t & g^- \end{matrix}}{\begin{bmatrix} \eta\tau^* & 1 \\ \eta & \omega \end{bmatrix}} \tag{18}$$

$$\mathbf{U}''_{dd} = \begin{matrix} t \\ g^+ \end{matrix}\overset{\begin{matrix} t & g^- \end{matrix}}{\begin{bmatrix} \eta\omega & 1 \\ \eta & \omega \end{bmatrix}}; \qquad \mathbf{U}''_{ll} = \begin{matrix} t \\ g^- \end{matrix}\overset{\begin{matrix} t & g^+ \end{matrix}}{\begin{bmatrix} \eta\omega & 1 \\ \eta & \omega \end{bmatrix}} \tag{19}$$

$$\mathbf{U}''_{dl} = \begin{matrix} t \\ g^+ \end{matrix}\overset{\begin{matrix} t & g^+ \end{matrix}}{\begin{bmatrix} \eta & \omega \\ \eta\omega & 1 \end{bmatrix}}; \qquad \mathbf{U}''_{ld} = \begin{matrix} t \\ g^- \end{matrix}\overset{\begin{matrix} t & g^- \end{matrix}}{\begin{bmatrix} \eta & \omega \\ \eta\omega & 1 \end{bmatrix}} \tag{20}$$

for any sequences of d and l asymmetric centers.†

It is noteworthy that the statistical weight matrices in the limit $\tau\omega = 0$, as given by the equations above, are independent of τ but not of ω. Thus, the condition $\omega = 0$ eliminates τ as a parameter, but ω remains when $\tau = 0$. Conformations exemplified by a g^- rotation in Fig. 4*a* or by g^+ in Fig. 4*b*, in which CH is in proximity to both the CH_2 and the R group three bonds removed, are therefore forbidden if $\omega = 0$, even though these conformations do not themselves entail four-bond overlaps. Hence, if $\omega = 0$, the parameter τ representing these conformations vanishes from the partition

† The same reduction of order holds for the initial statistical weight matrix entering into the generation of the partition function, as follows from the fact that only its first row is required (see Eq. III-24). Hence, the simplifications offered by Eqs. 18–20 are applicable to finite as well as to infinite chains.

function and from related products generated from the statistical weight matrices.

5. SYMMETRY RELATIONS AND THE PARTITION FUNCTION

We have observed above that the matrices $\mathbf{U}'$ and $\mathbf{U}''$ for mirror image or enantiomorphic structures (see Eqs. 14–17) are convertible one to the other by row and column interchanges. If $\mathbf{Q}$ is the elementary matrix

$$\mathbf{Q} = \begin{bmatrix} 1 & 0 & 0 \\ 0 & 0 & 1 \\ 0 & 1 & 0 \end{bmatrix} \tag{21}$$

which effects interchange of second and third rows or columns, then $\mathbf{U}_l' = \mathbf{Q}\mathbf{U}_d'\mathbf{Q}$, $\mathbf{U}_{ll}'' = \mathbf{Q}\mathbf{U}_{dd}''\mathbf{Q}$, and $\mathbf{U}_{ld}'' = \mathbf{Q}\mathbf{U}_{dl}''\mathbf{Q}$. Of course, $\mathbf{U}_d' = \mathbf{Q}\mathbf{U}_l'\mathbf{Q}$, etc., as follows from $\mathbf{QQ} = \mathbf{E}_3$. Only three statistical weight matrices therefore need to be specified. These may be the three matrices $\mathbf{U}'$, $\mathbf{U}_m''$, and $\mathbf{U}_r''$ defined as follows*:

$$\begin{aligned} \mathbf{U}' &= \mathbf{U}_d' \\ \mathbf{U}_m'' &= \mathbf{U}_{dd}'' \\ \mathbf{U}_r'' &= \mathbf{U}_{dl}''\mathbf{Q} \end{aligned} \tag{22}$$

where m and r denote *meso* and racemic, respectively. Statistical weight matrices other than $\mathbf{U}_d'$ and $\mathbf{U}_{dd}''$, defined explicitly by Eqs. 22, are

$$\begin{aligned} \mathbf{U}_l' &= \mathbf{Q}\mathbf{U}'\mathbf{Q} \\ \mathbf{U}_{ll}'' &= \mathbf{Q}\mathbf{U}_m''\mathbf{Q} \\ \mathbf{U}_{dl}'' &= \mathbf{U}_r''\mathbf{Q} \\ \mathbf{U}_{ld}'' &= \mathbf{Q}\mathbf{U}_r'' \end{aligned} \tag{23}$$

The partition function for a vinyl polymer conforming to formula A on p. 205 and consisting of x repeating units, or $n = 2x$ skeletal bonds, is given according to Eq. III-24 by

$$Z = \mathbf{J}^* \left[\prod_{(i/2)=1}^{(n/2)-1} (\mathbf{U}_i'\mathbf{U}_{i+1}'') \right] \mathbf{J} \tag{24}$$

* The designation of states for $\mathbf{U}'$ and $\mathbf{U}_m''$ when expressed in the 2×2 form applicable to the three-state scheme with $\tau\omega = 0$ is that given on the margins of $\mathbf{U}_d'$ in Eq. 18 and of $\mathbf{U}_{dd}''$ in Eq. 19. Inasmuch as $\mathbf{Q}$, operating as a postmultiplier, interchanges second and third columns, it follows from Eqs. 16 and 22 that

$$\mathbf{U}_r'' = \begin{matrix} & \begin{matrix} t & g^- \end{matrix} \\ \begin{matrix} t \\ g^+ \end{matrix} & \begin{bmatrix} \eta & \omega \\ \eta\omega & 1 \end{bmatrix} \end{matrix}$$

For chains conforming to the pattern of interactions treated in the preceding section, the $\mathbf{U}'$ and $\mathbf{U}''$ matrices may be taken from Eqs. 14–17 (or from Eqs. 18–20 if $\tau\omega = 0$). Alternatively, Eqs. 23 may be used for this purpose, $\mathbf{U}'$ and $\mathbf{U}''_m$ being defined by Eqs. 14, and $\mathbf{U}''_r$ by Eq. 16 with second and third columns exchanged as required according to Eq. 22, i.e.,

$$\mathbf{U}''_r = \begin{bmatrix} \eta & \tau\omega & \omega \\ \eta\omega & \tau\omega & 1 \\ \eta\omega & 0 & \omega \end{bmatrix} \tag{25}$$

In any event, the selection of matrices for generating the product in Eq. 24 must comply with the d and l configurations of the asymmetric centers throughout the chain.

Fulfillment of the latter stipulation is accomplished more efficiently through use of the alternative rendering of Z as follows:

$$Z = \mathbf{J}^*\left(\prod_{k=1}^{x-1} \mathbf{U}_k^{(2)}\right)\mathbf{J} \tag{26}$$

where $\mathbf{U}_k^{(2)}$ is the dyad matrix defined by

$$\mathbf{U}_k^{(2)} = (\mathbf{U}'_a\mathbf{U}''_{ab})_k \tag{27}$$

The characters, d or l, of the asymmetric centers of the kth dyad are denoted by a and b.

The various dyad matrices, four in number, are conveniently expressed in terms of *meso* and racemic dyad matrices defined as follows:

$$\begin{aligned} \mathbf{U}_m^{(2)} &= \mathbf{U}'\mathbf{U}''_m \\ \mathbf{U}_r^{(2)} &= \mathbf{U}'\mathbf{U}''_r \end{aligned} \tag{28}$$

Then

$$\begin{aligned} \mathbf{U}_{dd}^{(2)} &= \mathbf{U}'_d\,\mathbf{U}''_{dd} = \mathbf{U}_m^{(2)} \\ \mathbf{U}_{ll}^{(2)} &= \mathbf{U}'_l\,\mathbf{U}''_{ll} = \mathbf{Q}\mathbf{U}_m^{(2)}\mathbf{Q} \\ \mathbf{U}_{dl}^{(2)} &= \mathbf{U}'_d\,\mathbf{U}''_{dl} = \mathbf{U}_r^{(2)}\mathbf{Q} \\ \mathbf{U}_{ld}^{(2)} &= \mathbf{U}'_l\,\mathbf{U}''_{ld} = \mathbf{Q}\mathbf{U}_r^{(2)} \end{aligned} \tag{29}$$

Specification of two matrices, $\mathbf{U}_m^{(2)}$ and $\mathbf{U}_r^{(2)}$, suffices to define the four dyad matrices required by Eq. 26.

Now the character a of the first asymmetric center of dyad k must necessarily correspond to b for dyad $k - 1$. It follows from this requirement that, irrespective of the stereochemical sequence of the vinyl chain, the $\mathbf{Q}$'s will vanish from the expression for Z obtained by substituting Eqs. 29 into Eq. 26. The relations $\mathbf{J}^*\mathbf{Q} = \mathbf{J}^*$ and $\mathbf{QJ} = \mathbf{J}$ may be called upon to eliminate factors $\mathbf{Q}$ from terminal members of the serial product. It suffices therefore to replace Eq. 27 by

$$\mathbf{U}_k^{(2)} = \mathbf{U}_m^{(2)} \quad \text{or} \quad \mathbf{U}_r^{(2)} \tag{30}$$

with the choice depending on the character of the dyad. Only the two dyad matrices are required to develop the partition function for a vinyl polymer of any stereosequence. The artificial distinction between d and l centers in an absolute sense has vanished at this point from the expression for the partition function.

Equations 29 and their effectuation of the simplification enabling use of Eq. 30 for generating the partition function are not contingent upon special characteristics of the rotational potentials, apart from those dictated by the inherent symmetry of the vinyl chain. If in a given instance a greater number of rotational states should be required, the foregoing equations remain valid; the transformation $\mathbf{Q}$ must of course be revised to higher order. On the other hand, if the three-state scheme holds and it is legitimate to let $\tau\omega = 0$, then $\mathbf{U}_m^{(2)}$ and $\mathbf{U}_r^{(2)}$ may be formulated in 2×2 order from Eqs. 18–20.

Interconversion between the $\mathbf{U}$'s for enantiomorphic forms may be avoided altogether through adoption of the simple expedient of reckoning ϕ_i in the negative (i.e., left-handed) sense when bond i adjoins an asymmetric center designated l. The effect of this convention is equivalent to the interchange of g^+ and g^- states, but the result is accomplished without alteration of the required matrices. If $\mathbf{U}' = \mathbf{U}_d'$, $\mathbf{U}_m'' = \mathbf{U}_{dd}''$, and $\mathbf{U}_r'' = \mathbf{U}_{dl}''\mathbf{Q}$ (see Eq. 22) are chosen as the primary set, then the foregoing device permits their use, without rearrangement, for enantiomorphic forms. The generation of Z by Eq. 26 with the $\mathbf{U}_k^{(2)}$ given according to Eq. 30 follows at once. In the following section we shall expand upon the convention regarding the sign of ϕ in relation to the asymmetry of the unit.

For stereoregular chains the serial product of dyad matrices required to generate Z may be expressed succinctly as a power of the appropriate matrix $\mathbf{U}^{(2)}$. The partition function for an isotactic chain can be written

$$Z = \mathbf{J}^*(\mathbf{U}_m^{(2)})^{x-1}\mathbf{J} \tag{31}$$

and for a syndiotactic chain

$$Z = \mathbf{J}^*(\mathbf{U}_r^{(2)})^{x-1}\mathbf{J} \tag{32}$$

The partition function for a stereoregular chain in the limit $\omega = 0$ is of interest. Using Eqs. 18 and 19 for an isotactic chain, we find

$$\mathbf{U}_m^{(2)} = \eta \begin{bmatrix} 1 & \tau^* \\ 0 & 1 \end{bmatrix} \tag{33}$$

with equal eigenvalues $\lambda^{(2)} = \eta$. A preferred conformation was previously found to consist of a right-handed and of a left-handed helical segment. There are $x - 2$ such conformations corresponding to the various locations of the junction between the two helices. The statistical weight of each is $\eta^{x-1}\tau^*$. Additionally, there are two conformations comprising single helices, one being right-handed and the other left; the statistical weight of each is just η^{x-1}, there being no $t;t$ junction. Since all other conformations are eliminated by taking $\omega = 0$,

$$Z = (x - 2)\eta^{x-1}\tau^* + 2\eta^{x-1}$$

This result may be obtained directly from Eqs. 31 and 33, with $\tau^* = 1$ for the first dyad. It follows that

$$\lim_{x\to\infty} (Z^{1/x}) = \eta = \lambda^{(2)} \tag{34}$$

as required.

For a syndiotactic chain with $\omega = 0$

$$\mathbf{U}_r^{(2)} = \begin{bmatrix} \eta^2\tau^* & 1 \\ \eta^2 & 0 \end{bmatrix} \tag{35}$$

according to Eqs. 18 and 20, the explicit distinction between g^+ and g^- being abolished by the foregoing convention. The larger eigenvalue of $\mathbf{U}_{dl}^{(2)}$ is

$$\lambda_1^{(2)} = (\eta/2)(\eta\tau^* + \sqrt{4 + \eta^2\tau^{*2}}) \tag{36}$$

In the unlikely event that $\tau^* = 0$, we would have $\lambda_1^{(2)} = \eta$ and the syndiotactic chain would be forced to assume either the right- or the left-handed helical form listed for it in Table 1. Then $\lim_{x\to\infty} (Z^{1/x}) = \eta$, as for the isotactic chain. With increase in $\eta\tau^*$, the planar conformation becomes increasingly prevalent; for $\eta\tau^* = 1$,

$$\lambda_1^{(2)} = \eta(1 + \sqrt{5})/2$$

which emphasizes the fact that the partition function, regarded as a measure of the number of effective configurations of the chain molecule, is generally much greater for a syndiotactic chain than for its isotactic isomer.

The incidence of various bond conformations can be deduced by the methods presented in Chapter III.

6. MOMENTS OF VINYL CHAINS

The second moment $\langle M^2 \rangle$ for a vinyl chain consisting of $x = n/2$ units having any specified stereochemical configurations can be calculated according to Eq. IV-28, the appropriate $\mathscr{G}$ matrix being selected for each bond. Simplification would be possible if, in analogy to the formulation of the partition function, the result could be expressed in terms of dyad matrices $\mathscr{G}_m^{(2)}$ and $\mathscr{G}_r^{(2)}$. We shall explore this possibility. Written in terms of dyad matrices, Eq. IV-28 is

$$\langle M^2 \rangle = 2Z^{-1} \mathscr{J}^* \mathscr{G}_1 \left(\prod_{k=1}^{x-1} \mathscr{G}_k^{(2)} \right) \mathscr{G}_n \mathscr{J} \tag{37}$$

where $\mathscr{G}_k^{(2)}$ is the dyad matrix for the kth unit; i.e.,

$$\mathscr{G}_k^{(2)} = (\mathscr{G}_a' \mathscr{G}_{ab}'')_k \tag{38}$$

The "bond" matrices $\mathscr{G}_a'$ and $\mathscr{G}_{ab}''$ are defined according to Eq. IV-24, statistical weight matrices $\mathbf{U}'$ and $\mathbf{U}''$ of character corresponding to a and b being used. The matrices $\mathscr{G}_1$ and $\mathscr{G}_n$ for terminal bonds accord with previous specifications, i.e., $\mathbf{U}_1 = \mathbf{U}_n = \mathbf{E}_\nu$.

Pursuant to reduction of the number of dyad matrices $\mathscr{G}^{(2)}$ that will be required, we assign *left-handed coordinate systems for both skeletal bonds attached to an asymmetric center designated l*, a device introduced by Birshtein and Ptitsyn.[20] The left-handed reference frames are defined as before, but with the directions of their z axes reversed. In keeping with assignment of coordinate systems in this manner, rotation angles for the set of bonds adjoining l centers will be measured in the negative or left-handed sense. Hence, the convention on the signs of the rotation angles ϕ proposed in the preceding section is comprehended by the broader convention adopted here.

Some of the transformations $\mathbf{T}$, as well as the statistical weights $\mathbf{U}$, are affected by the introduction of left-handed reference frames. The transformation $\mathbf{T}_i'$ associated with the first bond of a dyad i, $i+1$ will connect reference frame $i+1$ to i within the given dyad (see Fig. 1). If the dyad is *dd*, then it will have its usual form $\mathbf{T}$ as expressed by Eq. I-25. If the dyad is *ll*, both reference frames are left handed. The reversal of directions of the z axes necessitates reversal of the signs of elements in the third row and again in the third column. Compliance with the stipulations above requires additionally that $\phi_i = \phi$ be replaced by $-\phi$. The matrix $\mathbf{T}$ is unaltered by these operations. This conclusion follows immediately from the mirror symmetry relating *ll* to *dd*. If, however, the dyad is *dl*, then the reference frames i and $i+1$ are of opposite hand. The relevant rotation angle ϕ is reckoned in the positive sense in this case, but the signs of elements in the third column of **T** must be reversed to obtain the appropriate transformation, which becomes

$$\mathbf{T}_* = \begin{bmatrix} \cos\theta & \sin\theta & 0 \\ \sin\theta\cos\phi & -\cos\theta\cos\phi & -\sin\phi \\ \sin\theta\sin\phi & -\cos\theta\sin\phi & \cos\phi \end{bmatrix} \tag{39}$$

If the dyad is *ld*, then the transformation operates from a right-handed to a left-handed reference frame; since bond i adjoins an l center, ϕ_i is to be taken in the left-hand sense. Signs of elements in the third row of **T** must be reversed and ϕ replaced by $-\phi$. The result is $\mathbf{T}_*$ given by Eq. 39; this follows directly from the mirror symmetry operation relating *ld* to *dl*.

Transformations required for matrices $\mathscr{G}''$ invariably relate two reference frames of the same hand. Hence, they are to be represented by **T**.

The $\mathscr{G}$ matrices for individual bonds are defined, subject to the foregoing convention, as follows†:

$$\begin{aligned} \mathscr{G}'_{dd} &= \mathscr{G}'_{ll} = \mathscr{G}(\mathbf{U}', \mathbf{T}, \mathbf{m}') \\ \mathscr{G}'_{dl} &= \mathscr{G}'_{ld} = \mathscr{G}(\mathbf{U}', \mathbf{T}_*, \mathbf{m}') \\ \mathscr{G}''_{dd} &= \mathscr{G}''_{ll} = \mathscr{G}(\mathbf{U}''_m, \mathbf{T}, \mathbf{m}'') \\ \mathscr{G}''_{dl} &= \mathscr{G}''_{ld} = \mathscr{G}(\mathbf{U}''_r, \mathbf{T}, \mathbf{m}'') \end{aligned} \tag{40}$$

where $\mathbf{U}'$, $\mathbf{U}''_m$, and $\mathbf{U}''_r$ are defined according to Eqs. 22, and $\mathbf{m}'$ and $\mathbf{m}''$ are the contributions (e.g., bond lengths) for the respective bonds of the dyad. The functions denoted on the right-hand side of each equation are defined by Eq. IV-24.

From the relations given by Eqs. 40 we obtain at once

$$\begin{aligned} \mathscr{G}_m^{(2)} &= \mathscr{G}(\mathbf{U}', \mathbf{T}, \mathbf{m}')\mathscr{G}(\mathbf{U}''_m, \mathbf{T}, \mathbf{m}'') \\ \mathscr{G}_r^{(2)} &= \mathscr{G}(\mathbf{U}', \mathbf{T}_*, \mathbf{m}')\mathscr{G}(\mathbf{U}''_r, \mathbf{T}, \mathbf{m}'') \end{aligned} \tag{41}$$

Through use of these definitions, the serial product in Eq. 37 can be generated from two dyad matrices. The procedure parallels the generation of Z, as set forth in the preceding section. Artificiality in the designation of asymmetric centers has vanished from the equations, and the procedure is applicable for any specified stereosequence.

† If the **U** matrices reduce to the 2×2 form of Eqs. 18–20, then the choice between $\mathbf{T}(\phi_{g+})$ and $\mathbf{T}(\phi_{g-})$ in formulating $\|\mathbf{T}\|$ for the construction of $\mathscr{G}$ is to be made in accordance with Eqs. 18–20 and 22. In the construction of $\mathscr{G}''_{dl} = \mathscr{G}''_{ld}$, the matrix $\mathbf{T}(\phi_{g-})$ is to occur as the second pseudoelement of $\|\mathbf{T}\|$ in order to be associated with the second column of $\mathbf{U}''_r$ (see the footnote on p. 223).

7. STEREOREGULAR CHAINS

If the chain is isotactic, i.e., if all dyads are *meso*, then

$$\langle M^2 \rangle = (2/Z)\mathscr{J}^*\mathscr{G}_1(\mathscr{G}_m^{(2)})^{x-1}\mathscr{G}_n\mathscr{J} \tag{42}$$

Similarly, for a syndiotactic chain

$$\langle M^2 \rangle = (2/Z)\mathscr{J}^*\mathscr{G}_1(\mathscr{G}_r^{(2)})^{x-1}\mathscr{G}_n\mathscr{J} \tag{43}$$

The mean-square value of the chain vector **r** can be calculated readily for stereoregular chains of any length through use of these equations in conjunction with Eqs. 26 and 30 for Z. The limiting value of the characteristic ratio, C_∞, may readily be obtained by successively squaring $\mathscr{G}_m^{(2)}$ or $\mathscr{G}_r^{(2)}$ until the ratio $\langle r_n^2 \rangle_0/nl^2$ converges.†

Results of calculations of C_∞ for isotactic vinyl chains[2] are given in Fig. 10. The supplements of both valence angles, θ' and θ'', were taken to be 68°. This value is greater by 2.5° than the estimate of Bunn and Holmes[8] based on the x-ray diffraction of oriented isotactic α-olefin polymers; it is less than the value (69°) found by Lide[7] for isobutane. Rotational states were chosen according to Eqs. 6 and 7, with $\Delta\phi = 0$, 20, and 30° as indicated in Fig. 10. The unabridged 3×3 statistical weight matrices, Eq. 14, were used. The parameter η was equated to unity; τ^*, being of little importance for infinite isotactic chains, was likewise set equal to unity. The calculated characteristic ratio rises rapidly with decrease in $\omega\tau$ except when ω is near unity. Rationalization of these results is straightforward: as ω decreases toward zero, all configurations but the preferred helix (3_1 helix for $\Delta\phi = 0$) are suppressed with the result that $\langle r^2 \rangle_0$ approaches proportionality to n^2, and C_∞ increases without limit. As ω is arbitrarily increased toward unity, the increasing participation of *gauche* states at random causes C_∞ to decrease.

Experimental values of C_∞ for typical isotactic polymers appear to occur in the range of 6–10; see Table II-1. The steep rise of C_∞ with $-\ln\omega$ over this range of Fig. 10 predicates a large negative temperature coefficient, contrary to experiments[21] showing $d\ln\langle r^2 \rangle_0/dT$ to be positive though small (Table II-2); it is certainly not much less than zero. Extensive calculations[2] fail to reveal any combination of parameters, ω, τ, and $\Delta\phi$, which resolve this striking incongruity with experiment. A calculated temperature co-

† The alternative method of Eqs. IV-43 and IV-44, involving the dominant eigenvalues of the appropriate recurrent products of statistical weight matrices, may be used to obtain C_∞ for stereoregular polymers. However, it is necessary to evaluate separate terms like the one occurring in Eq. IV-44 for each of the several sums over distinct sets of bond pairs present in these polymers. Three separate terms are required for both isotactic and for syndiotactic chains. The matrix squaring method is more straightforward and less cumbersome.

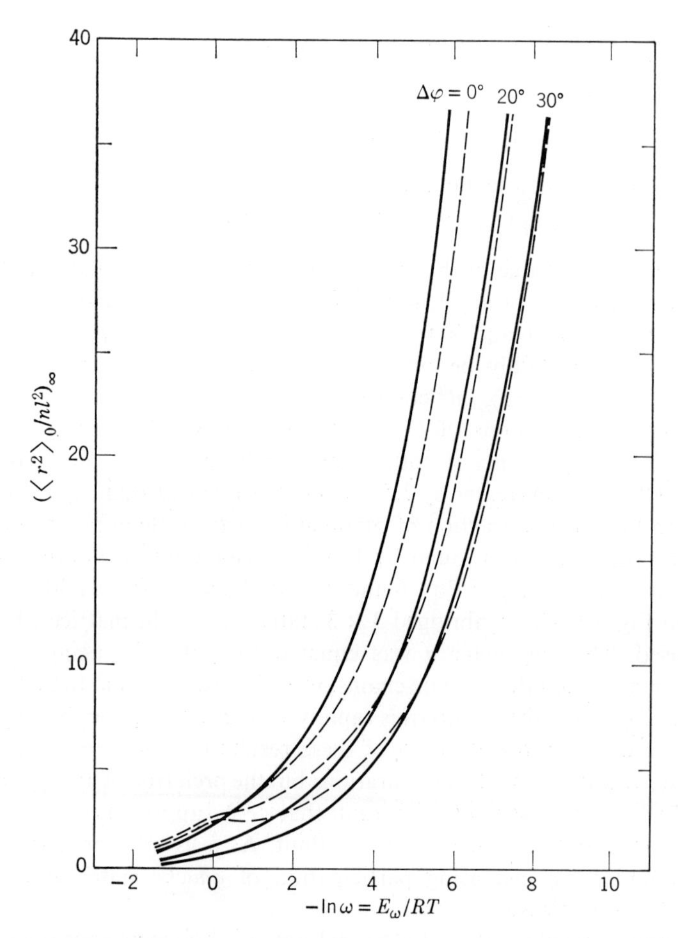

Fig. 10. The limiting value of the characteristic ratio C_∞ for a perfect isotactic chain calculated according to Eqs. 41 and 42 as a function of $-\ln \omega$, with $\eta = 1$ and $\Delta\phi = 0$, 20, and 30°, as indicated. Solid and dashed curves are for $\tau = 0.05$ and 0.5, respectively.[2]

efficient of zero can be deduced, with utter disregard of the chain structure and the steric interactions to which it is subject, by taking $\omega = \tau = 1$, or by letting $\omega = 0$. The former condition yields a characteristic ratio which is much too small (Fig. 10), and the latter causes $\langle r^2 \rangle_0$ to approach its value for full extension. Neither introduction of additional rotational states like those indicated by calculations on n-pentane (see Fig. V-4), nor any other reasonable revision of the conformational energy scheme effects a drastic alteration of the calculated results.

The rotation angles for bond pairs in the vicinity of the preferred state are subject to variation over ranges of as much as 20° along the valleys mapped in Figs. 8*b* and 9*b*, as noted earlier. Fluctuations in the values for ϕ_i, ϕ_{i+1} within this range for one unit should be sensibly independent of those for its neighbors. The resulting random variations may conceivably reduce the chain dimensions to some extent. Numerical calculations show,[2] however, that the randomness thus imparted is by no means sufficient to account for the reduction of C_∞ when its value is of the order of 10. If, for example, $\omega = 0$, these fluctuations could reduce C_∞ from infinity to about 100. They could not alone account for a major fraction of the large discrepancy between the rotational isomeric state calculations for a perfect isotactic chain and experimental observations on actual polymers.

In summary, examination of the structure of an isotactic chain shows unequivocally that a helical conformation, described approximately as *tg*, should be strongly preferred over all others. While the experimental characteristic ratio is by no means as large as estimates of conformation energies (E_ω and E_τ) would indicate, it is substantially greater than would be the case if the various conformations were permitted to occur more or less at random (i.e., if $E_\omega \cong E_\tau \cong 0$). Empirical adjustment of ω and τ to yield values of C_∞ in the range of observation would imply positive values of E_ω and E_τ which are much smaller than expected from analysis of the model, but nevertheless sufficiently greater than zero to require $d \ln \langle r^2 \rangle_0 / dT$ to be negative and large, contrary to experiments.

These considerations led to the postulation[2] of stereoirregular units in isotactic chains, the presence of which disrupt the predicted perpetuation of the preferred configuration throughout long sequences of units. The dimensions of the chains are consequently reduced. This hypothesis is examined in the following section devoted to the treatment of stereo-irregular chains. Before turning to vinyl polymers of that description, a few remarks on syndiotactic chains are in order.

For the small values of ω indicated by the severity of steric overlaps, the characteristic ratios for syndiotactic chains, calculated according to Eq. 43, are insensitive to ω. Hence, for illustrative calculations it suffices to let $\omega = 0$. The $\mathbf{U}$ matrices of 2×2 order given by Eqs. 18 and 20 may then be used to obtain $\mathscr{G}_r^{(2)}$ according to Eq. 41. The calculated dependence of C_∞ on $-\ln \tau^*$ for a syndiotactic chain is shown in Fig. 11[2] for two values of $\Delta\phi$. Starting at $\tau^* = 1$, the ratio decreases slightly with decrease in τ^*, but subsequently assumes a sustained increase with further decrease in τ^*. In the latter range, the syndiotactic chain approaches the conformation of its helical form, e.g., $ttg^+g^+ttg^+g^+$ etc., as pointed out above, and consequently C_∞ increases without limit. For τ^* near unity (as for $R = CH_3$), the characteristic ratio is calculated to be of the order of 10. In contrast to the isotactic chain, C_∞ increases[2] with $\Delta\phi$.

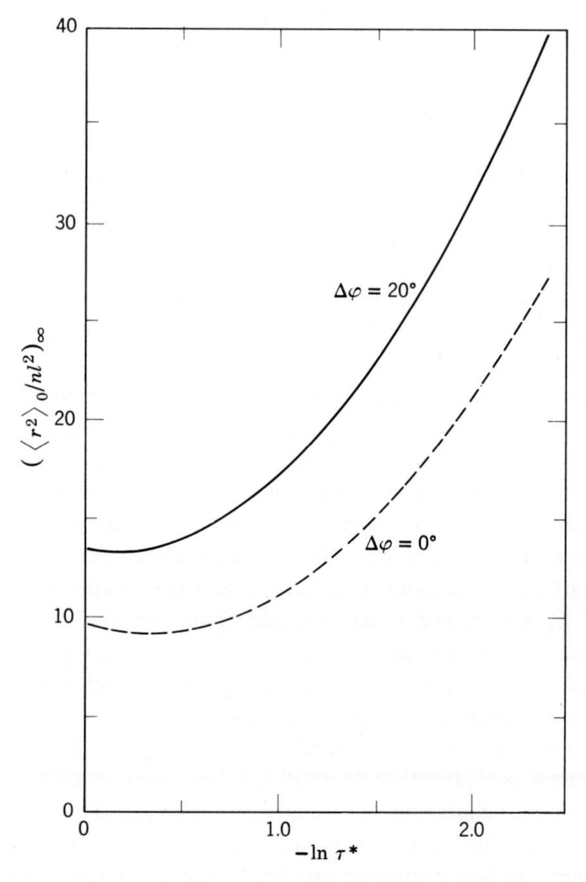

Fig. 11. The characteristic ratio C_∞ for a syndiotactic chain calculated according to Eqs. 41 and 43 as a function of $-\ln \tau^*$, with $\omega = 0$.[2]

8. AVERAGE DIMENSIONS OF STEREOIRREGULAR CHAINS

If the configurations of the asymmetric centers —CHR— do not occur in a regularly repeated succession throughout the chain, and in particular, if the chain departs from perfect isotactic or syndiotactic regularity through the presence of a few stereoirregular units which occur at random, then more elaborate computations are required. The chain must be treated as a copolymer (see Chap. IV, Sect. 7), and since the rotational states of neighbor units are interdependent, the **T** matrices cannot be independently averaged. Sequential products of them must be averaged over the chain as a whole, and this requires full specification of the stereochemical configurations of successive units in the chain. Inasmuch as a copolymer, or an atactic vinyl polymer, comprises a statistically distributed array of molecular species differing in the sequence of units, it is necessary to arrive at an average over a representative population of such species. As has been emphasized in Chapter IV (see p. 121), separate averaging of the serial product of the $\mathscr{G}$ matrices and of the serial product of the $\mathbf{U}$ matrices required for the partition function Z is not permitted, and to proceed in this manner may lead to serious error. Instead, characteristic ratios must be calculated from these respective products for individual molecular species, and then these ratios are to be averaged over the population of species in the system. Recourse to "Monte Carlo" generation of representative species as subjects on which to perform calculations is indicated in this situation (see p. 121).

Let us suppose that the conditions operative during synthesis of the collection of chains define a probability w_{iso} that the kth unit of any chain has the same stereochemical configuration as its predecessor; $w_{\text{syn}} = 1 - w_{\text{iso}}$ is the probability of a syndiotactic placement of the unit in question relative to the preceding unit. We assume the replication probability w_{iso} to be independent of the stereochemical configurations of preceding dyads. In this sense, the sequence of isotactic and syndiotactic dyads is assumed to be Bernoullian. Further, we assume that w_{iso} is independent of the serial index k, and that all molecules of the sample are subject to the same *a priori* replication probability w_{iso}.

A statistically representative set of molecules, each consisting of x units, may be specified by ordered sets of $x - 1$ random numbers ranging from 0 to 1. Those numbers within each set which are less than w_{iso} are taken to specify an isotactic dyad in which the d or l character of one unit is perpetuated in the next; those numbers exceeding w_{iso} denote syndiotactic dyads in which the —CHR— groups are of opposite character. The serial product

$$\prod_{k=1}^{x-1} \mathscr{G}_k^{(2)}$$

required in Eq. 37, and the product

$$\prod_{k=1}^{x-1} \mathbf{U}_k^{(2)}$$

required for calculation of Z (see Eq. 26) appearing in the denominator of Eq. 37 are computed for a given Monte Carlo species using the $\mathscr{G}^{(2)}$ defined by Eqs. 41 and the $\mathbf{U}^{(2)}$ defined by Eqs. 22 and 28, the *meso* and racemic characters of successive dyads being strictly observed. The characteristic ratio C_n for each of a number of Monte Carlo species is computed in this manner.

Calculations[2] showing the dependence of the characteristic ratio on $-\ln \omega$ for two typical Monte Carlo chains of 100 units, one generated with $w_{iso} = 0.90$ and the other with $w_{iso} = 0.50$, are shown in Fig. 12 for the several values of $\Delta\phi$ indicated. Other parameters were assigned as follows: $\eta = 1$, $\tau = 0.05$, and, assuming R to be CH_3 or an atom or group of equivalent effect, $\tau^* = 1$. The dashed curve was calculated for a four-state model simulating the separation of four-bond overlapped states (analogous to $g^{\pm}g^{\mp}$ states of polymethylene) into pairs of minima like those indicated by calculations for *n*-pentane and implied for higher *n*-alkanes (see Fig. V-4). The curves rise as expected to asymptotic levels with increase in $-\ln \omega$. The asymptote is reached at lower values of the abscissa for the four-state model, which is probably a more realistic representation. At the asymptotes the values of C_n are the same for the two models inasmuch as only the preferred conformations are permitted by either of them for $\omega = 0$. Although the significance of quantitative detail is dubious, we may note that, according to Fig. 12, an energy E_ω in the range of 3–5 kcal mole^{-1} is sufficient at $T = 300$–$400°$K to suppress conformations having major steric overlaps to the point where they contribute a negligible lowering of the characteristic ratio. The conformation of each isotactic sequence in the chain with $w_{iso} = 0.90$ then consists of the two helical segments, one right handed and the other left, which are permitted to coexist without occurrence of an unfavorable conformation (see Sect. 4). Whereas a large negative temperature coefficient is implied along the steep rise of the curve for lower values of $-\ln \omega$, the temperature coefficient obviously should fall to zero as the asymptote is approached. Other calculations[2] show the relationships depicted in Fig. 12 to be insensitive to τ for $\tau < 1$.

The dependence of the characteristic ratio on the replication probability w_{iso} is shown in Fig. 13. The calculations summarized in this figure were carried out[22] for $\tau^* = 1$ and $\omega = 0$ (see Eqs. 18–20). They are representative therefore of chains in which second-order overlaps are effectively suppressed and the affected states make no appreciable contribution to the lowering of the characteristic ratio. Three values of $\Delta\phi$ (see Eqs. 6 and 7), namely, 0, 10, and 20°, are represented. All calculations are for chains of 400 repeat units, which is a sufficient number for approximate convergence of C_n to its limiting value throughout the range $w_{iso} = 0$–0.95. Beyond the latter limit convergence is slowed; at $w_{iso} = 1.00$ the characteristic ratio is nonconvergent as pointed out above. Each point in the range $0 < w_{iso} \leq 0.95$ represents the average of results for ten Monte Carlo chains of 400 units, with the exception of those for $\Delta\phi = 0$ at $w_{iso} = 0.90$ and 0.95, where 20

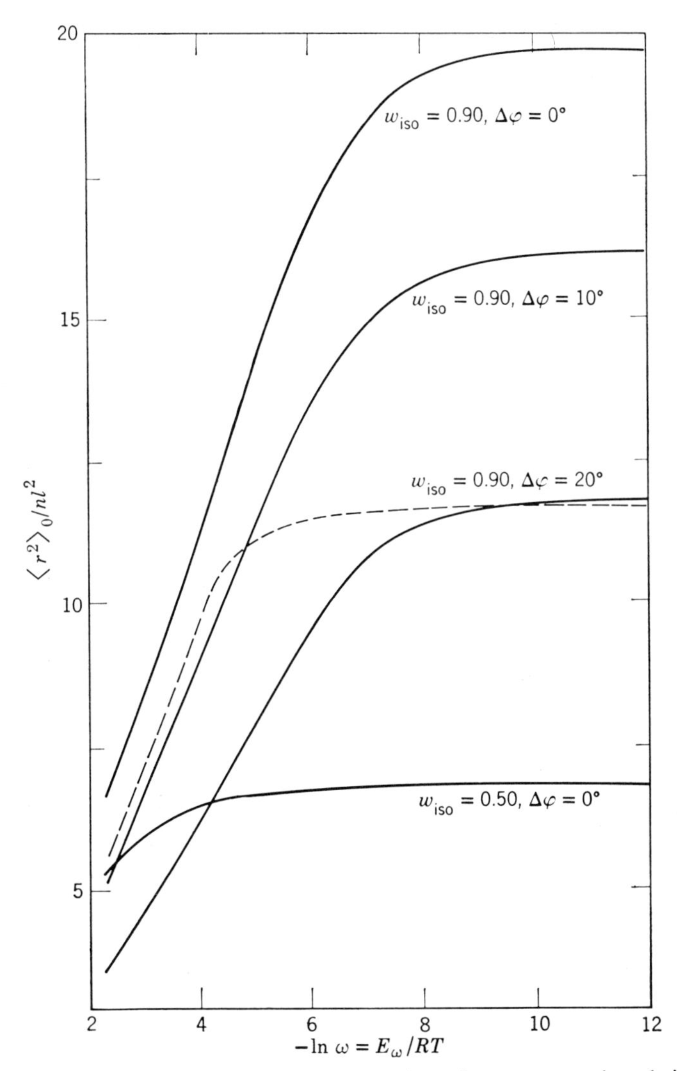

Fig. 12. The characteristic ratio versus $-\ln \omega$ for representative chains of 100 units generated with $w_{iso} = 0.90$ and 0.50. Calculations were carried out for $\eta = 1.00$, $\tau^* = 1.00$, $\tau = 0.05$, and $\triangle\phi$ as indicated. The dashed curve was calculated for a four-state model.[2]

chains were computed. Standard deviations, represented by error bars, increase as w_{iso} approaches unity owing to the diminished number of configurations available to a chain having few syndiotactic dyads, and hence the enhanced dependence on the numbers of them and their locations in a given Monte Carlo chain. The large uncertainty in the location of the

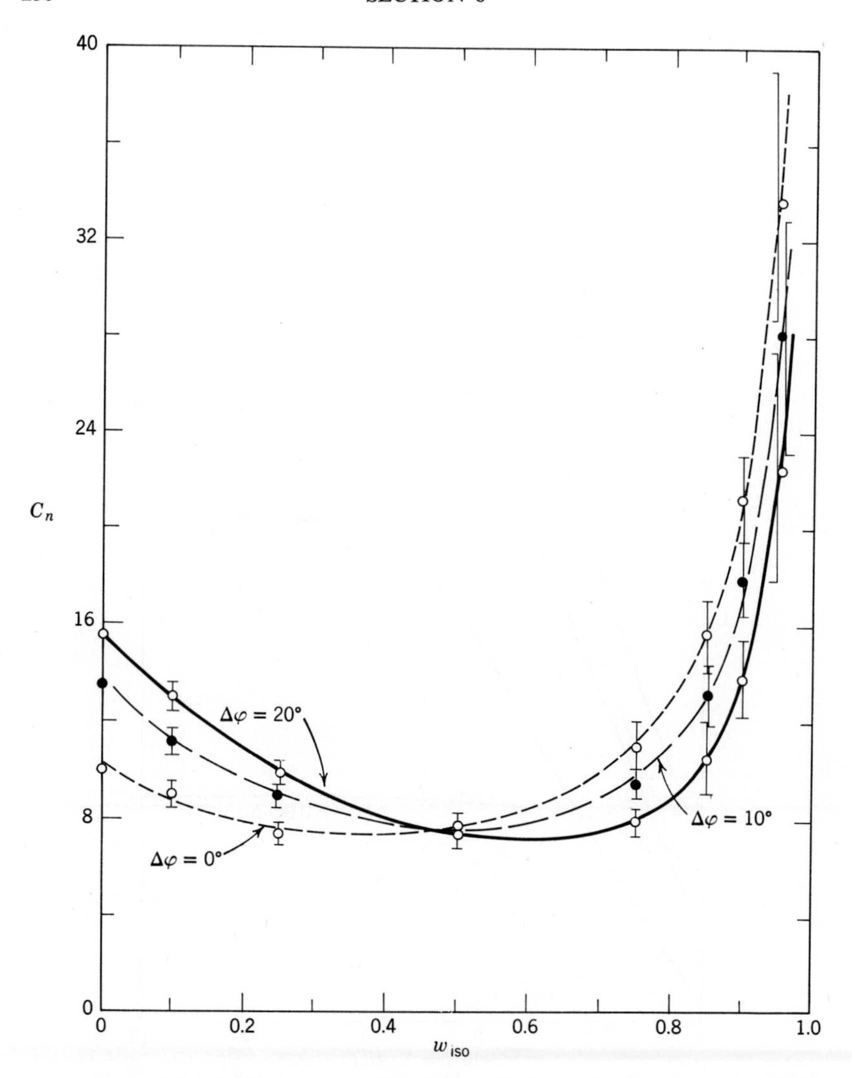

Fig. 13. Characteristic ratios calculated by Tonelli[22] for Monte Carlo chains of 400 units, with $\tau^* = 1$ and $\omega = 0$. The points for $0 < w_{iso} \leq 0.95$ represent averages for ten Monte Carlo chains (20 for $\triangle\phi = 0$ and $w_{iso} = 0.90$ and 0.95). Short dashed, long dashed, and solid curves, and the points associated with them, respectively, are for $\triangle\phi = 0$, 10, and 20°. Error bars show standard deviations.

points in this range notwithstanding, a sharp decrease in $\langle r^2 \rangle_0/nl^2$ with $1 - w_{iso}$ is clearly demonstrated. The decrease is more precipitous, the larger the value of $\Delta\phi$. Over the remainder of the diagram, the ratio is comparatively insensitive to w_{iso}; it increases somewhat from $w_{iso} = 0.50$

(atactic) to $w_{iso} = 0$ (syndiotactic). The effect of $\Delta\phi$ for syndiotactic chains is the reverse of its effect for isotactic chains, as is also apparent in Fig. 11.

According to these calculations, the experimentally observed values of C_∞ in the vicinity of 10 for isotactic polymers can be reconciled with $w_{iso} = 0.80$–0.90 for $\Delta\phi \cong 20°$, i.e., the presence of 10–20% syndiotactic dyads would account for experimental chain dimensions. Fluctuations from the preferred rotational states defined by the minima in the conformational energy would further decrease $\langle r^2 \rangle_0$. The diminution from this source must be comparatively small, however, as we have concluded in the preceding section.

Calculations[22] show the characteristic ratio to be lowered almost to the same degree by interposition of an occasional unit of one symmetry (e.g., *l*) in a chain consisting of very long sequences of units of the opposite character (e.g., *d*), the resulting stereosequence being $\cdots dddlddddldd \cdots$. The average chain dimensions are also subject to lowering by other types of stereoirregularity, e.g., by head-to-head placements of units as in

$$—CH_2—CHR—CHR—CH_2—CH_2—CHR—CH_2—CHR— \text{ etc.}$$

if such structures should be present.

The predominantly isotactic chain with $w_{iso} \geq 0.9$ can be treated, in crude approximation, as a succession of helical segments (see footnote, p. 165). The number of such segments in a very long chain is twice the number $(n/2)w_{syn}$ of isotactic sequences (see p. 218), and the average squared length of a segment measured along its axis is

$$\langle l_h^2 \rangle = (2l^*)^2 \langle x_h^2 \rangle \tag{44}$$

where $2l^*$ is the length of a structural unit (two bonds) projected on the helix axis, x_h being the number of units in a helical segment. Thus, $\langle x_h \rangle = 1/2w_{syn}$. For a random distribution of syndiotactic units and of the $t;t$ states within the isotactic sequences (a condition which should obtain when $\tau^* = 1$) the first and second moments of x_h are related according to

$$\langle x_h^2 \rangle = \langle x_h \rangle (2\langle x_h \rangle - 1)$$

Hence,

$$\langle x_h^2 \rangle = (1/2w_{syn})(w_{syn}^{-1} - 1) \tag{45}$$

If the helical sequences are of sufficient length to suppress lateral correlations of successive helical axes, so that they may be treated in the free rotation approximation, then the foregoing relations yield

$$C_\infty \cong 2(1 + \cos\theta_h)(1 - \cos\theta_h)^{-1}(l^*/l)^2(w_{syn}^{-1} - 1) \tag{46}$$

where θ_h is the angle between the directions of the axes of successive helical segments.

If bond angles are tetrahedral and the *gauche* states are situated at $\pm 120°$ (i.e., if $\Delta\phi = 0$), then the angle θ_h will be the tetrahedral angle. This result is easily ascertained both for the junction at the syndiotactic dyad between two isotactic sequences and for the right- and the left-handed helices joined by the conformation $\cdots(g^+t)(tg^-)\cdots$, which is permitted to occur within an isotactic sequence. It follows that $\cos\theta_h = \frac{1}{3}$. Under the same conditions $l^*/l = \frac{2}{3}$. Hence

$$C_\infty \cong (16/9)(w_{\text{syn}}^{-1} - 1) \tag{47}$$

If $w_{\text{iso}} = 1 - w_{\text{syn}} = 0.90$, the thus estimated value of C_∞ is 16, compared with the value of ca. 21 indicated by the curve for $\Delta\phi = 0$ in Fig. 13.

This procedure is obviously inadmissible for lower values of w_{iso}, for which the length of the average helical segment is small. The determination of θ_h and l^*/l when $\Delta\phi \neq 0$ is more difficult. In general, the exact methods used to obtain the results shown in Fig. 13 will be preferred for their greater generality and reliability. The approximate calculation affords an intuitive grasp of the physical basis for the results obtained by the exact methods, however.

If R is an articulated substituent like $—CH_2—CH_3$, a value of $\tau^* < 1$ must be expected. While τ^* is of little consequence in the isotactic range, it assumes importance for syndiotactic chains, as was pointed out in the preceding section (see Fig. 11). It remains to consider the effect of $\tau^* < 1$ on atactic chains. Characteristic ratios calculated[2] for Monte Carlo chains of 100 units generated with $w_{\text{iso}} = 0.50$ are plotted in Fig. 14 against

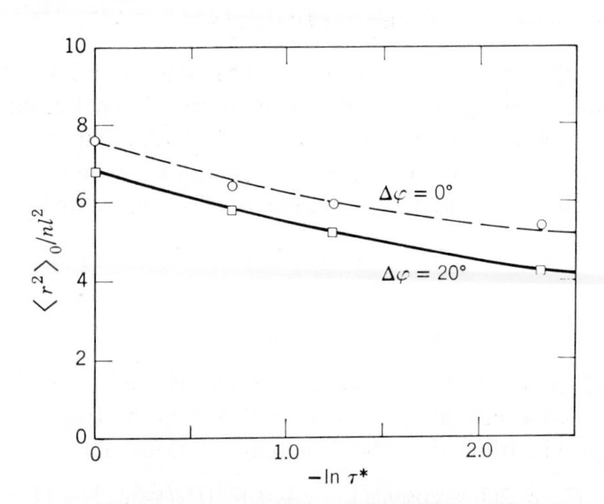

Fig. 14. The characteristic ratio for atactic chains of 100 monomer units as a function of $-\ln\tau^*$ for $\omega = 0$, $w_{\text{iso}} = 0.50$, and the values of $\Delta\phi$ indicated. Each point represents the average for five different Monte Carlo chains. (From Flory, Mark, and Abe.[2])

$-\ln \tau^*$. Each point is the mean of results for five such chains. The calculated mean dimensions diminish gradually as τ^* decreases. A positive temperature coefficient of $\langle r^2 \rangle_0$ is predicted since τ^* should increase with temperature. This feature has been suggested[2] as the basis for the positive temperature coefficients $d \ln \langle r^2 \rangle_0 / dT$ usually observed for vinyl polymers (see Table II-2).

9. EQUILIBRIUM STEREOCHEMICAL CONFIGURATIONS[23]

Consider a system consisting of N vinyl polymer molecules, each of which comprises exactly x units for a total of $n = 2x$ skeletal bonds, as represented by the formula A on p. 205. The x consecutive units may not necessarily be identical (i.e., their R groups may differ), but if this should be the case, we assume the several kinds of units to be ordered identically in the various molecules comprising the system. The stereochemical configurations of the —CHR— groups are not constrained to be the same either along a given chain or among the different molecules of the system. We shall in fact go further in asserting a catalyst to be present which effectuates reversible isomerization (i.e., epimerization) of the configurations of the —CHR— asymmetric centers. A state of stereochemical equilibrium is thus established.

The various species of molecules present in the system differ from one another only in the configurations of their asymmetric centers. Inasmuch as the molecules are otherwise identical, the frequency of occurrence of each species at equilibrium will be proportional to the configuration partition function Z for that species. Let $\mathscr{Z}^N$ be the configuration partition function appropriate for a canonical ensemble of N-molecule systems of this description. Then

$$\mathscr{Z} = \sum_{\{q\}} Z_{\{q\}} \tag{48}$$

where $\{q\} = dlld$ etc., or rmr etc., in terms of *meso* and racemic dyads.

The required sum of the products of the $\mathbf{U}_k^{(2)}$ in Eq. 26 over every stereoisomeric species may be generated by taking the serial product of

$$\mathscr{U}_k = (\mathbf{U}_m^{(2)} + \mathbf{U}_r^{(2)})_k \tag{49}$$

where $\mathbf{U}_m^{(2)}$ and $\mathbf{U}_r^{(2)}$ are defined by Eqs. 28. Thus,

$$\mathscr{Z} = \mathbf{J}^* \left(\prod_{k=1}^{x-1} \mathscr{U}_k \right) \mathbf{J} \tag{50}$$

The index k is retained to denote the kinds of units (apart from their asymmetries) occurring at locations k and $k+1$ in the chains of specified chemical constitution. Its present significance differs from usage above.

In formulating the partition function for the equilibrated system according to Eq. 50, we make no distinction between two enantiomorphic species such as *ddldl* etc. and *lldld* etc., and in fact only one term enters for each such pair. Full counting would therefore require a factor of 2 in Eq. 50. On the other hand, if we abandon the arbitrary labeling of one end of the chain (a suitable device for treating a chain of defined stereochemical sequence, but superfluous here), then division by a symmetry factor of 2 is appropriate for a vinyl chain molecule having the formula (A) assumed. It follows that Eq. 50 is the appropriate expression for $\mathscr{Z}$ if allowance is made for the twofold symmetry of the molecule about its center. For finite chains it is important to observe that $\tau^* = 1$ for the first dyad of the chain (see pp. 213 and 216).

If all units of the chains are identical chemically (apart from the configurations of their asymmetric centers), and we shall henceforth assume this to be the case, indexes k may be dropped except for the first unit, and we have

$$\mathscr{Z} = \mathbf{J}^* \mathscr{U}_1 \mathscr{U}^{x-2} \mathbf{J} \tag{51}$$

where $\mathscr{U}_1$ is written separately to allow for $\tau^* = 1$ therein.

If the distinction between $\mathscr{U}_1$ and $\mathscr{U}$ may be ignored, and it will always be legitimate to do so for sufficiently long chains, then the partition function $\mathscr{Z}$ may be expressed in terms of the eigenvalues λ_ζ of $\mathscr{U}$ as follows (see Appendix B and Chap. III)

$$\mathscr{Z} = \sum_{\zeta,\eta=1}^{\nu} A_{1\zeta} \lambda_\zeta^{x-1} B_{\zeta\eta} \tag{52}$$

where $\mathbf{A}$ is the matrix of eigenvectors of $\mathscr{U}$, and $\mathbf{B} = \mathbf{A}^{-1}$. For long chains the approximation

$$\mathscr{Z} \cong \lambda_1^{x-1} \tag{53}$$

may be adopted, where λ_1 is the largest eigenvalue of $\mathscr{U}$. We shall make use of this relationship later on.

The fraction of isotactic dyads in chains of any length is given by

$$f_{\text{iso}} = 1 - f_{\text{syn}} = (x-1)^{-1} \mathbf{U}_m^{(2)} \partial \ln \mathscr{Z} / \partial \mathbf{U}_m^{(2)} \tag{54}$$

Applying the methods of Chapter III, Section 6, we have

$$f_{\text{iso}} = (x-1)^{-1} \mathscr{Z}^{-1} [\mathbf{J}^* \mathbf{0}] \hat{\mathscr{U}}_1 \hat{\mathscr{U}}^{x-2} \begin{bmatrix} \mathbf{0} \\ \mathbf{J} \end{bmatrix} \tag{55}$$

where the $\mathbf{0}$ in the pre- and postmultiplying factors are of orders $1 \times \nu$ and $\nu \times 1$, respectively, and

$$\hat{\mathscr{U}} = \begin{bmatrix} \mathscr{U} & \mathscr{U}' \\ \mathbf{0} & \mathscr{U} \end{bmatrix} \tag{56}$$

where

$$\mathscr{U}' = \mathbf{U}_m^{(2)} \, \partial \mathscr{U} / \partial \mathbf{U}_m^{(2)} = \mathbf{U}_m^{(2)} \tag{57}$$

The matrix $\hat{\mathscr{U}}_1$ differs from $\hat{\mathscr{U}}$ through assignment of $\tau^* = 1$ in the former.

The f's may be regarded as *a priori* probabilities averaged over all dyads of the chain. Thus, f_{iso} is the analog of w_{iso} used in the preceding section. A different symbol is introduced here in acknowledgment of the fact that the succession of isotactic and syndiotactic dyads in the equilibrated polymers is neither Bernoullian nor Markoffian, in contrast to those described in Section 8. That they are not Markoffian will be evident from the equivalence of the conditional dependence of the character of a given dyad on its neighbors on both sides. Departures from Bernoullian and Markoffian distributions of units will be apparent in the numerical examples which are given below.

The fractions, or *a priori* probabilities, F_I, F_H, and F_S of triads which are isotactic (*ddd* alias *lll*), heterotactic (*ddl* alias *dll* and *lld* alias *ldd*), and syndiotactic (*dld* alias *ldl*) may be obtained by an extension of the foregoing procedure. Let a matrix be constructed from the statistical weights in such a way as to take account of the characters of *two* consecutive units. A matrix meeting this requirement is the following[23]:

$$\mathscr{W} = \begin{array}{c} \\ dd \text{ or } ll \\ dl \text{ or } ld \end{array} \!\!\begin{array}{c} \begin{array}{cc} dd \text{ or } ll & dl \text{ or } ld \end{array} \\ \begin{bmatrix} \mathbf{U}_m^{(2)} & \mathbf{U}_r^{(2)} \\ \mathbf{U}_m^{(2)} & \mathbf{U}_r^{(2)} \end{bmatrix} \end{array} \tag{58}$$

Rows are labeled to denote the characters of the two preceding units $k-2$ and $k-1$; column indexes refer likewise to units $k-1$ and k. Also let

$$\hat{\mathscr{W}}_1 = [\mathscr{W}_1 \quad \mathbf{0}] \tag{59}$$

and

$$\hat{\mathscr{W}}_I = \begin{bmatrix} \mathscr{W} & \mathscr{W}'_I \\ \mathbf{0} & \mathscr{W} \end{bmatrix} \tag{60}$$

where the $\mathbf{0}$ are of order $2\nu \times 2\nu$, and

$$\mathscr{W}'_I = \begin{bmatrix} \mathbf{U}_m^{(2)} & \mathbf{0} \\ \mathbf{0} & \mathbf{0} \end{bmatrix} \tag{61}$$

where the $\mathbf{0}$ are of order $\nu \times \nu$. Then the fraction of triads which are isotactic (i.e., *ddd* alias *lll*) at stereochemical equilibrium is

$$F_I = \mathscr{Z}^{-1}(x-2)^{-1}[\mathbf{J}^*\mathbf{0}]\hat{\mathscr{W}}_1 \hat{\mathscr{W}}_I^{x-2} \begin{bmatrix} \mathbf{0} \\ \mathbf{0} \\ \mathbf{J} \\ \mathbf{J} \end{bmatrix} \tag{62}$$

If the 2×2 forms of $\mathbf{U}_m$ and $\mathbf{U}_r$ are used (i.e., if $\nu = 2$), then $\hat{\mathscr{W}}_I$ is of order 8×8.

The *a priori* probabilities F_H and F_S for heterotactic and for syndiotactic triads are given by equations identical with Eq. 62 except for the replacement of $\hat{\mathscr{W}}_I$ by $\hat{\mathscr{W}}_H$ and by $\hat{\mathscr{W}}_S$, respectively, where

$$\hat{\mathscr{W}}_H = \begin{bmatrix} \mathscr{W} & \mathscr{W}'_H \\ \mathbf{0} & \mathscr{W} \end{bmatrix} \tag{63}$$

with

$$\mathscr{W}'_H = \begin{bmatrix} \mathbf{0} & \mathbf{U}_r^{(2)} \\ \mathbf{U}_m^{(2)} & \mathbf{0} \end{bmatrix} \tag{64}$$

and

$$\hat{\mathscr{W}}_S = \begin{bmatrix} \mathscr{W} & \mathscr{W}'_S \\ \mathbf{0} & \mathscr{W} \end{bmatrix} \tag{65}$$

where

$$\mathscr{W}'_S = \begin{bmatrix} \mathbf{0} & \mathbf{0} \\ \mathbf{0} & \mathbf{U}_r^{(2)} \end{bmatrix} \tag{66}$$

Obviously, $F_I + F_H + F_S = 1$. The incidence of various tetrads and of sequences of higher order could be found by elaboration of this procedure.

Equations 50, 51, and 54–66, inclusive, are applicable to chains of any length whose bonds are characterized by any (finite) number ν of rotational states. In the limit of large x, Eq. 53 expressing the partition function $\mathscr{Z}$ in terms of the largest eigenvalue λ_1 of $\mathscr{U}$ may be used, and the fraction f_{iso} of isotactic units at stereochemical equilibrium may be deduced from λ_1 and the secular equation of $\mathscr{U}$. We shall treat in particular the case in which it is permissible to take $\tau\omega = 0$, whereupon Eqs. 18–20 may be adopted to express the statistical weight matrices for individual bonds. Then

$$\mathbf{U}_m^{(2)} = \begin{bmatrix} (\eta^2\tau^*\omega + \eta) & (\eta\tau^* + \omega) \\ (\eta^2\omega + \eta\omega) & (\eta + \omega^2) \end{bmatrix} \tag{67}$$

$$\mathbf{U}_r^{(2)} = \begin{bmatrix} (\eta^2\tau^* + \eta\omega) & (\eta\omega\tau^* + 1) \\ (\eta^2 + \eta\omega^2) & (\eta\omega + \omega) \end{bmatrix} \tag{68}$$

and

$$\mathscr{U} = (1 + \omega)\begin{bmatrix} \eta(1 + \eta\tau^*) & (1 + \eta\tau^*) \\ \eta(\eta + \omega) & (\eta + \omega) \end{bmatrix} \tag{69}$$

Adopting Eqs. 67 and 68 for the dyad statistical weight matrices, we define the matrix

$$g\mathbf{U}_m^{(2)} + \mathbf{U}_r^{(2)} = \begin{bmatrix} (1 + g\omega)\eta^2\tau^* + (g + \omega)\eta & (1 + g\omega) + (g + \omega)\eta\tau^* \\ (1 + g\omega)\eta^2 + (g + \omega)\eta\omega & (1 + g\omega)\omega + (g + \omega)\eta \end{bmatrix} \tag{70}$$

where g is a factor introduced for reasons which will be apparent below. The secular determinant of this matrix is

$$\begin{aligned} \psi(g, \lambda) &= |g\mathbf{U}_m^{(2)} + \mathbf{U}_r^{(2)} - \lambda\mathbf{E}_2| \\ &= \lambda^2 - [(1 + g\omega)(\eta^2\tau^* + \omega) + 2(g + \omega)\eta]\lambda \\ &\quad + (g^2 - 1)(1 - \omega^2)(1 - \tau^*\omega)\eta^2 \end{aligned} \tag{71}$$

Let $\lambda_1(g)$ and $\lambda_2(g)$ be the eigenvalues obtained by equating Eq. 71 to zero, with $\lambda_1(g) > \lambda_2(g)$. For $g = 1$, these are the eigenvalues of the matrix $\mathscr{U}$; they are given by

$$\begin{aligned} \lambda_1 &= (1 + \omega)(2\eta + \eta^2\tau^* + \omega) \\ \lambda_2 &= 0 \end{aligned} \tag{72}$$

As will be apparent from the argument leading to Eq. III-62

$$\begin{aligned} f_{\text{iso}} &= [\partial \ln \lambda_1(g)/\partial g]_{g=1} \\ &= -\lambda_1^{-1}[(\partial\psi/\partial g)/(\partial\psi/\partial\lambda)]_{g=1,\psi=0} \end{aligned} \tag{73}$$

in the limit $x = \infty$. From Eq. 71 it follows that

$$f_{\text{iso}} = (2\eta + \eta^2\tau^*\omega + \omega^2)\lambda_1^{-1} - 2\eta^2(1 - \tau^*\omega)(1 - \omega^2)\lambda_1^{-2} \tag{74}$$

in the same limit. If $\omega = 0$,

$$\lambda_1 = \eta(2 + \eta\tau^*) \tag{75}$$

and

$$f_{\text{iso}} = 2(1 + \eta\tau^*)/(2 + \eta\tau^*)^2 \tag{76}$$

A similar treatment[23] of the *a priori* probabilities of various triads in the limit $x = \infty$ leads to the following results for the case $\omega = 0$:

$$F_I = (3\eta\tau^* + 2)/(\eta\tau^* + 2)^3 \tag{77}$$

$$F_H = 2[2(\eta\tau^*)^2 + 3\eta\tau^* + 2]/(\eta\tau^* + 2)^3 \tag{78}$$

$$F_S = 1 - [4(\eta\tau^*)^2 + 9\eta\tau^* + 6]/(\eta\tau^* + 2)^3 \tag{79}$$

The *a priori* probabilities for infinite chains are functions of $\eta\tau^*$ when $\omega = 0$, according to Eqs. 76–79.

It is of interest to compare Eqs. 76–79 for infinite chains with results obtained for short chains according to the methods of Eqs. 55–66, with $\omega = 0$ throughout. For $x = 2$, corresponding to a single dyad, we readily find

$$(f_{\text{iso}})_2 = 2\eta/(1 + \eta)^2 \tag{80}$$

and for $x = 3$ (two dyads, or one triad)

$$(f_{\text{iso}})_3 = (1 + \eta + \eta\tau^*)/(1 + \eta)(2 + \eta\tau^*) \tag{81}$$

$$(F_I)_3 = \eta(2 + \tau^*)/(1 + \eta)^2(2 + \eta\tau^*) \tag{82}$$

$$(F_H)_3 = 2(1 + \eta^2 + \eta^2\tau^*)/(1 + \eta)^2(2 + \eta\tau^*) \tag{83}$$

$$(F_S)_3 = \eta(2 + \eta^2\tau^*)/(1 + \eta)^2(2 + \eta\tau^*) \tag{84}$$

Generation of algebraic expressions for longer chains is tedious. Numerical results for chains of any length may readily be computed, however, by Eqs. 55–66.

Comparisons of numerical results for dimeric and trimeric molecules with those for indefinitely long chains are presented in Table 2 for three arbitrary sets of values of η and τ^*, with $\omega = 0$ as required for Eqs. 76–84 to apply. Results for $x = 2$ and 3 for the first choice of parameters suggest a random distribution of dyads without bias in favor of either isotactic or syndiotactic forms. The relations

$$\begin{aligned} F_I &= f_{\text{iso}}^2 \\ F_H &= 2f_{\text{iso}}\,(1 - f_{\text{iso}}) \\ F_S &= (1 - f_{\text{iso}})^2 \end{aligned} \tag{85}$$

which should hold if the dyads were independent (Bernoullian sequence distribution), are indeed obeyed for the populations of dimers and trimers in this example. This result is atypical; it does not hold for other values of η

Table 2

Equilibrium Stereoisomeric Composition[23]

($\omega = 0$)

Dyad composition		Triad composition			
x	f_{iso}	F_I	F_H	F_S	F_I/F_S
		$\eta = 1; \tau^* = 1$			
2	1/2				
3	1/2	1/4	1/2	1/4	1
∞	4/9	5/27	14/27	8/27	5/8
		$\eta = 2; \tau^* = 1$			
2	4/9				
3	5/12	1/6	1/2	1/3	1/2
∞	3/8	1/8	1/2	3/8	1/3
		$\eta = 2; \tau^* = 1/4$			
2	4/9				
3	7/15	1/5	8/15	4/15	3/4
∞	12/25	28/125	64/125	33/125	28/33

and τ^*, as the second and third examples show. Even for the first example, departures occur for larger values of x as indicated by the results for $x = \infty$. Equations 85 are generally incorrect for both finite and infinite chains. This is a consequence of the interdependence of the spatial configurations (or conformations) of neighboring units, which in turn precludes factorizing the statistical weight (i.e., Z) for the chain as a whole into separate contributions for its respective units, or dyads. The neighbor dependence of the conformational states causes the stereoisomeric state of a dyad to depend on those of its neighbors.

If the sequence of dyads were Markoffian, then the conditional probabilities for triads, e.g., F_I/f_{iso}, should be independent of the chain length x. Examination of the numerical results for each of the three examples illustrates the dependence of the conditional probabilities on x.

The smaller proportions of isotactic as compared with syndiotactic dyads and triads in long chains reflects the greater number of spatial configurations made possible by syndiotactic placements. The degree of preference depends of course on the values of the parameters. For $\tau^* = 0$, $(f_{iso})_\infty = (f_{syn})_\infty = \frac{1}{2}$, the two kinds of units being subject to equally severe conformational constraints as earlier pointed out. Values of $(f_{iso})_\infty$ and of the ratio F_I/F_S decrease with increase in τ^* and in η.

The stereochemical equilibrium is determined by the same rotational potentials, including interactions between nonbonded atoms and groups, as those affecting the occurrence of various conformations for a given stereochemical species. The same set of parameters serves to describe both the chemical and the conformational equilibria, as the foregoing theory demonstrates. This invites comparison[23] of experimental results for the two kinds of equilibria; the same values of the parameters should serve to represent both. Initial attempts in this direction using low molecular dimeric and trimeric analogs of vinyl polymers appear to be confirmatory.[12,23,24] The theory serves also to relate either stereochemical or conformational equilibria for low molecular homologs to equilibria in polymeric chains of any length. Future investigations along these lines should clarify the nature of conformational interactions in various vinyl polymers.

REFERENCES

1. G. Natta and F. Danusso, *J. Polymer Sci.*, **34**, 3 (1959); G. Natta, M. Farina, and M. Peraldo, *Makromol. Chem.*, **38**, 13 (1960); M. L. Huggins, G. Natta, V. Desreux, and H. Mark, *J. Polymer Sci.*, **56**, 153 (1962).
2. P. J. Flory, J. E. Mark, and A. Abe, *J. Am. Chem. Soc.*, **88**, 639 (1966).
3. D. W. Scott, J. P. McCullough, K. D. Williamson, and G. Waddington, *J. Am. Chem. Soc.*, **73**, 1707 (1951).
4. J. R. Thomas and W. F. Gwinn, *J. Am. Chem. Soc.*, **71**, 2785 (1949).
5. J. K. Brown and N. Sheppard, *J. Chem. Phys.*, **19**, 976 (1951).
6. A. Abe, *J. Am. Chem. Soc.*, to be published (paper on optical activities).
7. D. R. Lide, Jr., *J. Chem. Phys.*, **33**, 1519 (1960).
8. C. W. Bunn and D. R. Holmes, *Discussions Faraday Soc.*, **25**, 95 (1958).
9. T. Shimanouchi, *Pure Appl. Chem.*, **12**, 287 (1966).
10. T. Shimanouchi, M. Tasumi, and Y. Abe, *Makromol. Chem.*, **86**, 43 (1965); Y. Abe, M. Tasumi, T. Shimanouchi, S. Satoh, and R. Chûjô, *J. Polymer Sci. A-1*, **4**, 1413 (1966).
11. D. Doskočilová, J. Štokr, B. Schneider, H. Pivcová, M. Kolínský, J. Petránek, and D. Lím, *J. Polymer Sci. C*, **16**, 215 (1967); B. Schneider, J. Štokr, D. Doskočilová, S. Sýrkora, J. Jakes, and M. Kolínský, *J. Polymer Sci. C*, to be published.
12. A. D. Williams, unpublished.
13. Y. Fujiwara, S. Fujiwara, and K. Fujii, *J. Polymer Sci. A-1*, **4**, 257 (1966); S. Fujiwara, Y. Fujiwara, and T. Fukuroi, Intern. Symp. Macromol. Chem., Tokyo, Kyoto, 1966.
14. G. Natta, P. Pino, P. Corradini, F. Danusso, E. Mantica, G. Mazzanti, and G. Moraglio, *J. Am. Chem. Soc.*, **77**, 1708 (1955); G. Natta, *J. Polymer Sci.*, **16**, 143 (1955); G. Natta and P. Corradini, *Makromol. Chem.*, **16**, 77 (1955).
15. G. Natta and P. Corradini, *J. Polymer Sci.*, **34**, 21 (1959); *Nuovo Cimento Suppl.*, **15**, 9 (1960).

16. G. Natta, P. Corradini, and I. W. Bassi, *Nuovo Cimento Suppl.*, **15**, 68 (1960).
17. G. Natta and P. Corradini, *J. Polymer Sci.*, **20**, 251 (1956); G. Natta, I. Pasquon, P. Corradini, M. Peraldo, M. Pegoraro, and A. Zambelli, *Atti Accad. Nazl. Lincei Rend. Classe Sci. Fis. Mat. Nat.*, **28**, 539 (1960); G. Natta, *Makromol. Chem.*, **35**, 93 (1960); G. Natta, M. Peraldo, and G.Allegra, *ibid.*, **75**, 215 (1964).
18. N. P. Borisova and T. M. Birshtein, *Vysokomolekul. Soedin.*, **5**, 279 (1963).
19. P. DeSantis, E. Giglio, A. M. Liquori, and A. Ripamonti, *J. Polymer Sci. A*, **1**, 1383 (1963). A. M. Liquori, *J. Polymer Sci. C*, **12**, 209 (1966).
20. T. M. Birshtein and O. B. Ptitsyn, *Conformations of Macromolecules* (translated from the 1964 Russian ed., S. N. Timasheff and M. J. Timasheff), Interscience, New York, 1966, p. 157.
21. J. E. Mark and P. J. Flory, *J. Am Chem. Soc.*, **87**, 1423 (1965).
22. A. E. Tonelli, unpublished results.
23. P. J. Flory, *J. Am. Chem. Soc.*, **89**, 1798 (1967).
24. A. D. Williams, J. I. Brauman, N. J. Nelson, and P. J. Flory, *J. Am. Chem. Soc.*, **89**, 4807 (1967).

CHAPTER VII

Polypeptides, Proteins, and Analogs

Polypeptides command special attention not only because of their biological importance, but also because they possess a combination of structural features that is nearly unique. The chemical constitution of a polypeptide chain is shown in Fig. 1. It is characterized by a repeat unit which spans three skeletal atoms. These repeat units, or α-amino acid *residues* —NH—CHR—CO—, are numbered sequentially 0 to x as indicated by the numbers in parentheses opposite the α carbon (CHR) of each of the residues. The choice of terminal groups, amino (NH_2) and carboxyl (COOH), for the polypeptide represented in Fig. 1 is essentially arbitrary. Other end groups could of course be adopted with only minor alterations in the treatment to follow; these alterations would assume importance only for finite chains in any case.

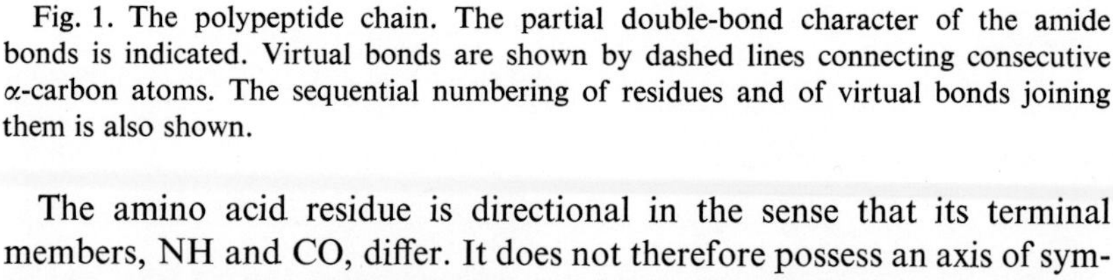

Fig. 1. The polypeptide chain. The partial double-bond character of the amide bonds is indicated. Virtual bonds are shown by dashed lines connecting consecutive α-carbon atoms. The sequential numbering of residues and of virtual bonds joining them is also shown.

The amino acid residue is directional in the sense that its terminal members, NH and CO, differ. It does not therefore possess an axis of symmetry, and the same dissymmetry is conferred on the chain as a whole. That is to say, rotation of the fully extended chain about an axis perpendicular to its length is nondegenerate, unlike the vinyl chains considered in the preceding chapter which are symmetric with respect to end-for-end rotation, except as this symmetry may be modified by the asymmetric centers in vinyl chains.

The second feature of note is the presence of an asymmetric center —CHR— in each α-amino acid residue, or structural unit, provided of course that R $\neq$ H. Because of the directional character of the amino acid residue, the configurations of its asymmetric centers are absolute and not subject to circumstances extraneous to the residue itself, as was observed to be the case for vinyl chains. In polypeptides of living organisms these centers are, with rare exceptions, of the L configuration.

The third and most distinctive structural characteristic of the polypeptide chain stems from the planarity of the amide group imparted by the partial double bonded nature of the amide linkage (see Chap. V, p. 182 *et seq.*). The system of skeletal bonds from one α carbon to the next, including also the C=O and N—H bonds of the intervening CONH group, is coplanar. This system of bonds and the atoms joined directly by them have been designated[1] as the *peptide unit*, which is not to be confused with the amino acid *residue*, or chemical repeat unit. Peptide unit i may be construed to include also the side chain R_i and the atom H_i attached to C_i^{α} (see Fig. 1). The chain comprises x such units.

Evidence showing the conformation of the amide linkage within a polymer chain to be exclusively *trans*, or virtually so, was presented in Chapter V, Section 12, in connection with treatment of polyamide chains of the nylon type. The same conformation appears to hold invariably for the amide groups in polypeptide chains consisting of residues —NH—CHR—CO— in which the NH groups are unsubstituted. We shall confine attention at the outset to residues of this type. In particular, we exclude for the present polypeptides containing the prolyl residue

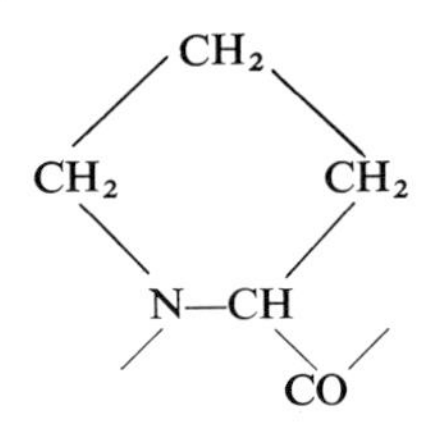

This residue manifests the capacity to occur in the *cis* as well as the *trans* conformation, depending on conditions. Polypeptides containing prolyl residues are treated toward the end of Section 2.

Uniform adherence of all peptide units to the planar *trans* form permits the spatial configuration of the polypeptide chain to be expressed in terms of hypothetical *virtual bonds* which join successive α-carbon atoms one to another. Virtual bonds are shown in Fig. 1 by dashed lines. To the extent that bond lengths and angles are fixed and the amide bond adheres to the *trans* conformation, the vector $\mathbf{l}_{u;i}$ representing virtual bond i is fixed in length and is rigidly embedded in the plane of the peptide unit which it spans. The orientation of a virtual bond thus defined relative to its neighbor is determined by the rotation angles about the two skeletal bonds flanking the α-carbon atom joining the virtual bond with its neighbor.

The virtual bond vectors fully describe the skeletal configuration. The polypeptide chain represented in Fig. 1 comprises x virtual bonds connecting successive α-carbon atoms serially numbered as indicated, commencing with carbon atom zero on the left. In accord with the serial numeration convention set forth in Chapter I, the ith virtual bond joins α carbons $i-1$ and i. We shall measure the end-to-end vector $\mathbf{r}$ from the zeroth α carbon to the xth. Hence

$$\mathbf{r} = \sum_{i=1}^{x} \mathbf{l}_{u;\,i} \tag{1}$$

Another property of the peptide unit to be noted is its large dipole moment (3.7 D). The fact that these moments recur in close proximity in the chain raises the possibility of a significant contribution to the conformational energy from dipole–dipole interactions. These interactions have been shown to affect the average dimensions of the random chain markedly.[2]

Poly(lactic acid) having the chemical repeat unit —O—CH(CH_3)—CO— will be considered also in the present chapter. It is an ester analog of the polypeptides. As was pointed out in Chapter V, pp. 182–3, the ester linkage, like the amide, is planar *trans*. The geometrical description of the poly(L-lactic acid) chain is therefore similar to its polypeptide analog. The dipole moment of the ester group (ca. 1.8 D) is much smaller however.

1. STERIC MAPS FOR POLYPEPTIDES

The structual geometry of the L-polypeptide chain is represented in Fig. 2. On the assumption that the peptide unit is planar *trans* and rigid, the angles ξ and η between virtual bonds and the single bonds N—C^{α} and C^{α}—C, respectively, are fixed. The spatial configuration of the chain is determined by the pairs of rotation angles φ_i, ψ_i about these single bonds flanking the α-carbon atom of each residue. Table 1 lists structural parameters describing the peptide unit, as compiled by Sasisekharan.[3*]

Of the various interactions affecting conformations of chains considered thus far, steric encounters between nonbonded atoms and groups assume foremost importance. Polypeptides are no exception to this rule. As will be readily apparent from even a cursory examination of models, the overlaps

* These parameters differ slightly from those given for the amide group in a polyamide chain of the kind treated in Chapter V; see Fig. V-27. The differences do not exceed experimental uncertainties.

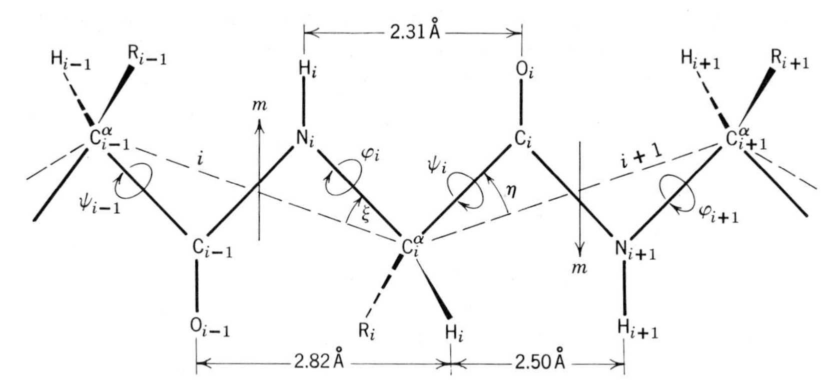

Fig. 2. Geometrical representation of a portion of a polypeptide chain in its planar, fully extended conformation for which bond rotation angles φ and ψ are zero. Other conformations are generated by alteration of these rotation angles. Successive amino acid residues are indexed $i-1$, i, and $i+1$. Virtual bonds connecting the α-carbon atoms of these residues are shown by light dashed lines. Group dipole moments are indicated by arrows **m**. Interatomic distances shown refer to the planar conformation of the chain backbone.

Table 1

Structural Parameters for Peptide Units[a]
(from Sasisekharan[3])

Bond lengths, Å		Bond angle and its supplement (θ), deg[c]
C^{α}—C,	1.53	$\angle C^{\alpha}CN$, 66
C—N,	1.32	$\angle CNC^{\alpha}$, 57
N—C^{α},	1.47	$\angle NC^{\alpha}C$, 70
C=O,	1.24	
N—H,	1.00	
C^{α}—C^{β},[b]	1.54	
C^{α}—H^{α},	1.07[d]	

[a] $\eta = 22.2°$; $\xi = 13.2°$ (see Fig. 2). l_u = length of virtual bond = 3.80 Å.

[b] C^{β} denotes the first carbon of the substituent R, assuming R to be —CH_2—R′.

[c] Angles for bonds to pendant atoms (O, H, R) may be assumed, in legitimate approximation, to subdivide equally the ranges available to them.

[d] Sasisekharan[3] quotes a lower value of 1.00 Å for the C—H bond length.

encountered in the course of varying one dihedral angle of the pair φ_i, ψ_i adjoining the ith α carbon are markedly dependent on the value assigned to the other angle of the pair. Critical inspection of molecular models reveals, however, that the interactions associated with rotations of one such interdependent pair are quite independent of the angles assumed by neighboring pairs.[2] In order to comprehend this important characteristic of the polypeptide chain more fully, we observe that if one pair of rotation angles φ_i, ψ_i is varied while values of zero are maintained for all others, atoms and groups preceding C^{α}_{i-1} (including R_{i-1}; see Fig. 2) do not for any values of φ_i and ψ_i come sufficiently close to atoms and groups beyond C^{α}_{i+1} (including R_{i+1}) for an appreciable interaction, repulsive or attractive, to occur between a member of one of these sets of atoms and groups and those of the other. If the φ_i, ψ_i pair is varied over the ranges not excluded by steric overlaps between atoms of the peptide units spanned by virtual bonds i and $i + 1$, and if at the same time the φ_{i+1}, ψ_{i+1} pair is permitted to vary correspondingly, no combination of the four rotations will be found which brings peptide unit i into interaction with unit $i + 2$. Interactions between units i and $i + 3$ can be brought about by similarly varying the sequence of three consecutive rotation pairs over their individually allowed ranges, but they occur only within a very small fraction of the associated angle space. Thus, the distinction is well defined between *short-range* and *long-range* interactions (see p. 33), which latter may be presumed to have been nullified by correcting experimental results to the Θ point (see Chap. II). In consequence of the planar, *trans* structure of the amide group, interactions of short range are virtually confined to those involving peptide units which are first neighbors.

The polypeptide chain exemplifies the case of "isolated" pairs of rotation angles, with members of the same pair strongly interdependent while those of different pairs are virtually independent. It suffices therefore to treat each such pair separately from others in the chain. The interactions associated with the ith pair of angles (involving atoms within the span from C^{α}_{i-1} to C^{α}_{i+1}, inclusive, but exclusive of atoms and groups beyond these limits) depend on the nature of the R group of the ith amino acid residue, and on the asymmetry of the α-CHR center of which it is a part; they do not depend on the substituents and asymmetries which characterize adjoining residues. We may therefore consider interactions for a given pair of peptide units (or the associated pair of virtual bonds) to be characteristic of the kind of amino acid residue at the junction of the pair, ignoring all others. We shall assume however that the residue considered is bonded to other α-amino acid residues on either side, or to equivalent chemical entities as required to complete two peptide units; it is *not* a terminal residue of the polypeptide chain, for example.

With the foregoing considerations in mind, Ramachandran and co-workers[3–7] have constructed diagrams to represent the permitted ranges over φ and ψ, which for a given residue are free of inordinate steric overlaps. A Ramachandran diagram, or steric map, is shown in Fig. 3 for L-alanine for which $R = CH_3$. Solid lines enclose regions within which none of the distances between nonbonded pairs of atoms is less than the value given in the second column of Table 2. Regions enclosed by dashed lines are those which comply with the somewhat smaller distances given in the last column of Table 2. These distances were chosen[3–6] on the basis of contact distances within molecules and between molecules in crystals. The "normally allowed" values are those found most frequently in crystals; values down to the "outer limits" are known but rare. The necessary calculations were carried out using the structural data in Table 1. It is

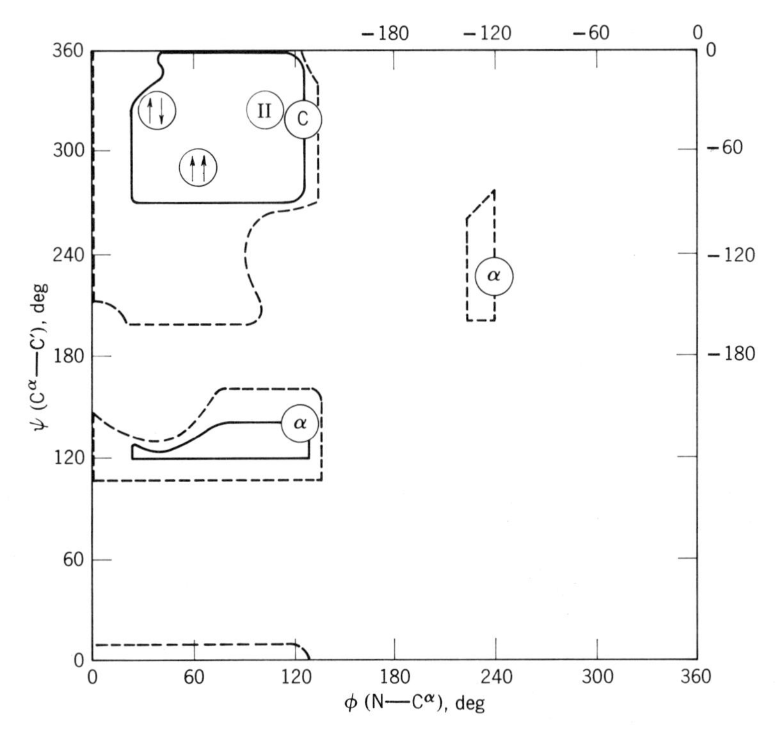

Fig. 3. Ramachandran steric map[4] for the L-alanyl residue. "Normally allowed" regions (Table 2) are enclosed by solid lines; those meeting the requirements set by the "outer limit" distances (Table 2) are enclosed by dashed lines. Conformations φ, ψ corresponding to several helices are indicated as follows: right- and left-handed α helices, (α), $\pm 122°$, $\pm 133°$[12,13]; parallel pleated sheet, (↑↑), $+62°$, $-68°$[13]; antiparallel pleated sheet, (↑↓), $38°$, $-35°$[13]; polyglycine-II (left-handed) and poly-L-proline-II, (II), $102°$, $-35°$[12]; collagen, (C), $123°$, $-40°$.[7,12]

Table 2

Minimum Contact Distances for Atom Pairs in Polypeptides According to Ramachandran and Co-workers[4]

	Minimum distance, Å	
Atom pair	Normally allowed	Outer limit
C · · · C	3.2	3.0
C · · · O	2.8	2.7
C · · · N	2.9	2.8
C · · · H	2.4	2.2
O · · · O	2.8[a]	2.7[a]
O · · · N	2.7	2.6
O · · · H	2.4	2.2
N · · · N	2.7	2.6
N · · · H	2.4	2.2
H · · · H	2.0	1.9

[a] Revised values for the O · · · O distances given by Ramakrishnan and Ramachandran[5] are lower by 0.1 Å.

apparent at once that the accessible domains for the L-alanine residue comprise only a small fraction of the total angle space. Corresponding diagrams for other α-amino acid residues have been computed by Ramakrishnan and Ramachandran[5] and by Leach, Némethy, and Scheraga.[8] The domain of angles permitted for glycine, for which R = H, is much greater than that shown for alanine in Fig. 3. Steric maps for residues such as leucine [R = $—CH_2CH(CH_3)_2$], glutamine (R = $—CH_2CH_2CONH_2$), and lysine (R = $—CH_2CH_2CH_2CH_2NH_2$) which have longer side chains but are of the form $—CH_2—R'$, differ little from those for alanine (R = $—CH_3$).[5,8] A side chain such as that of valine [R = $—CH(CH_3)_2$] which is branched at the β carbon diminishes the range of conformations available,[8] as should be expected.

The Ramachandran diagrams give a good account of observed conformations for residues in various peptides and in polypeptides whose structures have been determined by x-ray diffraction, including the residues in the proteins myoglobin[5] and lysozyme.[7,9] Several of them have conformations (φ, ψ) falling at the edge of the "outer limit" domain of the steric map, however. These maps are computed for fixed bond angles, and it must be borne in mind that bond angles are subject to appreciable deformation. Difficulty in accommodating a few of the residues within the protein structure can readily be resolved by altering bond angles by no more than 5–10°.[10,11] Allowed regions are considerably increased by relaxing the assignments of fixed bond angles to this extent. The energy for a deformation of this magnitude is only of the order of $2kT$ at ordinary temperature (Chap. I, p. 13). Thus, the boundaries of the so-called permitted regions of the steric domain should not be construed as rigid limits; they may be exceeded, but at some price of energy. Distortion of bond angles, or of the torsional angle about the C—N semidouble bond, offers the easiest means by which to transgress these limits.

Also designated in Fig. 3 are the conformations[5–7,12,13] of several of the more prominent polypeptide helices generated by repetition of a given pair of values of φ and ψ throughout all residues of the helical sequence. These helices fall into one or the other of two classes: (*1*) the right- and left-handed α helices approximating g^+g^+ and g^-g^-, respectively, and (*2*) the comparatively extended helices located in the large allowed region occupying the upper left-hand quadrant of the diagram. The latter are often referred to collectively as β forms. Whereas the α helix is stabilized by intramolecular hydrogen bonds,[12,14] the β forms are characterized by interchain hydrogen bonding, except in polyproline where the absence of an amide hydrogen deprives the residue of the capacity to form a hydrogen bond with another residue of the same kind. The Ramachandran diagram (Fig. 3) sanctions the conformation of the α helix, and it indicates a preference for the right-handed helix by an L residue such as L-alanine. This preference is well confirmed by experimental observations.

2. CONFORMATIONAL ENERGIES FOR POLYPEPTIDES

Treatment of the random coil configurations requires a more refined assessment of the conformations within the allowed regions. It is obvious that bond torsional potentials and London dispersion interactions (attractions) will render some of these regions more favorable than others. The resulting discriminations within a given allowed region and between competitive regions of the map can be expected to affect averages over the chain configurations, and these differences may be of critical importance. From this standpoint the requirements for analysis of the configurational statistics appear to be more stringent than for the mere assessment of allowed conformations. On the other hand, a sharp distinction between permitted regions and those which are disallowed is not required. Moreover, the effects of deformations of bond angles may be ignored, inasmuch as they are largely eliminated in the act of averaging over all configurations of the system, as has been pointed out previously (see p. 13).

The conformational energy associated with the interdependent pair of rotation angles φ and ψ can be estimated after the manner of the methods introduced by Chapter V (see pp. 131–133). A major difference arises however in consequence of the fact that the torsional potentials about the single bonds flanking C^{α} of the amino acid residue probably are fairly low[2]—on the order of 1.0 kcal mole^{-1} compared with ca. 3.5 kcal mole^{-1} for the C—C bond in the *n*-alkanes (see Table V-1). They are doubtless threefold, and minima probably occur at 0° (*trans*) for each.* The energy barriers separating them are not sufficiently large however to assure against their obliteration by other contributions to the conformational energy. Treatment of the configurational statistics in terms of discrete states located approximately at the minima in the bond torsional potentials is therefore inappropriate.

Adaptation of Eq. V-1 to expression of the conformational energy associated with bond rotations in residue i of a polypeptide chain yields[2,15]

$$E(\varphi, \psi) = (E_{\varphi}^{\circ}/2)(1 - \cos 3\varphi) + (E_{\psi}^{\circ}/2)(1 - \cos 3\psi) + \sum_{k,l} E_{kl}(\varphi, \psi) + E_C \tag{2}$$

Individual atoms are indexed k, l; the serial index i denoting the relevant bond or residue is omitted for simplicity. The summation term includes interactions for every nonbonded atom pair whose distance of separation depends on φ_i and ψ_i exclusively, i.e., atoms within the interval from C_{i-1}^{α} to C_{i+1}^{α}, inclusive. The term E_C represents the coulombic interactions between the adjoining pair of amide groups arising from their large electrical asymmetries as expressed for example by their dipole moments. The nonbonded interactions may be represented as in Eq. V-2 by the combination of an exponential term for the repulsion and a term in r^{-6} for the attraction between the atom pair. Alternatively,[15,16] an inverse power function may be used for the repulsion, in which case the pair interaction term takes the familiar form

$$E_{kl}(\varphi, \psi) = a_{kl}/r_{kl}^{m} - c_{kl}/r_{kl}^{6} \tag{3}$$

where the distances r_{kl} between atoms k and l are functions of φ and ψ, and a_{kl} and c_{kl} are parameters for the atom pair. If the exponent m is taken to be about 12 (Lennard-Jones potential), the results do not differ discernibly from those obtained using an exponential repulsive term (Buckingham potential) like that in Eq. V-2, with $b \cong 4.5$ Å^{-1}, provided of

* Evidence for these conclusions is reviewed in Chapter V, p. 185. See also ref. 2.

course that the parameters a_{kl} are determined in a consistent manner.[15] The c_{kl} may be evaluated from atomic polarizabilities as described in Chapter V, Section 2. Assignment of the remaining parameters, i.e., the a_{kl}, is critical. As in the treatment of n-alkanes and related molecules (Chap. V), they are given values arbitrarily selected to reproduce observed distances in crystals. It is important in assigning values to the a_{kl} to take account of interactions between nearby atoms other than the pair in contact,[15] as was pointed out in Chapter V, pp. 134–135. To this end the a_{kl} were assigned values such as to locate the minima in $E_{kl}(\varphi, \psi)$ at distances r_{kl} which are sums of atomic radii as follows:

$$r_{\mathrm{H}}^{\circ} = 1.3\,\text{Å},\ r_{\mathrm{C}}^{\circ} = 1.8\,\text{Å},\ r_{\mathrm{O}}^{\circ} = 1.6\,\text{Å},\ r_{\mathrm{N}}^{\circ} = 1.65\,\text{Å},\ \text{and}\ r_{\mathrm{CH_2}}^{\circ} = 1.95\,\text{Å}$$

These values were chosen[15] to be ca. 0.10 Å greater than the usual van der Waals radii for these atoms in order to correct for the effects of attractions imposed by other atoms in the molecule. They are consistent with the set of radii used for similar calculations on n-alkanes (see p. 134).

The coulombic term may be estimated on the basis of the interaction of two point dipoles situated at the midpoints of the amide bonds as indicated in Fig. 2, or by assigning partial electronic charges to the associated atoms. In the original calculations of Brant and Flory[2] the former method was used. The dipole moment of the amide group was assigned the value 3.7 D found by Meighan and Cole[17] for various alkyl-substituted amides. Computations were subsequently carried out according to the second procedure[15] by assigning partial electronic charges of $\pm 0.281e$ to H and N, respectively, of the N—H group, and $\pm 0.394e$ to C and O, respectively, of the carbonyl group, where e represents the magnitude of the charge of the electron. These four charges, arrayed as stated, reproduce the observed group dipole moment; they also correspond approximately to accepted bond dipole moments.

The effective dielectric constant which should be used in estimating the coulombic interactions by either method is problematical. Experiments on polypeptides are, by necessity, conducted in polar solvents, often in water. The multipolar array of charges, or dipoles, chosen to represent the electrical asymmetry of the structure is largely shielded from the surrounding bulk solvent of high dielectric constant by associated groups of the polypeptide chain, and possibly also by solvent molecules complexed (e.g., hydrogen bonded) to the amide group. Such complexed solvent molecules will not partake of the dielectric polarizability of free solvent (e.g., water) molecules. The effective dielectric constant for the coulombic interactions (multipolar and therefore of reduced range) between a neighboring pair of peptide units buried within a (solvated) polypeptide chain may therefore be expected to approximate more nearly the dielectric constant ϵ of the chain than that of the bulk solvent. Brant and co-workers[2,15] chose a value of 3.5 for ϵ. The absence of a significant dependence of the experimental characteristic ratio[18a] on the solvent medium and on the length of the R group suggests ineffectiveness of dielectric shielding of the coulombic interactions by the solvent. However, the dimensions calculated[2] for polypeptide chains are found to be fairly insensitive to the value chosen for the dielectric constant, provided that it falls within the range from about 2 to 10 (but not if it falls outside this range). Experiments are therefore indecisive on the authenticity of the numerical value of 3.5 chosen for ϵ for the calculations. For the same reason, the precise value assigned to ϵ is not important.

The Glycyl Residue

Maps of the conformational energy for the glycyl residue ($R = H$) calculated through the use of Eqs. 2 and 3 are shown in Figs. 4 and 5.[15] Energy contours are drawn at intervals of 1 kcal mole^{-1} relative to the lowest minima marked by crosses. Contours above 5 kcal mole^{-1} are not shown. The torsional energy barriers E_{φ}° and E_{ψ}° were taken to be 1.5 and 1.0 kcal mole^{-1}, respectively.[2,15] Figure 4 was calculated with neglect of the coulombic interaction energy, i.e., with $E_C = 0$. Figure 5 was obtained by adding to this energy surface the coulombic energy E_C calculated as the energy of interaction between the sets of four point charges specified above for each of two neighboring amide groups. The absence of an asymmetric center in the glycyl residue requires equivalence of energy, and statistical weighting, for rotations φ, ψ and $-\varphi$, $-\psi$. The diagrams consequently are symmetric with respect to reflection through their centers at $\varphi = \psi = 180°$.

Certain features of these conformational energy maps admit of straightforward interpretation. As is apparent at once from Fig. 2 (with $R = H$) or from a suitable model, a *cis* conformation about either of the pair of single bonds entails a very large overlap. Regions near either of the two lines $\varphi = 180°$ and $\psi = 180°$, which would bisect the diagram vertically and horizontally, respectively, are therefore strongly forbidden. The planar *trans* conformation $\varphi = \psi = 0$ is disfavored, although to a much lesser degree, due to the repulsive interaction between amide hydrogen H_i (attached to N_i) and O_i, these atoms being only 2.31 Å apart at $\varphi_i = \psi_i = 0$. Their mutual repulsion is readily diminished by a small rotation about either bond. Consequently, the energy in Fig. 4 diminishes in all directions from the origin (adjoined by quadrants at each of the four corners of the diagram), except near the diagonal $\psi = -\varphi$. This exception arises because the distance between the atoms in question is altered

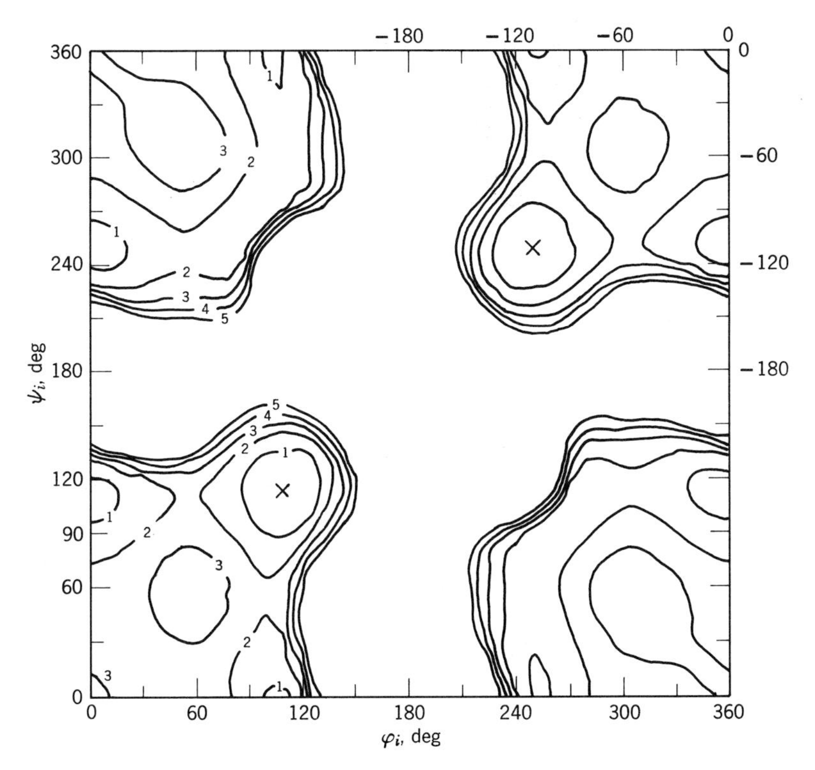

Fig. 4. Contour map of the conformational energy[15] of the glycyl residue calculated according to Eq. 2, with $E_C = 0$. Contours are shown at intervals of 1 kcal mole^{-1} relative to the lowest minima marked ×; contours above 5 kcal mole^{-1} are omitted.

imperceptibly by small rotations φ and ψ of equal magnitude and opposite sign. Small minima appearing along the margins of the figure are reminiscent of the $tg^{\pm}$ and $g^{\pm}t$ minima for the chains discussed in Chapter V. These minima are due in part to the inherent torsional potentials attributed to the single bonds. Repulsions between the H_i atoms attached to C_i^{α} and O_{i-1}, which are maximal at $\varphi_i \cong \pm 60°$ and diminish for $|\varphi| > 60°$, contribute also to the $g^{\pm}t$ minima along the abscissas. Interactions between the same H_i and the amide hydrogen H_{i+1} similarly affect the $tg^{\pm}$ minima along the ordinates. (The distances $O_{i-1} \cdots H_i$ and $H_i \cdots H_{i+1}$ are reduced, respectively, to 2.40 Å at $\varphi_i = 60°$ and to 2.08 Å for $\psi_i = -60°$; the corresponding distances for $\varphi_i = \psi_i = 0$ are shown in Fig. 2.) Incipience of the *cis* overlaps displaces the minima with respect to φ and ψ to values smaller than $|120°|$.

All of the repulsions cited above for the glycyl residue are relieved by simultaneously rotating *both* φ and ψ to angles $> |60°|$. Hence, the

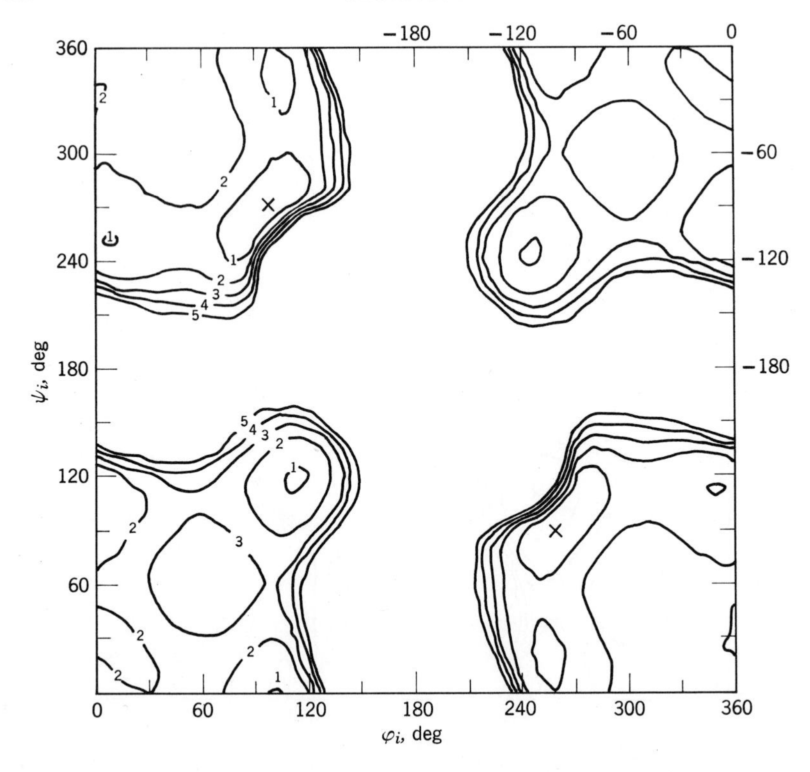

Fig. 5. Conformational energy map[15] for the glycyl residue calculated according to Eq. 2, with E_C estimated on the basis of point monopole charges (see text; $\epsilon = 3.5$). Contours are drawn at 1 kcal mole^{-1} intervals as in Fig. 4.

favored conformations in Fig. 4 are approximately g^+g^+ and g^-g^-. The g^+g^- state at 120°, −120° (and likewise its mirror image at −120°, 120°) is rendered unsatisfactory by overlaps not unlike those occurring in analogous conformations for other chains considered previously. However, a trend toward a minimum is evident in this region of Fig. 4. Instead of an actual minimum, a saddle occurs at about 95°, −95° and a trough extends from it in either direction.

The energy surface of the φ, ψ map is affected also by nonbonded attractions, but to a lesser degree. These are invariably small for any given pair, but, being of longer range, the number of them contributing in a given conformation is greater. For the same reason, the sum of these London dispersion energies is subject to a smaller variation with φ and ψ.

In the planar *trans* conformation shown in Fig. 2 the group dipole moments are favorably oriented antiparallel to one another, with each dipole approximately perpendicular to the line between them. At the minima in Fig. 4—which would correspond approximately to right- and left-handed α helices if perpetuated precisely for successive residues (compare Fig. 3)—they are very nearly parallel and inclined somewhat to the line between them. In this arrangement the dipole energy is positive (repulsive). Along the diagonal $\varphi + \psi = 0$, the dipole directions are predominantly antiparallel, but their distance apart decreases with increase in $|\varphi| = |\psi|$ from 0° to 180°. In the vicinity of this line the coulombic interaction remains favorable.

Figure 5 shows the conformational energy map[15] obtained by adding the coulombic energy term E_C, estimated through approximation of the electrical asymmetry by the array of partial charges specified on p. 257, to the energy surface depicted in Fig. 4. The negative coulombic energy along the line $\varphi + \psi = 0$ causes the saddle points in Fig. 4 to be replaced by minima near $\pm 100°$, $\mp 95°$, which may be designated approximately as g^+g^- and g^-g^+. The positive electrostatic energy in the g^+g^+ and g^-g^- regions raises the energy of the previous minima to a value above that of the new minima in Fig. 5. Although the new minima appearing in Fig. 5 are only a little lower in energy than the g^+g^+ and g^-g^- minima, the troughs of low energy occurring near the g^+g^- and g^-g^+ minima enhance the weighting of these regions. Thus, the $+ -$ and $- +$ quadrants of the energy surface represented in Fig. 7 account for 65% of the residue partition function at ordinary temperature, and the $+ +$ and $- -$ quadrants for 35%, according to calculations (see Eqs. 4 and 5 below) by Brant.[18b]

Approximation of the electrical asymmetry of the amide groups by point dipoles[2] as described earlier (p. 257) alters the conformational energy contours similarly but to a greater degree [15]; the g^+g^+ and g^-g^- minima are raised nearly 3 kcal mole^{-1} above the emergent g^+g^- and g^-g^+ minima. The array of partial charges is believed to offer a more reliable basis for estimation of the coulombic interaction between amide groups.

The L-Alanyl Residue

Diagrams computed similarly[15] for the L-alanyl residue are shown in Figs. 6 and 7. The former was computed according to Eq. 2, with $E_C = 0$; the latter includes the coulombic energy estimated on the basis of the array of partial charges. The methyl substituent R was treated as a spherical group having a van der Waals radius of 1.95 Å (see above). The results are considered to be applicable to other residues having side chains R of the type —CH_2—R′. The presence of the R group renders right- and left-handed rotations nonequivalent. The diagrams therefore are not symmetric. Positive rotations φ_i about N—C^α bond are preferred over negative ones owing to the repulsion between $(CO)_{i-1}$ and R_i (see Fig. 2). The energy in

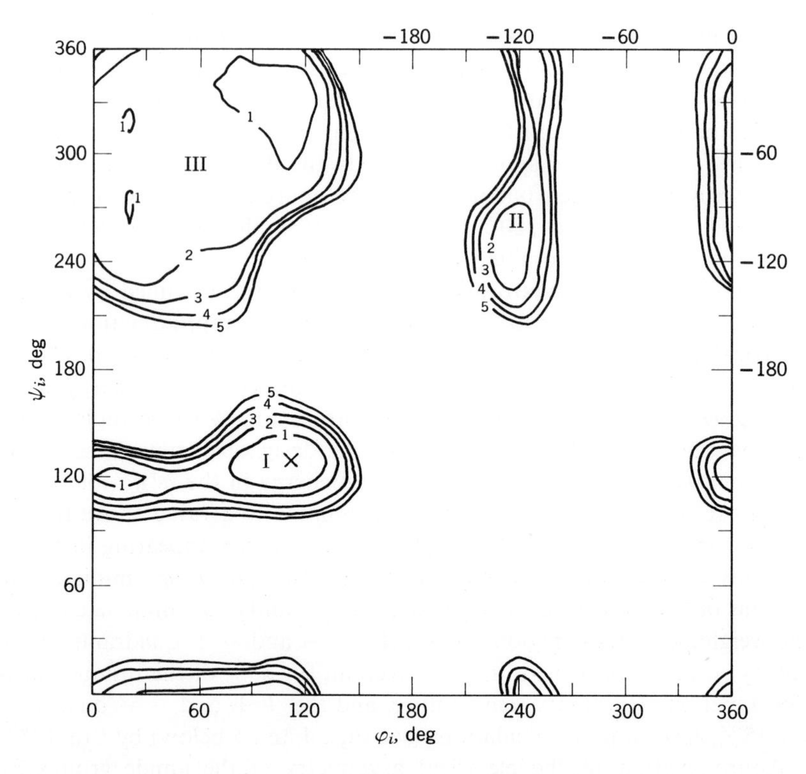

Fig. 6. Conformational energy map[15] for the L-alanyl residue calculated according to Eq. 2, with $E_C = 0$. The principal minima are labeled I, II, and III; the one of lowest energy is marked ×. Contours relative to the energy at this minimum are drawn as stated in the legend to Fig. 4.

the region near $\varphi = -60°$ is increased markedly on this account as may be seen by comparing Fig. 6 with Fig. 4. For rotations ψ_i of small magnitude about the C$^\alpha$—C bond, negative values are preferred over positive ones due to the interaction of $(NH)_{i+1}$ with R_i in Fig. 2; the region near $\psi = 60°$ is rendered virtually inaccessible by this repulsion. For larger $|\psi_i|$ the repulsions involving $(NH)_{i+1}$ are relieved, and the preference for negative rotations vanishes. In fact, the energy at $\varphi_i = 0°$, $\psi_i = 120°$, is calculated to be somewhat lower than at $\varphi_i = 0°$, $\psi_i = -120°$; at $\psi_i = 120°$, $(NH)_{i+1}$ is at a favorable distance from R_i, where London interactions lower the energy. In other respects, the rules enunciated above apply to the glycyl residue.

The $-+$ quadrant is disqualified almost in its entirety by the simultaneous interactions of R_i with $(CO)_{i-1}$ and with $(NH)_{i+1}$. In the $+-$ quadrant, on the other hand, these groups are well separated and the

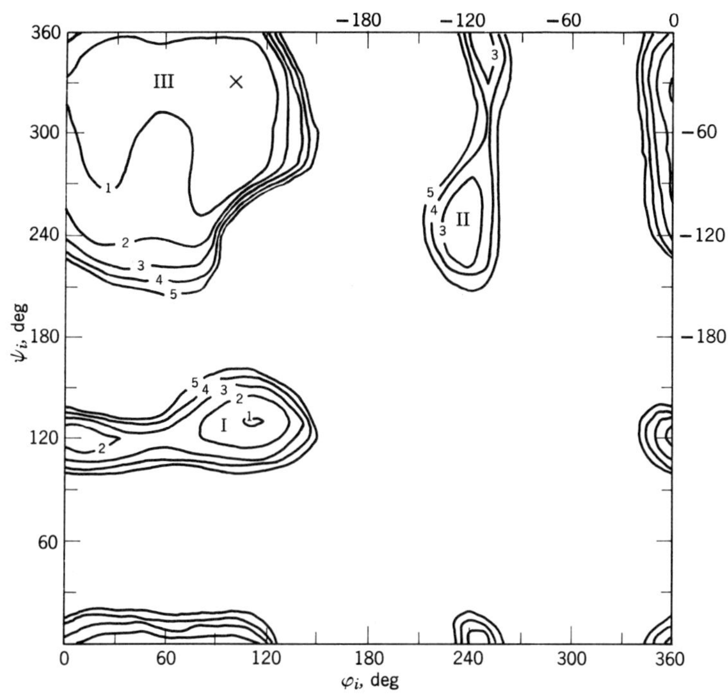

Fig. 7. Conformational energy map[15] for the L-alanyl residue calculated according to Eq. 2, with E_C estimated on the basis of point monopole charges (see text; $\epsilon = 3.5$ as in preceding calculations). The lowest minimum (×) occurs in region III (compare Fig. 6).

energy consequently is affected relatively little by the presence of the R group. The form of the lowest energy contours is altered compared with those in the same quadrant for the glycyl residue, but the contours (3 and 4 kcal mole^{-1}) delineate an accessible domain, designated III in Fig. 6, that resembles the corresponding region for the glycyl residue in Fig. 4.

Simultaneous rotations of the same sign are less sharply differentiated. The $--$ quadrant includes conformations in which $(CO)_{i-1}$ is near R_i; in the $++$ quadrant $(NH)_{i+1}$ approaches R_i. Principally because of the greater size of O compared with H, the former repulsion exceeds the latter. The energy at minimum I in Fig. 6 consequently is lower than at minimum II. The conformations of the right- and left-handed α helices lie close to these respective minima (see Fig. 3). The preference for the right-handed α helix usually exhibited by an L-polypeptide is thus seen to follow from rudimentary considerations of the asymmetry of rotations about the pair of bonds adjoining the α-carbon atom in an amino acid residue having the L configuration.

The diffuse region III in Fig. 6 is the domain in which the interactions encountered near the *tt* state, and cited in the foregoing account of the glycyl residue, are alleviated by rotations of both angles in their preferred directions. The repulsions associated with a g^+g^- pair delimit this region. Nevertheless, the accessible range in III considerably exceeds that of I, a feature manifested in the steric map shown in Fig. 3. Although the energies at the minima in I and III do not differ decisively, the former is calculated, with $E_C = 0$, to be lower than the latter. London dispersion interactions cited above, and operating between R_i and $(NH)_{i+1}$ when at the favorable distance dictated by this conformation (I), appear to be responsible for the difference.

Addition of E_C calculated by either of the two procedures moves the minimum from region I to III, as shown in Fig. 7. The latter domain includes the β conformations which, if repeated regularly throughout sequences of many units, would yield sheet-like structures wherein parallel chains are hydrogen bonded intermolecularly (see p. 255). As in the case of the glycyl residue, the calculated energy difference[15] between regions I and III is greater when E_C is estimated in the point dipole approximation, but, as previously indicated, this approximation probably overestimates the coulombic interactions. The large area of region III in Fig. 7 invests it with a considerable advantage over region I; the greater domain of III appears to be more significant than the small difference in energy in its favor. The residue partition function at 25°C comprises contributions from the principal domains as follows[18b] I, 5.6%; II, 1.0%; III, 93.4%.

Conformation energy maps for glycyl and L-alanyl residues which confirm the principal features shown in Figs. 5 and 7 have been reported by Scott and Scheraga.[16] They were calculated similarly, but by use of lower torsional potential barriers E_φ° and E_ψ°. The former barrier was shifted in phase by 60° from its representation in Eq. 2 above, the minima associated with rotation angle φ thereby being displaced to $\pm 60°$ and 180°. Differences between the energy maps thus calculated and the ones reproduced here are minor.

Diagrams for a D-alanyl residue are obtained by reflecting those of Figs. 6 and 7 through their centers; this operation is tantamount to reversing signs of rotation angles. These figures may be assumed to be representative also of other α-amino acid residues in which R is a longer, unbranched side chain of the kind $—CH_2—R'$. For bulkier side chains, additional interactions must be considered.[8]

Calculations of this nature, being mainly empirical in basis and approximate in numerical result, must be accepted with caution. Only the general features which persist when the parameters are varied over reasonable ranges, and when alternative procedures are compared, can be regarded as reliable. Steric overlaps are of overriding importance. The similarity of Figs. 6 and 7 to the steric map shown in Fig. 3 is apparent, and much of the content of the more detailed results can be gleaned from the simpler Ramachandran diagram. Thus, the latter (e.g., Fig. 3) gives clear indication of the preference for the right-handed α helix. The more elaborate calculations offer a measure of refinement needed for our purposes, however; the principal gradations of energy within the "allowed" regions of Figs. 6 and 7 can be seen to be genuine by critical examination of models. The effects of dipolar interactions, also important, are beyond the scope of the steric map. They materially affect the stabilities of helical forms of polyamides in general, as was first pointed out by Arridge and Cannon[19]; they also affect the random-coil configurational statistics of polypeptides.[2]

The Lactyl Residue

The poly(L-lactic acid) chain,

$$[\text{—O—}\underset{\displaystyle \text{CH}_3}{\underset{|}{\text{CH}}}\text{—CO—}]_x$$

is of interest as an analog of poly-L-alanine. Figure 8 displays the chain in its extended conformation. The ester group, like the amide, is planar and *trans*, as was pointed out in Chapter V, Section 12. Geometrical data for

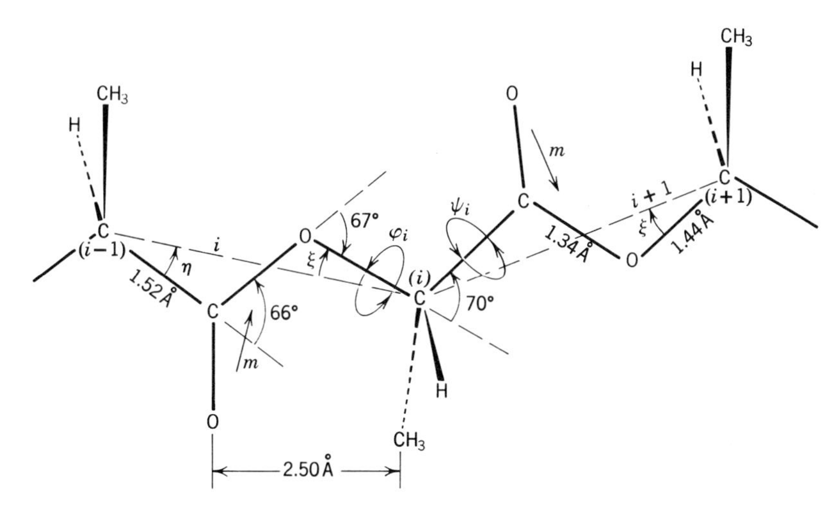

Fig. 8. The poly(L-lactic acid) chain in its extended, planar conformation. The dipole moment of the ester group (1.8 D) is shown as the vector **m**. Virtual bonds of length $l_u = 3.70$ Å spanning units i and $i + 1$ are shown by dashed lines. Geometrical data included in the diagram yield $\eta = 18.9°$ and $\xi = 19.9°$.

the ester group are included in Fig. 8. Two of the bond lengths quoted in Fig. 8 and used in calculations below differ insignificantly (by 0.01 Å) from those in Fig. V-27*b*. Principal structural differences relative to the peptide unit (compare Fig. 2) are the smaller valence angle (larger θ) by 10° at the oxygen atom as compared with $\angle CNC^{\alpha}$ (see Table 1), and the absence of the amide hydrogen atom present in the amino acid residue. The virtual bond is 0.10 Å shorter than that for the peptide unit, and the angles ξ and η differ somewhat (compare the structural data given in the legend to Fig. 8 with those in Table 1).

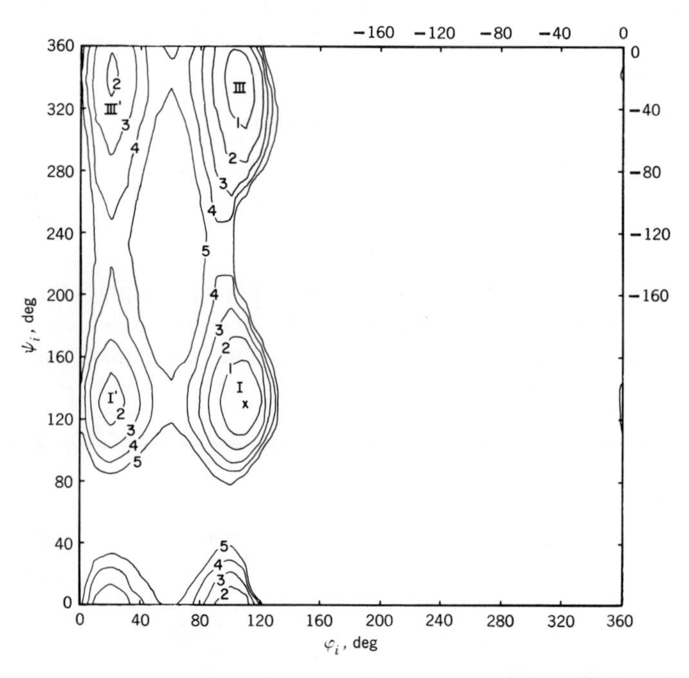

Fig. 9. Conformational energy map[20] for the L-lactyl residue calculated according to Eq. 2, with $E_C = 0$. Contours are drawn at 1 kcal mole^{-1} intervals as in Figs. 4–7 relative to the lowest minimum (I) marked ×.

The energy contour diagram calculated[20] according to Eqs. 2 and 3, with $E_C = 0$ and $E_{\varphi}^{\circ} = E_{\psi}^{\circ} = 1.0$ kcal mole^{-1}, is shown in Fig. 9. Nonbonded interactions were calculated using the same parameters employed in the computations presented above for peptide residues. Owing to the absence of the amide hydrogen, the energy is comparatively insensitive to ψ, except near $\psi_i = 60°$ where O_{i+1} in Fig. 8 is *cis* to $(CH_3)_i$. The energy in the range $\psi = 40$–$80°$ is excessive on this account. Absence of the amide hydrogen eliminates the repulsion mainly responsible for raising the energy of the glycyl and alanyl residues (see Fig. 4) in their planar *trans* forms.

The energy displays a much more marked dependence on φ, as is at once apparent in Fig. 9. The smaller valence angle at the oxygen atom of the lactyl unit as compared to the corresponding angle in the peptide unit is largely responsible for this feature of the diagram. Examination of the structural basis for the strong dependence on φ reveals large steric overlaps of $(CO)_{i-1}$, with $(CH_3)_i$ for small negative φ_i and with $(CO)_i$ for large negative φ_i. These repulsions raise the energy above 5 kcal mole^{-1} throughout both right-hand quadrants of Fig. 9. Even in the left-hand portion of the diagram, repulsions between these groups displace minima which otherwise would occur at $\varphi_i = 0$ and 120° to ca. 20 and 110°, respectively. Repulsion between $(CO)_{i-1}$ and H_i produces a ridge separating these minima and running throughout the range of ψ_i. Four minima labeled III′, III, I′, and I in Fig. 9 are thus explained. Minima III′ and I′ are about 1 kcal mole^{-1} greater in energy than minima III and I, respectively. The minimum I (approx. g^+g^+) is marginally lower in energy than III (approx. g^+t), in analogy to the energy functions calculated for glycyl and alanyl residues with $E_C = 0$.

The dipole moment of the ester group, 1.8 D, is only about half as large as that of the amide group, and it may be directed somewhat differently. Instead of being nearly parallel to the C=O bond as in the amides, it has been estimated to be oriented at an angle of about 20° with this bond (see Chap. V, p. 184). Bond moments suggest a representation of the electrical asymmetry as follows: $\pm 0.35e$ at the carbonyl carbon and oxygen atoms and $\pm 0.109e$ at the α carbon and skeletal oxygen atoms. This array of partial charges produces a dipole moment of 1.8 D and its direction makes an angle of 21° with the C=O bond. The electrostatic energy calculated on this basis using an effective dielectric constant of 3.0 is much smaller than the corresponding contribution for the polypeptides. The conformational energy including E_C thus estimated is shown in Fig. 10.[20] The coulombic energy contribution lowers the minimum III below I, but the difference is only about 100 cal mole^{-1}. These two regions are closely competitive. The two minima I′ and III′ near $\varphi = 20°$ are higher by about 1 kcal mole^{-1}.

The conformational energy map for the L-lactyl residue departs much more from the map for L-alanyl than might have been anticipated from the close similarity of their structures. Reasons for the differences are apparent from the foregoing examination of the chemical and geometrical differences in detail. The comparison of alanyl and lactyl residues underlines the importance of a source of reliable and sufficiently precise structural data.

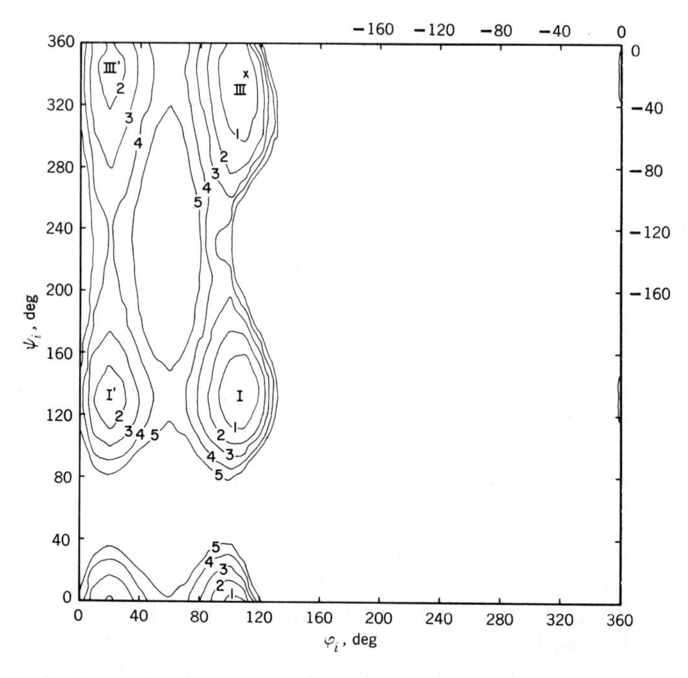

Fig. 10. Conformational energy map[20] for the L-lactyl residue calculated according to Eq. 2, with E_C estimated on the basis of monopole charges (see text).

The Prolyl Residue

The presence of *two* methylene substituents attached to the nitrogen of this residue (see the chemical formula on p. 249) obscures the grounds for preference for the *trans* over the *cis* configuration* about the amide bond, these forms being defined according to the disposition of the pendant bonds of the main chain skeleton. The configuration about the amide bonds is certainly planar; which of the two planar forms should be preferred, however, requires reexamination in this instance. Inspection of models readily reveals that the *cis* configuration engenders greater steric congestion than the *trans* when the prolyl residue is situated amongst other residues in a polypeptide chain. On this basis we may expect the *trans* form to be dominant. This expectation is partially confirmed in the case of the homopolymer, poly-L-proline[21,22]; the polymer chains in the apparently more common crystalline form, poly-L-proline-II, occur in the *trans*

* In the interests of clarifying the ensuing discussion, we shall refer to the *cis* and *trans* isomeric forms of the amide bonds as configurations rather than as conformations. Their forms will thus be distinguished from the conformations of the adjoining skeletal single bonds.

configuration.[23,24] However, another crystalline form, designated poly-L-proline-I, is known[21,22] and its x-ray diffraction pattern has been interpreted in terms of the *cis* configuration.[25] Moreover, two corresponding forms have been identified in solution.[21,22] They are readily distinguished by their optical rotations, which differ greatly. Poly-L-proline-II is the stable form in good solvents such as acetic acid, formic acid, and water; pyridine, and mixtures of acetic acid with an excess of either *n*-propanol or *n*-butanol, favor poly-L-proline-I. The two forms are interconvertible[26] with change of solvent composition, and the rate of the transformation from one form to the other may readily be followed by polarimetry. Further discussion of these forms of the homopolymer will be postponed pending consideration of the conformation of the *trans* residue when, as in a copolymer, it is isolated from the encumbrances imposed by neighboring prolyl residues.

An L-prolyl residue situated amongst residues of the kind —NH—CHR—CO— should be expected to occur preferentially in the *trans* form. The *cis* configuration of the amide bond compacts neighboring residues and limits their conformations, and for this reason its occurrence appears less likely. On the other hand, the *cis* form should not be ruled out altogether; in certain solvent media it appears to be preferred by the prolyl residues of gelatin.[27] The following analysis of the role of an isolated prolyl residue in a polypeptide chain will be concerned exclusively with the *trans* form, which doubtless is the more prevalent configuration.

An L-prolyl residue thus "isolated" within a chain of residues of the latter type is shown in Fig. 11. The pyrrolidine ring narrowly confines the

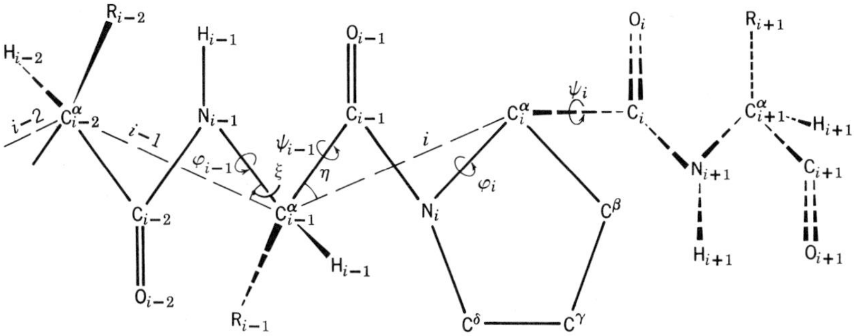

Fig. 11. Schematic diagram of a portion of an L-polypeptide chain containing an isolated *trans* L-prolyl residue. Virtual bonds $i-1$ and i are shown by light dashed lines. Dihedral angles φ and ψ specify rotations about N—C$^\alpha$ and C$^\alpha$—C bonds, as in preceding diagrams. The rotation angle $\varphi_i \cong 120°$ enforced by the pyrrolidine ring place the portion of the chain to the right of C_i^α out of the plane of preceding residues for which $\varphi = \psi = 0$.

rotation φ about the N—C$^{\alpha}$ bond of the prolyl residue, indexed i in Fig. 11. According to the x-ray crystallographic analysis of the structure of L-leucyl-L-prolyl-glycine carried out by Leung and Marsh,[28] $\varphi = 122°$ in the conformation adopted by this molecule in the crystalline state. A lower value, $\varphi = 102°$, was deduced by Sasisekharan[24] for the crystalline form of the *trans* homopolymer, poly-L-proline-II, from its x-ray diffraction. The former result may be presumed to approximate more closely the preferred value of φ for a prolyl residue isolated from other prolyl residues as in Fig. 11. In any event, the restriction on φ enforces departure of the chain skeleton from planarity, as the diagram in this figure illustrates.

The conformational energy for a *trans* L-prolyl residue isolated from other prolyl residues has been calculated by Schimmel[29] as a function of ψ according to Eqs. 2 and 3. The various parameters representing nonbonded interactions were assigned the same values used for the residues similarly treated above. Bond lengths and bond angles for the prolyl residue were taken from the work of Leung and Marsh[28]; the dihedral angle φ_i was assumed to be fixed at 122°. These calculations and the deductions that follow from them should hold also for a residue from a derivative of L-proline, e.g., hydroxy-L-proline. The coulombic interactions of the amide group were not included.

The results of these calculations[29] are shown in Fig. 12. The curve consists of two potential wells which correspond, respectively, to the minima labeled I and III in Figs. 6 and 7. In fact, Fig. 12 closely resembles

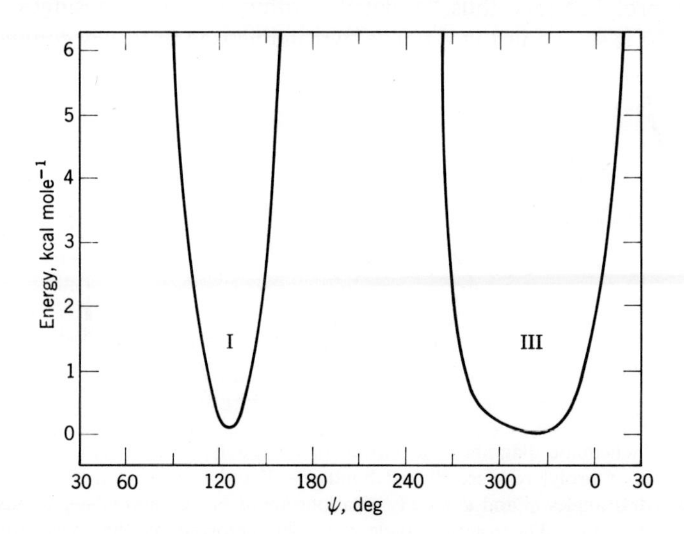

Fig. 12. Conformational energy calculated[29] for the isolated *trans* L-prolyl residue with $\varphi_i = 122°$. Minima I and III correspond to the regions so labeled in Figs. 6 and 7.

the section at $\varphi = 122°$ taken through the surfaces represented in these figures. The energies calculated for the minima I and III are almost identical. The minimum III is much broader than I, however. The dipole moment of the amide group of the prolyl residue is unknown, but is probably smaller than for other residues. Coulombic interactions may be presumed to lower minimum III somewhat but by no more than 1 kcal mole^{-1} relative to I. This revision would be within the probable limits of reliability of the calculations, which clearly indicate a preference for minimum III over I on account of the greater breadth of the former, and probably also in consequence of its lower energy. The isolated prolyl residue may therefore be likened to an alanyl residue with φ restricted to a value near 120°.

The effect of an isolated L-prolyl residue on the configuration of the polypeptide of which it is a part cannot, however, be dismissed with this summarizing statement.[29] The δ-methylene group of the pyrrolidine ring interposes steric repulsions which affect the rotation ψ_{i-1} of the residue *preceding* prolyl residue i in Fig. 11. Overlap of the δ-$(CH_2)_i$ group by $(NH)_{i-1}$ raises the energy for large values of $|\psi_{i-1}|$. If R_{i-1} is CH_3 or a larger group, the energy is rendered prohibitive throughout the range $0 < \psi_{i-1} < 180°$ by the overlap of R_{i-1} with δ-$(CH_2)_i$, in addition to the other interactions common to the residues of all α-amino acids.* Region I is therefore effectively excluded for the residue preceding an L-prolyl residue if the substituent R of that residue is substantially larger than hydrogen. If the preceding residue is glycyl, the effect of prolyl thereon is minor.[29]

The conformational energy calculated[29] for an L-alanyl residue (or other similar residue) preceding L-prolyl is shown in Fig. 13. The similarity of the contours to those in the upper half of Figs. 6 and 7 is apparent; energies in the region $0° < \psi < 240°$ are raised above 7 kcal mole^{-1} by the interactions involving the δ-CH_2 group of the pyrrolidine ring. This marked influence of a prolyl residue on its predecessor constitutes a departure from the rule enunciated on p. 252 to the effect that the conformational energy for residue i of a polypeptide chain depends only on the constitution of that residue and not on the characters of its neighbors. The prolyl residue does not impose constraints on its successor in the chain; its effect is operative only on the preceding residue.[29] It is to be noted also that the presence of an isolated *trans* prolyl residue does not predicate an exception to the rule that rotations within a given residue are independent of the rotations of neighboring residues.

* Similar steric effects on the preceding residue are revealed by the calculations of Mark and Goodman[30] on poly(N-methyl-L-alanine).

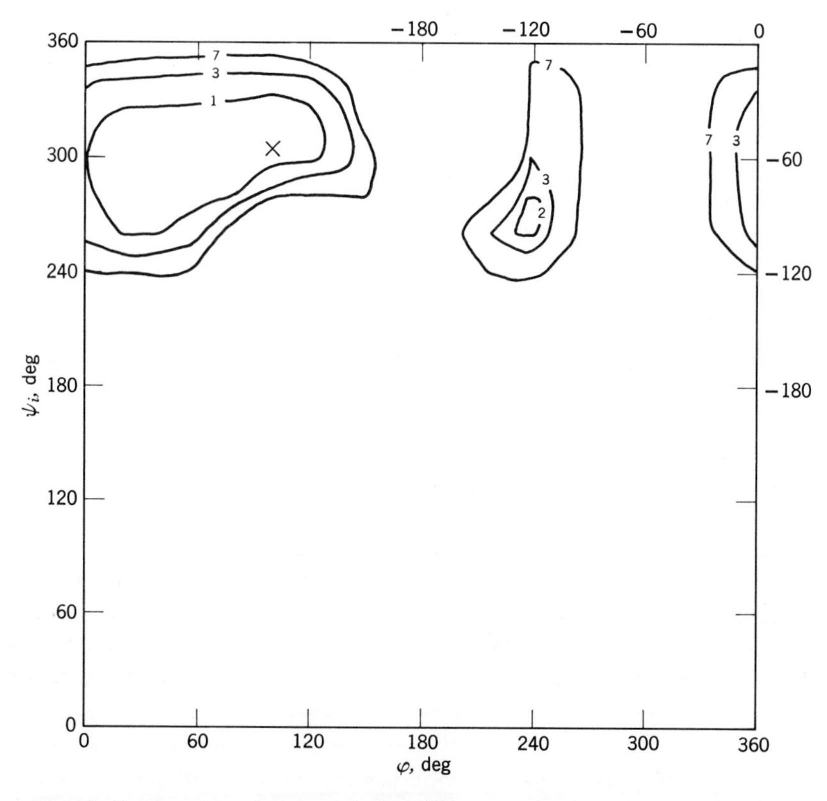

Fig. 13. Conformational energy calculated[29] for an L-alanyl residue which is followed in the polypeptide chain by *trans*-L-prolyl.

The effect of the L-prolyl residue of particular significance is its elimination of region I, including the right-handed α-helical conformation, from the conformations accessible to the preceding residue, unless that residue is glycyl. Thus, proline cannot occur *within* an α helix, except in conjunction with glycine.[29] It may however occupy the initial position of a right-handed α-helical sequence, inasmuch as it does not delimit the conformations of the following residue.

The influence of the prolyl residue on its predecessor is operative as well in a sequence of *trans* prolyl residues. Such a sequence is shown in Fig. 14. Again, conformation I is precluded for all but the last residue of the sequence. The conformational energy calculated as a function of ψ for a residue within a chain of L-prolyl residues very nearly duplicates the curve for minimum III in Fig. 11, the minimum I being eliminated. According to these calculations carried out by Schimmel[31] with φ assigned its value (102°) for the poly-L-proline-II helix,[24] the minimum occurs at

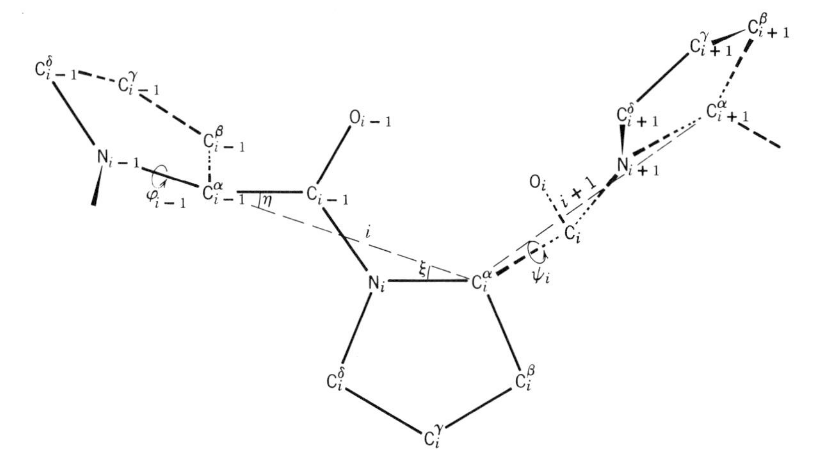

Fig. 14. The poly-L-proline chain with all amide bonds in the *trans* configuration.

$\psi = 304°$.* The region within which the energy is no more than 1 kcal mole^{-1} above the minimum is ca. $290° < \psi < 340°$, which may be compared with ca. $280° < \psi < 350°$ for the range of corresponding energy according to curve III in Fig. 12. The value of ψ deduced from the x-ray diffraction of crystalline poly-L-proline-II is 327°,[24] which is well within the range calculated to be of low energy.

The *cis* residues in poly-L-proline-I are subject to greater steric hindrances than the *trans* residues of poly-L-proline-II. One conformation free of excessive steric repulsion is accessible to poly-L-proline-I.[31,33] It occurs near $\psi = -15°$, and the admissible range of angle about this value is much smaller than for the residues of poly-L-proline-II. The acceptability of this conformation for a given residue of the all-*cis* chain is contingent, moreover, upon the adoption of the same conformation by neighboring residues on both sides. Thus, in exception to the rule asserted above for other residues, the interactions experienced by a *cis* prolyl residue in a polyproline-I chain are strongly dependent upon the conformations of neighboring residues.[31,33] Interactions occur which depend upon as many as three consecutive rotation angles ψ.

Cis residues in the allowed helical conformation with $\psi \cong -15°$ are tightly packed. The occurrence of I as the stable form under any circumstance implies that the groups forced into close contact interact attractively.[31]

* Liquori and co-workers[32] have calculated the conformational energy for a *trans* poly-L-proline chain in which the angles ψ for every residue are varied in concert, i.e., their calculations refer to the energy of a residue in the polyproline helices defined by various values of ψ. Their results correspond closely however to the calculations of Schimmel and Flory[31] for a single residue.

That form I should be preferred in poor solvents is consistent with this inference, inasmuch as the copious intrachain contacts in form I replace less favorable solvent–polymer interactions.

Examination of models for the poly-L-proline chain will show that realization of this favorable energy by a given *cis* residue requires its several neighbors to occur likewise in the *cis* configuration. Occurrence of a sequence of *cis* residues is therefore favored over dispersion of occasional *cis* residues in an otherwise *trans* polyproline chain. Additionally, isolated *cis* residues would obstruct the conformational freedom of adjoining *trans* residues. These circumstances are presumed to be responsible for the pronounced cooperativity of the I⇄II transformation,[31,33] as evidenced by its occurrence within a narrow interval of solvent composition.[21,34,35]

3. RANDOM-COIL STATISTICS OF HOMOPOLYMERIC CHAINS

The Partition Function

The partition function for a residue, formally defined as

$$z = \iint \exp\left[-E(\varphi, \psi)/RT\right] d\varphi d\psi \tag{4}$$

may be approximated as the sum of Boltzmann factors, taken at equal intervals of φ and ψ throughout their ranges. That is

$$z \cong \sum_{\varphi} \sum_{\psi} \exp\left[-E(\varphi, \psi)/RT\right] \tag{5}$$

Energies $E(\varphi, \psi)$ are furnished by the computed conformation energy represented by the contours in Figs. 4–7, 9, 10, 12, and 13. The average energy and configurational entropy per residue are

$$\langle E \rangle = z^{-1} \sum_{\varphi} \sum_{\psi} E(\varphi, \psi) \exp\left[-E(\varphi, \psi)/RT\right] \tag{6}$$

and

$$S = R(\ln z + \partial \ln z / \partial \ln T)$$

$$S = R \ln z + \langle E \rangle / T \tag{7}$$

The value of the entropy depends of course on the interval chosen for φ and ψ for the evaluation of the summations in these equations, i.e., the entropy is subject inevitably to an additive constant, and hence its absolute value has no significance. Separate contributions to the average energy $\langle E \rangle$ from the several terms in Eq. 2 may of course be determined according to Eq. 6, if desired.

Averaged Transformation Matrices

Pursuant to the evaluation of the characteristic ratio, let a coordinate system be affixed to each virtual bond. The X axis of this reference frame will be taken in the direction of the virtual bond, the Y axis will be placed in the plane of the peptide unit with its direction making an acute angle with the amide C⩦N bond of the same unit, and the Z axis will be taken in the direction required for completion of a right-handed orthogonal reference frame. Coordinate systems (xyz) for the various skeletal bonds will be defined in accordance with conventions previously used (see Chap. I, Sect. 8).

Let $\mathbf{R}(\tau, \rho)$ be the matrix which transforms the representation of a vector in one Cartesian reference frame to its representation in another, the former (initial) reference frame being brought into coincidence with the latter (final) frame by a rotation τ about the initial z axis followed by a rotation ρ about the final x axis. This matrix $\mathbf{R}(\tau, \rho)$ is given in Appendix B, Eq. B-4. The axes $X_{i+1}Y_{i+1}Z_{i+1}$ are related to the xyz system for the skeletal bond C$^\alpha$—C of residue i by rotations $\tau = -\eta$ and $\rho = -\psi_i$ (see Fig. 2). Hence, transformation from the former reference frame to the latter is effected by

$$\mathbf{R}(-\eta, -\psi_i) = \begin{bmatrix} \cos\eta & -\sin\eta & 0 \\ \sin\eta\cos\psi_i & \cos\eta\cos\psi_i & -\sin\psi_i \\ \sin\eta\sin\psi_i & \cos\eta\sin\psi_i & \cos\psi_i \end{bmatrix} \tag{8}$$

as follows from Eq. B-4. The transformation from this skeletal bond to the one preceding involves rotations θ^α and π-φ_i, and is given therefore by

$$\mathbf{R}(\theta^\alpha, \pi - \varphi_i) = \begin{bmatrix} \cos\theta^\alpha & \sin\theta^\alpha & 0 \\ \sin\theta^\alpha\cos\varphi_i & -\cos\theta^\alpha\cos\varphi_i & \sin\varphi_i \\ \sin\theta^\alpha\sin\varphi_i & -\cos\theta^\alpha\sin\varphi_i & -\cos\varphi_i \end{bmatrix} \tag{9}$$

Transformation from this skeletal bond (i.e., the N—C$^\alpha$ bond of residue i) to virtual bond i involves a single rotation $\tau = \xi$. It is represented therefore by

$$\mathbf{R}(\xi, 0) = \begin{bmatrix} \cos\xi & \sin\xi & 0 \\ -\sin\xi & \cos\xi & 0 \\ 0 & 0 & 1 \end{bmatrix} \tag{10}$$

The transformation $\mathbf{T}_i$ from the reference frame of virtual bond $i+1$ to that of virtual bond i is given by the product of these matrices as follows:

$$\mathbf{T}_i = \mathbf{T}(\varphi_i, \psi_i) = \mathbf{R}(\xi, 0)\mathbf{R}(\theta^\alpha, \pi - \varphi_i)\mathbf{R}(-\eta, -\psi_i) \tag{11}$$

Inasmuch as the bond rotational pairs are independent, the mean-square end-to-end distance $\langle r^2 \rangle_0$ and the average of the square of the radius of gyration $\langle s^2 \rangle_0$ may be calculated by resorting to Eqs. I-43 and I-45, respectively, with n replaced by x and l by the length l_u of the virtual bond. For very long chains, the limiting result given by Eq. I-44 may be used to obtained $\langle r^2 \rangle_0$; in this limit $\langle s^2 \rangle_0$ follows according to Eq. I-17. The average of $\mathbf{T}_i$ given by

$$\langle \mathbf{T}_i \rangle = \mathbf{R}(\xi, 0)\langle \mathbf{R}(\theta^\alpha, \pi - \varphi_i)\mathbf{R}(-\eta, -\psi_i)\rangle \tag{12}$$

is required for calculation of these quantities.

For the numerical evaluation of $\langle \mathbf{T} \rangle$ (see Eqs. I-30 and I-37), the product of matrices within angle brackets on the right-hand side of Eq. 12 may be taken at equal intervals of φ and ψ. Each product is multiplied by the Boltzmann factor, $\exp[-E(\varphi, \psi)/RT]$, of the conformational energy at the given φ and ψ. The sum of the resulting terms is then divided by the partition function z for the residue, z being evaluated according to Eq. 5 by summing over the same intervals. Multiplication of this quotient by $\mathbf{R}(\xi, 0)$ gives $\langle \mathbf{T} \rangle$.

Matrices computed for the several residues discussed above through choice of intervals of 30° for φ and ψ and use of the conformation energies mapped in Figs. 5, 7, and 10, respectively, are as follows:

$$\langle \mathbf{T}_{\text{gly}} \rangle = \begin{bmatrix} 0.36 & -0.077 & 0 \\ -0.092 & -0.037 & 0 \\ 0 & 0 & -0.12 \end{bmatrix} \tag{13}$$ [15]

$$\langle \mathbf{T}_{\text{L-ala}} \rangle = \begin{bmatrix} 0.51 & 0.20 & 0.59 \\ -0.046 & -0.61 & 0.21 \\ 0.65 & -0.23 & -0.30 \end{bmatrix} \tag{14}$$ [15]

$$\langle \mathbf{T}_{\text{L-lactyl}} \rangle = \begin{bmatrix} 0.33 & 0.066 & 0.051 \\ -0.216 & 0.021 & 0.170 \\ 0.75 & -0.34 & 0.021 \end{bmatrix} \tag{15}$$ [20]

The computation of $\langle \mathbf{T} \rangle$ could be refined by choice of an interval smaller than 30°, with consequent increase in the number of states. Exploratory calculations indicate,[2] however, that the value for the characteristic ratio would be little affected by reducing the interval below 30°. The refinement attainable in this manner is certainly within the limits of reliability of the conformational energy calculations.

The corresponding matrix for a *trans* L-prolyl residue situated in a poly-L-proline chain of identical residues is [31]

$$\langle \mathbf{T}_{\text{L-pro}} \rangle = \begin{bmatrix} 0.44 & 0.30 & 0.81 \\ -0.54 & -0.62 & 0.52 \\ 0.71 & -0.69 & -0.11 \end{bmatrix} \tag{16}$$

It was obtained from the sum taken at 10° intervals over the curve calculated for such a residue, which resembles curve III in Fig. 11 as described on pp. 272–273. The smaller interval was necessitated by the limited range of the potential wells for the prolyl residue.

The averaged transformation matrices for D residues may be obtained from those for the corresponding L residues by reversing the signs of φ and ψ in Eqs. 8 and 9. The effect of these operations is to multiply the third row and the third column of $\langle \mathbf{T}_L \rangle$ by -1, as may readily be proved. Thus, for example

$$\langle \mathbf{T}_{\text{D-ala}} \rangle = \begin{bmatrix} 0.51 & 0.20 & -0.59 \\ -0.046 & -0.61 & -0.21 \\ -0.65 & 0.23 & -0.30 \end{bmatrix} \tag{17}$$

The same follows on the grounds that residues of opposite symmetry are identically described in reference frames of opposite handedness, which differ only in the directions of their z axes (see p. 227). The equivalent transformation for the D residue is obtained therefore by reversing the signs of the z axes of both initial and final reference frames.

Average Dimensions of Random Coils

Limiting values of the characteristic ratios calculated from the averaged transformation matrices according to Eq. I-44 are given in Table 3. Results in the third column were obtained from $\langle \mathbf{T} \rangle$ matrices (not given above) representing the energy functions $E(\varphi, \psi)$ for $E_C = 0$ (see Figs. 4, 6, and 9). Those in the fourth column were calculated from the $\langle \mathbf{T} \rangle$ matrices given by Eqs. 13, 14, and 15, which represent energy functions $E(\varphi, \psi)$ with the coulombic contribution included (see Figs. 5, 7, and 10).

Table 3

Characteristic Ratios

	$C_\infty = (\langle r^2 \rangle_0 / x l_u^2)_\infty$			
	Free rotation	$E_C = 0$	E_C included	Exptl
Polyglycine[15]	1.93	1.79	2.16	
Poly-L-alanine[15]	1.93	2.97	9.27	9.0 ± 0.5[a]
Poly(L-lactic acid)[20]	1.92	1.24	2.13	2.1 ± 0.3[36]
Poly-L-proline-II[31]	1.86[b]	116		

[a] Average of results[18a] for analogs of poly-L-alanine having substituents —CH_2—R′, and not for poly-L-alanine itself; see text.

[b] Calculated with φ fixed at 102°.

Uncertainties pertaining to the electrical asymmetry of the prolyl residue discouraged estimation of E_C (see p. 270), but, inasmuch as all minima other than III are eliminated by the structural constraints on φ and by interactions between neighbor residues, the effect of E_C in this instance is not significant. Included for comparison in the second column are values calculated for free rotation about the bonds adjoining the α-carbon atoms, the ester and the amide bonds being rigidly fixed in the planar *trans* conformation.

The characteristic ratios calculated for the first three polymers with neglect of the coulombic energy E_C are low. They are even lower than for free rotation in two of the examples. Inclusion of dipolar interactions raises C_∞ in each instance, but to different extents. The explanations for these variable effects on the chain dimensions can be traced unambiguously to characteristics of the several kinds of chains, as we shall point out in due course. We note here that calculations relating to these different chains have been carried out on an equivalent basis, and in particular the same parameters for nonbonded interactions were used consistently throughout.

The characteristic ratio and the radius of gyration ratio $\langle s^2 \rangle_0 / x l_u^2$ calculated for poly-L-alanine are plotted against the degree of polymerization x in Fig. 15. The convergence is more protracted than for polymethylene as may be seen by comparison with Figs. V-9 and V-10. This is true notwithstanding the difference between the quantities represented on the abscissas: number of bonds in one case and number of virtual bonds in the other. The virtual bonds are joined by two single bonds susceptible to rotations, instead of one, and this in itself offers opportunity for a greater diversity of configurations as compared with polymethylene. The potentiality for enhanced flexibility thus imparted is suppressed in the case of poly-L-alanine through the intervention of steric and coulombic interactions (*cf. seq.*).

Experimental results leading to characteristic ratios for homopolymeric polypeptides are few. Values of C_∞ have been determined for polypeptides having side chains R as follows: $-CH_2COOCH_2C_6H_5$,[18a] $-CH_2CH_2COOCH_2C_6H_5$,[37] $-CH_2CH_2COO^-$,[18a] and $-CH_2CH_2CH_2CH_2NH_3^+$.[18a] Each should resemble $-CH_3$ in its steric interactions. For all four polypeptides, investigated in as many different solvents, $C_\infty = 9.0 \pm 0.5$ (see Table II-1).[18a] The calculated value for poly-L-alanine given in the last column of Table 3 stands in good agreement with these results. It is to be noted, however, that this value is rather sensitive to the choice of the van der Waals radius ascribed to the CH_3 group[2,15] (treated as spherical; see above). An increase in this radius by 0.05 Å would have raised the value of the characteristic ratio by about 10%.

Additional experimental evidence that is relevant in this connection

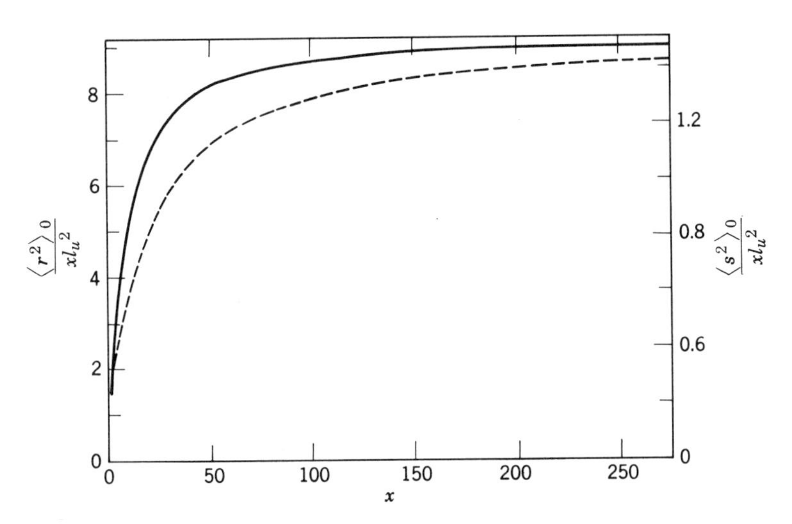

Fig. 15. The characteristic ratio (solid curve and left-hand ordinate scale) and the radius of gyration ratio $\langle s^2\rangle_0/xl_u^2$ (dashed curve and right-hand ordinate scale) calculated[2] as functions of x for poly-L-alanine using Eqs. I-43 and I-45, respectively. The upper boundary of the diagram is located at the asymptotes ($x = \infty$) for both curves.

comes from the dipole moments of the 14 diastereoisomeric di-, tri-, and tetra-alanines.[38] These have been treated successfully[39] according to the analysis above, the same averaged transformation matrices being used. Calculations show that the charges at the chain terminals, ^+H_3N— and —CO_2^-, make the preponderant contribution to the molecular dipole moment, which therefore reflects the root-mean-square of the end-to-end vector **r** and its dependence on the stereochemical configurations of the oligomeric polyalanines.[39]

The influence of E_C on the configurations of the polymers listed in Table 3 is most marked for poly-L-alanine. If E_C is disregarded, regions I and III of the conformational energy surface (Fig. 6) are competitive. Hence, intermixing of conformations from these two regions, which differ considerably, will occur abundantly; neither will be perpetuated consistently over sequences of appreciable length. When E_C is included (Fig. 7), region III is definitely preferred, with the result that successive residues tend to occur in conformations that are similar, though not identical.* These

* Even if every residue were constrained to assume a conformation within region III, the chain as a whole would not necessarily be described in terms of any simple geometrical form, helical or planar. The range of rotation angles permitted in this region is too great to constrain the chain to a uniform spatial configuration. A semblance to the β form, or some nearby conformation, would be discernible over the span of several units, but random variations within region III would dissipate all vestiges of order over long sequences.

conformations (III) moreover are in the vicinity of the fully extended form. The striking difference between the characteristic ratios calculated for poly-L-alanine with and without E_C is thus to be explained. It is significant that when E_C is ignored no reasonable combination of parameters, a_{kl}, in Eq. 3, raises the value calculated for the characteristic ratio above about 4.5,[2] which is well below experimental observations. Coulombic interactions, which cause region III to be favored over I, appear to be mainly responsible for the large characteristic ratio for polypeptides in which the steric effect of the side chain R resembles that of CH_3.

Equivalent electrostatic interactions operative in polyglycine affect its conformational energy similarly, but not so decisively; g^+g^- and g^-g^+ domains of the conformation map take precedence over the g^+g^+ and g^-g^- quadrants in the ratio of about 2 to 1 (see p. 261). The characteristic ratio is increased only marginally (Table 3) by the supersedure of one pair of domains by the other. Further calculations[15] show the characteristic ratio to be little affected by virtual suppression of the g^+g^+ and g^-g^- states. The circumstance of over-riding importance is the retention of *two* equally accessible states, and this is a consequence, of course, of the symmetry of glycine. The random occurrence of g^+g^- and g^-g^+ as the preferred conformations for successive residues suffices to maintain the characteristic ratio near its value for free rotation.

The characteristic ratio for polyglycine has not been determined by experiment owing to its insolubility in suitable media. Measurements on solutions of copolymers containing glycine,[40] discussed in the following section, afford indirect evidence supporting a low value of C_∞ for this polymer. Additionally, the dipole moments of oligomeric polyglycines $^{+}H_3NCH_2CO(\text{—}NHCH_2CO\text{—})_{x-1}NHCH_2CO_2^-$ determined by Wyman[41] for $x + 1 = 2$ to 7 are satisfactorily treated[39] by use of $\langle \mathbf{T}_{gly} \rangle$ given by Eq. 13.

The energy functions $E(\varphi, \psi)$ calculated for the L-lactyl residue, and shown in Figs. 9 and 10, display four minima. Trial calculations[36] show the average dimensions to be little affected by the subsidiary minima I′ and III′, which occur at energies approximately 1 kcal mole^{-1} above I and III. That is, omission of *both* I′ and III′ has a negligible effect on the values calculated for C_∞. This quantity is extraordinarily sensitive, however, to the difference between the energies of minima I and III.[36] It is for this reason that C_∞ calculated with E_C included is substantially greater than the result for $E_C = 0$, in spite of the small coulombic contribution in polymers of this residue. The ratio calculated with $E_C = 0$ is much smaller than the value for free rotation due to the frequent occurrence of the compact conformation I favored by the energy function calculated on this basis. Addition of E_C shifts the preference in favor of conformation

III. The close agreement with experiment achieved when E_C is included (see Table 3) must be regarded as somewhat fortuitous; the energy estimation is subject to an error greater than the calculated difference of ca. 100 cal mole^{-1} between minima I and III in Fig. 10. That the minimum III is in fact of lower energy than I by a small margin is confirmed by the temperature coefficient of C_{∞}, which is negative.[36]

The extraordinarily large characteristic ratio calculated for *trans* poly-L-proline[31] and given in Table 3 obviously is a consequence of the confinement of all residues to region III, and moreover to that portion of this region in which φ assumes a specified value, probably about 102°. (It will be recalled that the minimum I is excluded for an all-proline chain.) The breadth of the range of ψ, shown approximately by minimum III in Fig. 11, permits only limited variation of ψ from one residue to the next. Dimensions of short sequences of units consequently approximate those of the rigid Cowan-McGavin[23] helix (i.e., poly-L-proline-II; see Fig. 3). Thus, $\langle r^2 \rangle_0^{1/2}$ calculated for a sequence of 16 residues is 98% of its value for the rigid helix; for 128 residues, it is 80% of the length of the rigid helix.[31] In qualitative confirmation of these calculations, Steinberg *et al.*[26] concluded from the dilute-solution viscosity of a poly-L-proline-II consisting of 50 residues that its length approximates that of the Cowan-McGavin helix. Random coil behavior is indicated for chains of much greater length, but measurements from which to deduce the limiting value of the characteristic ratio are lacking. Calculations indicate convergence of $\langle r^2 \rangle_0 / x l_u^2$ to require in excess of 1000 units; for a chain of this length the ratio is approximately 90% of its limiting value.[31] The configurational character of the *trans*-poly-L-proline chain is well represented by the Porod-Kratky model (see Appendix G).

The polypeptides and their analogs include examples embracing a wide range of configurational characteristics. Some of them assume dimensions, and therefore a tortuosity, which approximate those for free rotation; others are semirigid chains of exceptionally large extension.

4. AVERAGE DIMENSIONS OF COPOLYPEPTIDES

A combination of circumstances peculiar to α-amino acid residues of the type —NH—CHR—CO— permits treatment of copolymers of such residues by methods of the utmost simplicity. These circumstances, already apparent in the treatment of the configurational statistics of homopolymeric polypeptides, are the following: (*1*) the skeletal configuration of the copolypeptide chain can be fully represented by a set of consecutively connected virtual bonds, one such bond being identified with each peptide unit; (*2*) the virtual bonds for all units are of the same fixed length l_u, adherence of each unit to the planar *trans* conformation being assumed (proline and hydroxyproline residues excluded); and (*3*) the conformational energy function associated with rotations (φ_i, ψ_i) at the skeletal bonds forming the junction between two peptide units (i and $i+1$) is determined by the character of only one residue, i.e., residue i. The last condition is most important, for it removes the necessity of treating residues pairwise (prolyl excepted, of course); the character of residue i alone determines the average transformation $\langle \mathbf{T}_i \rangle$ for unit i, regardless of its neighbors.

Let us first consider a *random* copolypeptide comprising two kinds of residues, a and b, meeting the specifications above. Generalization of the results to a greater number of co-units will be obvious. The average of the scalar product of virtual vectors j and i, these being separated by the series of residues $ba \cdots b$, will be

$$\mathbf{l}_u^T \langle \mathbf{T}_b \rangle \langle \mathbf{T}_a \rangle \cdots \langle \mathbf{T}_b \rangle \mathbf{l}_u$$

Repetition of the argument advanced in Chapter IV, Section 7, leads to the following expression for the matrix product over all ordered combinations of residues from i to $j-1$, each weighted according to its incidence in the random binary copolymer,

$$\overline{(\langle \mathbf{T}_i \rangle \langle \mathbf{T}_{i+1} \rangle \cdots \langle \mathbf{T}_{j-1} \rangle)} = \overline{\langle \mathbf{T} \rangle}^{j-i} \tag{18}$$

where

$$\overline{\langle \mathbf{T} \rangle} = w_a \langle \mathbf{T}_a \rangle + (1 - w_a) \langle \mathbf{T}_b \rangle \tag{19}$$

w_a being the fraction of units of kind a in the copolymer. In compliance with notation introduced in Chapter IV, the average over the various kinds of units of the copolymer is indicated by an overbar. It follows that the characteristic ratio for the copolypeptide may be calculated by resort to Eq. I-43, or in the case of an infinite chain to Eq. I-44, with $\overline{\langle \mathbf{T} \rangle}$ replacing $\langle \mathbf{T} \rangle$. Equation IV-74 with $\xi = 1$ may of course be used, but the procedure here outlined will be preferred for very long chains.

Results of calculations carried out in this manner for random copolymers of glycine with L-alanine are shown by the solid curve in Fig. 16, which is plotted against the mole fraction w_{gly} of glycine.[40]* The characteristic ratio decreases rapidly with w_{gly} near $w_{\text{gly}} = 0$; a small proportion of glycine exerts a disproportionately large effect. The curve decreases more gradually,

* The matrices $\langle \mathbf{T}_{\text{gly}} \rangle$ and $\langle \mathbf{T}_{\text{L-ala}} \rangle$ used[40] for the calculations shown in this figure differed somewhat from those given in Eqs. 13 and 14. The limiting ratio calculated for poly-L-alanine ($w_{\text{gly}} = 0$) on the basis is 8.9 instead of 9.27 as quoted in Table 3 and calculated using Eq. 14. The difference is unimportant.

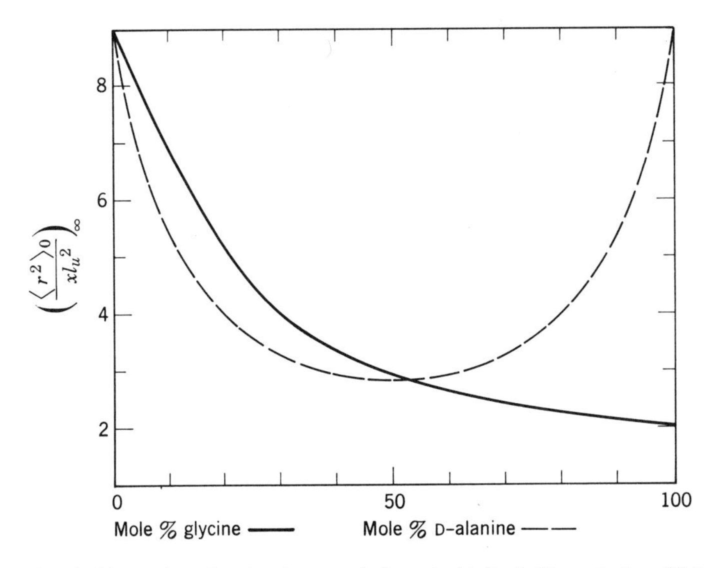

Fig. 16. Limiting values for the characteristic ratio $(\langle r^2 \rangle_0 / x l_u^2)_\infty$ calculated[40] for random copolymers of L-alanine and glycine (solid line) and for random copolymers of L- and D-alanine (dashed line).

but monotonically, with further increase in w_{gly}; a minor proportion of alanine exerts only a small effect on $\langle r^2 \rangle_0 / x l_u^2$ for polyglycine.

The dashed curve in Fig. 16 represents corresponding calculations for random copolymers of D and L residues of the alanyl type. A small proportion of the enantiomorphic alanyl residue is even more effective than an equal proportion of glycyl in lowering the characteristic ratio. For equimolar proportions ($w = 0.50$) the random-coil dimensions for the two copolymers of L-alanine are similar. The curve for D,L-alanine copolymers is necessarily symmetric about $w = \frac{1}{2}$.

Perpetuation of the quasi-helix generated by adherence of successive residues to conformations within the domain of region III of Fig. 7 for L-alanyl may be interrupted by interposition of either a glycyl or a D-alanyl residue. The glycyl residue presents two, equally favored minima. One member of this mirror image pair occurs in the general neighborhood of region III (compare Figs. 5 and 7), though displaced somewhat from it. The other represents a helix of opposite screw sense. If the glycyl residue chooses the former location (approximately g^+g^-), it introduces only a small departure from the preferred quasi-helicoidal form of the L-alanyl sequence. Choice of the latter (approximately g^-g^+) brings about a major disruption in the direction of the trajectory of the chain.

A D-alanyl residue offers one strongly favored conformational domain. This domain corresponds to III of Fig. 7, but it occurs in the opposite quadrant of the diagram. Whereas about a third of the glycyl residues are expected to adopt conformations resembling the preferred one for L-alanyl, this option is strongly forbidden for D-alanyl. The latter residue will usually assume the conformation of the quasi-helix of opposite sense of rotation which, as was noted above, represents a drastic departure from the preferred conformation for the L-alanyl residue. It is on this account that a small proportion of D-alanyl is much more effective than glycyl in reducing the characteristic ratio for poly-L-alanine.

Experimental results[40] tend to confirm these predictions. Twenty-five mole % of glycine randomly copolymerized with L-glutamic acid reduces the characteristic ratio to about 4 ± 0.5. A copolymer of D- and L-glutamic acids in the proportions of 2 to 3 exhibited a similar ratio.[40] A tendency to generate sequences of residues of the same kind in the copolymerization of *N*-carboxyanhydrides of the α-amino acids by which these copolymers were prepared would, if operative, raise the value of the characteristic ratio. Such tendencies do prevail under some conditions of polymerization,[42,43] but appear to be suppressed under others.[44,45]*

A particular class of polypeptides of interest in this connection consists of racemic copolymers of D and L residues in equal proportions copolymerized under conditions such that the probability of selection of one or the other residue depends on the one preceding it in the chain. The residue sequence is then Markoffian, as discussed on pp. 119–120 of Chapter IV. The matrix of sequence probabilities (see Eq. IV-75) is

$$\mathbf{w} = \begin{bmatrix} w_{\mathrm{DD}} & w_{\mathrm{DL}} \\ w_{\mathrm{LD}} & w_{\mathrm{LL}} \end{bmatrix} = \begin{bmatrix} w_{\mathrm{iso}} & 1 - w_{\mathrm{iso}} \\ 1 - w_{\mathrm{iso}} & w_{\mathrm{iso}} \end{bmatrix} \tag{20}$$

where w_{iso} is the probability of isotactic propagation, i.e., of addition of an amino acid of the same stereochemical configuration as its predecessor. Inasmuch as the symmetry of the zeroth α carbon (see Fig. 1) is irrelevant, we introduce equal *a priori* probabilities for amino acid residue 1, i.e., we let

$$\mathbf{w}_1 = [\tfrac{1}{2} \quad \tfrac{1}{2}] \tag{21}$$

The matrix of conditional probabilities, Eq. 20, applies to all succeeding residues. Matrices $\overline{\mathfrak{G}}$ defined by Eq. IV-78 with $\xi = 1$ may be formulated

* The effect of an occasional prolyl residue on the dimensions of a random-coil polypeptide consisting of a preponderance of L-alanyl residues should be similar to the effect of glycyl if the conformation I is competitive with III (Fig. 12) for the prolyl residue.[29] The preceding residue (alanyl) is restricted to region III (see Fig. 13), but the effect of this assignment is less consequential than the availability of conformation I for the isolated prolyl residue.

from Eqs. 20 and 21 and the matrices $\mathfrak{G}_D$ and $\mathfrak{G}_L$ for the respective alanyl residues. These latter matrices are obtained from the averaged transformation matrices $\langle \mathbf{T}_D \rangle$ and $\langle \mathbf{T}_L \rangle$ as prescribed by Eq. IV-60. Characteristic ratios may then be computed according to Eq. IV-79.* If one enantiomorph of the D,L pair should be present in excess of the other, the foregoing methods are applicable with nominal revisions.

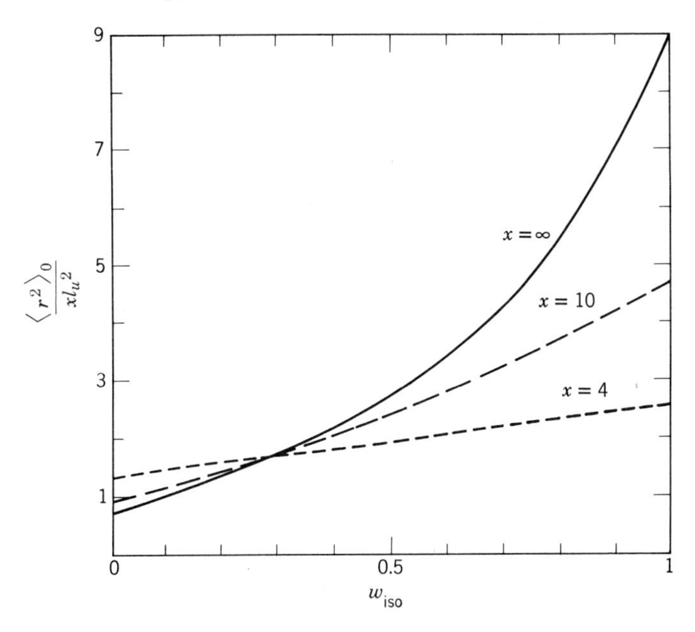

Fig. 17. Characteristic ratios calculated[40] for racemic copolypeptides of the polyalanine type as functions of the steroregularity. Degrees of polymerization are indicated for the respective curves.

Figure 17 presents results of calculations for racemic copolymers of alanine, or alanine analogs. The abscissa covers the range from the regularly alternating, or syndiotactic, copolymer of D and L residues, for which $w_{iso} = 0$, to the polymer having indefinitely long sequences of residues of the same isomer, for which $w_{iso} = 1$. For random copolymers $w_{iso} = \frac{1}{2}$. The solid line representing the characteristic ratio in the limit $x = \infty$ decreases with decreasing stereoregularity from 8.9 for all L- or all D-polyalanine to a value less than unity. Thus, the characteristic ratio predicted for the regularly alternating racemic copolymer is less than would obtain for a chain of freely jointed bonds each of length l_u. This result is unusual.

* The superficially different procedure used by Miller, Brant, and Flory[40] may be shown to be identically equivalent to the one outlined here.

The corresponding curves in Fig. 17 for finite chains show smaller variations with w_{iso}, as might be expected. The increase in the intercept at $w_{iso} = 0$ with decrease in x is unexpected, however. It denotes a *decrease* in the characteristic ratio with chain length, which is without precedent in other chains which have been treated heretofore.

The low value of the characteristic ratio for the alternating D,L copolymer finds explanation in the conformation map for L-alanine and its counterpart for the D residue.[40] If each residue is assigned its preferred conformation—region III in Fig. 7 for L and its inversion image for D—the resulting skeletal conformation will be found to approximate a helix of very low pitch. Strict adherence to helical form will not of course obtain owing to the breadth of the preferred domain III. The tendency toward a comparatively compact conformation which nevertheless prevails is responsible for the low values calculated for the average dimensions of the alternating copolymer.

5. COOPERATIVE CONFIGURATIONAL TRANSFORMATIONS; HELIX–COIL TRANSITIONS

The difficulties attending the coexistence of *cis* and *trans* prolyl residues in the same chain were pointed out in Section 2. Experimental evidence[34,35] was cited showing interconversion from one form to the other to occur, at equilibrium, within a narrow range of conditions. The variable conveniently subject to alteration in this instance is the solvent composition; the transformation of poly-L-proline between forms I and II can be brought about virtually in its entirety by a change of only a few percent in the composition of the solvent mixture. The occurrence of the transformation within a small range of values of an intensive thermodynamic variable is the mark of a cooperative process. Cooperativity of the polyproline transition is directly attributable to the interdependence of the configurations of the peptide bonds of neighboring units as discussed on p. 274.

This is not the first instance we have encountered of strong interactions affecting the conformations of neighboring units or bonds. In an isotactic vinyl chain of perfect stereoregularity (if such a chain were available) the preferred helical form (*tgtg*, etc.) undoubtedly would be perpetuated over many units. Occasional departures, at intervals of perhaps 100 units, from the preferred conformation would interrupt the helix of one sense (chirality) only to initiate another sequence of the opposite sense. The conformations of the two sequences are otherwise identical. Being related by mirror reflection, neither may be preferred over the other under any circumstances. It is in this respect that the poly-L-proline equilibrium between forms I and II differs from the "cooperative" correlation of the conformations over long sequences of units in a stereoregular vinyl chain. The two forms of poly-L-proline are unrelated by any symmetry operation; the difference between them is more profound than mere helical sense. Accordingly, one form will generally be preferred over the other, and their relative stabilities will depend on the intensive thermodynamic variables defining the environment.

The helix–coil transition in polypeptides has been exhaustively discussed as a process influenced by strong cooperative interactions.[46–49] In addition to being "sterically allowed," the α-helical conformation derives stability from dipolar interactions (*cf. seq.*) as well as from hydrogen bonds[14] formed between the carbonyl oxygen of one *peptide unit* (—CO—NH—CHR—) and the amide hydrogen of the third succeeding unit. These interactions are shown schematically in Fig. 18, hydrogen bonds

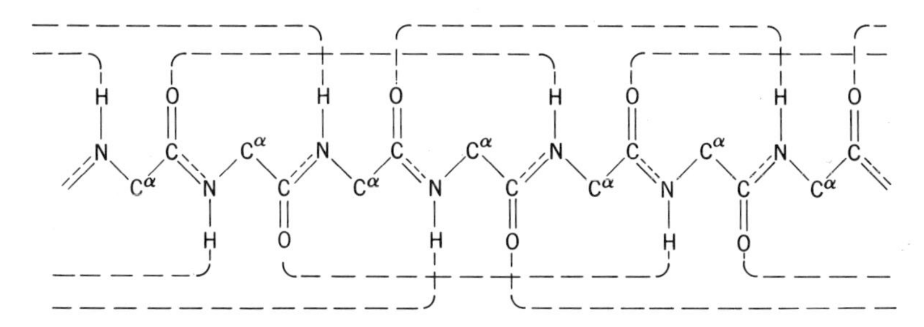

Fig. 18. Schematic diagram of the polypeptide chain with hydrogen bonds of the α-helical conformation indicated by dashed lines.

being indicated by dashed lines. The rotation angles φ and ψ for each of three successive residues must adopt the required values (see Fig. 3) within narrow limits if the participating atoms are to be brought into the relative positions required for effective hydrogen bonding on the one hand and avoidance of steric overlaps on the other.[16,50,51] This unique configuration has received scant attention in our discussion of the random coil configurations. Two valid reasons may be offered in justification for this apparent oversight. First, an α helix involves a series of residues identically conformed and cannot therefore be comprehended by φ and ψ for one residue alone. The single residue energies $E(\varphi, \psi)$ calculated above and applied to the statistical treatment of random-coil configurations of the chain in terms of its constituent residues are inapplicable in principle to the helix. Even an incipient helix requires four units to complete one turn. Second, the α helix occupies a very small portion of configuration space; hence, it may be set aside for separate consideration without detracting appreciably from the configuration space accessible to the random coil. It was legitimate, therefore, to treat the configurations which collectively comprise the random coil without taking cognizance of the specific configuration of the α helix.

Chiefly to be observed from the diagram in Fig. 18 is the vulnerability of three hydrogen bonds (and the concomitant dipolar interactions; *cf. seq.*) to rupture by alteration of the conformation of only one intervening residue—indeed, of only one of the rotation angles φ and ψ for a residue. Conversely, construction of an α helix from a random coil requires assignment of three residues to the narrowly defined conformation of the helix in order to form one hydrogen bond. The further assignment of one neighboring residue produces one additional hydrogen bond, etc. The initiation step requires the ordering of two units *before* the stabilizing effect by hydrogen bonds can be realized in the next assignment. This circumstance must obviously disfavor the generation of separate helical sequences. If conditions permit simultaneous occurrence of residues in both the helical and the random coil forms, the distribution of residues in the two forms must be such as will economize on the number of junctions between helical and coil sequences. In other words, the number of sequences will be diminished, the average number of units in a sequence being increased accordingly.

The foregoing qualitative description of the difficulty of initiation of an α-helical sequence, as compared with its perpetuation, is an over-simplification. Other factors are operative. We have pointed out above (see Sect. 2) that the assignment of φ and ψ for one residue to a conformation in the vicinity of region I of Figs. 6 or 7, which includes the conformation of the right-handed α helix, disposes the neighboring pair of dipole moments parallel to each other, and hence their mutual interaction is positive (i.e., repulsive). Repetition of the right-handed α-helical conformation for the next residue places second neighbor dipoles approximately parallel to one another, but inclined to the line between their centers. Their mutual energy turns out to be near zero.[51] Perpetuation of the same conformational assignment to a third residue places the dipole vectors for units separated by three residues approximately in collinear juxtaposition. The dipolar energy for their interaction is negative and large; of course, it may be overshadowed by hydrogen bonding. The sum of the dipolar interactions for a given residue with all succeeding residues of a long helix is negative. It is estimated to be -1.0 to -1.5 kcal per mole of residues.[2,19] This energy, representing the contribution of dipole–dipole interactions to the stabilization of a long α helix, is substantial. Realization of the full effect of stabilization of a given residue by dipolar interactions requires a helical sequence of about ten residues.[51] In other words, the deficit in dipolar energy associated with the terminus of a helical sequence extends over a range of about ten residues. This deficit makes an appreciable contribution to the difficulty of helix initiation. The usual attribution of the initiation factor solely to a deficit of entropy associated with the assignment of two residues to a narrowly restricted conformation, uncompensated by the formation of energetically favorable hydrogen bonds, is incorrect in failing to acknowledge other contributions, notably dipolar interactions, to the extra free energy required for initiation of a helical sequence.

Apart from specific details concerning the origin of effects at sequence termini, a correspondence of the helix ↔ coil and the polyproline I ↔ II interconversions will be apparent. Interruption of sequences of one type and initiation of the alternative form is disfavored by an adverse energy, or free energy, in each example. Consequently, sequences will tend to be long, and within a given molecule they will be few in number.

To consider first an extreme case, let the difficulty of juxtaposing two units in the alternative configurations be so great as to preclude altogether the occurrence of such neighbor pairs. Every chain molecule will then occur with all of its units in one form or the other. If for simplicity all molecules are assumed to consist of the same number x of units, and to be otherwise identical in chemical constitution as well, then the equilibrium between the two forms, present in a solution sufficiently dilute to eliminate effects of mutual interactions between different molecules, is simply described by

$$K_x = p_{\text{II}}/(1 - p_{\text{II}}) \tag{22}$$

where p_{II} is the fraction of units (equal to the fraction of molecules in this case) in form II, and of course $p_{\text{I}} = 1 - p_{\text{II}}$. Identification of p_{II} with form II in the polyproline transition is implied; in the case of a helix–coil transition, we shall associate p_{II} with the helical form. The interconversion may be thus represented as a simple chemical equilibrium characterized by an equilibrium constant K_x. If z_{I} and z_{II} denote the partition functions for residues in forms I and II, respectively, so that the corresponding molecular partition functions are $Z_{\text{I}} = z_{\text{I}}^x$ and $Z_{\text{II}} = z_{\text{II}}^x$, then

$$K_x = Z_{\text{II}}/Z_{\text{I}} = (z_{\text{II}}/z_{\text{I}})^x = s^x \tag{23}$$

where $s = z_{\text{II}}/z_{\text{I}}$ may be looked upon as the partition function for a residue in form II relative to unity for form I. In conformity with the notation introduced by Zimm and Bragg,[46] we shall use s in this sense throughout the ensuing discussion. In the language of thermodynamics,

$$s = \exp(\Delta F^{\circ}_{\text{II}\to\text{I}}/RT) \tag{24}$$

where $\Delta F^{\circ}_{\text{II}\to\text{I}}$ is the change in free energy per residue for the conversion of form II to form I, each in its standard state.

At the midpoint of the transformation, $\Delta F^{\circ}_{\mathrm{II}\to\mathrm{I}} = 0$, or $z_{\mathrm{II}}/z_{\mathrm{I}} = s = 1$, and $K_x = 1$. If x is very large, then a very small difference between z_{II} and z_{I} is translated into a large departure of K_x from unity. A small alteration of conditions, e.g., temperature or solvent composition, from those yielding $K_x = 1$ will shift the equilibrium strongly in favor of one form or the other form. The transition consequently occurs within a very narrow range. In the limit $x \to \infty$ it becomes infinitely sharp, and may be likened to a first-order phase transition in a three-dimensional system.

This description of the transformation rests on the arbitrarily asserted impossibility of juxtaposing the two conformations within the same chain molecule. In the examples cited and in other cases of interest as well, interruption of a sequence of units in one conformation and initiation of a series of units in the alternative form, though strongly disfavored, is not so difficult as to be suppressed altogether. We therefore relax the foregoing condition and take into consideration molecules in which sequences of the two kinds coexist.

Let ν be the number of sequences of type II. The number of type I will be $\nu - 1$, ν, or $\nu + 1$, depending on the terminal sequences; in a very long chain their number can be equated to ν in sufficient approximation. We assume that the molecular partition function may, in any event, be expressed as

$$Z_{x_{\mathrm{I}},x_{\mathrm{II}},\nu} = z_{\mathrm{I}}^{x_{\mathrm{I}}} z_{\mathrm{II}}^{x_{\mathrm{II}}} \sigma^{\nu}$$

where σ is the factor[46] ($\sigma \ll 1$) by which the partition function is diminished for each sequence of type II. Or, if the state I (e.g., random coil) is arbitrarily taken as the reference state with $z_{\mathrm{I}} = 1$

$$Z_{x_{\mathrm{II}},\nu} = s^{x_{\mathrm{II}}} \sigma^{\nu} \qquad (25)$$

It is to be observed that σ includes effects from *both* ends of the sequence, namely, those associated with the junctions I→II and II→I at the two terminals of a sequence of type II. Whether steric difficulties, the hydrogen bond deficit, and dipolar interactions are physically associated with one end or the other or jointly with both is immaterial insofar as the form of the partition function is concerned. In finite chains, however, account must be taken of the circumstances peculiar to terminal units of the chain molecule,[52] and also of the labilization of the terminal sequences of both types I and II when they occur at one end or the other of the chain molecule (*cf. seq.*). For very long chains these effects will obviously be subordinated, and we assume here that they may be neglected. The single parameter σ then suffices provided that σ may be assumed to be independent of the lengths of adjoining sequences. In the α-helix–coil transition, σ doubtless depends on the length of the helical sequence if it consists of fewer than about ten residues,[51] as preceding discussion implies. If the average sequence length is large, the variation of σ may be ignored.

Correspondence of $-RT \ln \sigma$ to the interfacial free energy between two macroscopic phases will be apparent. The helix–coil transition, for example, is analogous to the equilibrium between a crystal and its liquid. The factor σ is sometimes called the nucleation parameter.

The partition function which is the sum of $Z_{x_{II},v}$ over all combinations x_{II}, v is given by

$$Z = \mathbf{J}^* \mathbf{U}^x \mathbf{J} \tag{26}$$

where

$$\mathbf{U} = \begin{bmatrix} 1 & s\sigma \\ 1 & s \end{bmatrix} \tag{27}$$

With respect to the application of these equations to finite chains, we assume the interfacial effects for a sequence II with one or both of its termini coincident with the end of the chain to be the same as for an internal II sequence. Moreover, terminal I sequences are assumed to be free of special effects. These suppositions appear to be reasonable approximations for random-coil sequences in association with α-helical sequences; the latter approximation is somewhat in error for polyproline-I sequences.[31,52]

Solution of the characteristic equation of Eq. 27 gives

$$\lambda_{1,2} = [(1 + s) \pm \sqrt{(1 - s)^2 + 4\sigma s}]/2 \tag{28}$$

Diagonalization of **U** by the transformation $\mathbf{A}^{-1}\,\mathbf{U}\,\mathbf{A} = \Lambda$ (see Appendix B) and substitution in Eq. 26 leads to

$$Z = (1 - \lambda_2)(\lambda_1 - \lambda_2)^{-1}\lambda_1^{x+1} + (\lambda_1 - 1)(\lambda_1 - \lambda_2)^{-1}\lambda_2^{x+1} \tag{29}$$

For sufficiently long chains the approximation

$$Z \cong \lambda_1^x \tag{30}$$

is acceptable.

The eigenvalues λ_1 and λ_2 are plotted in Fig. 19 as a function of s for two values of σ. For $s < 1 - \sqrt{4\sigma}$, $\lambda_1 \cong 1$ and $\lambda_2 \cong s$; for $s > 1 + \sqrt{4\sigma}$, $\lambda_1 \cong s$ and $\lambda_2 \cong 1$. Within the range where $(1 - s)^2 \ll s\sigma$

$$\lambda_{1,2} \cong (1/2)(s + 1) \pm \sqrt{s\sigma} \tag{31}$$

and at $s = 1$

$$\lambda_{1,2} = 1 \pm \sqrt{\sigma} \tag{32}$$

The error involved in the approximation of Eq. 29 for the partition function

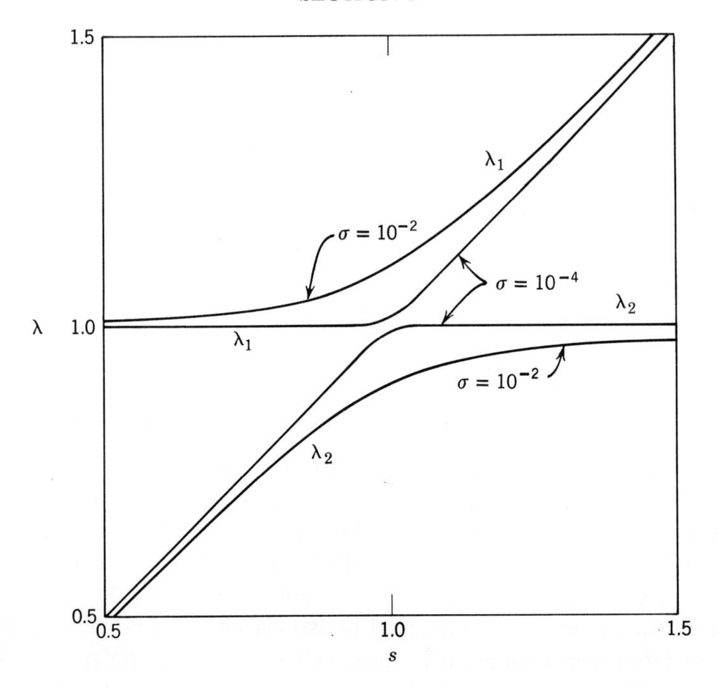

Fig. 19. Eigenvalues of **U** defined by Eq. 27 for the two values of σ indicated, plotted against s in the range $0.5 \leq s \leq 1.5$ (see Eq. 28).

by Eq. 30 obviously depends on the proximity of λ_2 to λ_1. It will be greatest at $s = 1$, where $\lambda_1 - \lambda_2 = 2\sqrt{\sigma}$ according to Eq. 32. Even here the approximation will be legitimate if $x \gg 1/2\sigma^{1/2}$, as may readily be verified.

The fraction of units in state II is given by

$$p_{\mathrm{II}} = x^{-1}\partial \ln Z/\partial \ln s \tag{33}$$

The general result applicable to chains of any length, apart from approximations in Eqs. 26 and 27 for the partition function, is obtained by differentiation of Eq. 29 according to Eq. 33. The resulting lengthy expression[46] is not reproduced here. The following development is limited to the case of long chains, i.e., chains for which $x \gg 1/2\sigma^{1/2}$, and use of Eq. 30 is therefore warranted. Then

$$p_{\mathrm{II}} = \partial \ln \lambda_1/\partial \ln s = (\lambda_1 - 1)/(\lambda_1 - \lambda_2) \tag{34}$$

In the case of a helix–coil transition, p_{II} represents the fraction of helical units, alternatively designated p_h. The average number of sequences of type II, or of type I, in a molecule is given by xp_σ where

$$p_\sigma = \partial \ln \lambda_1/\partial \ln \sigma = (\lambda_1 - 1)(1 - \lambda_2)/\lambda_1(\lambda_1 - \lambda_2) \tag{35}$$

and the average number of units in a type II sequence is

$$\langle y_{\text{II}} \rangle = p_{\text{II}}/p_\sigma = \lambda_1/(1 - \lambda_2) \tag{36}$$

At $s = 1$ we have (see Eq. 32)

$$p_{\text{II}} = 1/2 \tag{37}$$

and

$$\langle y_{\text{II}} \rangle = (1 + \sqrt{\sigma})/\sqrt{\sigma} \cong 1/\sqrt{\sigma} \tag{38}$$

Thus, $s = 1$ marks the midpoint of the transition, and the average length of a sequence of either kind is approximately $1/\sqrt{\sigma}$ for $\sigma \ll 1$. These relations are strictly applicable of course only to very long chains. The condition, $x \gg 1/2\,\sigma^{1/2}$, found earlier to be required to justify approximation of $Z^{1/x}$ by the larger eigenvalue of **U**, as in Eq. 30, reduces therefore to the assertion that the chain must be long enough to accommodate many sequences.

The fraction p_{II} of units in form II is shown in Fig. 20 calculated as a function of s according to Eq. 34, with $\sigma = 10^{-2}$ and $\sigma = 10^{-4}$. The range

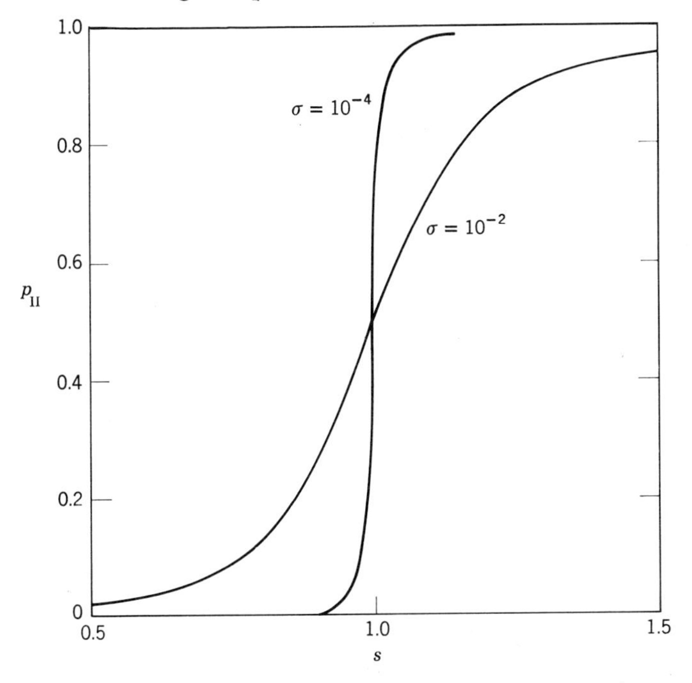

Fig. 20. Fraction p_{II} of helical units in the limit $x \to \infty$ calculated as a function of s according to Eq. 34. The values of σ correspond to those used in Fig. 19.

of s over that portion of the transition within which $\frac{1}{4} < p_{II} < \frac{3}{4}$ is approximately $(1 - \sqrt{s\sigma}) < s < (1 + \sqrt{s\sigma})$. The transition becomes sharper as σ decreases, but only in the limit $\sigma = 0$ does it become discontinuous. Inasmuch as σ must necessarily exceed zero for any real chain, the configurational transition must be continuous. We have already called attention to the analogy to a phase transition. The result here cited illustrates the axiom[53] that a phase transition in a one-dimensional system must take place without discontinuity even in the limit $x \to \infty$, unlike such transitions in two- or three-dimensional systems of unrestricted size.

Equation 24 relates s to the difference $\Delta F^{\circ}_{II \to I}$ between the free energies of units in forms I and II in their standard states at the specified temperature T. It follows that s should be a function of temperature. Specifically, it depenes on the change of $\Delta F^{\circ}/RT$ with temperature, i.e., on $-\Delta H^{\circ}_{m}/RT^2$ where $\Delta H^{\circ}_{m} \equiv \Delta H^{\circ}_{II \to I}$ is the enthalpy change associated with the process of "melting" the helix. Over a small range of temperature about the point T_m at which $s = 1$, we may take s to be linear in temperature. That is

$$s - 1 = -(\Delta H^{\circ}_{m}/RT^2)(T - T_m)$$

or

$$\Delta T = T - T_m = -(RT^2/\Delta H^{\circ}_{m})(s - 1) \tag{39}$$

Given the value of ΔH°_{m}, the abscissa may be rescaled in terms of temperature. For example, the curve in Fig. 20 for $\sigma = 10^{-4}$ approximates the results of Zimm, Doty, and Iso[54] for poly(γ-benzyl-L-glutamate) in an 80 : 20 mixture of dichloroacetic acid and ethylene dichloride. The temperature T_m at the midpoint of the transition for polymers of high molecular weight is 285°K; the helix is the more stable form at higher temperatures in this instance. From analysis of the dependence of p_{II} on temperature and chain length, Zimm, Doty, and Iso[54] arrived at a value of -900 cal per mole of residues for ΔH°_{m}. Thus, $RT^2/\Delta H^{\circ}_{m} = -180°$, and the points $s = 0.90$, 1.00, and 1.10 in Fig. 20 correspond to temperatures of 267, 285, and 303°K, respectively.

The average dimensions of polypeptide chains in the intermediate region between random coil and helix are readily calculated[55] by adaptation of methods developed in earlier chapters. To each residue which is in the random coil category we may assign the average transformation matrix $\langle \mathbf{T}_c \rangle$ evaluated as detailed in Section 3; the subscript is here appended in order to distinguish the coil state from the helix. Each helical residue is assigned the transformation $\mathbf{T}_h$ dictated by φ and ψ for the helical conformation. The matrix $\mathbf{U}$ given in Eq. 27 serves as the statistical weight matrix used to construct the $\mathscr{G}$ matrix of Eq. IV-24. In conformity therewith, we let

$$\|\mathbf{T}\| = \begin{bmatrix} \langle \mathbf{T}_c \rangle & \mathbf{0} \\ \mathbf{0} & \mathbf{T}_h \end{bmatrix} \tag{40}$$

Then $\langle r^2 \rangle_0$ may be calculated according to Eq. IV-28. Equations IV-54 and IV-56 similarly yield the radius of gyration.

Results of illustrative calculations[55] carried out in this manner for residues of the alanyl type are shown in Fig. 21 for chains consisting of the numbers of units indicated with each curve. A matrix closely resembling Eq. 14 for the alanyl residue was used for $\langle \mathbf{T}_c \rangle$. The comparatively large value of $\sigma = 3 \times 10^{-3}$ used throughout these calculations reduces the average sequence length, and hence sanctions manifestation of partial helicity even for short chains within the transition region. Calculations were carried out for various values of s as required to cover the entire range of $p_h \equiv p_{\mathrm{II}}$ from 0 to 1. The initial appearance of helicity is marked by a decrease in the characteristic ratio.[55,56] This is a consequence of the comparatively high extension of the random coil of poly-L-alanine on the one hand ($C_\infty \cong 9$), and of the low pitch of the α helix on the other. (The projection of the virtual bond vector, or of the unit, on the axis of the

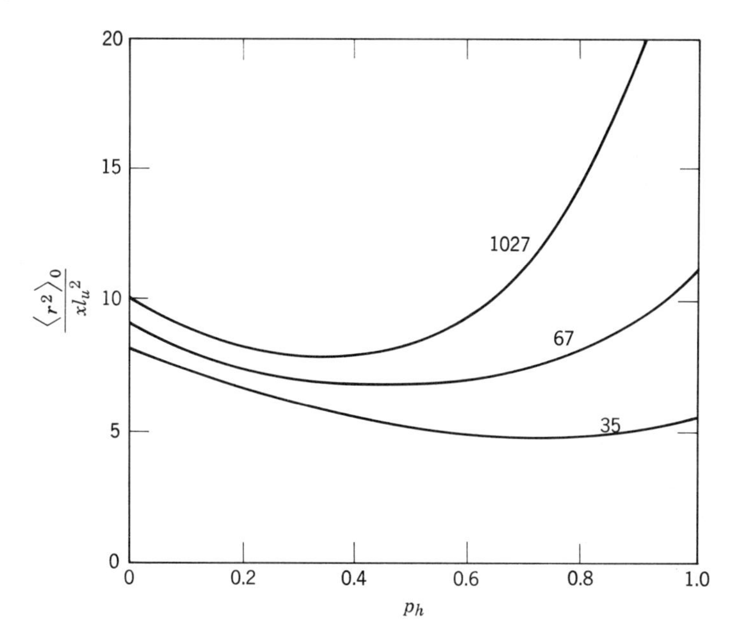

Fig. 21. The characteristic ratio plotted against the degree of helicity for poly-L-alanine chains of the degrees of polymerization x indicated. The fraction of helix p_h and $\langle r^2 \rangle_0$ were calculated[55] as functions of s using Eq. 34 for the former and Eqs. 27, 40, IV-24, and IV-28 for the latter; $\sigma = 3 \times 10^{-3}$ throughout.

α helix is 1.5 Å.) The curve passes through a minimum, then rises. The rise is steep if x is large inasmuch as the extension of the completely helical long chain greatly exceeds the mean extension of the random coil. For small chains ($x < \sim 60$), however, the value of $\langle r^2 \rangle_0$ for the random coil exceeds $\langle r^2 \rangle$ for the helix, as the lowest curve in Fig. 21 shows.

The theoretical treatment of configurational transitions according to Eqs. 25–27 above does not take account of details peculiar to specific examples. Thus, in the transition of a random-coil polypeptide to an α helix, embryonic intermediates comprising one or two residues in the helical conformation are ignored. The one or two units thus conformed do not acquire the advantage of stabilization by hydrogen bonds. Hence they may occur only in very low concentrations. More elaborate theories which take account of them have been developed,[46,57] but the refinements achieved by these theories are overshadowed by minor physical inaccuracies inherent in the model itself.

The helix–coil transition in polynucleotides, DNA and RNA, in which the helix consists of two strands, can be treated along similar lines. These helices derive their stability from interchain bonding (base pairing) and from "stacking" of bases one above the other along the same chain. The physical interpretation of s and σ differs accordingly, and the helix energy depends in a more complicated way on the base sequences in *both* strands.[58–60] In addition to revisions in the physical significance of the parameters in comparison to their interpretation for the single-strand α helix, partial "melting" within the two-strand helix produces "loops" of random coiled units.[59,60] Reentry of the participating chains into the next helical sequence introduces a constraint (equivalent to the requirement that the ends of a random coiled chain meet in space) which has no counterpart in the polypeptides considered above. Elaboration of the helix–coil transition theory obviously is required to take account of partial melting within the two-strand helix. Rigorous mathematical expression of the added constraint on midchain random-coil sections introduces insuperable complications.[59] The problem posed can be treated adequately, however, and without difficulty, by representation of the loop constraint in an approximate mathematical form that is compatible with the foregoing methodology.[60]

6. CONFIGURATIONAL TRANSITIONS IN PROTEINS

Most proteins are characterized by specific configurations, unique for a given protein. The configuration is essentially invariant for specimens of the protein isolated from different biological species. This holds true despite wide variations in the amino acid residues occurring at various positions in the chain for the corresponding protein from different species.[61,62] The fibrous proteins—collagen, keratin, myosin, and fibrinogen—occur in helical forms, the latter two probably as α helices. The globular proteins—serum albumin, myoglobin, lysozyme, carboxypeptidase, and chymotrypsin, for example—occur in conformations which are unique for each protein, but which are predominantly irregular in most of them. Myoglobin is exceptional in having a majority of its residues engaged in α-helical sequences. In other globular proteins the proportion of α helix is on the order of 20% or less.[61] A tendency for sequences of units to adopt extended conformations (region III) and to run approximately parallel to neighboring sequences similarly conformed is revealed by the x-ray crystallographic analysis of the structure of chymotrypsin,[63] a globular protein comprising 241 residues, and in lysozyme[64] (128 residues). In carboxypeptidase-A, consisting of 307 residues, this tendency is elaborated to a somewhat greater degree;[65] about 20% of the residues of the latter protein occur in an arrangement suggestive of a twisted pleated sheet.[65] With the exceptions noted, the majority of the residues of globular proteins occur in sequences which appear not to adhere to any repetitive conformational scheme; i.e., no systematic pattern is discernible in the conformations of successive residues. Most of the residues other than those occurring in α helices or pleated sheets probably are not involved in hydrogen bonds.

The elastic structure proteins are the only group of proteins in which the molecular configuration is not uniquely prescribed in the native state. The principal elastic protein is elastin, occurring as the main constituent of ligament, aortic tissue, and the walls of blood vessels, and as a minor component of skin. Others include resilin of the cuticle of arthropods[66] and the hinge-ligament protein of mollusks.[67] In fulfillment of their functions, the chains of these proteins occur as random coils even in the native state. The x-ray diffraction of elastin reveals only amorphous halos, and its elastomeric properties are those of a network of typical random-coiled chain molecules.[66–68]

Proteins having unique molecular configurations in the native state—and this includes all proteins with the exceptions noted in the preceding paragraph—are converted to random coils under suitable conditions, i.e., in solvents which interact favorably with amide linkages without being excessively hostile to hydrophobic side chains present in the given protein. When proteins are dispersed in dilute solutions in this "denatured" state, their disulfide crosslinkages having been severed by reduction, they exhibit the behavior typical of random-coiled polymers.[69] The transformation from native form to random coil is reversible under suitable conditions.

The helix–coil transition was for a time regarded as the prototype of the process of denaturation in proteins. Critical examination will show the correspondence to hold only in exceptional circumstances and not in general. Only in the case of a helical protein, e.g., collagen dispersed in a dilute solution, may the process of denaturation (i.e., conversion of the native form to the random coil) be identified with a helix–coil transition. Native collagen consists of triple-strand helices which can be dissolved intact. Their transformation in dilute solution may be likened to the melting of a one-dimensional crystal. The transformation is, quite literally, a helix–coil transition.[70]

In a fiber, or fibril, consisting of a bundle of many helical chains closely packed, the transformation is no longer one dimensional. Melting of chains across the section of the fibril must necessarily take place simultaneously.[71–74] The cooperative nature of the transformation relates not only to adjoining units of a given chain; neighboring chains are involved as well. Thus, the transition acquires three dimensionality.[75] The transition converts helices to random coils, to be sure, but the unimolecular (and hence one-dimensional) character generally denoted by the "helix–coil" designation does not hold. It appears to be sharp, or discontinuous, apart from limitations set by chemical and structural irregularities, and by the finite size of the fibril. This is in contrast to the one-dimensional helix–coil transition treated in the preceding section. In the case of β structures, intermolecular hydrogen bonds may be considered to enhance the intermolecular cooperativity. However, the conclusion asserted above holds equally in the absence of intermolecular bonding; geometrical constraints between neighbors suffice to impose cooperativity transverse to the axes of the chains in a fibril, or in a three-dimensional crystal.[75]

As pointed out above, the chain conformations in native globular proteins, with the exception of myoglobin, are largely devoid of order perpetuated regularly throughout a succession of units.[61] Short α-helical sequences often are discernible, and sequences of residues in extended (β-form) conformations may be evident,[63–65] but the preponderance of the units do not adhere repetitively to the same conformation. Yet, the configuration of a given native protein is unique; in the crystal consisting of many molecules, all chains adopt the same configuration within limits of resolution (1–2 Å) attainable from their x-ray diffraction patterns. The packing is dense. Essentially all of the space within the domain of the globular molecule is occupied by the residues; the void volume is negligible. Polar residues bearing ionic side chains occur on the exterior of the particle; nonpolar residues are located predominantly within. The configuration of the chain in the native state invariably disposes the residues in this manner. To put the matter in a different way, which is more to the point, residues of the two varieties are *so arranged in the chain sequence* to allow this condition to be fulfilled while at the same time permitting a high density of packing of residues in the globular state. In support of this point of view, inspection of the residue sequences for various specimens of the same protein[61,62] from different species shows that the variety of residue, polar or nonpolar, occupying a given position in the chain sequence nearly always is preserved. The externally situated polar residues perform the obvious function of providing an electrostatic double layer which assures solubility in aqueous media and stabilizes the dispersion against coagulation.

The thermal denaturation of a globular protein, chymotrypsinogen, dispersed in dilute aqueous solution is illustrated in Fig. 22, which is taken from the work of Brandts.[76] The change of the extinction coefficient with temperature affords a measure of the transformation from the native to the denatured form at the several pH's indicated in the legend. The process is reversible and the curves in Fig. 22 represent states of equilibrium.[76] The behavior illustrated in this figure is typical of globular proteins.[76,77] The transformation has a finite breadth; it occurs over a range of about 10°C. This observation suggests a correspondence to the helix–coil transition, the breadth of which is attributed to the presence of partially transformed molecules that persist at equilibrium within the transition interval. Careful analysis of the dependence of the degree of denaturation on temperature leads, however, to the conclusion that the process of denaturation occurs without intervention of intermediate states, under conditions of equilibrium at least.[76] A "two-state" model such as was first discussed in detail by Schellman[78] appears therefore to be applicable to the denaturation of globular proteins. In other words, the equilibrium is adequately represented in terms of only two species of molecules: the native protein and the fully denatured form. The latter may be identified, in light of evidence cited earlier, with the random coil form, modified by constraints imposed by such disulfide crosslinkages as may occur in a given protein.[76]

The two-state scheme[76–78] has been represented in the preceding section by Eqs. 22–24 for a simple chemical equilibrium. The small range of the transition is the direct consequence of the high molecular weight (e.g., $x = 240$), as was pointed out on page 290. The fact that it is of finite breadth is a manifestation of the coexistence of native and denatured molecules over the temperature interval where K_x changes from a negligibly small to a very large value. In support of the two-state treatment, the change of enthalpy ΔH° calculated from the dependence of the transformation on temperature (see Fig. 22) is confirmed[76] by direct calorimetric measurements.[79]

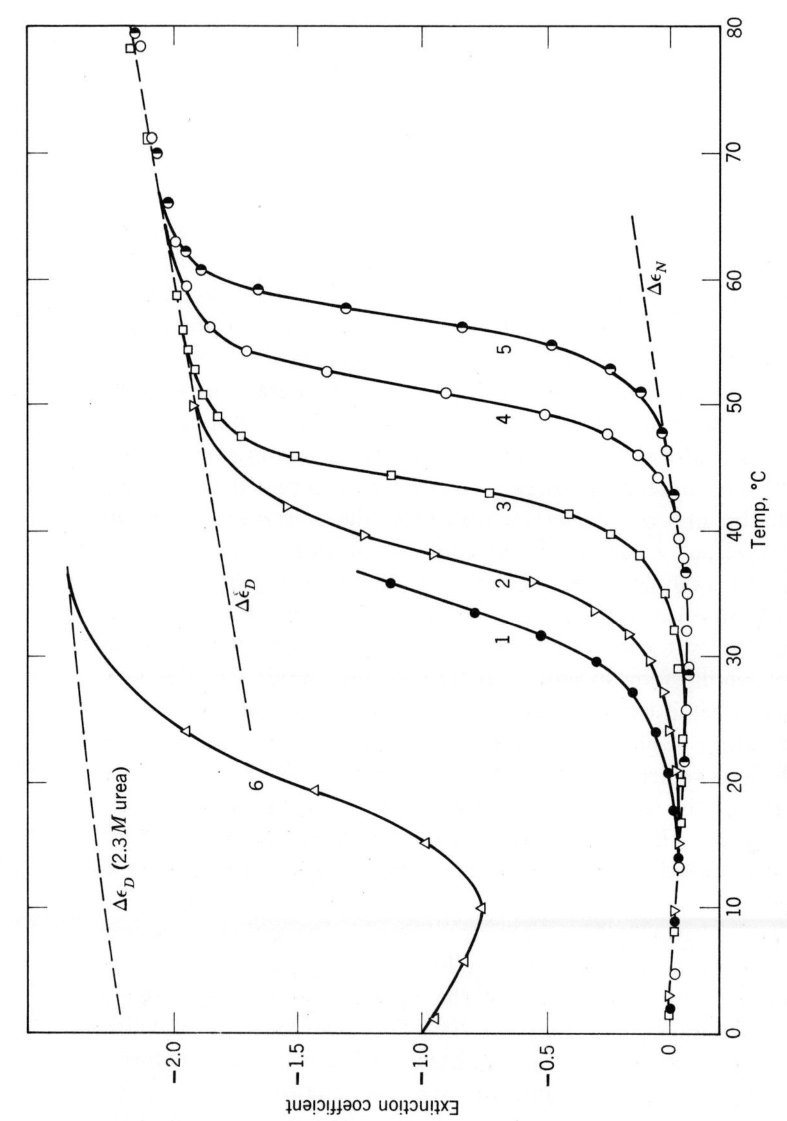

Fig. 22. Plot of the extinction coefficient at 293 mμ for dilute aqueous solutions of chymotrypsinogen against temperature at equilibrium. The several curves represent equilibria at the pH's indicated. The upper dashed curve represents the extinction coefficient for the denatured form as a function of temperature; the lower dashed curve refers to the native form. (From Brandts, 1964.[76])

For the foregoing reasons, the denaturation of a globular protein should not be likened to a helix–coil transition. To do so would imply the occurrence of intermediate states with partial retention (or formation) of the native conformation. There are of course additional compelling reasons for abandoning the analogy to the helix–coil transition: the molecule in the native state is largely nonhelical, it is globular and therefore three dimensional, and it is replete with interactions between residues other than those which are proximate in sequence along the polypeptide chain.

Synthetic analogs of globular proteins are unknown. The capability of adopting a dense globular configuration stabilized by self-interactions, and of transforming reversibly to the random coil, are characteristics peculiar to the chain molecules of globular proteins alone. The factors responsible for these exceptional properties invite inquiry, to which we now turn our attention.

At the outset we note that protein molecules occur in the form of random coils only in solvent media in which there is strong interaction of the amide and/or other polar groups with the solvent. Typically, as in water, this interaction involves solvation of these groups. The solvation layers are compatible of course with the surrounding solvent. They would be incompatible with a hydrophobic medium, or with hydrophobic side chains of the polymer. Their stabilizing effect is contingent therefore upon a large excess of solvent.

A possible alternative situation for the polypeptide chain is one in which its environment is provided by other peptide units, either from the same molecule as in a globular protein, or from like molecules as in a protein fiber. Contacts of hydrophobic groups with a hostile solvent can thus be avoided, or minimized. Polar side chains may, however, find such an environment less satisfactory. A polypeptide chain consisting of both polar and hydrophobic residues may conceivably achieve an optimum *modus vivendi* with its environment by adopting a compact configuration in which hydrophobic groups are placed in the interior, polar groups being relegated to the exterior where they may exist in contact with surrounding (aqueous) solvent. The two kinds of residues must occur, however, in a sequence which permits such an arrangement, as was pointed out above.

It is especially to be observed that the random coil and the globular, or native, states differ drastically and a compromise between them should be less favored than either of these extremes. Thus, the solvated groups of the random coil state are acceptable only if bulk solvent is in copious abundance; incorporation of solvated groups in the environment of the more hydrophobic parts of the molecule would be unsatisfactory.

Realization of effective hydrophobic interactions (and specific interactions between polar groups as well) requires close contact between the interacting groups. Hence, the globular state can be expected to be favorable only if solvent is expelled to the point where little or none is retained within the domain of the molecule. The propensity for dense packing of the polypeptide chain within the globular protein may thus be understood. The antipathy to partial packing accounts for the "all or none" character of the transition, which in this respect resembles a typical phase transition.

A necessary requirement for a configuration of the skeleton of a long chain molecule to be an admissible one is the absense of spatial overlaps between any pair of its units. Compliance with this condition can be achieved with *relatively* little difficulty in the random coil state where occupancy of the domain of the coil by units of the chain is quite low—on the order of a few per cent only. Even in the random coil, however, many otherwise acceptable configurations of a long chain are disqualified by spatial conflicts between units which are well separated in sequence along the chain. Long range interactions are multiplied profusely by increase of the density of units in the domain occupied by the molecule. With complete removal of solvent and consequent packing of the chain units to bulk density, the avoidance of spatial conflicts becomes acute. If the chain is very long, the problem bears a close resemblance to that encountered in a system consisting of *many* random coil molecules that are to be packed to high density without ordering their configurations.[75] An approximate treatment of the problem in the terms of the lattice model shows that, whereas the number of configurations available to an isolated random coil is exceedingly large, the fraction of these configurations which meet the requirements of nonoverlapping at bulk concentration of chains is exceedingly small.[80] Hence, the number of eligible configurations for the molecule under these circumstances may be small, and especially so if the chain is one like poly-L-alanine that is subject to rather severe conformational limitations.

With reference to a single polymer chain, the total number of configurations accessible to it suffers drastic reduction when induced by intramolecular attractions to adopt a compact, globular form owing to (*1*) the rejection of all configurations which pervade a volume in excess of the net volume of the molecule, and (*2*) exclusion of those of the remaining dense configurations which would enforce overlaps. The eligible configurations meeting these requirements will be few. If we impose the additional stipulation that polar side chains shall populate the surface of the globular arrangement, then the possibility of even one configuration meeting all requirements will be vanishingly small—unless of course the residues are arranged in a favorable sequence.

We are thus led to conclude that a polypeptide chain consisting of several hundred amino acid residues present in the proportions usually found in globular proteins, but assembled in random sequence, would rarely be found capable of being "folded," or collapsed, to a predominantly *irregular* configuration which is at once space filling, nonoverlapping, free of an excessive number of residue conformations of high energy, and so arranged as to place polar side chains on the exterior. The spatial requirements of side chains of internal residues must of course be met as well. Only by strategically fashioning the sequence of residues according to their special conformational capacities can a globular form be assured which does not entail an excessive commitment of energy. An assemblage of the same or an equivalent set of residues in random succession almost certainly would not yield a chain capable of assuming a globular conformation of low energy, according to the present line of reasoning. The necessity for a well-defined pattern in the sequence of residues is thus clear. In view of the stringency of the requirements for an acceptable globular configuration, it will not be surprising if a few of the residues are forced into conformations[6,7,9] with energies appreciably above the minima. Glycyl residues permit skeletal conformations departing from those for residues bearing substituents, and thus may enhance the configurational adaptability of the chain if suitably located in the chain sequence.

The foregoing arguments offer a rational basis for observations indicating the native conformation for the polypeptide chain of a given protein to be unique. This conformation is not one of many competitive possibilities which differ marginally in energy, such differences arising, for example, from alterations in the interactions between residues juxtaposed as neighbors in the competing conformations. Rather, its existence is an event of rare exception, allowed to occur only by virtue of a judicious selection of the residue sequence, which conduces to fulfillment of the several requisites, enunciated above, for an acceptable globular conformation. An alternative conformation also meeting these requisites would be exceedingly unlikely. The criteria for tolerance of a globular conformation are primarily steric, in the sense of accommodation of units which are remote in sequence along the chain as spatial neighbors without inflicting overlaps; specific interactions between such neighbors appear to play a subordinate role.[61]

Finally, it may be worth pointing out that the globular conformation of a native protein, or that part of it which is not helical, should not be likened to a random coil. The customary differentiation of helical and "random coil" portions of a native protein is a misapplication of the latter term, which can be quite misleading. From what has been said above it will be apparent that a globular conformation is highly specific and certainly not random. Such conformations are compact, whereas the random coil may occupy a domain many times the molar volume. The globular state should be differentiated from both the helix and the random coil. It may share some of the features of each, but its differences from both necessitate its description in separate terms.

REFERENCES

1. J. T. Edsall, P. J. Flory, J. C. Kendrew, A. M. Liquori, G. Némethy, G. N. Ramachandran, and H. A. Scheraga, *J. Biol. Chem.*, **241**, 1004 (1966); *J. Mol. Biol.*, **15**, 399 (1966); *Biopolymers*, **4**, 121, 1149 (1966).
2. D. A. Brant and P. J. Flory, *J. Am. Chem. Soc.*, **87**, 2791 (1965); **87**, 663 (1965).
3. V. Sasisekharan, in *Collagen*, N. Ramanathan, Ed., Interscience, New York, 1962, p. 39.
4. G. N. Ramachandran, C. Ramakrishnan, and V. Sasisekharan, *J. Mol. Biol.*, **7**, 95 (1963).
5. C. Ramakrishnan and G. N. Ramachandran, *Biophys. J.*, **5**, 909 (1965).
6. G. N. Ramachandran, C. M. Venkatachalam, and S. Krimm, *Biophys. J.*, **6**, 849 (1966).
7. C. M. Venkatachalam, and G. N. Ramachandran, in *Conformation of Biopolymers*, Vol. 1, G. N. Ramachandran, Ed., Academic Press, New York, 1967, pp. 83–105.
8. S. J. Leach, G. Némethy, and H. A. Scheraga, *Biopolymers*, **4**, 369 (1966).
9. D. A. Brant and P. R. Schimmel, *Proc. Natl. Acad. Sci. U.S.*, **58**, 428 (1967).
10. G. N. Ramachandran, C. Ramakrishnan, and C. M. Venkatachalam, *Biopolymers*, **3**, 591 (1965).
11. K. D. Gibson and H. A. Scheraga, *Biopolymers*, **4**, 709 (1966).
12. J. A. Schellman and C. Schellman, in *The Proteins*, 2nd ed., Vol. 2, H. Neurath, Ed., Academic Press, New York, 1964, p. 24.
13. T. Miyazawa, *J. Polymer Sci.*, **55**, 215 (1961).
14. L. Pauling, R. B. Corey, and H. R. Branson, *Proc. Natl. Acad Sci. U.S.*, **37**, 205 (1951); L. Pauling and R. B. Corey, *ibid.*, **37**, 729 (1951).
15. D. A. Brant, W. G. Miller, and P. J. Flory, *J. Mol. Biol.*, **23**, 47 (1967).
16. R. A. Scott and H. A. Scheraga, *J. Chem. Phys.*, **45**, 2091 (1966).
17. R. M. Meighan and R. H. Cole, *J. Phys. Chem.*, **68**, 503 (1964).
18a. D. A. Brant and P. J. Flory, *J. Am. Chem. Soc.*, **87**, 2788 (1965).
18b. D. A. Brant, private communication.
19. R. G. C. Arridge and C. G. Cannon, *Proc. Roy. Soc. (London)*, **A278**, 91 (1964); R. G. C. Arridge, *Proc. Phys. Soc. (London)*, **85**, 1157 (1965).
20. D. A. Brant and P. J. Flory, to be published.
21. L. Mandelkern, in *Poly-α-Amino Acids*, G. D. Fasman, Ed., Marcel Dekker, Inc., 1967, Chap. 13, p. 675.
22. J. P. Carver and E. R. Blout, in *Treatise on Collagen: Chemistry of Collagen*, Vol. 1, G. N. Ramachandran, Ed., Academic Press, London, 1967.
23. P. M. Cowan and S. McGavin, *Nature*, **176**, 501 (1955).
24. V. Sasisekharan, *Acta Cryst.*, **12**, 897 (1959).
25. W. Traub and V. Shmueli, *Nature*, **198** 1165 (1963).
26. I. Z. Steinberg, W. F. Harrington, A. Berger, M. Sela, and E. Katchalski, *J. Am. Chem. Soc.*, **82**, 5263 (1960).
27. A. Veis, E. Kaufman, and C. C. W. Chao, in *Conformation of Biopolymers*, Vol. 2, G. N. Ramachandran, Ed., Academic Press, New York, 1967, pp. 499–512.

28. Y. C. Leung and R. E. Marsh, *Acta Cryst.*, **11**, 17 (1958).
29. P. R. Schimmel and P. J. Flory, *J. Mol. Biol.*, **34**, 105 (1968).
30. J. E. Mark and M. Goodman, *J. Am. Chem. Soc.*, **89**, 1267 (1967); *Biopolymers*, **5**, 809 (1967). See also A. M. Liquori and P. De Santis, *ibid.*, **5**, 815 (1967).
31. P. R. Schimmel and P. J. Flory, *Proc. Natl. Acad. Sci. U.S.*, **58**, 52 (1967).
32. P. De Santis, E. Giglio, A. M. Liquori, and A. Ripamonti, *Nature*, **206**, 456 (1965); A. M. Liquori, P. De Santis, A. L. Kovacs, and L. Mazzarella, *ibid.*, **211**, 1039 (1966).
33. P. R. Schimmel, unpublished.
34. F. Gornick, L. Mandelkern, A. F. Diorio, and D. E. Roberts, *J. Am. Chem. Soc.*, **86**, 2549 (1964).
35. J. Engel, *Biopolymers*, **4**, 945 (1966); J. Engel, in *Conformation of Biopolymers*, Vol. 2, G. N. Ramachandran, Ed., Academic Press, New York, 1967, p. 483.
36. A. E. Tonelli and P. J. Flory, unpublished.
37. P. Doty, J. H. Bradbury, and A. M. Holtzer, *J. Am. Chem. Soc.*, **78**, 947 (1956).
38. J. Beacham, V. T. Ivanov, G. W. Kenner, and R. C. Sheppard, *Chem. Commun.*, **16**, 386 (1965).
39. P. J. Flory and P. R. Schimmel, *J. Am. Chem. Soc.*, **89**, 6807 (1967).
40. W. G. Miller, D. A. Brant, and P. J. Flory, *J. Mol. Biol.*, **23**, 67 (1967).
41. J. T. Edsall and J. Wyman, *Biophysical Chemistry*, Vol I, Academic Press, New York, 1958, Chap. 6.
42. E. R. Blout and M. Idelson, *J. Am. Chem. Soc.*, **78**, 3857 (1956).
43. R. D. Lundberg and P. Doty, *J. Am. Chem. Soc.*, **79**, 3961 (1957).
44. Y. Shalitin and E. Katchalski, *J. Am. Chem. Soc.*, **82**, 1630 (1960).
45. R. E. Nylund and W. G. Miller, *J. Am. Chem. Soc.*, **87**, 3537 (1965).
46. B. H. Zimm and J. K. Bragg, *J. Chem. Phys.*, **31**, 526 (1959).
47. L. Peller, *J. Phys. Chem.*, **63**, 1194, 1199 (1959).
48. J. H. Gibbs and E. A. DiMarzio, *J. Chem. Phys.*, **28**, 1247 (1958).
49. T. M. Birshtein and O. B. Ptitsyn, *Conformations of Macromolecules*, Interscience, New York, 1966, Chap. 9.
50. T. Ooi, R. A. Scott, G. Vanderkooi, and H. A. Scheraga, *J. Chem. Phys.*, **46**, 4410 (1967).
51. D. A. Brant, *Macromolecules* (in press).
52. G. Schwarz, *Biopolymers*, **5**, 321 (1967).
53. L. Landau and E. Lifshitz, *Statistical Physics*, Oxford University Press, Oxford, 1938, p. 232.
54. B. H. Zimm, P. M. Doty, and K. Iso, *Proc. Natl. Acad. Sci. U.S.*, **45**, 1601 (1959).
55. W. G. Miller and P. J. Flory, *J. Mol. Biol.*, **15**, 298 (1966).
56. K. Nagai, *J. Chem. Phys.*, **34**, 887 (1961).
57. S. Lifson and A. Roig, *J. Chem. Phys.*, **34**, 1963 (1961).
58. S. Lifson and B. H. Zimm, *Biopolymers*, **1**, 15 (1963); S. Lifson, *ibid.*, **1**, 25 (1963).
59. B. H. Zimm, J. Chem, Phys., **33**, 1349 (1960).
60. P. J. Flory and W. G. Miller, *J. Mol. Biol.*, **15**, 284 (1966).
61. R. E. Dickerson, in *The Proteins*, Vol. 2, H. Neurath, Ed., Academic Press, New York, 1964, Chap. 11.
62. E. Margoliash and A. Schejter, *Advan. Protein Chem.*, **21**, 113 (1966); W. M. Fitch and E. Margoliash, *Science*, **155**, 279 (1967).
63. B. W. Matthews, P. B. Sigler, R. Henderson, and D. M. Blow, *Nature*, **214**, 652 (1967).
64. D. C. Phillips, *Proc. Natl. Acad. Sci. U.S.*, **57**, 484 (1967).

65. G. N. Reeke, J. A. Hartsuck, M. L. Ludwig, F. A. Quiocho, T. A. Steitz, and W. N. Lipscomb, *Proc. Natl. Acad. Sci. U.S.*, **58**, 2220 (1967).
66. T. Weis-Fogh, *J. Mol. Biol.*, **3**, 520, 648 (1961).
67. R. M. Alexander, *J. Exp. Biol.*, **44**, 119 (1966); R. E. Kelly and R. V. Rice, *Science*, **155**, 208 (1967).
68. C. A. J. Hoeve and P. J. Flory, *J. Am. Chem. Soc.*, **80**, 6523 (1958).
69. C. Tanford, K. Kawahara, and S. Lapanje, *J. Biol. Chem.*, **241**, 1921 (1966); *J. Am. Chem. Soc.*, **89**, 729 (1967).
70. P. J. Flory and E. S. Weaver, *J. Am. Chem. Soc.*, **82**, 4518 (1960).
71. P. J. Flory, *Science*, **124**, 53 (1956); *J. Cellular Comp. Physiol.*, **49**, 175 (1957).
72. R. R. Garrett and P. J. Flory, *Nature*, **177**, 176 (1956); *J. Am. Chem. Soc.*, **80**, 4836 (1958).
73. L. Mandelkern, *Crystallization of Polymers*, McGraw-Hill, New York, 1964, pp. 163–164, 197–213. See also J. Mandelkern, *J. Gen. Physiol.*, **50**, 29 (1967).
74. H. A. Scheraga, *J. Gen. Physiol.*, **50**, 5 (1967).
75. P. J. Flory, *J. Polymer Sci.*, **49**, 105 (1961).
76. J. F. Brandts and R. Lumry, *J. Phys. Chem.*, **67**, 1484 (1963); J. F. Brandts, *J. Am. Chem. Soc.*, **86**, 4291 (1964); **87**, 2759 (1965); J. F. Brandts and L. Hunt, *ibid.*, **89**, 4826 (1967).
77. J. Hermans, Jr., and G. Acampora, *J. Am. Chem. Soc.*, **89**, 1543, 1547 (1967).
78. J. H. Schellman, *Compt. Rend. Trav. Lab. Carlsberg Ser. Chim.*, **29**, 223, 230 (1955); *J. Phys. Chem.*, **62**, 1485 (1958).
79. W. W. Forrest and J. M. Sturtevant, *J. Am. Chem. Soc.*, **82**, 585 (1960); R. Danforth, H. Krakauer, and J. M. Sturtevant, *Rev. Sci. Instr.*, **38**, 484 (1967).
80. P. J. Flory, *Proc. Roy. Soc.* (*London*), **A234**, 60 (1956).

CHAPTER VIII

The Statistical Distribution of Configurations

The characterization of a chain configuration by the chain vector $\mathbf{r}$ connecting the ends of the chain was introduced in Chapter I as an abridgement for the impossibly cumbersome designation of the spatial locations of all atoms, as would be required for the full specification of a configuration. The distribution $W(\mathbf{r})$ of $\mathbf{r}$ was briefly discussed, (Chap. I, Sect. 2), and it was pointed out that this distribution must be Gaussian in the limit $n = \infty$. The proof there offered for this assertion relied on arguments more intuitive than rigorous. Here we undertake a more penetrating analysis of the distribution $W(\mathbf{r})$ over the array of configurations of the chain molecule. Finite chains, for which $W(\mathbf{r})$ departs from its limiting Gaussian form, will assume main interest; infinite chains will appear as a limiting case. The distribution function $W(\mathbf{r})$ for finite chains eludes representation by any concise mathematical expression of closed form. In fact, the departure of $W(\mathbf{r})$ from the Gaussian function can only be approximated for any real chain.

The distribution function $W(\mathbf{r})$ is but a crude index of the range of variation among the many possible configurations of the chain. Other representations could be considered, e.g., the spatial density distribution of atoms, or segments, about their common center of gravity (see Chap. I, pp. 7–8). The range of this distribution is reflected in the radius of gyration s, which measures the root-mean-square distance of an atom from the center of gravity of the configuration. The distribution $w(s)$ of values of s for the array of configurations is touched upon briefly at the close of this chapter. We shall be concerned mainly however with the distribution of the chain vector $\mathbf{r}$. The distribution $W(\mathbf{r}_{ij})$ of the vector connecting skeletal atoms i and j of the chain, being closely related to $W(\mathbf{r})$, will not require separate treatment.

1. STATISTICAL THERMODYNAMIC RELATIONS FOR CHAIN MOLECULES[1,2]

The configuration partition function for a chain of n skeletal bonds free of constraints may be expressed with full generality as follows:

$$Z = (8\pi^2)^{-1} \int \cdots \int \exp(-E/kT)\, d\{\mathbf{l}\} \tag{1}$$

where the energy E is understood to be a function of the configuration $\{\mathbf{l}\}$. The factor $(8\pi^2)^{-1}$ normalizes Z with respect to the spatial orientations of the chain so that Z represents the internal configuration space only. The configuration partition function for the chain whose end-to-end vector is fixed at $\mathbf{r}$ is given by

$$\tilde{Z}_{\mathbf{r}} = (8\pi^2)^{-1} \int \underset{\mathbf{r}}{\cdots} \int \exp(-E/kT)\, d\{\mathbf{l}\}/d\mathbf{r} \tag{2}$$

The scalar variables of integration, $3(n-1)$ in number, represent the set defining the configuration subject to constancy of $\mathbf{r}$. Equation 2 is the analog of the conventional configuration integral for a canonical ensemble; $\mathbf{r}$ assumes the role of the volume. It will be apparent that

$$Z = \int \tilde{Z}_{\mathbf{r}}\, d\mathbf{r} \tag{3}$$

If bond lengths and bond angles are fixed, $\{\mathbf{l}\}$ may be replaced by $n-2$ bond rotation angles and three Eulerian angles χ, ψ, ω. Thus,

$$Z = (8\pi^2)^{-1} \int \cdots \int \exp(-E/kT) \sin\chi\, d\chi\, d\psi\, d\omega\, d\{\phi\} \tag{4}$$

$$= \int \cdots \int \exp(-E/kT)\, d\{\phi\} \tag{5}$$

where $\{\phi\}$ denotes the set of internal rotation angles $\phi_2, \phi_3 \cdots \phi_{n-1}$. The Eulerian angles may be construed to specify the orientation in space of the pair of bonds 1 and 2, the bond angle between them being fixed, or they may fix the direction of vector $\mathbf{r}$ and the rotation of the chain, in the given internal configuration, about $\mathbf{r}$ as an axis. Equation 2 for $\tilde{Z}_{\mathbf{r}}$ may be expressed similarly, provided that $n > 2$.

If the chain is subject to a constant force $\mathbf{f}$ acting on its ends, then the appropriate partition function is

$$Z_{\mathbf{f}} = (8\pi^2)^{-1} \int \cdots \int \exp[(\mathbf{f}\cdot\mathbf{r} - E)/kT]\, d\{\mathbf{l}\} \tag{6}$$

It is the analog of the statistical mechanical configuration integral for an isothermal–isobaric system of n particles. Obviously,

$$Z_{\mathbf{f}} = \int \tilde{Z}_{\mathbf{r}} \exp(\mathbf{f}\cdot\mathbf{r}/kT)\, d\mathbf{r} \tag{7}$$

For $\mathbf{f} = 0$, we have $Z_{\mathbf{f}} = Z$.

In light of the correspondence of Eq. 2 to the configuration integral for a canonical ensemble, we may take the Helmholtz free energy of the chain having its vector equal to $\mathbf{r}$ to be

$$F_{\mathbf{r}} = A(T) - kT \ln \tilde{Z}_{\mathbf{r}} \tag{8}$$

where $A(T)$ is a function of the temperature. It follows that the average *contractile* force acting on the chain ends fixed at the separation **r** is given by

$$\langle \mathbf{f} \rangle = \text{grad } F_{\mathbf{r}}$$
$$= -kT(\partial \ln \tilde{Z}_{\mathbf{r}}/\partial \mathbf{r})_T \tag{9}$$

In view of the spherical symmetry of $Z_{\mathbf{r}}$, this average force will be parallel to **r** with sign reversed.

If the chain is subject to a fixed external force **f**, then the average of the chain vector **r** will be

$$\langle \mathbf{r} \rangle = kT(\partial \ln Z_{\mathbf{f}}/\partial \mathbf{f})_T \tag{10}$$

Equations 9 and 10 call to mind the analogous statistical thermodynamic relations wherein p and V replace **f** and **r**, respectively.

2. THE CHAIN VECTOR DISTRIBUTION AND ITS FOURIER TRANSFORM

The distribution function for the chain vector **r** is given by

$$W(\mathbf{r}) = \tilde{Z}_{\mathbf{r}}/Z \tag{11}$$

as follows directly from the definitions of the configuration partition functions, or configuration integrals.[2] With greater generality, for a chain subject to a force **f** the distribution function is

$$W_{\mathbf{f}}(\mathbf{r}) = \tilde{Z}_{\mathbf{r}} \exp(\mathbf{f} \cdot \mathbf{r}/kT)/Z_{\mathbf{f}} \tag{12}$$

In order to realize full advantage of the more important of the foregoing relationships, it would be necessary to evaluate $\tilde{Z}_{\mathbf{r}}$ in closed form. In general, this configuration integral resists exact evaluation. Only for the freely jointed chain has it been put in exact form, and even for this rudimentary model the expression for $\tilde{Z}_{\mathbf{r}}$ is cumbersome (*cf. seq.*). On the other hand, $Z_{\mathbf{f}}$ is more readily susceptible to rationalization, as will be apparent below.

Let $G(\mathbf{q})$ be the Fourier transform of $W(\mathbf{r})$, as defined by

$$G(\mathbf{q}) = \int \exp(\tilde{\imath}\mathbf{q} \cdot \mathbf{r}) W(\mathbf{r})\, d\mathbf{r} \tag{13}$$

$$= (8\pi^2 Z)^{-1} \int \cdots \int \exp(-E/kT) \exp(\tilde{\imath}\mathbf{q} \cdot \mathbf{r})\, d\{\mathbf{l}\} \tag{14}$$

where $\tilde{\imath} = \sqrt{-1}$.* If $\tilde{\imath}\mathbf{q}$ is identified with $\mathbf{f}/kT$, then according to Eqs. 6 and 14

* The tilde is used to distinguish this symbol from the index i in equations appearing in Chapter IX.

$$Z_{\mathbf{f}} = ZG(-\tilde{\imath}\mathbf{f}/kT) \tag{15}$$

i.e., $Z_{\mathbf{f}}/Z$ and $W(\mathbf{r})$ are related by the Fourier transformation.

Inasmuch as $W(\mathbf{r})$ is spherically symmetric, Eq. 13 reduces to

$$G(\mathbf{q}) = \int_0^{\pi} \int_0^{\infty} \exp(\tilde{\imath}qr\cos\chi)W(\mathbf{r})2\pi r^2 \sin\chi \, d\chi \, dr$$

where χ is the angle between $\mathbf{r}$ and $\mathbf{q}$, and r and q are the scalar magnitudes of these vectors. Integration over χ gives

$$G(\mathbf{q}) = \int_0^{\infty} 4\pi r^2 W(\mathbf{r})(qr)^{-1} \sin(qr)\, dr \tag{16}$$

Through replacement of sin (qr) by its series expansion we obtain†

$$\begin{aligned} G(\mathbf{q}) &= \int_0^{\infty} (1 - q^2r^2/3! + q^4r^4/5! - \cdots)4\pi r^2 W(\mathbf{r})\, dr \\ &= 1 - (\langle r^2\rangle_0/3!)q^2 + (\langle r^4\rangle_0/5!)q^4 - (\langle r^6\rangle_0/7!)q^6 + \cdots \end{aligned} \tag{17}$$

a series in the even moments of the distribution $W(\mathbf{r})$. Following the treatment of Nagai,[3] we may factor $\exp(-q^2\langle r^2\rangle_0/6)$ from the series in Eq. 17 and thereby express $G(\mathbf{q})$ as‡

$$G(\mathbf{q}) = \exp(-v)[1 + g_2(2v) + g_4(2v)^2 + \cdots] \tag{18}$$

† Inclusion of the subscript zero on $\langle r^2\rangle$ and higher even moments of r is not obligatory here and elsewhere throughout the present section if, in accordance with previous usage, this subscript is construed to stipulate that the chain is unperturbed by the self-interactions generally associated with effects of excluded volume. The development in this section, being applicable to any spherically symmetric distribution, comprehends distributions distorted by such effects. Insofar as the subscript zero signifies absence of a force **f** acting on the ends of the chain, it is, of course, a necessary condition for the specification of the moments appearing in Eq. 17 and the sequel. In following sections of this chapter, as, for example, in the treatment of the freely jointed chain with all configurations of equal energy, the assumption of absence of perturbations by self-interactions is implicit.

‡ The procedure here departs from that of Nagai[3] through the use of $\langle r^2\rangle_0$ *for the chain of length n* in the exponential, and hence in Eq. 19 also. Nagai used the quantity $\langle r^2\rangle_* = n(\langle r^2\rangle_0/n)_{n\to\infty}$ instead of $\langle r^2\rangle_0$, and proceeded on the supposition that $\langle r^2\rangle_0$ is linear in n, i.e., that for finite chains it is permissible to take $\langle r^2\rangle_0 = \text{const} + n(\langle r^2\rangle_0/n)_\infty$. For sufficiently long chains this is legitimate,[4] but for small n it may not be (*cf. seq.*). We avoid this assumption. Hence, our definitions of the coefficients in the series in Eq. 18 (see Eqs. 20 and 21) differ somewhat from those of Nagai.[3] The equations given here are general and not contingent on any assumptions regarding the dependence of the moments on chain length.

where for convenience we introduce the variable

$$v = q^2\langle r^2\rangle_0/6 \tag{19}$$

to serve instead of q^2. The coefficients g_2, g_4, etc. may be evaluated by expanding the exponential in Eq. 18 in series, multiplying the two series, and comparing the result with Eq. 17. In this way the coefficients are found to be related as follows to the even moments of the distribution:

$$\begin{aligned}
g_2 &= 0 \\
g_4 &= -(1/2!2^2)(1 - 3\langle r^4\rangle_0/5\langle r^2\rangle_0{}^2) = -(1/2^2)(\Delta_4/2!) \\
g_6 &= -(1/3!2^3)[3(1 - 3\langle r^4\rangle_0/5\langle r^2\rangle_0{}^2) - (1 - 3^2\langle r^6\rangle_0/5\cdot 7\langle r^2\rangle_0{}^3)] \\
&= -(1/2^3)(\Delta_4/2! - \Delta_6/3!) \\
g_8 &= -(1/2^4)(\Delta_4/2!2! - \Delta_6/3! + \Delta_8/4!) \\
&\vdots
\end{aligned} \tag{20}$$

or

$$g_{2p} = 2^{-p}\left\{-\frac{\Delta_4}{2!(p-2)!} + \frac{\Delta_6}{3!(p-3)!} - \cdots \frac{(-1)^{p-1}}{p!}\Delta_{2p}\right\} \tag{21}$$

where

$$\Delta_{2p} = 1 - \left[\frac{3^p}{3\cdot 5\cdot 7\cdots(2p+1)}\right]\frac{\langle r^{2p}\rangle_0}{\langle r^2\rangle_0{}^p}, \qquad p > 0 \tag{22}$$

For a Gaussian distribution $W(\mathbf{r})$, the Δ_{2p} vanish, as may easily be shown. Hence, the g_{2p} equal zero for all $p > 0$, and $G(\mathbf{q})$ also is Gaussian. This is the limit for chains of sufficiently large n. For chains of finite length, the series in Eq. 18 may be taken to represent the departure of the Fourier transform $G(\mathbf{q})$ of the distribution $W(\mathbf{r})$ from a Gaussian representation.

Application of the Fourier integral theorem to Eq. 13 gives

$$W(\mathbf{r}) = (2\pi)^{-3}\int G(\mathbf{q}) \exp(-\bar{\imath}\mathbf{q}\cdot\mathbf{r})\, d\mathbf{q} \tag{23}$$

$$= (2\pi^2)^{-1}\int_0^\infty G(\mathbf{q})(qr)^{-1}\sin(qr)q^2\, dq \tag{24}$$

Substitution of Eq. 18 in Eq. 24 and integration over the series yields

$$\begin{aligned}
\mathscr{W}(\boldsymbol{\rho}) = \pi^{-3/2}[&(1 + 3\cdot 5g_4 + 3\cdot 5\cdot 7g_6 + \cdots) \\
&- (1 + 5\cdot 7g_4 + 5\cdot 7\cdot 9g_6 + \cdots)\rho^2 \\
&+ (1 + 7\cdot 9g_4 + 7\cdot 9\cdot 11g_6 + \cdots)\rho^4/2! \\
&- (1 + 9\cdot 11g_4 + 9\cdot 11\cdot 13g_6 + \cdots)\rho^6/3! + \cdots]
\end{aligned} \tag{25}$$

where

$$\mathscr{W}(\boldsymbol{\rho}) = (2\langle r^2\rangle_0/3)^{3/2}W(\mathbf{r}) \tag{26}$$

and $\rho^2 = 3r^2/2\langle r^2\rangle_0$. Thus, $\mathscr{W}(\boldsymbol{\rho})$ is the distribution function for the vector $\boldsymbol{\rho}$ of magnitude ρ and having the direction of $\mathbf{r}$.

As Nagai[3] has shown, the distribution function $\mathscr{W}(\boldsymbol{\rho})$, which is related to $W(\mathbf{r})$ according to Eq. 26, can be factored to give

$$\begin{aligned}\mathscr{W}(\boldsymbol{\rho}) = \pi^{-3/2}\exp(-\rho^2)[(1 + 3\cdot 5g_4 + 3\cdot 5\cdot 7g_6 + 3\cdot 5\cdot 7\cdot 9g_8 + \cdots)\\ - (20g_4 + 210g_6 + 2520g_8 + 34650g_{10} + \cdots)\rho^2\\ + (4g_4 + 84g_6 + 1512g_8 + \cdots)\rho^4 - (8g_6 + 288g_8 + \cdots)\rho^6\\ + \cdots]\end{aligned} \tag{27}$$

Jernigan[5] has shown that this result may be derived in a more direct manner by expanding $\mathscr{W}(\boldsymbol{\rho})$ in Hermite polynomials, a method developed by Grad[6] for the solution of the Boltzmann transport equation. The development of $\mathscr{W}(\boldsymbol{\rho})$ in this manner to obtain Eq. 27 is given in Appendix E.

The foregoing equations, Eqs. 16–27 inclusive, have great generality; they hold for any spherically symmetric density distribution function $W(\mathbf{r})$. They may actually suffer from being too general for their specific application to the distribution of the vector $\mathbf{r}$ for a chain molecule. The form of the particular distribution must be introduced entirely through its moments. The number of moments required for adequate specification may be excessive, a matter to which we shall devote attention in Section 6.

Nagai[7] has presented arguments generally applicable to chain molecules showing that g_4 in Eq. 27, as applied to the density distribution of chain vectors $\mathbf{r}$, is of order $1/n$ for large n, g_6 and g_8 are of order $1/n^2$, etc. The bonds of the chain may be subject to correlations of any kind whatever, provided only that the range of correlations is finite. Crudely stated, the finite flexibility of the chain is a sufficient condition to assure adherence of the coefficients g to these asymptotic orders in powers of $1/n$. It follows that for chains of sufficient length and at extensions such that $\rho^2 < O(n)$, i.e., at extensions $r \ll r_{max}$, all terms beyond unity of the series in Eq. 27 must vanish. Thus, $\mathscr{W}(\boldsymbol{\rho})$, and likewise $W(\mathbf{r})$, reduce to Gaussian distributions in this limit, and the conclusion to this effect tentatively asserted in Chapter I (pp. 6–7) is rigorously confirmed. Alternative proofs showing the distribution $W(\mathbf{r})$ to be Gaussian in the limit $n \to \infty$ have been offered by Moran[8] and by Tchen.[9]

Equation 27 expresses the departure of the distribution for a finite chain from the Gaussian function, $\pi^{-3/2} \exp(-\rho^2)$, in terms of the moments embodied in the g's. Even at low chain extensions where terms in ρ^2, ρ^4, etc., may be ignored, the portion of the first term which departs from unity, namely,

$$3 \cdot 5g_4 + 3 \cdot 5 \cdot 7g_6 + \cdots$$

could conceivably require many moments for adequate specification. [Although g_4 is generally negative, higher coefficients may be positive as well as negative (*cf. seq.*).] For greater extensions, other terms become significant, of course. Justification for truncation of the series representing each coefficient of a power of ρ^2 at its first or second term cannot be guaranteed in general. Particular cases must be examined individually. Nagai's series expansion, Eq. 27, is treated in greater detail in Section 6.

3. THE FREELY JOINTED CHAIN: EXACT TREATMENT[10–12]

The energy of the hypothetical freely jointed chain, having the lengths of its bonds fixed and bond angles unconstrained, is the same for all configurations. Hence, Eq. 14 for the Fourier transform of the chain vector distribution $W(\mathbf{r})$ reduces to

$$G(\mathbf{q}) = (8\pi^2 Z)^{-1} \int \cdots \int \exp(\tilde{\imath}\mathbf{q} \cdot \sum \mathbf{l}_i)\, d\mathbf{l}_1 \cdots d\mathbf{l}_n \tag{28}$$

$$= (4\pi)^{-n} \prod_i \int \exp(\tilde{\imath}\mathbf{q} \cdot \mathbf{l}_i)\, d\mathbf{l}_i$$

$$= \prod_i (ql_i)^{-1} \sin(ql_i) \tag{29}$$

If all bonds are of the same length l, which we henceforth assume to be the case, then

$$G(\mathbf{q}) = [\sin(ql)/ql]^n \tag{30}$$

$$= 1 - \left[\frac{n}{3!}\right] l^2 q^2 + \left[\frac{n(n-1)}{2(3!)^2} + \frac{n}{5!}\right] l^4 q^4$$

$$- \left[\frac{n(n-1)(n-2)}{3!(3!)^3} + \frac{n(n-1)}{3!5!} + \frac{n}{7!}\right] l^6 q^6$$

$$+ \cdots \tag{31}$$

By comparison of Eq. 17 with this series, additional terms included, we obtain

$$\langle r^2\rangle_0 = nl^2$$

$$\langle r^4\rangle_0 = \left[\left(\frac{5}{3}\right)n(n-1) + n\right]l^4$$

$$\langle r^6\rangle_0 = \left[\left(\frac{35}{9}\right)n(n-1)(n-2) + 7n(n-1) + n\right]l^6$$

$$\langle r^8\rangle_0 = \left[\left(\frac{35}{3}\right)n(n-1)(n-2)(n-3) + 42\,n(n-1)(n-2) + \frac{123}{5}\,n(n-1) + n\right]l^8$$

$$\langle r^{10}\rangle_0 = \left[\left(\frac{385}{9}\right)n(n-1)(n-2)(n-3)(n-4) + \left(\frac{770}{3}\right)n(n-1)(n-2)(n-3) + 341n(n-1)(n-2) + \left(\frac{253}{3}\right)n(n-1) + n\right]l^{10} \tag{32}$$

etc. for higher moments.

Application of the Fourier integral theorem, Eq. 23 or 24, to $G(\mathbf{q})$ as expressed by Eq. 30 yields

$$W(\mathbf{r}) = (2\pi^2 r)^{-1}\int_0^\infty \sin(qr)[\sin(ql)/ql]^n q\,dq \tag{33}$$

This exact expression for the freely jointed chain was adduced by Lord Rayleigh[10] in his classic investigations on random flights. Derivation by a method due to Markoff, and involving use of delta functions and their representation by Dirichlet integrals, has been given by Chandrasekhar.[11,12]

If n is large, the bracketed expression in the integrand in Eq. 33 may be approximated as follows:

$$[\sin(ql)/ql]^n = \left[1 - \frac{1}{3!}(ql)^2 + \cdots\right]^n \cong \exp(-nq^2l^2/6) \tag{34}$$

By substituting this exponential expression in Eq. 33, we obtain

$$\lim_{n\to\infty} W(\mathbf{r}) = (2\pi^2 r)^{-1}\int_0^\infty \exp(-nq^2l^2/6)\sin(qr)q\,dq$$

which, upon integration by parts, yields[12]

$$\lim_{n\to\infty} W(\mathbf{r}) = (3/2\pi nl^2)^{3/2}\exp(-3r^2/2nl^2) \tag{35}$$

Critical examination of Eq. 33 will show this result to be valid only if r is substantially less than nl. The physical necessity for this qualification is self-evident: r may not exceed nl, and indeed the integral in the exact equation, Eq. 33, is zero for $r \geq nl$. Equation 35 will be recognized as the Gaussian distribution previously expressed by Eq. I-7, and also by Eq. 27 above in the limit $n \to \infty$. The correspondence is established through the identification $\langle r^2 \rangle_0 = nl^2$ for freely jointed chains, to which the present considerations are restricted. The treatment above affords a rigorous proof that the distribution function $W(\mathbf{r})$ is Gaussian in the special case of a freely jointed chain, provided, of course, that $n \gg 1$ and that r is sufficiently less than nl.

For finite chains, the integration of Eq. 33 is tedious for all but very small values of n. Integrals for $n = 3$, 4, and 6 were obtained by Rayleigh.[10] For $n = 2$, 3, and 4 (see Appendix F) the integrals can be concisely expressed as follows:

$$W_2(\mathbf{r}) = 1/8\pi l^2 r, \qquad 0 < r/l < 2 \tag{36}$$

$$\begin{aligned} W_3(\mathbf{r}) &= 1/8\pi l^3, & 0 \leq r/l \leq 1 \\ W_3(\mathbf{r}) &= (16\pi l^2 r)^{-1}(3 - r/l), & 1 \leq r/l < 3 \end{aligned} \tag{37}$$

$$\begin{aligned} W_4(\mathbf{r}) &= (64\pi l^3)^{-1}(8 - 3r/l), & 0 \leq r/l \leq 2 \\ W_4(\mathbf{r}) &= (64\pi l^2 r)^{-1}(4 - r/l)^2, & 2 \leq r/l < 4 \end{aligned} \tag{38}$$

with $W_n(\mathbf{r}) = 0$ outside the ranges specified in each instance.

The integral of Eq. 33 for any value of n is the rather unwieldy expression

$$W_n(\mathbf{r}) = (8\pi r l^2)^{-1} n(n-1) \sum_{t=0}^{\tau} \frac{(-1)^t}{t!(n-t)!} \left[\frac{n - (r/l) - 2t}{2}\right]^{n-2} \tag{39}$$

where τ is specified by

$$[(n - r/l)/2] - 1 \leq \tau < (n - r/l)/2 \tag{40}$$

Equation 39 is due to Treloar.[13,14] He derived this relationship by a procedure differing from that employed here. Observing that the components of a system of unit vectors of random directions are uniformly distributed over ranges -1 to $+1$, Treloar derived his equation, Eq. 39, from the theory of random sampling as developed by Hall[15] and Irwin.[16] The equivalence of Treloar's result to integrals of Eq. 33 for lower values of n, as obtained earlier by Lord Rayleigh,[10] was pointed out by Volkenstein.[12] Proof that Eq. 39 is the general solution of Eq. 33 is given in Appendix F.

Treloar's result is most useful for lower values of n. The distribution for larger n may be obtained more readily by numerical evaluation of the integral in Eq. 33 with the aid of a digital computer. Results thus obtained are presented in Figs. 1–6 on pp. 322–325. Their discussion is deferred pending consideration of approximate distribution functions.

4. APPROXIMATE DISTRIBUTION FUNCTIONS FOR FREELY JOINTED CHAINS

The exact distribution of **r** for the freely jointed chain does not lend itself to simple mathematical expression, and its approximation by a Gaussian function, though asymptotically exact for sufficiently long chains at low extensions ($r \ll nl$), implies nonvanishing probabilities $W(\mathbf{r})$ for $r > nl$. The distribution function derived below overcomes this latter deficiency; that is, the function $W(\mathbf{r})$ which will be derived vanishes for $r \geq nl$ as required.

Two derivations will be given. These lead to identical results. The first is carried out according to well-established statistical mechanical procedures originally applied to the freely jointed chain by Kuhn and Grün[17,18]; the second relies upon a method due to James and Guth.[19] Both treatments in their original forms were subject to an error which will be amended below. Inasmuch as the distribution function obtained by these methods may be conveniently expressed in terms of the inverse Langevin function of the extension ratio r/nl, it has come to be known as the Langevin distribution, or the inverse Langevin distribution. These terms are obviously misleading. We shall adopt the term "$\mathscr{L}$* distribution," which retains association with past usage without perpetuating its false connotations.

In the treatment of Kuhn and Grün[17,18] the n bonds comprising the freely jointed chain are projected on an *arbitrary* axis, which we take to be the X axis of a (laboratory) Cartesian coordinate system. Let n_j represent the number of bonds having projections between l_{xj} and $l_{xj} + \delta l_{xj}$ on this axis. In absence of constraints (e.g., a force acting on the ends of a chain, "long-range" interactions between segments remote in sequence, etc.), the probability density of projections l_{xj} may easily be shown to be uniform throughout the permissible range $-l \leq l_{xj} \leq l$. Hence, the *a priori* probability of an orientation within any interval l_{xj} to $l_{xj} + \delta l_{xj}$ in this range is just $\delta l_{xj}/2l$. The set of numbers $\{n\} = n_1, n_2, \cdots, n_j \cdots$ for all orientation intervals will specify a distribution. The statistical weight $\Omega_{\{n\}}$ of such a distribution for the free chain may be written as the product of the *a priori* probabilities for individual bonds and the number of permutations of bonds amongst the states 1, 2, etc. Thus,

$$\Omega_{\{n\}} = n! \prod_j (\delta l_{xj}/2l)^{n_j}/n_j! \tag{41}$$

The set of numbers $\{n\}$ must comply with the conditions

$$\sum_j n_j = n \tag{42}$$

$$\sum_j n_j l_{xj} = x \tag{43}$$

if it is to be representative of a chain of n bonds having a projection x. The combined statistical weight $\Omega(x)$ for all distributions such that the projection of $\mathbf{r}$ on the x axis is x to $x + \delta x$ is given by the sum of all $\Omega_{\{n\}}$ meeting these conditions. Formally expressed, this statistical weight is

$$\Omega(x) = \sum_{\{n\}} n! \prod_j (\delta l_{xj}/2l)^{n_j}/n_j! \tag{44}$$

where the summation includes all distributions consistent with Eqs. 42 and 43.

The formidable task of evaluating this summation can be circumvented, provided that n is sufficiently large, by replacing the entire sum by its largest term—a familiar expedient in dealing with macroscopic statistical mechanical systems comprising 10^{20} or more entities. The maximum term is, of course, only a minute part of the entire sum, but it nevertheless suffices for the prescription of average properties of the system, with results which are exact in the limit $n \to \infty$. The validity of this device for a chain of only a few thousand bonds is another matter. Certainly, such a system is not sufficiently large for assertion of the laws of large numbers, on which the maximum term device depends. One may hope that the Stirling's approximations introduced below will compensate the error from the illegitimacy of application of the maximum-term approximation to the system under consideration, but it is difficult to formulate a logical analysis of the resultant of the opposing errors from these two sources. We adopt the maximum-term device therefore with due circumspection.

Letting $\Omega(x)$ be replaced henceforth by the maximum term of the sum in Eq. 44, we have

$$\Omega(x) = n! \prod_j (\delta l_{xj}/2l)^{n_j}/n_j! \tag{45}$$

where the n_j's now refer to the most probable distribution. Introduction of Stirling's approximations for the factorials—a step likewise justified only if $n_j \gg 1$ for all j, which is clearly not the case—yields

$$\ln \Omega(x) = \sum_j n_j[\ln(\delta l_{xj}/2l) - \ln(n_j/n)] \tag{46}$$

It remains to ascertain the most probable distribution. We turn to the familiar Lagrangian multiplier method for this purpose. By differentiation of Eqs. 46, 42, and 43

$$d \ln \Omega(x) = \sum_j [\ln (\delta l_{xj}/2l) - \ln n_j]\, dn_j \tag{47}$$

$$dn = \sum_j dn_j \tag{48}$$

$$dx = \sum_j l_{xj}\, dn_j \tag{49}$$

Equating each of these differential expressions to zero and combining the resulting equations after multiplying Eqs. 48 and 49 by arbitrary factors α and β/l, respectively, we obtain

$$\ln n_j - \ln (\delta l_{xj}/2l) - \alpha - \beta l_{xj}/l = 0$$

or

$$n_j = \exp(\alpha) \exp(\beta l_{xj}/l)(2l)^{-1}\, \delta l_{xj} \tag{50}$$

where α and β are Lagrangian multipliers which remain to be determined.

The condition expressed by Eq. 42 requires that

$$\begin{aligned} \exp(-\alpha) &= (2nl)^{-1} \int_{-l}^{l} \exp(\beta l_x/l)\, dl_x \\ &= (n\beta)^{-1} \sinh \beta \end{aligned} \tag{51}$$

the summation in Eq. 42 being replaced by the integral over l_x. Similarly, from Eq. 43

$$x = (2l)^{-1} \exp(\alpha) \int_{-l}^{l} \exp(\beta l_x/l) l_x\, dl_x$$

Substitution from Eq. 51 for exp (α) and execution of the integration gives

$$x = nl\mathscr{L}(\beta) \tag{52}$$

where $\mathscr{L}(\beta)$ is the Langevin function of β defined by

$$\begin{aligned} \mathscr{L}(\beta) &= \coth \beta - 1/\beta \\ &= \beta/3 - \beta^3/45 + \cdots \end{aligned} \tag{53}$$

Equation 52 may be written

$$\beta = \mathscr{L}^*(x/nl) \tag{54}$$

where $\mathscr{L}^*$ denotes the inverse Langevin function. The parameters α and β are determined by Eqs. 51 and 54.

According to Eqs. 51 and 50

$$n_j = (n\beta/\sinh \beta) \exp(\beta l_{xj}/l)(2l)^{-1}\delta l_{xj} \tag{55}$$

which, when substituted into Eq. 46, yields

$$\ln \Omega(x) = \sum_j n_j[\ln(\beta^{-1} \sinh \beta) - \beta l_{xj}/l]$$
$$= n \ln(\beta^{-1} \sinh \beta) - \beta x/l \qquad (56)$$

Inasmuch as the probability $w(x)\delta x$ of a projection x to $x + \delta x$ of the chain vector **r** on the arbitrary axis is proportional to $\Omega(x)$, we let

$$w(x) = cl^{-1}(\beta^{-1} \sinh \beta)^n \exp(-\beta x/l) \qquad (57)$$

where c is a numerical constant which normalizes $w(x)$.

In their original treatment Kuhn and Grün[17,18] chose the direction of the chain vector **r** as the "arbitrary" axis. They accordingly identified r with x in Eqs. 54–57. The end-to-end vector is not, however, an arbitrary axis. Components of **r** transverse to this axis are necessarily zero, and in this respect the axis obviously is unique. The sum of the projections of all bonds on an arbitrary axis must in general be less than r. Identification of the arbitrary axis with the chain vector **r** vitiates the treatment of Kuhn and Grün[17,18] by an error perpetuated throughout subsequent expositions[12,14] on the subject.

A method devised by Treloar[13] for use in another connection will serve to derive, from $w(x)$, the probability $4\pi r^2 W(\mathbf{r})$ of a scalar magnitude $r = |\mathbf{r}|$ per unit range of r, irrespective of the direction of the vector. Within a system of many identical chains in a random statistical array of configurations, let us consider those chains whose ends are separated by distances r to $r + dr$. Since the vectors **r** of these chains are uniformly distributed over all solid angles, their components x on the arbitrary axis occur with equal incidence, or uniform probability density, throughout the range $-r \leq x \leq r$. It follows that the decrease of $w(x)$ with x affords a measure of the contribution from chains of length $r = x$. That is, since chains of length r to $r + dr$ contribute $4\pi r^2 W(\mathbf{r})\, dr\, dx/2r$ to the fraction of chains $w(x)\, dx$ having components x to $x + dx$ if, and only if, $r > x$, it follows that

$$-(dw(x)/dx)_{x=r} = 2\pi r W(\mathbf{r}) \qquad (58)$$

Treating $w(x)$ given by Eq. 57 as an explicit function of β and x, we have therefore

$$W(\mathbf{r}) = -(2\pi r)^{-1}[(\partial w(x)/\partial\beta)(d\beta/dx) + (\partial w(x)/\partial x)]_{x=r}$$

Since $\partial w(x)/\partial\beta = 0$, as may be confirmed from Eqs. 57 and 54,

$$W(\mathbf{r}) = -(2\pi r)^{-1}(\partial w(x)/\partial x)_{x=r} \qquad (59)$$

Hence,

$$W(\mathbf{r}) = (A\beta/rl^2)(\beta^{-1} \sinh \beta)^n \exp(-\beta r/l) \qquad (60)$$

where $A = c/2\pi$ and β (see Eq. 54) is redefined as

$$\beta = \mathscr{L}^*(r/nl) \tag{61}$$

$$= 3(r/nl) + (9/5)(r/nl)^3 + (297/175)(r/nl)^5 + \cdots \tag{62}$$

For the alternative derivation[18,19] of the distribution function, Eq. 60, let one end of the freely jointed chain be acted upon by a force $\mathbf{f}$, the other being fixed. Its *average* extension is then given by Eq. 10. Confining attention to the average component in the direction of the force $\mathbf{f}$, and designating this component by x, we may cast Eq. 10 in the form

$$\langle x \rangle = kT(\partial \ln Z_{\mathbf{f}}/\partial f)_T \tag{63}$$

According to Eqs. 15 and 30

$$Z_{\mathbf{f}} = Z[\sinh(fl/kT)/(fl/kT)]^n \tag{64}$$

Differentiation of Eq. 64 and substitution in Eq. 63 yields

$$\langle x \rangle = nl\mathscr{L}(fl/kT) \tag{65}$$

The functional dependence of the average force $\langle f \rangle$ on x is of much greater interest than $\langle x \rangle$ as a function of f. If fluctuations about these averages could be ignored (i.e., if n were very great and r/nl much smaller than unity), then designation of averages could be abandoned and Eq. 65 could be rearranged to

$$f = (kT/l)\mathscr{L}^*(x/nl) \tag{66}$$

The work of extension to a length x in the direction of the force would be

$$(kT/l)\int_0^x \mathscr{L}^*(x'/nl)\,dx'$$

The probability of an extension x, taken to be proportional to the Boltzmann exponential of this energy, is on this basis[18]

$$w(x) = (c/l)\exp\left[-l^{-1}\int_0^x \mathscr{L}^*(x'/nl)\,dx'\right] \tag{67}$$

The exponent may be written

$$-n\int_0^\beta (1/\beta' - \beta' \operatorname{csch}^2 \beta')\,d\beta'$$

which, upon integration, reduces to

$$n \ln\left(\frac{\sinh \beta}{\beta}\right) - \left(\frac{\beta x}{l}\right)$$

where $\beta = \mathscr{L}^*(x/nl)$. Substitution of this expression for the exponent of Eq. 67 reproduces Eq. 57 of the preceding treatment. It will be apparent, by repetition of the previous argument, that an alternative to Eq. 60 for the density distribution function is

$$W(\mathbf{r}) = (A/rl^2)\mathscr{L}^*(r/nl) \exp\left[-l^{-1}\int_0^r \mathscr{L}^*(r'/nl)\,dr'\right]. \tag{68}$$

Equation 65 is a rigorous expression for the average *displacement* $\langle x \rangle$ as a function of a specified value of f/kT. The converse of this relationship, i.e., Eq. 66, is construed in the foregoing derivation to yield the average *force* as a function of x, when x is fixed. This transposition in the designation of averages is not generally justified. The error committed can be shown[20] to correspond to that involved in replacing the average quantity $\langle x \rangle$ by the most probable value of x when the chain is subject to a force of magnitude f. In the limit $n \to \infty$, these quantities converge and no error is introduced in replacing the one by the other. The invalidity of this step for finite n renders the treatment inexact.

Inasmuch as the result of the present derivation is identical with that obtained by the preceding method, the approximation here involved evidently corresponds to the resultant of the two approximations invoked in the previous derivation, namely, replacement of the partition sum by its maximum term and the use of Stirling's approximation for factorials.

The distribution function of Kuhn and Grün,[17,18] long accepted as the correct result by the methods detailed above, is

$$W(\mathbf{r}) = (A'/l^3)(\beta^{-1}\sinh\beta)^n \exp(-\beta r/l) \tag{69}$$

obtained from Eq. 57 through the illegitimate replacement of x by r, followed by normalization by the factor A'; β is defined by Eq. 61. The same result may be obtained alternatively through erroneous identification of x with r in Eq. 67. The correct result, Eq. 60 or Eq. 68, differs from Eq. 69 by the factoı $\beta l/r$; of course, the numerical constants A' and A differ as well. For small extensions, $\beta l/r \cong 3/n$ (see Eq. 62), and the forms of the two equations are equivalent. At higher extensions β increases much more rapidly than r, as will be apparent from the behavior of the Langevin function, or from the series expansion in Eq. 62. The two results differ markedly at extensions where r/nl approaches unity.

Equation 68 affords a convenient path to expression of the foregoing results in terms of their series expansions. Incorporating the series expansion for $\mathscr{L}^*(r/nl)$, Eq. 62, in Eq. 68, we obtain

$$\begin{aligned} W(\mathbf{r}) = (3A/nl^3)[1 &+ (3/5)(r/nl)^2 + (99/175)(r/nl)^4 \\ &+ \cdots] \exp\{-(3/2nl^2)r^2[1 + (3/10)(r/nl)^2 \\ &+ (33/175)(r/nl)^4 + \cdots]\} \end{aligned} \tag{70}$$

This expression reduces to the Gaussian function for small values of r relative to the maximum extension nl. Both of the series in this equation converge rapidly for $r/nl < \frac{1}{2}$.

The $\mathscr{L}^*$ distribution expressed alternatively by Eq. 60 or 68 vanishes for $(r/nl) > 1$, as required. The Kuhn and Grün expression, Eq. 69, vanishes likewise. Both of the derivations given for Eqs. 60 and 68 entail approximations, as we have taken pains to emphasize, which render the results strictly valid only in the limit $n \to \infty$. The impact of these approximations, which deprive the final results of rigor as applied to finite chains, is difficult to assess by mathematical analysis. We are obliged therefore to turn to comparisons of numerical results calculated by these approximate equations with those given by the alternative exact expressions, Eqs. 33 and 39.

Distribution functions $W_{10}(\mathbf{r})$ are shown in Fig. 1 for a freely jointed

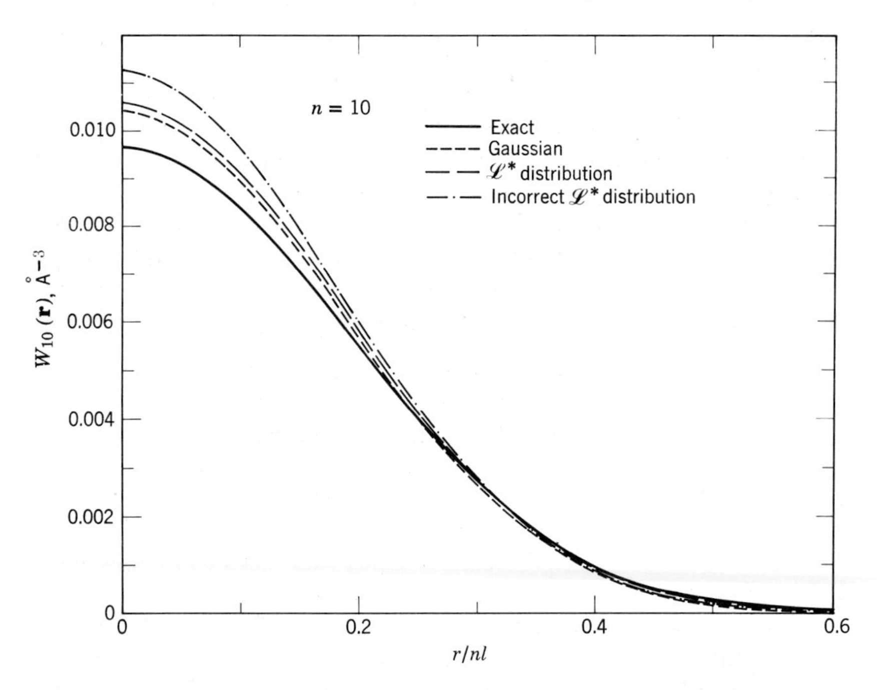

Fig. 1. Distribution functions $W(\mathbf{r})$ for the freely jointed chain consisting of ten bonds, each of unit length, i.e., $n = 10$ and $l = 1$. The curves are calculated as follows: exact, Eq. 33 or Eq. 39; Gaussian, Eq. I-7, with $\langle r^2 \rangle = nl^2$; $\mathscr{L}^*$ distribution, Eq. 60; incorrect $\mathscr{L}^*$ distribution, Eq. 69. Calculations carried out by Jernigan.[5,21]

chain of ten bonds ($n = 10$) of unit length. The four curves have been calculated[5,21] according to equations given above and identified in the legend. This manner of representing the distribution is most effective for the range of low extensions. Here the Gaussian function offers the best approximation to the exact distribution calculated from the Rayleigh equation, Eq. 33 or, equivalently, from Eq. 39. The $\mathscr{L}^*$ distribution in its amended form given by Eq. 60, or by Eq. 68, is displaced slightly above the Gaussian curve in this region. The curve calculated according to the incorrect $\mathscr{L}^*$ distribution function, Eq. 69, is seen to be in much greater error.

Radial distributions $4\pi r^2 W_n(\mathbf{r})$ are shown for chains of $n = 4$, 10, and 20 bonds in Figs. 2, 3, and 4, respectively, l being assigned a length of unity.[5,21]

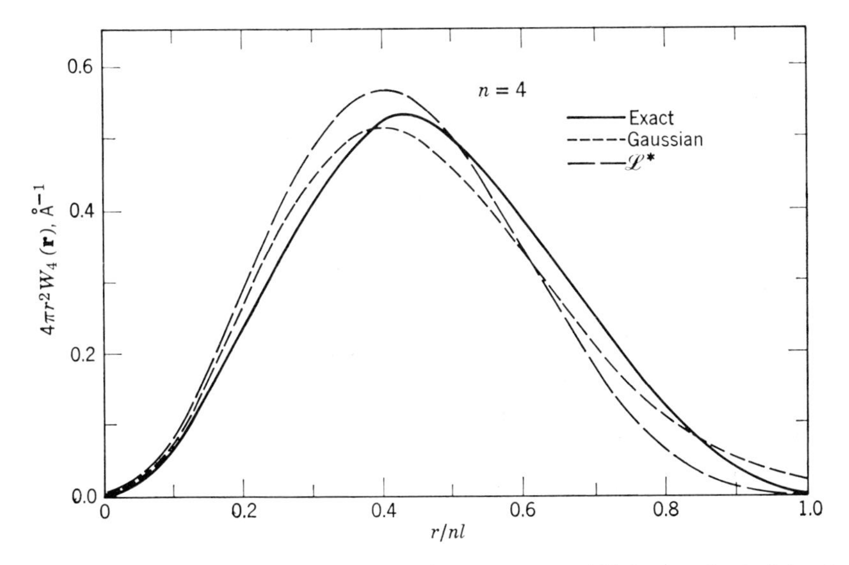

Fig. 2. Radial distribution functions $4\pi r^2 W(\mathbf{r})$ calculated[5,21] for the freely jointed chain with $n = 4$ (see legend to Fig. 1).

In each instance, the Gaussian function offers the best approximation up to extensions somewhat beyond the maxima in these curves. In Figs. 5 and 6 the range of high extensions is emphasized by plotting the logarithm of $W(\mathbf{r})$ against $(r/nl)^2$. The amended $\mathscr{L}^*$ distribution here is the preferred approximation. Use of Eq. 69 may entail an error of several orders of magnitude at high extensions, i.e., at $r/nl = 0.95$ and beyond. The Gaussian function fails in this range for obvious reasons pointed out earlier.

The amended $\mathscr{L}^*$ distribution, although a considerable improvement over its incorrect antecedent, offers no advantage over the Gaussian function except at values of r/nl approaching full extension. Throughout most of the range, the Gaussian function may actually be preferred.

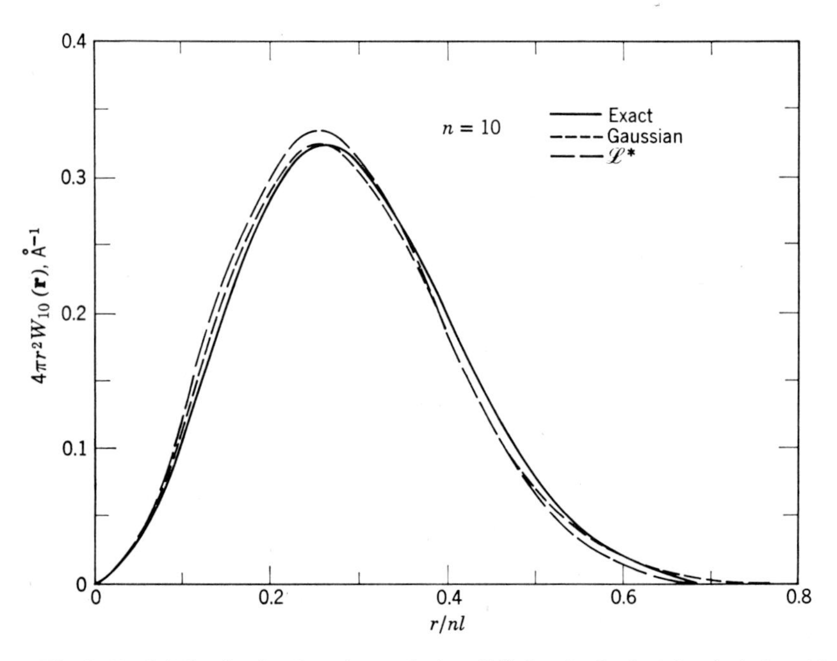

Fig. 3. Radial distribution functions calculated[5,21] for the freely jointed chain with $n = 10$ (see legend to Fig. 1).

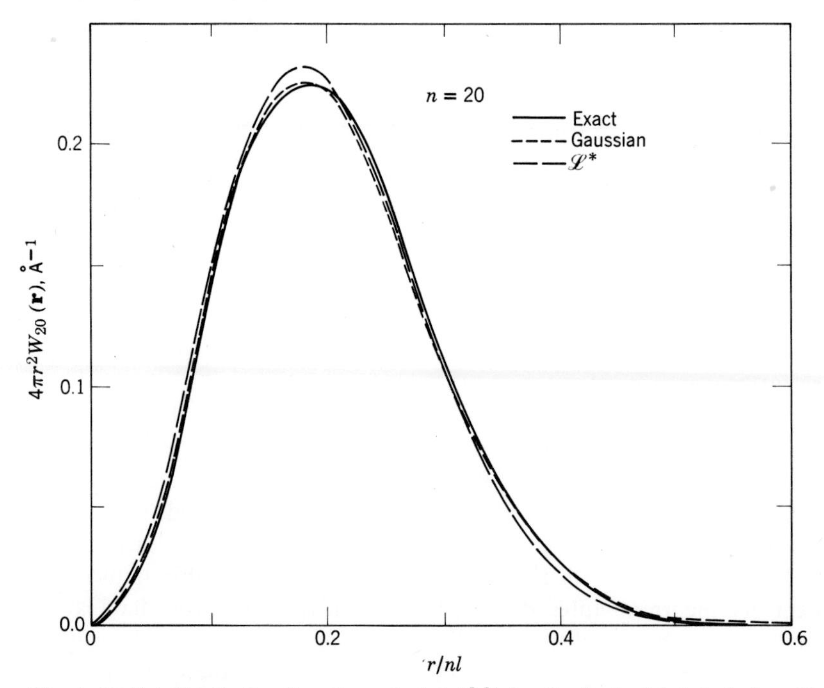

Fig. 4. Radial distribution function calculated[5,21] for the freely jointed chain with $n = 20$ (see legend to Fig. 1).

Nevertheless, the accuracy of the amended $\mathscr{L}^*$ distribution surpasses expectations warranted by the grave approximations involved in applying the maximum term method to chains of so few bonds, and likewise in

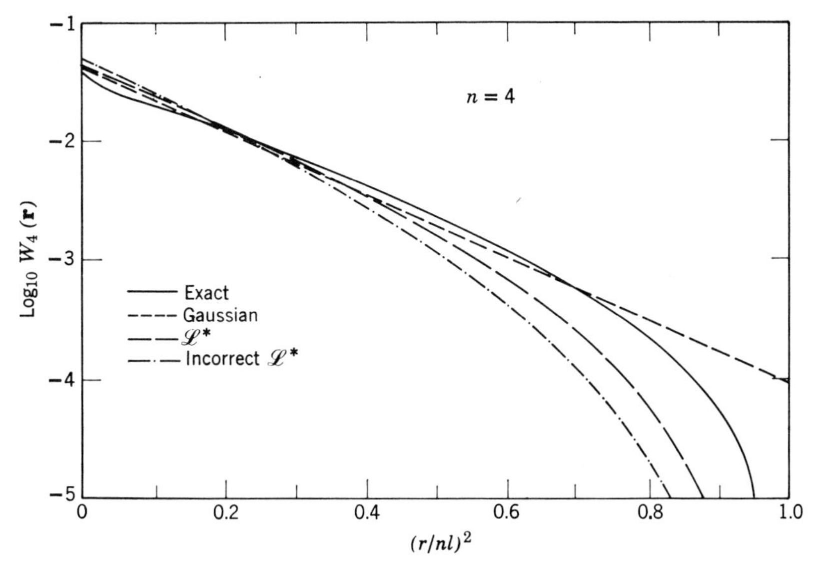

Fig. 5. Logarithms of the distribution functions $W(\mathbf{r})$ for the freely jointed chain of four bonds (see also Fig. 2) plotted against $(r/nl)^2$.[5,21]

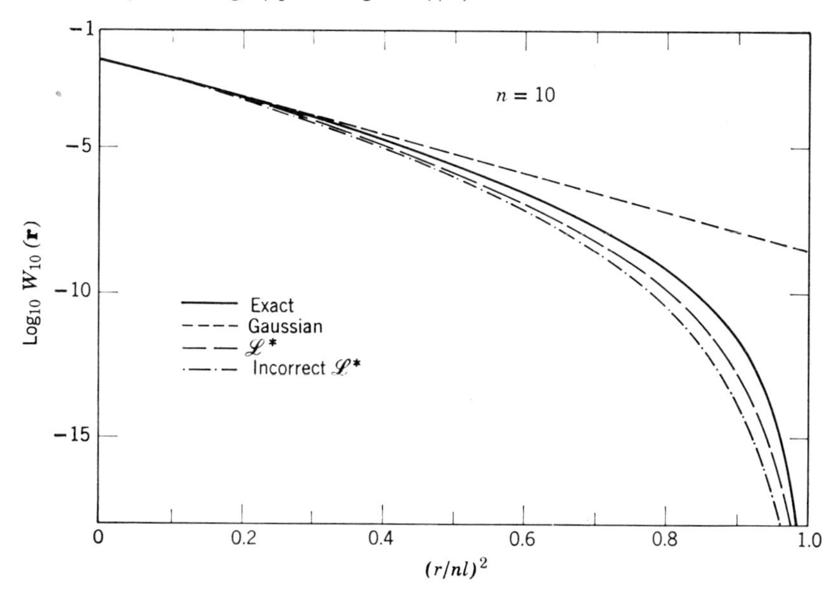

Fig. 6. Logarithms of the distribution functions for the freely jointed chain of ten bonds (see also Fig. 3) plotted against $(r/nl)^2$.[5,21]

using Stirling's approximations for the factorials $n_j!$ under the same circumstances. Fortuitous compensation of errors from these sources evidently occurs. In view of the forfeiture of rigor through these approximations, the $\mathscr{L}^*$ distribution function should perhaps be relegated to the status of an empirical representation of the distribution, the validity of which rests largely on numerical comparisons like those shown in Figs. 1–6.

The Gaussian function affords a generally serviceable approximation for freely jointed chains, except in the higher range of r/nl, where $4\pi r^2 W(\mathbf{r})$ for long chains falls to values so low as to be of little importance in most applications.

5. MOMENTS OF THE DISTRIBUTIONS FOR VARIOUS CHAINS, HYPOTHETICAL AND REAL

Exact distribution functions for finite real chains being unknown, the efficacy of various approximations cannot be subjected to direct test in the same way as the $\mathscr{L}^*$ distribution for the freely jointed chain was compared with numerical solutions of the Rayleigh equation. Moments of approximate distribution functions may be compared, however, with moments for finite, real chains deduced according to the methods of preceding chapters. Inasmuch as the moments and their ratios are functions of n, it is pertinent at the outset to explore the dependence of these quantities on chain length for various models and approximations.

The dependence of the characteristic ratio $C_n = \langle r^2 \rangle_0 / nl^2$ on n is shown in Fig. 7 for four hypothetical chains and for polymethylene (PM) at 140°C.[22] Parameters for the hypothetical chains have been chosen to match the limiting characteristic ratio $C_\infty = 6.87$ for the latter in order to provide a consistent basis for comparing the diverse chains. The chains specifically considered are the following:

(*1*) Equivalent freely jointed chain consisting of n' equivalent bonds of length l', with $n'l' = nl \cos(\theta/2)$ and $n'(l')^2 = nl^2 C_\infty$ (see Eqs. I-18). Thus, $\theta = 68°$ and $C_\infty = 6.87$ yield $l'/l = 6.87/0.829 = 8.29$ and $n'/n = (0.829)^2/6.87 = 0.100$.

(*2*) Freely rotating chain consisting of n bonds joined at angles $\theta = 41.8°$ (see Eq. I-20).

(*3*) Chain with rotations hindered by independent potentials; $\theta = 68°$ and $\langle \cos \phi \rangle = 0.514$ (see Eq. I-53).

(*4*) Porod-Kratky[23] chain, with $C_\infty = 6.87$, $L = nl \cos(\theta/2)$ (see Eqs. 71 and 72 below).

(*5*) Chain with neighbor-dependent rotational hindrances: specifically, a simple chain exemplified by PM at 140°C, with $\theta = 68°$, $\sigma = 0.54$, and $\omega = 0.088$, the rotational statistical weights being expressed by Eq. V-9.

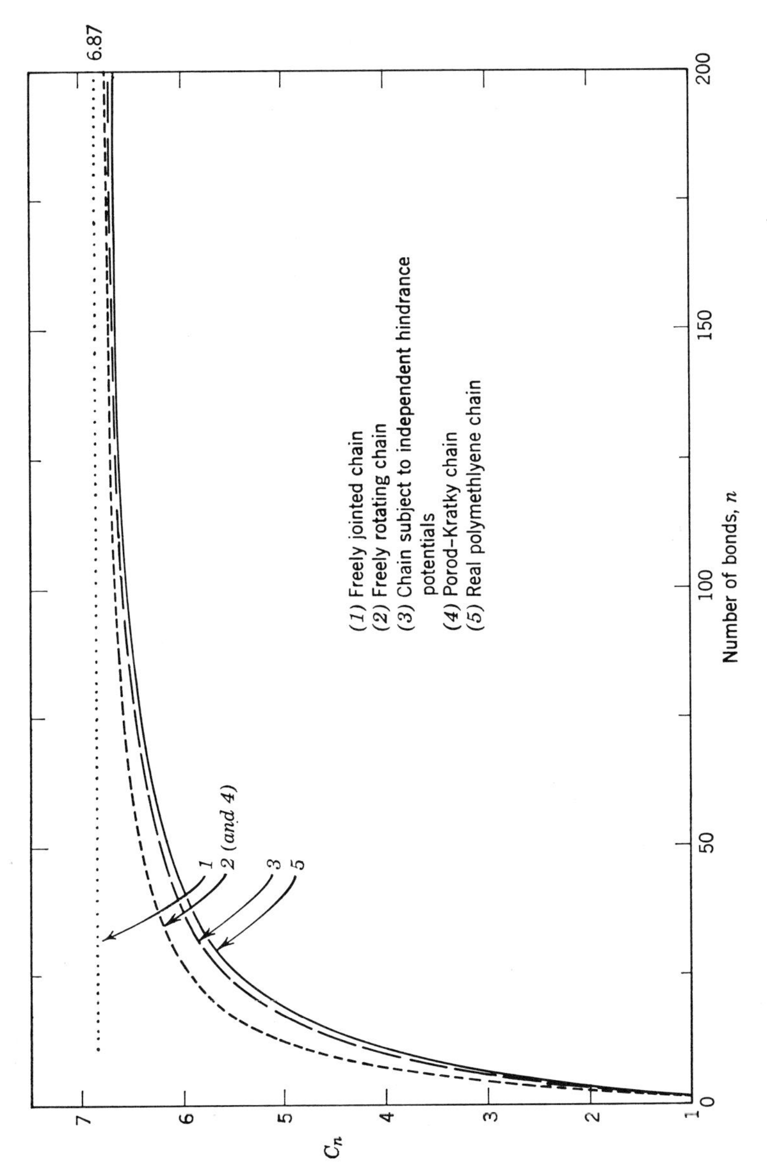

Fig. 7. Dependence of the characteristic ratio $\langle r^2 \rangle_0 / nl^2$ on n for the "real" polymethylene chain, curve (*5*), and the four model chains specified in the text, the parameters in each case being chosen to reproduce $C_\infty = 6.87$.[5,22] The curve (*4*) for the Porod-Kratky chain very nearly coincides with curve (*2*) for the freely rotating chain, and is not shown separately. The straight line (*1*) for the freely jointed chain is terminated at $n = 10$ in recognition of the correspondence of one equivalent bond of this model to ten real bonds of the PM chain; i.e., $n' = n/10$ (see text).

For the freely jointed chain, C_n is independent of n. The horizontal straight line shown dotted in Fig. 7 represents this chain. It is also the asymptote for the other curves, the coincidence of asymptotes having been achieved through the choices of parameters above. The curve (*5*) for the PM chain has been taken from Fig. V-9.

The distinguishing property of the Porod-Kratky[23] model chain (*4*) is the continuity of the direction of its contour in space. Its trajectory is described by a smooth curve whose direction changes at random, but in a continuous manner. The character of a given Porod-Kratky chain, manifested in its tortuosity, is prescribed by its persistence length a (see Chap. IV, p. 111) which enters as the sole parameter apart from the contour length of the chain, or its fully extended length, denoted by L. This "worm-like" chain bears a resemblance to the freely rotating chain (see p. 16 and Appendix G) for which

$$a = l \sum_{k=0}^{\infty} (\cos \theta)^k = l/(1 - \cos \theta) \tag{71}$$

It differs from the latter in that the bond length l and the angle θ between successive bonds are made to vanish. The limit is approached in such manner as to preserve the value of a. The characteristic ratio for the Porod-Kratky chain may be shown (see Appendix G) to depend on the length of the chain expressed by L through the relationship

$$C_n = C_\infty[1 - (a/L)(1 - e^{-L/a})] \tag{72}$$

where

$$C_\infty = (L/nl)(2a/l) \tag{73}$$

We may identify L with the length of the chain when it is fully extended; e.g., for a simple chain, for which all bond lengths l and all bond angle supplements θ are equal, $L = nl \cos(\theta/2)$. The ratio L/nl having been established, the value of C_∞ then serves to specify a through Eq. 73.

The several curves in Fig. 7 are qualitatively similar in form, the freely jointed chain (*1*) excepted of course. The Porod-Kratky chain (*4*) is indistinguishable in Fig. 7 from curve (*2*) for the freely rotating chain. The shape of curve (*3*) for the chain with independent and symmetric rotational potentials could be altered by arbitrarily changing θ while, at the same time, adjusting $\langle \cos \phi \rangle$ to maintain C_∞ at its chosen value. Wider variations are encountered among various real chains (*cf. seq.*). The dependences of the characteristic ratios on chain length are presented in a manner better suited for comparisons by the plots against $1/n$ in Fig. 8.[22] Again, similarities are more striking than differences.

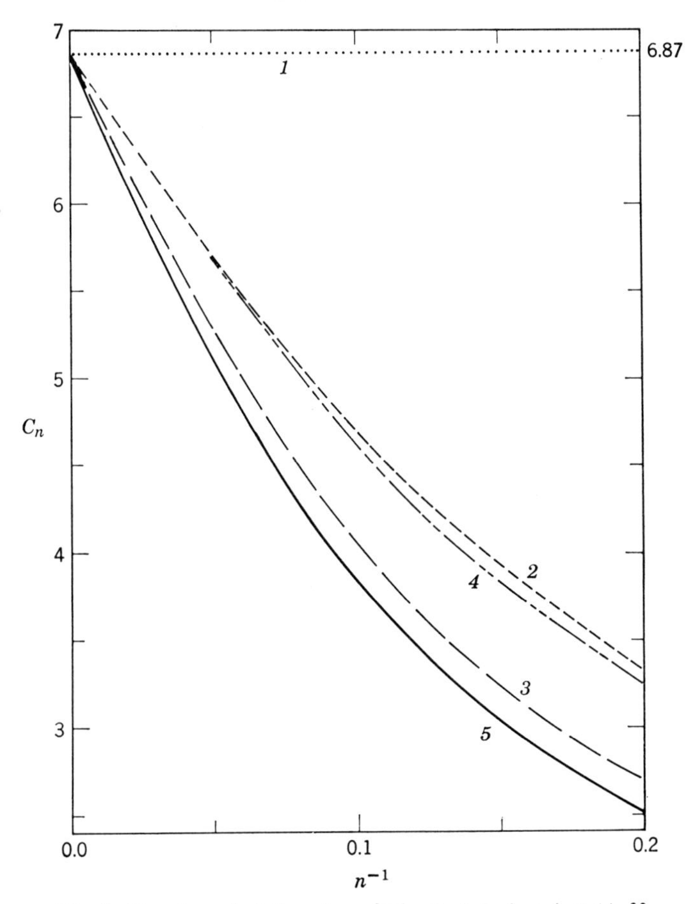

Fig. 8. The characteristic ratios of Fig. 7 plotted against $1/n$.[22]

In Fig. 9 the ratio $\langle r^4 \rangle_0 / \langle r^2 \rangle_0^{\,2}$ of the fourth moment to the square of the second moment of $\mathbf{r}$ is plotted against n for the various chains treated in Fig. 7, and designated (*1*) to (*5*) above. These calculations, like those above, were carried out by Jernigan.[5,22] The curve (*5*) for polymethylene is from Fig. V-11. Analogous plots of $\langle r^6 \rangle_0 / \langle r^2 \rangle_0^{\,3}$ vs. n are shown in Fig. 10 for the model chains, (*1*), (*2*), (*3*), and (*4*) for which the sixth moment of $\mathbf{r}$ has been computed.[5,22] It is to be noted with reference to the freely jointed chain that the scales on the abscissas in these figures represent the number of bonds in the real chain which is simulated, and not the numbers of number n' of virtual bonds in the model chain. The quantity plotted for the freely jointed chain is 10.0 n' (see above).

Although substantial differences are apparent among the curves in Fig. 9 at low values of n, those for the model chains (*2*), (*3*), and (*4*) depart

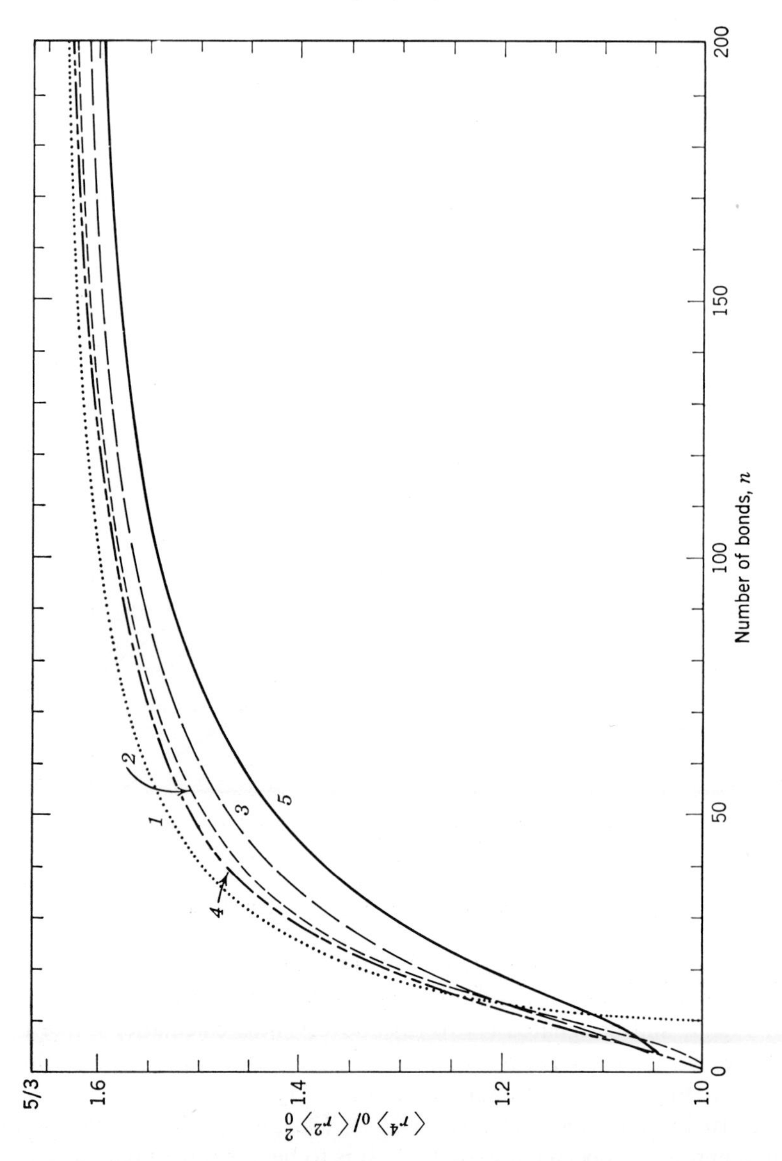

Fig. 9. The fourth moment of **r**, expressed as its ratio to the square of the second moment, plotted against the chain length.[5,22] The chains represented are the same as in Figs. 7 and 8 (see text). The upper margin represents the asymptote common to all curves.

systematically from curve (*5*) for the real chain by factors of about 1.25 to 1.5 in n over most of the range. The forms of the various curves are

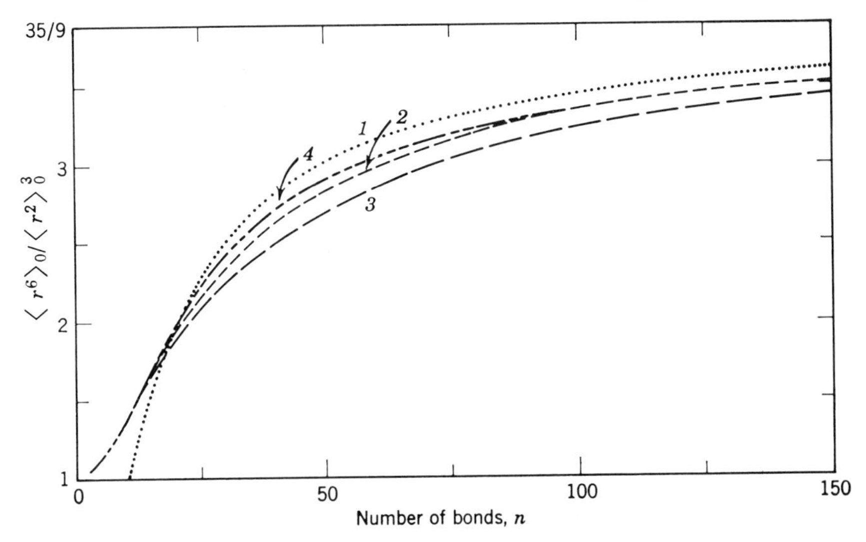

Fig. 10. The ratio of the sixth moment of **r** to $\langle r^2 \rangle_0{}^3$ plotted against n for the four model chains represented in Figs. 7–9.[5,22] The upper margin is the asymptote for all curves.

therefore similar. The extent to which this similarity between the real chain and the model chains persists for the higher moments remains unknown owing to the impracticality of calculation of $\langle r^6 \rangle_0$, $\langle r^8 \rangle_0$, etc. for real chains. Comparable differences between the model chains (*1*), (*2*), (*3*), and (*4*) are apparent in Fig. 10. This observation suggests that higher moments for the real chain may likewise resemble those for the model chains, (*2*) and (*3*) in particular. Significant differences between the various curves shown in Figs. 9 and 10 notwithstanding, the dependence of the moment ratios on chain length are generally similar for the several chains. Resemblance is, of course, favored by the arbitrary choice of parameters to reproduce the same value of C_∞ in each case. The main purpose of the comparisons is to test the suitability of model chains for estimating the trends of higher moments with n for real chains. This purpose is served by adjustment of the various curves to the same asymptote.

It would be misleading to assume the polymethylene chain to be representative of real chains in general. The relationships of C_n to n calculated for a variety of real chains have been shown in Chapters V and VII; see, for example, Fig. V-18 for polyoxyethylene, Fig. V-24 for polydimethylsiloxane, Fig. V-32 for poly(ethylene terephthalate), and Fig. VII-15 for poly-L-alanine. Among the examples cited, including polymethylene, the rate of convergence of C_n, or of C_x, to its asymptote seems to be correlated with the value of the asymptote, the approach being more protracted, the larger the value of C_∞. This correlation does not hold universally, however, as perusal of other examples would show. The dependence on chain length exhibited by the characteristic ratio of the second moment of **r** evades generalization; peculiarities of structure and of conformational

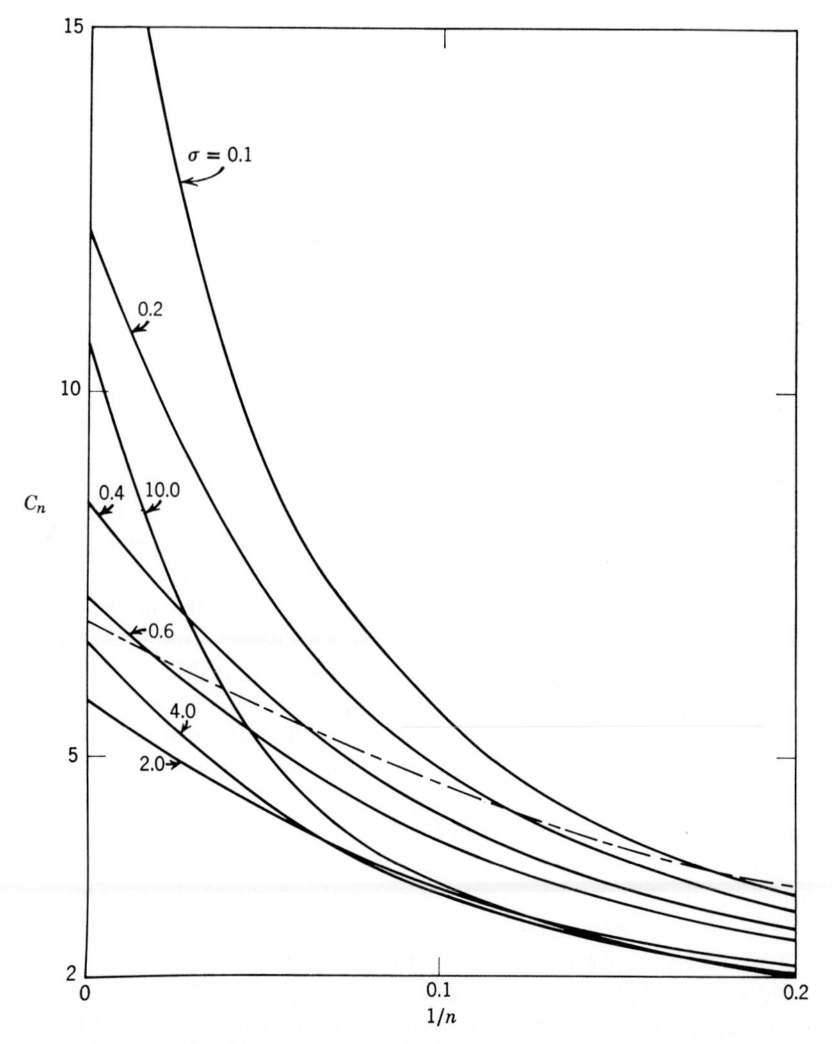

Fig. 11. The characteristic ratio calculated for several hypothetical, simple chains with interdependent neighbor rotations, plotted against the reciprocal chain length. The statistical weight matrix given by Eq. V-9, with $\omega = 0$, has been used for all calculations; $\theta = 68°$ and $|\phi_g| = 120°$. The various solid curves were calculated for the values of σ indicated. The dot-dashed curve, representing the Porod-Kratky "worm-like" chain, corresponds to curves (*4*) in Figs. 7 and 8.

interactions may affect the form of this relationship drastically. An extreme case is that of the alternating racemic polymer of DL-alanine for which C_x was calculated to *decrease* asymptotically with x (see pp. 285–286).

The calculated curves in Fig. 11, representing C_n as functions of $1/n$, are indicative of the variety to be anticipated in configurational characteristics of real chains, as manifested in the dependence of C_n on n [or of $\langle r_{hk}^2 \rangle_0/(k-h)l^2$ on $k-h$ for sequences h, k (see p. 147 and Fig. V-9)]. These curves have been calculated according to Eqs. IV-24 and IV-28, with Eq. V-9 representing the matrix of statistical weights for simple chains devoid of asymmetry. The values of σ are indicated with each curve, ω being equated to zero throughout. A bond angle supplement θ of 68° was used for all calculations and $|\phi_{g\pm}|$ was taken to be 120°. The curve for $\sigma = 0.6$ approximates polymethylene; polyoxymethylene* (POM) is simulated by $\sigma = 10$ in the approximation $\omega = 0$ (see Eqs. V-19). A considerable variation in the form of the functional dependence of C_n on n is illustrated by these curves. Linearity of C_n with $1/n$, sometimes assumed[3,24] as a basis for treating real chains and suggested by Eq. I-20 as a good approximation for freely rotating chains except at the smallest values of n, is obviously inaccurate for real chains. It holds only in the range of n for which C_n is near its limit, C_∞.

The dot-dashed line in Fig. 11 represents the curve for the Porod-Kratky chain in Fig. 7 replotted against $1/n$. Each of the solid curves in Fig. 11 could be approximated by the Porod-Kratky model through arbitrary adjustment of *both* a/L and C_∞ in Eq. 72, Eq. 73 being disregarded. Improvement gained in this manner within the range of Fig. 11 would be at the sacrifice of fit for small values of n. The Porod-Kratky function is not a generally satisfactory approximation for the second moment in relation to n.[22]

The family of curves for real chains ranges from those which prefer an extended form (e.g., the planar form of PM), with frequent departures therefrom, to those exhibiting strong preference for a relatively compact helicoidal conformation (e.g., POM), with comparatively infrequent (though drastic) departures from this conformation. It is not surprising that the dependence of C_n on n for these extreme types differs considerably. The DL-alternating copolymer of alanine, cited above, poses an even more striking contrast. Obviously, no simple empirical function can be found which will reproduce the dependence of C_n on n for all chains. To adopt a particular function or model, e.g., the Porod-Kratky model, for the purpose of representing polymeric chains in general is to ignore the structural and conformational characteristics peculiar to individual chains.

* Polyoxymethylene is not a simple chain but since $\theta_1 \simeq \theta_2$ (see Chap. V), it may be regarded as one in the present approximation that $\omega = 0$.

6. CHAIN VECTOR DISTRIBUTIONS FOR REAL CHAINS

Rigorous proof that $W(\mathbf{r})$ is Gaussian in the limit $n \to \infty$ for any real chain (of finite flexibility) has been given in Section 2. Reduction of $W(\mathbf{r})$ for the freely jointed chain to the Gaussian distribution in the same limit has been demonstrated in Section 3. Compelling, though intuitive, arguments were advanced earlier by Kuhn[25] in support of the assertion that the chain vector distribution must invariably converge to the Gaussian form in the limit $n \to \infty$. Thus, if the correlation between bonds vanishes with their distance apart, the gross properties of the real chain will be reproduced by its equivalent freely jointed analog, defined in Chapter I (p. 12), provided that n is sufficiently large. Since $W(\mathbf{r})$ for the freely jointed chain is Gaussian in the limit $n \to \infty$ (see Sect. 3), the same must hold for the real chain.

As noted above, the chain vector distribution function for finite real chains eludes exact mathematical representation. Departures of $W(\mathbf{r})$ from the Gaussian function may be estimated from Nagai's[3] series expression, Eq. 27, if a sufficient number of the higher moments $\langle r^{2p} \rangle_0$ are known. As was pointed out in Section 2, rapid convergence of this series is not guaranteed; inspection of individual cases is required. We first examine the convergence of the several series in Eq. 27 as applied to the freely jointed chain.

Table 1

Parameters for the Series Expansion of $W(\mathbf{r})$ and Its Fourier Transform $G(\mathbf{q})$ for the Freely Jointed Chain[22]

n	$g_4 \times 10^2$	$g_6 \times 10^3$	$g_8 \times 10^4$	$g_{10} \times 10^5$	$g_{12} \times 10^6$
2	−2.500	−2.38	0.443	2.70	2.81
5	−1.000	−0.382	0.325	0.291	0.0256
10	−0.500	−0.0952	0.1036	0.0427	−0.0066
25	−0.200	−0.0154	0.0175	−0.0006	−0.0072
70	−0.072	−0.0036	−0.0022	−0.0088	−0.0122

The coefficients g_4, g_6, etc. in Eq. 18 for the Fourier transform $G(\mathbf{q})$ of $W(\mathbf{r})$ are tabulated for the freely jointed chain for several values of n in Table 1. These coefficients have been calculated[22] according to Eqs. 20–22 from the moments given by Eq. 32 for the freely jointed chain. Distribution functions expressed in terms of $\rho = (3/2\langle r^2 \rangle_0)^{1/2} r$ according to Eq. 27 for $n = 5$ and 25 are as follows:

$$\mathscr{W}_5(\boldsymbol{\rho}) = \pi^{-3/2} \exp(-\rho^2)[(1 - 0.150 - 0.040 + 0.031 + 0.030 + 0.004 - \cdots) + (20.0 + 8.0 - 8.2 - 10.1 - \cdots)10^{-2}\rho^2 - (4.0 + 3.2 - 4.9 - \cdots)10^{-2}\rho^4 + (0.31 - 0.94 - \cdots)10^{-2}\rho^6 - \cdots] \quad (74)$$

$$\mathscr{W}_{25}(\boldsymbol{\rho}) = \pi^{-3/2} \exp(-\rho^2)[(1 - 0.0300 - 0.0016 + 0.0017 - 0.0000_6 - 0.0010 - \cdots) + (4.00 + 0.32 - 0.44 + 0.02 + \cdots)10^{-2}\rho^2 - (0.80 + 0.13 - 0.26 + \cdots)10^{-2}\rho^4 + (0.012 - 0.050 + \cdots)10^{-2}\rho^6 - \cdots] \quad (75)$$

Ultimate convergence of each of the series in parentheses is delayed by the occurrence of terms of both positive and negative signs, which do not alternate regularly. However, the terms in g_4 (see Eq. 27) appear to dominate higher terms in the coefficients of ρ^0, ρ^2, and ρ^4; contributions to $\mathscr{W}_{25}(\boldsymbol{\rho})$ from higher powers of ρ are small for $\rho < 2$, i.e., for r less than twice its value at the maximum in the radial distribution $4\pi r^2 W(\mathbf{r})$. The departure from the Gaussian distribution may therefore be reckoned in crude approximation by neglecting contributions from g_6, g_8, etc. in Eq. 27. On this basis,

$$\mathscr{W}(\boldsymbol{\rho}) \cong \pi^{-3/2} \exp(-\rho^2)[(1 + 15g_4) - 20g_4\rho^2 + 4g_4\rho^4] \quad (76)$$

This approximate equation is unacceptable for $\rho > 2$. Even for $\rho < 2$ its accuracy is difficult to ascertain. It may serve however to indicate the range of conditions under which neglect of terms beyond the first, unity, in this series is permissible.

Refinement through the introduction of additional terms in g_6, g_8, etc., which depend respectively on $\langle r^4\rangle_0$ and $\langle r^6\rangle_0$, on $\langle r^4\rangle_0$, $\langle r^6\rangle_0$, and $\langle r^8\rangle_0$, etc. (see Eqs. 20 and 21), is even more problematical. Terms in these parameters are of comparable magnitude, although smaller than terms in g_4. Owing to the peculiar pattern of their signs, significant refinement would require that account be taken of many more terms involving g_6, g_8 g_{10}, etc. The introduction of terms in g_6 only, for example, would not necessarily improve the result. In practice, therefore, Nagai's series is of limited value in refining the Gaussian approximation to the distribution $W(\mathbf{r})$. It may however afford an indication of the magnitude of the departure from the Gaussian distribution over the preponderance of the range of r. To this end, Eq. 76 may be useful. Improvement on this expression by the inclusion of terms depending on higher moments of $\mathbf{r}$ (i.e., on $\langle r^6\rangle_0$, etc.) is impracticable, even for the freely jointed chain.

The quantities Δ_6 and Δ_4 defined by Eq. 22 are plotted, one against the other, in Fig. 12 for those chains for which computation of $\langle r^6 \rangle_0$ is

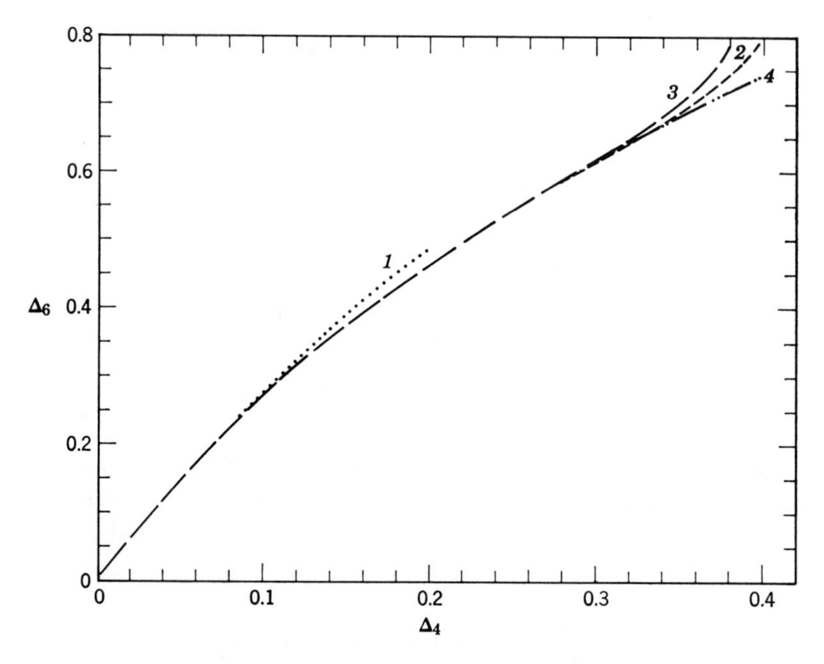

Fig. 12. Parameter Δ_6 (see Eq. 22) plotted against Δ_4 for the freely jointed chain (*1*), for the freely rotating chain (*2*), for the chain with independent hindrance potentials (*3*), and for the Porod-Kratky chain (*4*).[22] Curves (*2*) and (*3*) connect points of integral length $n \geq 2$; curve (*1*) similarly terminates at $n' = 2$. Curve (*4*) continues to $n = 1$, where all models coincide with $\Delta_4 = 2/5$ and $\Delta_6 = 26/35$.

feasible. The curves for the several model chains converge rapidly with increase of chain length (i.e., as they proceed toward the origin in Fig. 12). Throughout the significant range they are indistinguishable. Analogous calculations of the eighth moment, which would permit comparisons of the relationship of Δ_8 to Δ_4 for the various model chains, unfortunately are not available. The Nagai series in Eq. 27 depends on these quantities through Eqs. 20 and 21. If it is justified to infer that similarities like those revealed in Fig. 12 also hold for the higher quantities Δ_8, Δ_{10}, etc., as well, then it follows that the departure of $\mathscr{W}(\boldsymbol{\rho})$, and $W(\mathbf{r})$ from the Gaussian functions are related, approximately, to Δ_4 alone for these chains. This manner of comparing parameters governing the non-Gaussian character of $\mathscr{W}(\boldsymbol{\rho})$ eliminates n as an explicit variable, which obviously is essential since bonds for different models are not equivalent. Relationships like those shown in Fig. 12 avoid the inequity of comparing functions of n for different kinds of chains.*

The higher quantities Δ_6, Δ_8, etc. are not available for real chains. However, the relationships of $\langle r^2\rangle_0/nl^2$ to n (Figs. 7 and 8) and of $\langle r^4\rangle_0/\langle r^2\rangle_0^2$ to n (Fig. 9), are similar in *form* to the corresponding relationships for the model chains (*2*) and (*3*), apart from adjustment factors in the scale of n. This observation strongly suggests that the similarities observed in Fig. 12 hold also for real chains. On this basis, Δ_4, which is susceptible to calculation for real chains, emerges as a useful index of the departure of $\mathscr{W}(\boldsymbol{\rho})$, and likewise of $W(\mathbf{r})$, from the Gaussian form. It is suggested therefore that the bracketed expression in Eq. 76 should serve generally as an approximate indication of the severity of departures from the Gaussian distribution, high extensions excepted of course.

7. MOMENTS OF THE RADIUS OF GYRATION AND ITS STATISTICAL DISTRIBUTION

The radius of gyration s, defined by Eq. I-6, is uniquely specified for a given configuration of the chain. It differs in general for different configurations. The variance of s over the array of configurations of the chain will be much smaller than the variance of r, inasmuch as s is determined by the distances between all pairs of skeletal atoms (see Eq. I-6) and not merely by the distance between the terminal pair.

By methods applicable in the limit $n \to \infty$, Fixman[26] has evaluated even moments of the radius of gyration as follows:

$$\begin{aligned}
\langle s^2\rangle_0 &= \chi_1 \\
\langle s^4\rangle_0 &= \chi_1^2 + \chi_2 \\
\langle s^6\rangle_0 &= \chi_1^3 + 3\chi_1\chi_2 + \chi_3 \\
\langle s^8\rangle_0 &= \chi_1^4 + 6\chi_1^2\chi_2 + 4\chi_1\chi_3 + 3\chi_2^2 + \chi_4
\end{aligned} \tag{77}$$

where

$$\chi_p = \langle r^2\rangle_0^p B_p (2/3)^{p-1} 2^{2p-1}(p-1)!/(2p)! \tag{78}$$

and the B_p are the Bernoulli numbers, i.e., $B_1 = \frac{1}{6}$, $B_2 = \frac{1}{30}$, $B_3 = \frac{1}{42}$, and $B_4 = \frac{1}{30}$. The first of the Eqs. 77 corresponds to Eq. I-17. Ratios of these moments according to Eqs. 77 and 78 are given in the second column of Table 2. They are well reproduced by the following empirical distribution function[27]:

* One bond of the freely jointed chain is equivalent to about 15 to 20 bonds of the PM real chain when the two are compared at equal values of Δ_4. The correspondence is approximate and fails for very short chains.

$$w(s) = \text{const } s^6 \exp[-(7/2)\langle s^2\rangle_0^{-1}s^2] \tag{79}$$

as is shown by the entries in the last column of Table 2.

Table 2

Ratios of Moments of the Radius of Gyration[27]

Ratio	Exact value, Eq. 77	Calcd from Eq. 79
$\langle s^4\rangle_0/\langle s^2\rangle_0^2$	$\frac{19}{15} = 1.2667$	$\frac{9}{7} = 1.2857$
$\langle s^6\rangle_0/\langle s^2\rangle_0^3$	$\frac{631}{315} = 2.0032$	$\frac{99}{49} = 2.0204$
$\langle s^8\rangle_0/\langle s^2\rangle_0^4$	$\frac{1219}{315} = 3.8698$	$\frac{1287}{343} = 3.7522$

Equation 79 may be presumed therefore to approximate the actual distribution satisfactorily over most of its range. The factor s^6 has the effect of narrowing the distribution markedly in comparison with the Gaussian function for $W(\mathbf{r})$ in the same limit, $n = \infty$. The narrower range of the distribution of s is manifest also in the ratios of moments, e.g., $\langle s^4\rangle_0/\langle s^2\rangle_0^2$ is $\frac{19}{15}$ as compared with $\frac{5}{3}$ for the corresponding ratio for the Gaussian distribution.

REFERENCES

1. M. V. Volkenstein and O. B. Ptitsyn, *Zh. Tekhn. Fiz.*, **25**, 662 (1955).
2. M. V. Volkenstein, *Configurational Statistics of Polymeric Chains* (translated from the Russian ed., S. N. Timasheff and M. J. Timasheff), Interscience, New York, 1963, p. 454 *et seq.*
3. K. Nagai, *J. Chem. Phys.*, **38**, 924 (1963).
4. K. Nagai, *J. Chem. Phys.*, **44**, 423 (1966).
5. R. L. Jernigan, Ph.D. thesis, Stanford University, 1967.
6. H. Grad, *Commun. Pure Appl. Math.*, **2**, 331 (1949).
7. K. Nagai, *J. Chem. Phys.*, **40**, 2818 (1964).
8. P. A. P. Moran, *Proc. Cambridge Phil. Soc.*, **44**, 342 (1948).
9. C. M. Tchen, *J. Chem. Phys.*, **20**, 214 (1952).
10. Lord Rayleigh, *Phil. Mag.*, **37**, [6], 321 (1919).
11. S. Chandrasekhar, *Rev. Mod. Phys.*, **15**, 1 (1943).
12. M. V. Volkenstein, *Configurational Statistics of Polymeric Chains* (translated from the Russian ed., S. N. Timasheff and M. J. Timasheff), p. 165 *et seq.*
13. L. R. G. Treloar, *Trans. Faraday Soc.*, **42**, 77 (1946).
14. L. R. G. Treloar, *The Physics of Rubber Elasticity*, 2nd ed., Oxford Univ. Press, 1958, pp. 100–113.
15. P. Hall, *Biometrika*, **19**, 240 (1927).
16. J. Irwin, *Biometrika*, **19**, 225 (1927).
17. W. Kuhn and F. Grün, *Kolloid-Z.*, **101**, 248 (1942).

18. W. Kuhn and H. Kuhn, *Helv. Chim. Acta*, **26**, 1394 (1943); W. Kuhn, *ibid.*, **29**, 1095 (1946).
19. H. M. James and E. Guth, *J. Chem. Phys.*, **11**, 470 (1943).
20. P. J. Flory, C. A. J. Hoeve, and A. Ciferri, *J. Polymer Sci.*, **34**, 337 (1959).
21. R. L. Jernigan and P. J. Flory, to be published.
22. R. L. Jernigan and P. J. Flory, unpublished.
23. G. Porod, *Monatsh. Chem.*, **80**, 251 (1949); O. Kratky and G. Porod, *Rec. Trav. Chim.*, **68**, 1106 (1949).
24. K. Nagai, *J. Chem. Phys.*, **45**, 838 (1966).
25. W. Kuhn, *Kolloid-Z.*, **68**, 2 (1936); **87**, 3 (1939).
26. M. Fixman, *J. Chem. Phys.*, **36**, 306 (1962).
27. P. J. Flory and S. Fisk, *J. Chem. Phys.*, **44**, 2243 (1966).

CHAPTER IX

Optical Properties and Radiation Scattering

This final chapter is devoted to applications of methods developed earlier to the treatment of a selected set of properties of polymeric chains. Properties included are: (*1*) the angular dependence of radiation scattered by chain molecules dispersed in a dilute solution, for the full range of wavelength from x-rays to visible light; (*2*) the depolarization of light scattered at 90° from the incident beam; (*3*) the birefringence resulting from elastic deformation of a network of polymeric chains; and (*4*) the birefringence induced by an electric field, i.e., the Kerr effect. The treatment of strain birefringence is applicable also to dichroism in elastically deformed networks, with only nominal adaptation. Properties (*2*), (*3*), and (*4*) depend on a tensor quantity, the optical polarizability, and each is a manifestation of the anisotropy of that tensor. It will be necessary to sum tensor contributions of individual bonds or groups of the molecule, and to extract the configurational average of the tensor invariant relevant to the particular property.

The feasibility of relating the properties cited above, rigorously and unambiguously, to the structure of the chain molecule as demonstrated in this chapter presents the prospect of new methods of investigation and deeper insights into the properties of macromolecules. It is premature for adequately illustrative applications. These can confidently be anticipated for the near future, however. The procedures developed in full in this chapter may be suggestive of further developments which will enlarge the scope of the properties of polymeric chains susceptible to rigorous treatment.

An optical property not included below is the rotation of polarized light by macromolecules containing genuine asymmetric centers, as in vinyl polymers possessing side chains bearing asymmetric atoms. The steric constraints in vinyl polymer chains can be expected to restrict the conformation about asymmetric centers proximate to the chain skeleton, and thus to enhance the optical rotation. An elegant theory propounded by A. Abe has been remarkably successful in treating such effects.*

* See A. Abe, *J. Am. Chem. Soc.*, **90**, 2205 (1968).

1. RADIATION SCATTERING: DEPENDENCE ON ANGLE AND WAVELENGTH

Primary Relationships

Scattering of electromagnetic radiation by two elements, i and j, of a particle or a molecule is illustrated in the familiar way in Fig. 1. The

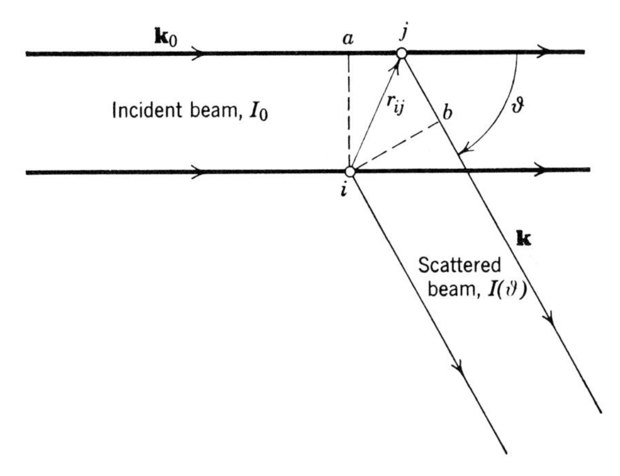

Fig. 1. Diagram depicting scattering of radiation at an angle ϑ from the incident beam.

intensities of the incident beam and of the radiation which is scattered at an angle ϑ from the direction of the incident beam are I_0 and $I(\vartheta)$, respectively, with $I(\vartheta) \ll I_0$, of course. Wave propagation vectors for the incident and scattered beams are $\mathbf{k}_0$ and $\mathbf{k}$; the angle between their directions is ϑ, and $|\mathbf{k}_0| = |\mathbf{k}| = 2\pi/\lambda$, where λ is the wavelength of the radiation in the scattering medium. The scattering elements i and j are assumed to be very small compared to the magnitude of the difference between the wave vectors, i.e., compared to the quantity μ defined by

$$\mu = |\mathbf{k}_0 - \mathbf{k}| = (4\pi/\lambda) \sin(\vartheta/2) \tag{1}$$

Fulfillment of this condition justifies treatment of the elements as point scatterers. They are separated by the vector $\mathbf{r}_{ij}$. The dashed lines ia and ib in Fig. 1 are normals to the incident and scattered rays for element j. We shall assume throughout that experimental conditions are so designed as to render secondary scattering negligible.

The difference between the path length for radiation falling on element j and scattered by it compared to that for radiation scattered by element i is given by the sum of the segments aj and jb. These segments are the projections of vector $\mathbf{r}_{ij}$ on the direction of the incident beam and on the reversed direction of the scattered beam, respectively. It follows that the difference in phase between the two scattered rays shown in Fig. 1 is given by

$$(\mathbf{k}_0 - \mathbf{k}) \cdot \mathbf{r}_{ij} = \boldsymbol{\mu} \cdot \mathbf{r}_{ij} \tag{2}$$

where $\boldsymbol{\mu} = \mathbf{k}_0 - \mathbf{k}$. The scalar magnitude of $\boldsymbol{\mu}$ has been defined in Eq. 1.

At the point of observation of the scattered radiation, the sum of the amplitudes of the rays scattered by elements i and j is expressed by

$$A_0 f_i \exp(\tilde{\imath}\omega t) + A_0 f_j \exp(\tilde{\imath}\boldsymbol{\mu} \cdot \mathbf{r}_{ij} + \tilde{\imath}\omega t)$$

apart from an arbitrary phase factor. Here A_0 is the amplitude of the incident beam of intensity $I_0 = A_0{}^2$, ω is the angular frequency of the radiation, t is the time, and $\tilde{\imath} = \sqrt{-1}$. The f's are scattering factors for the respective particles. The total amplitude of radiation scattered at angle ϑ by a molecule comprising $n + 1$ chain atoms, or groups, indexed 0 to n can be written

$$A(\vartheta) = A_0 \sum_{j=0}^{n} f_j \exp(\tilde{\imath}\boldsymbol{\mu} \cdot \mathbf{r}_{0j} + \tilde{\imath}\omega t) \tag{3}$$

where the zeroth element has been taken as origin, a role corresponding to that of element i in Fig. 1. The intensity of the radiation scattered by the molecule, being the product of the amplitude and its complex conjugate, is

$$I(\vartheta) = \text{const}\, I_0 \sum_{i,j} f_i f_j \exp(\tilde{\imath}\boldsymbol{\mu} \cdot \mathbf{r}_{ij}) \tag{4}$$

where the sum includes terms for every value of i and of j in the range $0 \le i, j \le n$. The chain molecules of main interest consist of sequences of identical elements, or they can usually be so represented in satisfactory approximation. The scattering factors* f therefore may be absorbed into the constant in Eq. 4 to give

$$I(\vartheta)/I_0 = \text{const} \sum_{i,j} \exp(\tilde{\imath}\boldsymbol{\mu} \cdot \mathbf{r}_{ij}) \tag{5}$$

The intensity ratio extrapolated to a scattering angle ϑ of zero is

$$I(0)/I_0 = \text{const}\,(n + 1)^2 \tag{6}$$

as follows from Eq. 5 in the limit $\boldsymbol{\mu} = \mathbf{0}$, where all terms of the sum are equal to unity. From a physical point of view, this result is a consequence of the vanishing of differences in phase for rays scattered by each of the

* The scattering factors for x-radiation may depend on the angle ϑ. At the small values of μ that will be of interest, however, these factors may be treated as constants with negligible error.

$n + 1$ elements as the scattering angle ϑ approaches zero. The dependence of the scattered intensity on ϑ and λ (through μ) for a fixed spatial configuration of the scattering elements comprising the chain molecule is suitably expressed by the ratio of Eq. 5 to Eq. 6, that is, by

$$I(\vartheta)/I(0) = (n+1)^{-2} \sum_{i,j} \exp(\tilde{\imath}\boldsymbol{\mu} \cdot \mathbf{r}_{ij}) \tag{7}$$

It will be observed that the vectors $\mathbf{r}_{ij}$ specify not only the internal configuration of the chain but its orientation in space as well.

The intensity ratio $I(\vartheta)/I(0)$ averaged without bias over all orientations of the molecule, treated for the present as a rigid body, is the integral of Eq. 7 over Eulerian angles χ, ψ, and ω. Denoting the resulting function by $P(\mu)$, we have

$$\begin{aligned} P(\mu) &\equiv [I(\vartheta)/I(0)]_{\text{av}} \\ &= (n+1)^{-2} \sum_{i,j} \int_0^{\pi} \int_0^{2\pi} \int_0^{2\pi} \exp(\tilde{\imath}\boldsymbol{\mu} \cdot \mathbf{r}_{ij}) \sin\chi \, d\chi \, d\psi \, d\omega / 8\pi^2 \end{aligned}$$

which upon integration yields the celebrated Debye scattering relation[1]

$$P(\mu) = (n+1)^{-2} \sum_{i,j} \sin(\mu r_{ij})/\mu r_{ij} \tag{8}$$

Or, by series expansion of $\sin(\mu r_{ij})$

$$P(\mu) = (n+1)^{-2} \sum_{i,j} (1 - \mu^2 r_{ij}^2/3! + \mu^4 r_{ij}^4/5! - \cdots) \tag{9}$$

The scattering function $P(\mu)$ depends, through μ, on the external variables: scattering angle ϑ and wavelength λ. It is a function also of the minimum set of *scalar* distances r_{ij} required to specify the internal configuration of the molecule. The function $P(\mu)$ is often designated as $P(\vartheta)$. The values of the functions are equal; they differ *mathematically* but are connected by the relation of μ to ϑ, as expressed by Eq. 1.

Equations 8 and 9 have been derived for a single particle or molecule. In a system of many particles, N in number, which we may assume to be identical replicas of the one considered above, the total scattered intensity will not, in general, be the sum of the intensities for the individual particles. Departure from additivity is a consequence of the correlation of phases of the radiation scattered by different particles. These correlations are manifested in interference or reinforcement, depending on the phase difference between the contributions from the N individual molecules to the combined scattered amplitude. If, however, the spatial distribution and orientation of the particles is random (as expressed, for example, in constancy of the pair density distribution function for all distances), then the phases are random and the intensities are additive. For a solution which is sufficiently dilute, this condition will be fulfilled. Then $P(\mu)$ of Eq. 8, or Eq. 9, becomes the appropriate expression for the angular distribution of radiation scattered by the macroscopic system. We shall assume throughout the following discussion that relevant experimental results refer to the condition of infinite dilution, realized for example by empirical extrapolation of measurements at finite concentrations.

Flexible Chains

For molecules which are nonrigid, and this, of course, is the case of interest here, it is necessary to average the terms of Eq. 8 or those of Eq. 9 over all configurations. We first introduce the required averages into Eq. 8, leaving Eq. 9 for later consideration. Equation 8 for the scattering function may be amended as follows:

$$P(\mu) = (n + 1)^{-2} \sum_{i,j} G_{ij}(\boldsymbol{\mu}) \tag{10}$$

where

$$G_{ij}(\boldsymbol{\mu}) = \langle(\mu r_{ij})^{-1} \sin(\mu r_{ij})\rangle \tag{11}$$

$$= \int_0^\infty (\mu r_{ij})^{-1} \sin(\mu r_{ij}) 4\pi r_{ij}^2 W(\mathbf{r}_{ij})\, dr_{ij} \tag{12}$$

$W(\mathbf{r}_{ij})$ being the distribution function for vector $\mathbf{r}_{ij}$ discussed in Chapter VIII.

Alternatively, we have directly from Eq. 7 that

$$P(\mu) = (n + 1)^{-2} \sum_{i,j} \int \exp(\tilde{\imath}\boldsymbol{\mu} \cdot \mathbf{r}_{ij}) W(\mathbf{r}_{ij})\, d\mathbf{r}_{ij} \tag{13}$$

The integrals in Eqs. 12 and 13 are equivalent expressions for the Fourier transform of $W(\mathbf{r}_{ij})$. Hence, drawing upon the results of the preceding chapter, we may identify $G_{ij}(\boldsymbol{\mu})$ with $G(\mathbf{q})$ defined by Eq. VIII-13 or by Eq. VIII-16, with $\mathbf{r}_{ij}$ and $W(\mathbf{r}_{ij})$ replacing $\mathbf{r}$ and $W(\mathbf{r})$. According to Eqs. VIII-18 and VIII-20,

$$G_{ij}(\boldsymbol{\mu}) = \exp(-\mu^2\langle r_{ij}^2\rangle/6)[1 - (1/8)(1 - 3\langle r_{ij}^4\rangle/5\langle r_{ij}^2\rangle^2)(\mu^2\langle r_{ij}^2\rangle/3)^2 + \cdots] \tag{14}$$

Coefficients of higher terms in $\mu^2\langle r_{ij}^2\rangle/3$ may be formulated from Eqs. VIII-18, VIII-20, VIII-21, and VIII-22.*

* Subscript zeros denoting the unperturbed state have been omitted from $\langle r_{ij}^2\rangle$ and $\langle r_{ij}^4\rangle$ in Eqs. 14–16. These equations are applicable to perturbed chains provided only that the distribution remains spherically symmetric. In later expressions, where $\langle r_{ij}^2\rangle$ is taken to be independent of n and of the location of the sequence i, j in the chain, strict adherence to rigor would dictate designation of the unperturbed state. Effects of excluded volume usually may be ignored in practice if actual moments $\langle r_{ij}^2\rangle$ are used instead of their unperturbed values (*cf. seq.*).

For values of $\mu^2\langle r_{ij}^2\rangle/3$ much in excess of 2, the magnitude of $G_{ij}(\boldsymbol{\mu})$ is rendered small by the exponential factor in Eq. 14. At this point the second term of the series in brackets in Eq. 14 is

$$-(\tfrac{1}{2})(1 - 3\langle r_{ij}^4\rangle/5\langle r_{ij}^2\rangle^2)$$

Its contribution is significant compared to unity only if the quantity in parenthesis is of the order of 0.1 or greater. For polymethylene (see Fig. 11 of Chap. V) $|i-j|$ must be less than about 25 bonds in order for $3\langle r^4\rangle/5\langle r^2\rangle^2$ to depart from unity by such an amount. For chain sequences of this length $\langle r_{ij}^2\rangle^{1/2}$ is only ca. 20 Å or less. Scattering experiments affording information on distances in this range must of course be carried out with x-rays, i.e., the required values of μ can be covered in practice only through the use of radiation in the x-ray range. The foregoing root-mean-square interunit distance of 20 Å is to be compared with a mean diameter of the polymethylene chain of about 5 Å. Obviously, the approximation of the scattering elements (CH_2, for example) as point centers when r_{ij} is as small as 20 Å is no longer valid. We conclude that for any combination of parameters validating the approximation of the scattering groups by point scatterers, the omission of higher terms of the series in Eq. 14 will certainly be justified for a random-coil chain molecule having a tortuosity comparable with, or not much less than, that of polymethylene. Only for very stiff chains, e.g., for poly-L-proline (see p. 281), may circumstances arise where terms higher than those specifically included in Eq. 14 will be needed.

In fact, for most flexible chains it will be justified, within the limitations of the point scattering approximation, to replace the series in brackets in Eq. 14 by unity under all conditions (i.e., for any feasible μ), so that

$$G_{ij}(\boldsymbol{\mu}) = \exp(-\mu^2\langle r_{ij}^2\rangle/6) \tag{15}$$

and

$$P(\mu) = (n+1)^{-2}\sum_{i,j}\exp(-\mu^2\langle r_{ij}^2\rangle/6) \tag{16}$$

This result was obtained by Debye[2] and it has been used by Kratky and Porod[3] in their treatment of x-ray scattering of chain molecules. These expressions are tantamount to representation of $W(\mathbf{r}_{ij})$ in the Gaussian approximation. It will be apparent from the foregoing analysis, and also from Chapter VIII, that conditions for compliance with Eqs. 15 and 16, i.e., for the Fourier transform of $W(\mathbf{r})$ to conform to the Gaussian function, are less stringent than for $W(\mathbf{r})$ itself to be Gaussian. We note also that truncation of the series in Eq. 14 for $G_{ij}(\boldsymbol{\mu})$ at its second term corresponds to expression of $W(\mathbf{r}_{ij})$ according to Eq. VIII-76, wherein coefficients beyond g_4 are omitted.

For sufficiently large values of $|i-j|$, it is legitimate to take $\langle r_{ij}^2 \rangle$ to be proportional to $|i-j|$, provided, of course, that the chain is not perturbed by volume exclusion and related interactions of long range. That is, under these conditions we may let

$$\langle r_{ij}^2 \rangle = \langle r_{ij}^2 \rangle_0 = C_\infty m l^2$$

where $m = |i-j|$. Substitution into Eq. 16 gives

$$P(\mu) = (n+1)^{-2} \sum_{i,j} \exp(-\mu^2 C_\infty m l^2/6) \tag{17}$$

In the range of the small values of μ accessible through light scattering experiments, the exponents of the terms in Eq. 16 are appreciable only for large values of $\langle r_{ij}^2 \rangle$, i.e., for large m. Hence, the segment pairs ij which contribute appreciably to the dependence of $P(\mu)$ on the scattering angle ϑ are those for which $m = |i-j|$ is large. For a sufficiently long chain molecule investigated over the range of small μ, adoption of Eq. 17, in which the foregoing proportionality of $\langle r_{ij}^2 \rangle$ to m is assumed, will therefore be legitimate. Replacement of the summation in Eq. 17 by an integral and adoption of other approximations valid for very large n lead at once to*

$$\begin{aligned} P(\mu) &= 2n^{-2} \int_0^n (n-m) \exp(-\mu^2 C_\infty m l^2/6)\, dm \\ &= (2/v^2)(v - 1 + e^{-v}) \end{aligned} \tag{18}$$

where

$$v = \mu^2 C_\infty n l^2/6 \tag{19}$$

For $v < \sim 3$,

$$P(\mu) = 1 - (v/3) + (3/4)(v/3)^2 - (9/20)(v/3)^3 + \cdots \tag{18'}$$

Equation 18 is due to Debye.[2] According to its derivation, it is conditional on (*1*) the suppression of perturbations by long-range interactions associated with the excluded volume effect, a condition fulfilled at the Θ

* Equation 17 can be written

$$P(\mu) = (n+1)^{-1} + 2(n+1)^{-2} \sum_{m=1}^{n} (n-m+1) \exp(-\mu^2 C_\infty m l^2/6)$$

Replacement of the summation by integration yields[4]

$$P(\mu) = (n+1)^{-1} + [2n/v^2(n+1)^2][(n-v)\exp(-v) + n(v-1)\exp(-v/n)]$$

which reduces to Eq. 18 for $n \gg 1$ and $v/n \ll 1$. Critical examination does not recommend this equation over Eq. 18; the error introduced in replacing summation by integration is partially compensated[4] by omission of terms of order $1/n$.

point where $\langle r^2 \rangle = \langle r^2 \rangle_0$, etc., and (*2*) the validity of the assumption of proportionality between $\langle r_{ij}^2 \rangle$ and $m = |i-j|$. The second condition requires not only that the chains be of great length, but also that μ be sufficiently small (ϑ small and/or λ large) so that $|i-j|$ for the significant terms in Eq. 16 is very large. It is important in this connection to observe that the approximation associated with condition (*2*) is much more drastic than, and unrelated to, the approximation of taking the Fourier transform of $W(\mathbf{r}_{ij})$ to be Gaussian. Thus, Eqs. 15 and 16 should be applicable over a much broader range of conditions than Eq. 17. Equation 18, *with* v *defined by Eq.* 19 (*cf. seq.*), is subject to the same limitations as Eq. 17. These equations are acceptable only for very long chains observed at small values of μ, i.e., in the light scattering range.

Molecular Scattering ($v < 3$)

At this juncture it is advisable to draw a distinction between scattering at small values of v such that all pairs of groups i, j contribute appreciably to the function $P(\mu)$, and scattering at larger v where the more remote pairs of the chain contribute negligibly thereto (see Eq. 16). Although the two regimes are not sharply differentiated, we may conveniently consider the former to hold for $v < \sim 3$ and the latter for $v > \sim 3$. The condition $v < 3$ obtains in light scattering experiments at all angles ϑ, provided that $\langle s^2 \rangle^{1/2}$ is no greater than about 0.15 λ, as follows from Eqs. 1 and 19 (see also Eq. 19″ below). We shall refer to this as the molecular scattering regime since all pairs of the entire molecule contribute appreciably to the scattering function. At much larger values of v, attained either by use of x-rays or through an increase in $\langle r^2 \rangle$ and $\langle s^2 \rangle$, e.g., by increasing n, the scattering is *submolecular* in the sense that the span of the chain over which pairs of groups contribute appreciably to the scattering function is much less than the total chain length n. For the present, our attention is confined to the range $v < 3$, which is inclusive of the usual optical scattering range, but may also include x-ray scattering by short chains or by longer chains at very small angles ϑ.

In the limit $n \to \infty$ the definition of v according to Eq. 19 is equivalent to

$$v = \mu^2 \langle r^2 \rangle_0 / 6 = \mu^2 \langle s^2 \rangle_0$$

Some of the error incurred in the application of Eq. 18 to finite chains subject to long-range perturbations may be alleviated by redefining v according to either

$$v = \mu^2 \langle r^2 \rangle / 6 \qquad (19')$$

or

$$v = \mu^2 \langle s^2 \rangle \qquad (19'')$$

where $\langle r^2 \rangle$ and $\langle s^2 \rangle$ refer specifically to the *finite chain* of length n, and not to $C_\infty nl^2$, as stipulated by Eq. 19 and implied by the formal derivation of Eq. 18. The revision is empirical, but obviously in the direction of diminishing the error arising from the failure of the perturbed, finite chains to comply with conditions (*1*) and (*2*) in the penultimate paragraph above. Equation 19′, incidentally, corresponds to the definition of v introduced in Chapter VIII (see Eq. VIII-19). The alternative definitions, Eqs. 19′ and 19″, are equivalent only for unperturbed chains in the limit $n \to \infty$.

Condition (*1*) above can be reasoned to have been effectively voided through the replacement of $\langle r^2 \rangle_0$ and $\langle s^2 \rangle_0$ by $\langle r^2 \rangle$ and $\langle s^2 \rangle$ in Eqs. 19′ and 19″.* The efficacy of defining v in terms of the value of $\langle r^2 \rangle$, or of $\langle s^2 \rangle$, for the finite chain as a device by which to relax condition (*2*) is less clear. Certainly, the smaller values of these quantities compared with $C_\infty nl^2/6$ will reduce the error incident upon application of Eq. 18 to finite chains, but the extent of the reduction is not immediately apparent.

Numerical calculations[4] for polymethylene chains of 10 and of 263 bonds show remarkably close agreement between $P(\mu)$ according to Eq. 18, with $v = \mu^2 \langle s^2 \rangle_0$ (Eq. 19″), and $P(\mu)$ calculated from Eq. 16 using values of $\langle r_{ij}^2 \rangle_0$ appropriate for a sequence of length $|i-j|$ for each term of the sum.† Definition of v according to Eq. 19′ is less satisfactory, although much to be preferred over Eq. 19. Thus, semiempirical adoption of Eq. 19″ for the definition of v appears to extend the range of Debye's Eq. 18 much beyond the limits set by the premises of its derivation.

A general expansion for $P(\mu)$ appropriate for the range of molecular scattering is indicated in Eq. 9. Adaptation of this equation to flexible chains by averaging the r_{ij}^2 and higher even powers of r_{ij} over all configurations gives

$$P(\mu) = 1 - (2/3!)S_2\mu^2 + (2/5!)S_4\mu^4 - (2/7!)S_6\mu^6 + \cdots \tag{20}$$

where

$$S_{2\kappa} = (n+1)^{-2} \sum_{0 \le i < j \le n} \langle r_{ij}^{2\kappa} \rangle \tag{21}$$

* Current treatments purporting to take account of the effect of the excluded volume perturbation on $P(\mu)$, or $P(\vartheta)$, are open to question as a consequence of the arbitrary manner in which the perturbation usually designated by α is assumed to depend on m. The assumed dependence of α on m^ε, where ε is a fractional quantity, is certainly wrong, and the use thereof entails a greater error than the approximation involved in replacing $\langle r^2 \rangle_0$ by $\langle r^2 \rangle$ in Eqs. 16, 19′, and 19″ in order to adapt them to perturbed chains.

† The distinction between $\langle r_{ij}^2 \rangle_0$ for a sequence within a chain of infinite length and $\langle r_n{}^2 \rangle_0$ for a finite chain of length $n = j - i$ can be ignored since the difference between them is negligible (see p. 148).

Obviously $S_2 = \langle s^2 \rangle$. Equation 20 follows also from Eqs. 10 and 11 by expansion of $(\mu r_{ij})^{-1} \sin(\mu r_{ij})$ in series. An alternative rendition of Eq. 20 for $P(\mu)$ is the following:

$$P(\mu) = 1 - (v/3) + (3S_4/20S_2{}^2)(v/3)^2 - (3S_6/280S_2{}^3)(v/3)^3 + \cdots \qquad (20')$$

where $v = \mu^2 S_2$ in keeping with Eq. 19″. The series in Eqs. 20 and 20′ ordinarily will converge for $v < 3$. In the limit $n \to \infty$, Eq. 20′ must converge to Eq. 18′, provided that long-range perturbations, which would vitiate Eqs. 18 and 18′ (but not Eqs. 20 or 20′), are not operative. Comparing Eqs. 18′ and 20′, we find in this limit

$$\begin{aligned} S_2 &= \langle s^2 \rangle_0 = (1/6)\langle r^2 \rangle_0 \\ S_4 &= 5\langle s^2 \rangle_0{}^2 = (5/36)\langle r^2 \rangle_0{}^2 \\ S_6 &= 42\langle s^2 \rangle_0{}^3 = (7/36)\langle r^2 \rangle_0{}^3 \\ S_{2\kappa} &= [(2\kappa+1)!/(\kappa+2)!]S_2^{\kappa} \end{aligned} \qquad (22)$$

Subscripts 0 have been included in Eqs. 22 as a reminder that these limiting relations are strictly valid only for unperturbed chains.

A particularly advantageous expansion of $P(\mu)$ is obtained by factoring the Debye expression, Eq. 18, from Eq. 20′, with the result[4]

$$\begin{aligned} P(\mu) = \left(\frac{2}{v^2}\right)(v - 1 + e^{-v})\Bigg\{1 - \frac{3}{4}\left[1 - \frac{S_4}{5S_2{}^2}\right]\left(\frac{v}{3}\right)^2 \\ - \frac{3}{4}\left[\left(1 - \frac{S_4}{5S_2{}^2}\right) - \frac{3}{5}\left(1 - \frac{S_6}{42S_2{}^3}\right)\right]\left(\frac{v}{3}\right)^3 - \cdots\Bigg\} \end{aligned} \qquad (23)$$

The departure from Debye's equation, modified by redefinition of v as stipulated above, is expressed by higher terms of the series in braces.

For radiation in the optical range of the spectrum, the reciprocal of the scattered intensity usually is the quantity considered. The relevant function is then the reciprocal of $P(\mu)$, given by

$$\begin{aligned} P^{-1}(\mu) = 1 + (v/3) + (1 - 3S_4/20S_2{}^2)(v/3)^2 \\ + (1 - 3S_4/10S_2{}^2 + 3S_6/280S_2{}^3)(v/3)^3 + \cdots \end{aligned} \qquad (24)$$

Equations 20′, 23, and 24 express the angular dependence of the scattered intensity as power series in $v/3$. The utility of each of these equivalent renditions of $P(\mu)$ under given circumstances depends, of course, on the convergence of the series, which, in turn, is dependent in first approximation on the magnitude of the quantity $v = S_2 \mu^2 = [4\pi \sin(\vartheta/2)]^2 \langle s^2 \rangle / \lambda^2$. Determination of the extent to which the third terms (i.e., the terms which are quadratic in v, and hence in $\sin^2 \vartheta/2$) contribute relative to the second terms of the several series requires evaluation of S_4 in addition to

$S_2 \equiv \langle s^2 \rangle$. The calculation of the latter quantity, $\langle s^2 \rangle_0$, for an unperturbed chain was treated in Chapter IV (see pp. 111–114). The same methods may be adapted to the calculation of S_4 for unperturbed chains, as we now show.

By steps analogous to those involved in arriving at Eq. IV-35 from Eq. IV-25, we have from Eq. IV-96

$$\langle r_{ij}^4 \rangle_0 = 4Z^{-1}\mathbf{J}^*\mathbf{U}_1^{(i)}[\mathbf{E}_\nu \mathbf{0} \cdots \mathbf{0}]\mathscr{K}_{i+1}^{(j-i)} \begin{bmatrix} \mathbf{0} \\ \vdots \\ \mathbf{0} \\ \mathbf{E}_\nu \end{bmatrix} \mathbf{U}_{j+1}^{(n-j)}\mathbf{J} \tag{25}$$

where $\mathscr{K}_i$ is defined by Eq. IV-95. Indices i and j here replace h and k in Eq. IV-35 (see also Fig. IV-1). Evaluation of S_4 defined by Eq. 21, with $\kappa = 2$, may be carried out in the same manner as was used to obtain Eq. IV-54 from Eq. IV-35. The result is

$$S_4 = 4(n+1)^{-2}Z^{-1}\mathscr{J}^* \begin{bmatrix} \mathbf{U} & [\mathbf{E}_\nu \mathbf{0} \cdots \mathbf{0}]\mathscr{K} & (l^4/4)\mathbf{U} \\ \mathbf{0} & \mathscr{K} & \mathscr{K}[\mathbf{0} \cdots \mathbf{0}\mathbf{E}_\nu]^T \\ \mathbf{0} & \mathbf{0} & \mathbf{U} \end{bmatrix}_1^{(n)} \mathscr{J} \tag{26}$$

where the statistical weight matrices $\mathbf{U}_1$ and $\mathbf{U}_n$ for the terminal bonds are to be set equal to the identity $\mathbf{E}_\nu$. The result pertains, of course, to the unperturbed chain.

The order of the square matrix is $27\nu \times 27\nu$. Simplification to the case of a chain in which bond rotations are independent is straightforward, and may be carried out by observing the correspondence of Eqs. IV-97 and IV-98 on the one hand to Eqs. IV-96 and IV-95 on the other.

Shown in Fig. 2 are illustrative calculations of $S_4/5S_2^2$ carried out[4] for polymethylene chains at about 140°C. Values of S_2 are from the calculations presented in Fig. V-10; S_4 was calculated according to Eq. 26 from the same set of parameters (see the legend to Fig. V-9), which have found extensive use for purposes of illustration in Chapters V and VIII. As expected, the approach of $S_4/5S_2^2$ to its asymptote, unity (see Eq. 22), is even more protracted than for $\langle r^4 \rangle_0/\langle r^2 \rangle_0^2$, shown in Fig. V-11. Although convergence with increase in n is slow, $S_4/5S_2^2$ exceeds 0.8 for all values of n beyond the sharp minimum at $n = 2$. Numerical values calculated[4] for ratios of higher sums, i.e., $S_6/42S_2^3$, etc., obtained by computing $S_{2\kappa}$ for every configuration of short PM chains up to $n = 10$, are likewise near their asymptotes (unity), even for values of n in this range.

It follows that the series for $P^{-1}(\mu)$ given by Eq. 24 converges more rapidly than the series for $P(\mu)$ in Eq. 20′. Coefficients in both of these series approach constant values with increase in chain length. The series in

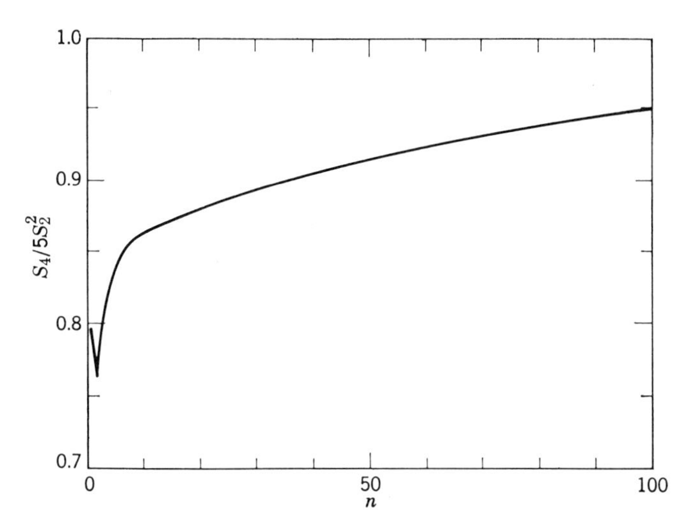

Fig. 2. The ratio $S_4/5S_2{}^2$ of sums $S_{2\kappa}$ defined by Eq. 21 plotted against the chain length for a polymethylene chain at ca. 140°C.[4] The parameters used in the calculations are those given in the legend to Fig. V-9.

Eq. 23 converges more rapidly than either Eq. 20′ or Eq. 24, and its coefficients vanish with increase in n. Terms beyond unity are small throughout the range $0 < v < 3$ for molecular scattering, even for comparatively short chains. Thus, the Debye equation, Eq. 18, with v redefined according to Eq. 19″, gains strong support for application much below the range of n implied to be legitimate by its original derivation and the associated definition of v according to Eq. 19. In the case of polymethylene chains, the definition $v = \mu^2\langle s^2\rangle$ validates the use of Eq. 18 as an excellent approximation, even for $n = 100$ bonds, where x-rays would be required to attain values of v such that $P(\mu)$ departs measurably from unity.

Submolecular Scattering ($v \gg 3$)

For large values of v the intensity of scattering $I(\vartheta)$ by the molecule becomes proportional to its length. Hence, the scattering when $v \gg 3$ is suitably discussed in terms of the intensity per scattering element, a quantity equal to $n^{-1}I(\vartheta)$ for large n and hence proportional to $nP(\mu)$. Under the conditions stated we have, according to Eq. 16,

$$nP(\mu) = n^{-1} \sum_{i,j} \exp\left(-\mu^2\langle r_{ij}^2\rangle/6\right)$$

$$= 2\sum_{m=1}^{\infty} \exp\left(-\mu^2\langle r_m^2\rangle/6\right), \qquad \lim n \to \infty \tag{27}$$

Thus, the scattered intensity per scattering element, or group, is independent of the chain length for very long chains. The physical basis for this result is apparent from the preceding discussion.

If $\langle r_m{}^2 \rangle$ were proportional to m throughout the range wherein $\mu^2 \langle r_m{}^2 \rangle /6$ is appreciably greater than zero, then the exponent in Eq. 27 could be replaced by $m\mu^2(\langle r^2 \rangle /n)_\infty/6 \equiv m\mu^2(\langle s^2 \rangle /n)_\infty$, and evaluation of the sum would yield

$$nP(\mu) = 2/\mu^2(\langle s^2 \rangle /n)_\infty \tag{27'}$$

This relation follows directly from the Debye equation, Eq. 18, in the limit of very large v. In general, however, the assumption of proportionality of $\langle r_m{}^2 \rangle$ to m is untenable under conditions such that $v \gg 3$. It is necessary instead, for evaluation of the sum in Eq. 27, to use values of $\langle r_m{}^2 \rangle$ calculated for each sequence of length m.

The function $n\mu^2P(\mu)$ computed[4] according to Eq. 27 from values of $\langle r_m{}^2 \rangle$ for finite sequences in polymethylene chains is plotted against μ in Fig. 3. The range covered is pertinent to scattering of x-rays. In contrast

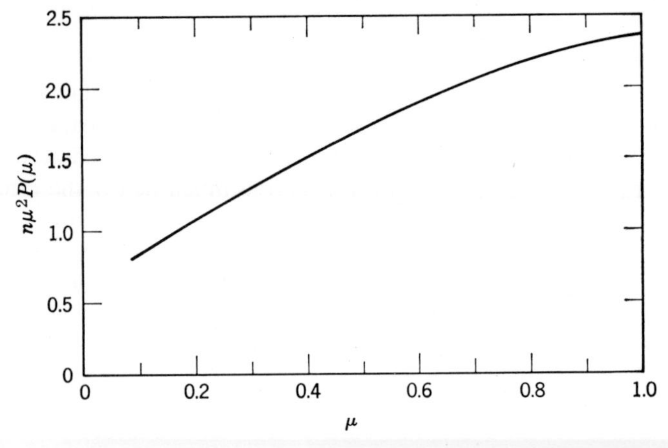

Fig. 3. The dependence of the scattering function $n\mu^2P(\mu)$ on μ as calculated[4] according to Eq. 27 for polymethylene chains when $v \gg 3$ with μ expressed in Angstrom units.

to predictions of Eq. 27′ according to which $n\mu^2P(\mu)$ should be constant, this function increases with μ for the example chosen. The curve is suggestive of the potentialities of x-ray scattering as a method for investigating the configurational characteristics of chain molecules. This approach is especially attractive as a means of "sampling" the configurations of comparatively short sequences within long chains.[5,6] An estimate of its utility must await more extensive studies on a variety of molecular chains.

2. DEPOLARIZATION IN OPTICAL SCATTERING

For physical reasons which are self-evident, light scattered at right angles to the incident beam is predominantly plane polarized. In fact, if the scattering molecule or particle is optically isotropic, i.e., if its polarizability is the same in all directions, then light scattered at 90° will be completely plane polarized.* The direction of polarization is such that the electric vector is perpendicular to the plane defined by the incident and scattered beams. This plane of polarization is denoted as vertical, and the plane of the incident and scattered beams as horizontal, as shown in Fig. 4. The

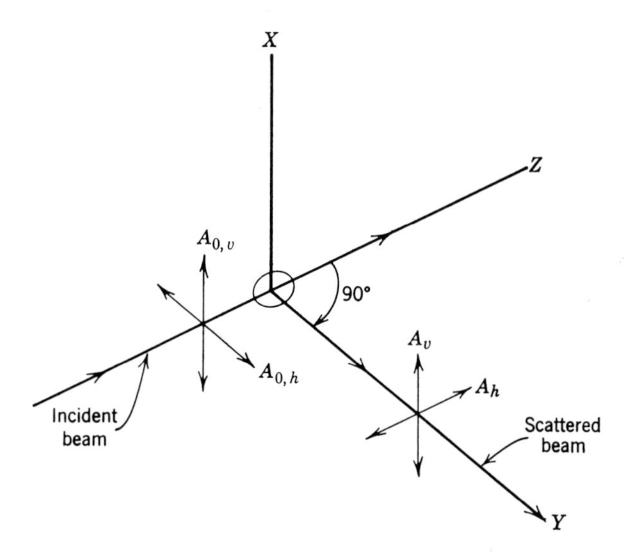

Fig. 4. Diagram showing the polarization of light scattered at 90° from the incident beam along the Z axis. Directions are shown for the vertical $A_{0,v}$ and the horizontal $A_{0,h}$ components of the incident amplitude, and for the vertical A_v and the horizontal A_h (depolarized) scattered components.

direction of polarization of the radiation scattered at 90° by isotropic particles is independent of the polarization of the incident beam; only the vertical component of the incident beam contributes.

In the event that the scattering particles are optically anisotropic, i.e., if their optical polarizabilities in different directions are unequal, then the

* Multiple scattering will contribute a depolarized (horizontal) component to the beam observed at any angle, including 90°, as follows from the fact that the angles of successive scattering events for a given ray will generally depart from 90°. Throughout the discussion in the text, multiple scattering is assumed to be suppressed.

scattered ray will include a horizontally polarized component as well as the main component, which is vertically polarized. This assertion will be violated by a given scattering particle only if it is oriented with one of the principal axes of its polarizability tensor parallel to the direction of propagation of the incident beam. For all other orientations (assuming an unpolarized incident beam), the optically anisotropic particle will contribute a horizontal component to the beam scattered at 90°. Situations of interest here are those in which the orientations of the molecules, or scattering particles, are unconstrained, all orientations being equally probable. Thus, the scattering medium is isotropic in the macroscopic sense, although its constituent molecules are optically anisotropic. It will be necessary to take the appropriate average of the depolarization of the scattered radiation over all orientations of the scattering particles.

General Formulation

Let the direction of propagation of the incident beam of amplitude $\mathbf{A}_0$ and of intensity $I_0 = A_0{}^2$ be taken along the Z axis of a laboratory reference frame, as illustrated in Fig. 4. Then $\mathbf{A}_0$ may be resolved into vertical and horizontal components taken respectively along the transverse axes X and Y of the same reference frame, as is indicated in Fig. 4. Rays scattered at an angle of 90° are propagated along the direction of the Y axis. We shall let $\boldsymbol{\alpha}$ be the polarizability tensor for the scattering particle expressed in a reference frame affixed to the particle. This reference frame may or may not be taken along the principal axes of $\boldsymbol{\alpha}$; the choice of axes will depend on circumstances discussed below. For a given orientation of the particle, its polarizability tensor in the laboratory reference frame (XYZ above) will be denoted by $\boldsymbol{\alpha}_L$. Then the amplitude of the ray scattered at 90°, comprising a vertical (X) and a horizontal (Z) component, will be given by

$$\mathbf{A} = \text{const}\begin{bmatrix} 1 & & \\ & 0 & \\ & & 1 \end{bmatrix}\boldsymbol{\alpha}_L \mathbf{A}_0 \tag{28}$$

where

$$\boldsymbol{\alpha}_L = \mathbf{R}^T \boldsymbol{\alpha} \mathbf{R} \tag{29}$$

$\mathbf{R}$ being the orthogonal matrix for the axis transformation from the laboratory reference frame to the set of axes in which $\boldsymbol{\alpha}$ is presented; i.e., $\mathbf{R}$ describes the orientation of the particle. The constant (equal to $4\pi^2/\lambda^2 a$, where a is the distance from scatterer to observer) is of no concern here.

For the case of plane-polarized incident radiation with its plane of polarization vertical,

$$\mathbf{A} = \text{const} \begin{bmatrix} 1 & & \\ & 0 & \\ & & 1 \end{bmatrix} \mathbf{R}^T \boldsymbol{\alpha} \mathbf{R} \begin{bmatrix} A_0 \\ 0 \\ 0 \end{bmatrix} \tag{30}$$

whence

$$\begin{aligned} A_v &= \text{const}\, A_0 \sum_{s,t} R_{s1} R_{t1} \alpha_{st} \\ A_h &= \text{const}\, A_0 \sum_{s,t} R_{s3} R_{t1} \alpha_{st} \end{aligned} \tag{31}$$

where v and h denote the vertically and horizontally polarized components, respectively, of the scattered radiation; s and t, enumerated 1, 2, 3, signify the coordinate axes of the Cartesian reference frame affixed to the scattering particle. The amplitudes A_v and A_h, and the intensities obtained as their squares, refer, of course, to the orientation specified by $\mathbf{R}$. The desired intensities are obtained by averaging over all orientations. For example,

$$I_v/I_0 = K \sum_{s,t,s',t'} \overline{(R_{s1} R_{t1} R_{s'1} R_{t'1})} \alpha_{st} \alpha_{s't'} \tag{32}$$

where K is a constant. The bar superscript denotes the average over all orientations, with uniform weighting. A similar expression holds for I_h/I_0. The averages are readily obtained by expressing $\mathbf{R}$ in terms of Eulerian angles and integrating over the angle space, those terms in the sum which vanish by symmetry being dropped at the outset.

The evaluation of the averages occurring in Eq. 32 is simplified if $\boldsymbol{\alpha}$ is expressed in canonical form, i.e., if it is presented in a reference frame whose axes coincide with the principal components of $\boldsymbol{\alpha}$. Then the only nonzero terms of the sum are those for which $s = t$ and $s' = t'$, and the quantities to be averaged impartially over all orientations are products $\cos^2 \chi_s \cos^2 \chi_{s'}$, where χ_s and $\chi_{s'}$ are the angles between the laboratory axis designated 1 and the principal axes s and s', respectively. For $s = s'$, the required average is $\overline{\cos^4 \chi_s} = \frac{1}{5}$; for $s \neq s'$ it is $\frac{1}{15}$. Evaluation of Eq. 32 in this manner leads at once to the well-known result[7,8]

$$I_v/I_0 K = \bar{\alpha}^2 + (4/45)\gamma^2 \tag{33}$$

where $\bar{\alpha}$ and γ are defined by[7–9]

$$\bar{\alpha} = (\alpha_1 + \alpha_2 + \alpha_3)/3 \tag{34}$$

$$\gamma^2 = (1/2)[(\alpha_1 - \alpha_2)^2 + (\alpha_1 - \alpha_3)^2 + (\alpha_2 - \alpha_3)^2] \tag{35}$$

the principal components α_1, α_2, α_3 of $\boldsymbol{\alpha}$ being designated by a single index. Thus $\bar{\alpha}$ is the mean polarizability; γ is a measure of the anisotropy of $\boldsymbol{\alpha}$.

Analogous treatment of the horizontal intensity ratio $I_h/I_0 = (A_h/A_0)^2$ (see Eq. 31) in terms of the principal components of $\boldsymbol{\alpha}$ yields[7–9]

$$I_h/I_0 K = (1/15)\gamma^2 \tag{36}$$

For the treatment which follows, more general definitions of $\bar{\alpha}$ and γ than the usual ones given by Eqs. 34 and 35[7,8] are required inasmuch as reduction of $\boldsymbol{\alpha}$ to canonical (i.e., diagonal) form for each configuration of the molecule (treated thus far as a rigid body) will be impracticable. Both $\bar{\alpha}$ and γ are invariants of the tensor $\boldsymbol{\alpha}$.[9] The former quantity is just

$$\bar{\alpha} = (1/3) \text{ trace } \boldsymbol{\alpha} \tag{37}$$

It will prove expedient to define γ in terms of the traceless tensor $\boldsymbol{\beta}$ formed from $\boldsymbol{\alpha}$ as follows[10]:

$$\boldsymbol{\beta} = \sqrt{3/2}(\boldsymbol{\alpha} - \bar{\alpha}\mathbf{E}) \tag{38}$$

If $\boldsymbol{\alpha}$ is expressed in canonical form, then γ^2 as defined by Eq. 35 is the sum of the squares of the elements of $\boldsymbol{\beta}$, as may easily be verified. This correspondence is expressed by

$$\gamma^2 = \text{trace } (\boldsymbol{\beta}^T\boldsymbol{\beta})$$

But the quantity on the right-hand side of this equation is an invariant for any second-order tensor $\boldsymbol{\beta}$. In other words, Eq. 38 must hold for $\boldsymbol{\alpha}$ expressed in any reference frame related to the principal axes of $\boldsymbol{\alpha}$ by a rotation. Since $\boldsymbol{\beta}$, like $\boldsymbol{\alpha}$, is symmetric, we have therefore*

$$\gamma^2 = \text{trace } (\boldsymbol{\beta}\,\boldsymbol{\beta}) = \sum_{s,\,t} \beta_{st}^2 \tag{39}$$

It follows from Eqs. 38 and 39 that[11]

$$\gamma^2 = (1/2)[(\alpha_{11} - \alpha_{22})^2 + (\alpha_{11} - \alpha_{33})^2 + (\alpha_{22} - \alpha_{33})^2] + 3(\alpha_{12}^2 + \alpha_{13}^2 + \alpha_{23}^2) \tag{40}$$

when $\boldsymbol{\alpha}$ is expressed in an arbitrary reference frame.

The "depolarization ratio" for vertically polarized incident light is

$$\rho_v \equiv I_h/I_v = 3\gamma^2/(45\bar{\alpha}^2 + 4\gamma^2) \tag{41}$$

* An alternative definition of γ^2 is

$$\gamma^2 = \text{trace } (\boldsymbol{\alpha}\,\boldsymbol{\alpha} - \text{adj } \boldsymbol{\alpha})$$

where adj $\boldsymbol{\alpha}$ is the adjoint matrix formed from $\boldsymbol{\alpha}$; i.e., adj $\boldsymbol{\alpha} = |\boldsymbol{\alpha}|\,\boldsymbol{\alpha}^{-1}$, where $|\boldsymbol{\alpha}|$ is the determinant of $\boldsymbol{\alpha}$. Equation 40 follows directly from this definition of γ^2.

A horizontally polarized incident beam contributes equal horizontal and vertical components to the beam scattered at 90°; the intensity of each component relative to I_0 is given identically by Eq. 36. The depolarization ratio for unpolarized incident radiation is therefore

$$\rho = 6\gamma^2/(45\bar{\alpha}^2 + 7\gamma^2) \tag{42}$$

For the formulation of $\boldsymbol{\alpha}$ for a chain molecule, treatment of which is the object of the present discussion, we adopt the valence optical scheme and the premises on which it is based.[12] Each bond is characterized by a polarizability tensor (usually cylindrically symmetric, but not required to be so by the following treatment) which is taken to be invariant to the configuration of the chain when this tensor is expressed in the coordinate system of the bond. The polarizability $\boldsymbol{\alpha}$ of the molecule as a whole is then the tensor sum of the contributions of its individual bonds. This tensor sum depends, of course, on the configuration of the chain as specified, for example, by the set of skeletal bond rotations $\{\phi\}$.

The validity of the valence-optical scheme for treating the optical anisotropy as manifested in scattering depolarization, in the Kerr effect (electro-optical effect), in streaming birefringence, or in the stress-optical coefficient for a cross-linked network, is open to question[12–14] from the standpoint of its theoretical foundations. On the other hand, experimental uncertainties, often regrettably large, may preclude a decisive empirical test. Effects of the internal field arising from induced polarization of other parts of the molecule, or of its neighbors in the liquid or solid states, have been cited[14,15] as major causes for departures from the additivity assumed in the valence-optical scheme. Such effects should be minimized when the molecules of interest occur in the pure liquid, or when in solution in an appropriate solvent. The main effect of the internal field may then be taken into account by treating the medium as an isotropic continuum characterized as is customary by an average optical polarizability, or by the refractive index.[16] The valence-optical scheme unquestionably stands in need of a more thorough test. It is adopted here for the purposes of the present section and the ones which follow without an attempt to pass judgment on its validity.

Usually it will be expedient to combine the contributions of individual bonds within the same group i to form a group polarizability tensor $\boldsymbol{\alpha}_i$. The groups must be so chosen as to satisfy the condition of invariance of $\boldsymbol{\alpha}_i$, when expressed in the reference frame of the skeletal bond with which group i is associated, to the skeletal configuration of the chain. Fulfillment of this condition requires that the mutual orientations of the bonds within a group be fixed relative to one another, as dictated by the geometry of the chain, i.e., by bond lengths and bond angles θ. It follows that bonds of a given group must be restricted to those associated with a single chain atom. For the polymethylene chain, the CH_2 group is an obvious choice; in addition to the two C—H bonds, it may be considered to include one or the other of the adjoining C—C bonds (*cf. seq.*). If in some other chain molecule the group bears a side chain capable of assuming various configurations, we shall assume that an average can be taken over these configurations independently of the configuration of the chain skeleton.*

Configurational Averages

The quantities $\bar{\alpha}$ and γ^2 appearing in Eq. 42 are to be constructed, in accordance with the principle of tensor additivity which is implied in the valence-optical scheme, as sums over contributions of the various bonds or groups comprising the chain. It will be required also to average these sums over all configurations of the chain. In accordance with notation used throughout, these averages may be designated by $\langle\bar{\alpha}\rangle$ and $\langle\gamma^2\rangle$. Since, however, the average polarizabilities $\bar{\alpha}_i$ of individual groups are both additive and independent of their orientations,

$$\langle\bar{\alpha}\rangle = \bar{\alpha} = \sum_i \bar{\alpha}_i \tag{43}$$

and the previous symbol $\bar{\alpha}$ may be retained. We therefore express the depolarization ratio for the chain molecule, with appropriate averaging over its configurations, as

$$\rho = 6\langle\gamma^2\rangle/(45\bar{\alpha}^2 + 7\langle\gamma^2\rangle) \tag{44}$$

The major task, that of evaluating $\langle\gamma^2\rangle$, the average of the anisotropy of $\boldsymbol{\alpha}$ as measured by γ^2, remains to be carried out. To this end, we undertake to formulate $\boldsymbol{\beta}$ (see Eq. 38) as the sum of $\boldsymbol{\beta}_1, \boldsymbol{\beta}_2, \ldots, \boldsymbol{\beta}_{n+1}$ representing the contributions of the respective groups, each being expressed in the reference frame of the skeletal bond of the same index. Then $\boldsymbol{\beta}$ for the molecule as a whole, expressed in the reference frame of the first bond, is

* Dependence of $\boldsymbol{\alpha}_i$ on the conformation of the skeletal bond i, or of bonds i and $i-1$, could be taken into account within the framework of present methods through revision of the generator matrix $\mathscr{P}$ (see Eq. 53) to accommodate different tensors $\boldsymbol{\gamma}_i$ (*cf. seq.*) for different rotational states. Our reasons for assuming invariance of group polarizabilities are not so much dictated by limitations of method as by the excessive information that would be required if a different $\boldsymbol{\alpha}_i$ (and hence $\boldsymbol{\gamma}_i$) were to be introduced for each bond conformation.

$$\boldsymbol{\beta} = \boldsymbol{\beta}_1 + \mathbf{T}_1\boldsymbol{\beta}_2\mathbf{T}_1{}^T + \mathbf{T}_1\mathbf{T}_2\boldsymbol{\beta}_2\mathbf{T}_2{}^T\mathbf{T}_1{}^T + \cdots$$
$$= \sum_{i=1}^{n+1} (\mathbf{T}_1^{(i-1)})\boldsymbol{\beta}_i(\mathbf{T}_1^{(i-1)})^T \tag{45}$$

where $\mathbf{T}_1^{(i-1)}$ represents the serial product comprising $i-1$ factors, in accordance with previous notation. Evaluation of tne average of this matrix by the multiplication methods extensively used in this book requires the factors comprising each term of this sum to be reordered in a manner such as to juxtapose transformations $\mathbf{T}$ and $\mathbf{T}^T$ of the same index. This end may be achieved[16] by first arranging the elements of $\boldsymbol{\beta}_i$ in a column, which we denote by $\boldsymbol{\gamma}_i$, the elements being ordered as follows:

$$\boldsymbol{\gamma}_i = \begin{bmatrix} \beta_{11} \\ \beta_{12} \\ \beta_{13} \\ \beta_{21} \\ \beta_{22} \\ \beta_{23} \\ \beta_{31} \\ \beta_{32} \\ \beta_{33} \end{bmatrix}_i \tag{46}$$

In keeping with previous practice, the serial index i applies to each element of the column. For a given configuration of the chain, the required quantity would be

$$\gamma^2 = \sum_{i,j}^{n+1} \boldsymbol{\gamma}_i{}^T\boldsymbol{\gamma}_j \tag{47}$$

if all $\boldsymbol{\gamma}_i$ were expressed in the same reference frame.

As may be proved directly, the following relationship due to Jernigan[16] holds for any three conformable matrices $\mathbf{A}$, $\mathbf{B}$, and $\mathbf{G}$:

$$(\mathbf{AGB})^C = (\mathbf{A} \otimes \mathbf{B}^T)\mathbf{G}^C \tag{48}$$

where the superscript C denotes the column formed from the elements of the matrix arranged in the order specified in Eq. 46. If $\mathbf{A}$, $\mathbf{G}$, and $\mathbf{B}$ are identified respectively with $\mathbf{T}_1^{(i-1)}$, $\boldsymbol{\beta}_i$, and $(\mathbf{T}_i^{(i-1)})^T$ of Eq. 45, then $\mathbf{G}^C$ corresponds to $\boldsymbol{\gamma}_i$, and

$$\boldsymbol{\gamma} \equiv \boldsymbol{\beta}^C = \sum_i (\mathbf{T}_1^{(i-1)} \otimes \mathbf{T}_1^{(i-1)})\boldsymbol{\gamma}_i \tag{49}$$

with γ_i expressed in the reference frame of bond i. Application of the theorem on direct products permits rearrangement to[16]

$$\gamma = \sum_i (\mathbf{T} \otimes \mathbf{T})_1^{(i-1)} \gamma_i \tag{50}$$

in which transformations $\mathbf{T}_i$ of the same index are conjoined within the same factor of the serial product. The average of γ^2 over all configurations required for Eq. 44 is

$$\langle \gamma^2 \rangle = \sum_{i=1}^{n+1} \gamma_i{}^T \gamma_i + 2 \sum_{1 \le i < j \le n+1} \gamma_i{}^T \langle (\mathbf{T} \otimes \mathbf{T})_i^{(j-i)} \rangle \gamma_j \tag{51}$$

This sum of matrix products is of the familiar form encountered in the treatment of the second moment of the chain vector $\mathbf{r}$ (see Chap. IV). (It bears a resemblance also to sum V evaluated in the treatment of the fourth moment[17] of $\mathbf{r}$; see pp. 122 and 123.) Thus, by replacing $\mathbf{m}$ in Eq. IV-24 by γ, and $\mathbf{T}$ by $\mathbf{T} \otimes \mathbf{T}$, we obtain[16] (see Eq. IV-28)

$$\langle \gamma^2 \rangle = 2Z^{-1} \mathscr{J}^* \mathscr{P}_1^{(n+1)} \mathscr{J} \tag{52}$$

where

$$\mathscr{P}_i = \begin{bmatrix} \mathbf{U} & (\mathbf{U} \otimes \gamma^T)\|\mathbf{T} \otimes \mathbf{T}\| & (\frac{1}{2})\gamma^2 \mathbf{U} \\ \mathbf{0} & (\mathbf{U} \otimes \mathbf{E}_9)\|\mathbf{T} \otimes \mathbf{T}\| & \mathbf{U} \otimes \gamma \\ \mathbf{0} & \mathbf{0} & \mathbf{U} \end{bmatrix}_i \tag{53}$$

The order of $\mathscr{P}_i$ is $11\nu \times 11\nu$.

The serial product $\mathscr{P}_i^{(n+1)}$ in Eq. 52 comprises $n + 1$ factors instead of the usual number n. This alteration is designed to take account of $n + 1$ group tensors γ_i, including those for the two terminal groups (*cf. seq.*). Of course, $\mathbf{T}_{n+1}$ is neither defined nor required inasmuch as only the final pseudo-column of $\mathscr{P}_{n+1}$ is used. The matrix $\mathbf{U}_{n+1}$ is to be equated to the identity $\mathbf{E}_\nu$. If the terminal groups are symmetric as in the n-alkane or poly-methylene chain, then $\mathbf{U}_1 = \mathbf{U}_n = \mathbf{E}_\nu$ as well, and the transformations $\mathbf{T}_1$ and $\mathbf{T}_n$ may be identically composed for each of the several rotational states of bonds 1 and n. The generality of the results embodied in Eqs. 52 and 53 will be apparent. Their application to the optical anisotropy of n-alkane chains, presented below, is illustrative.

As will be evident from the equations above, $\langle \gamma^2 \rangle$ must approach proportionality to n for flexible chains of sufficient length. According to Eq. 36, therefore, the depolarized scattering I_h/I_0 is additive in the number of units for large n. The total scattering, being proportional to $\bar{\alpha}^2$ (apart from the small effects of depolarization), increases as n^2. Hence, the depolarization ratio ρ (see Eq. 44) decreases toward zero as the length of the chain is increased without limit.

Treatment of Polymethylene Chains

The diagram of this chain shown in Fig. 5 will serve for the formulation of group polarizabilities. In order to facilitate the treatment of terminal groups, a terminal C—H bond at each end of the chain is arbitrarily chosen as a skeletal bond. With the definition of a zeroth bond in this manner, a Cartesian coordinate system is provided for bond 1 by application of the scheme used throughout for specifying coordinate systems for skeletal bonds. Methylene group $(CH_2)_i$ will include the two C_{i-1}—H_i bonds in addition to skeletal bond i joining carbons $i-1$ and i. Terminal group $(CH_3)_1$ is understood to include arbitrary skeletal bond H_0—C_0 in addition

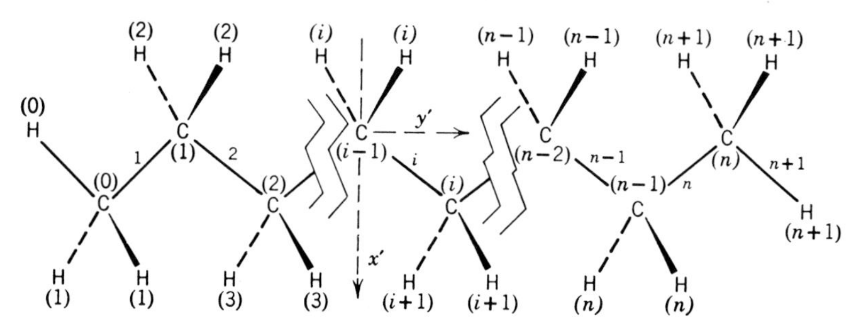

Fig. 5. Diagram of a polymethylene chain showing numeration of bonds and atoms. Axes x' and y', situated in the plane of skeletal bonds $i-1$ and i, and used for formulation of the polarizability tensor $\boldsymbol{\alpha}_i$ for CH_2 group i, also are shown.

to the two bonds C_0—H_1 and the C_0—C_1 bond. Terminal group $(CH_3)_{n+1}$ comprises the three C_n—H_{n+1} bonds only.

As before

$$\theta_i = \pi - \angle C_{i-1}C_iC_{i+1}, \qquad 0 < i < n$$

Terminal angles θ_0 and θ_n are similarly defined:

$$\theta_0 = \pi - \angle H_0C_0C_1$$

$$\theta_n = \pi - \angle C_{n-1}C_nH_{n+1}$$

and, of course, $\theta_0 = \theta_n$. Additionally, half-angles are introduced as follows:

$$\psi_i = (\tfrac{1}{2})\angle H_{i+1}C_iH_{i+1}$$

The equivalence of three of the bonds joined at the terminal carbon dictates that

$$\cos \psi_0 = \cos \psi_n = (\tfrac{1}{2})(1 + 3\cos^2 \theta_0)^{1/2} \tag{54}$$

The sum of the polarizability tensors for the two C_{i-1}—H_i bonds of the ith CH_2 group (with i in the range $1 < i < n + 1$) can be resolved along principal axes as follows (see Fig. 5): an axis x' along the bisector of the angle between these bonds, i.e., $\angle H_iC_{i-1}H_i$; a second axis y' likewise in the plane defined by C_{i-2}, C_{i-1}, and C_i; and a third axis z' perpendicular to these axes and in the plane of the two C—H bonds. The principal components of the polarizability tensor for these bonds are:

$$\begin{aligned} &2(\alpha_{\perp CH} + \Delta\alpha_{CH} \cos^2 \psi_{i-1}) \\ &2\alpha_{\perp CH} \\ &2(\alpha_{\perp CH} + \Delta\alpha_{CH} \sin^2 \psi_{i-1}) \end{aligned} \tag{55}$$

taken in the order in which the principal axes have been specified. Here $\Delta\alpha_{CH} = \alpha_{\parallel CH} - \alpha_{\perp CH}$ is the difference between the parallel and perpendicular components of the C—H bond polarizability. The serial subscripts ($i - 1$ for group i) in Eq. 55 are superfluous for all but terminal groups in view of the equivalence of all groups $i = 2$ to n, inclusive, of the polymethylene chain.

Subtraction of one-third of the trace of the tensor defined by Eq. 55 and multiplication by $\sqrt{\tfrac{3}{2}}$ as required by Eq. 38 gives[16]

$$(\boldsymbol{\beta}_{2CH})_i = \sqrt{\tfrac{3}{2}}\,\Delta\alpha_{CH} \begin{bmatrix} 2\cos^2 \psi_{i-1} - \tfrac{2}{3} & 0 & 0 \\ 0 & -\tfrac{2}{3} & 0 \\ 0 & 0 & 2\sin^2 \psi_{i-1} - \tfrac{2}{3} \end{bmatrix} \tag{56}$$

for the contribution of the two C—H bonds to the group tensor $\boldsymbol{\beta}_i$, expressed, of course, in the coordinate system having the axes identified above. This coordinate system is related to the one prescribed in the preceding section for the ith skeletal bond, i.e., C_{i-1}—C_i by a rotation about their common z axis through an angle ξ_{i-1} given by

$$\xi_{i-1} = (\pi - \theta_{i-1})/2, \qquad 1 < i < n + 1 \tag{57}$$

Transformation of Eq. 56 by this rotation and addition of the traceless tensor $(\boldsymbol{\beta}_{CC})_i$ for the skeletal bond yields[16]

$$\boldsymbol{\beta}_i = \sqrt{\tfrac{3}{2}} \begin{bmatrix} \frac{2}{3}\Delta\alpha_{CC} + 2(\cos^2 \xi \cos^2 \psi - \frac{1}{3})\Delta\alpha_{CH} & -2(\sin \xi \cos \xi \cos^2 \psi)\Delta\alpha_{CH} & 0 \\ -2(\sin \xi \cos \xi \cos^2 \psi)\Delta\alpha_{CH} & -\frac{1}{3}\Delta\alpha_{CC} + 2(\sin^2 \xi \cos^2 \psi - \frac{1}{3})\Delta\alpha_{CH} & 0 \\ 0 & 0 & -\frac{1}{3}\Delta\alpha_{CC} + 2(\sin^2 \psi - \frac{1}{3})\Delta\alpha_{CH} \end{bmatrix}_{i-1}$$

$$1 < i < n + 1 \tag{58}$$

for the contribution to $\boldsymbol{\beta}$ from internal methylene group i. The subscript $i-1$ appended to the matrix applies to angles ξ and ψ occurring therein.

The tensor $\boldsymbol{\beta}_1$ for the first group defined as above must include the contribution from bond H_0—C_0 in addition to those for the C_0—C_1 bond and for the two C_0—H_1 bonds comprising the $(CH_2)_1$ group. Hence[16]

$$\boldsymbol{\beta}_1 = (\boldsymbol{\beta}_{CH_2})_1 + \sqrt{\tfrac{3}{2}}\,\Delta\alpha_{CH}\begin{bmatrix} \cos^2\theta_0 - \frac{1}{3} & \sin\theta_0\cos\theta_0 & 0 \\ \sin\theta_0\cos\theta_0 & \sin^2\theta_0 - \frac{1}{3} & 0 \\ 0 & 0 & -\frac{1}{3} \end{bmatrix} \tag{59}$$

The term $(\boldsymbol{\beta}_{CH_2})_1$ representing the contribution of the first CH_2 group is given by Eq. 58 with $\psi = \psi_0$ (see Eq. 54) and with $\xi = \xi_0$ specified by

$$\cos\xi_0 = 2\cos\theta_0/(1 + 3\cos^2\theta_0)^{1/2} \tag{60}$$

The second term in Eq. 59 represents the contribution of the H_0—C_0 bond in the reference frame of bond 1.

The tensor for terminal group $n+1$ also may be constructed from Eq. 58 but in this case the contribution of the C—C bond is to be replaced by that of the C_n—H_{n+1} bond. It follows that[16]

$$\boldsymbol{\beta}_{n+1} = \sqrt{6}\,\Delta\alpha_{CH} \times \begin{bmatrix} \cos^2\xi_n\cos^2\psi_n & -\sin\xi_n\cos\xi_n\cos^2\psi_n & 0 \\ -\sin\xi_n\cos\xi_n\cos^2\psi_n & \sin^2\xi_n\cos^2\psi_n - \frac{1}{2} & 0 \\ 0 & 0 & \sin^2\psi_n - \frac{1}{2} \end{bmatrix} \tag{61}$$

where $\cos\psi_n$ is given by Eq. 54 and

$$\cos\xi_n = (1 - 3\cos^2\theta_n)/(1 + 3\cos^2\theta_n)^{1/2} \tag{62}$$

For normal alkane chains $\theta_{i-1} = 68°$ and $\psi_{i-1} \cong 54.5°$ (assuming $\angle$HCH $\cong$ $\angle$ HCC) for $1 < i < n+1$. The terminal groups may be taken to be tetrahedral, making $\cos\theta_0 = \cos\theta_n = \frac{1}{3}$ and $\cos\psi_0 = \cos\psi_n = \cos\xi_0 = \cos\xi_n = 1/\sqrt{3}$. Calculations[16] carried out on this basis are shown in Fig. 6. The statistical weight matrix **U** was the same as used for the polymethylene chain in the preceding section, apart from adjustment of the statistical weights to a temperature of 25°C (see pp. 142–146). The bond polarizabilities adopted for the calculations were those of Denbigh[18]: $\Delta\alpha_{CC} = 18.6 \times 10^{-25}$ cm^3 and $\Delta\alpha_{CH} = 2.1 \times 10^{-25}$ cm^3. Extensive calculations for other values of these quantities serve to establish the relationship[16]

$$(\langle\gamma^2\rangle/n)_\infty = 1.27(\Delta\alpha_{CC} - 1.95\,\Delta\alpha_{CH})^2 \tag{63}$$

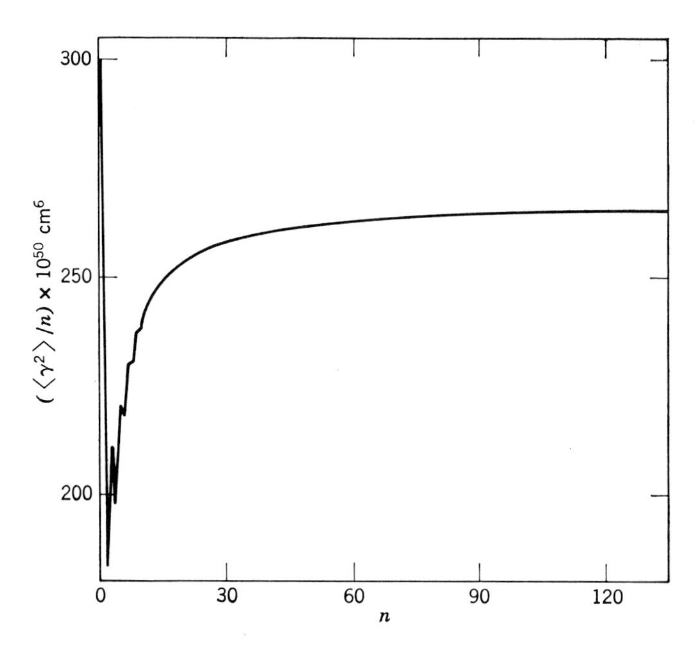

Fig. 6. Calculated ratio of the averaged tensor invariant $\langle\gamma^2\rangle$ to the chain length n plotted against n for polymethylene[16] (see text).

for the value of this ratio in the limit of sufficiently long chains. The convergence with chain length is not merely a function of the quantity $\Delta\alpha_{CC} - 1.95\,\Delta\alpha_{CH}$, however. The rapidity of the convergence varies greatly, depending on the separate values of the two bond anisotropies.

Experimental values[11,19] of $\langle\gamma^2\rangle$ obtained from depolarization of light scattered by various liquid n-alkanes in the range $n = 3$ to 31 are lower than those calculated (Fig. 6). The disparity is variable, reflecting irregularities in the experimental results, but may be as great as a factor of 2. The trend with chain length is compatible with the calculated curve shown in Fig. 6, however.

3. STRAIN BIREFRINGENCE IN AMORPHOUS POLYMERS

The treatment of the strain birefringence of an amorphous, nonglassy, polymeric material as an equilibrium property presupposes the presence of some kind of network structure. That is, the long-chain molecules comprising the polymer must be linked to one another at occasional points. Otherwise the anisotropy imposed by the strain would be dissipated by relaxation processes. These *cross-linkages* must be permanent, or at least of sufficient duration to last for the time interval of the experiment without being voided by rupture. The constituent of the network structure on which we focus attention is a *chain*, defined as that portion extending along a given molecular path from one cross-linked unit (or other kind of branched junction) to the next.[20] The term "chain" thus takes on a more specialized meaning within the context of this section. All chains so defined will be assumed to be of the same length n. The essentials of the results which follow would not be altered, however, if account were taken of the distribution of n for the collection of chains. Although the magnitude of n will not be restricted, we shall bear in mind that the chains ordinarily will be fairly long, i.e., $n > 100$. The amorphous network should then manifest the properties of a typical rubber or elastomer.

The property under consideration is a macroscopic one: it is the resultant of the combined contributions of many chains, or rather of the structural units comprising them. The optical polarizabilities of these units and their average spatial arrangement determine the birefringence of the system. The macroscopic birefringence differs in this respect from the depolarization of light scattered by free chain molecules, which, as discussed in the preceding section, depends on fluctuations of individual molecules from their average configurations. These configurations are essentially isotropic in the sense that $\langle \boldsymbol{\alpha} \rangle = \bar{\alpha}\mathbf{E}_3$ for very long chains. Fluctuations at the molecular level from this average condition, as measured by the mean-*square* anisotropy $\langle \gamma^2 \rangle$, are responsible for the depolarization. The network under strain possesses an average anisotropy, and the average polarizabilities $\langle \boldsymbol{\alpha} \rangle$ of its individual chains in general are not scalar tensors (i.e., not isotropic). The macroscopic polarizability may be formulated as the sum of averaged contributions from the various chains comprising the specimen.

In the unstrained state the system of chains in a network may be assumed to be isotropic; this must necessarily be true[21–23] for a network of Gaussian chains in the absence of a stress (apart from an isotropic pressure). Let the network be subject to a strain, which for simplicity we assume to be homogeneous (i.e., the same throughout the specimen). This strain may be specified by the *displacement gradient tensor* $\boldsymbol{\lambda}$, whose elements are defined by[23]

$$\lambda_{ts} = \partial x_t' / \partial x_s \tag{64}$$

where x_s is one of the Cartesian coordinates of a point in the elastic body when in the state of rest (i.e., the isotropic state), and x_t' is one of the coordinates of the same point after deformation. If axes are so chosen as to render $\boldsymbol{\lambda}$ diagonal, then its elements represent the three principal extension ratios λ_1, λ_2, and λ_3 which characterize the strain.

Let $W(\mathbf{r})$ be the distribution function for the vectors $\mathbf{r}$ for the chains of the network in the state of rest. For example, $W(\mathbf{r})$ might be just the unperturbed distribution which would obtain if all chains were free, i.e., severed from the cross-linkages connecting their termini, and subject to no constraints whatever.[22] The chain vector distribution in the state of rest in any case will be spherically symmetric, as we have pointed out above. Let $W'(\mathbf{r})$ be the distribution of vectors $\mathbf{r}$ after deformation. If the cross-linkages of the network are assumed to be displaced like points embedded in a homogeneous elastic continuum, then the distribution of vectors in the deformed state will be related to that in the state of rest by

$$W'(\mathbf{r}) = |\boldsymbol{\lambda}|^{-1} W(\boldsymbol{\lambda}^{-1}\mathbf{r}) \tag{65}$$

inasmuch as any vector $\mathbf{r}^0$ in the initial state is converted to $\mathbf{r} = \boldsymbol{\lambda}\mathbf{r}^0$ by the deformation; obviously, the converse of this relation holds as well, i.e., $\mathbf{r}^0 = \boldsymbol{\lambda}^{-1}\mathbf{r}$. The factor $|\boldsymbol{\lambda}|^{-1}$ i.e., the reciprocal of the determinant of $\boldsymbol{\lambda}$, takes account of the dilatation accompanying the deformation. Hence, W', like W, is the probability density *per unit volume* in the space of the chain vector $\mathbf{r}$, or of $\mathbf{r}^0$. (The unique relationship between a given vector before and after deformation implied by this formulation of Eq. 65 is not actually required; only the *distribution* of vectors is assumed to be deformed by Eq. 65.[20,22,23] This is a somewhat less stringent requirement that admits of fluctuations of vectors about mean positions, provided, however, that these fluctuations undergo the same deformation $\boldsymbol{\lambda}$.)

Those chains characterized by the same chain vector $\mathbf{r}$ may, of course, assume many configurations consistent with $\mathbf{r}$. These various configurations, taken collectively, will be disposed uniformly about $\mathbf{r}$. Hence, $\mathbf{r}$ may be treated as an axis of cylindrical symmetry with reference to this family of chains. (An individual chain in its instantaneous configuration will not, of course, be cylindrically symmetric about $\mathbf{r}$.) The averaged optical polarizability tensor representing these chains therefore assumes diagonal form if referred to axes one of which is parallel to $\mathbf{r}$, the other two being perpendicular to $\mathbf{r}$. Expressed as a matrix, the average polarizability in this reference frame is

$$\boldsymbol{\alpha}_r = \begin{bmatrix} \alpha_r & & \\ & (3\bar{\alpha} - \alpha_r)/2 & \\ & & (3\bar{\alpha} - \alpha_r)/2 \end{bmatrix} \tag{66}$$

where α_r is the average component of $\boldsymbol{\alpha}$ parallel to $\mathbf{r}$. Symmetry about $\mathbf{r}$ renders the two transverse components equal to each other, and hence equal to $(3\bar{\alpha} - \alpha_r)/2$ inasmuch as $3\bar{\alpha}$ is the sum of the diagonal elements of $\boldsymbol{\alpha}_r$ (see Eq. 37). The tensor $\boldsymbol{\alpha}_r$ could be defined equivalently as the average of $\boldsymbol{\alpha}$ over all configurations of one chain whose vector is fixed at $\mathbf{r}$, the chain being free of other constraints such as are imposed by the neighboring chains with which it is inextricably intertwined in the network.

The component of the averaged polarizability tensor $\boldsymbol{\alpha}_r$ along an arbitrary reference axis X with which $\mathbf{r}$ makes an angle ξ is

$$
\begin{aligned}
(\alpha_{xx})_{\mathbf{r}} &= \alpha_r \cos^2 \xi + [(3\bar{\alpha} - \alpha_r)/2] \sin^2 \xi \\
&= (\tfrac{1}{2})(\alpha_r - \bar{\alpha})(3 \cos^2 \xi - 1) + \bar{\alpha} \\
&= (\tfrac{1}{2})(\alpha_r - \bar{\alpha})(3x^2/r^2 - 1) + \bar{\alpha}
\end{aligned} \tag{67}
$$

where x is the component of $\mathbf{r}$ along this axis. Similarly

$$(\alpha_{yy})_{\mathbf{r}} = (\tfrac{1}{2})(\alpha_r - \bar{\alpha})(3y^2/r^2 - 1) + \bar{\alpha} \tag{68}$$

$$(\alpha_{zz})_{\mathbf{r}} = (\tfrac{1}{2})(\alpha_r - \bar{\alpha})(3z^2/r^2 - 1) + \bar{\alpha} \tag{69}$$

The difference between the polarizabilities along two axes, e.g., X and Y, is

$$(\alpha_{xx} - \alpha_{yy})_{\mathbf{r}} = (\tfrac{3}{2})(\alpha_r - \bar{\alpha})(x^2 - y^2)/r^2 \tag{70}$$

Or

$$(\alpha_{xx} - \alpha_{yy})_{\mathbf{r}} = \Delta\alpha_r(x^2 - y^2)/r^2 \tag{71}$$

where

$$\Delta\alpha_r = (\tfrac{3}{2})(\alpha_r - \bar{\alpha}) \tag{72}$$

is just the difference between α_r and the average transverse polarizability, $(3\bar{\alpha} - \alpha_r)/2$. Equations analogous to Eqs. 70 and 71 hold for other pairs of directions.

To proceed further, it is necessary to evaluate the average polarizability α_r of a chain along its chain vector of length r. Before undertaking the task of deriving general relationships suitable for evaluating α_r for any molecular chain, we may profitably consider the conventional treatment of the optical anisotropy in terms of a hypothetical freely jointed chain.

The Freely Jointed Chain[12,24,25]

Let such a chain be subject to the constraint requiring the projection of its chain vector $\mathbf{r}$ upon an arbitrary axis to be fixed. Let this component of $\mathbf{r}$ be specified by x. The arbitrary direction (X) becomes an axis of cylindrical symmetry for the collection of all configurations subject to the foregoing constraint. Then the number of bonds having orientations such that their projections on this axis are in the range l_x to $l_x + dl_x$ is

$$dn_x = (n\beta/\sinh \beta) \exp (\beta l_x/l)(2l)^{-1}\, dl_x \tag{73}$$

which follows from Eq. VIII-55 with minor alterations in notation. The average of the square of the cosine of the angle between any given bond and the axis X follows as

$$\begin{aligned}\langle (l_x/l)^2 \rangle &= \beta(\sinh \beta)^{-1} \int_{-l}^{l} \exp (\beta l_x/l)(l_x^2/2l^3)\, dl_x \\ &= 1 - 2\beta^{-1} \coth \beta + 2\beta^{-2} \\ &= 1 - 2\beta^{-1}\mathscr{L}(\beta) \end{aligned} \tag{74}$$

Inasmuch as all bonds of the freely jointed chain are equivalent, no serial index is required, and Eq. 74 may be construed as the average over all bonds, as well as over all configurations. Recalling the definition of β given by Eq. VIII-54, we have

$$\langle (l_x/l)^2 \rangle = 1 - 2(x/nl)/\mathscr{L}^*(x/nl) \tag{75}$$

where $\mathscr{L}^*$ is the inverse Langevin function.

In the original treatment of the problem by Kuhn and Grün[24] and in subsequent expositions as well,[12,25] the arbitrary axis X has been identified incorrectly with the chain vector **r**, giving $x = r$. The same error has been pointed out earlier in another connection (see Chap. VIII, p. 319 *et seq.*). Only in the limit of very long chains, and for $r/nl \ll 1$, may this identification be fully justified. Inasmuch as a rigorous treatment follows in due course, we tentatively replace x by r in Eq. 75, a step lacking rigor but abounding in precedent. On this basis

$$\langle \cos^2 \Phi \rangle \cong 1 - 2(r/nl)/\mathscr{L}^*(r/nl) \tag{76}$$

where Φ is taken to be the angle between a given bond and the chain vector **r**. Series expansion (see Eq. VIII-62) gives

$$\langle \cos^2 \Phi \rangle \cong \frac{1}{3} + \frac{2}{5}\left(\frac{r}{nl}\right)^2 + \frac{24}{175}\left(\frac{r}{nl}\right)^4 + \cdots \tag{77}$$

Assuming the bond polarizability tensor to be cylindrically symmetric (and scarcely any other circumstance is conceivable for a freely jointed chain), we let $a_\parallel$ and $a_\perp$ represent its components parallel and perpendicular, respectively, to the bond axis. Then

$$\alpha_r = n(a_\parallel \langle \cos^2 \Phi \rangle + a_\perp \langle \sin^2 \Phi \rangle) \tag{78}$$

and from Eq. 76

$$\alpha_r = n[a_\parallel - (a_\parallel - a_\perp)2(r/nl)/\mathscr{L}^*(r/nl)] \tag{79}$$

By substitution in Eq. 72

$$\Delta\alpha_r = n\ \Delta a[1 - (3r/nl)/\mathscr{L}^*(r/nl)] \tag{80}$$

$$= n\ \Delta a\left(\frac{3}{5}\right)\left(\frac{r}{nl}\right)^2\left[1 + \frac{12}{35}\left(\frac{r}{nl}\right)^2 + \frac{36}{175}\left(\frac{r}{nl}\right)^4 + \cdots\right] \tag{81}$$

where $\Delta a = a_{\|} - a_{\perp}$. For sufficiently long chains subject to relatively small extensions, only the first term will be required. Hence,

$$\Delta\alpha_r = (\tfrac{3}{5})\ \Delta a(r^2/nl^2) = \Gamma_2 r^2/\langle r^2\rangle_0 \tag{82}$$

where $\Gamma_2 = 3\Delta a/5$. In the general case of a real chain, Γ_2 will appear as the coefficient of the first term of a series in powers of $r^2/\langle r^2\rangle_0$. The higher coefficients will be designated Γ_4, etc. (see Eq. 100 and Appendix H). For real chains of sufficient length, it suffices to let $\Delta\alpha_r$ be proportional to r^2 for small extensions $r \ll r_{\max}$. Hence, with these qualifications Eq. 82 and those which follow are applicable not only to freely jointed chains but to real chains as well. The coefficient Γ_2, and likewise Δa, must, of course, be reinterpreted in terms of parameters of the real chain, a task undertaken in the following subsection.

According to Eqs. 71 and 82, the difference between the polarizabilities of the chain along axes X and Y is

$$(\alpha_{xx} - \alpha_{yy})_{\mathbf{r}} = \Gamma_2(x^2 - y^2)/\langle r^2\rangle_0 \tag{83}$$

and so forth for other pairs of axes. The difference between the polarizability components depends only on x and y, being otherwise independent of $\mathbf{r}$. This independence of $r = |\mathbf{r}|$ obtains only in the approximation whereby higher terms in Eq. 81 are neglected, as will be apparent from inspection of that equation.

The difference of polarizabilities in Eq. 83 averaged over all chains is obtained by replacing x^2 and y^2 by their averages. These depend, of course, on the strain specified by the deformation gradient $\boldsymbol{\lambda}$. This tensor has been defined above relative to the state of rest in which the network is isotropic. Let $\langle r^2\rangle_{\text{iso}}$ represent the mean-square value of r, averaged over all chains of the network, in this state of reference. (In general,[22,23] $\langle r^2 \rangle_{\text{iso}} \neq \langle r^2 \rangle_0$.*) Then, according to the definition of $\boldsymbol{\lambda}$

$$\overline{x^2} = \lambda_x{}^2 \langle r^2 \rangle_{\text{iso}}/3 \tag{84}$$

with corresponding relations for $\overline{y^2}$ and $\overline{z^2}$. Here λ_x is one of the principal extension ratios, the reference frame having been chosen so that its axes coincide with the principal axes of $\boldsymbol{\lambda}$. The differences between the averaged polarizabilities along these axes are given therefore by expressions[24,25] such as

$$\overline{\alpha_{xx} - \alpha_{yy}} = (\Gamma_2/3)(\langle r^2 \rangle_{\text{iso}}/\langle r^2 \rangle_0)(\lambda_x{}^2 - \lambda_y{}^2) \tag{85}$$

The principal axes of the polarizability tensor must coincide with those of $\boldsymbol{\lambda}$. Hence, with λ_x, etc. defined as above, $\bar{\alpha}_{xx}$ and $\bar{\alpha}_{yy}$ are principal components of $\boldsymbol{\alpha}$. Equations corresponding to Eq. 85 hold for other pairs of principal polarizabilities.

The refractive index $\tilde{n}$ may be related to the scalar polarizability α of a chain according to the Lorentz-Lorenz expression

$$\frac{\tilde{n}^2 - 1}{\tilde{n}^2 + 2} = \frac{4\pi}{3}\left(\frac{v}{V}\right)\alpha \tag{86}$$

where v is the number of chains in volume V. The refractive indexes along each of the principal axes of the strain may be obtained from the respective polarizabilities through use of this equation. Introducing approximations appropriate for the difference between two of the expressions on the left-hand side when the two refractive indexes differ only very slightly, we thus obtain[12,24,25]

$$\begin{aligned}\tilde{n}_x - \tilde{n}_y &= \left(\frac{2\pi}{9}\right)\left(\frac{v}{V}\right)\frac{(\tilde{n}^2 + 2)^2}{\tilde{n}}\overline{(\alpha_{xx} - \alpha_{yy})} \\ &= \left(\frac{2\pi v}{27V}\right)\left[\frac{(\tilde{n}^2 + 2)^2\Gamma_2}{\tilde{n}}\right]\left(\frac{\langle r^2 \rangle_{\text{iso}}}{\langle r^2 \rangle_0}\right)(\lambda_x{}^2 - \lambda_y{}^2)\end{aligned} \tag{87}$$

* The displacement gradient tensor $\boldsymbol{\lambda}$ was defined in reference 23 relative to the isotropic state of volume such that $\langle r^2 \rangle_{\text{iso}} = \langle r^2 \rangle_0$. That is, the factor $(\langle r^2 \rangle_{\text{iso}}/\langle r^2 \rangle_0)^{1/2}$ was incorporated into $\boldsymbol{\lambda}$. It is expedient here to retain this factor external to $\boldsymbol{\lambda}$ in order that it will appear explicitly in the equations to follow. The present $\boldsymbol{\lambda}$ corresponds to $\boldsymbol{\alpha}$ used elsewhere[23] (see also Eq. II–10). Adoption of the latter symbol to represent the polarizability precludes its use for the deviatory portion of the deformation in accordance with notation in reference 23.

In the case of an axially symmetric deformation, e.g., simple elongation, $\lambda_y = \lambda_z$, with x taken as the axis of symmetry. It is then convenient to let $\lambda_x = \lambda$, and if the deformation may be considered to occur at constant volume so that $\lambda_x \lambda_y \lambda_z = 1$, then $\lambda_y^2 = \lambda_z^2 = \lambda^{-1}$, and

$$\Delta\tilde{n} = \left(\frac{2\pi\nu}{27V}\right)\left[\frac{(\tilde{n}^2 + 2)^2\Gamma_2}{\tilde{n}}\right]\left(\frac{\langle r^2\rangle_{\text{iso}}}{\langle r^2\rangle_0}\right)\left(\lambda^2 - \frac{1}{\lambda}\right) \tag{88}$$

where $\Delta\tilde{n}$ is the difference between the refractive indices parallel and perpendicular to the axis of elongation.

According to the theory of rubber elasticity[23,25] of a network of Gaussian chains, the difference between two principal stresses is given by

$$\tau_x - \tau_y = (\nu kT/V)(\langle r^2\rangle_{\text{iso}}/\langle r^2\rangle_0)(\lambda_x^2 - \lambda_y^2) \tag{89}$$

By combination of this result[25] with Eq. 87

$$\tilde{n}_x - \tilde{n}_y = B(\tau_x - \tau_y) \tag{90}$$

where B is the stress-optical coefficient defined for the freely jointed chain by

$$\begin{aligned} B &= (2\pi\Gamma_2/27kT)(\tilde{n}^2 + 2)^2/\tilde{n} \\ &= (2\pi\,\Delta a/45kT)(\tilde{n}^2 + 2)^2/\tilde{n} \end{aligned} \tag{91}$$

Equation 90, asserting the birefringence to be proportional to the stress, expresses Brewster's law of strain birefringence. The constant B of proportionality is the stress-optical coefficient. For an axial deformation, we may take $\tau_x = \tau$ and $\tau_y = \tau_z = 0$. Equation 90 then simplifies to

$$\Delta\tilde{n} = B\tau \tag{92}$$

Real Chains

A more general treatment of the optical anisotropy of a system of real chain molecules subjected to stress was initiated by Gotlib[26] and Volkenstein.[12] Adopting their procedure, we consider the components of the polarizability tensor $\boldsymbol{\alpha}_i$ for group i relative to a coordinate system affixed to the chain vector $\mathbf{r}$. Groups are specified as prescribed in the preceding section. The total polarizability α_r along this vector being the quantity required as elaborated above, we seek the contribution to α_r from each element of $\boldsymbol{\alpha}_i$ averaged over all configurations of the chain for a specified value of r. Let Φ_{it} be the angle between $\mathbf{r}$ and a vector embedded in group i, this vector being indexed by t; specifically, t may designate one of the principal axes of $\boldsymbol{\alpha}_i$. Then, following Gotlib[26] and Volkenstein,[12] we assume that the average square of the cosine of this angle for the given value of r may be represented by a series in powers of r^2 as follows:

$$\langle \cos^2 \Phi_{it} \rangle_r = \frac{1}{3} + \beta_{2;\,it}\left(\frac{r^2}{\langle r^2 \rangle_0}\right) + \beta_{4;\,it}\left(\frac{r^4}{\langle r^2 \rangle_0{}^2}\right) + \cdots \tag{93}$$

The $\beta_{2;\,it}$, etc. are constant coefficients to be determined. The form of this expansion was suggested by Eq. 77 for the freely jointed chain. On the premise that the first two terms in Eq. 93 are adequate for a chain of sufficient length at small extensions, the constant $\beta_{2;\,it}$ may be evaluated by the following procedure.[12,26] Multiplying Eq. 93 by r^2 and averaging over all values of r for the chain free of constraints, Gotlib and Volkenstein obtained

$$\begin{aligned} \langle r^2 \cos^2 \Phi_{it} \rangle_0 &= \langle r^2 \langle \cos^2 \Phi_{it} \rangle_r \rangle_0 \\ &\cong \langle r^2 \rangle_0 / 3 + \beta_{2;\,it} \langle r^4 \rangle_0 / \langle r^2 \rangle_0 \end{aligned} \tag{94}$$

which is subject to whatever approximation may be entailed in the premises cited. (The subscript zero denotes the unperturbed chain, free of an external stress.) In keeping with these approximations, $\langle r^4 \rangle_0$ may be replaced by $(\frac{5}{3})\langle r^2 \rangle_0{}^2$, to which it converges in the limit $n \to \infty$. Then

$$\beta_{2;\,it} \cong 3\langle r^2 \cos^2 \Phi_{it} \rangle_0 / 5\langle r^2 \rangle_0 - \frac{1}{5} \tag{95}$$

and in this approximation

$$\langle \cos^2 \Phi_{it} \rangle_r \cong \frac{1}{3} + \frac{1}{5}\left[\frac{3\langle r^2 \cos^2 \Phi_{it} \rangle_0}{\langle r^2 \rangle_0} - 1\right] \frac{r^2}{\langle r^2 \rangle_0} \tag{96}$$

The coefficient of $r^2/\langle r^2 \rangle_0$ in Eq. 96 is confirmed as correct to terms of order $1/n$ by a rigorous analysis carried out by Nagai[27] using a Fourier transformation method. Further terms of order $1/n^2$ contributing to this coefficient were obtained by Nagai, who also derived the coefficient of $r^4/\langle r^2 \rangle_0{}^2$ in the expansion of $\langle \cos^2 \Phi_{it} \rangle_r$. A condensation of Nagai's derivation is given in Appendix H, which yields $\langle \cos^2 \Phi_{it} \rangle_r$ in higher approximation.

For completion of the task at hand, we adopt the approximation offered by Eq. 96 which, as has been pointed out, should be acceptable for long chains at extensions which are not too great. The importance of Eq. 96 lies in the fact that it relates a quantity, $\langle \cos^2 \Phi_{it} \rangle_r$, which eludes direct evaluation, to averages for the free chain. These averages can be evaluated by the methods used previously in other connections, as is demonstrated below.

The total average contribution of group i to the component α_r of the polarizability of the chain along its chain vector $\mathbf{r}$ is

$$\alpha_{r;\,i} = \sum_{t=1,2,3} \alpha_{it} \langle \cos^2 \Phi_{it} \rangle_r \tag{97}$$

where α_{it} is the principal component t of $\boldsymbol{\alpha}_i$, and the sum extends over the three components. Substitution of Eq. 96 into Eq. 97 yields

$$\begin{aligned}\alpha_{r;\,i} &\cong \bar{\alpha}_i + (\tfrac{1}{5}) \sum_t (3\alpha_{it}\langle r^2 \cos^2 \Phi_{it}\rangle_0/\langle r^2\rangle_0 - \alpha_{it})(r^2/\langle r^2\rangle_0) \\ &= \bar{\alpha}_i + (\tfrac{3}{5})(\langle \mathbf{r}^T\boldsymbol{\alpha}_i\mathbf{r}\rangle_0/\langle r^2\rangle_0 - \bar{\alpha}_i)(r^2/\langle r^2\rangle_0)\end{aligned} \tag{98}$$

Hence, in the present approximation we have from Eq. 72[27,28]

$$\Delta\alpha_r = (\tfrac{9}{10}) \sum_i (\langle \mathbf{r}^T\boldsymbol{\alpha}_i\mathbf{r}\rangle_0/\langle r^2\rangle_0 - \bar{\alpha}_i)(r^2/\langle r^2\rangle_0) \tag{99}$$

More precisely (see Appendix H)[27]

$$\Delta\alpha_r = \Gamma_2\, r^2/\langle r^2\rangle_0 + \Gamma_4\, r^4/\langle r^2\rangle_0{}^2 + \cdots \tag{100}$$

where Γ_2, Γ_4, etc. are coefficients that depend on $\sum_i \langle \mathbf{r}^T\boldsymbol{\alpha}_i\mathbf{r}\rangle_0/\langle r^2\rangle_0$, $\sum_i \langle(\mathbf{r}^T\boldsymbol{\alpha}_i\mathbf{r})(\mathbf{r}^T\mathbf{r})\rangle_0/\langle r^2\rangle_0{}^2$, etc., and also on the higher even moments of r. In the approximation that terms of order $1/n^2$, $1/n^3$, etc. may be omitted,

$$\begin{aligned}\Gamma_2 &\cong (\tfrac{9}{10}) \sum_i (\langle \mathbf{r}^T\boldsymbol{\alpha}_i\mathbf{r}\rangle_0/\langle r^2\rangle_0 - \bar{\alpha}_i) \\ &= (\tfrac{9}{10}) \sum_i \langle \mathbf{r}^T\hat{\boldsymbol{\alpha}}_i\mathbf{r}\rangle_0/\langle r^2\rangle_0\end{aligned} \tag{101}$$

where $\hat{\boldsymbol{\alpha}}_i$ is the traceless tensor

$$\hat{\boldsymbol{\alpha}}_i = \boldsymbol{\alpha}_i - \bar{\alpha}_i\mathbf{E}_3 \tag{102}$$

The calculation of the anisotropy of the polarizability in this approximation requires evaluation[29] of the sum of terms $\langle \mathbf{r}^T\hat{\boldsymbol{\alpha}}_i\mathbf{r}\rangle_0$ occurring in Eq. 101. These terms may be rearranged according to Eq. 48, and since they are scalars, each may be equated to its transpose. Thus,

$$\mathbf{r}^T\hat{\boldsymbol{\alpha}}_i\mathbf{r} = (\mathbf{r}^T \otimes \mathbf{r}^T)\hat{\boldsymbol{\alpha}}_i{}^C = \hat{\boldsymbol{\alpha}}_i{}^R(\mathbf{r} \otimes \mathbf{r}) \tag{103}$$

where $\hat{\boldsymbol{\alpha}}_i{}^C$ denotes the column array of the elements of $\hat{\boldsymbol{\alpha}}_i$ and $\hat{\boldsymbol{\alpha}}_i{}^R$ is the corresponding row form, i.e., the transpose of $\hat{\boldsymbol{\alpha}}_i{}^C$. Hence

$$\sum_i \mathbf{r}^T\hat{\boldsymbol{\alpha}}_i\mathbf{r} = \sum_i \Big(\sum_j \mathbf{l}_j{}^T\Big)\hat{\boldsymbol{\alpha}}_i\Big(\sum_k \mathbf{l}_k\Big) \tag{104}$$

$$= \sum_i \Big[\Big(\sum_j \mathbf{l}_j{}^T\Big) \otimes \Big(\sum_k \mathbf{l}_k{}^T\Big)\Big]\hat{\boldsymbol{\alpha}}_i{}^C \tag{105}$$

$$= \sum_i \hat{\boldsymbol{\alpha}}_i{}^R\Big[\Big(\sum_j \mathbf{l}_j\Big) \otimes \Big(\sum_k \mathbf{l}_k\Big)\Big] \tag{106}$$

Which of these three expressions is preferred will depend on the serial order of occurrence of i in relation to j and k.

Let the triple sum over i, j, k be separated into terms as follows:

$$\sum_i\sum_j\sum_k = \sum_{i=j=k} + 2\sum_{i=j<k}\sum + 2\sum_{j<k=i}\sum$$

$$+\sum_{i<j=k}\sum + \sum_{j=k<i}\sum$$
$$+2\sum_{i<j<k}\sum\sum + 2\sum_{j<i<k}\sum\sum + 2\sum_{j<k<i}\sum\sum$$

Proceeding on this basis, we obtain, through use of Eqs. 104, 105, and 106,

$$\begin{aligned}\sum_i \mathbf{r}^T\hat{\boldsymbol{\alpha}}_i\mathbf{r} = &\sum_i \hat{\boldsymbol{\alpha}}_i^R(\mathbf{l}_i\otimes\mathbf{l}_i) + 2\sum_{i<j}\hat{\boldsymbol{\alpha}}_i^R[\mathbf{l}_i\otimes(\mathbf{T}_i\cdots\mathbf{T}_{j-1}\mathbf{l}_j)]\\ &+2\sum_{j<i}[(\mathbf{l}_j^T\mathbf{T}_j\cdots\mathbf{T}_{i-1})\otimes\mathbf{l}_i^T]\hat{\boldsymbol{\alpha}}_i^C\\ &+\sum_{i<j}\hat{\boldsymbol{\alpha}}_i^R[(\mathbf{T}_i\cdots\mathbf{T}_{j-1}\mathbf{l}_j)\otimes(\mathbf{T}_i\cdots\mathbf{T}_{j-1}\mathbf{l}_j)]\\ &+\sum_{j<i}[(\mathbf{l}_j^T\mathbf{T}_j\cdots\mathbf{T}_{i-1})\otimes(\mathbf{l}_j^T\mathbf{T}_j\cdots\mathbf{T}_{i-1})]\hat{\boldsymbol{\alpha}}_i^C\\ &+2\sum_{i<j<k}\hat{\boldsymbol{\alpha}}_i^R[(\mathbf{T}_i\cdots\mathbf{T}_{j-1}\mathbf{l}_j)\otimes(\mathbf{T}_i\cdots\mathbf{T}_{k-1}\mathbf{l}_k)]\\ &+2\sum_{j<i<k}\mathbf{l}_j^T\mathbf{T}_j\cdots\mathbf{T}_{i-1}\hat{\boldsymbol{\alpha}}_i\mathbf{T}_i\cdots\mathbf{T}_{k-1}\mathbf{l}_k\\ &+2\sum_{j<k<i}[(\mathbf{l}_j^T\mathbf{T}_j\cdots\mathbf{T}_{i-1})\otimes(\mathbf{l}_k^T\mathbf{T}_k\cdots\mathbf{T}_{i-1})]\hat{\boldsymbol{\alpha}}_i^C\end{aligned}$$

By resort to the theorem on direct products, the various sums may be rearranged as follows so that quantities of the same serial index occur in the same factor:

$$\begin{aligned}\sum_i \mathbf{r}^T\hat{\boldsymbol{\alpha}}_i\mathbf{r} = &\sum_i \hat{\boldsymbol{\alpha}}_i^R(\mathbf{l}_i\otimes\mathbf{l}_i) + 2\sum_{i<j}\hat{\boldsymbol{\alpha}}_i^R(\mathbf{l}_i\otimes\mathbf{E}_3)\mathbf{T}_i\cdots\mathbf{T}_{j-1}\mathbf{l}_j\\ &+2\sum_{j<i}\mathbf{l}_j^T\mathbf{T}_j\cdots\mathbf{T}_{i-1}(\mathbf{E}_3\otimes\mathbf{l}_i^T)\hat{\boldsymbol{\alpha}}_i^C\\ &+\sum_{i<j}\hat{\boldsymbol{\alpha}}_i^R(\mathbf{T}_i\otimes\mathbf{T}_i)\cdots(\mathbf{T}_{j-1}\otimes\mathbf{T}_{j-1})(\mathbf{l}_j\otimes\mathbf{l}_j)\\ &+\sum_{j<i}(\mathbf{l}_j^T\otimes\mathbf{l}_j^T)(\mathbf{T}_j\otimes\mathbf{T}_j)\cdots(\mathbf{T}_{i-1}\otimes\mathbf{T}_{i-1})\hat{\boldsymbol{\alpha}}_i^C\\ &+2\sum_{i<j<k}\hat{\boldsymbol{\alpha}}_i^R(\mathbf{T}_i\otimes\mathbf{T}_i)\cdots(\mathbf{T}_{j-1}\otimes\mathbf{T}_{j-1})(\mathbf{l}_j\otimes\mathbf{E}_3)\mathbf{T}_j\cdots\mathbf{T}_{k-1}\mathbf{l}_k\\ &+2\sum_{j<i<k}\mathbf{l}_j^T\mathbf{T}_j\cdots\mathbf{T}_{i-1}\hat{\boldsymbol{\alpha}}_i\mathbf{T}_i\cdots\mathbf{T}_{k-1}\mathbf{l}_k\\ &+2\sum_{j<k<i}\mathbf{l}_j^T\mathbf{T}_j\cdots\mathbf{T}_{k-1}(\mathbf{E}_3\otimes\mathbf{l}_k^T)(\mathbf{T}_k\otimes\mathbf{T}_k)\cdots(\mathbf{T}_{i-1}\otimes\mathbf{T}_{i-1})\hat{\boldsymbol{\alpha}}_i^C\end{aligned} \tag{107}$$

We require the averages of these sums over all configurations. The combined result may be generated by multiplication of matrices, one for each skeletal bond, the procedure in this case being closely similar to that employed for generation of the fourth moment of r (Chap. IV, Sect. 8). Thus,[29]

$$\sum_i \langle\mathbf{r}^T\hat{\boldsymbol{\alpha}}_i\mathbf{r}\rangle_0 = 2Z^{-1}\mathscr{J}^*\mathscr{Q}_1^{(n+1)}\mathscr{J} \tag{108}$$

where

$$\mathcal{Q}_i = \begin{bmatrix} \mathbf{U} & (\mathbf{U}\otimes\mathbf{l}^T)\,\|\mathbf{T}\| & (\mathbf{U}\otimes\hat{\boldsymbol{\alpha}}^R)\,\|\mathbf{T}\otimes\mathbf{T}\| & \frac{1}{2}(\mathbf{U}\otimes\mathbf{l}^T\otimes\mathbf{l}^T)\,\|\mathbf{T}\otimes\mathbf{T}\| & \mathbf{U}\otimes[\hat{\boldsymbol{\alpha}}^R(\mathbf{l}\otimes\mathbf{E}_3)]\,\|\mathbf{T}\| & \frac{1}{2}\mathbf{U}[\hat{\boldsymbol{\alpha}}^R(\mathbf{l}\otimes\mathbf{l})] \\ \mathbf{0} & (\mathbf{U}\otimes\mathbf{E}_3)\,\|\mathbf{T}\| & \mathbf{0} & (\mathbf{U}\otimes\mathbf{E}_3\otimes\mathbf{l}^T)\,\|\mathbf{T}\otimes\mathbf{T}\| & (\mathbf{U}\otimes\hat{\boldsymbol{\alpha}})\,\|\mathbf{T}\| & \mathbf{U}\otimes[(\mathbf{E}_3\otimes\mathbf{l}^T)\hat{\boldsymbol{\alpha}}^C] \\ \mathbf{0} & \mathbf{0} & (\mathbf{U}\otimes\mathbf{E}_9)\,\|\mathbf{T}\otimes\mathbf{T}\| & \mathbf{0} & (\mathbf{U}\otimes\mathbf{l}\otimes\mathbf{E}_3)\,\|\mathbf{T}\| & \frac{1}{2}(\mathbf{U}\otimes\mathbf{l}\otimes\mathbf{l}) \\ \mathbf{0} & \mathbf{0} & \mathbf{0} & (\mathbf{U}\otimes\mathbf{E}_9)\,\|\mathbf{T}\otimes\mathbf{T}\| & \mathbf{0} & \mathbf{U}\otimes\hat{\boldsymbol{\alpha}}^C \\ \mathbf{0} & \mathbf{0} & \mathbf{0} & \mathbf{0} & (\mathbf{U}\otimes\mathbf{E}_3)\,\|\mathbf{T}\| & \mathbf{U}\otimes\mathbf{l} \\ \mathbf{0} & \mathbf{0} & \mathbf{0} & \mathbf{0} & \mathbf{0} & \mathbf{U} \end{bmatrix}_i \qquad (109)$$

Just as in the calculation of $\langle \gamma^2 \rangle$ entering into the treatment of depolarized scattering, $\mathbf{U}_{n+1} = \mathbf{E}_\nu$; if the terminal groups are symmetric, $\mathbf{U}_1 = \mathbf{U}_n = \mathbf{E}_\nu$ as well. Only the first row of $\mathscr{Q}_1$ is required; hence, elements of $\mathbf{T}_1$ which would depend on ϕ_1 need not be specified. Similarly, $\mathbf{T}_{n+1}$ is neither defined nor required.

Having evaluated the essential quantity according to Eq. 108, we may obtain Γ_2 from Eq. 101. Substitution into Eqs. 87–92 yields the strain birefringence in first approximation, appropriate for networks of long chains and for small deformations.[29]

The dependence of the dichroic ratio on the strain admits of similar treatment. Consider a macromolecule consisting of units having chromophoric groups which absorb radiation of a given wavelength. The quantity required is the average orientation of the transition moments of these chromophores with respect to the vector **r**. This quantity may be expressed as the average of the square of the cosine of the angle between the transition moment and **r**. Hence, the transition moment may be treated as a Cartesian tensor having one nonzero principal component. If its orientation relative to bonds of the chain skeleton is defined by the molecular structure, then the difference between the extinction coefficients parallel and perpendicular to **r**, this difference being analogous to $\Delta\alpha_r$ above, is given by an equation like Eq. 100. The coefficient Γ_2 may be obtained to terms of first order by use of Eqs. 101, 108, and 109.

4. ELECTRIC BIREFRINGENCE OF POLYMER CHAINS: THE KERR EFFECT

The birefringence induced by a homogeneous electric field acting on a system of linear polymer molecules, in a dilute solution, for example, has been treated rigorously to terms of first order by Nagai and Ishikawa.[28] Averages of the molecular quantities entering into these terms can be taken over all configurations of the chain,[30] as we proceed to show, by resort to the methods applied to the analysis of various configuration-dependent properties in earlier chapters and especially in this one. We first present Nagai and Ishikawa's[28] derivation of the relationship between $\Delta\alpha_F$, the difference between polarizabilities in the direction of the field and perpendicular to it, and the field strength F. The derivation proceeds along the lines of the treatment of the strain birefringence given in the preceding section. Essentially, a perturbation method is employed. The electric field acts on the molecule through its permanent electric moment $\boldsymbol{\mu}$ and also through its static polarizability, which we designate by $\boldsymbol{\alpha}'$ to distinguish it from the optical polarizability $\boldsymbol{\alpha}$ introduced in Section 2 and used further in Section 3. The action of the field on the molecular dipole moment $\boldsymbol{\mu}$ is quite analogous to the effect of the tension f which, as was shown in Section 3, produces a mean extension of the chain specified by $\boldsymbol{\lambda}$ and an average difference $\Delta\alpha_r$ in polarizabilities parallel and perpendicular to **f**, or to $\langle \mathbf{r} \rangle$. The role of the static polarizability $\boldsymbol{\alpha}'$ has no counterpart in the strain birefringence.

Let the electric field F act along the X axis. Then the energy of the molecule in a given configuration is

$$E_F = E - \mu_x F - (\alpha'_{xx}/2)F^2 \tag{110}$$

Here E is the internal energy determined by the set of internal bond rotation angles $\{\phi\}$ as previously identified. The second and third terms represent the electrostatic energy; μ_x and α'_{xx} are the components of $\boldsymbol{\mu}$ and of $\boldsymbol{\alpha}'$ along the X axis. The molecular partition function (see Eqs. VIII-4 and VIII-6) in the presence of the field is

$$Z_F = (8\pi^2)^{-1} \int \cdots \int \exp(-E_F/kT) \sin\chi \, d\chi \, d\psi \, d\omega \, d\{\phi\} \tag{111}$$

Substitution of Eq. 110 for E_F and series expansion of the exponential to terms in F^2 yields

$$Z_F \cong (8\pi^2)^{-1} \int \cdots \int [1 + (\mu_x F/kT) + (\tfrac{1}{2})(\mu_x F/kT)^2 + (\tfrac{1}{2})(\alpha'_{xx} F^2/kT)] \exp(-E/kT) \sin\chi \, d\chi \, d\psi \, d\omega \, d\{\phi\} \tag{112}$$

$$= Z[1 + (\tfrac{1}{2})\langle \mu_x{}^2 \rangle_0 (F/kT)^2 + (\tfrac{1}{2})\langle \alpha'_{xx} \rangle_0 (F^2/kT)] \tag{113}$$

where Z is the partition function, defined by Eq. VIII-4, in the absence of a field F. Angle brackets with the subscript zero again denote averages for the unperturbed chain ($F = 0$). The average over μ_x vanishes in Eq. 112. It follows also from the symmetry of space that $\langle \mu_x{}^2 \rangle_0 = (\tfrac{1}{3})\langle \mu^2 \rangle_0$ and that $\langle \alpha'_{xx} \rangle_0 = (\tfrac{1}{3})\langle \text{trace}\, \boldsymbol{\alpha}' \rangle = (\tfrac{1}{3})$ trace $\boldsymbol{\alpha}'$. Angle brackets are deleted in the last expression on the grounds that trace $\boldsymbol{\alpha}'$ is invariant to the chain configuration if $\boldsymbol{\alpha}'$ is additive in the group contributions $\boldsymbol{\alpha}_i'$. With substitution of these results in Eq. 113, we have

$$Z_F = Z[1 + (\tfrac{1}{6})\langle \mu^2 \rangle_0 (F/kT)^2 + (\tfrac{1}{6})(\text{trace}\, \boldsymbol{\alpha}')(F^2/kT)] \tag{114}$$

to the same approximations as above.

We require the difference $\Delta\alpha_F$ between the polarizabilities parallel and perpendicular to the field, and this difference must be averaged over all orientations and all internal configurations of the chain subject to the field F.

$$\Delta\alpha_F \equiv \langle \alpha_{xx} - \alpha_{yy} \rangle_F = (8\pi^2 Z_F)^{-1} \int \cdots \int (\alpha_{xx} - \alpha_{yy}) \exp(-E_F/kT) \sin\chi \, d\chi \, d\psi \, d\omega \, d\{\phi\} \tag{115}$$

Substitution of Eq. 110 for E_F and expansion of the exponential to terms in F^2 as above gives

$$\Delta\alpha_F \cong (8\pi^2 Z_F)^{-1} \int \cdots \int (\alpha_{xx} - \alpha_{yy})[1 + \mu_x F/kT + (\tfrac{1}{2})\mu_x{}^2(F/kT)^2 + (\tfrac{1}{2})\alpha'_{xx}(F^2/kT)]\exp(-E/kT) \times \sin\chi\, d\chi\, d\psi\, d\omega\, d\{\phi\} \tag{116}$$

$$= (\tfrac{1}{2})(Z/Z_F)[\langle(\alpha_{xx} - \alpha_{yy})\mu_x{}^2\rangle_0 (F/kT)^2 + \langle(\alpha_{xx} - \alpha_{yy})\alpha'_{xx}\rangle_0 F^2/kT] \tag{117}$$

Let $\alpha_1, \alpha_2, \alpha_3$ and $\alpha_1', \alpha_2', \alpha_3'$ denote the principal components of $\boldsymbol{\alpha}$ and of $\boldsymbol{\alpha}'$, respectively. We first take the averages of the quantities in angle brackets over all spatial orientations χ, ψ, ω, for a fixed internal configuration $\{\phi\}$. This average of the coefficient of the first term in square brackets in Eq. 117 is*

$$\langle(\alpha_{xx} - \alpha_{yy})\mu_x{}^2\rangle_{\{\phi\}} = \sum_{t=1,2,3} [\alpha_t \mu^2 \langle(\boldsymbol{\alpha}_t, \mathbf{X})^2(\boldsymbol{\mu}, \mathbf{X})^2\rangle_{\{\phi\}} - \langle(\boldsymbol{\alpha}_t, \mathbf{Y})^2(\boldsymbol{\mu}, \mathbf{X})^2\rangle_{\{\phi\}}] \tag{118}$$

where $(\boldsymbol{\alpha}_t, \mathbf{X})$ denotes the cosine of the angle between principal axis t of $\boldsymbol{\alpha}$ and the X axis of the laboratory reference frame, etc. Performance of the required averaging over χ, ψ, and ω yields

$$\langle(\alpha_{xx} - \alpha_{yy})\mu_x{}^2\rangle_{\{\phi\}} = (\mu^2/15)\sum_t \alpha_t\{[2(\boldsymbol{\alpha}_t, \boldsymbol{\mu})^2 + 1] - [2 - (\boldsymbol{\alpha}_t, \boldsymbol{\mu})^2]\}$$

$$= (\mu^2/15)\sum_t \alpha_t[3(\boldsymbol{\alpha}_t, \boldsymbol{\mu})^2 - 1]$$

$$= (\tfrac{1}{15})[3\boldsymbol{\mu}^T\boldsymbol{\alpha}\boldsymbol{\mu} - \mu^2 \text{ trace } \boldsymbol{\alpha}]$$

Hence

$$\langle(\alpha_{xx} - \alpha_{yy})\mu_x{}^2\rangle_0 = (\tfrac{1}{15})[3\langle\boldsymbol{\mu}^T\boldsymbol{\alpha}\boldsymbol{\mu}\rangle_0 - \langle\mu^2 \text{ trace } \boldsymbol{\alpha}\rangle_0] \tag{119}$$

A similar rationalization of the coefficient of the second term in the square-bracketed expression in Eq. 117 leads to

$$\langle(\alpha_{xx} - \alpha_{yy})\alpha'_{xx}\rangle_{\{\phi\}} = (\tfrac{1}{15})\sum_{s,t} \alpha_s \alpha_t'[3(\boldsymbol{\alpha}_s, \boldsymbol{\alpha}_t')^2 - 1]$$

$$\langle(\alpha_{xx} - \alpha_{yy})\alpha'_{xx}\rangle_0 = (\tfrac{1}{15})[3\langle\text{trace } (\boldsymbol{\alpha}\boldsymbol{\alpha}')\rangle_0 - \langle(\text{trace } \boldsymbol{\alpha})(\text{trace } \boldsymbol{\alpha}')\rangle_0] \tag{120}$$

* Departing from the convention followed throughout the text, we use angle brackets in Eq. 118 and several of the succeeding equations to denote averages performed over spatial orientations only, the chains being fixed in the configuration $\{\phi\}$ as noted.

Substitution of Eqs. 119 and 120 into Eq. 117 yields Nagai and Ishikawa's result[28]

$$\Delta\alpha_F = (\tfrac{1}{30})\{[3\langle \boldsymbol{\mu}^T \boldsymbol{\alpha}\boldsymbol{\mu}\rangle_0 - \langle \mu^2 \text{ trace } \boldsymbol{\alpha}\rangle_0](F/kT)^2 + [3\langle \text{trace } (\boldsymbol{\alpha}\boldsymbol{\alpha}')\rangle_0 - \langle (\text{trace } \boldsymbol{\alpha})(\text{trace } \boldsymbol{\alpha}')\rangle_0](F^2/kT)\} \tag{121}$$

The factor Z/Z_F has been replaced by unity, as is consistent with the omission of terms of higher order. Equation 121 may be written alternatively as follows:

$$\Delta\alpha_F = (\tfrac{1}{10})[\langle \boldsymbol{\mu}^T \hat{\boldsymbol{\alpha}}\boldsymbol{\mu}\rangle_0 (F/kT)^2 + \langle \text{trace } (\hat{\boldsymbol{\alpha}}\hat{\boldsymbol{\alpha}}')\rangle_0 F^2/kT] \tag{122}$$

where $\hat{\boldsymbol{\alpha}} = \boldsymbol{\alpha} - \bar{\alpha}\mathbf{E}_3$ and $\hat{\boldsymbol{\alpha}}' = \boldsymbol{\alpha}' - \bar{\alpha}'\mathbf{E}_3$.

From the Lorentz-Lorenz relation we have (see Eq. 86) for the difference between refractive indices parallel and perpendicular to the field,

$$\Delta\tilde{n} = \frac{(\tilde{n}^2+2)^2}{6\tilde{n}} \frac{4\pi}{3} N_A \left(\frac{c}{M}\right) \Delta\alpha_F = \frac{(\tilde{n}^2+2)^2}{6\tilde{n}} K \left(\frac{c}{M}\right) F^2 \tag{123}$$

where c is the concentration in grams per cubic centimeter, M is the molecular weight of the polymer, N_A is Avogadro's number, and K is the molar Kerr constant, i.e.,

$$K = (4\pi/3) N_A \, \Delta\alpha_F / F^2 \tag{124}$$

From Eq. 122 we have[30]

$$K = (2\pi N_A / 15kT)[\langle \boldsymbol{\mu}^T \hat{\boldsymbol{\alpha}}\boldsymbol{\mu}\rangle_0 (kT)^{-1} + \langle \text{trace } (\hat{\boldsymbol{\alpha}}\hat{\boldsymbol{\alpha}}')\rangle_0] \tag{125}$$

The quantity appearing in the first term in Eq. 125 is the analog of $\langle \mathbf{r}^T \hat{\boldsymbol{\alpha}} \mathbf{r}\rangle_0$ treated in the preceding section, and it can be evaluated according to Eqs. 108 and 109, with group dipole moments $\mathbf{m}_i$ replacing bond vectors $\mathbf{l}_i$. The second term in brackets in Eq. 125 can be written

$$\langle \text{trace } (\hat{\boldsymbol{\alpha}}\hat{\boldsymbol{\alpha}}')\rangle_0 = \langle \hat{\boldsymbol{\alpha}}^R \hat{\boldsymbol{\alpha}}'^C \rangle_0 \tag{126}$$

where the superscripts R and C denote row and column forms of $\hat{\boldsymbol{\alpha}}$ and $\hat{\boldsymbol{\alpha}}'$, in accordance with previous usage (see Eq. 103). It will be apparent that

$$\langle \hat{\boldsymbol{\alpha}}^R \hat{\boldsymbol{\alpha}}'^C \rangle_0 = Z^{-1} \mathscr{J}^* \mathscr{A}_1^{(n+1)} \mathscr{J} \tag{127}$$

where

$$\mathscr{A}_i = \begin{bmatrix} \mathbf{U} & (\mathbf{U} \otimes \hat{\boldsymbol{\alpha}}^R)\|\mathbf{T} \otimes \mathbf{T}\| & (\mathbf{U} \otimes \hat{\boldsymbol{\alpha}}'^R)\|\mathbf{T} \otimes \mathbf{T}\| & (\hat{\boldsymbol{\alpha}}^R \hat{\boldsymbol{\alpha}}'^C)\mathbf{U} \\ \mathbf{0} & (\mathbf{U} \otimes \mathbf{E}_3)\|\mathbf{T} \otimes \mathbf{T}\| & \mathbf{0} & \mathbf{U} \otimes \hat{\boldsymbol{\alpha}}'^C \\ \mathbf{0} & \mathbf{0} & (\mathbf{U} \otimes \mathbf{E}_3)\|\mathbf{T} \otimes \mathbf{T}\| & \mathbf{U} \otimes \hat{\boldsymbol{\alpha}}^C \\ \mathbf{0} & \mathbf{0} & \mathbf{0} & \mathbf{U} \end{bmatrix}_i \tag{128}$$

The statistical weight matrix $\mathbf{U}_{n+1}$ is to be represented by the identity $\mathbf{E}_\nu$; replacement of $\mathbf{U}_1$ and $\mathbf{U}_n$ by $\mathbf{E}_\nu$ may also be appropriate if the terminal group is symmetrical. Equations 127 and 128 are formulated for chains in which the succession of units from one end to the other is differentiable from the reverse direction, i.e., for chains devoid of end-for-end symmetry. If the chain is structurally symmetrical in this respect, the third pseudorow and third pseudocolumn may be stricken from Eq. 128; a factor of 2 must then be introduced in Eq. 127.

The Kerr constant K may be calculated by use of Eqs. 125, 127, and 128, in conjunction with Eqs. 108 and 109 of the preceding section, provided that the necessary data are available. Besides the geometrical parameters required to define the transformations $\mathbf{T}$ and the information needed to formulate the statistical weights in $\mathbf{U}$, the group dipole moments $\mathbf{m}_i$ and the anisotropic parts of the polarizability tensors $\boldsymbol{\alpha}_i$ and $\boldsymbol{\alpha}_i'$ for groups making up the chain must be known.

In keeping with treatments of other properties of chain molecules by the foregoing methods, no approximations have been introduced beyond the adoption of the rotational isomeric state model and the assumption of additivity of group polarizability tensors, taken to be invariant to the chain configuration when expressed in the coordinate reference frame affixed to the associated skeletal bond.

REFERENCES

1. P. Debye, *Ann. Physik*, **46**, 809 (1915).
2. P. Debye, *J. Phys. Chem.*, **51**, 18 (1947).
3. O. Kratky and G. Porod, *Rec. Trav. Chim.*, **68**, 1106 (1949).
4. P. J. Flory and R. L. Jernigan, *J. Am. Chem. Soc.*, **90**, 3128 (1968).
5. S. Heine, O. Kratky, G. Porod, and P. J. Schmitz, *Makromol. Chem.*, **46**, 682 (1961); O. Kratky, *Pure Appl. Chem.*, **12**, 483 (1966).
6. A. Peterlin, *J. Polymer Sci.*, **47**, 403 (1960); see also, Proc. Interdisciplinary Conf. Electromagnetic Scattering, Potsdam, N. Y., 1962, p. 357 (1963).
7. S. Bhagavantam, *Scattering of Light and the Raman Effect*, Chemical Publishing Co. Inc., Brooklyn, N. Y., 1942, pp. 27–35.
8. H. C. Van de Hulst, *Light Scattering by Small Particles*, Wiley, New York, 1957, pp. 79–81.

9. E. V. Chalam, *Proc. Indian Acad. Sci.*, **15A**, 190 (1942).
10. R. P. Smith and E. M. Mortensen, *J. Chem. Phys.*, **32**, 502 (1960).
11. C. Clement and J. Bothorel, *J. Chim. Phys.*, **61**, 878, 1262 (1964).
12. M. V. Volkenstein, *Configurational Statistics of Polymeric Chains* (translated from the Russian ed., S. N. Timasheff and M. J. Timasheff), Interscience, New York, 1963, Chap. 7.
13. J. Powers, D. A. Keedy, and R. S. Stein, *J. Chem. Phys.*, **35**, 376 (1961).
14. R. L. Rowell and R. S. Stein, *J. Chem. Phys.*, **47**, 2985 (1967).
15. K. S. Pitzer, in *Advances in Chemical Physics*, Vol. 2, I. Prigogine, Ed., Interscience, New York, 1959, pp. 79–81.
16. R. L. Jernigan and P. J. Flory, *J. Chem. Phys.*, **47**, 1999 (1967); see also, R. L. Jernigan, Ph.D. thesis, Stanford University, 1967.
17. P. J. Flory and R. L. Jernigan, *J. Chem. Phys.*, **42**, 3509 (1965).
18. K. G. Denbigh, *Trans. Faraday Soc.*, **36**, 946 (1940).
19. H. A. Stuart, in *Landolt-Bornstein Zahlenwerte und Funktionen*, 6th ed., Vol. II, Part 8, Springer-Verlag, Berlin, 1962, p. 816.
20. P. J. Flory, *Principles of Polymer Chemistry*, Cornell Univ. Press, Ithaca, N. Y., 1953, Chap. XI.
21. A. S. Lodge, *Rheology of Elastomers*, Pergamon Press, London, 1958, p. 70.
22. P. J. Flory, *J. Am. Chem. Soc.*, **78**, 5222 (1956).
23. P. J. Flory, *Trans. Faraday Soc.*, **56**, 722 (1960); **57**, 829 (1961).
24. W. Kuhn and F. Grün, *Kolloid-Z.*, **101**, 248 (1942).
25. L. R. G. Treloar, *The Physics of Rubber Elasticity*, 2nd ed., Clarendon Press, Oxford, 1958, Chap. X.
26. Yu. Ya. Gotlib, M. V. Volkenstein, and E. K. Byutner, *Dokl. Akad. Nauk SSSR*, **99**, 935 (1954); Yu. Ya. Gotlib, *Zh. Tekhn. Fiz.*, **27**, 707 (1957).
27. K. Nagai, *J. Chem. Phys.*, **40**, 2818 (1964).
28. K. Nagai and T. Ishikawa, *J. Chem. Phys.*, **43**, 4508 (1965).
29. P. J. Flory, R. L. Jernigan, and A. E. Tonelli, *J. Chem. Phys.*, **48**, 3822 (1968).
30. P. J. Flory and R. L. Jernigan, *J. Chem. Phys.*, **48**, 3823 (1968).

APPENDIX A

The Theorem of Lagrange

The theorem of Lagrange,[1-3] first published in 1783, related the center of gravity for a system of masses to the distances between their centers taken pairwise. Equation I-6 is the special case of that theorem as applied to a system of particles of equal mass.[2] The derivation for this special case, simplified by the use of vector algebra unknown in Lagrange's time, follows.[3]

Let $\mathbf{s}_i$ be the vector from the center of gravity to chain atom i, and let $\mathbf{r}_{0i}$ be the vector from the zeroth atom to the ith of the chain in its specified configuration. Then

$$\mathbf{s}_i = \mathbf{s}_0 + \mathbf{r}_{0i} \tag{1}$$

$\mathbf{s}_0$ being the vector leading from the center of gravity to the zeroth atom. The square of the radius of gyration is by definition (see Eq. I-5)

$$s^2 = (n+1)^{-1} \sum_0^n s_i^2 = (n+1)^{-1} \sum_0^n \mathbf{s}_i \cdot \mathbf{s}_i \tag{2}$$

Substitution of Eq. 1 into Eq. 2 gives

$$\begin{aligned} s^2 &= (n+1)^{-1} \sum_0^n (\mathbf{s}_0 + \mathbf{r}_{0i}) \cdot (\mathbf{s}_0 + \mathbf{r}_{0i}) \\ &= s_0^2 + 2(n+1)^{-1}\mathbf{s}_0 \cdot \left[\sum_1^n \mathbf{r}_{0i}\right] + (n+1)^{-1} \sum_1^n r_{0i}^2 \end{aligned} \tag{3}$$

Inasmuch as

$$\sum_0^n \mathbf{s}_i = 0$$

we have from Eq. 1

$$\mathbf{s}_0 = -(n+1)^{-1} \sum_1^n \mathbf{r}_{0i} \tag{4}$$

and

$$s_0^2 = (n+1)^{-2} \sum_{i=1}^n \sum_{j=1}^n \mathbf{r}_{0i} \cdot \mathbf{r}_{0j} \tag{4'}$$

Substitution of these results in Eq. 3 yields

$$s^2 = (n+1)^{-1} \sum_{1}^{n} r_{0i}^2 - (n+1)^{-2} \sum_{i=1}^{n} \sum_{j=1}^{n} \mathbf{r}_{0i} \cdot \mathbf{r}_{0j} \tag{5}$$

Now by the law of cosines

$$\mathbf{r}_{0i} \cdot \mathbf{r}_{0j} = (r_{0i}^2 + r_{0j}^2 - r_{ij}^2)/2$$

By substitution into Eq. 5, we find

$$\begin{aligned} s^2 &= (n+1)^{-2} \left[\sum_{i=1}^{n} r_{0i}^2 + \frac{1}{2} \sum_{i=1}^{n} \sum_{j=1}^{n} r_{ij}^2 \right] \\ &= (\tfrac{1}{2})(n+1)^{-2} \sum_{i=0}^{n} \sum_{j=0}^{n} r_{ij}^2 \\ &= (n+1)^{-2} \sum_{0 \leqslant i < j \leqslant n} r_{ij}^2 \end{aligned} \tag{6}$$

which is Lagrange's result given by Eq. I-6 of the text.

REFERENCES

1. J. L. Lagrange, *Oeuvres* (*Paris*), **5**, 535 (1870).
2. R. A. Sack, *Nature*, **171**, 310 (1953).
3. H. Lamb, *Statics*, 3rd ed., Cambridge Univ. Press, 1928, p. 166.

APPENDIX B

The Axis Transformation Matrix $\mathbf{T}_i$ Relating Coordinate Systems of Consecutive Skeletal Bonds and Diagonalization of its Average for Bonds Subject to Independent Rotational Potentials

Consider a unit vector having its direction parallel to the axis x_{i+1} in Fig. I-3. Its components in the Cartesian reference frame i, as defined on p. 20, will be found to be

$$\begin{bmatrix} \cos\theta_i \\ \sin\theta_i \cos\phi_i \\ \sin\theta_i \sin\phi_i \end{bmatrix}$$

where they are presented in the order x_i, y_i, z_i. Similarly, the components in reference frame i of a unit vector along axis y_{i+1} in Fig. I-3 are

$$\begin{bmatrix} \sin\theta_i \\ -\cos\theta_i \cos\phi_i \\ -\cos\theta_i \sin\phi_i \end{bmatrix}$$

and those of a unit vector along z_{i+1} are

$$\begin{bmatrix} 0 \\ \sin\phi_i \\ -\cos\phi_i \end{bmatrix}$$

It follows that any vector $\mathbf{v}$ with components v_x, v_y, v_z in the coordinate system $i+1$ has components in coordinate system i given by

$$\begin{aligned} v_x' &= v_x \cos\theta_i + v_y \sin\theta_i \\ v_y' &= v_x \sin\theta_i \cos\phi_i - v_y \cos\theta_i \cos\phi_i + v_z \sin\phi_i \\ v_z' &= v_x \sin\theta_i \sin\phi_i - v_y \cos\theta_i \sin\phi_i - v_z \cos\phi_i \end{aligned}$$

The same result is expressed succinctly in matrix notation as follows:

$$\mathbf{v}' = \mathbf{T}_i \mathbf{v} \qquad (1)$$

where $\mathbf{T}_i$ is the orthogonal matrix

$$\mathbf{T}_i = \begin{bmatrix} \cos\theta_i & \sin\theta_i & 0 \\ \sin\theta_i \cos\phi_i & -\cos\theta_i \cos\phi_i & \sin\phi_i \\ \sin\theta_i \sin\phi_i & -\cos\theta_i \sin\phi_i & -\cos\phi_i \end{bmatrix} \tag{2}$$

appearing in the text as Eq. I-25, and $\mathbf{v}$ and $\mathbf{v}'$ are expressed as column vectors, e.g.,

$$\mathbf{v} = \begin{bmatrix} v_x \\ v_y \\ v_z \end{bmatrix} \tag{3}$$

The matrix $\mathbf{T}_i$ used as in Eq. 1 effects the transformation required to convert the components of a vector presented in the Cartesian reference frame $i + 1$ to its components in reference frame i. The form of transformation $\mathbf{T}_i$ is contingent upon the definition of reference frames for consecutive bonds, e.g., bonds i and $i + 1$, adopted in Chapter I.

The transformation represented by $\mathbf{T}_i$ and defined by Eq. 2 may be formulated in more general terms as follows. Consider two Cartesian reference frames xyz and $x'y'z'$ corresponding to $x_{i+1}\, y_{i+1}\, z_{i+1}$ and $x_i y_i z_i$, respectively. Let them be so related that the former can be brought into coincidence with the latter (or rendered parallel to it) by execution of two rotations as follows: (*1*) a rotation of xyz about its z axis through the angle τ that renders x coincident with (or parallel to) x', and (*2*) the subsequent rotation of xyz through an angle ρ about the x axis shared in common by the two reference frames as a result of (*1*). (The complete Eulerian transformation, which would involve rotation about a third axis noncoplanar with the other two, is not required for the purposes at hand.) The angles τ and ρ are measured in the right-handed sense. The matrix $\mathbf{R}(\tau, \rho)$ that will transform a vector $\mathbf{v}$ represented in xyz to its representation $\mathbf{v}'$ in $x'y'z'$ comprises the resultant of the operations (*1*) and (*2*) carried out in the order specified. It follows that

$$\mathbf{R}(\tau, \rho) = \begin{bmatrix} 1 & 0 & 0 \\ 0 & \cos\rho & \sin\rho \\ 0 & -\sin\rho & \cos\rho \end{bmatrix} \begin{bmatrix} \cos\tau & \sin\tau & 0 \\ -\sin\tau & \cos\tau & 0 \\ 0 & 0 & 1 \end{bmatrix}$$

$$= \begin{bmatrix} \cos\tau & \sin\tau & 0 \\ -\sin\tau \cos\rho & \cos\tau \cos\rho & \sin\rho \\ \sin\tau \sin\rho & -\cos\tau \sin\rho & \cos\rho \end{bmatrix} \tag{4}$$

with $\mathbf{v}' = \mathbf{R}\mathbf{v}$. Alternatively, $\mathbf{R}$ may be looked upon as the transformation that rotates a vector $\mathbf{v}$ into $\mathbf{v}'$, the coordinate frame remaining fixed. In these terms, $\mathbf{v}$ is rotated through angles $-\tau$ and $-\rho$ about z and x, respectively, and in the order stated.

In the special case of the right-handed reference frames $i+1$ and i affixed to consecutive skeletal bonds of a chain molecule according to the conventions introduced on p. 20 (see Fig. I-3) and used throughout succeeding chapters, $\tau = \theta_i$ and $\rho = \pi - \phi_i$. Systematic definition of reference frames according to these conventions entails a reversal of the directions of y and z axes for successive bonds when $\theta_i = 0$ and $\phi_i = 0$. It is on this account that ρ is displaced by 180° from $-\phi_i$, the negative value of the bond rotation angle. Thus, ϕ_i corresponds to $\pi - \rho$ and

$$\mathbf{T}_i = \mathbf{R}(\theta_i, \pi - \phi_i) \tag{5}$$

This identification may be verified by substitution of Eq. 4 into Eq. 5 to obtain Eq. 2 above.

The result of averaging the elements of $\mathbf{T}_i$ over all angles ϕ_i subject to a symmetric rotational potential which is independent of neighboring rotations is given by Eq. I-46. Omitting the serial subscript index, we may write this as

$$\langle \mathbf{T} \rangle = \begin{bmatrix} \alpha & \beta & 0 \\ \beta\eta & -\alpha\eta & 0 \\ 0 & 0 & -\eta \end{bmatrix} \tag{6}$$

where $\alpha = \cos\theta$, $\beta = \sin\theta = \sqrt{1-\alpha^2}$, and $\eta = \langle \cos\phi \rangle$. This matrix is not orthogonal, as will be apparent, for any value of $\eta^2 < 1$; i.e., if used in Eq. 1 it would contract the vector $\mathbf{v}$ as well as change its orientation.

Let us assume that $\langle \mathbf{T} \rangle$ can be converted to diagonal form by a similarity transformation*

$$\mathbf{A}^{-1}\langle \mathbf{T} \rangle \mathbf{A} = \begin{bmatrix} \lambda_1 & 0 & 0 \\ 0 & \lambda_2 & 0 \\ 0 & 0 & \lambda_3 \end{bmatrix} \equiv \mathbf{\Lambda} \tag{7}$$

That is, we postulate the existence of a matrix $\mathbf{A}$ which will accomplish this result. Without entering into the conditions which are necessary to assure that $\langle \mathbf{T} \rangle$ is reducible in this manner, we note that the similarity transformation is tantamount to alteration of the adjoining reference frames $i+1$ and i, each by the same transformation $\mathbf{A}^{-1}$. Hence, from a geometrical standpoint, transformation according to Eq. 7 recasts the operand, i.e., matrix $\langle \mathbf{T} \rangle$ in this case, in the coordinate system in which it assumes diagonal form. For later convenience we rewrite this equation as follows:

$$\mathbf{B}\langle \mathbf{T} \rangle \mathbf{A} = \mathbf{\Lambda} \tag{7'}$$

*The procedure for diagonalization of this matrix is set forth in detail mainly for purposes of illustration. It was first applied to the present problem by Oka,[1] and independently by Benoit[2] and also by Volkenstein and Ptitsyn.[3,4]

where $\mathbf{B} = \mathbf{A}^{-1}$, i.e., $\mathbf{AB} = \mathbf{E}$ where

$$\mathbf{E} = \begin{bmatrix} 1 & 0 & 0 \\ 0 & 1 & 0 \\ 0 & 0 & 1 \end{bmatrix}$$

is the identity matrix.

Premultiplication of Eq. 7 or 7′ by $\mathbf{A}$ gives

$$\langle \mathbf{T} \rangle \mathbf{A} = \mathbf{A}\boldsymbol{\Lambda}$$

which may be separated into three vector equations

$$\langle \mathbf{T} \rangle \mathbf{A}_k = \mathbf{A}_k \lambda_k, \qquad k = 1, 2, 3 \tag{8}$$

where the $\mathbf{A}_k$ are the column *eigenvectors*

$$\mathbf{A}_k = \begin{bmatrix} A_{1k} \\ A_{2k} \\ A_{3k} \end{bmatrix}; \qquad k = 1, 2, 3 \tag{9}$$

These eigenvectors of $\langle \mathbf{T} \rangle$ have the property (see Eq. 8) of being altered only by scalar factors λ_k when they are premultiplied by the parent matrix $\langle \mathbf{T} \rangle$. The λ_k are the *eigenvalues*, or *characteristic values*, of this matrix.

Similarly, postmultiplication of Eq. 7′ by $\mathbf{B} = \mathbf{A}^{-1}$ gives

$$\mathbf{B}\langle \mathbf{T} \rangle = \boldsymbol{\Lambda}\mathbf{B}$$

or

$$\mathbf{B}_k^* \langle \mathbf{T} \rangle = \lambda_k \mathbf{B}_k^*, \qquad k = 1, 2, 3 \tag{10}$$

where

$$\mathbf{B}_k^* = [B_{k1} \quad B_{k2} \quad B_{k3}] \tag{11}$$

are the *eigenrows* of $\langle \mathbf{T} \rangle$. Since $\mathbf{AB} = \mathbf{E}$, the eigenvectors and eigenrows as here defined satisfy the condition of mutual normalization and orthogonality

$$\mathbf{B}_j^* \mathbf{A}_k = \delta_{jk}$$

δ_{jk} being the Kronecker delta, which is equal to unity for $j = k$ and to zero for $j \neq k$.

Equation 8 can be written

$$(\langle \mathbf{T} \rangle - \lambda_k \mathbf{E})\mathbf{A}_k = 0 \tag{8′}$$

It represents three homogeneous linear equations in the components of $\mathbf{A}_k$. For $\langle \mathbf{T} \rangle$ defined by Eq. 6 these equations are

$$\begin{aligned}(\alpha - \lambda_k)A_{1k} + \beta A_{2k} &= 0\\ \beta\eta A_{1k} - (\alpha\eta + \lambda_k)A_{2k} &= 0\\ (\eta + \lambda_k)A_{3k} &= 0\end{aligned} \tag{12}$$

In order for such a set of homogeneous equations to be solvable, it is a necessary and sufficient condition that the determinant of their coefficients shall vanish. That is,

$$|\langle \mathbf{T} \rangle - \lambda_k \mathbf{E}| = 0 \tag{13}$$

where the determinant of a matrix is indicated by enclosure within vertical lines. By introducing Eq. 6 into Eq. 13, we obtain

$$\begin{vmatrix} \alpha - \lambda & \beta & 0 \\ \beta\eta & -\alpha\eta - \lambda & 0 \\ 0 & 0 & -\eta - \lambda \end{vmatrix} = 0 \tag{14}$$

or

$$[\lambda^2 - \alpha(1-\eta)\lambda - \eta](\eta + \lambda) = 0 \tag{15}$$

β having been eliminated by use of $\alpha^2 + \beta^2 = 1$. The quantity on the left-hand side of Eqs. 13 and 14 is the *secular determinant*, and Eqs. 13, 14, and 15 are expressions of the *secular equation*, or *characteristic equation*, of $\langle \mathbf{T} \rangle$. The eigenvalues of $\langle \mathbf{T} \rangle$ are obtained as the roots of the characteristic equation. We thus find

$$\begin{aligned}\lambda_{1,2} &= (1/2)\left[\alpha(1-\eta) \pm \sqrt{\alpha^2(1-\eta)^2 + 4\eta}\right]\\ \lambda_3 &= -\eta\end{aligned} \tag{16}$$

where the plus and minus signs distinguish λ_1 from λ_2, respectively.

The set of homogeneous Eqs. 12 may now be solved for A_{1k}, A_{2k}, and A_{3k} in terms of the eigenvalues. For $k = 1$ or 2, the first and second of Eqs. 12 can be shown to be identical according to Eq. 15 or 16. That is,

$$A_{2k} = [(\lambda_k - \alpha)/\beta]A_{1k} = [\beta\eta/(\alpha\eta + \lambda_k)]A_{1k}$$

The second equality can be verified directly from Eq. 15 or 16. The third of Eqs. 12 requires $A_{3k} = 0$ for $k = 1, 2$. For $k = 3$, Eqs. 12 require $A_{13} = A_{23} = 0$, and A_{33} is unspecified, since its coefficient in the third Eq. 12 is zero.

As will be obvious from Eqs. 8, 8′, or 12, solution of the set of homogeneous linear equations for the elements of the vector $\mathbf{A}_k$ must admit of the arbitrary assignment of one of them. Accordingly, let $A_{11} = A_{12} = A_{33} = 1$. The foregoing results yield on this basis

$$\mathbf{A} = \begin{bmatrix} 1 & 1 & 0 \\ (\lambda_1 - \alpha)/\beta & (\lambda_2 - \alpha)/\beta & 0 \\ 0 & 0 & 1 \end{bmatrix} \tag{17}$$

Its inverse is readily found to be

$$\mathbf{A}^{-1} = \mathbf{B} = (\lambda_1 - \lambda_2)^{-1} \begin{bmatrix} \alpha - \lambda_2 & \beta & 0 \\ \lambda_1 - \alpha & -\beta & 0 \\ 0 & 0 & \lambda_1 - \lambda_2 \end{bmatrix} \tag{18}$$

The matrix **B** may bc obtained alternatively by rearranging Eq. 10 to

$$\mathbf{B}_k^*(\langle \mathbf{T} \rangle - \lambda_k \mathbf{E}) = 0, \qquad k = 1, 2, 3 \tag{19}$$

and treating the homogeneous equations in the eigenrow elements in similar fashion. The validity of these results can be confirmed by substituting them in Eq. 7′.

The foregoing reduction of the matrix $\langle \mathbf{T} \rangle$ to diagonal form is presented principally for purposes of illustration of standard methods which find more important applications in other connections, as for example in reduction of statistical weight matrices **U** (Chap. III). These methods are developed more fully in appropriate texts.

REFERENCES

1. S. Oka, *Proc. Phys. Math. Soc. Japan*, **24**, 657 (1942).
2. H. Benoit, *J. Chim. Phys.*, **44**, 18 (1947); H. Benoit and P. M. Doty, *J. Phys. Chem.*, **57**, 958 (1953).
3. M. V. Volkenstein and O. B. Ptitsyn, *Dokl. Akad. Nauk SSSR*, **78**, 657 (1951); *Zh. Fiz. Khim.*, **26**, 1061 (1952).
4. See also M. V. Volkenstein, *Configurational Statistics of Polymeric Chains* (translated from the Russian ed., S. N. Timasheff and M. J. Timasheff), Interscience, New York, 1963, Chap. 4.

APPENDIX C

An Alternative Reduction of the Configuration Partition Function

Inasmuch as Z is a scalar quantity, it is equal to its own trace, the trace of a matrix being defined as the sum of its diagonal elements. That is, from Eq. III-27

$$Z = \text{trace}\,(\mathbf{J}^*\mathbf{U}^{n-2}\mathbf{J}) \tag{1}$$

Now the trace is unaltered by a cyclic permutation of its argument. Hence,

$$Z = \text{trace}\,(\mathbf{U}^{n-2}\mathbf{J}\mathbf{J}^*) \tag{2}$$

The equivalence of this result to Eq. III-27 can be verified directly.

Substitution of Eq. III-28′ for $\mathbf{U}$ and a further cyclic permutation gives

$$Z = \text{trace}\,(\mathbf{\Lambda}^{n-2}\mathbf{B}\mathbf{J}\mathbf{J}^*\mathbf{A}) \tag{3}$$

Equations III-31 and III-32 can be obtained from Eq. 3 by reference to Eq. III-25 for $\mathbf{J}^*$ and $\mathbf{J}$.

It is customary in treating one-dimensional Ising problems[1–3] to introduce so-called cyclic boundary conditions whereby bond $n-1$ is assigned the same rotational state as bond 2. This requirement categorically eliminates all terms in Eq. III-27 except those occurring in diagonal elements of $\mathbf{U}^{n-2}$. These may be summed as the trace, giving just

$$Z \cong \text{trace}\,(\mathbf{U}^{n-2}) = \sum_{\zeta=1}^{\nu} \lambda_{\zeta}^{n-2} \tag{4}$$

which resembles the exact expression Eq. III-31, but with the coefficients Γ_{ζ} set equal to unity.

We have avoided the use of cyclic boundary conditions in the text, inasmuch as it is physically extraneous and mathematically unnecessary to do so. Resort to this artifice in the limit of very large n where none but the largest eigenvalue needs to be retained yields, of course, the correct asymptotic result, Eq. III-34.

REFERENCES

1. H. A. Kramers and G. H. Wannier, *Phys. Rev.*, **60**, 252 (1941).
2. G. F. Newell and E. W. Montroll, *Rev. Mod. Phys.*, **25** 353 (1953).
3. T. M. Birshtein and O. B. Ptitsyn, *Conformations of Macromolecules* (translated from the Russian ed., 1964, S. N. Timasheff and M. J. Timasheff), Interscience, New York, 1966, Chap. 4.

APPENDIX D

Macrocyclization Equilibrium[1,2]

Consider a system consisting of a preponderance of bifunctional monomer units M, i.e., units constrained to enter into bonds with two, and only two, other monomer units. The system may also contain a very small proportion of monofunctional units that provide chain ends for linear species. Following Jacobson and Stockmayer,[1] we consider the processes

$$\mathrm{M}_y\text{—} \rightarrow \mathrm{M}_{y-x}\text{—} + \text{—}\mathrm{M}_x\text{—} \tag{1}$$

$$\text{—}\mathrm{M}_x\text{—} \rightleftharpoons \text{c-}\mathrm{M}_x \tag{2}$$

where c-M_x denotes the cyclic, x-meric species. For simplicity in developing the following argument, the terminal groups of various acyclic species, including the hypothetical intermediate —M_x—, may be regarded as free radicals. Their nature is immaterial to the argument. The resultant of the sum of Eqs. 1 and 2, i.e.,

$$\mathrm{M}_y\text{—} \rightleftharpoons \mathrm{M}_{y-x}\text{—} + \text{c-}\mathrm{M}_x \tag{3}$$

is the chemical process whose equilibrium constant is sought.

The proportion of the cyclic species consisting of x units in equilibrium with linear chains is related to the probability of coincidence of the ends of a sequence of x units, and hence to the statistics of the spatial configurations of the chains under consideration. Cyclization equilibrium constants for larger rings may serve in principle as measures of the statistical configuration of chain molecules, or what is loosely referred to as chain "flexibility." This method of approach is applicable at degrees of polymerization much below the range where the familiar light scattering and viscosity methods are effective. Moreover, it is one of the few methods capable of yielding information on the configurations of chains in the bulk state. The theory presented below follows closely the original theory of Jacobson and Stockmayer,[1] cast, however, in terms of real chains rather than in terms of the freely jointed model. Also to be taken into account is the requirement of correspondence of the directions of the terminal bonds[2] (i.e., the "free valences" of —M_x—), and the consequences of this requirement will be pointed out.

The standard molar free-energy change for process 1 above can be written

$$\Delta G^\circ_{(1)} = \Delta H^\circ_{(1)} - RT \ln (4\pi/N_A\, \sigma_a\, \delta\mathbf{r}\, \delta\omega) \tag{4}$$

where N_A is Avogadro's number, and $\Delta H^\circ_{(1)}$ is the standard state heat of dissociation according to Eq. 1; σ_a is the symmetry number for acyclic or chain species ($\sigma_a = 2$ for PDMS chains); $\delta\mathbf{r}$ is the volume element within which termini of the two separated acyclic species must meet in order to reestablish a bond via the reverse process, and $\delta\omega$ is, similarly, the permitted range of solid angle for one terminal bond of the fragment (diradical) —M_x— relative to the direction of its partner on the other fragment. Thus, two conditions are required for occurrence of the reverse of reaction 1: (*1*) the termini must be situated in juxtaposition, and (*2*) the directions of terminal bonds, here considered for convenience to be free radicals, must be collinear within the range $\delta\omega$. Since both the locations and the orientations of the dissociated species are uncorrelated, the probabilities of fulfillment of these respective conditions are $N_A\,\delta\mathbf{r}$ and $\delta\omega/4\pi$, with all species present in standard state concentrations of one mole per unit volume. Equation 4 follows at once from these considerations.

In process 2 an intramolecular bond is formed which is equivalent to the one severed in process 1. Unless the ring is so small as to induce strain, $\Delta H^\circ_{(2)} = \Delta H^\circ_{(1)}$. The termini of —$M_x$— must meet within the same ranges $\delta\mathbf{r}$ and $\delta\omega$ previously defined. The probability of fulfillment of condition (*1*) above is $W_x(\mathbf{0})\delta\mathbf{r}$ where $W_x(\mathbf{r})$ is the function expressing the distribution of the end-to-end vector $\mathbf{r}$, per unit range in $\mathbf{r}$, for a chain of x units. If x is sufficiently large, the directions of the bonds when condition (*1*) is fulfilled will remain uncorrelated; hence, the probability that condition (*2*) also is met will be $\delta\omega/4\pi$ as before. Then

$$\Delta G^\circ_{(2)} = -\Delta H^\circ_{(1)} - RT \ln\,[W_x(\mathbf{0})\delta\mathbf{r}\delta\omega\sigma_a/4\pi\sigma_{Rx}] \qquad (5)$$

where σ_{Rx} is the symmetry number of an x-meric ring. From Eqs. 4 and 5 we have

$$\Delta G^\circ_{(3)} = -RT \ln\,[W_x(\mathbf{0})/N_A\sigma_{Rx}] \qquad (6)$$

and the equilibrium constant for reaction 3 is

$$K_{(3)} \equiv K_x = W_x(\mathbf{0})/N_A\,\sigma_{Rx} \qquad (7)$$

These are the equations of Jacobson and Stockmayer.[1] They showed further that if the reactivity of a terminal functional group can be considered to be independent of the length of the chain, then the cyclization constant K_x is given by

$$K_x \equiv [M_{y-x}—][\text{c-}M_x]/[M_y—] = [\text{c-}M_x]/p^x \qquad (8)$$

where p is the extent of reaction of functional groups for the acyclic constituents. Alternatively, p may be defined as the ratio of the concentrations of acyclic species of sizes x and $x - 1$. If the average chain length of acyclic species greatly exceeds x, then $p^x \cong 1$ and we have to a sufficient approximation[1]

$$K_x = W_x(\mathbf{0})/N_A \sigma_{Rx} \cong [\text{c-M}_x] \qquad (9)$$

Cancellation of $\delta\mathbf{r}$ and $\delta\omega/4\pi$ on combining Eqs. 4 and 5 to obtain Eq. 6 is contingent upon equivalence of the application of conditions (*1*) and (*2*) above to intramolecular cyclization (i.e., to reaction 2) and to intermolecular combination (i.e., to the reverse of reaction 1). In particular, fulfillment of condition (*1*) in reaction 2 must not vitiate the probability of compliance with (*2*). This implies in effect that the relative orientation of the terminal bonds of —M_x— must remain random (i.e., uncorrelated) when these bonds are forced into close proximity. It will be readily apparent that this independence of condition (*2*) on (*1*) must fail when x is small. If, for example, x approximates the minimum size for ring closure without strain, the proximity of the termini is conducive to a regular polygonal conformation, or to the puckered, nonplanar analog dictated by the preference for bond staggering. The conditional probability of simultaneous fulfillment of condition (*2*) is thus much greater than $\delta\omega/4\pi$. This circumstance is responsible for the facility with which cyclic compounds are formed from units of the minimum size for an unstrained ring.[2] In the PDMS series, for example, the cyclic tetramer occurs in an abundance several times that predicted by calculation according to Eq. 9 (*cf. seq.*).

A more elaborate analysis would be required to pursue the contingency of condition (*2*) on (*1*) as x increases beyond its value for the optimally favored rings which, depending on the homologous series, usually comprise six to eight bonds. For sufficiently large rings, these two conditions obviously will operate independently, whereupon the Jacobson-Stockmayer[1] equations, Eqs. 6 to 9, can be asserted to be rigorous, apart from the limitations stemming from inaccuracies in the specification of $W_x(\mathbf{0})$ for a finite chain.

For chains of a length sufficient to comply with the conditions shown to be necessary for validation of Eqs. 6 to 9, $W_x(\mathbf{r})$ should be Gaussian to an adequate approximation (see Chap. VIII). Thus,

$$W_x(\mathbf{0}) = (3/2\pi\langle r_x^2 \rangle)^{3/2} \qquad (10)$$

where $\langle r_x^2 \rangle$ is the mean-square end-to-end length averaged over all configurations of the real chain of size x. Substitution of Eq. 10 into Eq. 7 yields[2]

$$K_x = (3/2\pi\langle r_x^2 \rangle)^{3/2} N_A^{-1} \sigma_{Rx}^{-1} \qquad (11)$$

We shall identify $\langle r_x^2 \rangle$ with the unperturbed mean-square end-to-end length $\langle r_x^2 \rangle_0$ as determined, for example, on a dilute solution of the polymer in a Θ solvent. Expansion of the chain molecule due to the exclusion of self-intersections becomes negligible in concentrated solutions and vanishes in the undiluted bulk polymer.

Application of these equations to polydimethylsiloxane (PDMS) is illustrative. In this case, $\sigma_{Rx} = 2x$. Hence, Eq. 11 can be written

$$K_x = (3/\pi)^{3/2}/2^{5/2}\langle r_x^2\rangle_0^{3/2}N_A x \quad (12)$$

or

$$K_x = (3/\pi)^{3/2}/2^4 l^3 C_x^{3/2} x^{5/2} N_A \quad (13)$$

where $C_x = \langle r_x^2\rangle_0/2xl^2$ is the characteristic ratio for the x-mer.

Cyclization equilibrium constants K_x, calculated[2] according to Eq.13 and expressed in moles per liter, are represented by the solid line in Fig. 1.

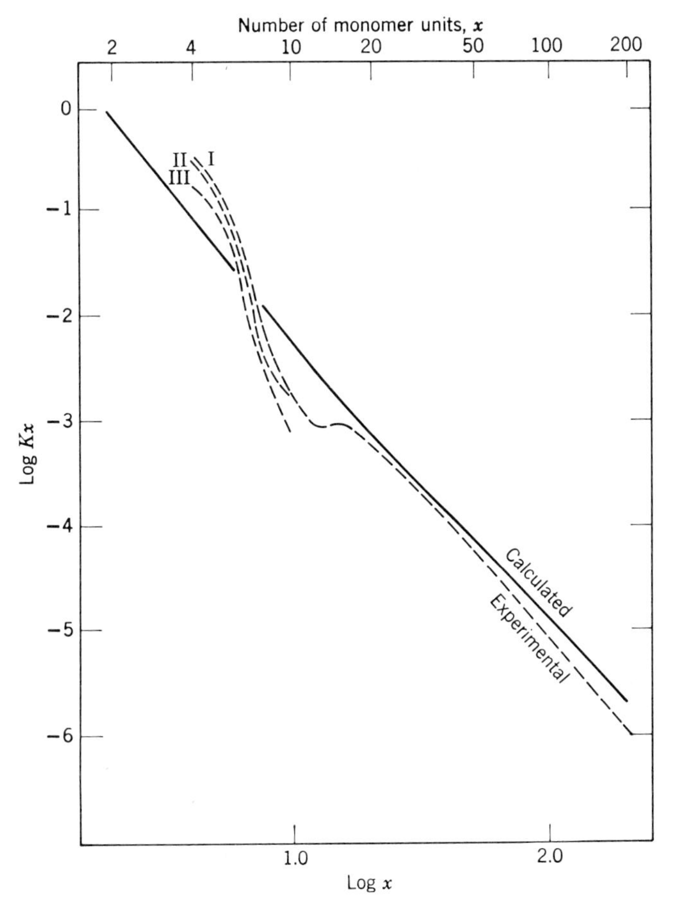

Fig. 1. Macrocyclization equilibrium constants for polydimenthylsiloxane. The dashed curves represent experimental results as follows: I, Brown and Slusarczuk[3]; II, Hartung and Camiolo[3]; III, Carmichael and Winger.[5] The solid curve was calculated by Semlyen[2] from Eq. 13 and values of $\langle r_x^2\rangle_0$ for PDMS of various degrees of polymerization x.

Other curves represent experimental results from the sources indicated. The most extensive set of results are those of Brown and Slusarczuk[3] covering the range from $x = 4$ to over 200 for PDMS equilibrated in toluene solution at a concentration of 222 g/liter and at a temperature of 110°C. Gel permeation chromatography was the principal analytical method. The results of Hartung and Camiolo[4] for 25–75% by weight PDMS in xylene and of Carmichael and Winger[5] for the undiluted polymer, also shown in Fig. 1, cover the lower range of x only. They tend to support Brown and Slusarczuk's data.

The latter authors[3] refer to less extensive experiments carried out on PDMS equilibrated in the absence of diluent. Cyclic species comprise a smaller fraction of the total polymer under these conditions as must be expected, but in the range $x > 11$, the concentrations of macrocyclic siloxanes agree within ca. 20% with those for polymer equilibrated in toluene solutions. Thus, the equilibrium constants K_x found by Brown and Slusarczuk in solution and in the bulk polymer are in substantial agreement.

The calculated curve[2] shown in Fig. 1 agrees remarkably well with the results of Brown and Slusarczuk for higher degrees of polymerization. The differences between calculated and experimental curves for $x > 15$ probably do not exceed the experimental error. The absolute values of K_x are reproduced without adjustment of arbitrary parameters.

The cyclic tetramer occurs at a concentration several times that calculated from $\langle r_4^2 \rangle_0$. This departure from the theoretical curve can be plausibly explained as a manifestation of the ease with which condition (*2*) above is fulfilled when condition (*1*) is met by a chain of this length. The minimum near $x = 10$–12 is not comprehended, however, by the foregoing considerations. Detailed analysis of the conformations of linear and of cyclic species in this range may be required for its explanation. It is to be borne in mind that $W_x(\mathbf{0})$ may not be represented with sufficient accuracy by the Gaussian approximation for short chains, and this may contribute in some degree to the disparity between the calculated and observed values of K_x for small rings. The error from this approximation would be especially serious with respect to a quantitative treatment of the tetramer.

REFERENCES

1. H. Jacobson and W. H. Stockmayer, *J. Chem. Phys.*, **18**, 1600 (1950).
2. P. J. Flory and J. A. Semlyen, *J. Am. Chem. Soc.*, **88**, 3209 (1966).
3. J. F. Brown and G. M. J. Slusarczuk, *J. Am. Chem. Soc.* **87**, 931 (1965).
4. H. A. Hartung and S. M. Camiolo, Abstracts of papers of the Meeting of the American Chemical Society, Washington, D.C., March, 1962.
5. J. B. Carmichael and R. Winger, *J. Polymer Sci., A*, **3**, 971 (1965).

APPENDIX E

Expansion of the Chain Vector Distribution Function in Hermite Polynomials[1]

Let Hermite polynomials be defined as follows[2]:

$$H_\nu(x) = (-1)^\nu e^{x^2}\, \partial^\nu(e^{-x^2})/\partial x^\nu \tag{1}$$

Several of the polynomials thus defined are:

$$\begin{aligned} H_1(x) &= 2x \\ H_2(x) &= 2(2x^2 - 1) \\ H_3(x) &= 2^2x(2x^2 - 3) \\ H_5(x) &= 2^3x(4x^4 - 20x^2 + 15) \\ H_7(x) &= 2^4x(8x^6 - 84x^4 + 210x^2 - 105) \end{aligned} \tag{2}$$

These functions, when multiplied by $e^{-x^2/2}$, obey the orthogonality condition

$$\int_{-\infty}^{\infty} e^{-x^2} H_\nu(x) H_\xi(x)\, dx = 2^\nu \pi^{1/2} \nu!\, \delta_{\nu\xi} \tag{3}$$

where $\delta_{\upsilon\xi}$ is the Kronecker delta.

The distribution function $\mathscr{W}(\boldsymbol{\rho})$ of Chapter VIII, p. 312, may be expanded in terms of the Hermite polynomials as follows[1]:

$$\mathscr{W}_n(\boldsymbol{\rho}) = \pi^{-3/2} e^{-\rho^2} \sum_{\nu=0}^{\infty} h_\nu(n) \rho^{-1} H_{\nu+1}(\rho) \tag{4}$$

$$= \mathscr{W}_\infty(\boldsymbol{\rho}) \sum_{\nu=0}^{\infty} h_\nu(n) \rho^{-1} H_{\nu+1}(\rho) \tag{5}$$

The subscripts n and ∞ are here applied to designate explicitly the length of the chain. The coefficients $h_\nu(n)$ may be derived by the usual procedure involving the multiplication of Eq. 4 by $\rho H_{\nu+1}(\rho)$ and integration. We thus obtain by use of the orthogonality condition, Eq. 3, that

$$\int_{-\infty}^{\infty} H_{\nu+1}(\rho) \mathscr{W}(\boldsymbol{\rho}) \rho\, d\rho = h_\nu \pi^{-1} 2^{\nu+1} (\nu+1)! \tag{6}$$

or

$$h_\nu = [1/2^{\nu+3}(\nu+1)!] \int_{-\infty}^{\infty} \rho^{-1} H_{\nu+1}(\rho) \mathscr{W}(\boldsymbol{\rho}) 4\pi\rho^2\, d\rho \tag{7}$$

Taking $\mathscr{W}(\boldsymbol{\rho})$ to be an even function of ρ, we observe that the integral vanishes if ν is an odd integer. This follows from the fact that $\rho^{-1}H_{\nu+1}$ is an odd function of ρ for $\nu = 1, 3, 5$, etc. Hence, $h_\nu = 0$ for odd integral values of ν. For even ν, the half-range integral may be used with the result that

$$h_\nu = [1/2^{\nu+2}(\nu+1)!] \int_0^\infty \rho^{-1} H_{\nu+1}(\rho) \mathscr{W}(\boldsymbol{\rho}) 4\pi\rho^2 \, d\rho$$
$$= [1/2^{\nu+2}(\nu+1)!] \langle \rho^{-1} H_{\nu+1}(\rho) \rangle \tag{8}$$

Substitution of the Hermite polynomials of odd index into Eq. 8 and the identification $\rho^2 = 3r^2/2\langle r^2 \rangle$ gives at once

$$h_0 = \tfrac{1}{2}$$
$$h_2 = \left(\frac{1}{2^2 \cdot 3!}\right)(2\langle \rho^2 \rangle - 3) = 0$$
$$h_4 = \left(\frac{1}{2^3 \cdot 5!}\right)(4\langle \rho^4 \rangle - 20\langle \rho^2 \rangle + 15)$$
$$= \left(\frac{1}{2^5 \cdot 2!}\right)\left(\frac{3\langle r^4 \rangle}{5\langle r^2 \rangle^2} - 1\right) \tag{9}$$
$$h_6 = \left(\frac{1}{2^4 \cdot 7!}\right)(8\langle \rho^6 \rangle - 84\langle \rho^4 \rangle + 210\langle \rho^2 \rangle - 105)$$
$$= \left(\frac{1}{2^7 \cdot 3!}\right)\left(\frac{3^2\langle r^6 \rangle}{5 \cdot 7\langle r^2 \rangle^3} - \frac{3^2\langle r^4 \rangle}{5\langle r^2 \rangle^2} + 2\right)$$

etc.

Subscript zeros on the $\langle r^{2p} \rangle$ are omitted for the reasons given in Chapter VIII (see the footnote on p. 310).

A correspondence between the h_{2p} and the g_{2p} defined in Eqs. VIII-20 and VIII-21 is apparent. In fact,

$$h_{2p} = (-1)^p g_{2p}/2^{p+1} \tag{10}$$

Substitution of Eq. 10 and the Hermite polynomials, Eq. 2, into Eq. 4 gives Eq. VIII-27,[1] which was first obtained by Nagai[3] using the lengthier Fourier transformation procedure. The latter procedure has been adopted in the text in order to make available the Fourier transform of $\mathscr{W}(\mathbf{r})$ for subsequent use.

REFERENCES

1. R. L. Jernigan, Ph.D. thesis, Stanford University, 1967.
2. H. Margenau and G. M. Murphy, *The Mathematics of Physics and Chemistry*, 2nd ed., Van Nostrand, New York, 1956, pp. 122–126.
3. K. Nagai, *J. Chem. Phys.*, **38**, 924 (1963).

APPENDIX F

The Chain Vector Distribution for Freely Jointed Chains

Equation VIII-33 for the chain vector distribution function for a freely jointed chain consisting of n bonds each of length l may be written

$$W(\mathbf{r}) = (2\pi^2 r l^2)^{-1} I_n(\sigma) \qquad (1)$$

where $\sigma = r/l$ and

$$I_n(\sigma) = \int_0^\infty (\sin \sigma y \sin^n y) y^{-(n-1)}\, dy \qquad (2)$$

with y replacing ql. We first consider these functions for small values of n.

For $n = 2$, we obtain through use of a standard trigonometric identity

$$I_2(\sigma) = (1/2) \int_0^\infty [\cos(\sigma y - y) - \cos(\sigma y + y)] y^{-1} \sin y\, dy$$

Now

$$\int_0^\infty (\cos ay \sin y) y^{-1}\, dy = \begin{cases} \pi/2, & a^2 < 1 \\ \pi/4, & a^2 = 1 \\ 0, & a^2 > 1 \end{cases}$$

Hence

$$\begin{aligned} I_2(\sigma) &= \pi/4, \qquad 0 < \sigma < 2 \\ &= \pi/8, \qquad \sigma = 2 \\ &= 0, \qquad \sigma = 0 \text{ or } \sigma > 2 \end{aligned} \qquad (3)$$

For $n = 3$

$$I_3(\sigma) = (1/2) \int_0^\infty [\cos(\sigma y - y) - \cos(\sigma y + y)](\sin^2 y) y^{-2}\, dy$$

Partial integration, with $y^{-2}\, dy$ treated as the differential, yields

$$\begin{aligned} I_3(\sigma) = (1/2) \int_0^\infty \{&2[\cos(\sigma y - y) - \cos(\sigma y + y)] \cos y \\ &- [(\sigma - 1) \sin(\sigma y - y) - (\sigma + 1) \sin(\sigma y + y)] \sin y\} y^{-1} \sin y\, dy \end{aligned}$$

$$= (1/4) \int_0^\infty [(3 - \sigma) \cos(\sigma y - 2y) + 2\sigma \cos(\sigma y) - (3 + \sigma)\cos(\sigma y + 2y)] y^{-1} \sin y \, dy$$

$$\begin{aligned} &= \pi\sigma/4, && 0 \leqslant \sigma \leqslant 1 \\ &= \pi(3 - \sigma)/8, && 1 \leqslant \sigma \leqslant 3 \\ &= 0, && \sigma \geqslant 3 \end{aligned} \tag{4}$$

Equations 3 and 4, when substituted into Eq. 1, lead to Eqs. VIII-36 and VIII-37.

Partial integration of Eq. 2 in the same manner as above yields for any value of n

$$I_n(\sigma) = (n - 2)^{-1} \int_0^\infty [\sigma \cos \sigma y \sin y + n \sin \sigma y \cos y] y^{-(n-2)} \sin^{n-1} y \, dy$$

$$= [2(n - 2)]^{-1} \int_0^\infty [(n + \sigma) \sin(\sigma y + y) + (n - \sigma)\sin(\sigma y - y)] \times y^{-(n-2)} \sin^{n-1} y \, dy$$

Hence

$$2(n - 2) I_n(\sigma) = (n + \sigma) I_{n-1}(\sigma + 1) + (n - \sigma) I_{n-1}(\sigma - 1) \tag{5}$$

The solution of this recursion relation is

$$I_n(\sigma) = (\pi/4) n(n - 1) \sum_{t=0}^{\tau} \left[\frac{(-1)^t}{t!(n - t)!}\right] \left[\frac{n - \sigma - 2t}{2}\right]^{n-2} \tag{6}$$

where

$$(n - \sigma - 2)/2 \leqslant \tau < (n - \sigma)/2$$

This result may be confirmed by substitution of Eq. 6 into Eq. 5. Verification, though tedious, is straightforward. Equation 6, when substituted into Eq. 1, gives Treloar's[1] equation, Eq. VIII-39. The correspondence of his result to the equation of Lord Rayleigh,[2] Eq. VIII-33, is thus established.

REFERENCES

1. L. R. G. Treloar, *Trans. Faraday Soc.*, **42**, 77 (1946); see also *The Physics of Rubber Elasticity*, 2nd ed., Oxford Univ. Press, 1958, p. 97 *et seq.*
2. Lord Rayleigh, *Phil. Mag.* [6], **37**, 321 (1919).

APPENDIX G

The Porod-Kratky Chain[1,2]

This hypothetical model for a chain molecule incorporates the concept of continuous curvature of the chain skeleton, the direction of curvature at any point of the trajectory being random.[1,2] It is frequently referred to as the worm-like chain. The model has had particular appeal for representing stiff chains, but its use has not been restricted to them alone.

The freely rotating chain, comprising bonds joined at fixed angles θ, serves as the starting point for definition of the Porod-Kratky chain. The Porod-Kratky chain is more closely related to the freely rotating chain than to others treated in detail in the text. The average projection of the kth bond of a freely rotating chain on the direction of the first bond is $l'\alpha^{k-1}$, where l' is the bond length and $\alpha = \cos\theta'$, with θ' denoting the fixed angle between successive bonds. Primes are used in order to distinguish bonds and bond angles of the freely rotating, model chain from the corresponding quantities for the real chain it is intended to represent. The average sum of projections of n' of these bonds on the direction of the first bond is[1,2]

$$\overline{X}_1 = \mathbf{r}\cdot(\mathbf{l}_1/l_1) = l'\sum_{k=0}^{n'-1}\alpha^k \tag{1}$$

where $\mathbf{l}_1/l_1$ is the unit vector on the first bond.

If the chain is made indefinitely long, then $\overline{X}_1$ becomes the *persistence length* defined in Chapter IV (see p. 111) as the sum of the average projections of all bonds $i = 1$ to ∞ on the direction of the first bond. Denoting the persistence length by a, we have from Eq. 1

$$a = l'/(1-\alpha) \tag{2}$$

The persistence length for a real chain is determined by its structure and by hindrances to bond rotations. It is directly related to the characteristic ratio C_∞ (see Eq. IV-48).

Let the freely rotating chain of finite length considered above be subdivided into shorter and shorter bonds in such a way as to maintain the constancy of the *contour length* L and of the persistence length a at their predetermined values. Continuation of the subdivision to the limit $l' = 0$ and $n' = \infty$ yields the Porod-Kratky[1,2] chain of continuously varying direction. In this limit $1 - \alpha$ also vanishes, and it does so in the manner required by Eq. 2 to maintain a at its specified value. Equation 2 may be replaced by

$$a = \lim_{l' \to 0} (-l'/\ln \alpha) \tag{3}$$

Through the introduction of a thus defined into Eq. 1, we obtain

$$\bar{X}_1 = l' \sum_{k=0}^{n'-1} \exp(-kl'/a) \tag{4}$$

$$= \int_0^L \exp(-K/a)\, dK \tag{5}$$

$$= a[1 - \exp(-L/a)] \tag{6}$$

with $L = n'l'$.

Replacement of $\bar{X}_1$ by $\langle r^2 \rangle_0$ as the relevant variable may be accomplished by relating differential quantities as follows[1,2]:

$$d\langle r^2 \rangle_0 = 2\langle \mathbf{r} \cdot d\mathbf{r} \rangle = 2\bar{X}_1\, dL \tag{7}$$

The latter relation may be verified readily by considering the increment dL to be added at the beginning of the chain. Then the magnitude of $d\mathbf{r}$ is dL and its direction coincides with X_1. Equation 7 follows at once. Substitution of Eq. 6 for $\bar{X}_1$ into Eq. 7 and integration yields[1,2]

$$\langle r^2 \rangle_0/L = 2a[1 - (a/L)(1 - e^{-L/a})] \tag{8}$$

In the limit $L \to \infty$

$$(\langle r^2 \rangle_0/L)_\infty = 2a \tag{9}$$

Thus, the representation of the hypothetical Porod-Kratky model chain depends on two parameters a and L. In order to proceed further, it is necessary to establish a correspondence between these quantities and characteristics of the real chain. This step is attended by a degree of arbitrariness. A reasonable course is the following. Let L be identified with the length r_{max} of the real chain when fully extended. The ratio L/nl is thereby established. Note that $L = r_{max}$ is less than nl, in general, owing to valence angle restrictions, these angles being assumed to be fixed.* Second, we so choose a as to establish coincidence between C_∞ for the model and for the real chain. Thus, the limiting value of the characteristic ratio for the model chain is given according to Eq. 9 by

$$C_\infty \equiv (\langle r^2 \rangle_0/nl^2)_\infty = (L/nl)(2a/l) \tag{10}$$

*The specification of L in this manner is not without complications in some cases. If valence angles θ differ for successive skeletal atoms, then the most highly extended conformation may be nonplanar and its precise geometrical description is not immediately obvious.

where l and n are the bond length and number of bonds, respectively, for the *real* chain.* The Porod-Kratky relation, Eq. 8, for the second moment of **r** may be expressed in like terms as follows:

$$\langle r^2 \rangle_0 / nl^2 = C_\infty[1 - (L/a)^{-1}(1 - e^{-L/a})] \tag{11}$$

where L is understood to be related to n according to Eq. 10.

For short chains, or very stiff ones, Eq. 8 may be expanded in the following series:

$$\langle r^2 \rangle_0 / L = L[1 - (1/3)(L/a) + (1/12)(L/a)^2 - \cdots], \qquad (L/a) < 1 \tag{12}$$

For long, or "flexible," chains

$$\langle r^2 \rangle_0 / L \cong 2a[1 - (L/a)^{-1}], \qquad (L/a) \gg 1 \tag{13}$$

or

$$\langle r^2 \rangle_0 / nl^2 \cong C_\infty[1 - (L/a)^{-1}], \qquad (L/a) \gg 1 \tag{14}$$

In this limit, the characteristic ratio is linear in L^{-1}, or in n^{-1}.

Expressions for the fourth and sixth moments of **r** for the Porod-Kratky chain have been derived by Hermans and Ullman[3] and by Heine, Kratky, and Porod.[4] Benoit and Doty[5] have derived the following expression for the unperturbed radius of gyration of this model chain:

$$\langle s^2 \rangle_0 / L = (a/3)\{1 - (3a/L)[1 - 2(a/L) + 2(a/L)^2 - 2(a/L)^2 e^{-L/a}]\} \tag{15}$$

or

$$\langle s^2 \rangle_0 / nl^2 = (\langle s^2 \rangle_0 / nl^2)_\infty [1 - 3(a/L) + 6(a/L)^2 - 6(a/L)^3(1 - e^{-L/a})] \tag{16}$$

REFERENCES

1. G. Porod, *Monatsh. Chem.*, **80**, 251 (1949).
2. O. Kratky and G. Porod, *Rec. Trav. Chim.*, **68**, 1106 (1949).
3. J. J. Hermans and R. Ullman, *Physica*, **18**, 951 (1952).
4. S. Heine, O. Kratky, and G. Porod, *Makromol. Chem.*, **44**, 682 (1961).
5. H. Benoit and P. Doty, *J. Phys. Chem.*, **57**, 958 (1953).

*According to Eq. IV-48

$$(\langle r^2 \rangle_0 / nl^2)_\infty = (2a - l)/l$$

for a real chain. This expression differs, of course, from that for the Porod-Kratky model chain as expressed by Eq. 10. The difference arises in part from the inclusion of the first step (bond) of the real chain which by definition is coincident with the direction of $\bar{X}_1$.

APPENDIX H

The Average Orientation of a Vector within a Chain of Specified End-to-End Vector r

The following analysis[1] is addressed to a molecular chain whose end-to-end vector $\mathbf{r}$ is specified within a laboratory reference frame XYZ. The "molecule" under consideration may, for example, be the portion of a network extending from one cross-linkage to the next, as is explained more fully in Chapter IX, Section 3. In any event, the chain is subject only to the constraint imposed by stipulation of the vector $\mathbf{r}$ separating its ends.

We focus attention on a unit vector $\mathbf{v}_{it}$ affixed to the ith skeletal bond or unit. If this vector is identified with the direction of one of the principal components of the optical polarizability tensor $\boldsymbol{\alpha}$, the results here obtained are applicable to analysis of the strain birefringence (see Chap. IX, Sect. 3). Identification of $\mathbf{v}_{it}$ with the transition moment for excitation of group i by absorption of radiation of a given wavelength provides the basis for treatment of the dichroic ratio. Finally, the preferential orientation of a given bond by extension of the chain may be obtained by identifying the unit vector with the direction of the bond. Whatever the identification of $\mathbf{v}_{it}$ may be, we seek the average square of its projection on chain vector $\mathbf{r}$ as a function of r. The relationship will be obtained as a series in even powers of r, by resort to a procedure developed by Nagai.[1]

Letting $(\mathbf{v}_{it}, \mathbf{X})$ denote the cosine of the angle between $\mathbf{v}_{it}$ and axis X of the fixed reference frame, we have

$$\langle(\mathbf{v}_{it}, \mathbf{X})^2\rangle_r = \tilde{Z}_{\mathbf{r}}^{-1} \int \underset{\mathbf{r}}{\cdots} \int (\mathbf{v}_{it}, \mathbf{X})^2 \exp(-E/kT) \sin\chi \, d\chi \, d\psi \, d\omega \, d\{\phi\}/8\pi^2 \, d\mathbf{r} \tag{1}$$

where $\tilde{Z}_{\mathbf{r}}$ is the configuration partition function for a chain of specified $\mathbf{r}$ (see Eq. VIII-2); χ, ψ, and ω are Eulerian angles, and $\{\phi\}$ is the set of internal, skeletal bond rotations; these and other symbols carry the definitions given in Chapter VIII, Section 1.

The Fourier transform of the integral in Eq. 1 is

$$H_{it}(\mathbf{q}) = \int e^{\tilde{\imath}\mathbf{q}\cdot\mathbf{r}}\tilde{Z}_{\mathbf{r}}\langle(\mathbf{v}_{it}, \mathbf{X})^2\rangle_r \, d\mathbf{r}$$

$$= (8\pi^2)^{-1}\int\cdots\int(\mathbf{v}_{it}, \mathbf{X})^2 \exp(-E/kT)\exp(\tilde{\imath}\mathbf{q}\cdot\mathbf{r}) \times \sin\chi \, d\chi \, d\psi \, d\omega \, d\{\phi\} \tag{2}$$

Pursuant to the evaluation of this expression, it will be helpful to define a new Cartesian coordinate system, xyz, with the z axis taken parallel to vector $\mathbf{q}$, and the x axis in the Xz plane; the direction of the x axis will be chosen to make an acute angle with X. The Eulerian angles are conveniently defined as in Fig. 1. That is, χ and ψ are the polar and azimuthal angles, respectively, locating $\mathbf{r}$ with reference to $\mathbf{q}$ (i.e., z) as the polar axis; ω measures the rotation of the plane defined by $\mathbf{v}_{it}$ and $\mathbf{r}$ from the plane of $\mathbf{r}$ and z. Further, let Φ_{it} be the angle made by $\mathbf{v}_{it}$ with $\mathbf{r}$, and let τ denote the angle between $\mathbf{q}$ and the fixed axis X (not shown in Fig. 1), i.e.,

$$\cos\tau \equiv (\mathbf{q}, \mathbf{X}) = q_1/q \tag{3}$$

where q_1 is the projection of $\mathbf{q}$ on X, and $q = |\mathbf{q}|$. Then a unit vector along the X axis is expressed in the coordinate system xyz by

$$\mathbf{X}/X = \begin{bmatrix} \sin\tau \\ 0 \\ \cos\tau \end{bmatrix} \tag{4}$$

The unit vector $\mathbf{v}_{it}$ expressed in the same coordinate system is

$$\mathbf{v}_{it} = \begin{bmatrix} \sin\chi\cos\psi\cos\Phi_{it} + (\cos\chi\cos\psi\cos\omega - \sin\psi\sin\omega)\sin\Phi_{it} \\ \sin\chi\sin\psi\cos\Phi_{it} + (\cos\chi\sin\psi\cos\omega + \cos\psi\sin\omega)\sin\Phi_{it} \\ \cos\chi\cos\Phi_{it} - \sin\chi\cos\omega\sin\Phi_{it} \end{bmatrix} \tag{5}$$

The cosine of the angle between these two unit vectors is

$$(\mathbf{v}_{it}, \mathbf{X}) = \sin\tau[\sin\chi\cos\psi\cos\Phi_{it} + (\cos\chi\cos\psi\cos\omega - \sin\psi\sin\omega)\sin\Phi_{it}] + \cos\tau[\cos\chi\cos\Phi_{it} - \sin\chi\cos\omega\sin\Phi_{it}] \tag{6}$$

Substitution of Eq. 6 into Eq. 2 and integration over the Eulerian angles at fixed internal configuration $\{\phi\}$ leads to

$$H_{it}(\mathbf{q}) = \frac{1}{3}\int\cdots\int\left\{\frac{\sin qr}{qr} + \left(\frac{1}{2}\right)(3\cos^2\tau - 1)(3\cos^2\Phi_{it} - 1) \times \left[\frac{\sin qr}{qr} + \frac{3\cos qr}{(qr)^2} - \frac{3\sin qr}{(qr)^3}\right]\right\}e^{-E/kT}\,d\{\phi\} \tag{7}$$

$$= \frac{1}{3}\int\cdots\int\left\{\left[1 - \frac{(qr)^2}{3!} + \frac{(qr)^4}{5!} - \frac{(qr)^6}{7!} + \cdots\right]\right.$$

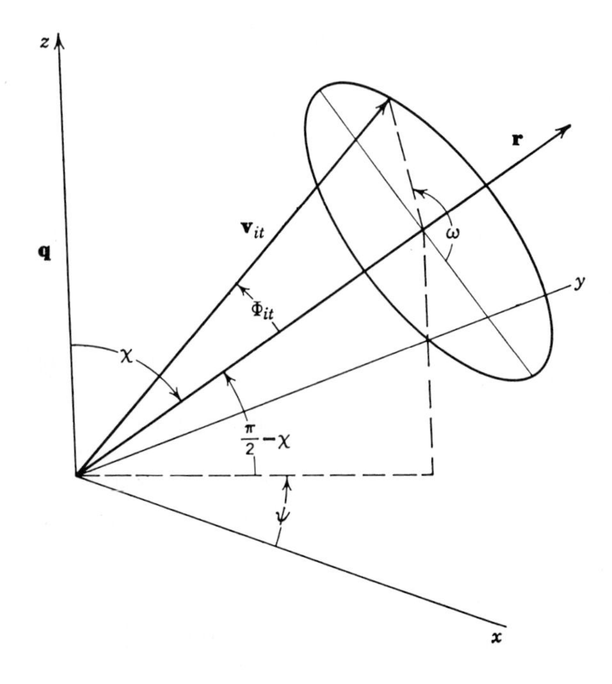

$$-\frac{1}{2}(3\cos^2\tau - 1)(3\cos^2\Phi_{it} - 1)$$

$$\times\left[\frac{(qr)^2}{5\cdot 3} - \frac{(qr)^4}{7\cdot 5\cdot 3!} + \frac{(qr)^6}{9\cdot 7\cdot 5!} - \cdots\right]\Bigg\}e^{-E/kT}\,d\{\phi\} \tag{8}$$

The result of integration over the internal configuration variables may be expressed in terms of the configurational averages of the various quantities, i.e.,

$$H_{it}(\mathbf{q}) = \frac{Z}{3}\Bigg\{\left[1 - \frac{\langle r^2\rangle_0 q^2}{3!} + \frac{\langle r^4\rangle_0 q^4}{5!} - \cdots\right]$$

$$+\left[\frac{1}{30}(3\langle r^2\cos^2\Phi_{it}\rangle_0 - \langle r^2\rangle_0) - \frac{1}{420}(3\langle r^4\cos^2\Phi_{it}\rangle_0 - \langle r^4\rangle_0)q^2\right.$$

$$\left.+\frac{1}{15120}(3\langle r^6\cos^2\Phi_{it}\rangle_0 - \langle r^6\rangle_0)q^4 - \cdots\right](q^2 - 3q_1^2)\Bigg\} \tag{9}$$

where q_1 is defined by Eq. 3 and Z is the partition function for the unconstrained chain (see Eqs. VIII-1 and VIII-5). Through the multiplication of the two series in brackets in Eq. 9 by the series expansion of $\exp(\langle r^2\rangle_0 q^2/6)$, Eq. 9 can be rearranged to

$$H_{it}(\mathbf{q}) = \frac{Z}{3}\exp\left(\frac{-\langle r^2\rangle_0 q^2}{6}\right)\left\{1 + \left[\eta_2\left(\frac{\langle r^2\rangle_0}{3}\right) + \eta_4\left(\frac{\langle r^2\rangle_0}{3}\right)^2 q^2 + \eta_6\left(\frac{\langle r^2\rangle_0}{3}\right)^3 q^4 + \cdots\right](q^2 - 3q_1^2) + \left[g_4\left(\frac{\langle r^2\rangle_0}{3}\right)^2 q^4 + g_6\left(\frac{\langle r^2\rangle_0}{3}\right)^3 q^6 + \cdots\right]\right\} \tag{10}$$

where

$$\eta_2 = \frac{1}{10}\left(\frac{3\langle r^2 \cos^2 \Phi_{it}\rangle_0}{\langle r^2\rangle_0} - 1\right)$$

$$\eta_4 = \frac{1}{20}\left[\left(\frac{3\langle r^2 \cos^2 \Phi_{it}\rangle_0}{\langle r^2\rangle_0} - 1\right) - \frac{3}{7}\left(\frac{3\langle r^4 \cos^2 \Phi_{it}\rangle_0}{\langle r^2\rangle_0^2} - \frac{\langle r^4\rangle_0}{\langle r^2\rangle_0^2}\right)\right]$$

$$\eta_6 = \frac{1}{80}\left[\left(\frac{3\langle r^2 \cos^2 \Phi_{it}\rangle_0}{\langle r^2\rangle_0} - 1\right) - \frac{6}{7}\left(\frac{3\langle r^4 \cos^2 \Phi_{it}\rangle_0}{\langle r^2\rangle_0^2} - \frac{\langle r^4\rangle_0}{\langle r^2\rangle_0^2}\right) + \frac{1}{7}\left(\frac{3\langle r^6 \cos^2 \Phi_{it}\rangle_0}{\langle r^2\rangle_0^3} - \frac{\langle r^6\rangle_0}{\langle r^2\rangle_0^3}\right)\right] \tag{11}$$

etc.

and g_4, g_6, etc. are defined by Eqs. VIII-20. Subscripts *it* applicable to the η_{2p} have been omitted for simplicity.

By the Fourier integral theorem,

$$\langle(\mathbf{v}_{it}, \mathbf{X})^2\rangle_r = (2\pi)^{-3}\tilde{Z}_{\mathbf{r}}^{-1} \int H_{it}(\mathbf{q}) \exp(-\tilde{\imath}\mathbf{q}\cdot\mathbf{r})\, d\mathbf{q} \tag{12}$$

Our purposes will be served if the chain vector $\mathbf{r}$ introduced in Eq. 12 is located along the fixed axis X. Then $(\mathbf{v}_{it}, \mathbf{X}) = \cos \Phi_{it}$, and $q_1 = (\mathbf{q}\cdot\mathbf{r})r^{-1}$ defines the polar angle of $\mathbf{q}$ with respect to $\mathbf{r}$, which now assumes the role of a fixed axis. Substitution of Eq. 10 into Eq. 12 and integration over the azimuthal angle gives

$$\langle\cos^2 \Phi_{it}\rangle_r = (1/12\pi^2)(Z/\tilde{Z}_{\mathbf{r}}) \int_0^\infty \int_{-q}^{q} [1 + \eta_2(\langle r^2\rangle_0/3)(q^2 - 3q_1^2) + \eta_4(\langle r^2\rangle_0/3)^2(q^2 - 3q_1^2)q^2 + \cdots + g_4(\langle r^2\rangle_0/3)^2 q^4 + \cdots] \exp(-\tilde{\imath}q_1 r) \times \exp(-\langle r^2\rangle_0 q^2/6) q\, dq\, dq_1 \tag{13}$$

The integrations are straightforward. The result carried to terms in r^4, with omission of terms involving parameters beyond h_4 and g_4, is

$$\langle \cos^2 \Phi_{it} \rangle_r = \frac{1}{3}\left(\frac{3}{2\pi\langle r^2\rangle_0}\right)^{3/2}\left(\frac{\tilde{Z}_{\mathbf{r}}}{Z}\right)^{-1}\left[\exp\left(-\frac{3r^2}{2\langle r^2\rangle_0}\right)\right.$$
$$+ \eta_2\left(\frac{6r^2}{\langle r^2\rangle_0} - \frac{9r^4}{\langle r^2\rangle_0^2} + \cdots\right)$$
$$+ \eta_4\left(\frac{42r^2}{\langle r^2\rangle_0} - \frac{81r^4}{\langle r^2\rangle_0^2} + \cdots\right) + \cdots$$
$$\left.+ 15g_4\left(1 - \frac{7r^2}{2\langle r^2\rangle_0} + \frac{189r^4}{40\langle r^2\rangle_0^2} - \cdots\right) + \cdots\right] \tag{14}$$

Substitution of

$$\frac{\tilde{Z}_{\mathbf{r}}}{Z} = W(\mathbf{r}) \cong \left(\frac{3}{2\pi\langle r^2\rangle_0}\right)^{3/2} \exp - \left(\frac{3r^2}{2\langle r^2\rangle_0}\right)\left[1 + 15g_4\left(1 - \frac{2r^2}{\langle r^2\rangle_0} + \frac{3r^4}{5\langle r^2\rangle_0^2}\right)\right]$$

according to Eqs. VIII-11, VIII-26, and VIII-76, leads to

$$\langle \cos^2 \Phi_{it} \rangle_r - \frac{1}{3} = \beta_{2;it} r^2/\langle r^2\rangle_0 + \beta_{4;it}(r^2/\langle r^2\rangle_0)^2 + \cdots \tag{15}$$

which is Eq. IX-93, with

$$\beta_{2;it} = 2(\eta_2 + 7\eta_4 + \cdots)(1 - 15g_4 + \cdots) \tag{16}$$

$$\beta_{4;it} = 6(10\eta_2 g_4 - \eta_4 + 85\eta_4 g_4 - \cdots) \tag{17}$$

These equations reproduce, essentially,* the results of Nagai[1] with omission of terms in r^6 which he included in part. Nagai has shown[1] that η_2 (in our notation) is of order n^{-1} and that η_4 (and also η_6) is of order n^{-2}. Similarly, g_4 is of order n^{-1} for large n. Hence, Eqs. 15, 16, and 17 include all terms of orders n^{-1} and n^{-2}. These relations should therefore be adequate for chains of sufficient length subject to small relative extensions, e.g., for r^2 not much greater than $\langle r^2\rangle_0$.

Finally, we consider specifically the application of the foregoing results to strain birefringence.[2] Proceeding in the manner of Eqs. IX-97 to IX-101 and identifying the unit vector $\mathbf{v}_{it}$ with a principal axis of the polarizability tensor $\boldsymbol{\alpha}_i$ for group i, we find for the difference between the polarizabilities parallel and perpendicular to $\mathbf{r}$

*Included in Eqs. 16 and 17 are terms in $\eta_4 g_4$ omitted by Nagai.[1]

$$\begin{aligned}\Delta\alpha_r &= (3/2)(\alpha_r - \bar{\alpha})\\ &= (3/2)\sum_i\sum_t \alpha_{it}(\langle\cos^2\Phi_{it}\rangle_r - 1/3)\\ &= \Gamma_2(r^2/\langle r^2\rangle_0) + \Gamma_4(r^2/\langle r^2\rangle_0)^2 + \cdots\end{aligned} \tag{18}$$

where

$$\Gamma_2 = (\gamma_2 + 7\gamma_4 + \cdots)(1 - 15g_4 + \cdots) \tag{19}$$

$$\Gamma_4 = 3(10\gamma_2 g_4 - \gamma_4 + 85\gamma_4 g_4 - \cdots) \tag{20}$$

and

$$\gamma_2 = (9/10)\sum_i \langle \mathbf{r}^T\hat{\boldsymbol{\alpha}}_i\mathbf{r}\rangle_0/\langle r^2\rangle_0 \tag{21}$$

$$\gamma_4 = (9/20)\left[\sum_i \langle \mathbf{r}^T\hat{\boldsymbol{\alpha}}_i\mathbf{r}\rangle_0/\langle r^2\rangle_0 - (3/7)\sum_i \langle(\mathbf{r}^T\hat{\boldsymbol{\alpha}}_i\mathbf{r})(\mathbf{r}^T\mathbf{r})\rangle_0/\langle r^2\rangle_0^2\right] \tag{22}$$

where $\hat{\boldsymbol{\alpha}}_i$ is the traceless tensor $\boldsymbol{\alpha}_i - \bar{\alpha}_i\mathbf{E}_3$ defined by Eq. IX-102. The parameter we here designate by γ_2 is subject to evaluation according to Eqs. IX-108 and IX-109. A much more elaborate scheme would be required to evaluate the second sum of γ_4.

REFERENCES

1. K. Nagai, *J. Chem. Phys.*, **40**, 2818 (1964).
2. P. J. Flory, R. L. Jernigan, and A. E. Tonelli, *J. Chem. Phys.*, **48**, 3822 (1968).

Glossary of Principal Symbols and Conventions

1. Mathematical Conventions

$\langle\ \rangle$	The statistical mechanical average of the quantity enclosed, taken over all configurations of the chain
$\langle\ \rangle_0$	The corresponding average for the chain unperturbed by interactions of long range, or by external constraints
$\mathbf{b}$	A vector, or its representation as a column matrix
$\mathbf{b}^T$	The transposed, or row form, of $\mathbf{b}$
$\lvert\mathbf{b}\rvert$	The scalar magnitude of $\mathbf{b}$
$[b_{\alpha\beta}]$	The matrix $\mathbf{B}$ comprising the array of elements $b_{\alpha\beta}$
$\mathbf{B}^T$	The transpose of a matrix $\mathbf{B}$
$\lvert\mathbf{B}\rvert$	The determinant of $\mathbf{B}$
$\lVert\mathbf{T}\rVert$	The pseudodiagonal matrix defined as in Eq. IV-6
$\mathbf{A}\otimes\mathbf{B}$	The direct matrix product of $\mathbf{A}$ with $\mathbf{B}$
$\mathbf{B}^R$	The matrix $\mathbf{B}$ expressed as the row $[b_{11}, b_{12}, \ldots, b_{1\nu}, b_{21}, \ldots, b_{2\nu}, \ldots, b_{\nu\nu}]$
$\mathbf{B}^C$	The corresponding column, being the transpose of $\mathbf{B}^R$
$\mathbf{B}_i^{(m)}$	Serial product commencing with $\mathbf{B}_i$ and consisting of m factors; i.e., $\prod_{h=i}^{i+m-1} \mathbf{B}_h = \mathbf{B}_i \mathbf{B}_{i+1} \cdots \mathbf{B}_{i+m-1}$

2. English and German Letter Symbols

a	Persistence length
Å	Angstrom unit, 10^{-8} cm
$\mathbf{A}_0$, $\mathbf{A}$	Vector amplitudes of incident and scattered radiation
A_0, A	Corresponding scalar magnitudes
A_v, A_h	Amplitudes of horizontal and vertical components of scattered radiation
$\mathbf{A}$	Matrix of eigenvectors $\mathbf{A}_k$ (column eigenvectors)
A_{jk}, $A_{\zeta\eta}$, etc.	Elements of $\mathbf{A}$
$\mathbf{B}$	Matrix of eigenrows $\mathbf{B}_k^*$ mutually normalized to the $\mathbf{A}_k$; e.g., $\mathbf{B} = \mathbf{A}^{-1}$
B_{jk}, $B_{\zeta\eta}$, etc.	Elements of $\mathbf{B}$
B	Stress-optical coefficient
c	Concentration (weight/volume)

C_n	Characteristic ratio $\langle r^2 \rangle_0 / nl^2$ for a chain of n bonds
C_∞	C_n in limit $n \to \infty$
$\mathbf{D}$	Matrix of statistical weights for interactions of first order, in diagonal array
$E, E\{\phi\}, E_{\text{tors}}(\phi)$, etc.	Energy (per mole), energy associated with the configuration $\{\phi\}$, torsional energy for rotation ϕ, etc.
E°	Torsional energy barrier
$\mathbf{E}, \mathbf{E}_\nu$	Matrix identity, identity of order ν
f	Tensile force acting on a polymer chain or on a polymer network
f_{II}, f_h	Fraction of units occurring in the conformation (state) II, or in the helical conformation (h)
F	Electric field intensity
g, g^+, g^-	*Gauche* conformations; used also as subscripts denoting these conformations
$G(\mathbf{q})$	Fourier transform of $W(\mathbf{r})$
$g_2, g_4, \ldots, g_{2p} \cdots$	Coefficients in the series expansion of $G(\mathbf{q})$
$\mathrm{G}_{ij}(\boldsymbol{\mu})$	Fourier transform of $W(\mathbf{r}_{ij})$
$\mathscr{G}_i$	Matrix for bond i, used to generate $\langle M^2 \rangle_0$ or $\langle r^2 \rangle_0$
$\mathscr{G}_k^{(\xi)}$	Serial product of $\mathscr{G}$ matrices for the kth repeat unit consisting of ξ bonds
$\mathfrak{G}_i$	Matrix, analogous to $\mathscr{G}_i$, for a bond subject to an independent rotational potential
$\mathfrak{G}_k^{(\xi)}$	Matrix, analogous to $\mathscr{G}_k^{(\xi)}$, for the sequence of ξ bonds of the kth repeat unit, with rotational independence for bonds in separate units
h, i, j, k	Serial indexes of bonds, or of repeat units (subscripts)
$\tilde{\imath}$	Imaginary, $\sqrt{-1}$
$I_0, I(\vartheta)$	Radiation intensities, incident and scattered at angle ϑ, respectively
$\mathbf{J}$	The column vector $[1, 1, \ldots, 1]^T$ of order $\nu \times 1$
$\mathbf{J}^*$	The row vector $[1, 0, \ldots, 0]$ of order $1 \times \nu$
$\mathscr{J}$	Column vector $[0 \cdots 0, 1]^T \otimes \mathbf{J}$ comprising a sequence of $(\mu - 1)\nu$ zeros followed by ν elements of unity
$\mathscr{J}^*$	Row vector $[1, 0 \cdots 0] \otimes \mathbf{J}^*$ comprising one element of unity followed by $\mu\nu - 1$ zeros
$\mathbf{k}_0, \mathbf{k}$	Wave propagation vectors associated with incident and scattered radiation
k	Serial index for the kth repeat unit of the chain; or, serial index for atom k or bond k (used in conjunction with h for hk pair)

k	Difference $\|j-i\|$ between ordinal numbers (serial indexes) of bonds j and i; hence, length of a sequence
k	Boltzmann constant
K_x	Equilibrium constant for configurational (e.g., helix–coil) transition in a chain of x units (Chap. VII), or for cyclization (Appendix D)
K	Molar Kerr constant (Chap. IX)
$\mathscr{K}_i$	Matrix for bond i used to generate $\langle M^4 \rangle_0$ or $\langle r^4 \rangle_0$
$\mathfrak{K}_i$	Analog of $\mathscr{K}_i$ for ith bond of a chain subject to independent rotations (see $\mathfrak{S}_i$)
$\mathbf{l}$, $\mathbf{l}_i$	Bond vector
$\{\mathbf{l}\}$	Set of all bond vectors for the chain of n bonds in a given configuration
$l = \|\mathbf{l}\|$	Magnitude of $\mathbf{l}$, i.e., the bond length
$\overline{l^2}$	Mean-square bond length averaged over all (n) bonds of the chain
$\mathscr{L}(\beta)$	Langevin function, $\mathscr{L}(\beta) = \coth \beta - 1/\beta$
$\mathscr{L}^*(\gamma) = \beta$	Inverse Langevin function
L	Macroscopic length of sample (Chap. II)
L	Contour length of the Porod-Kratky model chain (Chap. VIII and Appendix G); usually identified with $r_{\max}$
m	Difference $\|j-i\|$ between ordinal numbers for bonds j and i (Chap. IX)
m	Subscript denoting *meso* steroechemical configuration of a dyad in a vinyl polymer chain
$\mathbf{m}_i$	Vector moment associated with a given bond or group i; in particular, a bond or group dipole moment
$\mathbf{M} = \sum \mathbf{m}_i$	Vector moment for the molecule as a whole in a specified configuration
$\langle M^2 \rangle$	Average square of $\|\mathbf{M}\|$
M	Molecular weight
M_b	Molecular weight per skeletal bond
n	Number of skeletal bonds in the chain molecule
$n_{\xi\eta}$	Number of skeletal bond pairs in states ξ and η
n_t, etc.	Numbers of *trans* bonds, etc.
$\tilde{n}$	Refractive index
$p(\phi_i)$, $p_{\eta;i}$	*A priori* probability that bond i occurs in the rotational state specified by ϕ_i or by η
$p_{\zeta\eta;i}$	*A priori* probability that bonds $i-1$ and i occur in states ζ and η, respectively
$\mathbf{P}$, $\mathbf{P}_i$	The matrix $[p_{\zeta\eta}]$ of *a priori* probabilities

$P(\vartheta)$, $P(\mu)$	Radiation scattering functions expressing the dependence of the scattered intensity on the scattering angle ϑ, or on $\mu = \lvert \mathbf{k}_0 - \mathbf{k} \rvert = (4\pi/\lambda)\sin(\vartheta/2)$
$\mathscr{P}_i$, $\mathscr{Q}_i$	Generator matrices for bond i, used in treating optical anisotropies
$q_{\zeta\eta;i}$	Conditional probability for bond i in state η, given that bond $i-1$ is in state ζ
$\mathbf{Q}$, $\mathbf{Q}_i$	The matrix $[q_{\zeta\eta}]$ of conditional probabilities
$\mathbf{q}$, q	Variable of the Fourier transform $G(\mathbf{q})$ of $W(\mathbf{r})$
$\mathbf{r}$, r	The chain vector, i.e., the vector connecting the ends of the chain in a given configuration, and its magnitude $r = \lvert \mathbf{r} \rvert$
$\mathbf{r}_{ij}$, r_{ij}	The vector connecting skeletal atoms i and j, and its magnitude
$d\mathbf{r}$	The volume element (e.g., $dx\,dy\,dz$) about the terminus of vector $\mathbf{r}$
r_{max}	The value of r for the configuration of maximum extension
$\langle r^2 \rangle$	The statistical mechanical average of r^2 over all configurations of the chain
$\langle r^2 \rangle_0$	Same for the unperturbed chain
r	Subscript denoting racemic stereochemical configuration of a dyad in a vinyl polymer chain
R	Gas constant
$\mathbf{R}$, $\mathbf{R}(\tau, \rho)$	An orthogonal matrix representing rotations τ and ρ about orthogonal axes (see Appendix B)
$\mathbf{s}_i$, $\mathbf{s}_k$	Vector from the center of gravity of the configuration to the ith chain atom, or the kth unit
s	Radius of gyration in a specified configuration (see Eqs. I-5 and I-6)
$\langle s^2 \rangle$, $\langle s^2 \rangle_0$	Squared radius of gyration averaged over all configurations, and the same for the unperturbed chain (see Eq. I-11)
$\mathscr{S}_i$	Matrix for bond i, used to generate $\langle s^2 \rangle_0$
$\mathfrak{S}_i$	Analog of $\mathscr{S}_i$ for the ith bond of a chain of independent bonds
s, t	Cartesian coordinate indexes
t	Index indicating the type of structural unit
t	*Trans* conformation; used also as a subscript
T	Absolute temperature (°K)
$\mathbf{T}_i$	Transformation from the coordinate system of bond $i+1$ (or virtual bond $i+1$) to that of bond i (or virtual bond i); usually specified by Eq. I-25
$\lVert \mathbf{T}_i \rVert$	Pseudodiagonal array of transformation matrices $\mathbf{T}_i(\phi_\eta)$ for the several rotational states η for bond i (see Eq. IV-6)

$u_{\zeta\eta;i}$	Statistical weight for bond i in state η, given that bond $i-1$ is in state ζ
$\mathbf{U}_i = [u_{\zeta\eta;i}]$	Matrix of statistical weights for bond i
$v = q^2\langle r^2\rangle_0/6$	See also alternative definitions according to Eqs. IX-19, 19′, and 19″ as adapted to Rayleigh scattering of radiation
V	Volume
$\mathbf{V}$	Matrix of statistical weights for interactions of second order
$w(x)$, $w(y)$, $w(z)$	One-dimensional distributions of x, y, and z
$\mathbf{w}$	Matrix of probabilities w_{ab} that a unit of kind b succeeds one of kind a in a Markoffian copolymer
w_a	Fraction of units of kind a
$W_{ij}(\mathbf{r}_{ij})$ or $W(\mathbf{r}_{ij})$	Distribution functions for the vector $\mathbf{r}_{ij}$ in three dimensions
$W(\mathbf{r})$	Corresponding distribution function for the end-to-end vector
$\mathscr{W}(\boldsymbol{\rho})$	Distribution function for $\boldsymbol{\rho} = (3/2\langle r^2\rangle_0)^{1/2}\mathbf{r}$
x, y, z	Cartesian components of chain vector
x, y, z, and X, Y, Z	Axes of cartesian reference frames
x	Number of repeat units in a chain molecule
y_t, etc.	Number of bonds in a sequence of *trans* bonds, etc.
z, z_i, z_k	Partition function for one bond, for bond i, or for structural unit k, in cases where Z for the chain as a whole is factorable
Z	Partition function for the chain as a whole
$Z_{\mathbf{f}}$	Partition function for the chain subject to a force $\mathbf{f}$
$\tilde{Z}_{\mathbf{r}}$	Partition function for the chain with vector $\mathbf{r}$ specified
$\mathscr{Z}$	Sum of the partition functions Z for all stereochemical configurations of the asymmetric centers in a vinyl polymer chain

3. Greek Letter Symbols

$\boldsymbol{\alpha}$	Optical polarizability tensor for the molecule as a whole
$\hat{\boldsymbol{\alpha}}$	Traceless polarizability tensor, $\boldsymbol{\alpha} - \bar{\alpha}\mathbf{E}_3$
$\boldsymbol{\alpha}_i$	Optical polarizability for the group associated with skeletal bond i
$\bar{\alpha}$	Average polarizability; $\bar{\alpha} = (1/3)$ trace $\boldsymbol{\alpha}$
$\Delta\alpha_{\mathbf{r}} = (3/2)(\alpha_{\mathbf{r}} - \bar{\alpha})$	Difference between polarizabilities parallel ($\alpha_{\mathbf{r}}$) and perpendicular to end-to-end vector $\mathbf{r}$
$\Delta\alpha_F$	Difference between polarizabilities parallel and perpendicular to the electric field $\mathbf{F}$
$\boldsymbol{\alpha}'$, $\boldsymbol{\alpha}'_i$	Analogous static polarizabilities
α, β, ζ, η	Indexes for rotational states; also $\alpha = \cos\theta$, $\beta = \sin\theta$, and $\eta = \langle\cos\phi\rangle$
β	$= \mathscr{L}^*(x/nl)$ or $\beta = \mathscr{L}^*(r/nl)$

$\boldsymbol{\beta}$ $= \sqrt{3/2}\,(\boldsymbol{\alpha} - \bar{\alpha}\mathbf{E}_3)$ (Chap. IX)

$\boldsymbol{\gamma} = \boldsymbol{\beta}^C$ Column form of $\boldsymbol{\beta}$

γ^2 $=$ trace $(\boldsymbol{\beta}\,\boldsymbol{\beta})$

Γ_ζ Coefficient in the expression (Eq. III-32) of Z as a polynomial in λ_ζ^{n-2}

Γ_2, Γ_4, etc. Coefficients in the series expansions of $\Delta\alpha_r$ in powers of $r^2/\langle r^2\rangle_0$ (see Eq. IX-100)

η, $[\eta]$ Viscosity, and intrinsic viscosity

$\eta, \tau, \sigma, \psi, \omega$ Factors entering into the elements of the statistical weight matrices **U**

θ, θ_i Supplement of the valence angle, or same at skeletal atom i

ϑ Angle between scattered and incident beams of electromagnetic radiation

λ_1, etc. Eigenvalues, especially of the statistical weight matrix **U**, with λ_1 taken to be the largest

$\boldsymbol{\Lambda}$ $=$ diag $(\lambda_1, \lambda_2$, etc.)

λ Wavelength

$\boldsymbol{\lambda}$ Displacement gradient tensor (Chap. IX), with components $\lambda_x, \lambda_y, \lambda_z$

$\boldsymbol{\mu}$ Molecular dipole moment

$\boldsymbol{\mu} = \mathbf{k}_0 - \mathbf{k}$ Difference between wave propagation vectors for incident and scattered radiation, with $\mu = |\boldsymbol{\mu}| = (4\pi/\lambda)\sin(\vartheta/2)$

ν Number of rotational isomeric states for a given bond

ν Number of chains in volume V (Chap. XI)

ξ Number of bonds in a repeat unit of the chain

ρ Depolarization ratio

$\boldsymbol{\rho}$ $= (3/2\langle r^2\rangle_0)^{1/2}\mathbf{r}$

ϕ, ϕ_i (or φ, φ_i) Rotation angle about bond i, measured in right-handed sense relative to the *trans* planar conformation

$\{\phi\}$ (or $\{\varphi\}$) Set of all skeletal bond rotation angles for the chain molecule

Φ, Φ_{it} Angle between a given bond, or vector (it) affixed to a given bond, and **r** (Chap. IX and Appendix H)

χ, ψ, ω Eulerian angles

ω Angular frequence (Chap. IX)

$\Omega_{\{\phi\}}$ Statistical weight of the configuration $\{\phi\}$

Author Index

Numbers in parentheses are reference numbers and show that an author's work is referred to although his name is not mentioned in the text. Numbers in *italics* indicate the page on which the full reference appears.

Subject Index

THE

CHRISTIAN'S DAILY WALK,

IN HOLY

SECURITY ... CE.

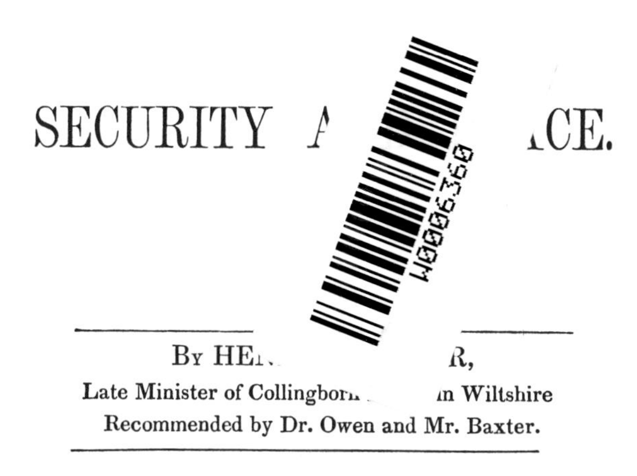

By HE... R,

Late Minister of Collingborn ... in Wiltshire

Recommended by Dr. Owen and Mr. Baxter.

Thine ears shall hear a voice behind thee, saying, This is the Way, walk ye in it.—
Isaiah xxx. 21.

Harrisonburg, Virginia

SPRINKLE PUBLICATIONS

1984

Sprinkle Publications
P. O. Box 1094
Harrisonburg, Virginia 22801

CONTENTS.

CHAPTER IV.

OF RELIGIOUS FASTING.

CHAPTER V.

OF THE LORD'S DAY, OR CHRISTIAN SABBATH.

CHAPTER VI.

CHAPTER VII.

OF WALKING WITH GOD ALONE.

CHAPTER VIII.

OF KEEPING COMPANY.

CHAPTER IX.

RULES FOR OUR RELIGIOUS CONDUCT IN PROSPERITY.

CHAPTER X

DIRECTIONS FOR WALKING WITH GOD IN ADVERSITY.

CHAPTER XI.

OF UPRIGHTNESS.

CHAPTER XII.

OF LAWFUL CARE, AND FREEDOM FROM ANXIOUS CARE.

CHAPTER XIII.

OF THE PEACE OF GOD.

CHAPTER XIV.

OF THE IMPEDIMENTS OF PEACE.

CHAPTER XV.

CONCERNING FALSE FEARS.

CHAPTER XVI.

MEANS TO ATTAIN THE PEACE OF GOD.

RECOMMENDATION

BY THE

REV. DR. OWEN.

It is now above thirty years ago since I first perused the ensuing treatise. And although until upon this present occasion I never read it since; yet the impression it left upon me in the days of my youth, have (to say no more) continued a grateful remembrance of it upon my mind. Being, therefore, unexpectedly, upon this new edition, desired, by him concerned therein, to give some testimony unto its worth and usefulness; I esteem myself obliged so to do, by the benefit I myself formerly received by it. But considering the great distance of time since I read it, and hoping perhaps that there might be, since that time, some little improvements of judgment about spiritual things in my own mind; I durst not express my thoughts concerning it, until I had given it another perusal: which I have now done. I shall only acquaint the reader, that I am so far from subducting my account, or making an abatement in an esteem thereof, that my respect unto it, and valuation of it is greatly increased; wherein also I do rejoice, for reasons not here to be mentioned. For although, perhaps, some few things might be expressed in different words or order, yet there is generally that soundness and gravity in the whole doctrine of the book, that weight and wisdom in the directions given in it for practice, that judgment in the resolution of doubts and objections, that breathing of a spirit of holiness, zeal, humility, and the fear of the Lord, in the whole; that I judge and am satisfied therein, that it will be found of singular use unto all such as in sincerity desire a compliance with his design; namely, such a walking with God here, that he may come to the enjoyment of him hereafter. I know, that in the days wherein we live, there are other notions esteemed higher or more raised, and those otherwise expressed with more elegancy of words, and pressed with more appearing strenuous ratiocinations than those contained in this book, wherewith the generality of professors seem to be more taken and satisfied. But for my part, I must say, that I do find in this, and some other practical discourses of the worthy ministers of the age past, that authority and powerful evidence of truth, arising from a plain transferring of the sacred sense of the Scripture in words and expressions suited to the experience of gracious, honest, and humble souls, that the most accurate and adorned discourses of this age do not attain or rise up unto. Such, I say, is this discourse; the wisdom and ability of whose author discover themselves from first to last, not in expressing his mind "with enticing words of man's wisdom," but in evident deduction of all his useful directions from express testimonies of Scripture, in such a way as to give light unto them, without intercepting the influence of their authority on the minds and consciences of the readers. I shall therefore say no more, but that if those into whose hands this book shall come, be not either openly or secretly enemies unto the whole design of it, as being "alienated from the life of God through the ignorance that is in them," or be not possessed with prejudices against the simplicity of the gospel, and that strictness of obedience it requireth; they will find that guidance, direction, and spiritual advantage, as having their faith, love, and obedience, increased and improved thereby; which will issue in the praise of God's grace, that ought to be the end of all our writing and reading in this world.

JOHN OWEN.

Feb. 24, 1673–4.

RECOMMENDATION

BY THE

REV. RICHARD BAXTER.

READER, I take it for some dishonour of our age, that such a book as this should need any man's recommendation, to procure its entertainment, having been so long known and so greatly approved by the most judicious and religious ministers and people, as it hath been; even to be to practical Christians, the one instead of many, for the ordering of their daily course of life, and securing their salvation and well-grounded peace. And though I know that there are some few words, especially about *perseverance*, of which all good Christians are not fully of one mind, (and I never undertake to justify *every word*, in my own books, or any others, while we all confess that we are not absolutely infallible;) yet I must say, (without disparagement to any man's labours,) that I remember not any book which is written to be the daily companion of Christians, to guide them in the practice of a holy life, which I prefer before this: I am sure, none of my own. For so sound is the doctrine of this book, and so prudent and spiritual, apt and savory the directions, and all so fully suited to our ordinary cases and conditions, that I heartily wish no family might be without it; and many volumes (good and useful) are now in religious people's hands, which I had rather were all unknown than this. And I think it of more service to the souls of men, to call men to the notice and use of such a treasure, and to bring such old and excellent writings out of oblivion and the dust, than to encourage very many who overvalue their own, and to promote the multiplication of things common and undigested, to the burying of more excellent treatises in the heap.

Reader, if thou wilt make this book (after the sacred Scripture) thy daily counsellor, and monitor and comforter, I am assured the experience of thy own great advantage, and increase of wisdom, holiness, and peace, will commend it to thee more effectually than my words can do.

Read, love, and practise that which is here taught thee, and doubt not of thy everlasting happiness.

RICHARD BAXTER.

Jan. 16*th*, 1673–4.

THE EPISTLE

TO THE READER.

The searching out of man's true happiness has exercised the wits and pens of many philosophers and divines with a different success.

1. Some, by a mistake of the end, have erred about the means. All their enterprises have ended in vanity and vexation, whilst they have caught at the shadow of fruit in a hedge of thorns, and have neglected the tree itself, whence the fruit might have been gathered with more certainty and less trouble. Man's natural corruption has so darkened his understanding, Eph. ix. 18, that in vain have the wisest men sought the happiness, which without the help of God's word and Spirit, they could never find, Acts xvii. 27. And his spiritual appetite and taste is so distempered, that he can judge of the chief good no better than a sick man can do of the best of meats.

2. Others, Eph. i. 18, having the eyes of their understanding enlightened, and their senses exercised to discern both good and evil, Heb. v. 12, have concluded, that man's true happiness consists in the soul's enjoyment of God by a holy conformity, and sweet communion with him, through Christ Jesus.

For what else is true happiness than the enjoyment of the chief good? And that God is the chief good, appears in this, that all the properties, which exalt goodness to the highest perfection, are in God only. For he is the most pure, John i. 5; perfect, universal, primary, unchangeable, communicative, desirable, and delightful good, Gen. i. 31; the efficient, pattern, and utmost end of all good, Gen. i. 27; without whom there is neither natural, moral, nor spiritual good in any creature, 1 Peter i. 16. Prov. xvi. 4. Matt. xix. 17. Our conformity to him, the apostle Peter expresses, when he says, that the saints are "made partakers of the divine nature," 2 Peter i. 4; that is, "they are renewed in the spirit of their mind, and have put on the new man, which after God, is created in righteousness, and true holiness," Eph. iv. 23, 24. So that they have, 1. A new light in their understanding, Col. iii. 10, that they know God, not only as Creator, but as Redeemer also of the world, John xvii. 3; and whilst they "behold, as in a mirror, the glory of the Lord, with open face, they are changed into the same image from glory to glory, as by the Spirit of the Lord." This knowledge is begun in this life, in the knowledge of faith, Isa. liii. 11, and shall be perfected in the life to come, in the knowledge of sense, Rom. viii. 24. This is, in a glass; that shall be face to face, 1 Cor. xiii. 9, 12. Secondly, they have a new life in their will and affections; that is, they have dispositions and inclinations in their hearts, conformable to the directions of God's holy word. This the apostle Paul intended, when he said to the Romans, that they had obeyed from the heart, the form of doctrine, whereunto they were delivered, Rom. vi. 17; that is, the word is as a mould whereinto being cast, they are fashioned according to it. Hence it is, that the saints are said to be "sealed with the Holy Spirit," Eph. i. 13, because as the seal leaves its print upon the wax, so the Spirit makes holy impressions in the soul: this is called the "writing of the law in our hearts," Jer. xxxi. 32; in allusion whereunto the apostle compares the hearts of believers to tables, 2 Cor. iii. 2, 3; and their affections or conversation to an epistle, which is said to be read and understood of all men when they walk as examples of the rule, 2 Cor. iii. 2.

3. Hence it is, that godliness hath a self-sufficiency joined with it, 1 Tim. vi. 6. Because the Christian is now in communion with God, whose face when a man beholds in righteousness, he shall be satisfied with his image, Psa. xvii. 15. Hence comes that peace of conscience, joy unspeakable and glorious, and that holy triumph and exultation of spirit, which you may observe in the apostle Paul and others, Rom. v. 1. 1 Peter i. 8. Rom. viii. 25.

Having briefly shewed what this conformity, and communion with God is, I will add one or two more words to make it manifest, that only those are truly happy who are in this estate. For, 1. Man's utmost end is, that it may be perfectly well with him, which he can never attain unto without communion with God, who is the Father of spirits, and the best of goods. Other things are desired as subordinate to this. The body is for the soul, as the matter for its form, or the instrument for its agent. Human wisdom and moral virtues are desired, not for themselves, but for the fruit that is expected by them, as glory, pleasure, and riches. Worldly and bodily pleasures, excessively desired, are as drink in a fever, or dropsy: better it is to be without the malady than to enjoy that remedy. Riches are desired not for themselves, but for the conveniences of life. Life is not so much desired for itself as for the enjoyment of happiness, which when a man has sought in the labyrinth of earthly vanities, after much vexation and disquietude of spirit, he must conclude, that it is only in that truest and chief good, which is the fountain whence true delight first flows, and the object, wherein finally it rests.

Secondly, That is man's happiness, in the possession and enjoyment whereof, his heart rests best satisfied. So far a man is from true happiness as he is from full contentment in that which he enjoys. The bee would not sit upon so many flowers, if she could gather honey enough from any one, neither would Solomon have tried so many conclusions, if the enjoyment of any creature could have made him happy. Would you know the cause why so many (like Ixion) make love to shadows and leave the substance, or (that I may speak in a better phrase) Jer. ii. 13, forsake the fountain of living water, and dig to themselves broken cisterns that will hold no water? Briefly, it is because man, who in his pride would have seen as much as God, is now become so blind that he sees not himself, Gen. iii. 5. For if men knew either the disposition of their souls by creation, or the indisposition of their souls by corruption, they would easily escape this delusion. 1. The soul is a spiritual substance, whose original is from God, and therefore its rest must be in God; as the rivers run into the sea, and as every body rests in his centre. The noblest faculties are abased, not improved; abused, not employed; vexed, not satisfied; when they are subjected to these inferior objects, as when Nebuchadnezzar fed among beasts, Dan. iv. 29;

or, as when servants rode on horseback, and masters walked like servants on the ground, Eccles. x. 7.

2. Consider the soul as it is in this state of corruption; nothing can now content it, but that which can cure it. The soul is full of sin, which is the most painful sickness; hence the prophet compares wicked men to the raging waves of the sea, that is never at rest, whose waters cast up mire and dirt, Isa. lvii. 1. What will you do to comfort him that is heart-sick? Bring him the choicest delicates, he cannot relish them; compass him about with merry company and music, it is tedious and troublesome to him; bring him to a better chamber, lay him on an easier bed: all will not satisfy him. But bring the physician to him, then he conceives hopes; let the physician cure him of his distemper, and then he will eat coarser meat, with a better stomach, and sleep on a harder bed, in a worse chamber, with a more cheerful and contented heart.

Just so it is with a guilty conscience, though he is not always sensible of it. What comfort can his friends give him, when God is his enemy? What delight can he take in his stately buildings, or frequent visits, who may expect, even this night, to have his soul required of him, and be made a companion with devils? Luke xii. 20. What is a golden chain about a leprous person, or the richest apparel upon a dead carcass? Or, what comfort will a costly banquet yield to a condemned malefactor, who is just going to execution? Surely no more than Adam found, when he had sinned in the garden, Gen. iii. 10, or than Haman had, when Ahasuerus frowned on him in the banquet, Esth. vii. 6—8. On the other side, let a man be at peace with God, and, in a sweet communion enjoy the influence of heavenly graces and comforts in his soul, he can rejoice in tribulation, Rom. v. 3 sing in prison, Acts xvi. 25, solace himself in death, Psa. xxiii. 4, and comfort his heart against principalities and powers, tribulation and anguish, height and depth, things present and things to come, Rom. viii. 38, 39. This true happiness, which all men desire, (but most miss it, by mistaking the way conducing to it,) is the subject-matter of this book. Here you may learn the right way of peace, Rom. iii; how a man may do every day's duty conscientiously, and bear every day's cross comfortably. Receive it thankfully, and read it carefully.

'But this course is too strict.'

In bodily distempers we account that physician the wisest and best, who regards more the health than the will of his patient. The carpenter squares his work by the rule, not the rule by his work. O miserable man, what an antipathy against truth is in thy cursed corrupted nature, which had rather perish by false principles, than be saved by receiving and obeying the truth! But secondly, as it is strict, so it is necessary, and in that case, strictness does not blunt, but sharpen the edge of industry to duty, therefore, saith our Saviour, Strive to enter in at the strait gate, Luke xiii. 34; that is, therefore strive to enter because the the gate is strait. Bradford well compared the way of religion to a narrow bridge, over a large and deep river; from which, the least turning awry is dangerous. We see into what a gulf of misery Adam plunged himself, and his posterity, by stepping aside from God's way. Therefore forget not these rules of the apostle: "Walk circumspectly, and make straight paths to your feet, lest that which is lame be turned out of the way," Eph. v. 15. Heb. xii. 13.

2. 'But many of God's children attain not to this strictness, yet are saved.'

It is true; though all God's children travel to one country, yet not with equal speed; they all shoot at one mark, yet not with the same dexterity. Some difference there is in the outward action, none in their inward intention; some inequalities there are in the event, none in the affection. In degrees there is some disparity, none in truth and uprightness. All that are regenerate are alike strict in these five things, at least. First, they have but one path or way wherein they all walk, Isa. xxxv. 8. Secondly, they have but one rule to guide them in that way which they all follow, Gal. vi. 15, 16. Thirdly, all their eyes are upon this rule, so as they are not willingly ignorant of any truth, 2 Peter iii. 5. Nor do they suppress, or detain any known truth in unrighteousness, Rom. i. 18, but they stand in the ways, and ask for the old path, which is the good way, Jer. vi. 16. Fourthly, they all desire, and endeavour to obey every truth, Luke i. 9, not only to walk in all the commandments of God without reproof, before men, Heb. xiii. 19, but also in all things, to live honestly and uprightly, before God, Gen. xvii. 1. Fifthly, if they fall by temptation, Gal. vi. 1, (as a member may, by accident, be disjointed) yet they are in pain till they be set right again. If they stumble, through infirmity (as sheep may slip into a puddle) yet they will not lie down, and wallow in the mire, which is the property of swine. If they are sometimes drawn aside by violent temptations, or step aside by mistake, yet they will not walk on in the counsel of the wicked, Psa. i. 1, nor will any way of wickedness, (that is, a constant, or daily course in any one sin) be found in them; they are so far from perverting the right ways of God, Acts xiii. 10, (that is, speaking evil of what is good) that they will justify God in condemning themselves, and subscribe to the righteousness of his word, praying that their ways might be directed to keep his statutes, Psa. cxix. 5.

To conclude, laying aside all cavils, beg of God a teachable disposition, and make the best profit of the labours of this faithful servant of Jesus Christ. For the matter of this book, use it as thy daily counsellor; learn to write by this copy. I mean, stir up the gifts of God that are in thee, to become more profitable to others, both in presence, by discourse, and in absence, by writing.

The Christian and intelligent reader shall find in this, some things new, other things expressed in a new manner, all digested in such a method, with such brevity and perspicuity, as was necessary to make the book a vade mecum, or pocket companion, especially profitable to the poor and illiterate.

I will here stop, wishing thee (candid and serious reader) to consider that an account must be given of what thou readest, as well as of what thou hearest, and therefore, to join prayer with thy reading, that spiritual wisdom and strength may be increased in thee, for the practice of what thou learnest. So I commend the book to thy reading; and thee, and it to God's blessing.

Thine in the Lord Jesus,

JOHN DAVENPORT.

THE

CHRISTIAN'S DAILY WALK.

CHAPTER I.

OF WALKING WITH GOD IN GENERAL.

THE INTRODUCTION.

Beloved friend, observing your forwardness and zeal in seeking to know how you might please God, and save your soul; I thought it would be acceptable and profitable to you, if I should, by the infallible rule of God's word, direct you how, with most certainty, speed, and ease, you might attain to this your holy aim. Wherefore, considering that most of God's children make their lives unprofitable and uncomfortable, by troubling themselves about many things, Luke x. 40, 41; and that too much in things less needful, by caring and fearing what shall befall them and theirs hereafter, with respect unto this present life; that you may obtain that one thing needful, Luke x. 42, and contain yourself within your own line and calling, I exhort you heedfully to apply yourself to do each present day's work with Christian cheerfulness, and to bear each present day's evil with Christian patience.

SECT. 1. WALKING WITH GOD DESCRIBED.

The best and surest way to please God, and gain a cheerful quiet heart in the way to heaven is, to walk with God in uprightness, (through faith in Jesus

Christ,) being careful in nothing: but in every thing, by prayer and supplication, with thanksgiving, to make your request known unto God, which if you do, the peace of God, which passeth all understanding, shall so establish your heart and mind, in and through Christ Jesus, that you may live in a heaven upon earth, and may be joyous and comfortable in all estates and conditions of life whatsoever.

That you should walk with God in uprightness, is commended to you in the cloud of examples, of Enoch, Gen. v. 22, 24; Noah, Gen. vi. 9; Job, Job i. 1; David, 1 Kings ix. 4; Zacharias and Elizabeth, Luke i. 6: with many others, renowned in Scripture; and is commanded to Abraham, and, in him, to all the faithful, Gen. xvii. 1.

To live by faith (which is, to frame your heart and life according to the will of God revealed in his word) and to walk with God, are all one. Enoch was said to have walked with God, Gen. v. 24; what was this else, but to rest and believe on God, whereby he pleased him? Heb. xi. 5, 6. For according to what we live, according to that we are said to walk, Colos. iii. 7. The moral actions of man's life are fitly resembled by the metaphor of walking, which is a moving from one place to another. No man, while he liveth here, is at home in the place where he shall be, Heb. xiii. 14. There are two contrary homes, to which every man is always going, either to heaven, or to hell. Every action of man is one pace or step whereby he goeth to the one place or the other. The holiness or wickedness of the action is the several way to the place of happiness, or place of torment.

So that God's own children, while they live in this world as pilgrims and strangers, are but in the way, not in the country which they seek, which is heavenly, Heb. xi. 3–16.

This life of faith and holiness, 1 Thess. i. 9, 10, what is it, but a going out of a man's self, and a continual returning to God (by Christ Jesus) from the way of sin and death, and a constant perseverance in all those acts of obedience which God hath ordained to be the way, for all his children to walk in, unto eternal life? Eph. ii. 10.

A godly life is said to be a walking with God in respect of four things that concur thereunto.

First, Whereas by sin we naturally are departed from God, Isa. liii. 6, and gone away from his ways which he has appointed for us, Rom. iii. 12, we, by the new and living way of Christ's death and resur rection, Heb. x. 20, and by the new and living work of Christ's Spirit, are brought near to God; and are set in the ways of God, by repentance from dead works, and by faith towards God in Christ Jesus; which are the first principles of true religion, Heb. vi. 1, and the first steps to this great duty of walking with God. Now, to believe and to continue in the faith, is, to walk in Christ, Col. ii. 6, 7, therefore to walk with God.

Secondly, The revealed will of God is called God's way, because in it God doth as it were display the secrets of his holy Majesty, to shew his people their way to him, and so bring them nigh unto himself; as the inspired Psalmist speaks: Righteousness shall go before him, and shall set us in the way of his steps, Psa. lxxxv. 13. Now this way of righteousness, revealed in the sacred scriptures, is the rule of a godly life: He who walketh according to God's law, is said to walk before God, (compare 1 Kings viii. 25, with 2 Chron. vi. 16.) So that he who walketh according to God's will in the various changes and conditions of life, keeping himself to this rule, walketh with God.

Thirdly, He that liveth a godly life, walketh after the Spirit, not after the flesh. He is led by the Spirit of God, Rom. viii. 1—14, having him for his guide; wherefore in this respect also he is said to walk with God, Gal. v. 16.

Fourthly, He that walketh with God, sees, by the eye of faith, God present with him in all his actions; seriously thinking of him upon all occasions, remembering him in his ways, Isa. lxiv. 5: setting the Lord always before him, as David did, Psa. xvi. 8; seeing him that is invisible, as Moses did, Heb. xi. 27; doing all things, as Paul did, as of God, in the sight of God, 2 Cor. ii. 17. Now he who so walketh that he always observeth God's presence, and keepeth him still in his view in the course of his life, not only with a general and habitual, but, as much as he can, with an actual intention to please and glorify God, this man may be said to walk with God.

Thus you may know when you walk with God: (1.) When you daily go on to repent of sins past, believe in Jesus Christ for pardon, and believe his word for direction. (2.) When you walk not according to the will of man, but of God. (3.) When you walk not after the flesh, but after the Spirit. (4.) When you set God before you, and walk as in his sight, then you walk with, before, after, and according to God: for all these are understood in one sense.

That you may walk with God, consider these arguments further to convince and induce you:

SECT. 2. REASONS WHY CHRISTIANS SHOULD WALK WITH GOD.

First, You are commanded to walk as Christ walked, 1 John ii. 6; and it concerns you so to do, if you would approve yourself to be a member of his body: for it is monstrous, nay, impossible, that the head should go one way, and the body another. Now our Saviour himself observed all these methods of walking with God, justifying faith and repentance only excepted, because he was without sin.

Secondly, It is all which the Lord requireth of you, for all his love and goodness showed unto you, in creating, preserving, redeeming, and saving you. For what doth the Lord require of you, but to do justly, and to love mercy, and to walk humbly with your God? Micah vi. 8.

Thirdly, If you walk with God, and keep close to him, you will be sure to go in the right way, in that good old way, Jer. vi. 16, which is called the way of holiness, Isa. xxxv. 8; in a most straight, Prov. iii. 17, most sure, and (to a spiritual man) most pleasant way, the paths of which are peace; the very happiness and rest of the soul, Jer. vi. 16. God teacheth his children to choose this way, Isa. xlviii. 17, Psa. lxxxv. 13, Psa. xxxvii. 23. And if they happen to err, or to doubt of their way, they shall hear the voice of God's Spirit behind them, saying, This is the way, walk in it, Isa. xxx. 21.

Fourthly, If you walk with God, you shall walk safely; Prov. iii. 24, Psa. xxxvii. 24; you will not need to fear, though ten thousand set themselves against you, Psa. iii. 5, 6; for his presence is with you, and for you. His holy angels encamp about you, Psa. xxxiv. 7; and while you walk in his ways, they are charged to support you, Psa. xci. 11, 12, lest you should receive any harm.

Fifthly, When you walk with God (though you be alone, separate from all other society) you still walk with the best company, even such whereof there is most need, and best use. While God and you walk together, you have an advantage above all that walk not with him; for you have a blessed opportunity of that holy acquaintance with God, which is expressed Job xxii. 21—30. You have opportunity to speak unto him, praying with assurance of a gracious hearing. Abraham and his faithful servant made use of their walking with God for these purposes, Gen. xxiv. Is it not a special favour that the most high God, whose throne is in heaven, should condescend to walk on earth with sinful man? nay, rather to call up man from earth to heaven, to walk with him? Phil. iii. 20, Colos. iii. 2. It would be therefore shameful ingratitude not to accept this offer, and not to obey this charge.

Sixthly, To set the Lord always in your sight, is an excellent preservative and restraint from sin. With this shield Joseph did repel and quench the fiery darts of the temptations of his designing mistress, Gen. xxxix. 9. For who is so foolish, and shameless, as wilfully to transgress the just laws of a father, king, and judge, knowing that he is present, and observes him with detestation if he so do?

Seventhly, To set the Lord always before you, Psa. cxix. 168, is an excellent remedy against spiritual sloth and negligence in duties, and it is a sharp spur to quicken, and make you diligent and abundant in the work of the Lord. What servant can be slothful and careless in his master's sight? And what master will keep a servant that will not observe him, and do his commands, while he himself looketh on?

Eighthly, Walking with God in manner aforesaid, doth exceedingly please God, Heb. xi. 5. It also pleases God's holy angels, 1 Cor. xi. 10. It pleases God's faithful ministers, 3 John ver. 3, and doth please and strengthen all the good people of God, Psa. cxix. 74, with whom you do converse. It is to walk worthy of God in all well pleasing, Colos. i. 9, 10.

Ninthly, Thus walking with God, you shall be assured of God's mercy and gracious favour. He keepeth covenant and mercy with all his servants, that walk before him with all their heart, 1 Kings viii. 23. When you do thus walk in the light, you have a gracious fellowship with God, and the blood of Jesus Christ cleanseth you from all sin, 1 John i. 7. There is no condemnation to you who thus walk, Rom. viii. 1. Your flesh, when you die, shall rest in hope. For to them that set God before them, he doth show the path of life, which will bring them into his glorious presence, where are fulness of joys, and pleasures for evermore, Psa. xvi. 11.

Any one of these motives, seriously thought upon by an humble Christian, is enough to persuade him to this holy walking with God.

Notwithstanding, it is sad to consider, how few there are who walk thus. For most men seek not after God, God is not in all their thoughts, Psa. x. 4; they walk in the vanity of their minds, Eph. iv. 17, after their own lusts, 2 Peter iii. 3; the lust of the flesh, the lust of the eye, and the pride of life, 1 John ii. 16; walking according to the course of this world, according to the will of Satan, the prince of the power of the air, Eph. ii. 2, the spirit that now worketh in the children of disobedience; who refuse to return, or to call themselves into question concerning their ways, though God doth wait and hearken for it; no, not so much as to say, What have we done? Jer. viii. 6; but every one runneth to his course, as the horse rusheth into the battle.

Now concerning all that walk thus contrary unto God, God hath said, that he will set his face against them, and punish them seven times, Lev. xxvi. 21—28; even with many and sore plagues. And if yet they will walk contrary to him, he will walk contrary to them in fury, and punish them seven times more for their sins. And if yet they will walk in impenitency, notwithstanding God's offer of mercy to them in Christ, Paul could not speak of such with dry eyes, but peremptorily pronounceth that their end is destruction, Phil, iii. 18, 19.

Weigh well, therefore, these premises; compare the way, wherein you walk with God, with all other ways; compare this company with all other company, and the issues and end of this way with the issues and end of all other ways: and the proper choice of your walk will easily and quickly be made.

Thus much may be said in general of walking with God.

SECT. 3. WALKING WITH GOD, TO BE CONSTANT AND UNIVERSAL.

The commandment to walk with God is indefinite, without limitation, therefore must be understood to be a walking with him in all things, and that at all times, in all companies, and in all changes, conditions, and estates of your life, whatsoever. To walk with God in general and at large is not sufficient.

You are not dispensed with for any moment of your life: but all the days of your life, and each day of your life, and each hour of that day, and each minute of that hour; you must pass the time, 1 Peter

i. 17, the whole time of your dwelling here in fear; even all the day long, saith Solomon, Prov. xxiii. 17. You must endeavour to have a conscience void of offence always, Acts xxiv. 16. You must live the rest of your life, 1 Peter iv. 2, not to the lusts of men, but to the will of God; taking heed lest at any time there be in you an evil heart of unbelief, in departing from the living God, Heb. iii. 12.

1. For this end Christ did redeem you from the hands of your enemies, that you might serve him in holiness and righteousness (which is the same with walking with God) all the days of your life without fear, Luke i. 74, 75.

2. The end of the instructions of God's word, which is the light of your feet in this walking, is, that it be bound upon your heart continually, to lead, keep, and converse with you at all times, Prov. vi. 21, 22.

3. The lusts of your own heart, and your adversary the devil lie always upon the advantage to hinder you in, or divert you from, this godly course, 1 Peter v. 8, so that, upon every intermission of your holy care to please God, they take their opportunity to surprise you.

4. You are accountable unto God for losing and mis-spending all that precious time wherein you do not walk in his ways, Eph. v. 16.

5. Besides, he that hath much work to do, or that is in a long journey, or is running a race for a wager, hath no need to lose any time. If you be long obstructed in your Christian work and race, by sin and sloth, you will hardly recover your loss but with much sorrow, with renewed faith, and with more than ordinary repentance.

Wherefore when you awake in the night, or in the morning, and while you are employed in the day, and when you betake yourself to sleep at night, you must, as David, have thoughts on God, and set him always before you, Psa. xvi. 8. Acts ii. 25. When I awake, I am still with thee, saith he, Psa. cxxxix. 18; and in the night he remembered God, Psa. lxiii. 6; and his hope and meditation was on God's word, Psa. cxix.

147, 148. And Isaiah (in the person of all the faithful) saith, With my soul have I desired thee in the night, yea, with my spirit within me will I seek thee early. Isa. xxvi. 9.

CHAPTER II

OF BEGINNING THE DAY WITH GOD.

SECT. 1. HOW TO AWAKE WITH GOD.

In the instant of awaking let your heart be lifted up to God with a thankful acknowledgment of his mercy to you. For it is he that giveth his beloved sleep, Psa. cxxvii. 2; who keepeth you both in soul and body while you sleep, Prov. vi. 22; who reneweth his mercies every morning, Lam. iii. 22, 23. For, while you sleep, you are as it were out of actual possession of yourself, and all things else. Now, it was God that kept you, and all that you had, and restored them again, with many new mercies, when you awaked.

2. Arise early in the morning (if you be not necessarily hindered) following the example of our Saviour Christ, John viii. 2, and of the good matron in the Proverbs, Prov. xxxi. 15. For this will usually much conduce to the health of your body, and the prosperity, both of your temporal and spiritual state; for hereby you will have the day before you, and will gain the best, and the fittest times for the exercises of religion, and for the works of your calling.

3. In the time between your awaking and arising (if other suitable thoughts offer not themselves) it will be useful to think upon some of these: I must awake from the sleep of sin, to righteousness, Eph. v. 14. 1 Cor. xv. 34; as well as out of bodily sleep, unto labour in my calling. The night is far spent, the day

is at hand, I must therefore cast off the works of darkness, and put on the armour of light, Rom. xiii. 11, 12, 13. I must walk honestly as in the day. I am, by the light of grace and knowledge, to arise and walk in it, as well as by the light of the sun to walk by it. Think also of your awaking out of the sleep of death, and out of the grave, 1 Cor. xv. 55; at the sound of the last trumpet, 1 Thess. iv. 16; even of your blessed resurrection unto glory, at the last day. It was one of David's sweet thoughts (speaking to God) When I awake, I shall be satisfied with thy likeness, Psa. xvii. 15.

4. When you arise, and dress yourself, lose not that precious time (when your mind is freshest) with impertinent and fruitless thoughts, as is the custom of too many to do. This is a fit time to think upon the cause why you have need of apparel; namely, the fall and sin of your first parents, which from them is derived to you. For before their fall, their nakedness was their comeliness, Gen. i. 31, and seeing it, they were not ashamed, Gen. ii. 25. It will likewise be to good purpose to consider what the wise providence of God hath appointed to be the substance of your apparel. The rinds of plants, the skins, hair, or wool of brute beasts, and the bowels of the silkworn; the very excretions and superfluous apparel of unreasonable creatures. Which, as it doth magnify the wisdom, power, and goodness of God, in choosing, and turning such mean things to such excellent use: so it should humble and suppress the pride of man. For what man in his senses would be proud of the badge of his shame, even of that apparel, for which (under God) he is beholden even to plants and beasts?

Now also is a good time to call to mind what rules are to be observed, that you may dress yourself as becometh one that professeth godliness: namely, 1. That your apparel, for matter and fashion, do suit with your general and special calling, 1 Tim. ii. 9, 10, and with your estate, sex, and age, Deut. xxii. 5.

2. That your apparel be consistent with health and comeliness, 1 Cor. xi. 14, 15. 1 Cor. xii. 23.

3. That you rather go with the lowest, than with the highest of your state and place.

4. That the fashion be neither strange, immodest, singular, nor ridiculous, Zeph. i. 8.

5. That you be not over curious, or over long, taking up too much time in putting it on.

6. Neither the making nor wearing of your apparel, must savour of pride, lightness, curiosity, lasciviousness, prodigality, or base covetousness, Isa. iii. 18—24 But it must be such as becometh holiness, wisdom, and honesty, and such as is well reported of, Phil. iv. 8. 1 Cor. xi.

7. Follow the example of those of your rank and means, who are most sober, most frugal, and most discreet.

While you dress yourself, it will be seasonable and profitable also, by this occasion, to raise your thoughts, Rev. iii. 18; and fix them upon that apparel which doth clothe and adorn your inward man, 1 Peter iii. 4, which is spiritual, and of a divine matter, which never is out of fashion, which never weareth out, but is always the better for the wearing. Think thus: If I go naked without bodily apparel, it will be to the shame of my person, and to the hazard of my health and life. But how much more will the filthy nakedness of my soul appear to the eyes of men, of angels, and of God himself, Rev. iii. 17. Rev. xvi. 15. Exod. xxxii. 25, whose pure eyes cannot abide filthiness, Hab. i. 13, whereby my soul will be exposed to most deadly temptations, and my whole person to God's most severe judgments; except I have put on, and do keep on me the white linen of Christ's spouse, the righteousness of the saints, Rev. xix. 8, that is, justification by faith in Christ, and sanctification by the Spirit of Christ?

And because every day you will be assaulted with the world, the flesh, and the devil, you will do well to consider whether you have put on, and do improve your coat of mail, that complete armour, prescribed Eph. vi. 11–18.

When you use your looking-glass, James i. 23, 24,

25, and by experience and that it serveth to discover, and to direct you how to reform whatever is uncomely, and out of order in your body; you may hereby remember the necessity and admirable use of the glass of God's word, and gospel of Christ, both read and preached, for the good of your soul. For, this being understood and believed, doth not only show what is amiss in the soul, and how it may be amended; but in some measure will enable you to amend; for, it doth not only show you your own face, but the very face and glory of God in Christ Jesus, which by reflection upon you, will, through the Spirit, work on you a more excellent effect than on Moses's face in the mount, Exod. xxxiv. 29, 30, which yet was so glorious, that the people could not endure to behold it. For by this glory of God, which by faith you behold in the word, you will be changed into the same image, from glory to glory, even as by the Spirit of the Lord, 2 Cor. iii. 18.

Concerning these things which I have directed to be thought upon, when you arise, and put on your apparel in the morning, and those which I shall direct when you put off your apparel at night; my meaning is not to urge them as necessary, as if it were sin to omit any of these particulars: but to be used, except better come in place, as most convenient.

SECT. 2. OF BEGINNING THE DAY WITH GOD, BY RENEWED FAITH AND REPENTANCE.

(1.) WHEN you are thus awake, and are risen out of your bed, that you may walk with God the remainder of the day, it will be needful that you first renew your peace with God, by faith in Jesus Christ; and then endeavour to show your dutifulness and gratitude to God, by doing those works of piety, equity, mercy, and sobriety, which may any way concern you that day. For how can two walk together, except they be agreed? Amos iii. 3. And how can any walk with God, if he be not holy in all his conversation? You have as much cause to beware of him, and to obey his voice, and not provoke him who goeth before you in the wilderness of this world, to guide and bring you to his heavenly kingdom, Exod. xxiii. 20, 21, 22, as the Israelites had to beware of him who went before them to keep them in the way, and to conduct them unto the earthly Canaan, the place which he had promised and prepared for them. It was for this, that Joshua told the people, that except they would fear the Lord, and serve him in sincerity, and put away their strange gods, they could not serve God, Josh. xxiv. 14—19; they could not walk with him. For he is (saith he) a holy God: he is a jealous God: he will not forgive your transgressions, nor your sins.

(2.) For this cause (if unavoidable necessity hinder not) begin the day with solemn prayer and thanksgiving, Psa. xcii. 1, 2. Psa. lxxxviii. 13. Before which (that these duties may be the better performed) it will be convenient, if you have time, that you prepare yourself by meditation, Lam. iii. 40, 41. Job xi. 13; the matter whereof should be an inquiry into your present state, how all things stand between God and you; how you have behaved since you last prayed and renewed your peace with God; what sins you have committed, what graces and benefits you want, what fresh favours God has bestowed on you, Psa. cxvi. 1—13, since last you gave him this tribute of thanks; and how much praise and thanks you owe to him also for the continuance of former blessings. Think also what employments you shall have that day, in which you may need his special grace and assistance. Consider likewise what ground and warrant you have to approach to the throne of grace, to ask pardon, and to hope for favour and help of God. Upon these considerations, you must seriously and faithfully endeavour, in the strength of Christ (without whom you can do nothing) to reform whatsoever you find to be amiss, Job xi. 14; flying unto, and only relying upon God's mercy in Christ; to acknowledge him in all things; and that you will now seek grace and help of him, whereby you may walk as in his sight in all well pleasing, all that day.

To assist you therein, do thus:

First, Lay a strict charge upon your conscience to deal impartially, plainly, and fully, in this examination and judging of yourself.

Secondly, You should be so well acquainted with the substance and meaning of God's holy law, Deut. vi. 8, 9, that you may be able to carry in your head a catalogue or table of the duties required, and vices forbidden, in each commandment; whereby you may try your obedience past, and may set before you a rule of life for time to come.

Thirdly, (lest the calling to mind the multitude and greatness of your sins should make you despair of God's favour) You should be so well instructed in the Christian faith, and in the principal promises of the gospel, that you may be able also quickly to call them to mind, for the strengthening of your faith and hope in God. The form of sound words in the gospel, 2 Tim. i. 13, should be familiar unto you for these purposes.

All these need not take up much time: you will find it to be time well redeemed. For, first, by such preparation you will keep yourself from that rude and irreverent thrusting yourself into God's holy presence, whereof you are warned in the Scriptures, Eccles. v. 1, 2.

Secondly, When by this means your heart is well humbled, softened, and set right towards God, so that you can say, you regard no iniquity in your heart, Psa. lxvi. 18. John ix. 31; and when hereby you have called in your thoughts from straggling, and have gotten composedness of mind, and inward strength of soul (without which the arrow of prayer can never fly home to the mark) then you may approach into God's special presence with more faith and boldness; you shall be more able to utter before him apt confessions, lawful requests, and due thanksgivings, with more understanding, more humbly, more feelingly, more fervently, and with more assurance of a gracious hearing, (all which are requisite in prayer) than you could ever possibly be able to do without such preparation.

Thirdly, This due preparation to prayer doth not only fit you to pray; but is an excellent furtherance to an holy life. For it maketh the conscience tender and watchful, by the daily exercise of the knowledge of the precepts and threats of the law, and the precepts and promises of the gospel; and it being enforced to examine, accuse, judge, and pass sentence, and do a kind of execution upon you for your sin; smiting your heart, and wounding itself with godly fear, grief, and shame, (a work to which the conscience is loth to come, till it needs must;) wherefore, to prevent all this trouble and smart, it will rather give all diligence in other acts which are more pleasing; namely, it will direct you in the ways to God, check and warn you before hand, lest you should sin; to the end that when you come to examine yourself again, it might find matter, not of grieving and tormenting, but of rejoicing and comforting your heart, which is the most proper, and most pleasing work of a sanctified conscience, 2 Cor. i. 12. He that knoweth that he must be at much pains to make himself whole and clean, when he is wounded and defiled, will take the more heed lest he wound and defile himself.

Fourthly, This due preparation to prayer, by examining, judging, and reforming yourself, doth prevent God's judging you; for when you judge yourself, you shall not be judged of the Lord, saith the apostle, 1 Cor. xi. 31.

(3.) Being rightly prepared, you must draw near into God's special presence, falling low at his footstool, Psa. xcv. 6, representing him to your thoughts as one who is in himself, and of himself, the only heavenly, all-knowing and almighty Majesty, Matt. vi. 9, now become your loving and merciful Father, through Christ his Son your Lord: then you must pour out your soul before him in confessing your sins, 1 Sam. i. 15; and in making your desires (through the Spirit) known unto him in the name of Christ, for yourself and others, in all lawful petitions and supplications, with thanksgiving, Phil. iv. 6; and all this with understanding, 1 Cor. xiv. 15, with the intention and full bent of the soul, James v. 16, and expectation of being heard, Mark xi. 24, in due time and measure, and in the best manner.

SECT. 3. FURTHER DIRECTIONS CONCERNING PRAYER.

UNTO the directions both for preparation to prayer, and concerning prayer itself, take these cautions.

First, If it may be, omit neither the one nor the other, and let them be the first work after you are up, Psa. v. 3. But if that cannot be, because of some necessary hinderance, yet perform them so soon as you can, and as well as you can; though you can do neither, either so soon, or so well as you would, yet omit them not altogether. Break through all seeming necessities, which will daily come in your way, to hinder and thrust out these duties. The devil, knowing that nothing doth undermine and overthrow his kingdom more than these duly performed; knowing also that the spiritual performance of them is tedious to corrupt nature, will thrust upon you seeming necessities, so many, and so often, that if you be not watchful to gain, and to take time, breaking through all such hinderances as are not truly necessary; you will often, by the circumvention of the flesh and of the devil, be brought to an omission of preparation, or of prayer, or both. Upon which will follow similar temptations, together with a proneness to the like neglect, and a greater indisposition to these duties afterward.

Secondly, Lay not too great a task upon yourself in this preparation to prayer; I mean, so much as will take up more time than the works of your calling, and other needful affairs, will permit; but contrive and husband your time so, that every lawful business may have its own time, Eccles. iii. 1. God has subordinated the works of your general and particular calling in such sort, that, usually, the one shall not obstruct the other.

If, through taking up too much time in preparation to prayer, and in prayer, either of them grow necessarily tedious and burthensome; Satan will circumvent you by this means, causing you out of a true weariness of too much (even before you are aware) to omit them altogether.

Thirdly, Whereas when you prepare yourself to pray, and when you do pray, it is lawful to think of your worldly business, to the end that you might pray for direction and for good success therein (for you may ask your daily bread,) Matt. vi. 11; you must take heed, when you think of these things, that your thoughts be not worldly through distempers and distractions about the same, Luke xii. 29. For these will abate your spirituality and fervour in prayer, and will shut the ears of God against your prayer.

SECT. 4. SIGNS OF WORLDLY-MINDEDNESS IN DEVOTION, AND REMEDIES AGAINST IT.

If you desire to know the signs and remedies of distempers and distractions about worldly things in your preparation for holy duties; by distempers, I mean, inordinate trouble about the means; and by distractions, I mean, a vexing trouble about success.

I. As to the signs of it. You may know that your mind is distempered with worldliness (even in thinking on lawful business) when you prepare yourself to prayer, and at other seasons, by these marks:

1. When (except in case of necessity in their apparent danger) your worldly affairs are first in your thoughts to be the matter of your meditation. For thoughts how to hallow God's name, and how his kingdom may come, and how you may do his will, should usually be in your mind, before those that concern your daily bread.

2. When they interpose themselves, interrupt, and

jostle out those good thoughts whereon you were thinking, before you have thought of them sufficiently.

3. When your thoughts of worldly business are with greater intention of mind, than the thoughts of things spiritual and heavenly.

4. When they last longer than such as immediately concern the glory of God, and the good of your soul: or hold you too long upon them.

5. You may know it by the ends which you propose to yourself in your thoughts of worldly business; are the ends you propose, only, or chiefly, that you may prevent poverty, or that you may have wherewith to satisfy your natural desires? If you propose not other, and more spiritual ends, your thoughts of them at that time, are worldly: But if your thoughts of your worldly business, be to the end that you may lay them to the rule of God's word, that you may not offend him in your labour and care about them; or that you might crave God's direction and blessing upon your said care and labour, you being spiritual in thoughts of worldly business, then your thoughts of lawful business are not distempered with worldliness.

II. To remedy these distempered thoughts; 1st, Let a sound and clear judgment discern what is good, what is bad; also what is best, and what is least good; preferring things spiritual, heavenly, and eternal, incomparably before those which are earthly and temporal. Make those best things your treasure, Matt. vi. 21; then your heart will be chiefly set, and your thoughts will chiefly run on them; and you will be moderate in thinking of those things which are less needful.

2. Do as a wise counsellor at law, or as a master of requests, who must hear many clients, and receive and answer many petitions. Consider whose turn it is, and what is the most important suit; and despatch them first. Let thoughts of worldly business be shut out, and made to stand at the door, till their turn come to be thought upon, and let the more excellent, and more needful be despatched first.

3. If thoughts of the world will imprudently intrude themselves, and will not be kept out; rebuke them sharply; give them no hearing, but dishearten them, and rebuke the porter and keeper of the door of your heart; that is, smite, wound, and check your conscience, because it did not check and restrain them.

4. In all lawful business, insure yourself fully and sufficiently to intend that one thing which you have in hand for the present, Eccles. ix. 10; and at all times restrain wandering thoughts as much as may be. Let your reason get such power over the fancy, that you may be able to think of what you please, when you please. You will say, "to a fickle mind this is hard, if not impossible." To this I answer, if you would not nourish and entertain evil, flying, and unseasonable thoughts when they arise; and would (as often as they offer themselves) be much displeased with them, and with yourself for them; then in time you will find it possible, and not exceedingly hard to think of what good things you would, and not of what evil things you would not.

5. Lastly, When the time of thinking and doing of your worldly business is come, then think thereof sufficiently, and to good purpose; for then they will be the less troublesome in thrusting themselves in out of place, because it is known that in their place, they shall be fully regarded. Idleness and improvidence about these things, puts a man into straits many times, and into distempers about his worldly business, more than needs, or else would be.

You would also know when your thoughts of success in your worldly affairs are evil, together with a remedy against them.

To think, that, if you be not prudent and diligent in your calling, and that if God do not bless your diligence, you may do the works of your calling in vain, and may expect ill success; thus to think is lawful and useful. For it will excite in you a resolution to be frugal and diligent; and when you have done all you can, these thoughts also will quicken you to prayer unto God for success. But if your thoughts of thriving, or not thriving, be other than these, and bring forth other effects; namely, if desires of success drive you to think of using unlawful means, from doubting that you cannot so soon, or so certainly, or not at all, speed by the use of lawful only: if it make you full of anxiety and fear, that though you use what good means you can, all will be in vain: if you be yet doubtful and take anxious thought about what you shall eat, what you shall drink, and what you shall put on, or how you and yours shall live another day, then your thoughts about success in wordly business, are worldly and distracted.

I shall speak to this sin with its remedy more fully when I write against taking care in any thing.

Yet for the present, know: All the fruit you will reap from unbelieving fears and distrusts, doubts of success, &c., will be nothing else, but a further degree of vexation of heart. For all the anxiety in the world cannot bring good success. Besides, nothing provokes the Lord to give ill success sooner, than when you nourish distrustful care.

Secondly, Consider the power and faithfulness of God, who has taken care of the success of your labour upon himself: commanding you not to care but to cast all the care on him, 1 Peter v. 7. If you would rest upon this, you might be secure of good success in your outward state, even according to your desire; or else God will more than recompense the want thereof, by causing you to thrive, and to have good success in spiritual things, which is much better, and which you should desire much more.

4. A fourth caution to be observed in your preparation to prayer, and in prayer, is, be not slight and formal herein, which is, when cursorily and out of custom only you call your sins, your duties, God's favours, and his promises, into a bare and fruitless remembrance. For if the heart be not seriously affected with anger, fear, grief, and shame for sin; and if it be not affected with a thankful acknowledgment of being beholden to God for his favours; moreover, if it be not affected with hope and confidence in God

at the remembrance of his blessed promises; and if withal, the heart be not gained to a renewed resolution to reform what is faulty, and to cry earnestly to God for grace and mercy; and for the time to come to endeavour to live a godly life; all your preparation is nothing. Nay, this slight and fruitless calling of sin and duty to remembrance, and no more, is a great emboldener and strengthener of sins; and a great weakener and quencher of the Spirit. For sins are like to idle vagrants, and lawless subjects; if officers call such before them, and, either say nothing to them, or only give them threatening words, but do not smite them and make them smart, they grow ten times more bold, insolent, and lawless. Good thoughts are like to dutiful servants and loyal subjects, such as are ready to come at every call, and offer themselves to be employed in all good services; now if such be not entertained with suitable regard, if they be not cherished in their readiness, they (like David's people) return disheartened, and their edge to future service is taken off, 2 Sam. xix. 3. Besides, this cursory performing of holy duties, is the highway to a habit of hypocrisy, that accursed bane of all that is good.

5. My last caution is, that if in your meditations, and in your prayers, you find a dulness and want of spirituality, I would have you to be humbled in the sense of your impotency and infirmity; yet, be not discouraged nor give them over, but rather betake yourself to these duties with more diligence and earnestness. When you want water, (your pump being dry) you by pouring in a little water, and much labour in pumping, can fetch water; so, by much labouring the heart in preparation, and by prayer, you may recover the gift of prayer, Luke xi. 13. And, as when your fire is out, by laying on fuel, and by blowing the spark remaining, you kindle it again: so by meditation you stir up the grace that is in you, 2 Tim. i. 6, and by the breath of prayer, may revive and inflame the spirit of grace and prayer in you. Yet, if you find that you have not time to prepare by meditation; or having done so, if you find a confusion and distraction in

your meditation, then it will be best to break through all hinderances, and without further preparation attend to the duty of prayer, only with premeditation of God to whom, and of Christ by whom, through the Spirit, you must pray.

If for all this you do not find satisfaction in these holy exercises; yet give them not over: for God is many times best pleased with your services, when, through an humble sense of your failings, you are displeased with yourself for them. Yet more, if when you have wrestled and striven with God and your own heart in prayer, you are forced to go halting away, with Jacob, Gen. xxxii. 25—31, in the sense of your infirmities; yet be not discouraged, for it is a good sign that you have prevailed with God as Jacob did, Gen. xxxii. 28.

God uses, when he is overcome by prayer, to work in them that do overcome some sense of weakness, to let them know, that they prevail with him in prayer, not by any strength of their own or by any worthiness of their prayer, when they have prayed best; but from the goodness of God's free grace, from the worthiness of Christ's intercession, by whom they offer up their prayers, and from the truth of his promise made unto them that pray. If it were not thus, many, when they have their hearts' desire in prayer, would ascribe all to the goodness of their prayers, and not to the free grace of God; and would be proud of their own strength, which, in truth, is none at all.

CHAPTER III.

DIRECTIONS FOR WALKING WITH GOD, IN THE PROGRESS OF THE DAY.

SECTION 1.—When you have thus begun the day in prayer by yourself, seeking peace with God through Jesus Christ, and craving his gracious presence to be with you, and for you, that day, you must then conscientiously, according to the nature of the day (be it one of the six days, or the Lord's day) apply yourself to the business of that day, whether it be in acts of religion, or of your personal calling, or in any other works belonging unto you, as you are superior or inferior in family, church, or commonwealth; doing all as in God's sight.

And because all lawful business is sanctified by the word and prayer, 1 Tim. iv. 5, and it is part of your calling (if you are master of a family) to govern your children and servants in the fear of God, and to teach them to live godly; therefore it is your duty to take the fittest time in the morning to call them together, and pray with them; before which prayer, it will be profitable to read the Scripture in order, with due reverence, taking all opportunities, in fit times, to instruct them in the principles of religion, often pressing the word upon them, Deut. vi. 7.

If it be a working day, with cheerfulness and diligence, attend to the work of your particular calling. For whosoever hath no calling whereby he may be profitable to the society of man in family, church, or commonwealth; or having a lawful calling doth not follow it, he lives inordinately, 2 Thess. iii. 10, 11. God never made any man for play or to do nothing. And whatever a man doth, he must do it by virtue of his Christian calling, receiving warrant from it, else he cannot do it in faith, without which no man can please God, Heb. xi. 6. Besides, whosoever is called to Christianity, has no way to heaven but by walking with God in his personal and particular calling, as well as in his general calling, 1 Cor. vii. 17—24.

1. That you may do this, be sure that the thing whereabout you labour, either with head or hand, be lawful and good.

2. Be diligent and industrious, Eph. iv. 28; for the sluggard and idle person desireth, but has nothing; but the diligent hand maketh rich, Prov. x. 4.

3. Let there be truth, plainness, and equity in all your dealings with men, Prov. x. 4. Circumvent and defraud no man, 1 Thess. iv. 6. Make not your own gain the weight and measure to trade by. I will propose to you sealed weights and rules, according to which, you must converse with all men.

(1.) Consider your neighbour's good as well as your own. Weigh impartially with yourself what proportionable advantage (in common estimation) your neighbour is like to have for that which you receive of him. For you must love your neighbour as yourself, Matt. xxii. 39. In whatsoever you have to do with men, you must not look only to your own advantage, but to the benefit also of your neighbour, Phil. ii. 4. Observe therefore the royal law, the standard of all equity in this kind: Whatsoever you (with a rectified judgment and honest heart) would that men should do unto you, do you even so unto them: for this is the law and the prophets, Matt. vii. 12.

(2.) Be watchful that you let not slip your opportunities of lawful advantage, Prov. vi. 6—8; and take heed lest in these evil times you be circumvented by fraud and falsehood, and be ensnared by unnecessary suretyship, Prov. xi. 15, xxii. 26, vi. 1—6.

Whereas in every calling there is a mystery, and for the most part each calling and condition of life has its special sin or sins, which the devil, and custom, for gain or credit sake amongst evil men, have made to seem lawful; yea have put a kind of necessity upon it, which cannot be shunned without exposing a man's self to censure; look narrowly therefore by the light of God's word, and by experience, to find out that or those sins, and then be as careful to avoid them.

SECT. 2. CONCERNING SUPERIORS AND INFERIORS.

THERE are other works also, such as concern you as you are a superior, and in authority; or as you are inferior, and subject, either in family, church, or commonwealth; in doing which you must act for the glory of God, following the directions of his word and Spirit.

I. As you are a superior, 1st. Walk worthy of all honour and due respect, behaving yourself in your place with such holiness, wisdom, gravity, justice, and mercy; and observing such a medium between too much rigour and remissness, between straining your authority too far, and relaxing it too much, that those under your charge may have cause to fear and love you, Lev. xxv. 43.

2. Wait on your office, and be watchful over your charge with all diligence and faithfulness; using all good means to direct and preserve them in the duties of godliness and honesty, 1 Tim. ii. 2, which is the only end why God has set you over them. The means are, (1.) Go before them in good example. Examples of superiors have a kind of constraining power, working strongly and insensibly upon inferiors. (2.) Pray with and for them, Job i. 5. (3.) Command only things lawful, possible, and convenient, and only those to which the extent of your authority from God and man doth allow you. (4.) As much as in you lies, procure for them the means, and put them upon the opportunities of being, and of doing good, Exod. xx. 8—10. (5.) Prevent likewise and remove all occasions of their being, and of doing evil. (6.) Protect and defend them, according to your power from all wrongs and injuries. (7.) When they do well, encourage them, by letting them see that you take notice as readily of their well-doing, as of their faults, Psa. ci. 6; and so far as is fit, let them have the praise and

fruit of their well-doing, Prov. xxxi. 31. (8.) When they do evil, rebuke them more or less, according to the nature of their fault: but never with bitterness, Col. iii. 19—21, Eph. vi. 9, by railing at, or reviling them, in terms of disdain and contempt. There should be always more strength of reason in your words to convince them of their sin, and to make them see their danger, and to know how to be reformed, than heat of anger, in uttering your own displeasure. (9.) If admonitions and words will reclaim them, then proceed not to correction and blows; but if they regard not your reproofs, then according to the nature of the fault, and condition of the person, and the limits of your authority, you must, in mercy to their soul, give them sufficient but not excessive punishment, Prov. xxix. 15—19. (10.) When you have done thus, and have waited a convenient time for their amendment, but find none; when they thus declare themselves to be rebellious, you must seek the help of higher authority, Deut. xxi. 18—21.

That you may govern according to these directions: consider well and often, first, that those whom you govern, are such whom you must not oppress, neither may you rule over them with rigour, Lev. xxv. 39—43; because they now are, or may be heirs of the same grace together with you, 1 Peter iii. 7.

Secondly, Remember often, that you have a Superior in heaven, Eph. vi. 9, Col. iv. 1; that you are his servant and deputy, governing under him; that all your authority is from him, and that, at last, a time will come when you must give account to him of your government.

II. As you are under authority, Exod. xx. 12. (1.) You must honour and reverence all whom God has set over you. (2.) You must obey them, Eph. v. 24, and vi. 1—5, Heb. xiii. 17, in all such their lawful commands as are within the compass of their authority and commission, and that with fidelity, and singleness of heart, for the Lord's sake, 1 Pet. ii. 13, 14, Eph. vi. 5, 6. (3.) You must submit to their reproofs, corrections, and just restraints with patience, without murmuring, or answering again or resisting, Tit. ii. 9. For if you do not submit to the powers that be ordained of God, or if you resist them, Rom. xiii. 2; you rebel against God, and resist the ordinance of God: which whoso doth, shall receive to himself damnation or judgment. But if you, not only for wrath, but chiefly for conscience to God, Rom. xiii. 5, do submit yourselves to every ordinance of man, 1 Peter ii. 13, 14, doing therein the will of God from the heart, Eph. vi. 6—8, then, whether men requite you or not, you shall be sure of the Lord to receive the reward of the inheritance, Col. iii. 24, for thus obeying men, you serve the Lord Christ.

SECT. 3. CONCERNING BODILY REFRESHMENT AND RECREATIONS.

THE constitution of man's soul and body is such that they cannot long endure to be employed, and stand bent with earnestness upon any thing, without relaxation and convenient refreshment.

(1.) The whole man is refreshed by eating and drinking: in which you must be, first, holy, secondly, just, thirdly, temperate.

1. It was their sin, who fed themselves without all fear of God, Jude 12. Meats and drinks are not sanctified to a man, if he be not pure and holy, Tit. i. 15, 1 Tim. iv. 4, 5; and if they be not received with prayer and thanksgiving.

2. You must not eat bread of deceit, Prov. xx. 17, 2 Thess. iii. 12, nor ill-gotten food: every man must eat his own bread. God would have no man to eat the bread of wickedness, nor yet drink the wine of violence, Prov. iv. 17.

3. Moreover, you must not eat and drink for gluttony and drunkenness, Rom. xiii. 13, Prov. xxiii. 20, 21, to please the palate, and to gorge the appetite; but for health and strength, Eccles. x. 17.

(2.) A man when he is weary may be refreshed likewise by variety and interchange of the duties of his particular and general calling. And the best recreation to a spiritual mind, when it is weary of worldly employments, is to walk into Christ's garden, Cant. iv. 12—15, and v. 1; and there, by reading and meditating, Psa. xciv. 19, singing of Psalms and holy conference, Col. iii. 16, you may solace yourself with the sweet comforts of the Holy Spirit, and enliven your heart with joy in God, even joy in the Holy Ghost · and a delight in the commandments and word of God Psa. cxix. 14, 16, 24. These are the most profitable, most ravishing, and most lasting delights of all other. And by how much the soul is of a more spiritual, heavenly constitution, by so much more it will content and satisfy itself in these delights.

Yet since bodily and natural delights are part of our Christian liberty, therefore (taking heed that you abuse not your liberty) you may, when you have need, recreate yourself with them. Now that you may innocently enjoy recreation, follow these directions:

1. The matter of your recreation must be of a common nature, and of things of indifferent use. Things holy are too good, and things vicious are too bad, to be sported or played with.

2. Recreations must be seasonable for time; not on the Lord's day, in which time God forbiddeth all men to seek their own pleasures, Isa. lviii. 13. Usually, diversions must be used, not before, but after the body or mind has been thoroughly employed in honest business. Not over long, to the expense and loss of your precious time, which you should study to redeem, not to trifle away, Eph. v. 16.

3. Recreations must always be inoffensive, 1 Cor. xvi. 14; such as do no harm to yourself, or to your neighbour. If your diversions do impeach or hazard your own, or your neighbour's life, estate, or comfortable living, they are unlawful.

4. Recreations must be moderate, not sensual or brutish; looking at no higher or further end than earthly delights. For as he that eateth and drinketh that he may enlarge his appetite, to eat and drink yet

more; so he who sporteth that he may sport, is brutish and sensual. It is very Epicurism: God has threatened that he who loves sport, shall be a poor man, Prov. xxi. 17; and he that loves wine and oil, shall not be rich.

5. Whatsoever your diversions be, you must so recreate the outward man, that you be no worse, but rather better in the inward man. For God hath set such a blessed order in all lawful things, that the meanest being, lawfully used, shall not hinder, but assist us in the best things.

6. In all recreations you must propose the right end; the nearest and immediate end is to revive your weary body, and to quicken your dull mind; but your highest and principal end is, that with this refreshed body and quickened spirit, you may the better serve, and glorify God, 1 Cor. x. 31; that whether you eat or drink, or whatsoever you do else, all may be done to the glory of God.

This may serve for direction how you should walk with God upon any of the six days, except there be special cause of setting a day apart for holy use, as, for fasting and prayer.

CHAPTER IV.

OF RELIGIOUS FASTING.

SECT. 1. THE NATURE OF, AND REASONS FOR, RELIGIOUS FASTING.

THE fast which I mentioned in the former chapter, of which I am now to treat, is a religious fast; which is, sanctifying a day to the Lord by a willing abstinence from meat and drink, from delights and worldly labours, that the whole man may be more thoroughly humbled before God, and more fervent in prayer.

This fast has two parts; the one, outward, the chastening the body; the other, inward, the afflicting of the soul; under which are contained all those religious acts which concern the setting of the heart right towards God, and the seeking help of God for those things, for which the fast is intended.

Take fasting strictly for bodily abstinence, so it is an indifferent thing, and is no part of God's worship: but take it as it is joined with the inward part, and is referred to a religious end, being a profession of an extraordinary humiliation; and it is a great assistance to a man's spiritual and reasonable service of God, giving a stronger and speedier wing to prayer, which must always go with it, Ezra viii. 23. Psa. xxxv. 13; so it is more than an ordinary worship.

It has the name from the outward part; Mark ix. 29. 1 Cor. vii. 5. Acts xiii. 3; it being most sensible; but has its excellency and efficacy from the inward, being that for which the outward is observed.

A fast is called public, when a whole state, or when any one public congregation doth fast.

Private, when one alone, one family, or some few together, do fast.

Public and private fasts have their warrant from the New Testament, as well as from the Old; which shows that religious fasts were not peculiar to the Jews; but are a Christian duty, belonging to all fitly qualified for them.

In the sacred Scriptures we have manifold examples of private fasts; and examples and commandment for public ones.

Our Lord and Saviour said, that his disciples after his departure from them should fast, Matt. ix. 15. Matt. vi. 16, 17, and gives directions unto all concerning private fasts. The apostle speaks of husbands and wives abstaining from conjugal embraces, that they might give themselves to fasting and prayer, 1 Cor. vii. 5. And we have repeated examples of the apostles, and primitive Christians, for religious fasts, Acts xiii. 2, 3. Acts xiv. 23. All which prove fasting to be a Christian duty.

The case of a person's self, or family, the church, or commonwealth, may be such, that ordinary humiliation and prayer will not suffice. For as there were some devils that could not be cast out, but by fasting and prayer, Mark ix. 29; so it may be that such hardness of heart may be grown upon a person; or some sinful lusts may have gotten so much strength, that they will not be subdued; some evils, private and public, (1 Sam. vii. 5, 7. Judges xx. 18, 23, compared with verse 26) which cannot be prevented or removed; some special graces and blessings, which shall not be obtained or continued, but with the most importunate seeking of God, by fasting and prayer.

Reasons for fasting.

Fasting is contrary to that fulness of bread, which makes both body aud soul more disposed to vice, and, indisposed to religious duties, through drowsiness of head, heaviness of heart, dulness and deadness of spirit. Now these being removed, and the dominion of the flesh subdued by fasting, the body will be brought into subjection to the soul, and both body and soul to the will of God, more readily than otherwise they would be.

A day of fasting is a great assistance to the soul, for the better performing of holy duties, such as meditations, reading, and hearing the word, prayer, examining, judging, and reforming a person's self; both because his spirits are better disposed, when he is fasting, to serious devotion; and the mind being so long taken wholly off from the thoughts, cares, and pleasures of this life, he may be more intent and earnest in seeking of God.

"Fasting is an open profession of guiltiness before God, and an expression of sorrow and humiliation; being a real acknowledgment of man's unworthiness, even of the common necessaries of this present life."

But it is not enough that the body be chastened, if the soul be not also afflicted, Isa. lviii. 5; because, it is else but a mere bodily exercise, which profits little; nay, it is but an hypocritical fast, abhorred and condemned by God; frustrating a chief end of the fast, which is, that the soul may be afflicted.

Afflicting the soul works repentance; another chief end, and companion of fasting; for godly sorrow causes repentance, never to be repented of, 2 Cor. vii. 10.

When the soul is afflicted, and heavy laden with sin, then a man will readily and earnestly seek after God, even as the sick do to the physician for health, and as a condemned man to the king for a pardon. In their affliction (saith God) they will seek me diligently, Hosea v. 15. If this be true of the outward, then much more of inward affliction.

The afflicted soul is a fit object of God's mercy; to him doth God look that is poor, and of a contrite spirit, Isa. lxvi. 2, that trembles at his word; yea, the bowels of his fatherly compassion are troubled for him, Jer. xxxi. 20, who is troubled and ashamed for his sin.

Moreover, upon a day of humiliation (if a man deal sincerely) this affliction of his soul drives him quite out of himself to seek help of God in Christ; and makes him endeavour to bring his soul into such good frame, that he may truly say he doth not regard iniquity in his heart, Psa. lxvi. 18, and that his unfeigned purpose is, and endeavour shall be, to keep a good conscience toward God and man alway. Whence follows boldness, and assurance, through Christ Jesus, that God will be found of him, John xv. 7, and that in God's own time, and in the best manner, he shall have all his holy desires fulfilled.

Who are to observe religious fasts.

All whom lawful authority enjoins, are to keep a public fast, Joel i. 14, so far as health will permit.

These only may keep a private fast:

First, Such as are of understanding; else how can they search out their ways, judge themselves, or pray? In public fasts, if authority think fit, little children may be caused to fast, that the parents, and others of understanding, may (as by objects of misery) be stirred up to a more thorough humiliation; but, in private, children and idiots are to be exempted.

Secondly, Novices and unexperienced Christians are not usually to fast in private; such were Christ's disciples, Matt. ix. 14, 16, 17. Luke v. 33, 34, 35, &c. When exception was taken at our Saviour, because they fasted not, he excuses them, not only that it was unseasonable to fast in a time of joy, while he, the bridegroom, was with them: but because they were not able to bear so strong an exercise, they being like old vessels and old garments, which would be made worse rather than better by the new wine, or new cloth of fasting. Strong physic is good, but not for babes. There is not the same reason why they may fast in private as in public, because the minister by teaching them, and by praying with them, and for them, taketh from them the greatest part of the burden of the fast in private.

Thirdly, All such as are not in their own power, are not to keep a private fast, when those under whose power they are shall expressly contradict it. For the husband might disallow the vow of his wife, even that wherewith she had bound herself to afflict her soul by fasting, Numb. xxx. 3—8, 13. Wherefore none may fast against the will of those who have full power to command their service and attendance.

When, and how long, fasts are to be observed.

Public fasts are to be kept as often as authority shall see cause.

Private, as often as a man shall have more than ordinary cause of seeking unto God, either for others or himself, 2 Sam. xii. 16. Neh. i. 4; for removing or preventing imminent judgments from the church and commonwealth, or for procuring their necessary good, Dan. ix. 3; for subduing some headstrong lust; for obtaining some necessary grace, or special blessing, Acts x. 30; for preparing himself for some special service of God, or the like.

Though I cannot but justly complain of Christians seldom fasting; yet I dare not allow you to make this extraordinary exercise of religion to be ordinary and common: for then it will soon degenerate into mere form or superstition: but wish you to observe it as you shall have special occasion, and when ordinary seeking of God is not likely to prevail.

It is indifferent which of the six days you set apart for fasting. Let it be as shall best suit with your occasions. As for the Lord's day, though it cannot be denied but that if the present necessity require, you may fast upon that day, neither can I utterly deny servants, and such as are under the power of others (if they can have no other time) sometimes to make choice of that day; yet because the Sabbath is a day of Christian cheerfulness, and fasting is somewhat of the nature of a free-will offering, I think you will do best to set such a day apart to yourself for fasting, which is more your own, and not the Lord's day.

The Scripture has not determined how long a continued fast should be kept. We have examples that some have fasted a longer time, as three days, Esther iv. 16; some a shorter, but none less than one day, Judges xx. 26. In hotter countries they could, without injury to health, abstain from food longer than we can who live in a colder; but I think the body cannot usually be sufficiently afflicted through want of food in less time than one day.

Thus I have proved religious fasting to be a Christian duty. And have showed what it is; who should and may fast, when and how long. It remains that I show you how you may keep a fast acceptable to God, and profitable to yourself, which is the principal thing to be regarded in a fast. And this I do the rather, because many well affected Christians have professed that they would gladly set about the duty, but ingenuously confess that they know not how to do it, and (in particular) how to be intent and spiritually employed, for want of matter, for a whole day together. But of this in the next section.

SECT. 2. DIRECTIONS FOR THE KEEPING A RELIGIOUS FAST.

By way of preparation to a religious fast, do thus: Take but a moderate supper the night before: for if a man glut himself over night, he will be more unfit for the duty of humiliation the next day, and it differs in effect little from breaking of fast next morning.

When you commend yourself to God alone by prayer that night (as every good Christian doth) then set yourself in a special manner to seek the Lord, as the saints of God; in the beginning of their fasts, have done; 2 Chron. xx. 3. Dan. x. 12; proposing to yourself the end of your intended fast; remembering this, that if the chief occasion and end be your own private good, that you forget not others, nor the public; or if it be the public, yet mind also your own private; for until you have renewed your own peace with God, your fasting and praying will prevail little for the public. And God having joined the public with our private good in prayer, we must not disjoin them in our fasting. Resolve with yourself, to the utmost of your power, to keep a religious fast unto God, according to his will; for this cause in your prayers add serious petitions to God for his grace to assist you therein.

When you awake that night, let not your thoughts be upon worldly business, much less upon any evil thing; but let them be holy, such as may tend to the assistance of the holy duties of the next day.

Also, if necessity hinder not, arise early on the day of your fast. It is most agreeable to a day of fasting, whereon your flesh is to be subdued, that you allow not yourself so much sleep as at other times. It is probable, that for this cause some lay on the ground, others in sackcloth, in the nights of their fasts, 2 Sam. xii. 16; not only to express, but to assist their humiliation, by keeping them from sleeping over much, or over sweetly, Joel i. 13.

When the day is come, be strict in observing the outward fast. To this end,

First, Forbear all meat and drink, Esth. iv. 16. Luke v. 33, until the set time of the fast be ended, which usually is about supper-time. A general council in the primitive church decreed, that total abstinence should be observed until evening prayer was ended. In case of necessity, that is, when total abstinence will indeed disable you from attending to the chief duties of that day, you may eat or drink; for in such cases God will have mercy rather than sacrifice; but then it must be a small refreshment, and that not of a dainty kind; only such and so much as may remove the impediment to the spiritual performance of the duties of that day.

Secondly, Abstain from all other worldly delights, as (so far as will stand with comeliness) from fine apparel, Exod. xxxiii. 4, 5, 6; from all recreations and pleasant music. Isa. lviii. 3: from the marriage bed, and the like, 1 Cor. vii. 5. Joel ii. 16.

Thirdly, Abstain from all worldly labour, as upon a Sabbath day, Isa. lviii. 3; for worldly business, and the cares thereof, do distract the thoughts, and hinder devotion, as well as worldly delights; and a ceasing from these gives a full opportunity to holy employments the whole day. Therefore the Jews were commanded to sanctify a fast, Joel ii. 26. And that yearly fast, called the day of atonement, Lev. xxiii. 27—32, was, upon peril of their lives, to be kept by a forbearance of all manner of work. Now although the ceremonials of that day are abolished in Christ, yet, forbearing work, as well as meat and drink (being of the substance and morality of a fast) remains to be observed in all truly religious fasts.

Thus much for the outward fast: you must be as strict in observing the inward.

Begin the day with prayer, according as I directed you to do every day; but with more than ordinary preparation, with fervency and faith, praying for God's special grace, to enable you to sanctify a fast that day according to the commandment.

Then apply yourself to the main work of the day, which has these parts: (1.) Unfeigned humiliation; (2.) Reformation, together with reconciliation; and (3.) Earnest invocation.

The soul is then humbled, the heart broken and truly afflicted, when a man is become vile in his own eyes, through consciousness of his own unworthiness, and when his heart is full of grief and anguish through fear of God's displeasure; and with godly sorrow and holy shame in himself, and anger against himself for sin. These affections excited do much afflict the heart.

This deep humiliation is to be wrought, partly by awakening your conscience through a sight of the law, and apprehensions of God's just judgments due to you for the breach of it, which will break your heart; and partly by the gospel, raising your mind to an apprehension and admiration of the love of God to you in Christ, which will melt your heart, and cause you the more kindly to grieve, and to loath yourself for sin, and also to entertain hope of mercy, whence will follow reconciliation, reformation, and holy calling upon God by prayer.

To work this humiliation, there must be,

First, Examination, to find out your sins.

Secondly, An accusation of yourself, with due aggravation of your sins.

Thirdly, Judging and passing sentence against yourself for sins.

Sin is the transgression of the law, and revealed will of God; wherefore for the better finding out of your sins, you must set before you God's holy law, for your light and rule, Psa. cxix. 105. And if you have not learned, or cannot remember the heads of the manifold duties commanded, or vices forbidden; then get some catalogue or table, wherein the same are set down to your hand, which you may read with serious consideration and self-inquiry, fixing your thoughts most upon those particular sins whereof you find yourself most guilty.

If you do not meet with one more fit for your purpose, then use the following table.

But expect not herein an enumeration of all particular sins and duties, which would require a volume; but of those which are principal and most common; by which, if your conscience be awakened, it will bring to your remembrance other sins and omissions of duty, not mentioned in the table, of which you may be guilty.

The first table of the law concerns the duties of love and piety to God, the performance whereof tends immediately to the glory of God, and mediately to the salvation and good of man.

1st. The first commandment respects the loving, serving, and glorifying the only true God, as your God, Exod. xx. 2, 3.

Examining yourself by this (and so in the other commandments) think thus with yourself: Do I know and acknowledge the only true God to be such an one as he has revealed himself in his word and works, namely, One only infinite, immaterial, immutable, incomprehensible spirit, and everlasting Lord God; having being and all-sufficiency in and from himself; one who is absolutely full of all perfections, and incapable of the least defect; being wisdom, goodness, omnipotence, love, truth, mercy, justice, holiness, and whatsoever is originally and of itself excellent; the only Potentate, King of kings, Lord of lords, of whom, through whom, and to whom are all things: the Father, Son, and Holy Ghost, God blessed for ever? Amen.

Do I believe his word, in all things related, commanded, promised, and threatened therein? and that his holy and wise Providence is in all things? Have I him and his word in continual remembrance?

Do I esteem and exalt God in my heart above all, so that it humbly adores him at the very mention and thought of him; judging myself to be nothing in mine own eyes, yea, esteeming all creatures to be nothing in comparison of him?

Have I given religious worship to him only? Have I believed in him, and in him only? Have I sworn by him as there has been cause, and by him alone? Have I prayed unto him, and him alone? And sought to obtain help of him only by such means as he has appointed; giving the glory and thanks of my being and well-being, and of all other things which are good, unto him?

Is my conscience so convinced of the truth and authority of God, that it holds itself absolutely bound to obey him in all things, so that it does incite to that which is good, restrain from that which is evil, encourage me in well-doing, and check me when I do ill?

Is my will resolved upon absolute and unfeigned obedience; to do whatsoever God commands, to forbear whatsoever he forbids, to subscribe to whatsoever he does, as well done; and have I borne patiently, all that, which either by himself or by any of his creatures, he has inflicted upon me?

Have mine affections been so for God, that I have loved him with all my heart, loving nothing more than him, nothing equal to him? Do I hate every thing that is contrary to him? Hath my confidence been only in him, and my expectation of good from him? Have my desires been to him, and from him, longing above all things to have communion with him? Has it been my greatest fear to offend him, or to be separated from him? Has it been my greatest grief and shame that I have sinned against him? Have I rejoiced in God as my chief good? Has mine anger risen against whatsoever I saw contrary to his glory? Have I been zealous for God? And have I made him the utmost end of all mine actions?

Hath my whole outward man, as tongue, senses, and all other active powers of my body, been employed in the service of the true God, and yielded obedience to his will?

Or, contrariwise, am I not guilty of denying of God in word, in works, or at least in heart? questioning the truth of his being, and of his word, denying his providence, power, or some other of his divine attributes? Have I not been ignorant of God, and of his will, and erroneous and misbelieving, if not heretical in my conceptions concerning God the Father, Son, or Holy Ghost?

Have I not been over curious in prying into the nature and secret counsels of God, beyond the rule of the revealed will of God? Have I not put myself, or any other creature in the place of God, through pride preferring, and resting upon my own way and will before God's, or by making myself mine utmost end, professing God and his religion, only to serve my own designs, or by seeking to the creature, (as to angel, saint, devil, or witch,) instead of the Creator?

Have I not been forgetful of God, and of his will? Is not my conscience impure, blind, deluded, or seared; and my will perverse, obstinate, impatient, and murmuring against God, and full of dissimulation?

Have I not set my affections upon the world, rather than upon God, loving that which is evil, hating that which is good, yea, God himself, if not directly, yet in his holiness, shining in his ordinances and in his children, or as he is a severe inflicter of punishment? fearing man more than God, trusting in the creature, making something besides God my chief joy? Have I not presumed when I had cause to despair, and despaired after that I had cause to hope? Have I not tempted God many ways? And have I not in the matters of God been either cold, lukewarm, or blindly or presumptuously zealous?

Has there not been a proneness in my whole outward man, to rebel against God?

2d. The second commandment concerns all such worship of God, which he only has appointed; whereby he communicates himself to man, and man again makes profession of him: forbidding, under one kind, all such as are not by him ordained, Exod. xx. 4—6.

Think thus: have I worshipped God in spirit and

truth, in all the kinds and parts of his worship, public and private, ordinary or extraordinary; as, by hearing, reading, and meditating of his word; by praying, praising, and giving thanks to him; by a right use of his sacraments, baptism, and the Lord's supper; and by religious fasting, religious feasting, and making of vows, according as I have had special occasion? And have I done what has been in my power for the maintaining and promoting of God's true worship; and have I, according to my place, executed aright, or submitted unto the government and discipline of the church of God?

Or, besides the omission of the former duties, am I not guilty, some way or other, of idol worship, conceiving of God in my mind, or representing him to my sense, in the likeness of any creature?

Have I not added to or detracted from, any part of God's worship? Have I not run into the appearances and occasions of idolatry, as, by presence at idol service, by marriage and needless familiarity with idolatrous persons? At least, is not my heart guilty of not hating, but rather lingering after, idolatrous worship? Have I not been guilty of superstition or will-worship, &c.

3d. The third commandment concerns the glory of God's holy name, shining forth in his titles, attributes, religion, word, ordinances, people, or any thing that has in it any signatures of his holiness or excellency; forbidding the taking of it in vain, and that in all words or actions, religious or common, Exod. xx. 7.

Have I glorified God, by answering my holy profession, with an holy and unblamable conversation; by performing all holy duties with due preparation, knowledge and devotion, also by thinking and speaking of the names and holy things of God with holy reverence; and in particular by fearing an oath?

Or, have I not caused the name, religion, and people of God, to be ill thought of, and dishonoured by my evil course of living, or at least by committing some gross sin? Am I not guilty of rash, unprepared, heedless, forgetful, and fruitless reading, hearing, receiving the sacraments, or performances of any other part of the worship of God?

Have I not thought or spoken blasphemously or contemptuously of God, or any of the things of God? Have I not used the name of God needlessly, rashly, wickedly, or falsely in swearing, or lightly in my salutations, admirations, or otherwise in my ordinary discourse?

Have I not abused the name of God, his Scriptures, his ordinances and creatures, using them for other purposes than he allows, as, for sports, charms, or any sorcery, luxury, or the like? Have I not passed by the great works of God's power, mercy, and judgments, (especially of his redeeming love in Christ Jesus,) without due observation and acknowledgment of God therein?

4th. The fourth commandment concerns the ordinary solemn time of the service and worship of God, requiring that the seventh day (now our Lord's day) be kept as an holy rest, Exod. xx. 8—11.

Have I upon the six days remembered the Lord's day, that I might despatch all my worldly business, and prepare my heart, that when it came I might keep an holy Sabbath to the Lord, according to the commandment? Did I, according as my health would permit, rise early on that day?

Have I performed my daily (both morning and evening) exercises of religion alone, and with my family, that day in prayer?

Have I caused all under my authority, according to my power, to rest from all manner of works and worldly recreations; also myself not only from the labour of my body, but of my mind in all worldly business; except about the things that concern common honesty, and comeliness, works of mercy, and such works of necessity as could not be done before, or let alone till afterwards?

Have I always prepared my heart before I went

into the house and presence of God, by meditation of God's word and works, and in particular by examination and reformation of my ways, by prayer, thanksgiving and holy resolution to carry myself as in God's presence, and to hear and obey whatsoever I should be taught out of the word of God?

Have I caused my family to go with me to the church? And did I with them come in due time, and, being there, stay the whole time of prayer, reading, and preaching of the word, singing of psalms, receiving and administering the sacraments, even that of baptism, when others are baptized; and did I attend diligently, and join with the minister and the rest of the congregation in all those holy exercises?

Did I spend the day, after the morning and evening prayers, sermons, or catechizing, in meditation, and (as I had opportunity) in conference and repetition of what I had heard? Also in visiting the sick, and other works of mercy; and so, from the beginning to the end of the day, have been employed in holy thoughts, words, and deeds, and all this with spiritual delight?

Or, am I not guilty of forgetting the Lord's day before it came, and of neglecting and profaning it when it came? as by mere idleness, or by taking opportunity of leisure from the business of my calling to be licentious in company keeping, &c., or by reserving that day for journeys, idle visits, and for the despatch of worldly business?

Have I not been careless of the service of God, frequenting it no oftener than law, or very shame did compel me?

Have I not been careless whether my servants or children, did keep the Sabbath or no? And when I was at church, did I not idle away the time, by gazing about or by sleeping, or by worldly thoughts?

Have I not bought, sold, spoken of, or done other works forbidden to be done, spoken, or contrived upon that day?

Have I not, under the name of recreation, sought mine own pleasure, using sports and games, which cause the mind to be more indisposed to the due performance of holy duties than honest labours do, to which they are subordinate, and with them forbidden to be done that day?

Has not the strict observance of the Sabbath been at least tedious to me, so that I could have wished that it had been gone long before it was ended?

5th. The second table concerns duties of love and righteousness towards man, the performance whereof tends immediately to the good of man; but mediately to the proof of his being truly religious, and to the glory of God.

God made man not to be alone, nor to be only for himself; therefore, for the greater good of mankind, he has endued men with variety of gifts, and degrees of place, some excelling others, both in family, church, and commonwealth; yet so as each is excellent in his gift and place, even the meanest made worthy of respect from the greatest, because of his usefulness for the common good: even as the least member of the natural body is truly useful and to be respected as well, though not so much, as the most honourable.

Now when each member in the body politic does acknowledge the several gifts and mutual use one of another, according to their place, then is there a sweet harmony in the society of man, and there is a sure foundation laid of all good offices of love between man and man.

Wherefore, in the first place, God in the fifth commandment, Exod. xx. 12, provides that the order which he had set amongst men, should inviolably be observed; requiring all inferiors, under the name of children, to honour their superiors, that is, to acknowledge that dignity and excellency which is in them, showing it in giving due respect unto their persons and names; implying that all superiors should walk worthy of honour, and that they should mutually shew good respect to their inferiors, seeking their good, as well as their own.

Concerning this fifth commandment, think thus: do I live in a lawful calling? And have I walked worthy my general calling of Christianity, and discharged my particular calling, and employed the gifts which God gave me, for the good of the society, of man in family, church, or commonwealth?

Have I honoured all men, for that they were made after the image of God, and have yet some remains thereof; are capable of having it renewed, if it be not renewed already; and because they are or may be useful for the common good of man; using them with all courteousness and kind respect; excepting when, and wherein, they have made themselves vile by open wickedness; so that it will not stand with the glory of God, the good of others, or of themselves, or with the discharge of my place, to show them countenance? Have I showed my due respect to others, in praying to God, and, as there has been cause, in giving him thanks for them?

Have I conceived the best, that in charity I might, of others? And by love have I endeavoured, according to my place, to cure their grosser evils, and to cover their infirmities? And have I to my power promoted my neighbour's good name and reputation, and have I been contented, nay desirous, that he should be esteemed as well, nay, better than myself? And have I, both in his life-time, and after his death, given him the honour of common humanity, as in common civilities at least, and in comely burial, so far as any way it did belong to me, and in maintaining his injured reputation? &c.

Have I, being superior to others in gifts of any kind, as learning, wit, wealth, strength, &c., employed those gifts to the honour of God, and the good of man, more than others?

As I am beyond others in years, am I superior to them in gravity, good counsel, and good example?

As I am above others in authority, do I acknowledge that it is not originally in me, but derived to me from God, and have I held it, and used it for him;

keeping within the due limits thereof, governing with wisdom and moderation; procuring the good of their bodies and souls, so far as lay in me; commanding only things lawful and convenient; encouraging them in well-doing, by commendation and rewards; preventing evil as much as I could, and restraining it in them by seasonable and due reproofs, according to the quality of the offence, and of the person, when fairer means would not prevail?

As I am an equal; have I esteemed others better than myself, and striven in honour to prefer them?

As I am below others in gifts and age, have I in word and gesture, showed them due reverence, and thankfully made use of their good parts and experiences?

As I am under authority, whether in family, church, or commonwealth, have I submitted myself to all my governors, reverencing their persons, obeying readily all those their lawful commandments, which are within the compass of their authority to enjoin me? Have I received their instructions, and borne patiently and fruitfully their reproofs and corrections?

Or do I not live without a lawful calling? or idly or unprofitably in it? Have I not buried or abused my talent and place, to the hurt rather than the good of myself and others?

Have I not been high-minded, esteeming better of myself than there was cause, seeking after the vain applause of men?

Have I not despised others? Yea, those who were good, yea, my superiors? Showing it by my irreverent gestures, and by my speeches to them, and of them? Have I not, some way or other, detracted from, and diminished the credit of others, or, at least, envied their due estimation?

As I am a superior, have I not carried myself insolently, lightly, or dissolutely?

As I am under authority, have I not carried myself stubbornly and undutifully?

6th. God having set an order in human society, does next provide for the life and safety of the person of man, who must keep this order, and make this society, by forbidding, in the sixth commandment, whatsoever may take it away, or impair it.

Have I had a care of mine own health, in a sober use of meat, drink, labour, sleep, recreation, physic, or whatever else is apt to promote health, and to prevent disease?

Have I been, or am I meek, patient, long-suffering, easy to be appeased, apt to forgive, full of compassion, kind, merciful; showing all these in soft speeches, gentle answers, courteous behaviour, requiting evil with good, comforting the afflicted, relieving the needy, peace-making, and by doing all other offices of love, which might tend to my neighbour's safety or comfort?

Or, have I not wished myself dead, or neglected the means of my health? Have I not impaired it, by surfeits, by excessive labour or sports, by fretting and over grieving, or by any other means? And have I not had thoughts of doing myself harm?

Have I not been angry unadvisedly, maliciously, and revengefully? showing surly gestures and behaviour, as sour looks, shaking the head or hand, gnashing the teeth, stamping, mocking, railing, cursing, quarreling, smiting, hurting, or taking away the life of man any way, without God's allowance?

Have I not been a sower of discord, or some way or other been an occasion of the discomfort, if not the death of others?

7th. The seventh commandment concerns chastity, whereby God provides for a pure propagation and conservation of mankind; forbidding all bodily pollution, under the name of adultery, Exod. xx. 13.

Have I been modest, sober, shamefaced, possessing my body in chastity, shutting mine eyes, and stopping mine ears, and restraining my other senses from all objects and occasions of lust? bridling my tongue from lascivious speeches; forbearing all manner of obsceneness and wantonness; abstaining from self-pollution, fornication, or any other natural or unnatural defilement of my body, either in deed or desire?

And being married, was I wise in my choice? and have I kept the marriage-bed undefiled, through a sanctified, sober, and seasonable use thereof?

Or, am I not guilty of manifold acts of uncleanness; at least of immodest looks, touches and embraces; of wanton speeches, gestures, apparel, and behaviour?

Have I not run into the manifold occasions of adultery and uncleanness; as idleness, gluttony, drunkenness, choice of such meats, drinks, or any other things that provoke lust; effeminate dancing, frequenting wanton company, or of unseasonable conversing with the other sex alone?

8th. The eighth commandment concerns the preservation of man's goods, the means of his comfortable maintenance in this life, forbidding all injuries and wrongs, under the name of stealing.

Have I a good title to the things which I possess, as by lawful inheritance, gift, reward, contract, or any other way which God allows? Have I been industrious and faithful in my calling, frugal and provident? Have I done that for which I have received pay or maintenance from others; and have I given to every man his own, whether tribute, wages, debts, or any other dues?

Or, have I not got my living by an unlawful calling? or have I impoverished myself and mine by idleness, luxurious and unnecessary expenses? by gaming, unadvised suretiship, or otherwise? Have I not withheld from myself or others, through covetousness, that which should have been expended?

Have I not gotten or kept my neighbour's goods, by fraud, oppression, falsehood, or by force, and made no restitution? Have I not some way or other impaired my neighbour's estate?

9th. The ninth commandment concerns truth of speech; the great means of intercourse between man and man, and of preserving the rights, and redressing the disorders of human society; forbidding all falsehood of speech, under the name of bearing false witness, Exod. xx. 14.

Have I at all times, in all things spoken the truth from my heart? giving testimony, in public or private, by word or writing, of things concerning mine own or neighbour's name, credit, life, chastity, goods, or in any matters of speech between me and others, whether in affirming, denying, with or without oath, or in any bare reports, promises, or in any other way?

Or am I not guilty of telling lies jestingly, officiously, or perniciously? Have I not raised, spread, or received false reports of my neighbour? Have I not spoken falsely in buying and selling; also in commending by word or writing unworthy persons, dispraising the good, in boasting of myself, or flattering of others?

Have I not given false evidence, used equivocations, or concealed the truth which I should have spoken, or perverted it when I did speak it?

10th. The tenth commandment concerns contentment with a man's own condition; the foundation of all order and justice amongst men; forbidding the contrary, namely, coveting that which is not his, Exod. xx. 15.

Am I contented with mine own condition, as, with my place which I hold in family, church, or commonwealth, with husband or wife, house or estate? Can I heartily rejoice in the prosperity of others, even when they are greater, happier, wiser, or better than myself?

Or have I not been full of discontent with my condition, coveting after something or other which was my neighbour's? at least by actual concupiscence, in multitude of evil and envious thoughts, arising from the law of my members, though my will hath contradicted them?

RULES FOR SELF-EXAMINATION FROM THE GOSPEL OF CHRIST.

Besides the breaches of God's holy law, have I not been guilty of many sins peculiarly against the gospel of our Lord Jesus Christ? Such as opposition to, and hatred of Christ, and his cause; being incensed against him, and his method of salvation; or vilifying his gospel by word or writing, Isa. xlv. 25.

Scepticism and gross infidelity, from a disinclination to conviction; and not impartially, in the fear of God, weighing the evidences in proof of the heavenly mission of our Lord and Saviour, John v. 39.

Unsound faith; not extended to all the revealed truths and duties of the gospel; either through culpable ignorance, strong prejudice, resolving to believe no further than I can comprehend, or may be consistent with the quiet of my conscience in an evil course, John iii. 19—21. Or has it been a mere notional and historical faith? However extended to all the doctrines, duties, promises, and threatenings of the gospel; yet not attended with heart-impressions, humbling the soul, making me poor in spirit at the feet of Christ; seeking the glory of God and the Redeemer, and my own salvation, as my chief business, Gal. vi. 15. Has it been such a faith that does not purify the heart, Acts xxiv. 18; that worketh not by love, Gal. v. 6; that unites not the soul to Christ; so as to crucify the flesh with the affections and lusts, Gal. v. 24; that directs not the whole conversation by the will and example of our acknowledged Lord and Master; not resting by faith in his promises, in all seasons of adversity and prosperity, 1 John ii. 6; that moderates not fear and hope concerning things present and temporal, by looking to Jesus, and things eternal, 2 Cor. iv. 13. Heb. xii. 12; that does not trust and rely upon Christ alone (in the prescribed way) for justification and salvation; submitting unto the righteousness of God in him, Rom. x. 3, 4.

Impenitency; not being seriously affected with an

humbling sense of the odious nature of sin; not searching out my offences, but hiding and extenuating them, Rom. ii. 3, 4. Not abasing myself for my sins (so many and aggravated) against all the love of the Father, the grace of the Son, and the strivings of the Holy Spirit, 2 Cor. vi. 2. No resolved and vigilant forsaking of sin, and bringing forth fruits meet for repentance, Matt. iii. 8.

Despair of God's mercy in Christ Jesus, saying, There is no hope, Jer. ii. 25.

Presumption, and turning the grace of God into lasciviousness, Jude 4; continuing in sin, that grace may abound, Rom. vi. 1, 2.

Making light of Christ, not esteeming him as the pearl of great price, and being willing to part with all to obtain it, Matt. xiii. 45, 46.

Slighting the benefits of redemption, Luke xiv. 16—20; such as peace with God through the blood of Christ; the gift of the Holy Spirit, as sanctifier; meetness for, and a title to the kingdom of heaven; and communion with God in the way to it.

Undervaluing the means of salvation; the holy Scriptures, secret prayer, public worship, the sacraments, &c. and not being spiritual in, if attendant upon them, John iv. 23. Heb. x. 25.

Great coldness and indifference about the honour of the sacred name into which I was baptized; and all the peculiar doctrines of the gospel, Phil. iii. 3.

No joyful progress in the works of faith and labours of love, to the full assurance of hope, even where faith is unfeigned, Phil. iii. 12—15.

Inconstancy and fickleness in the service of God, with the natural consequences thereof, despondency, diffidence, and the spirit of bondage again to fear, Gal. v. 7. Rom. viii. 15.

Slavish fear and cowardice, 2 Tim. i. 7.

Declensions in the love of Christ and the fruits of holiness; and growing conformity to the world, luxury, gaiety, pastimes, &c. with increasing inattention to the soul's immortality, the approach of death and eternity, the coming of the Lord, the resurrection and judgment-day, heaven's joys or hell's horrors, Rom. ii. 1—10. 2 Peter iii. 14.

Upon the whole,—How shall man be just with (or justify himself before) God? If he contend with him, he cannot answer him one of a thousand, Job ix. 2, 3. So that every mouth must be stopped, since all the world is become guilty before God. Being justified (if ever) freely by his grace, through the redemption that is in Jesus Christ; whom God hath set forth (in the most illustrious manner) to be a propitiation, through faith in his blood, &c. Rom. iii. 19—27.

Beware, therefore, lest that come upon you which is spoken of in the prophets: Behold, ye despisers, and wonder, and perish, Acts xiii. 40, 41. Examine yourselves, whether ye be in the faith; prove your own selves; know ye not your own selves, how that Jesus Christ is in you, except ye be reprobates? 2 Cor. xiii. 5.

SELF-JUDGING FOR SIN.

THE EVIL NATURE AND EFFECTS OF SIN.

Thus having by God's holy law found out your sins, you must arraign and accuse yourself, as it were at the bar of God's tribunal; representing your sins to your mind as they are, in their heinousness and mischievousness, according to their several aggravations.

First, Consider sin in its nature. It is a moral evil, an irregularity in the soul and actions, and enmity to God the chief good; it is the worst evil, worse than the devil and Satan, he had not been a devil but for sin; worse than hell, which, as it is a torment, is caused by sin, and is only contrary to the good of the creature, whereas sin itself is contrary to the good of the Creator. It is such a distemper of the soul, that the Scripture calls it wickedness of folly, even foolishness of madness, Eccles. vii. 25.

Secondly, Considering from whence sin in man had its original, even from the devil, John viii. 44. Gen. iii. who is the father of it; it came, and comes from hell, James iii. 15; therefore is earthly, sensual, devilish. Whensoever you sin, you do the lusts of the devil.

Thirdly, Consider the nature of the law whereof sin is a transgression: a law most perfect, most holy, just, and good, Rom. vii. 12. Gal. iii. 21. Rom. viii. 3; which would have given eternal life to the doers of it, had it not been for this cursed sin.

Fourthly, Consider the person against whom sin is committed, whom it highly offends and provokes; it is God, Psa. li. 4, to whom you owe yourself and all that you have; who made, Acts xviii. 28, and does preserve you, and yours; who, though you have sinned, desires not your death, Ezek. xxxiii. 11, nor afflicts you willingly; but had rather that you should humble yourself, repent, and live; who, that you might be saved, gave his only begotten Son to death, to ransom you, John iii. 16; who, by his ministers, makes known his word and good-will towards you, making proclamation, that if you will repent and believe, you shall be saved; yea, entreats you by his ministers to be reconciled to him, 2 Cor. v. 20. It is that God, who is rich in goodness, forbearance, and long-suffering, 2 Peter iii. 9, waiting when you will return, that you may live; who, on the other hand, if you despise this his goodness, and shall continue in your sin, thereby provoking the eyes of his glory, Isa. iii. 8, is a terrible and revengeful God, Nahum i. 2; who, if you still err in heart, and will not walk in his ways, has sworn in his wrath, that you shall not enter into his rest, Heb. iii. 11; who in his wrath is a consuming fire, Heb. xii. 29; and is ready and able to destroy body and soul in the eternal vengeance of hell-fire, Matt. x. 28.

Fifthly, Consider sin in the evil effects of it, namely, it brought a curse upon the whole creation, Gen. iii. 17. Rom. viii. 20, for man's sake; whereby the creatures are become defective, and oftentimes unserviceable, nay, hurtful to you. From your sins come all manner of diseases and afflictions that ever befell you. This your sin (until it be repented of and pardoned) makes you hateful to God, Psa. xi. 5, separates between you and God, Isa. lix. 2, causing him to withhold good things from you, Jer. v. 25, and to inflict evil upon you, even in this life. It defiles the whole man, Tit. i. 15, and every renewed act of sin does strengthen the body of sin, and works a decay of grace in you, though you be regenerate. And if it be gross iniquity, if it does not benumb and sear your conscience, yet it will wound it, and break the peace thereof, if it be tender; vexing it as motes do your eye, or thorns your feet, Psa. li. causing terrors and doubtings of salvation; God's withdrawing his favour and loving countenance from you; and, if you be not in Christ, it will in the end bring upon you everlasting damnation, Matt. xxv. 46. Rev. xxi. 8.

Sixthly, Consider the ransom for sin, who paid it, and what was paid; consider Christ Jesus, who he was, and what he did and suffered to take away your sin. He, the only Son of God, very God, did veil his glory for a time, and left heaven to dwell in the tabernacle of human flesh, taking upon him the estate of a servant, Phil. ii. 6—8; he was poor, despised of men, Isa. liii. 3; persecuted from the manger to the cross; made to shed tears abundantly; yea, so tormented with the sense of God's wrath for your sin, that for very anguish he did sweat as it were drops of blood, Luke xxii. 44. He was accused, condemned, spit upon, mocked, buffeted, and scourged by wicked men, Matt. xxvii. 1—31; made to bear his own cross, till for very faintness he could bear it no longer, Mark xv. 21; then he was crucified between thieves, dying the most accursed death, Gal. iii. 13; and, which to him was more than all the rest, he, in his human apprehension, was forsaken of God, crying out, My God, my God, why hast thou forsaken me? Matt. xxvii. 46.

Now you may be assured, that if the justice of God could have been satisfied, and your sin expiated and done away by a less price, Jesus Christ, his only Son, would never have been caused to pour out his soul a sacrifice for your sin, Isa. liii. 10, 12.

This looking (by the eyes of your faith) upon Christ whom you have pierced, Zech. xii. 10, will at once show you the greatness and hatefulness of your sin, which required such an infinite ransom; and the infinite love of God in Christ towards you, even when you were his enemy; in providing for you a sure remedy, which will free you from both the guilt and power of this sin. The thoughts hereof will (if any thing will) even melt the heart into godly sorrow for sin, and withal, give hope (in the use of means) of mercy and forgiveness.

That the former aggravations may be more pressing, observe these directions:

1st. You must consider sin in particulars, one after another; for generals leave no impressions; therefore David cries out of his bloody sin in particular, 2 Sam. xxiv. 10. Psa. li. 14.

2d. You must judge the least sin to be damnable, James i. 15, until it be pardoned, and repented of in particular, if known unto you; at least in general, if not known.

3d. The greater any sin is, Heb. x. 29, the greater you must judge the guilt and punishment to be.

4th. Sins committed long since, unrepented of, and the punishments deserved but deferred, are to be judged to be as near, lying at the door, Gen. iv. 7, and exposing you to condemnation, as if committed at the present; so that you may look for God's hand to be upon you this present moment. They, like the blood of Abel, or sins of Sodom, cry as loud to God for vengeance now, as the first day they were committed; nay, louder, because they are aggravated by impenitency, and by the abuse of God's long-suffering, Rom. xxiv. 5.

5th. Your humiliation must, in your endeavour, proportion your guilt of sin, Ezra ix. x. 1—14; Matt. xxvi. 75; the greater the guilt, the greater the humiliation.

Know, therefore, that sins against God, of the first table, all things considered, are greater than those of the second, 1 Sam. ii. 25. Matt. xxii. 37, 38.

The more grace hath been offered you by the gospel, Matt. xi. 21—24; and the more means you have had to know God and his will, the greater is your sin, if you be ignorant, impenitent, and disobedient.

The number of sins, according as they are multiplied, do increase the guilt and punishment, Isa. lix. 12, 13. Ezek. xvi. 51.

The more bonds are broken in sinning, as, committing it against the law of God, of nature, and nations, Jude 10. Jer. xxxiv. 18, against conscience, promises, and vows; the greater the sin and punishment.

All these things known and considered, now judge yourself, 1 Cor. xi. 31; pass a condemnatory sentence against yourself: whence will, through the grace of God, follow affliction of soul. Now you will see that you are base and vile, and that you may justly fear God's judgments; now you will see cause to be grieved, ashamed, yea, even confounded in yourself, and to conceive an holy indignation against yourself.

You will now think thus: Ah! that I should be so foolish, so brutish, so mad, to commit this, to commit these sins (think of particulars) to break so holy a law, to offend, grieve, and provoke so good and so great a majesty! So ill to requite him, Deut. xxxii. 6; so little to fear him, vile wretch that I am! That I should commit not only sins of common frailty, but gross sins, many and oft against knowledge, conscience, &c. (but still mind particulars.) Jesus Christ my Saviour shed his precious blood for me, to redeem me from my vain conversation, and do I yet again and again transgress, O miserable man that I am! What am I in myself, at best, but a lump of sin and pollution, not worthy to be loved, but worthy to be destroyed; one that may justly look to have my heart hardened, or my conscience terrified, and that, if God be not infinitely merciful, he should pour upon me all his plagues. Wherefore remembering my doings that they are not good, but abominably evil, I loath myself for mine abominations, Ezek. xxxvi. 31, and abhor myself, and repent, as in sackcloth and ashes, Job xlii. 6.

Now set upon the work of reformation and of reconciliation; general or particular, as you find there is need. It is not enough to search out and consider your ways, nor yet to lament them; if withal you do not turn again unto the Lord, Lam. iii. 40—42. Psa. cxix. 59. Zeph. ii. 1—3; and turn your feet unto his testimonies; and withal seek grace and forgiveness.

The gospel opens a way, and affords means to attain both, through the commands and promises thereof, in the doctrine of faith and repentance.

Now, therefore, bring yourself to the gospel; try yourself thereby, first, whether your first faith and repentance were sincere: then set upon reforming, and getting pardon of particular and later offences.

But learn to put a difference between the commands of the gospel and of the law; the law exacts absolute obedience; the gracious gospel does, through Christ, accept of the truth of faith and repentance, so that there be an endeavour after their perfection.

It would be too long to show you at large the signs of unfeigned faith and repentance. I will, for the present, only say this:

Have you been truly humbled for sin? and through the promises and commandment of the gospel, which bids you believe, have you conceived hope of mercy, relying on Christ for it? And thereupon have had a true change in your whole man, so that you make God your utmost end, and receive the Lord Jesus as your only Saviour; and, out of hatred of sin, and love unto Christ and his ways, have a will in all things to live honestly, Heb. xiii. 18; and to keep always a good conscience towards God and man, Acts xxiv. 16; desiring the sincere milk of the word, to grow by it, 1 Peter ii. 2; loving the brethren, 1 John iii. 14. Psa. xvi. 3; desiring and delighting in communion with them? Then you may be confident that your first faith, repentance, and new obedience were sound.

If upon trial you find that they were not sound, then you must begin now to repent and believe; it is not yet too late.

6th. Concerning reformation and obtaining of pardon and power of your particular sins, do thus:

1st. Consider the commandment which bids you to repent and amend, Ezek. xxxiii. 11, Rev. ii. 5.

2d. Consider the commandment which bids you to come unto Christ, when you are weary and heavy laden with your sins, Matt. xi. 28; believing that through him they shall be pardoned and subdued, Mic. vii. 18, 19. To this end,

3d. Consider that Christ, has fully satisfied for such and such a sin, yea, for all sin, 1 John ii. 1, 2; and that you have many promises of grace and forgiveness, 1 John i. 9; yea, a promise that God will give you grace to believe in him, that you may have your sins forgiven, Heb. x. 15—17.

4th. Consider that there is virtue and power in Christ's death and resurrection, Phil. iii. 10, John i. 16, applied by faith, Acts xv. 9, 1 Peter i. 21, 22, through his Holy Spirit, for the mortifying the old man of sin, and quickening the new man in grace; as well as merit to take away the guilt and punishment of your sin.

5th. Improve this power of Christ in you unto an actual breaking off your sins, and living according to the will of Christ, which is done by mortifying that old man of sin, and by strengthening the new and inner man of grace, Col. iii. 5, Rom. xii. 2, Eph. iii. 16.

In mortifying your sin, do thus:

1st. Take all your sins, especially your bosom sins, those to which the disposition of your nature, and condition of your place does most incline you, your strongest and most prevailing sins, and with them the body of corruption in you, the original and fountain of sin, Psa. li; smite at them, strike at the very root, arraign them, condemn them in yourself, bring them to the cross of Christ, and nail them thereunto, Col. i. 20, ii. 10—16; that is, believe that, not only in respect of their guilt, but also of their reigning power (through faith in his precious sacrifice and intercession) they shall be crucified with him, dead, and buried, Rom. vi. as is lively signified to you in your baptism. When you see that your old man is crucified with Christ, that the body of sin may be destroyed, you will take courage against sin, and will refuse to serve it, since by Christ you are freed from the dominion of it. When you thus by faith put on the Lord Jesus Christ, Rom. xiii. 14, you shall not fulfil the lusts of the flesh.

2d. Grieve heartily for your sins, James iv. 9, Job xlii. 6, 2 Cor. vii. 10; conceive deadly hatred against them, and displeasure against yourself for them. These, like a corrosive, will eat out the life and power of sin.

3d. Make no provision for the flesh, to fulfil the lusts of it, Rom. xiii. 14; but be sober in the use of all worldly things, 1 Cor. vii. 29—31; this, by little and little, will starve sin.

4th. Avoid all objects and occasions of sin, Job xxxi. 1, Prov. xxiii. 20—31; yea, abstain from the appearance of it, 1 Thess. v. 22; this will disarm sin.

5th. When you feel any motion unto sin, whether it arise from within, or come from without, resist it speedily and earnestly, by the sword of the Spirit, the word of God, 1 Peter. v. 9, Acts viii. 20, as your Saviour did, Matt. iv. 4, and as Joseph did, Gen. xxxix. 9; for which cause it must dwell plentifully in you, Col. iii. 16. Thus you shall kill sin.

That you may strengthen the inner man by the Spirit, whereby you may not only mortify the deeds of the flesh, but bring forth the fruits of the Spirit, do thus:

1st. Apply Christ, risen from the dead, to yourself particularly, Rom. iv. 24, 25, vi. 4; believing that God by the same power quickens you, and raises you together with Christ, Eph. ii. 5, 6; to walk in newness of life; reckoning yourself now to be alive unto God, Rom. vi. 9—11; being dead unto sin, and become the servant of righteousness. This believing in Christ, embracing and relying upon him, as set forth in the precious promises of the gospel, 2 Peter i. 4, does draw virtue from Christ into your heart, and does more and more incorporate you into him; and by it, he, by his Spirit, dwells in you, Eph. iii. 17, whereby of his life and grace, John i. 12—16, you receive life and grace; and so you are made partaker of the divine nature, shunning the corruption which is in the world through lust.

2d. Affect your heart with joy unspeakable, and with peace in believing, Rom. xv. 13, considering that you are justified through our Lord Jesus Christ, Rom. v. 1—3, Phil. iv. 4. This joy of the Lord, Neh. viii. 10, as a cordial, will exceedingly strengthen grace in the inner man.

3d. Take heed of quenching or grieving the Spirit, 1 Thess. v. 19—21, but nourish it by the frequent use of holy meditation, prayer, hearing and reading the word, receiving the sacrament, by a Christian communion with such as fear God, Acts i. 12—14, ii. 42—46, iv. 32, 33, and by attending to the motions of the Spirit of God; which you shall know to be from it, when the thing whereunto it moves is, both for matter and circumstance, according to the Scripture, the word of the Spirit. This is to be led of the Spirit; and this will be to walk in the Spirit, and then you shall not fulfil the lusts of the flesh, Gal. v. 16—18.

There remains yet one principal work wherein consists the chief business of the day of your fast, for which all hitherto spoken makes way, and by which, with the former means, you may attain to true reformation of yourself, and reconciliation with God; which is invocation and earnest prayer to God, in the name of Christ, through the Holy Ghost, 1 Sam. vii. 6, Neh. i. 4, &c., ix. 5, &c., Dan. ix. 3, &c., in particular, large and hearty confessions and complaints against yourself for your sins, asking forgiveness, making known your holy resolutions, asking grace, and giving thanks that God is at peace with you, having given Christ for you and to you, (upon your believing in him) and that he has given you a mind to know him, 1 John v. 20, and the power of his resurrection; with other first-fruits of the Spirit, which is the earnest of your inheritance, Eph. i. 13, 14.

Let this solemn and more than ordinary seeking of

God by prayer alone, be twice at least, in the day of your fast, besides your ordinary prayers in the morning and evening; and having thus obtained peace with God, through faith in Christ Jesus, you may, nay ought to pray for the good, or against the evil, which was the occasion of the fast, Ezra viii. 23, 2 Chron xx. 3—6, &c.

But in praying you must in fervency of spirit cry mightly, Jonah iii. 8; striving and wrestling in prayer.

The extraordinary burnt-offerings and sin-offerings, besides the sin-offering of the atonement, to be offered on the solemn day of the fast, Numb. xxix. 7—12, under the law (which, as I told you, in the morality of it, is the standard of religious fasts,) does show, that a fast must be kept in manner as has been said; for hereby we prepare and sanctify ourselves, and seek to God in Christ; hereby we by faith lay hold on Christ, the only true sacrifice for sin; and hereby we do by him draw nigh to God, and in token of thankfulness do give ourselves to be a whole and living sacrifice, holy and acceptable to God, which is our reasonable service, Rom. xii. 1.

For your greater and more thorough humbling of yourself, and further exercise of your faith in God, and love to your brethren and church of God, something yet is to be added.

You must represent to your thoughts also the sins and evils that are already upon, or hanging over the head of your family and nearest friends, and of the town, country, or kingdom where you live, together with their several aggravations; lay them to heart, Psa. cxix. 136—158, Jer. ix. 1, xiii. 17; considering that they by sinning do dishonour God your Father, and do bring evil upon the souls and bodies of those whom you should love as well as yourself: and it is a thousand to one but that you are involved in their sins, and become accessary, if not by example, counsel, permission, or concealment, yet in not grieving for them, in not hating them, and in not confessing and disclaiming them sufficiently before God. These also bring common judgments upon church and state, which you should prefer before your own particular interest, and wherein you may expect to share a part.

You must therefore affect your heart with these thoughts, and mourn for your own first, and then for the abominations of your family, town, country, and kingdom, Ezek. ix. 4; for the sins of princes and nobles, Neh. ix. 34; for the sins of ministers and people. And not only for present sins of the land, but for the sins long since committed, whereof it has not yet repented, Dan. ix. 5, 6; rivers of waters should run down from your eyes, Psa. cxix. 136; at least sighs and groans should rise from your heart, Ezek. ix. 4, Jer. ix. 1, because others as well as yourself have forgotten God's law, and have exposed themselves to his destroying judgments. Do all this so, that you may pour out your heart like water to the Lord in their behalf, Lam. ii. 18, 19.

This is to stand in the breach, Exod. xxxii. 11—15, Psa. cvi. 23; the prayer of a righteous man avails much, if it be fervent, James v. 16, 17, though he have infirmities. If it should not take good effect for others, yet your tears and sighs shall do good to yourself, Ezek. xiv. 14; it causes you to have God's seal in your forehead, Ezek. ix. 4; you are marked for mercy. God will take you from the evil to come, Isa. lvii. 1, or will make a way for you to escape, Jer. xxxix. 16, or will turn the hearts of your enemies to you, Jer. xxxix. 12; or, if you smart under the common judgment, it shall be sanctified to you: and if you perish bodily, yet, when others that cannot live, and are afraid to die, are at their wits' end, you shall be able, in the consciousness of your godly sorrow for your own and others' sins, to welcome death as a messenger of good tidings, and as a gate to everlasting happiness.

If it be a public fast, all these things before mentioned are to be done alone, both before and after the public exercises, at which time you must join in public hearing the word read and preached, and in prayer with more than ordinary attention and fervency.

If you fast with your family, or with some few, let

convenient times be spent in reading the word of God, or some good book, or sermons, which may be fit to direct and quicken you for the present work; also in fervent prayer: the other time alone, Matt. vi. 18, let it be spent as I have showed before.

If some public or necessary occasion, such as you could not well foresee or prevent, when you made choice of your day of private fast, happen to interrupt you; I judge that you may attend those occasions, notwithstanding your fast. But do it thus: if they may be despatched with little ado, then despatch them, and after continue your fast; but if you cannot, I think that you had better be humbled that you were hindered, break off your fast, and set some other day apart instead thereof; even as when a man is necessarily hindered in his vow. Numb. vi. 9—12.

The benefits of religious fasting.

The benefit that will accrue to you by religious fasting, will be motive enough to a frequent use of it, as there shall be cause.

1st. It was never read or heard of, that a fast was kept in truth, according to the former directions from the word, but it either obtained the particular blessing for which it was kept, or at least a better, to him that fasted. Judges xx. 26—35, 1 Sam. vii. 6—10, Ezra viii. 23, 2 Chron. xx. 3—22, Jonah iii. 7—10.

2d. And besides those advantages, thus fasting will put the soul into such good frame, into such an habit of spiritual-mindedness, that (as when against some special entertainment, a day has been spent in searching every corner in a house, to wash and cleanse it) it will be kept clean with common sweeping a long time after.

I do acknowledge that some have fasted, and God has not regarded it, Isa. lviii. 3; yea, he tells some before-hand, that if they fast, he will not hear their cry, Jer. xiv. 12. But these were such who fasted not to God, Zech. vii. 5—12; they only sought themselves; they would not hearken to his word; there was no putting away of sin, or loosing the bands of wickedness, &c., Isa. lviii. 6; no mortification of sin, no renewing their covenant with God. Now, unless we do join the inward with the outward, we may fast, but the Lord sees it not, Isa. lviii. 3—5; we may afflict ourselves, but he takes no notice; we may cry and howl, but cannot make our voice to be heard on high. But when God sees the works of them that fast, that turn from their evil way, Jonah iii. 10; yea, that they strive to turn and seek him with all their heart, then he will turn to them; his bowels of compassion does yearn towards them; and I will have mercy on them, saith the Lord, Jer. xxxi. 18—20.

After the time of the fast is ended, eat and drink but moderately. For, if you then over-indulge yourself, it will put your body and soul both out of order.

Secondly, Your fast being ended, hold the strength which you got that day as much as you can; keep your interest and holy acquaintance which you have obtained with God, and the holy exercises of religion. Though you have given over the exercises of the day, yet unloose not the bent of your care and affections against sin, and for God. It is a corruption of our nature, and it is a policy of Satan to help it forward, that (like some unwise warriors, when they have gotten victory over their enemies) we grow full of presumption and security, by which the enemy takes advantage to re-collect his forces, and coming upon us unlooked for gives us the foil, if not the overthrow. We are too apt, after a day of humiliation, to fall into a kind of remissness, as if then we had gotten the mastery; whereas, if Satan fly from us, if sin be weakened in us, it is but for a season, Luke iv. 13, and but in part; and, especially if we stand not upon our watch, Satan will take occasion to return, and sin will revive in us, Matt. xii. 43—45.

I will add a few cautions touching this excellent but neglected duty of fasting.

1st. The body, although it must be kept under, 1 Cor. ix. 27, Col. ii. 23; yet it must not be destroyed with fasting. It must not be so weakened as to be disabled to perform the works of your ordinary calling.

2d. In private fasts, you must not be open, but as private as conveniently you may, Matt. vi. 16.

3d. Separate not the inward from the outward work in fasting, Isa. lviii. 6, 7.

4th. Think not to merit by your fasting, as papists do.

5th. Presume not that presently upon the work done, God must grant every petition, as hypocrites do, that say to him, We have fasted, and thou dost not regard it, Isa. lviii. 3. You may and must expect a gracious hearing upon your unfeigned humiliation, Matt. xxi. 22; but as for when and how, you must wait patiently: faith secures you of good success, 1 John v. 14, but neither prescribes unto God how, Isa. xl. 13, nor yet does it make haste, Isa. xxviii. 16; but waits his leisure, when in his wisdom he shall judge it most seasonable.

CHAPTER V.

OF THE LORD'S DAY, OR CHRISTIAN SABBATH.

On the Sabbath or Lord's day, you must remember to keep it holy, according to the commandment, Exod. xx. 8—11, xxxv. 2, 3. For this cause consider,

(1.) The divine institution of the Lord's day, or Christian Sabbath.

First, Put a difference between this and the other six days, even as you put a difference between the bread and wine in the sacrament, and that which is for common use. And that because it is set apart for holy use, by divine institution. For as the seventh day, from the beginning of the creation, until the day of Christ's blessed resurrection; so our Lord's day, which is the day of the resurrection, is by divine institution moral. The commandment to keep an holy rest upon the seventh day, after the six days of work (which is the substance of the fourth commandment) remains the same. And this Adam (no doubt by the instinct of uncorrupted nature, which desires a time for God's honour and solemn worship) knowing that God finished the creation in six days, and rested on the seventh, would have observed; yet it was requisite that the particular day should be by institution, for natural reason could not certainly tell him which day. The Lord of the Sabbath therefore limited it unto the seventh from the creation, until Christ's resurrection, and then removed it to the day we keep, which is the first.

Now it appears, that it was the will of our Lord and Saviour Christ, that we should, since his resurrection, keep for our Sabbath that first day of the week; forasmuch as he arose on that day, John xx. 1—19, and appeared divers times on this our Lord's day to his disciples before his ascension; and did on this day, being the day of pentecost, Acts ii. 1—4, fill his disciples with the gifts of the Holy Ghost, they being assembled together; all which gives a pre-eminence to this day, and a probability to the point.

But inasmuch as the apostles, 1 Cor. xi. 1, who followed Christ, and delivered nothing but what they received from Christ, 1 Cor. xi. 23. xiv. 37, did observe this day as a Sabbath, 1 Cor. xvi. 1, 2; what can this argue but a divine institution of this day? The apostle Paul might have chosen any other day, for the people to assemble to hear the word, and receive the sacrament: but they assembled to receive the sacrament, and to hear the word, upon the first day of the week, which is our Lord's day, Acts xx. 6, 7. Now the approved practice of the apostles, and of the church with them, recorded in Scripture, carries with it the force of a precept.

Moreover, the Spirit of God honours this day with the title of the Lord's day, Rev. i. 10, as he does the communion with the title of the Supper of the Lord, 1 Cor. x. 21. xi. 20. What does this argue but as they both have reference to Christ, so they are both appointed by Christ? The Spirit of Christ knew the mind of Christ, who thus named this day.

(2.) Directions for the religious observance of the Lord's day.

Secondly, Being convinced of the holiness of this day, the better to keep it holy when it comes, you must,

1st. On the week day before the Sabbath, or Lord's day, remember it, Exod. xx. 8, 9, to the end that none of your worldly business be left undone, or put off till then; especially upon Saturday, you must prepare for it. Then you must put an end to the works of your calling; and do whatsoever may be well done beforehand, to prevent bodily labour even in your necessary actions, that, when the day comes, you may have less occasion of worldly thoughts, less incumbrance and distractions; and may be more free, both in body and mind, for spiritual exercises.

2d. You yourself, and, as much as in you lies, all under your authority, must rest upon this day, Exod. xxiii. 12; xxxiv. 21, the space of the whole day of four-and-twenty hours, from all manner of works, except those which have true reference to the present day's works of piety, mercy, and true necessity, Matt. xii. 1—13, not doing your own ways, not finding your own pleasures, nor speaking your own words Isa. lviii. 13.

2d. It is not enough that you observe this day as a rest, but you must keep a holy rest. Which that you may do, you must, on your awaking in the morning, make a difference between it and other days, not thinking on any worldly business more than will serve for a general providence, to preserve you from great hurt or loss. Both in your lying awake, and rising in the morning, make use of the former directions, showing you how to awake and rise with God. Rise early, Psa. xcii. 2, if it will consist with your health, and not hinder your fitness for spiritual exercises through drowsiness afterward, that you may show forth God's loving-kindness in the morning. Double your devotions on the Lord's day, as the Jews did their morning and evening sacrifice on the Sabbath day, Numb. xxviii. 3, 9, 10. Prepare yourself for the public holy services by reading, by meditation, Eccles. v. 1, 2; and by putting away all filthiness, James i. 21. 1 Peter ii. 1, 2; that is, repenting of every sin; and casting away the superfluity of naughtiness; that is, let no sin be allowed or suffered to reign in you. Then pray for yourself, and for the minister, Eph. vi. 18—20, that God would give him a mouth to speak, and you an heart to hear, as you both ought to do. All this, before you shall assemble for public worship. Being thus prepared, bring your family with you to the church. Join with the minister and congregation Set yourself as in the special presence of God, following the example of good Cornelius, Acts x. 33, with all reverence attending and consenting; saying Amen with understanding, faith, and affection, to the prayers uttered by the minister; believing, Heb. iv. 2, and obeying, James i. 22, whatsoever is by him commanded you from God. Afterward, by meditation, and by conference, Acts xvii. 11, 12; and if you have opportunity, by repetitions, call to mind, and wisely and firmly lay up in your heart what you have learned, Psa. cxix. 11. The like care must be had before, at, and after, the evening exercise.

The nature and design of baptism, and the Lord's Supper, &c.

1. If baptism be administered, stay, Ezek. xlvi. 10, and attend unto it, (1.) To honour that holy ordinance with the greater solemnity. (2.) And in charity to the persons to be baptized, joining with the congregation in hearty prayer for them, and in a joyful receiving them into the communion of the visible church. (3.) Also in respect of yourself. For hereby you may call to mind your own baptism, in which you did put on Christ, Gal. iii. 27, which also does lively represent the death, burial, and resurrection of Christ, together with your crucifying the affections and lusts, Gal. v. 24; being dead and buried with him unto sin, and rising with him to newness of life, Rom. vi. 3—5, and to hope of glory, Col. i. 27. ii. 11—13; understanding clearly that the blood and Spirit of Christ, Heb. ix. 14. x. 22, signified by water, doth cleanse you

from the guilt and dominion of sin to your justification and sanctification, Matt. iii. 11. Tit. iii. 5. 1 John i. 7. Remembering, moreover, that, by way of sealing, Gen. xvii. 11. Rom. iv. 11, your baptism did in particular exhibit and apply to you that believe, Christ with all the benefits of the covenant of grace ratified in his blood; minding you also of this, that it does not only seal God's promises of forgiveness, grace, and salvation to you; but that also it seals and binds you to the performance of your promise, and vow of faith and obedience, which is the branch of the covenant to be performed, according as was professed, on your part.

Recourse to your baptism is an excellent strengthener of your weak faith, 1 Peter iii. 21, and an occasion of renewing of your vow, you having broken it: and of resisting temptations, considering that they are against your promise and vow in baptism.

Directions for the right attendance on the Lord's Supper.

2. When there is a communion, receive it as oft as, without interrupting the order of the church, you may. But be careful to receive it worthily, 1 Cor. xi. 27.

It is not enough that you be born within the covenant, and that you have been baptized; but you must have knowledge of the nature of the sacrament of the Lord's Supper, 1 Cor. xi. 23. Rom. iv. 11; both that it is of divine institution, and that it is a sign and seal of the righteousness of faith, signifying to you, by the breaking and giving of the bread, and by pouring out and delivering the wine, 1 Cor. xi. 26, the meritorious sacrifice of the Lord Jesus Christ, in whom the covenant of grace is established, 2 Cor. i. 20, 21. Heb. vii. 22; presenting also, and sealing unto you, by the elements of bread and wine, the very body and blood of Christ, with all the benefits of the new covenant, of which you receive indeed livery and seisin in the act of receiving by faith, whereby you also grow into a nearer union with Christ your head, and communion with all his members your brethren, 1 Cor. x. 16, 17.

Besides, there must be a special preparation by

examining yourself, 1 Cor. xi. 28, and renewing your peace with God before you receive, according to the directions before given, chap. v. sect. 2. Also make your peace, at least be at peace, and in charity with your neighbour, by a hearty acknowledging your fault so far as is fit, and making recompence, if you have done him wrong, Matt. v. 23, 24; and by forgiving, and forbearing revenge, if he hath done you wrong, Col. iii. 13.

In the act of administering and receiving, join in confession and prayers, and attend to the actions of the minister when he breaks the bread, pours out the wine, and by blessing sets it apart for holy use, 1 Cor. x. 16; by faith behold Christ, in representation, wounded, bleeding, and crucified before your eyes for you; looking upon him whom your sins condemned and pierced to the death, rather than his accusers, and those who nailed him to the cross; who, though malicious, were but instruments of that punishment which God, with other tokens of his wrath, did execute upon him, though in himself a Lamb without spot, justly for your sin, he being your surety.

This looking upon him whom you have pierced, Zech. xii. 10, should partly dissolve you into a holy grief for sin: but chiefly, considering that by this his passion he hath made full satisfaction for you, and also seeing what blessings God and Christ himself, by the hand of his minister, giving Christ's body and blood sacramentally, do signify and seal unto you, it should raise your heart to a holy admiration of the love of God and of Christ, and it should excite you, in the very act of taking the bread and wine, to a reverend and thankful receiving of this his body and blood by faith, discerning the Lord's body, 1 Cor. xi. 29; gathering assurance hereby that now all enmity between God and you is done away, if you are believers indeed; and that you by this, as by spiritual food for life, shall grow up in him, with the rest of his mystical body, unto everlasting life.

1st. After that you have received, until you be to join in public praise and prayers, affect your heart with joy and thankfulness in the assurance of the pardon of all your sins, and of salvation by Christ; and that more than if you being a bankrupt, should receive an acquittance sealed of the release of all your debts, and with it a will and testament wherein you should have a legacy of no less than a kingdom, sealed with such a seal as gives clear proof of the fidelity, ability, and death of the testator; or than if, having been a traitor, you shall receive a free and full pardon from the king, sealed with his own seal, together with an assurance that he has adopted you to be his child, to be married to his son, the heir of the crown. This is your case, when by faith you receive the bread and wine, the body and blood of the Lord. Think thus, therefore, with joy and rejoicing in God. O! how happy am I in Christ my Saviour! God, who has given him to death for me, and also given him to me, how shall he not with him freely give me all things? Rom. viii. 32; even whatsoever may pertain to life, godliness, and glory, 2 Peter i. 3. Who shall lay any thing to my charge? Who, or what can separate me from the love of Christ? &c. Rom. viii. 33—39.

2d. Resolve withal upon a constant and an unfeigned endeavour to perform all duties becoming one thus acquitted, thus redeemed, pardoned, and advanced; and this in token of thankfulness; even to keep the covenant required to be performed on your part; undoubtedly expecting whatsoever God has covenanted and sealed on his part.

3d. Join in public praise and prayer heartily, and in a liberal contribution to the poor, if there be a collection.

4th. After the sacrament, if you feel your faith strengthened, and your soul comforted, nourish it with all thankfulness.

If not, yet, if your conscience can witness that you endeavoured to prepare as you ought, and to receive as you ought, be not discouraged, but wait for strength and comfort in due time. We do not always feel the benefit of bodily food presently, but stirring of humours and sense of disease is sometimes rather occasioned;

yet in the end being well digested, it strengthens; so it is often with spiritual food. Corruption may stir, and temptations may arise more upon the receiving than before; especially since Satan, if it be but to vex a tender-hearted Christian, will hereupon take occasion to tempt with more violence. But if you resist these, and stand resolved to obey, and to rely upon God's mercy in Christ, this is rather a sign of receiving worthily; so long as your desires and resolutions are strengthened, and you thereby are made more carefully to stand upon your watch. Endeavour in this case to digest this spiritual food by further meditation, improving that strength you have, praying for more strength, remembering the commandment, which bids you to be strong; and you shall be strengthened, Eph. vi. 10. Dan. x. 19.

5th. Lastly, if you find yourself worse indeed, or do feel God's heavy hand in special manner upon you, 1 Cor. xi. 30, following upon your receiving, and your conscience can witness truly that you came not prepared, or that you did wilfully and carelessly fail in such or such a particular in receiving, it is evident you did receive unworthily. In which case you must heartily bewail your sin, confess it to God, 1 John i. 9; ask, and believe that he will pardon it, through Christ Jesus, upon your sincere faith and repentance, 1 John ii. 1, 2, and take heed that you offend not in that kind another time.

(3.) Upon the Lord's day you must likewise be ready to visit and relieve the distressed, 1 Cor. xvi. 2.

Take some time this day to look into your past life, and chiefly to your walking with God the last week, as being freshest in memory, and be sure to let no old scores of sin remain between God and you.

Last of all, on every opportunity, take good time to consider God's works; what they are in themselves, what they are against the wicked, what they are to the church, and to yourself and to yours, and, in particular, take occasion from the day itself, to think fruitfully of the creation, of your redemption, sanctification, and of your eternal rest and glory to come.

For God in his holy wisdom has set such a divine mark upon this our Lord's day, that at once it reminds us of the greatest works of God, which either conduce to his glory, or his church's good; as, of the creation of the world in six days, when he rested the seventh, which specially is attributed to the Father; and of man's redemption by Christ, of whose resurrection this day is a remembrance, which is specially attributed to the Son; also of our sanctification by the Spirit, for that the observation of the Sabbath is a sign and means of holiness, which work is specially attributed to the Holy Ghost; lastly, of your and the church's glorification, which shall be the joint work of the blessed Trinity, when we shall cease from all our works, Heb. iv. 9, 10, and shall rest, and be glorious with the same glory which our head Christ has with the Father, to whom be glory for ever and ever, Amen. Do all these with delight, Psa. xcii. Isa. lviii. 13, raising up yourself hereby to a greater measure of holiness and heavenly-mindedness.

(4.) Motives to keep holy the Lord's day.

Do all this the rather because there is not a clearer sign to distinguish you from one that is profane, Exod. xxxi. 13, than this, of conscientiously keeping holy the Lord's day. Neither is there any ordinary means of gaining strength and growth of grace in the inward man, like this, of due observing the Sabbath, Ezek. xx. 12. For this is God's great mart, or fair-day for the soul, on which you may buy of Christ, wine, milk, bread, marrow, and fatness, Isa. lv. 1—4, gold, white raiment, eye-salve, Rev. iii. 18, even all things which are necessary, and which will satisfy, and cause the soul to live. It is the special day of proclaiming and sealing of pardons to penitent sinners, Acts ii. 38. It is God's special day of publishing and sealing your patent of eternal life. It is a blessed day, sanctified for all these blessed purposes, Exod. xx. 11.

Now, lest this so strict observance of the Lord's day, in spending the whole day in holy meditation, holy exercises, and works of mercy, excepting only necessary repasts, should be thought, as it is by some, to be merely Jewish, or only the private opinion of some zealots, more nice than wise; know, that as the fourth commandment is of moral obligation, there is the same reason for the strict observance of it, as any other divine precept, as against idolatry, murder, fornication, &c. And the taking away of the morality of the fourth commandment, and unloosing the conscience from the immediate bonds of God's command to observe a day for his solemn worship, overthrows true religion, and the power of godliness, and opens a wide gap to atheism, profaneness, and all licentiousness; as daily experience proves in those persons and places, by whom and where the Lord's day is not holily and duly observed.

CHAPTER VI.

DIRECTIONS HOW TO END THE DAY WITH GOD.

When you have walked with God from morning until night, whether on a common day, a day of fasting, or on the Lord's day, according to the former directions; it remains that you conclude the day well, when you would give yourself to rest at night. Wherefore,

First, Look back and take a strict view of your whole carriage that day past. Reform what you find amiss; and rejoice, or be grieved, as you find you have done well or ill, as you have advanced or declined in grace that day.

Secondly, Since you cannot sleep in safety, if God, who is your keeper, Psa. cxxi. 4, 5, do not wake and watch for you, Psa. cxxvii. 1; and though you have God to watch when you sleep, you cannot be safe, if he that watches be your enemy; wherefore it is very convenient, that at night, you not only conclude the day with your family, by reading some Scripture, and by prayer; but you must alone renew and confirm your peace with God by faith and prayer, and with like preparations thereto, as you received directions for the morning; commending and committing yourself to God's tuition by prayer, Psa. iii. 4, 5. xcii. 2, with thanksgiving, before you go to bed. Then shall you lie down in safety, Psa. iv. 8.

All this being done, yet while you are putting off your apparel, when you are lying down, and when you are in bed, before you sleep, it is good that you commune with your own heart, Psa. iv. 4. If other good and fit meditations offer not themselves, some of these will be seasonable:

1. When you see yourself without your apparel, consider what you were at your birth, and what you shall be at your death, when you put off this earthly tabernacle (if not in the mean time, as concerning your outward estates:) how that you brought nothing into this world, nor shall carry any thing out, 1 Tim. vi. 7; naked you came from your mother's womb, and naked shall you return, Job i. 21. This will be an excellent means to give you sweet content in any thing you have, 1 Tim. vi. 8, though never so little; and in the loss of what you have had, Job i. 21, though never so much.

2. When you lie down, you may think of lying down in your winding-sheet, and in your grave. For besides that sleep, 1 Cor. xi. 30, and the bed do aptly resemble death and the grave, Isa. lvii. 2, who knoweth when he sleeps, that ever he shall awake again to this life?

3. You may think thus also: if the sun must not go down upon my wrath, Eph. iv. 26, lest it become hatred, and so be worse ere morning; then it is not safe for me to lie down in the allowance of any sin, lest I sleep not only the sleep of natural death, Psa. xiii. 3, but of that death which is eternal; for who knows what a night will bring forth? Now, it is a high point of holy wisdom, Deut. xxxii. 29, upon all opportunities, to think of, and to prepare for your latter end.

4. Consider likewise, that if you walk with God in

uprightness, your death unto you is but to fall into a sweet sleep, an entering into rest, a resting on your bed for a night, Isa. lvii. 2, until the glorious morning of your happy resurrection.

5. Lastly, If you possibly can, fall asleep with some heavenly meditation, Then will your sleep be more sweet, Prov. iii. 21—25, and more secure, Prov. vi. 21, 22; your dreams fewer, or more comfortable; your head will be fuller of good thoughts, Prov. vi. 22; and your heart will be in a better frame when you awake, whether in the night, or in the morning.

Thirdly, Being thus prepared to sleep; you should sleep only so much as the present state of your body requires; you must not be like the sluggard, to love sleep, Prov. xx. 13; neither must you sleep too much: for if you do, that (which being taken in its due measure, is a restorer of vigour and strength to your body, and a quickener of the spirits) will make the spirits dull, the brain sottish, and the whole body inactive and unhealthy. And that which God has ordained for a furtherance, through your sin shall become an enemy to your bodily and spiritual welfare, Prov. vi. 6—11. Thus much of walking with God in all things, at all times.

CHAPTER VII.

HOW TO WALK WITH GOD ALONE.

SECTION 1.—There is no time wherein you will not be either alone or in company, in either of which you must walk in all well-pleasing, as in the sight of God.

Rules concerning Solitude.

1. Concerning being alone. First, Affect not too much solitude; be not alone, except you have just cause, namely, when you separate yourself for holy duties, and when your needful occasions do withdraw you from society; for in other cases, two are better

than one (saith Solomon) and woe be to him that is alone, Eccles. iv. 9, 10.

2. When you are alone, you must be very watchful, and stand upon your guard, lest you fall into manifold temptations of the devil; for solitariness is Satan's opportunity, Gen. iii. 1, xxxix. 11, 2 Sam. xi. 2, Matt. iv. 1; which he will not lose, as the manifold examples in Scripture, and our daily experience, do witness. Wherefore you must have a ready eye to observe, and a heart ready bent to resist all his assaults. And it will now the more concern you to keep close to God, and not lose his company; that through the weapons of your Christian warfare, you may by the power of God's might quit yourself, and stand fast, Eph. vi. 10, &c.

3. Take special heed, lest when you be alone, you, yourself, conceive, devise, or indulge any evil, to which your nature is then most prone.

And beware in particular, lest you commit alone, by yourself, contemplative wickedness, Mic. ii. 1, Psa. xxxvi. 4, Matt. v. 28; which is, when by feeding your fancy, and pleasing yourself in covetous, lustful, revengeful, ambitious, or other wicked thoughts, you act that in your mind and fancy, which, either for fear, or shame, you dare not, or for want of opportunity or means, you cannot act otherwise.

4. When you are alone, be sure that you are well and fully exercised about something that is good, either in the works of your calling, or in reading, or in holy meditation or prayer. For whensoever Satan does find you idle, and out of employment in some or other of those works which God has appointed, he will take that as an opportunity to use you for himself, and to employ you in some of his works, Matt. xii. 44. But if you keep always in your place; and to some or other good work of your place, you are under God's special protection, as the bird in the law was, while she sat upon her eggs or young ones, keeping her own nest; in which case no man might hurt her, Deut. xxii. 6.

I have already showed how you should behave

yourself as in God's sight, both in prayer, and in the works of your calling, I will say something for your direction concerning reading and meditation.

SECT. 2. OF READING.

Besides your set times of reading the holy Scriptures, you will do well to gain some time from your vacant hours, that you may read in God's book, and in the good books of men.

How to read profitably.

First, When you read any part of the word of God, you must put a difference between it and the best writings of men, preferring it far before them. To this end, (1.) Consider it in its properties and excellencies. No word is of like absolute authority, holiness, truth, wisdom, power, and eternity, Psa. xix. 7—11. (2.) Consider this word in its ends and good effects. No book aimeth at God's glory, John v. 39, 2 Cor. iii. 18, and the salvation of man's soul, Rom. xv. 4, James i. 21, like this; none concerns you like to this. It discovers your misery by sin, together with the perfect remedy, Rom. iii. 23, 24. It proposes perfect happiness unto you, Isa. lv. 1—3, affording means to work it out in you, and for you, Rom. i. 16, 1 Thess ii. 13. It is mighty, through God to prepare you for grace, 2 Cor. x. 4, 5. It is the immortal seed to beget you unto Christ, 1 Peter i. 23. It is the milk and stronger meat to nourish you up in Christ, 1 Peter ii. 2, Heb. v. 13, 14. It is the only soul-physic (through Christ Jesus) to recover you, 2 Tim. i. 13, and to free you of all spiritual evils. By it Christ gives spiritual sight to the blind, hearing to the deaf, speech to the dumb, strength to the weak, health to the sick: yea, by it he does cast out devils, and raise men from the death of sin (through faith) as certainly as he did all those things for the bodies of men by the word of his power, while he lived on the earth, John v. 25. This book of God does contain those many rich legacies bequeathed to you in that last will and testament of God, sealed with the blood of Jesus Christ our Lord, Heb. ix. 15—18. It is the magna charta, and statute-book of the kingdom of heaven, Isa. viii. 20. It is the book of privileges and immunities of God's children, Rom. vi. 14—22, 1 John v. 13. It is the word of grace, which is able to build you up, and give you an inheritance amongst all them that are sanctified, Acts xx. 32. For it will make you wise to salvation, through faith in Christ Jesus, making you perfect, thoroughly furnished unto all good works, 2 Tim. iii. 15—17.

Whenever therefore you hear this word preached, and when at any time you read it, you must receive it not as the word of man, but (as it is in truth) the word of God, then it will work effectually in you that believe, 1 Thess. ii. 13.

Secondly, When you read this word, lift up the heart in prayer to God for the spirit of understanding and wisdom, Psa. cxix. 18, that your mind may be more and more enlightened, and your heart more and more strengthened with grace by it. For this word is spiritual, containing the great counsels of God for man's salvation, and which is as a book sealed up, Isa. xxix. 11, 12, in respect of discovery of the things of God in it, 1 Cor. ii. 10, 11, to all that have not the help of God's Spirit; so that none can know the inward and spiritual meaning thereof, powerfully, and savingly, but by the Spirit of God.

Thirdly, Read the word with a hunger and thirst after knowledge and growth in grace by it, 1 Peter ii. 2, with a reverent, humble, teachable, and honest heart, Luke viii. 15; believing all that you read; trembling at the threatenings and judgments against sinners; rejoicing in the promises made unto, and the favours bestowed upon the penitent, and the godly; willing and resolving to obey all the commandments.

Thus if you read, blessed shall you be in your reading, Rev. i. 3; and blessed shall you be in your deed, James i. 25.

Who must read the Scriptures.

The holy Scriptures are thus to be read of all, of every sort and condition, and of each sex; for all are commanded to search the Scriptures, John v. 32; as well the laity as the clergy; women as well as men, Acts xvii. 11, 12; young as well as old, 2 Tim. iii. 15; all sorts of all nations, Isa. xxxiv. 1—16. Rev. i. 3. For though the Spirit of God is able to work conversion and holiness immediately without the word, as he does in those infants that are saved, yet in adult persons the Holy Ghost will not, where the word may be had, work without it as his instrument, Luke xvi. 29—31; using it as the hammer, plough, seed, fire, water, sword,* or as any other instrument to pull down, build up, plant, purge, or cleanse the souls of men. For it is by the word both read, Rev. i. 3, and preached, that Christ does sanctify all that are his, John xvii. 17; that he may present them to himself, and so to his Father without spot or wrinkle, a church most glorious, Eph. v. 26, 27.

And whereas it is most true, that those who are unlearned and unstable, 2 Peter iii. 16, do wrest not only hard Scriptures, but all other also, to their destruction; yet let not this, as papists would infer, cause you to forbear to read; any more than, because many surfeit and are drunk by the best meats and drinks, you do forbear to eat and drink.

To prevent misunderstanding and wresting of Scriptures to your hurt, do thus: (1.) Get and cherish a humble and honest heart, resolved to obey what you know to be God's will. If any man will do his will, saith Christ, he shall know of the doctrine, whether it be of God, John vii. 17. (2.) Get a clear knowledge of the first principles of the Christian religion, and believe them stedfastly. And endeavour to frame your life according unto those more easy and known Scriptures, whereon these first principles of the oracles of God are founded; for these give light, even at the first entrance, unto the very simple, Psa. cxix. 130. This do, and you shall neither be unlearned in the mysteries of Christ, nor yet unstable in his ways. (3.) Be much in hearing the word interpreted, by learned and faith-

* They are Scripture metaphors.

ful ministers, Isa. viii. 20. (4.) If you meet with a place of Scripture too hard for you, presume not to frame a sense to it of your own head; but take notice of your ignorance, admire the depth of God's wisdom, suspend your opinion, and take the first opportunity to ask the meaning of some or other, of those whose lips should preserve knowledge, Mal. ii. 7.

Motives to read Scripture.

Let no colourable pretence keep you from diligent reading of God's book; for hereby you will be better prepared to hear the word preached. For it lays a foundation for preaching, Acts viii. 28, 34, 35; leading the way to a better understanding thereof, and more easily preserving it in memory; also, to enable you to try the spirits and doctrines delivered, Acts xvii. 11. 1 John iv. 1. 1 Thess. v. 21; even to try all things, and to cleave to that which is good.

How to read men's writings profitably.

1. In reading men's writing, read the best, or at least those by which you can profit most.

2. Read a good book thoroughly, and with due consideration.

3. Reject not hastily any thing you read, because of the mean opinion you have of the author. Believe not every thing you read, because of the great opinion you have of him that wrote it. But, in all books of faith and manners, try all things by the Scriptures, Isa. viii. 20. Matt. xxii. 29, 31. Receive nothing upon the bare testimony or judgment of any man, any further than he can confirm it by the canon of God's holy word, Luke x. 26, or by evidence of reason, or by undoubted experience; provided always, that what you call reason and experience, be according unto, not against the word of God. If the meanest speak according to it, then receive and regard it: but if the most judicious in your esteem, yea, if he were an angel of God, should speak or write otherwise, refuse and reject it, Gal. i. 8.

Thus much for private reading.

Only take this caution. You must not think it to

be sufficient that you read the Scriptures and other good books at home in private, when by so doing you neglect the hearing of the word read and preached in public. For God has not appointed, that reading alone, or preaching alone, or prayer, or sacraments should singly and alone save any man, where all, or more than one of them may be had; but he requires the joint use of them all in their place and time. And in this variety of means of salvation, God has in his holy wisdom ordained such order, that the excellency and sufficiency of one shall not, in its right use, keep any from, but lead him unto a due performance of the other; each serving to make the other more effectual to produce their common effect, namely, the salvation of man's soul.

Indeed, when a man is necessarily hindered by persecution, sickness, or otherwise, that he cannot hear the word preached, then God does bless reading with an humble and honest heart, without hearing the word preached. But where hearing the word preached, is either contemned or neglected, for reading sake, or for prayer sake, Prov. xxviii. 9, or for any other good private duty, there no man can expect to be blessed in his reading, or in any other private duty, but rather cursed. Witness the evil effects, which by experience we see do issue from thence, viz. self-conceitedness, singularity in some dangerous opinions; and schism, and too often a falling away into damnable heresies and apostasy.

SECT. 3. OF MEDITATION.

When you are alone, then also is a fit season for you to be employed in holy meditation. For according to a person's meditation such is he. The liberal man devises liberal things; the covetous man the contrary, Isa. xxxii. 8. The godly man studies how to please God, the wicked how to please himself.

In meditation, the mind or reason of the soul fixes

itself upon something conceived or thought upon for the better understanding thereof, and for the better application of it to itself for use.

The distinct acts and parts of meditation.

(1.) In meditating aright, the mind of man exercises two kinds of acts; the one direct upon the thing meditated; the other reflex upon himself, the person meditating. The first is an act of the contemplative part of the understanding; the second is an act of conscience. The end of the first is to enlighten the mind with knowledge: the end of the second is to fill the heart with goodness. The first serves, I speak of moral actions, to find out the rule whereby you may know more clearly what is truth, what is falsehood, what is good, what is bad; whom you should obey, and what manner of person you should be, and what you should do, and the like. The second serves to direct you how to make a right and profitable application to yourself, and to your actions, of the rule.

In this latter are these two acts: First, an examination, whether you and your actions be according to the rule, or whether you come short, or are swerved from it, giving judgment of you, according as it finds you.

The second is a persuasive and commanding act, charging the soul in every faculty, understanding, will, affections, yea, the whole man, to reform and conform themselves to the rule, that is, to the will of God, if you find yourself not to think and act according to it: which is done by confessing the fault to God with remorse, praying for forgiveness, returning to God by faith and repentance, and reforming the heart and life through new obedience. This must be the resolution of the soul. And all this a man must charge upon himself peremptorily, commanding himself with sincere desire and fixed endeavour to conform to it.

When you meditate, join all these three acts, else you will never bring your meditation to a profitable issue. For if you only muse and study to find out what is true, what is false, what is good, what is bad, you may gain much knowledge of the head, but little goodness to your heart. If you only apply to yourself that whereon you have mused, and no more; you may, by finding yourself to be a transgressor, lay guilt upon your conscience, and terror upon your heart, without fruit or comfort; but if to these two, you lay a charge upon yourself to follow God's counsel concerning what you should believe and do, when you have offended him; if you also form an upright design, through God's grace, to be such an one as you ought to be, and to live such a life hereafter as you ought to live; then unto science you add conscience, and to knowledge, you join practice, and will find the comfortable and happy effects thereof. Observe David's meditations, and you will find they came to this issue. His thoughts of God and of his ways, made him turn his feet unto God's testimonies, Psa. cxix. 59. The meditation of God's benefits made him resolve to take the cup of salvation, and call upon the name of the Lord, and to pay his vows, Psa. cxvi. 12—14. When he considered what God had done for him, and thence inferred what he should be to God again, he saith to his soul, My soul, and all that is in me, praise his holy name, Psa. ciii. 1—3. When in his meditation he found that it was his fault to have his soul disquieted in him through distrust, he charges it to wait on God, Psa. xlii. 5—11, and raises up himself unto a holy confidence. I will meditate on thy precepts, saith he. What, is that all? No, but he proceeds to this last act of meditation, and saith, I will have respect unto thy ways, Psa. cxix. 15, 16, 106.

Rules for meditation.

(2.) God's holy nature, attributes, word, works, also what is duty, and what is sin; what you should be, and do; what you are, and what you have done, what are the miseries of the wicked, and what are the happiness and privileges of the righteous, are fit subjects of meditation.

(3.) That which must settle your judgment, and be the rule to direct you what to hold for true and good, must be the canon of God's word rightly understood, 2 Peter i. 19, and not your own reason or opinion: nor yet the opinions or conceits of men; for these are false and crooked rules.

Cautions about the matter of meditation.

(4.) In seeking to know the secrets and mysteries of God and godliness, you must not pry into them further than God hath revealed; for if you wade therein further than you have sure footing in his holy word, you will presently lose yourself, and be swallowed up in a maze and whirlpool of errors and heresies. These deep things of God must be understood with sobriety, Psa. cxxxi. 1; Rom. xii. 3, according to that clear light which God has given you by his word.

(5.) When sin happens to be the matter of your meditation, take heed lest while your thoughts dwell upon it, though your intention be to bring yourself out of love with it, it steal into your affections, and work in you some secret liking to it, and so circumvent you. For the cunning devices of sin are undiscoverable, Eccles. vii. 24, 26, 28, and you know that your heart is deceitful above all things, Jer. xvii. 9. Wherefore, to prevent this mischief; (1.) As sin is not to be named, Eph. v. 3, but when there is just cause; so it is not to be thought upon, but upon special cause, namely, when it shows itself in its motions and evil effects, and when it concerns you to try and find out the wickedness of your heart and life. (2.) When there is cause to think of sin, represent it to your mind as an evil, the greatest evil, Gen. xxxix. 9, most loathsome and abominable to God, and most hateful and hurtful to yourself. Whereupon you must raise your heart to a holy detestation of it, and resolution against it. (3.) Never stand reasoning or disputing with it, as Eve did with Satan, Gen. iii. 2, 3; but without any indulgence of it, you must do present execution upon it, by sheathing the word of God, Matt. iv. 4, 7, 10, the sword of the Spirit, into the heart of it; and by mortifying of it through the help of his Spirit, Rom. viii. 13. And if you would dwell long in meditating upon any subject, make choice of matter more pleasant and less infectious

(6.) It is necessary that you be skilful in this first part of meditation, for hereby you find out, who is to be adored, who not; what is to be done, what not; what you should be, what not. But the life of meditation lies in the reflex acts of the soul, whereby that knowledge which was gotten by the former act of meditation, doth reflect and return upon the heart, 2 Chron. vi. 37; causing you to apply to yourself what was proposed; whence also you are induced to endeavour to form your heart and life according to that which you have learned it ought to be.

This, though it be most profitable, yet, because it is tedious to the flesh, is most neglected. Wherefore it concerns you who are instructed in the points of faith and holiness, to be most conversant in this when you are alone, whether it be when you are engaged in the common business of life, or retirement for solemn worship.

(7.) You should therefore be well read in the book of your conscience, as well as in the Bible, 1 Cor. xi. 28, 31. 2 Cor. xiii. 5. Commune often with it, and it will fully acquaint you with yourself, and with your estate, through the light of God's Holy Spirit. It will tell you what you were and what you now are; what you most delighted in, in former times, and what now. It will tell you what straits and fears you have been in, and how graciously God delivered you; what temptations you have had, and how it came to pass, that sometimes you were overcome by them; and how, and by what means you overcame them. It will show what conflicts you have had between flesh and Spirit, and what was the issue thereof, Psa. lxxvii. 1—13; whether you were grieved and humbled when sin got the better, and whether you rejoiced and were thankful when God's grace restrained you, or gave you the victory. Your conscience being set on work, will call to remembrance your oversights; and the advantages which you gave to Satan and to the lusts of your flesh, that you may not do the like again. It will remember you by what helps and means, through God's grace, you prevailed and got a conquest over some sin, that you may use the same another time. If you thus diligently observe the passages and conflicts of your Christian race and warfare, your knowledge will be an experimental knowledge; which, because it is a knowledge arising from the frequent proof of that whereof you were taught in the word; it becomes a more fixed, perfect, and fruitful knowledge than that of mere contemplation.

It is only this experimental knowledge that will make you skilful in the duties and trials of the Christian life. Take a man that hath only read much of husbandry, physic, merchandise, policy, &c., who hath gotten into his head the notions of all these, and makes himself believe that he hath great skill in them; yet one that hath not read half so much, but hath been of long practice, and of great experience in these, as far excells him in husbandry, physic, trading, &c. as he excells one that is a mere novice in them. Such difference there is between one that hath only a superficial knowledge of Christianity, without experimental observation; and him that is often looking into the records of his own conscience, carefully observing the workings of his own heart, and God's dispensations towards him.

The experience which by this means you will obtain, of God's love, truth, and power; of your enemies' falsehood, wiles, and methods; of your own weakness without God, and of your strength by God to withstand the greatest lusts, and strongest temptations; yea, of an ability to do all things through Christ that strengthens you, will beget in you, faith and confidence in God, and love to him, watchfulness and circumspection, lest you be overtaken with sin; with such degrees of humility, wisdom, and Christian courage, that no opposition shall daunt you, nor shake your confidence in Christ Jesus.

Where do you read of two such champions as David and Paul? 1 Sam. xvii. 36. 2 Tim. i. 12. iv. 7, 17, 18. And where do you find two that recorded, and made use of their experiences of God's truth and goodness, like these?

Wherefore, next to God's book, which gives light and rule to your conscience, read often the book of your conscience. See what is there written for or against you. When you find that your heart and life are according to the rule of God's word, hold that fast to your comfort; but, wherein you find yourself not to be according to this rule, give yourself no rest, until in some good measure, at least in endeavour, you do live according to it.

I have insisted the more largely on this point of meditation, because of the great necessity and profitableness of it; many of God's people omit it, because they know not how to do it; and because they know not their need, nor yet the benefit which they may receive from it.

Motives to meditation.

(9.) The necessity and use of meditation will appear, if you consider, 1. That reading, hearing, and transient thoughts of the best things leave not half that impression of goodness upon the soul, which they would do, if they might be recalled, and fixed there by serious thought. Without this meditation, the good food of the soul passes through the understanding, and either is quite lost, or is like raw and undigested food, which doth not nourish those creatures that chew the cud, till they have fetched it back and chewed it better. Meditation is instead of chewing the cud. All the outward means of salvation do little good in comparison, except by meditation they are thoroughly considered, and laid up in the heart.

2. The great usefulness of meditation appears in that, (1.) It does digest, ingraft, and turn the spiritual knowledge gained in God's word and ordinances, into the very life and substance of the soul, changing and fashioning you according to it, so that God's will in his word and your will become one, choosing and delighting in the same things. (2.) Meditation fitteth for prayer, nothing more. (3.) Meditation also promotes the practice of godliness, nothing more. (4.) Nothing does perfect and make a man an understanding Christian more than this. (5.) Nothing does make

a man more know and enjoy himself with inward comfort, nor is a clearer evidence that he is in a state of happiness, than this. For in the multitude of my thoughts within me (saith David to God) thy comforts delight my soul, Psa. xciv. 19. And he does by the Spirit of God pronounce every man blessed, that does thus meditate in God's law day and night, Psa. i. 2.

CHAPTER VIII.

OF COMPANY IN GENERAL. RULES CONCERNING IT.

When you are in company, of what sort soever, you must amongst them walk with God.

Directions relating hereunto are of two sorts. First, showing how you should behave towards all: Secondly, how towards good or bad company.

First, In whatsoever company you are, your conversation in word and deed must be such, as may procure (1.) Glory to God, Matt. v. 16. (2.) Credit to religion, 1 Tim. vi. 1. (3.) All mutual, lawful, content, help, and true benefit to each other, Gen. ii. 18. For these are the ends, first, of society; secondly, of the variety of the good gifts that God has given unto men to do good with, 1 Cor. xii. 7—25.

To attain these ends, your conversation must be, 1. Holy; 2. Humble; 3. Wise; 4. Loving.

First, It must be holy, 1 Peter i. 15; you must, as much as in you is, prevent all evil speech and behaviour, which might else break forth, being careful to break it off, if it be already begun in your company. Suffer not the name and religion of God, nor yet your brother's name be traduced, or evil spoken of; but in due place and manner vindicate each. Be diligent to watch, and improve all fit opportunities of introducing pious and useful conversation; even whatsoever may tend to the practice and increase of godliness and honesty.

Secondly, Your conversation must be humble. You must give all due respect to all men, according to their several places and gifts; reverencing your betters, submitting to all in authority over you, 1 Peter ii. 17, Eph. v. 21; esteeming others as better than yourselves, in honour preferring them before you, Phil. ii. 3; condescending unto, and behaving respectfully towards, those of meaner rank, Rom. xii. 16.

Thirdly, You must be wise and discreet in your carriage towards all, and that in divers particulars.

(1.) Be not too open, nor too reserved; not over suspicious, 1 Cor. xiii. 7, nor over credulous, John ii. 24. Jer. xl. 14—16. For the simple believes every word, but the prudent looks well to his going, Prov. xiv. 15.

(2.) Apply yourself to the several conditions and dispositions of men in all indifferent things, so far as you may, without sin against God, or offence to your brother, becoming all things to all men, 1 Cor. ix. 19—23; suiting yourself to them in such a manner, that if it be possible, you may live in peace with them, Rom. xii. 18, and may gain some interest in them, to do them good.

But far be it from you to do as many, who under this pretence, are for all companies; seeming religious with those that are religious; but profane and licentious with those that are profane and licentious; for this is carnal policy, and damnable hypocrisy, and not true wisdom.

(3.) Intermeddle not with other men's business, 1 Thess. iv. 11, but upon due and necessary occasion.

(4.) Know when to speak, and when to be silent, 1 Tim. v. 13. How excellent is a word spoken in season! Eccles. iii. 7. As either speech or silence will make for the glory of God, and for the cause of religion, and good one of another, so speak, and so hold your peace, Prov. xv. 23, xxv. 11.

(5.) Be not hasty to speak, Prov. xxix. 11, nor be much in speaking, Prov. xvii. 27, Eccles. x. 14, but only when just cause shall require; for as it is shame and folly to a man to answer a matter before he hears it, Prov. xviii. 13, so is it for any to speak before his time and turn, Job xxxii. 4—6. Likewise consider, that in the multitude of words there wants not sin; but he that refrains his lips is wise, Prov. x. 19.

(6.) Be sparing to speak of yourself or actions, to your own praise, except in case of necessary apology, 2 Cor. xii. 11, and defence of God's cause maintained by you, and in the clearing of your wronged innocency, or needful manifestation of God's power and grace in you; but then it must be with all modesty, giving the praise unto God, Phil. iv. 12, 13. Neither must you cunningly hunt for praise, by debasing or excusing yourself and actions, that you may give occasion to draw forth commendations of yourself from others. Thus seeking of applause, argues pride and folly. But do praiseworthy actions, seeking therein the praise of God, that God may be glorified in you, then you shall have praise of God, Rom. ii. 29, whatever you have of man. However, follow Solomon's rule: Let another praise thee, not thine own mouth; a stranger, and not thine own lips, Prov. xxvii. 2.

(7.) As you must be wise in your carriage towards others, so you must be wise for yourself; which is to make a good use to yourself of all things that occur in company. Let the good you see, be matter of joy, and thankfulness to God, and improved for your own imitation, Rom. xii. 9. Let the evil you see, be matter of grief and humiliation, and a warning to you, lest you commit the like, since you are made of the same mould that others are, and are liable to the same temptations. If men report good of you to your face, repress those speeches as soon and as wisely as you can, giving the praise of all things to God, Gen. xli. 15, 16, Acts xi. 23; knowing that this is but a temptation and a snare, Prov. xxvii. 14, and a means to breed self-love, pride, and vain-glory in you. If this good report be true, bless God that he has enabled you to deserve it, and study by virtuous living to continue it. If this good report be false, endeavour to make it good by being hereafter answerable to the report.

(8.) If men report evil of you to your face, be not so much inquisitive who raised it, or how to confute them, or clear your reputation amongst men; as to make a good use of it to your own heart before God.

For you must know, this evil report does not rise without God's providence, 2 Sam. xvi. 11. If the report be true, then see God's good providence; it is that you may see your error and failings, that you may repent. If the report be false, yet consider, if you have not run into the appearance and occasions of those evils. Then say, though this report be false, yet it comes justly upon me, because I did not shun the occasions and appearances. This should humble you, and cause you to be more circumspect in your ways. But if neither the thing reported be true, nor you have given occasion for it, yet see God's wise and good providence; not only in discovering the folly and malice of evil men, who raise and take up an evil report against you without cause: but in giving you warning to look to yourself, lest you deserve thus to be spoken of. And how do you know, but that you should have fallen into the same, or the like evil, if by these reports you had not been forewarned? Make use therefore of the railings and revilings of an enemy, 2 Sam. xvi. 10—12; though he be a bad judge, yet he may be a good remembrancer; for you shall hear from him those things, of which flatterers will not, and friends, being blinded, or over indulgent through love, do never admonish you.

Fourthly, Your conversation amongst all must be loving: you should be kind and courteous towards all men, Tit. iii. 2. Do good to all, according as you have ability and opportunity, Gal. vi. 10. Give offence willingly to none, 1 Cor. x. 32. Do wrong to no man, 1 Cor. vi. 1—8, either in his name, life, chastity, or estate, or in any thing that is his; but be ready to forgive wrongs done to you, Col. iii. 13, and to take wrong, rather than to revenge, or unchristianly to seek your own vindication. As you have calling and opportunity, do good to the souls of your neighbours; exhort and encourage unto well-doing, 1 Thess. v. 14. If they show not themselves to be dogs and swine, Matt. vii. 6; that is, obstinate scorners of good men, and contemners of the pearl of good counsel, you must, so far as God gives you any interest in them, admonish and inform them with the spirit of meekness and wisdom, Lev. xix. 17. With this cloak of love you should cover and cure a multitude of your companions' infirmities and offences, 1 Peter iv. 8. In all your behaviour towards him, seek not so much to please yourself as your companion, in that which is good to his edification, Rom. xv. 2.

(1.) Speak evil of no man, Tit. iii. 2; nor yet speak the evil you know of any man, except in these or the like cases. (1.) When you are thereunto lawfully called by authority. (2.) When it is to those whom it concerns, to reform and reclaim him of whom you speak, and you do it to that end, 1 Cor. i. 11. (3.) When it is to prevent certain damage to the soul or estate of your neighbour, Acts xxiii. 16, which would ensue, if it were not by you thus discovered. (4.) When the concealment of his evil may make you guilty and accessory. (5.) When some particular remarkable judgment of God is upon a notorious sinner for his sin, then, to the end that God may be acknowledged in his judgments, and that others may be warned, or brought to repent of the same or like sin, you may speak of the evils of another, Psa. lii. 6, 7. But this is not to speak evil, so long as you do it not in envy and malice to his person, nor with aggravation of the fault more than is cause, nor yet to the judging of him as concerning his final estate.

(2.) When you shall hear any in your company speak evil of your neighbour, by slandering, whispering, or tale-bearing, whereby he detracts from his good name; you must not only stop your ears at such reports, but must set your speech and countenance against him, like a north wind against rain, Prov. xxv. 23.

(3.) When you hear another well reported of, let

it not be grievous to you, as if it detracted from your credit; but rejoice at it, inasmuch as God has enabled him to be good, and to do good; all which makes for the advancement of the common cause of religion, wherein you are interested; envy him not therefore his due praise.

(4.) Detract not from any man's credit, either by open backbiting, Psa. xv. 3, or by secret whispering, Prov. xvi. 28, or by any cunning means of casting evil aspersions, whether by way of pitying him, or otherwise: as, He is good or does well in such and such things; but, &c. This *but* mars all.

(5.) And, in a word, in all speeches to men, and communications with them, your speech must be gracious, Col. iv. 6, that which is good to the use of edifying, that it may minister grace, not vice, to the hearers. It must not be profane, nor any way corrupt, Eph. iv. 29, as defiled with oaths, curses, or profane jests; it must not be flattering, Job xvii. 5, nor yet detracting; not bitter, not railing, censorious, or injurious to any man, Eph. iv. 31. It must not be wanton, lascivious, and filthy, Eph. v. 3, 4. Col. iii. 8. It must not be false, Col. iii. 9; no, nor yet foolish, idle, and fruitless; for all evil communication does corrupt good manners, 1 Cor. xv. 33. And we must answer for every idle word which we speak, Matt. xii. 36. Besides, a man may easily be discerned of what country he is, whether of heaven, or of the earth, by his language; his speech will betray him.

(6.) There is no wisdom or power here below, can teach and enable you to do all, or any of the forementioned duties. This wisdom and power must be had from above, James iii. 13—18. Wherefore, if you would in all companies carry yourself worthy the gospel of Christ:

First, Be sure that the law of God, and the power of grace be in your heart, else the law of grace and kindness cannot be in your life and speech, Psa. xxxvi. 30, 31. Prov. xxxi. 26. You must be endued, therefore, with a spirit of holiness, humility, love, gentleness, long-suffering, meekness, and wisdom;

else you can never converse with all men as you ought to do. For such as the heart is, such the conversation will be. Out of the evil heart come evil thoughts and actions, Matt. xv. 19; but a good man, out of the good treasure of his heart brings forth good things, and according to the abundance of the heart the mouth speaketh, Matt. xii. 34, 35. A man must have the heart of the wise, before the tongue can be taught to speak wisely, Prov. xvi. 23.

Secondly, You must resolve before-hand, as David did, to take heed to your ways, that you sin not with your tongue; and that you will keep your mouth as with a bridle, Psa. xxxix. 1. Before your speech and actions, be well advised; weigh and ponder in the balance of reason, all your actions and words, before you vent them.

Thirdly, Let no passion of joy, grief, fear, anger, &c. get the head, and exceed their limits. For wise and good men, as well as bad, when they have been in any of these passions, have spoken unadvisedly with their lips, Job iii. 3, 23. Psa. cvi. 32, 33. Mark ix. 5, 6. Jonah iv. 8, 9. Mark vi. 22, 23. And experience will teach you, that your tongue never runs before your wit so soon, as when you are over-afraid, over-grieved, over-angry, or over-joyed.

Fourthly, You must be much in prayer unto God, before you come into company, that you may be able to order your conversation aright; let your heart also be lifted up often to God when you are in company, that he would set a watch before your mouth, and keep the door of your lips, and that your heart may not incline to any evil thing, to practise wicked works with men that work iniquity, Psa. cxli. 3, 4; and that he would open your lips, that your mouth may show forth his praise, Psa. li. 15; and that you may speak as you ought to speak, knowing how to answer every man, Col. iv. 6; for the tongue is such an unruly evil, that no man, but God only, can tame and govern it, James iii. 8.

SECT. 2. CAUTIONS AND DIRECTIONS CONCERNING EVIL COMPANY.

WHEN company is evil or sinful, if you may choose, come not into it at all, Prov. i. 15; xxiii. 20; Psa. xxvi. 4, 5. For keeping evil company will (1.) Blemish your name. (2.) It will expose you often to many hazards of your life and state, 1 Kings xxii. 29—32; 2 Chron. xviii. 31, and xxii. 6—9; Gen. xiv. 11, 12. And (3.) you are always in danger to be corrupted by the contagion and infection of it, Prov. xxii. 24, 25.

By bad company, I do not only understand, seducers, and such as are openly profane or riotous; but also such civil men, who yet remain mere worldings, and all lukewarm professors, who are strangers to the life and power of religion. For although the sins of these latter do not carry such a manifest appearance of gross impiety and dishonesty, as those of open blasphemers, drunkards, adulterers, and the like; yet they are not less dangerous. Your heart will quickly rise against these manifest enormous evils: but the other, by reason of their unsuspected danger, through that tolerable good opinion which, in comparison, is had of them, will sooner ensnare and infect you, by an insensible chilling of your spirits, and by taking off the edge of your zeal towards the power of godliness; and so, by little and little, draw you to a remissness and indifferency in religion, and to a love of the world.

If you shall think, that by keeping evil company, you may convert them, and draw them to goodness; be not deceived; it is presumption so to think. Has not God expressly forbidden you such company? Prov. xxiii. 20. If you be not necessarily called to be in sinful company, you may justly fear that you shall be sooner perverted, Psa. cvi. 34, 35, and made evil by their wickedness, than, that they should be converted and made good by your holiness.

Secondly, When by reason of common occasions in respect of the affairs of your calling, generally, or in particular, in church, commonwealth, and family, you cannot shun ill company; (1.) Be specially watchful that your conversation be honest, 1 Thess. iv. 12, unblamable and harmless, Phil. ii. 15, 16; Eph. v. 15, 16; even with a dove-like innocency, Matt. x. 16; that by your good example, they may without the word be brought to love the power and sincerity of that true religion which you profess, 1 Peter iii. 1. However, give no advantage to the adversay to speak evil, 1 Tim. v. 14, either of you, or of your religion; but, by a holy life, stop the mouths of ignorant and foolish men, 1 Peter ii. 15; or if they will notwithstanding speak against you, let your holy life shame all that blame your good conversation in Christ Jesus, 1 Peter iii. 16. (2.) Be wise as serpents, Col. iv. 5; Matt. x. 16. Walk cautiously, lest they bring you into temporal evils and inconveniences: but especially lest they infect you with their sin; for a little leaven will quickly leaven the whole lump, 1 Cor. v. 6.

That you may not be infected by that ill company which you cannot avoid, use these preservatives: (1.) Be not high minded, Rom. xi. 20; but fear, lest you do commit the same or the like sin; for you are of the same nature, and are subject to the same, or the like temptations. He that sees his neighbour slip and fall before him, had need to take heed lest he himself fall, 1 Cor. x. 12. (2.) Your soul, like that of righteous Lot, must be vexed daily with seeing and hearing their unlawful deeds, 2 Peter ii. 8, Psa. cxix. 136—158. (3.) Raise your heart to a sensible loathing of their sin; yet have compassion on the sinner, Jude 22, 23; and so far as you have opportunity, admonish him as a brother, 2 Thess. iii. 14, 15. (4.) When you see or hear any wickedness, lift up your heart to God, and before him confess it, and disclaim all liking of it, Psa. cxx. 5, 6; pray unto God to keep you from it, and that he would forgive your companion his sin, and give unto him grace to repent of it.

Lastly, Though you may converse with sinful company, when your calling is to be with them, in a common and colder kind of fellowship, by a common love, whereby you wish well to all, and would do good to all; yet you must not converse with them with such special and intimate Christian familiarity and delight, Psa. xvi. 3, as you do with the saints that are excellent. Thus do, and the Lord can and will keep you in the midst of Egypt and Babel, as he did Joseph and Daniel, if he call you to it.

Thirdly, As soon as possibly you can, depart out of their company, when you find not in them the lips of knowledge, Prov. xiv. 7, or when they any way declare that they have only a form, but deny the power of godliness, 2 Tim. iii. 5. From such turn away, saith the apostle. And so use the preservatives prescribed, or any other, as prudence shall direct, that you depart not more evil, or less good, than when you came together.

SECT. 3. DIRECTIONS FOR CHRISTIAN FELLOWSHIP.

Now concerning good company, or Christian fellowship, First, highly esteem it, Psa. xvi. 3, and much desire it. For you should love the brotherhood, 1 Peter ii. 17, however the world scoff at it; and forsake not the fellowship, Heb. x. 25, or the company of the godly, as the manner of some is; but, with David, as much as may be, be a companion with them that fear God, Psa. cxix. 63.

Secondly, When you are in good company, you must express all brotherly love; improving your time together for your mutual good, chiefly in the increase of each other's faith and holiness, Rom. i. 11, 12; provoking one another to love, and to good works, Heb. x. 24.

Then is your Christian love of the right kind, (1.) When you love them out of a pure heart fervently, 1 Peter i. 22; which is, when you love them, because they are brethren, partakers of the same faith and spirit of adoption, Heb. iii. 1; having the same Father, and being of the same household of faith with you, Gal. vi. 10. (2.) When you love them not only with a love of humanity, as they are men, for so you should love all men, even your enemies; nor yet only with a common love of Christianity, wherewith you love all professing true religion, though actually they show little fruit and power thereof; but with a special love, 1 Peter i. 22, iv. 8; for kind, spiritual; and for degree, more abundant. Therefore it is called brotherly kindness, Rom. xii. 10, and a fervent love, distinct from charity, or a common love, 2 Peter i. 7.

Where this love is, it will unite hearts together, like Jonathan's and David's, 1 Sam. xviii. 1, making you to be of one heart and soul, Acts iv. 32. It will make you enjoy each other's society with spiritual delight, Psa. xvi. 3. It will make you to sympathize with one another; and to bear each other's burthens, Gal. vi. 2. It will make you to communicate in all things communicable, with gladness, and singleness of heart, Acts ii. 46, as you are able, and that with a special love, Gal. vi. 10, beyond that which you show to them which are not alike excellent. Yea, it is so entire and so ardent, that you will not hold your life to be too dear, to lay down for the common good of the brethren, 1 John iii. 16.

When therefore you meet with those that fear God, improve the communion of saints, not only by communicating in natural and temporal good things as you are able, and as there is need; but especially in the communion of things spiritual, edifying yourselves in your most holy faith, Jude 20, 1 Thess. v. 11, by holy speech and conference, and in due time and place, in reading the holy Scriptures and good books, and by prayer, and singing of psalms together, Col. iii. 16.

That your singing may please God, and edify yourself and others, observe these rules:

1. Sing as in God's sight, and, in matter of prayer and praise, speak to God in singing, Psa. xxx. 4.

2. The matter of your song must be spiritual, either indited by the Spirit, or composed of matter agreeing thereunto, Col. iii. 16.

3. You must sing with understanding, 1 Cor. xiv. 15.

4. You must sing with judgment, being able in private to make choice of psalms suitable to the present time and occasion; and both in private and public to apply the psalm sung to your own particular case, only taking heed that you do not apply the imprecations made against the enemies of Christ and his church in general, to your enemies in particular; also endeavour to confirm your faith, and incline your will and affections according to the subject of your psalmody, whether you sing the prophecies of Christ, his promises, threats, commands, mercies, or judgments, &c.

5. You must make melody to the Lord in your heart, Col. iii. 16; which is done (1.) By preparing and setting the heart in tune, Psa. lvii. 7. It must be an honest heart. (2.) The heart must be lifted up, Psa. xxv. 1. (3.) The mind intent, 1 Cor. xiv. 15. (4.) The affections lively, Psa. xxxiii. 3, Rev. xiv. 3; the heart believing, and, in the matter of praise and thanks, joyous, Psa. lxxxiv. 1.

6. Lose not your short and precious time, with idle compliments, worldly discourses, or talking of other men's matters and faults, 1 Tim. v. 13; nor yet in a barren and fruitless hearing and telling of news, Acts xvii. 21, out of affectation of strangeness and novelty. But let the matter of your talk be, either of God, or of his word and ways, wherein you should walk; or of his works of creation, preservation, redemption, sanctification, and salvation; of his judgments which he executes in the world, and of his mercies showed towards his people: or matter of Christian advice, either of the things of this life, or of that which is to come. Impart also each to other the experience and proofs you have had of God's grace and power, in your Christian warfare. And as there shall be cause, exhort, admonish, and comfort one another, 1 Thess. v. 11—14.

To do all these well, will require special godly wisdom, humility, and love. If these three be in you, and abound, your society will be profitable: The strong will not despise the weak, neither will the weak judge the strong, Rom. xiv. 1—3. You will be far from putting a stumbling block, or an occasion to fall in your brother's way, but you will follow after the things which make for peace, and things wherewith you may edify one another, 1 John ii. 10, Rom. xiv. 13—19. You will then bear with each other's infirmities, Rom. xv. 1—3, and not seek to please yourself, but your neighbour, for his good to edification.

You must first be wise to make choice, not only of such matter of speech as is good and lawful, but such as is fit, considering the condition and need of those before whom you speak. In proposing questions, you must not only take heed that they be not vain, foolish, and needless, 2 Tim. ii. 23; such as engender strife, Tit. iii. 9; and do minister and multiply questions, rather than godly edifying, 1 Tim. i. 4; but you must be careful that they be fit and pertinent, both in respect of the person to whom they are proposed, and in respect of the person or persons before whom they must be answered.

Some men have special gifts for one purpose, some for another. Some for interpreting Scripture; some for deciding of controversies; some for discovering Satan's methods and enterprises; some are excellent for comforting and curing afflicted and wounded consciences; some are better skilled, and more exercised in one thing than in another. And some also of God's dear children, as they are not able to bear all exercises of religion, Matt. ix. 15—17; so neither are they capable of hearing and profiting by all kind of discourses of religion, Heb. v. 11—13. If this were wisely observed, Christian conference would be much more useful, than usually it is.

Secondly, You must be lowly minded, and of an humble spirit, not presuming above your gifts and calling, Rom. xii. 3. When you speak of the things of God, be reverent, serious, and sober, keeping yourself within the line, 2 Cor. x. 13, both of your calling, and the measure of that knowledge and grace which God has given you; speaking positively and confidently only of those things which you clearly under stand, and whereof you have experience, or sure proof. Think not yourself too good to learn of any, Acts xviii. 26; neither harden your neck against the admonitions and reproofs of any. If you have an humble heart, you will do as David did, when he was admonished and advised by a woman. He saw God in it, and blessed him for it, he received the good counsel, and blessed her that gave it. Now blessed be God who has sent thee to meet me this day, said he, and blessed be thy advice, and blessed be thou who hast kept me this day from coming to shed blood, &c. 1 Sam. xxv. 32, 33.

Thirdly, There will be need of the exercise of much fervent love and charity, even amongst the best. For as Satan has malice against all good company and good conference, he will infuse matters of difference and discord. And because the best men differ in opinion, though not in fundamentals, yet in ceremonies, and less necessary points of religion; and forasmuch as they all have infirmities, and, while the remains of corrupt nature are in them, are subject and apt to mistake and misconstrue one another's actions and speeches, you will need that this bond of love be strong, that it be not broken asunder by any of these, or other such means; but that you remain strongly and sweetly knit together in the unity of the Spirit, through this bond of peace, Eph. iv. 3.

I especially recommend this Christian society in brotherly love, 1 John iii. 14; because, 1. There is nothing gives a more sensible evidence of conversion, and translation from death to life than this. 2. Nothing does more assist the increase and power of godliness in any place or person, than this, Acts ii. 44—47; iv. 32, 33. For, let it be observed, though there be never such an excellent minister in any place; you will see little improvement in grace amongst the people, until many of them become of one heart; showing it by consorting together in Christian fellowship, in the communion of saints. 3. Nothing brings more sensible joy, comfort, and delight, next to communion with God in Christ, than the actual communion of saints and love of brethren, Psa. xvi. 3; Acts ii. 46, 47. It is the beginning of that happiness on earth, which shall be perfected in heaven. It is for kind, the same, only differing in degree.

And, to conclude this subject, after you have been in company, good or bad, it will be worth your while to examine how far you have hindered any evil in others, and have preserved yourself from evil: how far you have endeavoured to do good to others, and how much you have gained in knowledge, serious affection, zeal, or any other good grace, by your company; and according as you find, let your conscience reprove or comfort you.

CHAPTER IX.

RULES FOR OUR RELIGIOUS CONDUCT IN PROSPERITY.

When at any time you prosper in any thing, and have good success, that you may therein walk according to God's word,

First, Take heed of committing those sins to which the nature of man is most addicted, when his heart is satiated with prosperity.

Secondly, Be careful to produce those good fruits, which are the principal ends why God gives good success.

1st. The sins especially to be watched against, are, (1.) Denying of God, Prov. xxx. 9; by forgetting him and his ways, Deut. vi. 12; departing from him, Deut. xxxii. 15; when you are waxen fat like Jeshurun; taking the more licence to sin, Job xxi. 14, by how much you prosper the more in the world. (2.) Ascribing the praise of success to yourself or to second causes, Dan. iv. 30; sacrificing to your own net, Hab. i. 15, 16. (3.) High-mindedness, 1 Tim. vi. 17; thinking too well of yourself, because you have that which others have not, and despising and thinking too meanly of those, who have not what you possess, 1 Cor. xi. 22. (4.) If riches increase, or if you thrive in any other earthly thing, set not your heart thereon, Psa. lxii. 10; 1 Tim. vi. 17, either in taking too much delight therein, Job xxxi. 25, or in trusting thereto. Holy Job, and good David, were in some particulars overtaken with this fault. When Job was prospered, he entertained this secure conceit, that he should die in his nest, and multiply his days as the sand, Job xxix. 18; and David in his prosperity said, he should never be moved, Psa. xxx. 6. But the Lord by afflictions taught them both to know by experience, how vain all earthly things are to trust unto, and ingenuously to confess their error.

2d. I reduce the good effects or fruits, which are the principal ends why God gives good success, unto these two heads: (1.) Professed praise and thankfulness to God. (2.) Real proofs of the said thankfulness, in well using and employing this good success for God.

SECT. 2. MOTIVES TO PRAISE AND THANKFULNESS.

First, Praise and thank God. For, (1.) It is the chief and most lasting service and worship, which God has required of you. (2.) It is most due, Psa. xxix. 2, and due to him only: he only is worthy, Rev. iv. 11; v. 12; for of him are all things, Rom. xi. 36, and he is called the God of praises. (3.) It is the end why God declares his excellency and goodness, both in his word and works, Prov. xvi. 4, that it may be matter of praise and thanksgiving; also why he has given man a heart to understand, and a tongue to speak, that for them, and with them, as by apt instruments, they might acknowledge his goodness and excellency: thinking and speaking to his praise and glory. Wherefore David, speaking to his heart, or tongue, or both, when he would give thanks, saith, Awake, my glory, and I will give praise, Psa. lvii. 8, and cviii. 1, compared with Psa. xvi. 9, and Acts ii. 26. (4.) There is not any service of God more beneficial to man, than to be thankful, 1 Tim. iv. 4; for it makes those gifts of God, which are good in themselves, to be good to you, and they are the best preservatives of good things to you; nay, thankfulness for former blessings, are real requests for further favours, as well as the best security you enjoy, Phil. iv. 6, 7; for God will not withdraw his goodness from the thankful.

This praise and thanksgiving is a religious service, wherein a man makes known to God, that he acknowledges every good thing to come from him, and that he is worthy of all praise and glory, for the infinite excellency of his wisdom, power, goodness, and all his other holy and blessed attributes, manifested in his word and works; and that he is beholden to God for all that he has had, now has, and which he still hopes to enjoy.

Praise and thanksgiving go together, and do differ only in some respect. The superabundant excellency in God, showed by his titles and works, is the object of praise, 1 Chron. xxix. 11—13 ; Psa. viii. 1—9. The abundant goodness of God, showed in his titles and works, to his church, to you, or to any person or thing to which you have reference, is the object and matter of your thanks, 1 Chron. xxix. 14.

(2.) Directions for thanksgiving.

These following things, concerning praise and thanksgiving, are needful to be known and observed:

First, Who must give praise and thanks? Namely, you, and all that have understanding and breath, must praise the Lord, Psa. cl. 6.

Secondly, To whom are praise and thanksgiving due? Only to God, Psa. l. 13. Not unto us, not unto us, saith the church, but unto thy name give glory, Psa. cxv. 1.

Thirdly, By whom must this sacrifice of thankfulness be offered? Even by Christ only Eph. v. 20; Heb. xiii. 15, the only high-priest of our profession, out of whose golden censer our prayers and praises ascend, and are acceptable to God as incense, Rev. viii. 3, 4.

Fourthly, For what must we praise God and give him thanks? We must praise him in all his works, be they for us, or against us; we must thank him for all things, spiritual and temporal, wherein he is any way good unto us, Eph. v. 20.

Fifthly, With what must we praise and thank him? Even with our souls, and all that is within us, and with all that we have, Psa. ciii. 1. We must praise and thank God with the inward man; praise him with the spirit, and with the understanding, 1 Cor. xiv. 15; praise him with the will; praise and thank him with all the affections, with love, desire, joy, and gladness; praise him with the whole heart. We must likewise praise him with the outward man, both with tongue and hands, Psa. xxxv. 28; our words and our deeds must show forth his praise. When our thanks are cordial and real, then they make a good harmony and sweet melody, most pleasant in the ears of God.

Sixthly, When must we give thanks? Always, morning, noon, evening, at all times; as long as we live and have any being, we must praise him, Eph. v. 20. Psa. lv. 17. cxix. 164. civ. 33.

Seventhly, How much? We must praise and thank him abundantly, Psa. xlviii. 1. We must endeavour to proportion our praise to his worthiness and goodness: As we must love him, so we must thank him, with all our soul, and with all our strength.

(3.) The evil of unthankfulness, and dissuasives against it.

There is no sin more common than unthankfulness, Luke xvii. 17, 18; for scarce one out of ten gives thanks to God for his benefits; and those who do give thanks, besides many errors in thanksgiving, do not thank God for one mercy in twenty. Many in distress will pray, or cry and howl at least, Hosea vii. 14, as they of old, for corn and oil; but who returns proportionable praises to his prayers? Whereas the Christian should be oftener in thanks than in prayers, Psa. lix. 10, because God prevents our prayers with his good gifts a thousand ways.

Take heed therefore that you be not unthankful. It is a most base, hateful, and damnable sin. For he that is unthankful to God, is (1.) A most dishonest and disloyal man: he is injurious to God, in detaining from him his due, in not paying him his tribute. (2.) He is foolish and improvident for himself; for by not paying his tribute of thankfulness, and doing this homage, he forfeits all that he has into the Lord's hands, Deut. xxviii. 47, 48. Hosea ii. 8, 9; which forfeiture many times he takes: But if he does not presently take the forfeiture, it will prove worse to the unthankful in the end. For prosperity, without a thankful heart, always increases sin, Hosea iv. 7. Rom. i. 21, and prepares a man for greater destruction. The more such an one thrives, the more does pride, hard-heartedness, and many other evil lusts grow in him. This unthankfulness is the highway to be given over to a reprobate sense, Rom. i. 21—29. Such prosperity always proves a snare, and ends in utter ruin, Psa. lxix. 22. For the prosperity of fools shall destroy them, Prov. i. 32. And when the wicked prosper, it is but like sheep put into fat pastures, that they may be prepared for the slaughter, Jer. xii. 1—3. An unthankful man is, of all men, most unfit to go to heaven. Heaven can be no heaven to him: for there is praising of God continually. Now to whom thanksgiving and singing of the praises of God is tedious, to him heaven cannot be joyous.

(4.) It does concern you, therefore, that you be much and often in thanksgiving and praise unto God. To this end, attend to these directions; 1. Stir up your heart to holy resolution and longing desire so to do. 2. Beware of, and remove impediments to thankfulness. 3. Improve all the means of gaining such a frame of mind.

First, Consider that gratitude and thankfulness is the best service, being the end of all other worship; and is God's due; and is the end why God gives matter and means by which, and for which we should be thankful; and that nothing is more beneficial than thankfulness, nor any thing more mischievous than unthankfulness, as has been already showed. Consider also, that hearty and constant thankfulness is a testimony of uprightness; it does excellently become the upright to be thankful, Psa. xxxiii. 1. It is all the homage, and all the service which God requires at your hands, for all the good that he bestows on you. It is pleasant and delightful, Psa. cxlvii. 1. It is possible and easy through the grace of God's Spirit. It is a small matter, to what God might exact; even as an homage-penny or pepper-corn. Thankfulness does elevate and enlarge the soul, making it fruitful in good works, beyond any other duty. For the thankful man, with David, is often consulting with himself what he shall render to the Lord for all his benefits to him, Psa. cxvi. 12. Lastly, This spiritual praise and thanks to God by Christ, is the beginning of heaven upon earth, being part of that communion and fellowship which saints and angels have with God above. It is that everlasting service which endures for ever.

(5.) Impediments to thankfulness.

Not only stir up your soul to this great duty of praise and thanksgiving, but carefully shun all the impediments thereunto. Amongst many, take heed especially of these: (1.) Ignorance. (2.) Pride. (3.) Forgetfulness. (4.) Doubting of God's love. (5.) Undue affection to the benefits received, especially to such as are temporal.

First, If you are ignorant of the excellency and worth of God's good gift, or if you misprize things, preferring natural, temporal, or common gifts, before spiritual, eternal, and special graces, peculiar to God's children, you will either give no thanks at all; for who can give thanks for that which he esteems worth little or nothing? or if you do give thanks, it will be preposterous, giving thanks for temporal blessings sooner, and more than for spiritual and eternal. Moreover, though you do know each good gift according to its due value, yet if, through ignorance, you mistake the giver, you will bestow your thanks upon men, and second causes, but not on God, who is the giver of every good and perfect gift, James i. 17.

Secondly, If you be proud and highly conceited of your own worth and good deservings, you will expect matters otherwise than God will think fit to give, as Naaman did, before he was cleansed, 2 Kings v. 11, 12; and when you miss of your expectation, you will be so far from thanks, that you will murmur and complain.

Thirdly, Though you know the worth of the gift, and do acknowledge the giver; and also think yourself unworthy of the gift; yet, if you have not these in actual remembrance; if you have forgotten them, and they be out of mind, how can you be duly thankful? Therefore, when David calls upon himself to be thankful, he saith, Forget not all his benefits, Psa. ciii. 2.

Fourthly, Suppose that you know well the worth of the gift, and do judge yourself unworthy of it, and remember well that you received it of God; yet if through misbelief and doubting of God's love, you think that God does not give it to you in love and mercy, but in wrath, as he gave Israel a king, Hosea xiii. 11; your heart will sink, and be so clogged with this fear, that you cannot raise it up to praise and thankfulness, for any gift which you conceive to be so given.

Fifthly, Suppose that you are free from all the former impediments; yet if you be too eagerly affected with the gift, you will in a kind of over joyousness be so taken up with it, that, as little children, when their parents give them sweet-meats, or such things as they most delight in, fall to eating of the sweet-meat, and run away for joy, before ever they have shown any sign of thankfulness, so you will easily be overtaken in this kind, and neglect God that gave it.

(6.) Helps to thankfulness.

The helps to thankfulness, are most of them directly contrary to the former hindrances; of which, take these;

First, Get sound knowledge of God, and of his infinite excellencies, Psa. viii. 1, 9, and absoluteness every way, Matt. vi. 13. Rom. xi. 36, and of his independency on man, or any other creature; whence it is that he needeth not any thing that man hath, Psa. l. 12, 15. 1 Chron. xxix. 14—16, or can do; neither can he be beholden to man. But know, that you stand in need of God, Acts xiv. 17. xvii. 28, and must be beholden to him for all things. Know, also, that whatsoever God doth, by whatever means it be, he doth it from himself, Isaiah xliii. 25. Hosea xiii. 4, induced by nothing out of himself, being free in all that he doth. Know likewise, that whatsoever was the instrument of your good, God was the author of both the good and the instrument, James i. 17.

Next, get a clear understanding of the full worth and excellent use of God's gifts, both common and special. Wealth, honour, liberty, health, life, senses, reason, &c., considered in themselves, and in their use, will be esteemed to be great benefits; but if you consider them in their absence, when you are sensible of poverty, sickness, and the rest; or if you be so blessed, that you know not the want of them; then if you considerately and humbly look upon the poor, base, imprisoned, captive, sick, deaf, blind, dumb, distracted, &c. putting yourself in their case, Heb. xiii. 3, you will say that you are unspeakably beholden to God for these corporal and temporal blessings.

But chiefly learn to know, and consider well the worth of spiritual blessings. One of them, the peace of God, passeth all understanding, Phil. iv. 7. To enjoy the gospel upon any terms, to have salvation, such a salvation as is offered by Christ, to have faith, hope, love, and other the manifold saving graces of the Spirit, though but in the least measure, in the very first seed of the Spirit, though no bigger than a grain of mustard seed, Luke xvii. 6, with never so much outward affliction, is of such inestimable value and consequence, that it is more than eye has seen, or ear has heard, or ever entered into the heart of man, 1 Cor. ii. 9. For besides that the least grace is invaluable in itself, it is also the evidence of better gifts, namely, that God has given you his Spirit, has given you Christ, and in him has given himself, a propitious and gracious God, and with himself has given you all things, Rom. viii. 32. When you know God aright, and his gifts aright, knowing all things in God, and God in all things, then you will be full of praises and thanks.

Secondly, Be humble and base in your own eyes, 1 Chron. xxix. 13, 14. Let all things be base in your eyes, in comparison of God, account them worthless and helpless things, without him, Psa. cxlvi. 1, 3; xxxiii. 16, 17. Judge yourself to be, as indeed you are, less than the least of God's mercies, Gen. xxxii. 10. For what are you of yourself, but a compound of dust and sin, unworthy any good, deserving of all misery? You stand in need of God, but not he of you; it is of his mercy that you are not consumed, Lam. iii. 22. When you are thus sensible of your own need, and that help can come only from God, and that you are worthy of no good thing, then you will be glad and thankful at heart to God for any thing. An humble man will be more thankful for the least mercy, than a proud man will for the greatest.

Thirdly, Frequently reflect upon the infinite excellencies of God and his great benefits. Commune with your soul, and cause it to represent lively to your thoughts, what God is in himself, what to his church, and to you, how precious his thoughts are to you-ward, Psa. cxxxix. 17. Consider often what God has done, and what he will do for your soul, Psa. xl. 5. Call to mind with what variety of good gifts he enriches his church, and has blessed you: and you will find that they will pass all account and number. When also you consider that God is free in all his gifts to you, who are unworthy the least of them; if you would thus dwell upon these, and such like thoughts, they would excite in you a holy rapture and admiration, causing you to break out, with David, into these, or in the like praises; O Lord, our Lord, how excellent is thy name in all the earth! Psa. viii.

1. I thank thee, I praise thee, I devote myself, as my best sacrifice to thee, Rom. xii. 1. I will bless thy name for ever and ever.

Fourthly, Be persuaded of God's love to you in these good things, which he gives unto you: First, He loves you as his creature, and if only in that respect he does preserve you, and do you good, you are bound to thank him. Secondly, You know not but God may love you with a special love to salvation, 1 John iii. 16; 1 Tim. ii. 4. God's revealed will professes as much, for you must not meddle with that which is secret. I am sure he gives all-sufficient proof of his love, making offers of it to you; and which you are daily receiving the tokens of, both in the means of this life, and that which is to come. Did not he love you when out of his free and everlasting good will towards you, he gave his Son to die for you, that you, believing in him, should not die, but have everlasting life? John iii. 16. What though you are yet in your sins, does he not command you to return to him? Hosea xiv. 2—4; and has he not said, he will love you freely? What though you cannot turn to him, nor love him as you would, yet apply by humble faith to the Lord Jesus Christ, as your only Saviour and great physician, and endeavour in the use of all good means, to be and do as God will have you; then doubt not but that God does love you; and patiently wait, till you see it in the performance of all his gracious promises unto you.

(7.) Signs to know when God gives good things in love.

If you would consider things aright, you may possibly know with certainty, that the good things you have received of God, are bestowed in love to you. I will only ask these questions: Have God's mercies excited you to labour more diligently to please him well in all things? Have you had a will to be thankful upon the consideration thereof? Or, if you find a defect and barrenness herein, has not this unfruitful and unthankful receiving of blessings from God, been a great burden and grief of heart to you? If so, this is an evident sign that God gave those good things to you in love, because this holy and good effect is wrought in you by them. Again, do you love God? Would you love God, and his ways, and ordinances yet more? This proves that God loves you; for no man can love God, till God has first loved him, 1 John iv 10—19. Likewise, do you love the children of God? Then certainly you are God's child, and are beloved of God, 1 John iv. 7. By these things you have proof of your calling and election, that you are now translated from death to life, 1 John iii. 14. So that, though God may give you some things in anger, as a father gives correction, yet he never gives any thing in hatred and in wrath, as he does to his enemies. All things work together for good to them that love God, Rom. viii. 28; therefore whatsoever he gives to such, is in love.

Fifthly, Prefer the honour and glory of God, before and above all things that may be beneficial to yourself; prefer likewise the kindness and love of God in the gift, far above the gift itself; then you will never be so taken up with the enjoyment of the gift, as to forget to give praise and thanks to the giver.

Sixthly, Unto the former helps, add this: lay a holy command upon your soul, and strictly charge yourself to be thankful; and, since you have such good reason for it, make no excuses against it, but say, with David, Bless the Lord, O my soul, and all that is within me, bless his holy name, &c. Psa. ciii. 1.

Lastly, To all other means, join earnest prayer to God, to give you a thankful heart. It is not all the reasons you can allege for it, nor all the moral persuasions you can propose to yourself, can effect it, though these be good means, yea, God's means; yet if you go about to raise your heart to it, in the power of your own might, all will be vain. For as you cannot pray but by God's Spirit, so neither can you give thanks but by the same Spirit. Therefore say, as David did: Renew, O Lord, a right spirit in me; and open my lips, that my mouth may show forth thy praise, Psa. li. 10—15.

SECT. 2. OF THE REAL PROOFS OF GRATITUDE AND THANKFULNESS TO GOD.

It is not enough to profess and utter praise and thanks to God; but you must give real proof thereof.

First, By devoting and giving yourself to God, Rom. xii. 1; to be at the will of him, who is your sovereign Lord, who gives you all that you have, who is always giving unto you, and always doing you good; paying your vows to him that performs his promises to you, Psa. cxvi. 14. Let it appear that you acknowledge him to be such an one as you say in your praises, and that you stand obliged and beholden to him indeed, as you say in your thanks, in that both in the frame of your heart, and the conduct of your life, you behave towards him as one who only is excellent, who only is God, who is your God, the God of your life and salvation; and that, in all holy service. For thanks-living is the best way of thanksgiving, and it is a divine saying, The good life of the thankful, is the life of thankfulness. Wherefore let every new mercy quicken your resolution to persevere and increase in well-doing, serving God so much the more with gladness of heart, because of the abundance of all things, Deut. xxviii. 47.

Secondly, Do good with those blessings, which God gives you. For every good gift is given to a man to profit withal, 1 Cor. xii. 7; not only himself, but every member of that body, whereof he is part. Whatsoever good gift God has given you, whether temporal or spiritual, it must be employed to God's glory, and to your neighbour's good, as well as to your own, as you have opportunity. If riches, (and the same rule will serve for health, strength, wisdom, skill, &c.) be given to you, you must honour God therewith, Prov. iii. 9; and as God prosper you in any thing, you must communicate to them that need, 1 Cor. xvi. 2, as to the poor, sick, weak, simple, and ignorant. If God give knowledge, faith, spiritual wisdom, ability to pray, or any other of his rich graces, you must not hoard them up, and keep them reserved for your own private benefit; but you must communicate them to others, and improve them for the promoting their spiritual good, and edifying them in faith, hope, and love.

By communicating your good and common gifts of God in this sort, you make yourself friends with them, Luke xvi. 9, against a day of need; and when you honour God, and do good with the talents which God puts into your hand, then you make the best improvement of them. He who thus walks with God in prosperity, shall certainly find him to be his sure friend in adversity, and when he shall be put out of his stewardship at death, then he shall be received into the everlasting habitations, Luke xvi. 9. When, the more you prosper, the better you desire and endeavour to be, and do more good, this is an infallible proof of true thankfulness, and is an evident sign that you walk with God in prosperity as he would have you.

Give all diligence therefore, to learn this lesson, how to be full, and how to abound, Phil. iv. 12; but know, it can be learned no where but in Christ's school, and can never be practised but by Christ's strength. This is it which the apostle had learned, and said, he was able to do it through Christ that strengthened him, Philip. iv. 12, 13. It is a most needful and high point of learning, to be instructed and to know, every where, and in every thing, how to be full and how to abound: of the two, it is more rare and difficult, than to know how to be abased, and to suffer want, which shall be the subject of the next chapter.

CHAPTER X.

DIRECTIONS FOR WALKING WITH GOD UNDER AFFLICTIONS

EVERY day will bring forth its evil and cross, Matt. vi. 34, whether lighter and ordinary, or more heavy and extraordinary. The first sort rises partly from the common frailties of the persons with whom you converse, and partly from your own; as from pride and peevishness, and suspicion of evil, &c. Such as discourtesies from those of whom you expected kindness; imperiousness, and too much domineering of superiors; sullenness, negligence, and disregard from inferiors; awkwardness and perverseness in the persons and things with which you have to do.

(1.) Rules concerning lighter crosses.

First, Lay not these to heart, make them not greater than they be through your impatience, as many do, who, upon every light occasion of dislike, cast themselves into such a state of vexation and discontent, that all the blessings the yenjoy, are scarcely observed, or can make their lives comfortable. Whereas wisdom should prevent, and love and prudence should cover and pass by most of these; seeing, as if you saw not: or if you will give way to any passion at these, let it be with hatred of their and your sin, which is the cause of these, and all other crosses.

Secondly, These should cause you to pity, and pray for them that give you this offence, and for yourself, who many times without cause take offence. You may if need require, show your dislike, and admonish the offender, provided you do it with meekness of wisdom, James iii. 13; but learn hereby to warn yourself, that you give not the like offence.

(2.) Directions how to bear all afflictions well.

But whether your crosses and afflictions be imaginary only, or real; whether from God immediately, or from man, whether light or heavy, follow these directions: 1. Be not transported with passion and anger, like proud Lamech, Gen. iv. 23, 24, and froward Jonas, Jonah iv. 7—9. 2. Be not overwhelmed, or sullen with grief, like covetous Ahab, 1 Kings xxi. 4, and foolish Nabal, 1 Sam. xxv. 37. But, 3. Bear them patiently. 4. Bear them cheerfully and thankfully. 5. Bear them fruitfully.

Remedies against sinful anger.

To help you, that passion and heat of anger kindle not, or at least break not out beyond due bounds,

First, Convince your judgment thoroughly, that passion and rash anger is forbidden and hated of God, Matt. v. 22; Eccles. vii. 9. It is a fruit of the flesh, Gal. v. 20. A work of the devil, James iii. 14, 15. Bred and nourished by pride, Prov. xxi. 24, folly, Prov. xiv. 29, and self-love, Jonah iv. 1—3. Also, that it surprises all the powers of right reason, putting a man beside himself, causing him to abuse his tongue, hands, and the whole man; making him like a fool, to cast firebrands at every thing which crosses him, and that not only against his neighbour and dearest friends, 1 Sam. xx. 30—33, but against God himself, Jonah iv. 9. Consider likewise that it makes a man unfit to pray, 1 Tim. ii. 8, to hear the word, 1 Peter ii. 1; James i. 19, or to perform any worship to God; and unfit to speak, or hear reason, or to give or receive good counsel. God forbids his children the company of the froward, Prov. xxii. 24, and saith, that such an one abounds in transgression, Prov. xxix. 22; and that there is more hope of a fool than of him, Prov. xxix. 20. Wherefore he must needs be exposed to all the just judgments of God, Prov. xix. 19, temporal and eternal. For which cause, fix in your mind such an abhorrence of this vice, that you may beware and shun it with all caution.

Secondly, Observe watchfully when anger begins to kindle and stir in you, and before it flame and break forth into speech or behaviour, set your reason at work, to prevent or restrain it. Nay set faith at work, having in readiness, upon your mind, such pertinent scriptures as these: Be angry, but sin not, Eph. iv. 26; and, anger rests in the bosom of fools, Eccles. vii. 9. Shall I then sin against God? Shall I thus play the fool?

Rules to know when anger is sinful.

You sin in your anger, first, when it is without cause; as when neither God is dishonoured, nor your neighbour or yourself indeed injured; when it is for trifles, and only because you are crossed in your will and desire, and the like; but chiefly when you are angry with any for well doing, 1 Kings xxii. 24—26. Secondly, Though you have cause, yet if it extinguish your love to the person with whom you are angry; so that you neglect the common and needful offices thereof. Thirdly, When it exceeds due measure, as when it is over much, and over long. Fourthly, It is sinful when it brings forth evil and unseemly effects, such as neglect, or ill perfomance of any duty to God or man; also when it breaks out into loud, clamorous, or reviling speeches, or into churlish, sullen, or indecent behaviour, or when it is attended with any injurious act.

Thirdly, If you cannot keep anger from rising within you, yet be sure that you bind your tongue and hand to good behaviour. Make a covenant with them, and charge them not to show it, nor partake with it any further than considerate reason, and good conscience shall advise you, Psa. xxxix. 1. Set a law to yourself, Psa. cxli. 3, that you will not chide, nor strike while you are in the heat of anger. If there be cause of either, defer it until you have more government over yourself. If you say, that "If you do them not in your heat, you shall not do them at all;" I answer, that, in saying so you discover a great deal of folly and weakness. I am sure you never do them well in passion. And conscience of duty should lead you to chiding and correcting when there is cause, not passion: for, in it, you serve and revenge yourself upon the party, but not God.

Fourthly, Both before, and when you are angry,

see God, by the eye of your faith, as present with you, in hearing and looking upon you, Psa. xi. 4, 5. This will make you peaceable and quiet, causing you not only to hold your hands and tongue, as you find by experience you use to do, when some reverend friend is present; but this will calm and abate the inward heat and passion of your mind.

Fifthly, If you feel your corruption and weakness to be such, and the provocation to anger so great, that you fear you cannot contain yourself, then, if it be possible, avoid all occasions of anger, and remove yourself, in a peaceable and quiet manner, from the person, object, or occasion thereof. And at all times shun the company of an angry man, as much as your calling will give you leave, lest you learn his ways, Prov. xxii. 24, 25.

Sixthly, Howsoever it may happen that anger kindles in you, and breaks out; be sure that you subdue it before it grow into hatred of him with whom you are angry. For this cause let not the sun go down upon your wrath, Eph. iv. 26; you know not what hatred it may grow into before morning. And the best means that I know to subdue it, is, if you find your heart to rise against any, pray heartily to God for him in particular, for his good, Matt. v. 44; to this you are commanded. And be so far from seeking revenge, that you force yourself to be loving and kind, showing all good offices of love with wisdom, as you shall have occasion; overcoming evil with good, Rom. xiii. 17—21. Pray also to God for yourself, that he would please to subdue this passion in you. This act of love to him with whom you are angry, performed before God, in whose sight you dare not dissemble, will excellently quench wrath, and prevent hatred against him, and will give proof between God and your conscience that you love him.

If, pleading for yourself, you shall say, "It is my natural constitution to be choleric, and flesh and blood will have their course;" know, this is to nourish your passion. Know also, it is a wicked and hateful constitution of body, which came in with the fall. And flesh and blood shall not inherit the kingdom of God, 1 Cor. xv. 50. Say not, "I am so crossed and provoked, never "any the like;" for Christ was more injured and more provoked than you, and yet never was in a passion, 1 Peter ii. 23; Heb. xii. 2, 3. And you provoke God a thousand times more every day, yet he is patient with you. Say not, "It is such a headstrong passion, that it is impossible to bridle and subdue it;" for, I can assure you, that by using means, these prescribed, if you also do often and much abase yourself before God for your passion and folly, and daily repent thereof, and watch over yourself, you may, of most hasty and passionate, become most meek and patient before you die. I have seen it in old men whose age in itself giveth advantage to peevishness and forwardness, who were exceedingly passionate in their youth, yet through the grace of God, by constant conflict against this vice, have attained to an admirable degree of meekness.

2. The cure of worldly grief.

Next, as carnal anger, so worldly grief must be avoided in all sorts of crosses. For, by it, you repine against God, fret against men, and make yourself unfit for natural, civil, and spiritual duties, 1 Kings xxi. 4; and if it be continued, it works death, 2 Cor. vii. 10.

The best remedy against worldly sorrow for any affliction, is to turn it into godly sorrow for sin, which is the cause of all our troubles. This will work repentance to salvation, never to be repented of, 2 Cor. vii. 10; and will drive you to Christ, in whom, if you believe, you will have joy and comfort; even such joy unspeakable as will dispel and dry up both this and all other griefs whatsoever, 1 Peter i. 6, 8. For godly sorrow does always, in due time, end in spiritual joy.

(3.) The nature of Christian patience.

In the third place, I proceed to show the nature of Christian patience. By patience, I do not mean a stoical senselessness, or dull stupidity, like that of Issachar, Gen. xlix. 14, 15; nor yet a counterfeit patience, like Esau's, Gen. xxvii. 41, 42, and Absalom's, 2 Sam. xiii. 13, 22; nor a mere civil or moral patience, which wise heathens, to free themselves from vexation, and for vain glory and other ends, attained unto; nor yet a profane patience, Rev. ii. 2, of men insensible of God's dishonour or afflicting hand; nor a patience per-force, when the sufferer is merely passive, because he cannot relieve himself: but a Christian holy patience, wherein you must be sensible of God's hand, and when you cannot but feel an unwillingness in nature to bear it, yet, for conscience towards God, you do submit to his will, and that voluntarily, with an active patience, causing yourself to be willing to bear it so long as God shall please; after the example of Christ, Matt. xxvi. 39, 42, Not my will, but thine be done. The excellency of Christ's sufferings was not in that he suffered, but in that he was obedient in his sufferings. He was obedient to the death, Phil. ii. 8. So likewise no man's suffering is acceptable, if he be not active and obedient in suffering.

This patience is a grace of the Spirit of God, wrought in the heart and will of man, through believing, and applying the commandments and promises of God to himself; whereby, for conscience sake towards God, 1 Peter ii. 19, he does submit his will to God's will, quietly bearing, without bitterness and vexation, all the labour, changes, and evil occurrences which befal him in the whole course of his life, whether from God immediately, or from man: as also waiting patiently for all such good things as God has promised, but yet are delayed and unfulfilled.

(4.) Motives to Christian patience under adversities.

To induce you to get, and to show forth this holy patience, know, that you have great need of it, Heb. x. 36, and that in these respects:

First, You are but half a Christian, you are imperfect, and want a principal grace in the Christian life, if you want patience: thus James argues, implying that he who will be entire, James i. 4, and want nothing to make him a Christian, must have patience. This passive obedience is greater than active; it is more excellent, and more difficult to obey in suffering, than to obey in doing.

Secondly, You cannot have a sure possession of your soul without patience; in your patience possess ye your souls, saith our Saviour, Luke xxi. 19. A man without patience, is not his own man: he has not power to rule over his own spirit, Prov. xxv. 28, nor yet of his own body. The tongue, hands, and feet of an impatient man will not be held in by reason. But he that is patient, enjoys himself, and has rule over his spirit, Prov. xvi. 32; no affliction can put him out of possession of himself.

Thirdly, There are many oppositions and hinderances in your Christian race and warfare, that without patience to suffer, and to wait, Rom. viii. 25, you cannot possibly bring forth good fruit to God, nor hold out your profession of Christianity to the end; but will decline, and give over before you have enjoyed the promise, Heb. x. 36. Therefore you are bid to run with patience the race that is set before you, Heb. xii. 1. And the good ground is said to bring forth fruit with patience, Luke viii. 15; and the faithful are said through faith and patience to inherit the promises, Heb. vi. 12.

Fourthly, Patience works experience, Rom. v. 5, without which no man can be an established Christian; this experience being of the highest use to confirm the soul of a Christian in the greatest difficulties. This must be said of the necessity, together with the benefit of patience, that you may love it, and may desire to have and show it.

(5.) Means to gain Christian patience under afflictions.

By what means you may attain it, follows:

First, Spend those passions on your lusts, which war in your members, which are exercised on other objects; fall out with them, and mortify them, Col. iii. 5; for nothing makes a man impatient, so much as his lusts do, both because they will never be satisfied, and it is death to a man to be crossed in them; and because the fulness of lusts causes a guilty conscience, whence follows impatience and troublesome vexation upon every occasion, like the raging sea, which with every wind does foam and rage, and cast up nothing but mire and dirt, Isaiah lvii. 20. And, as James saith, Whence are wars and fightings, James iv. 1, so I say of all other fruits of impatience, but from your lusts which war in your members? Take away the causes of impatience, then you have made a good advance towards gaining Christian patience.

Secondly, Lay a good foundation of patience by being humble and low in your own eyes, through an apprehension that you are less than the least of God's mercies, and that your greatest punishments are less than your iniquities have deserved, Ezra ix. 13. As Christians abound in humility so will they abound in patience, witness the examples of Abraham, Moses, Job, David, and others.

Thirdly, Labour to gain and improve the Christian graces of faith, hope, and love: all and either of these calm the heart, and keep it steady in adversity. For besides that, they quiet the heart in the assurance of God's love in Christ: For being justified by faith, we have peace with God, rejoice in hope, Rom. v. 1, 3, 4; whence proceed joy and patience in tribulation. And who can be impatient with him whom he loves with all his heart and strength? These graces also furnish the Christian with an ability of spiritual reasoning and disputing with a disquieted soul, whereby it may be happily composed, and brought to possess itself in patience under any adversity.

Wherefore the fourth means of patience is, to do as David did. Whensoever you find your heart begin to fret and be impatient, you must, before passion or grief has got the mastery over you, ask your soul what is the matter; and why it is so disquieted within you, Psa. xlii. 11. This do seriously, and your heart will quickly represent to you such and such afflictions aggravated by many circumstances of distress. All which you must answer by the spiritual reasoning of your faith, founded on the word of God, whereby you may quiet your heart, and put your grief to silence.

Whatsoever the affliction be that may trouble you, you may be furnished with reasons why you should be patient, either (1.) From God that sent it: (2.) From yourself, on whom it lies: (3.) From the nature and use of the affliction itself: (4.) By considering the evils of impatience: (5.) By comparing the blessings you have, and are assured that you shall have, with the crosses you have, especially if patiently endured. From all these considerations you will see reason why your heart should be quiet under the greatest afflictions.

First, Consider well, that whatsoever the trouble and cross be, and whosoever be the instrument of it, either in the sense of evil, or in the want of good promised, it comes from God your Father, (1.) Who does all things according to the wisdom and counsel of his own will; (2.) Who doth afflict with most tender affection; (3.) Who corrects and afflicts in measure; (4.) Who has always holy purposes and ends in all afflictions, directing them for your good.

1. Consider that it was God who did it. There is no evil, that is of punishment, in a city, which the Lord has not done, saith Amos, Amos iii. 6; 2 Sam. xvi. 10.—It is the Lord, let him do what seems him good, saith Eli, 1 Sam. iii. 18. I opened not my mouth, saith David, because thou, Lord, didst it, Psa. xxxix. 9. The Lord gave, and the Lord has taken away; blessed be the name of the Lord, saith Job i. 21. Hosea vi. 1; 1 Sam. ii. 6. 7.

2. All this God does to his children with a fatherly affection, in much love and pity, Heb. xii. 5, 6. He has your soul still in remembrance, while you are in adversity, Psa. xxxi. 7. Yea, he bears some part of the burden with you: for, speaking after the manner of man, he saith, that in all the afflictions of his children he is afflicted, Isaiah lxiii. 9. He delights not in afflicting the children of men, Lam. iii. 33, much less his own children.

If you ask, Why then does he afflict, or why does he not ease you speedily? I ask you, why a tender-hearted father, being a surgeon, who is grieved and troubled at the pain and anguish, which he himself causes his child to feel by necessary operation, does notwithstanding apply the burning irons, and suffer those plasters to afflict him for a long time? You will say, Sure the wound or malady of the child required it, and that else it could not be cured. This is the case between God and you: God's heart is tender, and yearns towards you, when his hand is upon you: therefore bear it patiently.

3. God afflicts you in measure, Isaiah xxvii. 8; fitting your affliction for kind, time, and weight, according to the strength of grace which he has already given you, or which certainly he will bestow upon you. He does never lay more upon you, than what you shall be able to bear, 1 Cor. x. 13, and will always with the cross and temptation, make a way to escape. The husbandman will not always be ploughing, Isaiah xxiii. 24, 25, and harrowing of his ground, but only gives it so much as it hath need of, or as the nature or situation of the soil requires. So likewise he threshes his divers sorts of grain, with divers instruments, according as the grain can endure them: the fitches are not threshed with a threshing instrument, neither is the cart-wheel turned about upon the cummin: bread-corn is bruised, because he will not ever be threshing it, nor break it with the wheel of his cart, nor bruise it with his horsemen, Isaiah xxviii. 26—28. If the husbandman do all this by the discretion wherewith God has instructed him; can you think that God, who is wonderful in counsel, and excellent in working, Isaiah xxviii. 29, will plough and harrow any of his ground, or thresh any of his corn, above that which is fit, and more than his ground and corn can bear? Should not his ground and corn therefore be patient at such tillage, and at such threshing?

4. God's end in afflicting, is always his own glory in your good; as, to humble you, and to bring you to a sight of your sin, to break up the fallow ground of your heart, that you may sow in righteousness and reap in mercy, Hosea x. 12, to harrow you, that the seed of grace may take root in you. All God's afflic-

tions are to remove impediments of grace. By this, saith Isaiah, shall the iniquity of Jacob be purged, and this is all the fruit, to take away his sin, Isa. xxvii. 9. All the ploughing is but to kill weeds, and to fit the ground for seed; all the threshing and winnowing is but to sever the chaff from the corn; and all the grinding and bolting by afflictions, is but to sever the bran from the flour, that God's people may be a pure offering, acceptable to him, Isa. lxvi. 20. Or else he afflicts, that his children might have experience of his love and power in preserving and delivering them, or that they might have the exercise, proof, and increase of faith, hope, Rom. v. 4, love, and other principal graces, which serve for the beautifying and perfecting of a Christian. God does judge his children here, 1 Cor. xi. 32, that they may repent, and be reformed, that they may not be condemned with the world. God's end in chastening you, will be found to be always for your good, that you shall be able to say, It was good for me to be afflicted, Psa. cxix. 67, 71. For it is that you may be partakers of his holiness, Heb. xii. 10, 11, and accordingly of his glory and happiness. Bear therefore all afflictions patiently, for they are for your good.

If this be your cross and trouble, that you want many of the graces and good gifts of God which he has promised, know also that this deferring to give graces and comforts, is of God, not out of neglect or forgetfulness of you; but of wise and good purposes towards you; as to inflame your desires more and more after them; and, that you should seek them in a better manner. It is likewise to try your faith and hope, whether you will do him that honour, as to wait and rest upon his bare word. When you are fit for them, you shall have them. You must therefore charge your heart yet to wait patiently for them, considering the faithfulness and power of God that promised: and that all the promises of God are, yea and amen in Christ, 2 Cor. i. 20. He is wise, true, and able to fulfil them in the due time, and in the best manner; for faithful is he that has promised and will fulfil it: and yet a little while, and he that shall come, will come, and will not tarry, Heb. x. 23, 37.

Secondly, When the soul begins to be disquieted, consider how unworthy you are of any blessing, how worthy you are of all God's curses, yea, of eternal damnation in hell; and that justly, because of the sins of your nature, of your heart, and of your life. When you do thus, your heart will be quiet and contented, you will say with the church, whatsoever your trouble be, I will bear the indignation of the Lord, for I have sinned against him, Micah vii. 9. He who acknowledges that he has deserved to be hanged, drawn, and quartered, for an offence against the king, if the king will be so merciful that he shall escape only with a severe whipping, to remember him of his disloyalty, though he smart terribly with those lashes, yet in his mind he can bear them patiently and submissively. If you think thus, "I deserve more punishment in this kind, nay, in any other more grievous than this; my punishment is less than mine iniquities deserve, Ezra ix. 13, for I might have been long since despairing in torments, and past all means and hope of salvation; but I live, and have time and means to make a good use of my afflictions;" these thoughts will cause you to say, Why do I, who am a living man, complain for the punishment of my sin, Lam. iii. 39, which is so much lighter than my desert? And you will say, with the church, in all your distresses, It is God's mercy it is not worse. It is God's mercy I am not utterly consumed, Lam. iii. 19—22.

Thirdly, When your soul begins to be impatient under afflictions, whether in soul, body, or estate; consider the nature and use of them. To the eye of sense they are evil as poison, hurtful and dangerous, Heb. xii. 11; but to the eye of faith, they are good and useful, as physic, most healthful to the soul, and saving, 2 Cor. iv. 16—18. God, the skilful physician, has quite altered the nature of crosses to his children; he that brings light out of darkness, so orders afflictions, that they become good antidotes and preservatives against sin, and good purgatives of sin, Isa.

xxvii. 9. The sting and curse of the cross, which remains to the wicked, is by Christ's patient suffering, and God's mercy, taken quite away out of the afflictions of believers. Afflictions to the godly, are not properly punishments, serving to pacify God's wrath for sin; but are only chastisements to remove sin, and are exercises of graces, and means of holiness. For they serve either to prevent evil, or to reform it; either to prepare way for grace, to quicken and increase grace, or to discover and give proof of it. God is a wise and skilful refiner, he knows how to purge his gold, by casting it into the fire of affliction, 1 Peter i. 7; which fire is not the same to the dross, that it is to the gold; it consumes the dross, but refines the gold, that it may be fit to be made a vessel of honour. Fire serves to try gold, as well as to purge it; for pure gold, though it remain in the fire many days, the fire cannot waste it; when it is once pure, it will hold its weight still for all the burning. Hence it is that the Psalmist saith, It is good for me that I have been afflicted, that I might learn thy statutes, Psa. cxix. 67, 71; and the apostle saith, All things work together for good to them that love God, Rom. viii. 28. He is a froward and foolish person, who, being sick of a deadly disease, does not patiently and cheerfully bear the gripings and sickness of stomach, when he knows this sickness, caused by bitter physic, is for his health.

You will say, if you could find that your afflictions did you any good, you should not only be patient, but cheerful under them.

I answer, Whatsoever you feel, faith in God's word will tell you, that they now do you good, and hereafter you shall feel the benefit of it. The benefit of physic is not always felt the day you take it, but chiefly when the physic has done working. The chief end why God tries and purges you by afflictions, is, that he may humble you, and prove you, to do you good at your latter end, Deut. viii. 15, 16. You should therefore be patient in the mean time.

Fourthly, If yet your heart remain disquieted, because of your affliction; consider with yourself, what harm impatience will do you, compared with the good that will follow a patient enduring of it. For, besides that it deprives you of your right understanding, and makes you to forget yourself, as I have said, even to forget your duty both to God and man; it is the readiest means to double and lengthen the affliction, not to abate it, and take it off. That parent who intended to give a child but light correction, if he be impatient and rebellious under it, is hereby more incensed, and does punish him more severely. But if, in any affliction, you do patiently submit yourself under God's mighty hand, 1 Peter v. 6, besides the ease and quiet it gives to the soul, and experience and hope which it produces in you, it is the readiest means of seasonable deliverance out of it; for then God will exalt you in due time. God is wise, and too strong to be overcome by any means, but by fervent prayer and humble submission to his will, Hosea xii. 4.

Fifthly, If yet your mind be disquieted within you at any crosses; that you may quiet your soul, you must not, as most do, only consider the weight and number of your crosses, together with their several aggravations; but withal seriously think upon the manifold mercies and favours of God, both in the evils you have escaped, and in the benefits which you have received, and do now enjoy, and which, through Christ, you have cause to hope to receive hereafter. But amongst all his mercies, forget not this one, which you have already, God has given Christ unto you, whereby he himself is yours, as your all-sufficient portion. Now, if you have Christ, you have with him, all things also which are worth the having, Rom. viii. 32.

When you have thus weighed impartially blessings and mercies against crosses, you will tell me, that for one cross, you have a hundred blessings, yea, a blessing in your crosses, Psa. cxix. 71, and you will say, that this one mercy of being in Christ, alone weighs up all crosses, and makes them as light as nothing· giving you so much matter of joy and thankfulness even in the midst of affliction, that you can neither have cause nor time to be impatient, or to repine at any affliction, but to rejoice even in your tribulations, Rom. v. 1—3.

And as for the time to come, when you think upon all your crosses and sufferings of this present time, yet reckon, that they are not worthy to be compared with the glory that shall be revealed in you, Rom. viii. 18. For they are but short for time, and light for weight, being compared with the everlasting weight of glory which they will work for you, if you endure them patiently. I will say nothing of the shortness and lightness of your afflictions, in comparison of the far more intolerable and eternal weight of torments in hell, which you escape: and in comparing afflictions with glory, I will point out to you only the apostle's gradation; you shall have, for affliction, glory, 2 Cor. iv. 17; for light affliction, weight of glory; for short affliction, an eternal glory; for common and ordinary affliction, excellent glory. And although it might be thought that he had said enough, yet he adds degrees of comparison; yea, goes beyond all degrees, calling it more excellent, far more excellent: for thus he saith: Our light affliction, which is but for a moment, works for us a far more excellent and eternal weight of glory. Indeed, you must not look at the things which are seen with the eye of sense, 2 Cor. iv. 18, but at things which are not seen, which are spiritual and eternal, seen only by the eye of faith.

You will say, if you did but bear afflictions for Christ, then you could rejoice in hope; but you ofttimes suffer afflictions justly for your sin.

I answer, Though this place principally points to suffering for Christ's cause, yet it is all one, in your case, if you bear afflictions patiently for his sake. A man may suffer afflictions for Christ two ways: First, When he suffers for his religion and for his cause. Secondly, When a man suffers any thing which God lays upon him, quietly, for Christ's will and commandment sake. This latter is more general than the former, and the former must be comprehended in this latter; else the former suffering for Christ's cause, if it be not in love and obedience, 1 Cor. xiii. 3, and for Christ's sake, out of conscience to fulfil his will, is nothing: whereas he that endures patiently God's just punishment for sin for Christ's sake, endeavouring to submit his will to the will of Christ, this man suffers, that is, patiently endures affliction for Christ, though he never suffer for profession of Christ; and, if such an one were put to it, he would readily suffer for Christ's cause. And such afflictions as these, thus patiently endured, work also this excellent weight of glory, as well as the other.

By these and the like reasonings of faith, you may possess your soul in patience, as David and others have done, Psa. xlii. xliii. by casting anchor on God, and on his word, fixing their stay and hope in him. Let the issue of your reasoning be this: I will wait on God, and yet, for all the causes of distress, praise him who is the health of my countenance, and my God. Thus David quieted his heart, 1 Sam. xxx. 6, when he heard tidings that his city Ziklag was burnt, and that his wives and all that he had, together with the wives and children of all his soldiers, were carried captive; and when he saw that his soldiers began to mutiny, and heard them speak of stoning him, he encouraged himself in the Lord his God. And good Jehoshaphat, in his desperate condition, cast anchor here, saying, O our God, we know not what to do, but our eyes are on thee, 2 Chron. xx. 12. Thus by the exercise of your hope in God, the heart may be wrought unto much patience and quietness in all distresses.

A further means of grace is, observe the patience of others, as of the prophets and faithful servants of God, who are recorded in Scripture, and left as examples of suffering affliction, and of patience. We count them happy that endure, saith James; you have heard of the patience of Job, and have seen the end of the Lord; that the Lord is very pitiful, and of tender mercy, James v. 10, 11. But especially represent to your thoughts the patience of your head and Saviour Jesus Christ, whom you pierced by your sins who as a lamb, dumb before the shearer, opened not his mouth, Isa. liii. 7. Now, if you would consider him who is the author and finisher of your faith; who endured such contradiction of sinners, Heb. xii. 1—3, &c. and such intolerable anguish of soul, when he wrestled with his Father's wrath; then you would not be wearied nor faint in your minds, when you are under any affliction. If with Christ you set the joy before you, you will be able to endure the cross, and despise the shame of all persecution for well doing, and so run that race which is before you with patience, that you shall in the end sit down with Christ at the right-hand of the throne of God, Heb. xii. 2; Rev. iii. 21.

Sixthly, and lastly, Pray much for patience, waiting patiently for it, James i. 4, 5; and without doubt, the God of patience and consolation, who has commanded it, who sees that you have need of it, and who has promised to give you all your petitions which you make according to his will, will surely give you patience.

(6.) Of bearing afflictions thankfully and fruitfully.

To bear adversity and afflictions well, it is not enough that you bear them patiently, because you deserve them, and because they come from God; but you must bear them thankfully, Lam. iii. 22, 23; Job i. 21; cheerfully and comfortably, Rom. v. 3, because they are, as you have heard, for your good, Psa. cxix. 71; Lam. iii. 27. We do not only patiently endure the hand of the surgeon, and the prescriptions of the physician, but we thank them, pay them, and are glad of their recipes, though they put us to pain. Count it exceeding joy, saith James, when you fall into divers temptations, knowing this, that the trying of your faith works patience, &c. James i. 2, 3.

Last of all, unto patience and thankfulness, you must add fruitfulness and growth of grace, Psa. cxix. 67—71; this should be the fruit of all crosses and afflictions, that with David you may be better for them, and that you may, with Job, come out of them as gold refined and purged from dross, Job xxiii. 10.

Therefore God chastens you as he did Jacob: This is all the fruit, to take away your sin, Isa. xxvii. 9, and that you should be partaker of his holiness, Heb. xii. 10. Be better, therefore, for crosses; then God has his end, when, after his ploughing, harrowing, and threshing of you, he shall reap the harvest of well doing, which he reaps not so much for himself, as for you; for the ground that brings forth fruit meet for him that dresses it, receives blessing from God, Heb. vi. 7. All good works are "treasured up in heaven for the doers of them."

When you have learned this lesson also, how to be abased and to suffer need, as well as how to be full and to abound, Phil. iv. 11, 12, with all the fore-mentioned directions, how at all times, and in all things, to walk with God, you will prove yourself to be a good proficient in the school of Christ, one that has walked to good purpose before God; showing that you are neither barren or unfruitful, in the knowledge of our Lord Jesus Christ, 2 Peter i. 5—8.

Thus much concerning the outward frame of your life and conversation, according to which you must walk with God. The inward truth and life of all this, which is, doing all in uprightness, remains to be spoken unto, and is as follows.

CHAPTER XI.

OF UPRIGHTNESS.

The sum of this head is contained in this, that in your whole walking with God, you must be upright. Both these, to walk with God, and to be upright, are joined in this precept: Walk with me, and be perfect, or upright, Gen. xvii. 1. He speaks not of an absolute perfection of degrees, in the fulness of all graces, which is only aimed at in this life, towards which the Christian, by watchfulness and diligence, may come nearer and nearer; but is never attained until we come to heaven, amongst the spirits of just men made perfect, Heb. xii. 23. He speaks here of the perfection of parts, and of truth and grace in every part, expressing itself in unfeignedness of will and endeavour; which is uprightness.

SECT. 1. THE NECESSITY OF UPRIGHTNESS IN RELIGION.

THAT you should be sincere and upright, read Joshua xxiv. 14; 1 Chron. xxviii. 9. And the apostle tells you, that since Christ Jesus, your passover, is slain, you must keep the feast, which shadows forth the whole time of our life here, with the unleavened bread of sincerity and truth, 1 Cor. v. 7, 8. The examples of Noah, Gen. vi. 9; Job, Job i. 1; Nathaniel, John i. 47; with many others in the Scriptures, are therefore written, that of them you may learn to be upright.

There is special reason why you should be upright:

First, Your God with whom you walk, is perfect and upright, Matt. v. 48; he is truth; he loves truth in the inward parts, Psa. li. 6; all his works are done in truth; and there was no guile ever found to be either in the mouth, hand, or heart, of your head Christ Jesus, 1 Peter ii. 21, 22. Now, you should please God, and be like your Father, and your head Christ Jesus, following his steps.

Secondly, It is to no purpose to do that which is right in God's sight, in respect of the matter of your actions, if in the truth and disposition of your soul you be not upright therein. For the best action, void of uprightness, is but like a well-proportioned body without life and substance. And that is counted as not done at all to God, Zech. vii. 5, 6, which is not done in uprightness. This exception is taken against Amaziah's good actions; it is said, He did that which was right in the sight of the Lord, but he did it not in uprightness, he did it not with a perfect heart, 2 Chron. xxv. 2.

Thirdly, The best actions, without uprightness, do not only lose their goodness; but in God's account are esteemed abominable evils. Such were the prayers and sacrifices of the hypocritical Jews, Isa. i. 13, 14, lxvi. 3. For God judges such actions, and such services, to be mere flattery, lying, and mocking him to his face, Psa. lxxviii. 34, 36, 37.

Now, because there is none so ready to presume that he is upright, as the hypocrite, saying, with Ephraim, In all my labours they shall find no iniquity in me, that were sin, Hosea xii. 8; and because there are none so ready to doubt whether they be upright, as are the tender-hearted and sincere: so it was with David, when he prayed to have a right spirit renewed in him, Psa. li. 10: it will be needful and useful that I show you what uprightness is, and by what infallible signs you may know whether you be upright or no.

SECT. 2. THE DESCRIPTION OF UPRIGHTNESS.

Christian uprightness, for of that I speak, is a saving grace of the Holy Ghost, wrought in the heart of a man rightly informed in the knowledge of God in Christ, whereby his soul stands so entirely and sincerely right towards God, that in the true disposition, bent, and firm determination of his will, he would, in every faculty and power of soul and body, approve himself to be such an one as God would have him to be, and would do whatsoever God would have him to do, and all as God would have him, and that, for and unto God, and his glory.

The author of this uprightness is God's sanctifying Spirit.

The common nature of it, wherein it agrees with other graces, is, it is a saving grace; it is peculiar to them that shall be saved, for only they are endued with it; but it is common to all, and every one, who is effectually called.

The proper seat of this grace is the will.

The fountain in man, from whence, through the special grace of the Holy Ghost, it springs, is sound knowledge of God and of his will, concerning those things which the will should choose and refuse; and from faith in Christ Jesus, through whom every believer does, of his fulness, receive this grace to be upright. Hereby Christian uprightness differs from that uprightness, which may be in a mere natural, superstitious, and misbelieving man, for even such may be unfeigned in their actions in their kind, both in actions civil and superstitious, doing that which they do, in their ignorance and blindness, without dissimulation either with God or man. This Paul did before his conversion, he did as he thought he ought to do, Acts xxvi. 9.

The form and proper nature of uprightness, is the good inclination, disposition, and firm intention of the will to a full conformity with God's will, and that not in some faculties and powers of man, or in some of his actions, but the Christian would be universally sincere in all his parts, and in all things; he would be, and do, as God would have him to be and do, making God's will, revealed in his word and works, to be his will, and God's glory to be his end.

This holy uprightness expresses itself in these three things:

First, It shows itself in a well-grounded and unfeigned purpose and resolution to cleave to the Lord, and to make God's will to be his will, Acts xi. 23; Psa. cxix. 57, 106. This is an act of the will, guided and concluded from sound judgment.

The second act is, an unfeigned desire and longing of the heart to attain this good purpose and resolution, willing or desiring in all things to live honestly, Heb. xiii. 18, and to live worthy of the Lord in all well-pleasing; longing, with David, after God's precepts, Psa. cxix. 40. This is an act of the affection of desire, a motion of the will, drawing and exciting a man forward, giving him no rest, until he have obtained, at least in some good measure, his said purpose.

Thirdly, Uprightness shows itself in a true endeavour and exercise, according to the strength and measure of grace received, to be, and to do according to the former resolutions and desires. Such was the apostle's endeavour to have always a conscience void of offence towards God and towards men, Acts xxiv. 16. This endeavour is an act of the whole man. All and every active power of soul and body, as there shall be use of them, are employed in unfeigned endeavour.

Now concerning endeavour, know, there are who think they endeavour sufficiently, when they do not; others that they do not, when yet they do. The first, if they, to the sluggard's longing and wishing, do join an outward conformity to the means of grace, as to hearing the word, praying now and then, and receiving the sacraments; and if they do some things which may be done with little labour and difficulty; and if to these they add some slight essays to abstain from sin, and to do well, think they endeavour much; whereas, if they do no more, all is to little purpose.

For, to endeavour, is to exercise the head with study how, Acts xxiv. 16, and the heart with will and desire, and the hand and tongue, and the whole outward man, to do their utmost, putting to their whole strength, their whole skill, and their whole will, to subdue sin, and to be strengthened in grace, and built up more and more in knowledge, faith, and holiness, removing or breaking through every hindrance, shunning all occasions of evil, or whatsoever may strengthen sin, and seeking after and embracing all opportunities and means to be strengthened in the inward man, Phil. iii. 11, 14. If one means will not be sufficient, if there be others to be used, they will find out and use them also; if they cannot attain their good purposes at once, they will try again and again. They who endeavour indeed, not only seek to obtain their ends, but they strive in seeking; as hard students, as good warriors and wrestlers, and as those who run in a race do; so that they may obtain that which they study, fight, wrestle, and run for, 1 Cor. ix. 24—27. It is not a bare wishing or *woulding* for a fit, or a cold and common seeking; but an earnest striving to enter in at the strait gate, Luke xiii. 24, that gives admittance into the way of holiness, and into the kingdom of heaven. It is a studying and exercising a man's self as in a matter of life and death; and as a wise man would do for a kingdom, where there is possibility and hope of obtaining it.

Others, who indeed endeavour to keep a good conscience toward God and man, yet because they cannot bring into act always that which they labour for, or because they see oversights, neglects, or some weakness in their endeavours, think that they endeavour to no purpose. Whereas, if they do what they can, according to the strength of grace received, or according to the condition or state wherein they are, which is sometimes better, sometimes worse; if they see their failings in their endeavours, and bewail them, and do ask pardon, resolving by God's grace to strive to do better; this is true endeavour, this is that which God for Christ's sake does accept of, Mark xiv. 8. For since endeavour is a part of our holiness, you must not think that it will be perfect in this life; if it be true, you must thank God, for he will accept of that.

A man's endeavour may be as true, and as much, when yet he cannot perform what he endeavours to do, as it is at other times, when with the endeavour he has also ability to perform. As you may see in natural endeavours. The same man being well in health, if he fall, and break not his arms or legs, endeavours to get up, and readily does it: but if he be weak, or if falling he breaks his arms and legs, he also has a will and desire to rise, and strives earnestly to help himself, but cannot do it effectually, and in that case, he is fain to lie until he see help coming: then he will call, and entreat help, and when one gives him the hand, though he cannot rise of himself, yet he will lift up himself, as well as he can. Does not this man, in his latter condition, as truly endeavour as he did in his former? So it is with a spiritual man in his spiritual endeavours. If he essay to do what he can, and call to God for his help, and when he has it, is glad, and willing to improve it, this is the true endeavour, which, concurring with the two former acts, purpose and desire, gives proof of uprightness.

There is a twofold uprightness; the one of the heart and person, the other of the action. I have described the uprightness of the person. And then an action is upright, when a man does not dissemble, but means, as he saith, intending as much as is pretended, whether it be in actions toward God or man. The first is, when the heart of man agrees with, and in the intention thereof, is according to the will of God. The second is, when the outward act agrees with, and is according to the heart of him that does it.

SECT. 3. RULES TO JUDGE OF UPRIGHTNESS.

THAT you may rightly judge whether you are upright or no; first, take certain rules for direction, to rectify your judgment; then observe the marks of uprightness.

First, Uprightness being part of sanctification, is not fully perfect in any man in this life; but is mixed with some hypocrisy, conflicting one against the other. It has its degrees, sometimes more, sometimes less; in some things more, in some things less, according as each part prevails in the opposition, and according as the Christian grows or decays in other principal and fundamental graces.

Secondly, A man is not to be called an upright man, or an hypocrite, because of some few actions wherein he may show uprightness or hypocrisy: for an hypocrite may do some upright actions, in which he does not dissemble, though he cannot be said to do them in uprightness; as Jehu destroyed the wicked house of Ahab, and the idolatrous priests of Baal, with all his heart, 2 Kings x. And the best man may do some hypocritical and guileful actions, as, in the matter of Uriah, David did, 1 Kings xv. 5. It is not the having of hypocrisy that denotes an hypocrite, but the reigning of it, which is, when it is not seen, confessed, bewailed, and opposed.

A man should judge of his uprightness rather by his will, bent, and the inclination of his soul, and good desires, and true endeavours to well doing in the whole course of his life, than by this or that particular act, or by his power to do. David was thus esteemed a man according to God's own heart, no otherwise; rather by the goodness of the general course of his life, &c., than by particular actions: for in many things he offended God, and polluted his soul, and blemished his reputation.

Thirdly, Although uprightness is to be judged by the inward frame of the heart towards God, yet wheresoever uprightness is, it will show itself in men's actions in the course of their lives, James ii. 18. Only observe this, that in judging your actions, you must not judge them so much by the greatness of the quantity, as by the soundness and goodness of the quality. If it be good in truth, according to the measure of grace received, God accepts it in Christ, 2 Cor. viii. 12. She has done what she could, saith our Saviour, Mark xiv. 8. A little sound and true fruit, though weak in comparison, is far better than many fair blossoms, yea, than plenty of grapes, if they be wild and sour.

SECT. 4. PARTICULAR MARKS OF UPRIGHTNESS.

That you may conceive more distinctly, and better remember the signs of uprightness, I reduce them to these heads. They are taken, 1. From universality of respect to all God's will. 2. From a special respect to such things as God requires especially. 3. From a will and desire to please God in one place as well as another; in secret as well as openly. 4. From a constancy of will to please God at one time, as well as another. 5. From the true causes from whence good actions flow. 6. From the effects that follow well doing. 7. From the effects that follow evil doing 8. From the conflict which shall be found between uprightness and hypocrisy.

First, The upright man is universal in his respect to the whole will of God, Psa. cxix. 6.

(1.) In an unfeigned desire and endeavour to know what manner of man he ought to be, and what he ought to do, Psa. cxix. 33, 34. He would know and believe any one part of God's will, so far as it may concern himself, as well as another; threats as well as promises, commandments as well as either; and that not some, but all the threats, all the promises, and all the commandments; coming to the light readily, that his deeds may be made manifest, John iii. 21.

He is willing to know and believe what he should do, as well as what he should have, and hope for. But the hypocrite does not so, he winks with his eyes, and is willingly ignorant of that sin, which he would not leave, Matt. xiii. 15. 2 Peter iii. 5, and of that duty which he would not do, and of that judgment which he would not feel. He is willing to know the promises of the gospel, but willingly ignorant of the precepts of the gospel, and of the conditions annexed to the promises.

(2.) His universal respect to God s will is not only to know but to do, and to submit unto it in all things, willing to leave and shun every sin; willing to do every thing which he knows to be his duty; willing to bear patiently, thankfully, and fruitfully, every correction wherewith the Lord does exercise him. He dislikes sin in all. He loves grace and goodness 'n all. He would keep a good conscience in all acts of religion towards God, Acts xxiv. 16, and in all acts of righteousness and sobriety towards and amongst men. He would forbear not only those sins to which his nature is not so much inclined, or to which his condition in life affords not so many temptations; but those to which his nature and condition of life most carry him; he will cross himself in his dearest lust, especially his formerly beloved sin; his own sin, as David calls it, Psa. xviii. 23. Neither does he endeavour to abstain from those vices which may bring loss, and are out of credit; which human laws punish, and all men cry out against: but such as, through the iniquity of the times, are in countenance with the greatest, and practised by most; the forbearance whereof may threaten and procure danger and discredit, Dan. iii. 18; Acts iv. 19; the doing whereof may promise and promote much worldly gain and honour. Moreover, the upright man does not only strive to do those holy and virtuous actions, which are in credit, and for his advantage in the world; but those also which may expose to disgrace and loss even of his life and livelihood, Dan. vi. 10. He would abstain as well from less evils, even from appearance of evils, 1 Thess. v. 22, as from gross sins, and would so do the greater things of the law, as not to leave the other undone, Matt. xxiii. 23. But the hypocrite does not so, Mark vi. 20—27; there is some sin he will not leave, some duty he will not do, &c. Follow the opposition.

Secondly, An upright man is known by this: where God has laid a special charge, there he will have a first and special respect to it; as, to seek the kingdom of God and his righteousness, Matt. vi. 33; that one thing necessary, Luke x. 42; and to show a special love to the household of faith, Gal. vi. 10; Psa. xvi. 2; to be first and most at home, reforming himself, pulling the beam out of his own eye, Matt. vii. 5: to be most zealous for matter of substance in religion, and less in matter of ceremony and circumstance, Matt. xxiii. 23. Lastly, his chief care will be to apply himself to a conscientious discharge of the duties of his particular calling, Luke iii. 10—14; 1 Thess. iv. 11; knowing that a man has no more conscience nor goodness in truth, than he has will and desire in it to show the works of his particular place and calling. The hypocrite is contrary in all these. Matt. vii. 3—5; 1 Tim. v. 13; 2 Thess. iii. 11.

Thirdly, The upright man endeavours to approve himself to God, as well in secret, as openly; as well in the inward man, as in the outward; as well in thought, as in word and deed. But it is quite otherwise with the hypocrite; if he may seem good to men, it is all he cares for, Matt. vi. 2, &c.

Fourthly, The upright man is constant; his will is that he might always please God, Acts xxiv. 16. He does as much endeavour to approve himself to God in prosperity as in adversity, and even then studies how to be able to hold out before God, if his state should alter. I do not mean such a constancy as admits of no intermission or obstructions in his Christian course. A constant running spring may be hindered in its course for a time, by damming it up; yet the spring will approve itself to run constantly, for it will be still thrusting to get through, or to get under; or, if it can do none of these, it will raise itself in time, according to its strength, and get over all hinderances, and will bear down all before it, and run with a more full streams afterwards, by as much as it was before interrupted: so it is with an upright man. But the hypocrite's religion is by fits and starts; as he calls not on God at all times, Job xxvii. 10, so it is with all other his goodness, it is but as the seed in stony ground, and amongst thorns, Matt. xiii. 21, 22, and as morning dew, Hosea vi. 4, it endures but for a season.

Fifthly, An upright man is known by the causes from which all his good actions spring, and to which they tend.

(1.) That which causes the upright man to endeavour to keep a good conscience alway, is an inward principle and power of grace, causing him through faith in Christ, John xv. 2, 5, in and from whom, as the root of all grace, he brings forth fruit; and from love and fear of God, 1 Cor. ix. 16—18; 2 Cor. v. 14; Gen. xlii. 18; and from conscience of the commandment, to do the will of God, 2 Cor. ii. 17. Not only fear of wrath, and hope of reward, causes him to abstain from evil and do good; but chiefly love of God, and conscience of duty.

Now, if you would know when you obey out of conscience of the commandment, and from love of Christ; consider, 1. Whether your heart and mind stand bent to obey every of God's commandments which you know, as well as any, and that because the same God who has given one, has given all, James ii. 11. If so, then you obey out of conscience. 2. Consider what you do, or would do, when Christ, and his true religion, and his commandments go alone, and are separated from all outward credit, pleasure, and profit. Do you, or will you then cleave to Christ, and to the commandment? Then love of Christ, fear of God, and conscience of God's command, was, and is the true cause of your well-doing; especially if you choose and endeavour this, when all these are by the world clothed with peril and contempt. 3. Consider whether you can go on in the strict course of godliness alone, and whether you resolve to do it though you shall have no company, but all or most go in the way of sin, and also persuade you thereunto. When you will walk with God alone, and without other company, this shows that your walking with God is for his sake. So walked Noah, Gen. vii. 1, and Elijah, 1 Kings xix. 13, as he thought.

But the cause of an hypocrite's well-doing is only goodness of nature, or good education, or mere civility, or some common gifts of the Spirit, or self-love, slavish fear, or the like. See this in Ahab's repentance, 1 Kings xxi. 27, in Jehu's zeal, 2 Kings x. 16, and Joash's goodness, 2 Chron. xxiv. Ahab's humiliation was only from a slavish fear of punishment. The zeal of Jehu was only from earthly joy and carnal policy; for had it been in zeal for God, he would as well have put down the calves at Dan and Bethel, as slain the priests of Baal. And the goodness of Joash was chiefly for Jehoiadah's sake, whom he reverenced, and to whom he esteemed himself beholden for his kingdom, and not for God's sake. For the Scripture saith, that after Jehoiadah's death, the princes solicited him, and he yielded to them, aud fell to idolatry; added this also, he commanded Zechariah the high-priest, Jehoiadah's son, to be slain, because, in the name of the Lord, he reproved him for his sin, 2 Chron. xxiv. 6, 17, 18, 20, 21.

(2.) The upright man's actions, as they come from

a good beginning, so they are directed to a good end, 1 Cor. x. 31, namely, the pleasing of God, and the glory of his name, as his direct, chief, and utmost end; not that a man might not have respect to himself, and to his neighbour also, proposing to himself his own and his neighbour's good, as one end of his actions, sometimes; but these must not be proposed either only, or chiefly, or as the ultimate end, but only as they are subordinate to those chief ends, and are the direct means to promote God's glory. For so far as a man's health and welfare, both of body and soul. lie directly in the way to glorify God; he may in that respect aim at them in his actions. Our Saviour Christ, in an inferior and secondary respect, aimed at his own glory, and at the salvation of man, in the work of man's redemption, when he said, Glorify thy Son; and prayed, that his church might be glorified, John xvii. 1, &c. Here he had respect unto himself, and unto man; but when he says, "that thy Son may glorify thee," here he made God's glory his utmost end, and the only mark which for itself he aimed at.

The upright man's aim at his own, and at his neighbour's good, is not for themselves, as if his desire ended, and was terminated there; but in reference to God, the chief good, and the highest end of all things.

Indeed, such is God's wisdom and goodness, that he has set before man evil and good; evil, which follows upon displeasing and dishonouring him by sin, that man might fear and avoid sin; good, and recompence of reward, which follows upon faith and obedience; that he might hope, and be better induced to believe and obey. This God did, knowing that man has need of all reasonable helps to deter him from evil, and to allure him to good. Now, God having set these before us, we may and ought for these good purposes to set them before ourselves. Yet the upright man stands so fully and only to God, that, so far as he knows his own heart, he is thus resolved, that if there were no fear of punishment, nor hope of reward; if there were neither heaven nor hell, he would endeavour to please and glorify God, even out of that duty he owes to him, and from that high and awful estimation which he has of God's sovereignty, and from that entire love which he bears unto him. He that habitually in doing of common and earthly business, though they concern his own good, has a will to do them with a heavenly mind, and to a heavenly end, certainly stands well, and is uprightly resolved, although, in temptations and fears, he does not always feel the said resolution.

But the hypocrite does not so: he only or chiefly aims at himself, Matt. vi. 2, 5, 16, and in his aim serves himself in all that he does. If he look to God's will and glory, as sometimes he will pretend, he makes that but the bye, and not the main end, 2 Kings x. 16; he seeks God's will and glory not for itself, but for himself: not for God's sake, but for his own. Thus did Jehu.

Sixthly, An upright man may know he is upright, by the effects that follow upon his well doing.

(1.) His chief inquiry is, and he does observe, what good comes by it, and what glory God has had, or may have, rather than what earthly credit and benefit he has gotten to himself, Philip. i. 12—20. Or if this latter thrust in itself before the other, as it will ofttimes in the best, he is greatly displeased with himself for it. The hypocrite not so; all that he inquires after, and is pleased with, after he has done a good deed, is, what applause it has amongst men, &c.

(2.) When an upright man has done a praise-worthy action, he is not puffed up with pride, and high conceit of his own worth, glorying in himself; but he is humbly thankful unto God; thankful, that God has enabled him to do any thing with which he will be well pleased, and accept as well done; humble and low in his own eyes, because of the manifold failings in that good work, and because he has done it no better; and because whatsoever good he did, it was by the grace and power of God, not by any power of his own. Thus David showed his uprightness in that solemn thanksgiving when he said, But who am I, and what is my people, that we should be able to offer so willingly after this sort? &c. 1 Chron. xxix. 13, 14.

But it is otherwise with the hypocrite: for either he ascribes all the glory of his good work to himself; or if he seem to be thankful, it is with a proud thankfulness, like that of the pharisee, Luke xviii. 11, accompanied with disdain of others, who in his opinion do not so well as himself.

(3.) The upright man having begun to do well, does not set down his rest there; but strives to do more, and to be better: he, with the apostle, forgets what is behind, looking to that which is before, not thinking that yet he has attained to what he should do, Philip. iii. 13—15. So many as are indeed perfect and upright do thus.

But the hypocrite, if he have some flash of common illumination, and some little taste of those things which concern the kingdom of heaven, and has attained to a form of godliness, thinks that he has enough, and needs nothing. So did Laodicea, Rev. iii. 17.

Seventhly, The upright man and the hypocrite are distinguished by their different affections and carriages, after that they have fallen into a sin, for in many things we sin all, James iii. 2. As the upright man did not commit his sin with that full consent of will, which the hypocrite may do, and often does; but always with some reluctance and opposition of will, though not always felt and observed, insomuch that he can say, It was not he, but sin that dwelt in him, Rom. vii. 15—17; so after he is fallen into sin, when his sin is made known to him, he does not hide, excuse, or defend his sin, Job xxxi. 33, or if he do, it is but seldom in comparison, and but faintly and not long, his conscience smiting him when he does it, or quickly after it, Job xl. 3, 4, xlii. 3—6.

An upright man will not be much or long angry with any, who admonish him of his sin, yea, though an enemy, by malicious railing, call his sin to remembrance, as Shimei did to David, 2 Sam. xvi. 10—12; even therein he can see God, and can for the most part abstain from revenge, and will stir up his heart to godly sorrow for his sin. But if any, like Abigail, 1 Sam. xxv. 32, 33, shall, in wisdom and love, admonish him, he blesses God that sent him or her; he blesses and makes good use of the admonition, and blesses the admonisher, and takes it for a special kindness. Thus David, a man according to God's own heart, as he displayed human frailties in his many and great falls; so he gave clear proof of his uprightness, sooner or later, by his behaviour after his falls. He could say, and his repentance did prove it, that though, to his grief and shame, sometimes he departed from God; yet he did not wickedly depart from God. Psa. xviii. 21. Though upright men be transgressors, yet they are not wicked transgressors, Psa. lix. 5; there is a great difference between these two. And though there be evil in their actions, yea, in some of them filthiness, and grievous iniquity, yet in their filthiness is not lewdness, Ezek. xxiv. 13, as God complains of Judah, that is, they are not obstinate and rebellious, standing out against the means of purging and reclaiming them. For when God does correct them by his word or providence, they are willing to reform whatever is discovered to be amiss, Job xlii. 6.

Moreover, although the upright man may be often drawn into a way that is not good, and often, through his weakness and heedlessness, falls into a state that is not good; yet he does not set himself in a way which is not good, Psa. xxxvi. 2—4, nor yet, like the swine, delight to wallow and lie in it. When an upright man is fallen, and has recovered out of his spiritual swoon, when he is come to himself, he is like a man sensible of his bones broken or out of joint; he is not well, nor at quiet, nor his own man, until he has confessed his sin, repented of it, asked pardon and grace, and renewed his peace with God. An upright man is likewise like the needle of the mariner's compass, which may, by violent motion, sometimes swerve to the west, or to the east; but stands steady no way but towards the north, and if it be truly touched with a loadstone, has no rest but in that one point; so an upright man may, through boisterous temptations, and strong allurements, oftentimes look towards the pleasure, gain, and glory of this present world: but because he is truly touched with the sanctifying Spirit of God, he still inclines towards God; and has no rest until his mind is steadily fixed on Christ and heaven.

But it is not so with the hypocrite. He is in each particular directly contrary. I leave the full and particular application thereof to yourself.

Eighthly, You will find the most evident mark of uprightness from your sense of hypocrisy in yourself, and from your conflict with it, Gal. v. 17. The upright man is sensible of too much hypocrisy and guile in his heart, Psa. li. 10. Yea, so much, that oftentimes he makes it a question whether he have any uprightness; and, until he has brought himself to due trial by the balance of the sanctuary, the word and gospel of Christ, he fears he is still a hypocrite. But there is nothing which he would oppose more, nothing which he complains of, or prays to God more against, than this hypocrisy, nor is there any thing he longs after, labours and prays for more, than that he may love and serve the Lord in sincerity, 1 Cor. ix. 26, 27. All this plainly shows, that this man would be upright, which hearty desire so to be, is uprightness itself.

The hypocrite contrariwise, neglects to observe his guile and false heartedness in religion: or if he can see it, he is not much troubled at it, but suffers it to reign in him: and as he boasts of his good actions, so likewise of his good heart, and good meaning in all that he does, except when his lewdness and hypocrisy are discovered to his face; flattering himself in his own eyes, till his iniquity is found to be hateful, Psa. xxxvi. 2.

Before I leave this, I will answer a question or two, concerning judging of uprightness by these marks.

(1.) Whether an upright man can at all times discern his uprightness, by these or any other marks?

Ordinarily, if he will impartially compare himself with these evidences, he may. But sometimes it so happens that he cannot; namely, in the case of spiritual desertions, when God, for his neglect of keeping his peace with him, is hidden from him for a time, and when in his displeasure he looks angrily, and writes bitter things against him. Likewise, when he is in some violent and prevalent temptation, and thereby cast into a kind of spiritual swoon, and in such like cases. But a man must not judge himself to be dead, because when he is asleep, or in a swoon, he has no feeling, or sense of life.

(2.) Whether is it necessary that a man should find all these marks of uprightness in him, if he be upright?

No. Although, if he were in a condition to judge and try himself thoroughly, he might find them all in him; yet if he find most, or but some of these, he should comfort himself with those, until he find the rest.

Take heed therefore that you do not as many, who when they hear and see many signs given of this, or any other needful grace, if they cannot approve themselves by all, will make a question whether they have the grace or no. One may give you twenty signs of natural life, as seeing, hearing, talking, breathing, &c. What though you cannot prove yourself by all? Yet if you know you feel, or breathe, or move, you know you are alive by any one of them.

(3.) What is to be done when you cannot find that you are upright, whereas heretofore sometime you did hope that you were?

Do not presently conclude you are an hypocrite; but look back unto former proofs of uprightness. And though you have, for the present, lost your evidence and assurance of heaven, yet give not over your possession of what you have had, nor your hope. A man that has once had possession of house and lands, if his estate be questioned, will seek out his evidence; and, suppose that he has laid aside, or lost his evidence thereof, yet he is not such a fool as to give over his possession, or his right; but will seek till he find his evidences, or if he cannot find them, will search the records, and get them from thence. So must you in this case; you must seek for evidence again, Psa. li.

12. However cleave fast to God and to his promises, Acts xi. 23; Job xiii. 15, 16; frequently renew your acts of faith on the Lord Jesus Christ, and continue to persevere in the ways of godliness as you are able, and you shall not be long, before you shall know that you are upright: or if you attain not to this, yet be sure the Lord will know you to be his, 2 Tim. ii. 19, though you do not so certainly know that he is yours. But more of this when I shall speak of peace of conscience.

But in trying my uprightness, I find many of the signs of hypocrisy in me. I do not find myself to be so universal in my respect to all God's commandments as I should; I do not hate all sins alike; I find myself inclined to one sin more than another, and I am readier to neglect some duty than other; I cannot so thoroughly seek God's kingdom as I should; I am readier to find fault with others, than to amend my own conduct, &c. I find that I am not so constant as I ought to be in holy duties, and I have too much respect to myself in all that I do, and too little to God's glory. In reading all the notes of hypocrisy, except the last, I find hypocrisy, nay, much hypocrisy to be in me; must I not therefore judge myself to be a hypocrite?

No. For truth of uprightness may be in the same person, in whom there is sense of much hypocrisy; nay this, to feel hypocrisy with dislike, is the certain evidence of truth of uprightness. Indeed, if you felt not thus much, you might fear you were not upright. All that you have said, if it be true, only proves that you have hypocrisy remaining in you, and that you feel it. You must remember I told you, that not the having, but the reigning of hypocrisy, makes an hypocrite. Besides, a man may have an universal respect to all God's commandments, and yet not an equal respect to all. If you see and bewail your sin, and fight against your hypocrisy when you feel it, assure yourself you are no hypocrite.

(4.) What if a man finds indeed by these notes of hypocrisy, that it does reign in him?

He must know that he is for the present hated of God, and in a damnable state, yet his state is not desperate. If the hypocrite forsake his hypocrisy, and become upright, he shall not die for his hypocrisy; if this be true of a sinner's forsaking all sin, then, it is true of this in particular, of forsaking his hypocrisy; but in the uprightness wherein he lives, he shall live, Ezek. xviii. 21, 22. What Christ said to hypocritical and lukewarm Laodicea, Rev. iii. 19, that I say to all such: they must be zealous, they must amend, and be upright; hypocrisy is as pardonable as any other sin to him who is penitent, and believes in Christ Jesus, Isa. i. 11, 16, 18.

By this which I have written, you may plainly see, (1.) That you ought to be upright; (2.) What it is to be upright; (3.) Whether you be upright or no. It concerns you therefore to hate and avoid hypocrisy, and to love and embrace sincerity. Which that you may do, make use of the motives and means which follow in the next sections.

SECT. 5. DISSUASIVES FROM HYPOCRISY, AND MOTIVES TO UPRIGHTNESS.

If you would abandon hypocrisy, consider the dissuasives; taken from the evils and mischiefs that accompany it where it reigns, and how troublesome and hateful it is, where it does not reign.

The evils of hypocrisy, where it reigns.

First, Hypocrisy takes away all the goodness of the best actions. They are good only in name, not in deed. The repentance and obedience of a hypocrite is none, because it is feigned; his faith is no faith, because it is not unfeigned; his love is no love, because it is not from a pure heart, without dissimulation, 1 Tim. i. 5. Judge the same of all other graces and good actions of a hypocrite.

Secondly, All the good actions of a hypocrite are, together with himself, wholly lost, Luke xiii. 25; Matt.

vii. 22; xxv. 11, 12. Such is preaching, hearing, praying, almsgiving, &c.

Thirdly, Hypocrisy, in whom it reigns, does not only take away all goodness from the best gifts and actions, and cause the loss of all reward from God, but it poisons and turns the best actions into most loathsome and abominable sins, Isa. lxvi. 3; insomuch that in those good works wherein the hypocrite seems to make haste to heaven, he still runs post to hell. For such allowed hypocrisy is worse than professed wickedness, Rev. iii. 15. It is so odious in God's sight, that for it he will plague those in whom it rules with his severest judgments. For the hypocrisy of men professing the truth, brings the name, religion, and best services of God into disgrace and contempt, Rom. ii. 24, and causes the best actions and best men to be suspected. For such as have not spiritual wisdom to judge rightly, stumble thereat, and forbear the exercises of religion, and the company of those that are religious, ignorantly judging all who profess that religion to be alike. Besides, hypocrisy is high treason against God; for it is a gilding over, and setting the king's stamp upon base metal. It is tempting and mocking of God to his face, Psa. lxxviii. 36; a sin so abominable, that his holy justice cannot endure it.

God's just judgments upon hypocrites.

Fourthly, God's judgments on such hypocrites are manifold. For this cause God gives them over to believe lies, 2 Thess. ii. 10, 11; Heb. vi. 5, 6; even popery, or any other damnable error or heresy. Hence it is that God gives them up many times to fall from seeming goodness to real wickedness, and from one evil to another, Luke viii. 18, even unto final apostasy, Heb. x. 25, 26. And at last, when God takes away a hypocrite's soul, he is sure not only to lose his hope, which adds much to his hell, Job xxvii. 8, but to be made to feel that which he would not fear, being ranked with those sinners, who shall be punished with the greatest severity in the eternal vengeance of hell fire, Matt. xxiv. 51. For after that a hypocrite has played the civil and religious man for a while upon the stage of this world, his last act, when his life is ended, is to be, indeed, and to act to the life, the part of an incarnate and tormented devil. He shall have his portion with the devil and his angels, Matt. xxv. 41. When fearfulness has surprised the hypocrites; who shall dwell with the devouring fire? who shall dwell with everlasting burnings? saith the prophet, Isa. xxxiii. 14. Happy were it for them, if this warning might effectually awaken them out of this damning security.

The evils of hypocrisy, though it doth not reign.

Consider likewise; that hypocrisy does much harm, even where it does not reign, and that more or less, according as it is more or less mortified.

For, first, it brings the soul into a general consumption of grace; no sin more so. Secondly, It blinds the mind, and insensibly hardens the heart; no sin more. Thirdly, It makes a man formal and careless in the best actions. Fourthly, It causes fearful sins, and decays of grace. Fifthly, It deprives a man of peace of conscience in such a manner, that a spiritual physician can hardly suggest any hope or comfort to him on whose conscience lies the guilt of hypocrisy; yea, hardly to him that does but fear he is guilty; for he refuses the comfort of his good affections and actions, saying, All that I did was but in hypocrisy. Lastly, Besides that, it brings many temporal judgments; it causes a man to lose many of his good works done in hypocrisy, 2 John 8; 1 Cor. iii. 15, though, through God's mercy, he lose not himself, because he is still found in Christ, and Christ's spirit of uprightness reigns in him.

Motives to uprightness.

Now to induce you to love uprightness, and to labour after it; consider the good which accompanies uprightness: First, temporal and outward; but Secondly, and chiefly, that which is spiritual and eternal.

First, Uprightness has the promise of this life, 1 Tim. iv. 8. It is a means to keep off judgments, Psa. xci. 9, 10, 14, or in due time to remove them.

If affliction like a dark night overspread the upright, for their correction and trial for a time, yet light is sown for them, and in due time will arise unto them, Psa. xxxiv. 9, 10. Moreover, this uprightness does not only provide well for a person's self; but if any thing can procure a blessing to his children, and his children's children, uprightness will, Prov. xx. 7. The Holy Ghost saith, the generation of the upright shall be blessed, Psa. cxii. 2.

Secondly, The spiritual blessings which belong to the upright, are manifold.

1. The upright man is God's favourite, even his delight, Prov. xi. 20.

2. He is hereby assured of his salvation, Psa. xv. 1, 2. For although an upright man fall into many grievous sins, yet presumptuous sins, Psa. xix. 13, shall not reign over him; he shall be kept from the great transgression; he shall never sin the sin unto death; yea, he shall be kept from the dominion of every sin.

3. By uprightness a man is strengthened in the inward man; it being that girdle which buckles and holds together the chief parts of the Christian armour, Eph. vi. 14. Nay, it is that which gives efficacy to every piece of that armour: it strengthens the back and loins, yea, the very heart of him that is begirt with it.

4. He that is upright, is sure to have his prayers heard, Jer. xxix. 13, and to be made able to profit by the word of God, and by all his holy ordinances. Do not my words, saith God, do good to him that walks uprightly? Micah ii. 7.

5. The upright man's services to God in prayer, hearing, receiving sacraments, &c. though performed with much weakness and imperfections, shall, through Christ, be accepted of God, 2 Chron. xxx. 18—20. Nay, where there is not power, the will of the upright man is taken for the deed, 2 Cor. viii. 12; and where there is power and deed both, even there the uprightness and readiness of the will is taken for more than the deed, according to that commendation of them, whc were said not only to do, but to be willing a year

ago, 2 Cor. viii. 10. For many may do good things, who yet do them not with an upright will and ready mind.

6. The upright man has always matter of boldness before men. He can make an apology and defence for himself against the slanders of wicked men, and against the accusations of Satan, Acts xxiii. 11; xxiv. 14—16, who are ready upon every slight occasion, to reproach him as a hypocrite, and say, that all which he does is but in hypocrisy: but he can give all the lie, who charge him with dissimulation or hypocrisy. He knows more of his hypocrisy than they can tell him; he finds fault with, and accuses himself for it, more than they can do: yet this he can say, he allows it not, he hates it, and his heart is upright towards God. He cares not though his adversary write a book against him, Job xxxi. 35, 36; xix. 23—25. He has his defence; if men will receive it, they may; if not, he dares to appeal to heaven. For his record is on high. He has always a witness both within him, and in heaven for him, Job xvi. 19; 2 Cor. i. 12.

7. Uprightness is an excellent preventer and cure of despair, arising from accusations of conscience; even of a wounded spirit, of which Solomon saith, Who can bear it? For either it keeps it off, Job xxvii. 5, 6; or, if it be wounded, this uprightness in believing, and in willingness to reform and obey, is a most sovereign means to cure and quiet it, or at least will allay the extremity of it.

Not but an upright man may have trouble of mind, and that to some extremity; but he may thank himself for it, because he will not see and acknowledge that uprightness which he has, and does not properly apply it, or cherish it; which if he would do, there is nothing, next to the precious blood of Jesus Christ, would answer the charges of his accusing conscience, or bring feeling comfort to his soul, sooner or better.

8. The upright man has a holy boldness with God. When Abimelech could say, In the integrity of my heart, and innocency of my hands have I done this,

Gen. xx. 5, he had boldness to expostulate and reason his case with God. An upright man in his sickness, or in any other calamity, yea, at all times, when he needs God's help, can be bold to come before God, notwithstanding his sin that remains in him, his original sin, and his many actual transgressions. So did Hezekiah, upon his death bed, as he thought, saying, Remember, O Lord, I beseech thee, how I have walked before thee in truth, and with a perfect heart, and have done good in thy sight, Isa. xxxviii. 3. So did Nehemiah, saying, Remember me, O my God, concerning this, and spare me according to the greatness of thy mercy, Neh. xiii. 22. This uprightness gives boldness with God, but without all presumption of merit, as you see in good Nehemiah.

9. Lastly, Whatsoever the upright man's beginning was, and whatsoever his changes have been in the times that have gone over him, both in the outward and inward man, in his progress of Christianity; mark this, his end shall be peace, Psa. xxxvii. 37. The last and everlasting part which he shall act indeed, and to the life, is, everlasting happiness, Prov. xxviii. 18.

And, to contract all these motives into a short, but final sum, The Lord is a sun and shield; the Lord will give grace and glory: no good thing will he withhold from them that walk uprightly, Psa. lxxxiv. 11.

SECT. 6. MEANS TO SUBDUE HYPOCRISY, AND PROMOTE UPRIGHTNESS.

It remains now that you should know by what means you may abate and subdue hypocrisy; and may get, keep, and increase this grace of uprightness.

First, You must, by a due and serious consideration of the evils of hypocrisy, and advantages of uprightness, fix in your heart, by the help of Christ, a loathing and detestation of the one, and an admiration, love, and longing desire of the other; with a sincere purpose of heart, by the grace of God, to be upright. This must first be wrought, for until a man stand thus affected, and resolved against hypocrisy, and for up rightness, he will take no pains to be free from the one, nor yet to obtain the other.

Secondly, You must be sensible of that hypocrisy which yet is in you, and of the want of uprightness, though not altogether, yet in great part. For no man will be at the pains to remove that disease whereof he thinks he is sufficiently cured though he judge it to be never so dangerous; nor yet to obtain that good of which he thinks he has enough already, though he esteem it never so excellent.

Hitherto both in the motives and means, I have endeavoured to gain the will: to will and resolve to be upright, and to be willing to use all good means to be upright. Now those means that will effect it follow.

Thirdly, Do your best to root out those vices that beget and nourish hypocrisy; and to plant in their room those graces which produce and strengthen uprightness.

The chief vices are ignorance and unbelief, self-love, pride, and an irresolved and unsettled heart, unstable and not firmly resolved what to choose, whereby it wavers and is divided between two objects, dividing the heart between God, and something else, Zeph. i. 5. either false gods, a man's self, or the world; whence it is, that the Scriptures call a hypocrite a man that has a heart and a heart, one that is double-minded, James iv. 8.

The graces are, a right knowledge of God and of his will, and faith in him; self-denial, humility, and lowly-mindedness; stability, and singleness of heart towards God.

For, the more clear light you can get into your mind, the more truth you will have in your will. And when you can so deny yourself, that you can quite renounce yourself, and first give yourself to Christ, and unto God, 2 Cor. viii. 5, 10, 11, then there will follow readiness of mind, and heartiness of will, to do whatsoever may please God. Also, the more humility you have in your mind, the more uprightness you will have in your heart. For while the soul is lifted up, that man's heart is not upright in him, saith God, Hab. ii. 4. Lastly, when your eye is single, and your heart one, and undivided, you will not allow yourself to be in part for God, and in part for mammon, Matt. vi. 22—24, in part for God, and in part for your lusts, whether of the flesh, or of the world, or of the pride of life; you will not give your name and lips to God, and reserve your heart for the world, the flesh, or the devil; but, by your will, God shall be all in all unto you.

Fourthly, If you would be in earnest and in truth against sin, and for goodness, you must represent sin to your thoughts as the most hurtful, hateful, and most loathsome thing in the world, Gen. xxxix. 9; and must represent the obeying and doing of God's will unto your mind, as the best and most profitable, most amiable, most sweet and excellent thing in the world, Psa. xix. 7, 8, 11; cxix. 72. Hereby you may affect your heart with a thorough hatred and loathing of sin, and with an hearty love and delight in God's commandments, Psa. cxix. 97. If you do thus, you cannot choose but shun sin, and follow after that which is good, not in pretence only, but in deed and in truth, with all your heart, For, a man is always hearty against what he truly hates, and for what he dearly loves.

Fifthly, If you would be sincere, and do all your actions for God's glory, and for his sake, you must, by the light of God's word and works, fully inform and persuade yourself of God's sovereignty and absoluteness, and that, because he is the first absolute and chief good, he must needs be the last, the absolute and chief end of all ends. For he, that is the Alpha, must needs be the Omega, of all things, Rev. i. 8. Since all things are of God, Rom. xi. 36, and since he made all things for himself, Rev. iv. 11, therefore you should, in all things you do, be upright, intending God's glory as your principal and ultimate end in all things, 1 Cor. x. 28, 31.

Sixthly, Consider often and seriously, that how close and secret soever hypocrisy may lurk, yet it cannot be hid from the eyes of God, with whom you have to do, and before whom you walk, Heb. iv. 12, 13, who will bring every secret thing to judgment, Eccles. xii. 14. Wherefore take continual notice, that you are in the sight of God that made your heart, Psa. xciv. 9—11, who requires truth of heart, Psa. li. 6, and who perfectly knows the guile or truth of your heart. This will much further your uprightness; for who can dare to promote and dissemble in the presence of his Lord and judge, who knows his dissimulation better than himself?

Seventhly, Unite yourself more and more strongly unto your head Christ Jesus, by faith and love; continually renounce your own wisdom, righteousness, and strength, that you may every day be more and more united unto him. Grow daily in faith and hope in him, from whence you shall more and more partake of his fulness, Phil. iii. 8, &c., even grace for grace, John i. 12, 16. For the measure of your uprightness, will usually be in proportion to your faith. For in proportion as the branch partakes more of the vine, so it draws more virtue and bears more good fruit, John xv. 5.

Eighthly, You must, with an holy jealousy of the deceitfulness of your heart, examine yourself often; not only of what you have done, and now do, but of the motives and ends of your religious actions; as was before directed in the marks of uprightness. Lay yourself often to the rule of uprightness, that is, the will of God, and finding yourself defective, study and labour to amend, and be upright, and that to the utmost of your power.

Ninthly, Excuse that measure of uprightness which you have, and be more thankful for the little you have, than discouraged as many are, because they have no more. If you find yourself upright, be abundantly thankful, and resolve to keep and increase it by all means. Keep your heart thus with all diligence, Prov. iv. 23; then, as all other graces, so this of uprightness will increase in the using.

Tenthly, and lastly, Use the means of all means,

tne catholicon for all graces, which is prayer. Think not to gain uprightness by the power of your own might: but in the sense of your insufficiency, repair often to God by prayer; even to him who made your heart, in whose hands your heart is, who best knows the crooked windings and turnings of your heart, who only can amend and rectify your heart; who, because he delights in an upright heart, and has commanded you to seek it in the humble use of his means, will assuredly give it. Thus prays David; Renew, O Lord, a right spirit within me; Psa. li. 10; and, Let my heart be sound in thy statutes, Psa. cxix. 80.

CHAPTER XII.

OF LAWFUL CARE, AND OF FREEDOM FROM ANXIOUS CARE.

When you have thus exercised a holy care to walk with God in uprightness, according to the foregoing directions, it remains that you free yourself of all other care, and that you rest holily secure in God: enjoying your most blessed peace with him, according to the divine direction, Be careful for nothing, &c. Phil. iv. 6, &c.

The care which is commanded, and carefulness which is forbidden, differ thus:

Lawful care is an act of wisdom, whereby after that a person has rightly judged what he ought to do, what not, what good he is to pursue, and what evil is by him to be shunned, or removed; he, accordingly with more or less intention and eagerness of mind, as the things to be obtained or avoided are greater or less, is careful to find out, and diligent to use lawful and fit means for the good, and against the evil, and that with all circumspection; that he may omit nothing which may assist him, nor commit any thing that may hinder him in his lawful designs; which, when he has done, he rests quiet, and cares no further; casting all care of success upon God to wnom it belongs, expecting a good issue upon the use of good means, yet resolving, howsoever, to submit his will to God's will, whatsoever the success shall be.

Sinful care is an act of fear and distrust, exercising not only the head, but chiefly the heart, to the disquietude and disturbance thereof, causing a person inordinately, and anxiously to pursue his desires, perplexing himself with doubtful and fearful thoughts about success.

Lawful care may be called a provident care, and care of the head.

Carefulness may be called a distrustful care, or a care of the heart.

This provident care is not only lawful, but necessary; for without it, a man cannot possibly be secure, nor have reasonable hope of good success.

This provident care is commended to you in the examples of the most wise and industrious brute creatures, Prov. vi. 6—11; and in the examples of the most prudent men.

As of Jacob's care of his safety, how to escape the rage of his brother Esau, Gen. xxxii. xxxiii; of Paul's care of the churches, 2 Cor. xi. 28; of the Corinthians' care and study to reform themselves, 2 Cor. vii. 11; of the good noble woman's care to entertain the Lord's prophet, 2 Kings iv. 10; of the good housewife's care of well ordering and maintaining her family, Prov. xxxi. 10, &c. The same good examples you have in the care of godly unmarried men and women, 1 Cor. vii. 32, 34, how to please God, and that they might be holy both in body and soul; and of Mary, who cared for that one thing needful, Luke x. 42.

Moreover, you are commanded this provident care, namely, to study to be quiet, to be no busy-body, not idle, 1 Thess. iv. 11; but to labour in a lawful calling the thing that is good, Ephes. iv. 28. Also, to walk honestly towards them who are without, 1 Thess. iv.

12; to endeavour so to walk towards God's people that you keep the unity of the spirit in the bond of peace, Ephes. iv. 3; to provide for your own, 1 Tim. v. 8; to give diligence to make your calling and election sure, 2 Pet. i. 5—10; to study to maintain good works, Tit. iii. 8. But amongst all, you are commanded chiefly to seek the kingdom of God, and his righteousness, Matt. vi. 33, as the best means to free you from all unlawful cares.

The properties of lawful care are these:

SECT. 1. DESCRIPTION OF LAWFUL CARE.

First, The seat wherein lawful care resides, is the head; for that is the seat of understanding, wisdom, and discretion; but carefulness is chiefly seated in the heart.

Secondly, Godly care is always about good and lawful things. It has a good object, and good matter to work upon, and be conversant about: proposing always some good thing to be the end, which it would attain. It is not a care about evil, as how to make provision for the flesh to fulfil the lusts thereof, Rom. xiii. 14; like Ahab's and Jezebel's carefulness for Naboth's vineyard and life, 1 Kings xxi. nor yet like Absalom's carefulness, how to usurp his father's kingdom, 2 Sam. xv. nor like Haman's, how to destroy the Jews, Esther iii. 9; neither is it like the carefulness of those of whom Solomon speaks, who cannot sleep unless they do mischief, Prov. iv. 16.

Thirdly, This holy provident care makes choice only of lawful means, to obtain this lawful end. David had care of his own life; therefore he got intelligence from Jonathan of Saul's evil purposes towards him, 1 Sam. xx. 1. He did fly and hide himself from Saul, but would by no means lay violent hands upon his anointed lord and king; though he had fair opportunities, and strong solicitations to kill him, he falling twice into his power, and was earnestly called upon by his servants to dispatch him, 1 Sam. xxiv. 3—7; xxvi. 7—25.

Observe likewise Jacob's care to save himself, and all that he had, from the fury of his brother Esau, Gen. xxxii. xxxiii. He used only fit and lawful means. For though a man's intention be ever so good, and the thing cared for be good, yet if the means to get it be unlawful, the care is evil. To care how to provide for yourself and yours, is in itself good and needful; but so to care, that you run to unjust and indirect means, makes it evil. To care how to be saved, is an excellent care, but when you seek to attain it by ways of your own, or of other men's inventions; as by idolatrous worship, and voluntary religion, or looking to be saved by your own works, by purgatory, Pope's pardons and indulgences, as the papists do; this is a most sinful carefulness, Col. ii. 18, &c. To care how to bring glory to God is the best care; but if to procure it, you use lying for God, or any other unlawful means, it is an unholy care, Rom. iii. 7, 8.

Fourthly, This laudable holy care is a full and impartial care, even of all things belonging to a person's condition. It is not such a care of the body and estate, as causes neglect of the soul: neither is it such a care of the soul, as is attended with neglect of the body, life, estate, or name, 1 Tim. v. 23. It is not such a care of the private, as to neglect the public good, or of the public, so as to neglect the private. It extends itself to whatsoever God has committed to our care; both for ourselves and others. Those who care only for themselves, and for the things only of this life, sin in their care. Likewise those who seem to care only how to please God, and to save their souls, yet weakly or carelessly neglect their bodies, and affairs of their families belonging to their place, or the common good of others in church or commonwealth, all these are partial, and do sin in their care. All worldings and self-loving men offend in the first kind. All superstitious and indiscreetly devout men offend in the second kind: also all such who for devotion sake neglect the necessary duties of their particular calling.

Fifthly, Lawful care is a discreet and well ordered care; it puts difference between things more or less good, and between things necessary or not necessary, between things more necessary and less necessary. In all things it would keep first due order, then due measure.

1. Caring most for God's glory, as Moses and Paul did, who cared more for the glory of God than for their own lives, honours, and welfare. Exod. xxxii. 12, 32; Rom. ix. 3. Next, it cares for that one thing needful, how the soul may be saved in the day of the Lord, Luke x. 42. As any thing is best, or more needful for the present, that is cared for first and chiefly, Matt. vi. 33. If all cannot be cared for, the less worthy things, the less necessary for the present, and those things to which we are least bound, should be omitted.

2. As lawful care does, through discretion, keep due order, so it keeps due measure, seeking spiritual and heavenly things with more diligence and zeal than those that are temporal and earthly, 2 Pet. i. 5; caring for the things of this life with great moderation, without eagerness and greediness of desire; always proportioning the care to the goodness and worth of that which is to be cared for. Now because the world is to be loved and used as if we loved and used it not, 1 Cor. vii. 31, 32, it being of little worth in comparison; therefore the cares about it in comparison of the best and most necessary things must be, as if you cared not.

SECT. 2. SIGNS OF IMMODERATE CARE.

Cares of the things of this life are inordinate and immoderate, 1. When they will not give men leave to take the comforts and natural refreshments of this life, Eccles. v. 12, as sleep, meat, and drink, and other needful and lawful things; but especially when they hinder them from the exercise, profitable use, or due performance of religious duties, Matt. vi. 21; xiii. 22; xxii. 5; Ezek. xxxiii. 31, 32.

2. When they are first and chief in a man's thoughts; the mind always running upon them.

3. When they cause a man, out of his eager haste to be rich and to enjoy the world, to use unlawful and indirect means, Prov. xxviii, 20—22, or to engage in dealing and trading beyond his skill, stock, and means well to manage the same.

4. When they cause a man so to mind his worldly business, that he thinks nothing well done, or safe, if his eye or hand be not in it, and if it be not in his own custody; although there is cause why others should be used, and entrusted with it.

Lastly, This holy laudable care is confined within its due measure and bounds, as well as fixed upon its proper objects. It knows its due limits, how far to go, and where to stay: namely, when it has chosen a lawful object, and has found out and used lawful means, 2 Tim. i. 12, and applies itself to one thing as well as another, in due order and measure, it stays there, caring no further; but waits patiently God's pleasure for good success, Psa. xxxvii. 7, casting all care of event and success upon God by prayer and supplication, with thanksgiving.

SECT. 3. THE DUTY OF QUIET TRUST IN GOD.

By what has been said, you may see that although you may and must take thought about many things, according to the directions there given; yet you must, as the apostle says, be careful for nothing, with an anxious, perplexing care.

This is now the matter to be insisted on, viz. That God would have none of his servants and his children to be inordinately careful about any thing; nor yet when in obedience to his commandment, and due observance of his providence, they have diligently used lawful means for the attainment of all lawful things, that they should distress themselves at all about the issue or success. He would not that they should suffer their minds to hang in doubtful suspense and fear about them; but would that they should commit their ways unto him, and trust in him, Psa. xxxvii. 5, whether it be in the matter of their souls, or bodies, of the things of this life, or of that which is to come. God frees them from all carefulness, and would that they should free themselves from it too.

God would have you use all good means for this life, but without taking thought for to-morrow about what you shall eat, what you shall drink, what you shall put on, or what shall become of you and yours another day, Matt. vi. 25—34. He would not have you be so distrustful of him, as to take the care of futurity, the care of success from him, upon yourself, perplexing your heart with doubt and fear till you find it, Luke xii. 22—29. But his will is, that when you have done what you can, with a cheerful and ready mind, you should leave the whole matter of good, or ill success to his care, Psa. lv. 22; 1 Peter v. 7.

In like manner, God would have you to use means to save your soul; but when you have so done, and continue so to do, he would have you care no further. He would not have you to doubt and fear that all shall be in vain, and to no purpose, Psa. lxxiii. 13; or that you shall not be saved notwithstanding. He would not that you should discourage and enfeeble your heart by taking thought about the issue of any trials and temptations that may befall you, before they do come, Matt. x. 19; xxiv. 6, for that is vain; nor yet when they do come, for that is needless.

In such cases, you need only to serve God's providence in the use of the present means of salvation, gaining as much grace and strength as you can against such times, improving that grace and strength which you have in such times of trial: but as to success, either how much grace and comfort you shall have, or when you shall have it, and whether you shall hold out in the time of trial, or be saved in the end, you must not indulge doubtful and distrust ful cares; but must trust God with these things also.

For our Saviour prohibits his disciples all trouble, that might arise through fear of ill success in their Christian course, John xiv. 1, 27. And Paul eases himself of this trouble and fear, committing his soul, and the issue of all his trials unto God, saying, I know whom I have believed, and am persuaded that he is able to keep that which I have committed unto him against that day; 2 Tim. i. 12; iv. 6, 8, 18. He is confident in God for good success in his whole Christian warfare; so should you be.

SECT. 4. REASONS AGAINST ANXIOUS CARE, AND FOR QUIET TRUST IN GOD.

Now to dissuade you from all carefulness, and to persuade you to rest secure in God, concerning the particular events of all actions, and touching the final event and good success of your Christian profession, consider these reasons: (1.) Showing, why you should not care eagerly and inordinately for earthly things; (2.) Why you should not take doubtful or distrustful thought about any thing, whether earthly or heavenly.

(1.) Seriously consider, that all earthly things are of little worth, very fading and transitory, 1 John ii. 17, likened, when they are at best, to the flower of grass, Isa. xl. 6; James i. 10, 11. Wherefore they cannot be worthy of your anxious thought, or careful perplexity about them. It is extreme folly for man, being endued with reason, to set his mind upon that, which is little or nothing worth, nay, which as Solomon calls riches, is not, Prov. xxiii. 5; which is but of short continuance, and only for bodily use, while he has it: which also is given, by God, unto the wicked, even to his enemies, rather than unto the godly, Psa. xvii. 13, 14.

(2.) Inordinate care of earthly things is exceedingly hurtful: for besides that it breeds many foolish and hurtful lusts, which drown men in perdition, 1 Tim. vi. 9, it does hinder the care of things spiritual and heavenly. It causes persons either not to come at all to the means of salvation, Matt. xxii. 5; Luke xiv. 18, &c., or if they come to the word, prayer, sacraments, good company, and good conference, to depart without spiritual profit, Matt. xiii. 22; Ezek. xxxiii. 31. It will cause a man to err from the faith, 1 Tim. vi. 10, and to be altogether unfit for death, and unprepared for his latter end. For when any one part draws more nourishment to itself than it ought, some other parts must needs be hindered in their growth; and when the strength of the ground is spent in nourishing weeds, tares, or corn of little worth, the good wheat is obstructed in its growth, choked, or starved, Phil. iii. 19. "He whose cares are too much about the earth, his cares will be too little about the things of heaven."

Why a man should not be careful about success in any thing.

Next, consider the reasons, why you must not indulge any anxious care about success in your lawful endeavours, any more than by prayer to commend them to God.

First, Because it is to usurp upon God's peculiar right, God's divine prerogative, taking his sole and proper work out of his hands; for care of success, and of what shall be hereafter, is proper to God, 1 Peter v. 7.

Secondly, It is a vain and fruitless thing, when you have diligently used lawful means for any thing, to take thought for success, Psa. cxxvii. 2. For who can by taking thought add any thing to his stature? Luke xii. 25, 26; or make one hair white or black? Matt. v. 36. Understand the same of all other things.

Thirdly, Every day brings its full employment with it, together with its crosses and griefs, Matt. vi. 34; so that you will have full work enough for your care, to endeavour to do the present day's work holily; and to bear each present day's affliction fruitfully and patiently; you have little reason therefore to perplex your heart with taking thought of future events, or of what shall be to-morrow.

Fourthly, It is altogether needless to take thought about the success of your actions, for success is cared for already by God, Matt. vi. 26, 30, 32; one whose care is of more use and consequence than yours can be. You are cared for by one, who loves you better than you can love yourself, who is wise and knows what is best for you, and what you most need, better than yourself; who is always present with you, and is both able and ready to do exceeding abundantly for you, above all that you can ask or think, Eph. iii. 20; even God, who cares for meaner creatures than you are, who also is your God, your heavenly Father, of whose care you have had happy experience, who in times past cared for you, when you could not care for yourself, who has kept you in, and from your mother's womb; who, if you are believers indeed, ordained you to salvation before you had a being, Eph. i. 4; who in due time gave his only begotten Son for you, and to you, Rom. viii. 32, as appears in that now he has given you faith and hope in him, and love to him. It is your God and Father who has commanded, that for the present, and for the future, you should cast your care and burthen on him, 1 Peter v. 7; Psa. lv. 22; having made many gracious promises, that he will care for you, that he will sustain you, and that he will bring your desire to pass, Psa. xxxvii. 5. What wise man then will encumber himself with needless cares?

Fifthly, Carefulness or anxious thoughts about success, proceed from base and cursed causes; namely, ignorance of God, and unbelief and distrust of God, in whomsoever this sin reigns; hence it was that the heathen abounded in this sin, Matt. vi. 32. And by how much this carefulness is indulged by any, though it reign not, by so much he may be said to be of little sound knowledge, and of little faith, Matt. vi. 30.

Sixthly, Carefulness, and doubtful suspense about success in your lawful endeavours, be it whether you or yours shall prosper, or whether you shall profit by the means of grace, or whether you shall be saved in the end, does produce many dangerous and mischievous effects:

1. It will cause you to neglect the proper use of the means of this life, or of that which is to come, 2 Kings vi. 33, according as you doubt of success in either, or if you neglect them not utterly, yet you will have no heart to go about them. For as those that needlessly intermeddle with other person's business, usually neglect their own, so you will be apt to leave your own work undone, when you take God's work out of his hands; and who is he that will take pains about that which he fears will be to no purpose, or labour lost?

2. You will be ready to use unlawful means for any thing when you doubt of success from lawful, Gen. xii. 11—13; xvi. 2; xxvii. 5, 19.

3. Taking thought does divide, distract, overload, and consume the heart and spirits; nothing more.

4. You can never be thankful to God for any thing whereof you fear that you shall have no good success.

5. This anxious thought and distressing fear about success, will deprive you of the comfort of all those good things you have had, and which now you do enjoy.

6. Nothing will bring ill success upon you sooner than unbelieving and distrustful fears about futurity. For when any person shall, notwithstanding the experience he has had, or might have had, of God's power, love, care, and truth of his promises, yet distrustfully care so far, as not to content himself with his own work, so far as prudent care leads him; but also will take God's work, and the burthen of his work upon himself, caring about success, which only belongs to God, and which God only can do, and bear; this folly and presumption does so much provoke God, that it causes him, out of his wise justice to cease caring for such an one, leaving him to his own care, and to his wit, friends, or any other earthly help, to make him by woful experience see and feel, how little any,

or all these, without God, can avail him. Nay, it causes God not only to withdraw his own help, but the help of all things whereon such a man does rely; and what is more, causes them instead of being for him, to be utterly against him. Is it not just with God, that whosoever will not be beholden to God to bear their burthen, but will take it up and bear it themselves, should be made to bear it alone, and to the distress and disquietment of their own hearts?

Wherefore all these things considered, I return to the exhortation, or conclusion before proposed, viz. Commit thy ways unto the Lord, and trust in him. Cast all your care on God; be careful in nothing, Psa. lv. 22.

O! how happy are we Christians, if we did but know, or knowing would enjoy our happiness! We are cared for in every thing that we need, and that can be good for us; we may live without taking thought, or care in any thing. Our work is only to study and endeavour to please God, walking before him in sincerity, and with a perfect heart; then we may cleave to him, and rest on him both for our bodies and souls without fear or distraction, 1 Cor. vii. 35. God is all-sufficient, and all in all to such; he is known by his name Jehovah to such, Exod. vi. 3; even to the being the accomplisher of his promises to them. If we shall wisely and diligently care to do our work, we, serving so good and so able a Master, need not take thought about our wages. If we would make it our care to obey and please so good, and so rich, and bountiful a Father; we need not be careful for our maintenance here, in our minority and non-age; nor yet for our eternal inheritance, when we shall come to full age. We in this holy security and freedom from carefulness, if we are not wanting to ourselves, might live in an heaven upon earth; and that not only when we have means, for even then our security is in God, not in the means, but when to the eye of flesh we have no means: for God is above, and more than all means.

SECT. 5. MEANS TO ATTAIN QUIETING CONFIDENCE IN GOD, AND FREEDOM FROM PERPLEXING CARES.

THAT you may leave anxious caring, and be brought to cast all your care on God,

(1.) Deny yourself, and your own wisdom, Prov. xxiii. 4; be not wise in your own conceit, nor presumptuous of your wit, skill, or means.

(2.) Get sound knowledge, faith, hope, and confidence in God, Rom. viii. 32; live by faith, for the preservation both of body and soul, Heb. x. 38, 39. Get not only faith in his promise, but in his providence also. When you shall see no way or means of gaining the good you desire, or of keeping you from the evil which you fear, or of delivering you from the evil you feel, then call to mind, not only the promises of God, viz. I am with you, Joshua i. 5; I will not leave nor forsake you, Heb. xiii. 5; all things work together for good, Rom. viii. 28; and many such like; but believe also that God will provide means to bring to pass what he has promised, though yet you see not how. When you can say, with faithful Abraham, God will provide, Gen. xxii. 8, it will cast out fear and doubt. But if, with Abraham, Gen. xii. 11; xvi. 2, you believe God's promises in the main, but not God's providence in the means; you will then be tempted to seek out, and use unlawful means to obtain the thing promised, as he did; or faint in waiting, as many others have done. For we see the like in David, 1 Sam. xxvi. 10, 11, when he had faith in God's providence, he could say of Saul, The Lord shall smite him, or his day shall come to die, or he shall descend into the battle and perish. The Lord forbid that I should stretch forth my hand against the Lord's anointed. But when he doubted of God's providence, then he saith, I shall now perish one day by the hand of Saul, 1 Sam. xxvii. 1.

(3.) Give all diligence to make your calling and election sure; for when you know assuredly that God is your heavenly Father, and Christ Jesus your Re deemer, and that you are of his family, having your name written in heaven, you then will easily free your heart from being troubled with fear and restless care, John xiv. 1, 2; being sure that your heavenly Father and Saviour does and will provide for you.

(4.) Lastly, you must often renew your acts of faith on God, his promises and providence, casting all your care on him; making your request known to God by prayer and supplication, for what you would have; being heartily thankful for what you have had, now have, and hope to have hereafter. Then the peace of God which passes all understanding, Phil. iv. 6, 7, shall keep your heart and mind from vexing thoughts, and heart-distressing fears, and that, in and through Christ Jesus; of which peace I intend next to speak.

CHAPTER XIII.

OF PEACE WITH GOD.

THE NATURE AND EXCELLENCY OF PEACE WITH GOD.

THAT you may be persuaded to walk before God in uprightness, in all well-pleasing, and to live without taking anxious thought about any thing, casting your care on God according to the former directions; God has assured you that peace shall be upon you, Gal. vi. 16, even that peace of God which passes all understanding, which shall keep your heart and mind through Christ Jesus, Phil. iv. 6, 7, if you thus do.

Peace and quiet is most desirable. All things that have motion desire it as their perfection: bodily things enjoy it by their rest in their places; reasonable things enjoy this peace in the quiet of their mind and heart, when they have their desires satisfied, being freed from such opposition as might disquiet them.

Peace is a true agreement and concord between persons or things, whereby not only all enmity is laid aside, and all injuries are forborne; but all amity is entered into, and all readiness of communicating and doing good to each other is showed.

Natural peace is of great price, and very much to be desired, for the exceeding great benefit which it brings to the body, family, and state. But the peace of which I am to speak, which is promised to all who walk with God according to the rule of faith, and of the new creature, Gal. vi. 15, 16, casting their care on God, exceeds all other peace, as far as the soul, heaven, and eternity exceed the body, the earth, and a moment of time. Which will easily appear, if you shall observe by what motives and arguments the Holy Ghost does commend, and set this forth unto you, Phil. iv. 7. It has its commendation above all other peace in three respects:

First, In respect of the excellency of the person, with whom and from whom it is, namely, God; therefore it is called the peace of God, Phil. iv. 7. It is so called, (1.) Because it has God for its object; it is a peace with God. (2.) Because God by his Spirit is the author of it: it is peace from God, a peace which God gives; such a peace which the world neither can, nor will give, John xiv. 27.

Secondly, This peace is commended in respect of the unspeakable and inconceivable goodness and worth that is in it. It passes all understanding; and this it does, not only because unsanctified men are mere strangers to it, and understand it not; but because regenerate men, to whom it belongs, and in whom it is, even they when God gives them any lively feeling of it, find it to be such a peace as they could not imagine it to be before they felt it. For they cannot so distinctly and fully conceive the transcendent excellency of it, as by any means fitly to describe it. It rather takes up the mind into an holy rapture, unto admiration of what it sees, and of what it perceives is yet to be known, than possibly can be distinctly and fully comprehended or expressed by mind or tongue. It is with them that feel it in any special degree, as it was with the queen of the South, when she saw Solomon's wisdom, 1 Kings x. 4—7 She had a great opinion of Solomon's wisdom, by what she had heard; but when she saw it, she was stricken with such admiration, that it is said she had no more spirit in her; his wisdom was so much beyond her expectation, that she breaks out into words of admiration, saying the half was not told her of Solomon's wisdom, it exceeds the fame thereof; so does the peace of God, being, like the dimensions of the love of Christ, the root thereof, and like the ravishing joy of Christians, the fruit thereof, surpassing all full and distinct knowledge, and all means of full and clear expression, Eph. iii. 18, 19; being, as the Holy Ghost also saith, unspeakable, 1 Peter i. 8. This peace is included amongst those other graces and gifts accompanying the gospel, which are such as eye has not seen, nor ear heard, nor have entered into the heart of man, so as clearly to perceive them, or fully to express them, 1 Cor. ii. 9.

Thirdly, This peace is commended in respect of the excellent effect thereof, which is a proof that it passes understanding, namely, it keeps the heart and mind, in and through Christ Jesus.

This is an excellent and most useful effect on man's behalf; for it supplies the place and office of a castle or strong garrison, 2 Cor. xi. 32, as the original signifies, to keep the principal forts of the soul from being surprised or annoyed, either by invasion from without, or by insurrection from within.

The parts of man, which are kept by this peace of God, are the heart and mind; by heart is meant the will and affections; by mind, the power of thinking and understanding. For true peace of God does fill the heart with such joy, patience, hope, and comfort in believing, that it keeps it from heart-vexing grief, fear, distrust and despair. It likewise fills the mind so full of apprehension of God's favour, fidelity, and love, that it makes it rest secure in God, and delivers it from distress of mind, or anxious cares about any thing; keeping out the dominion of all perplexing and distrustful thoughts.

The strength which this peace has, whereby it keeps the heart and mind as with a garrison, is impregnable. It is derived from Christ, it has it in and from Christ: the text saith, through Christ, that is, through the power of Christ's Spirit. For as we are kept by faith, from which this peace springs, as with a strong garrison, by the power of God to salvation, 1 Peter i. 5; so, by the same power of Christ, our hearts and minds are kept by the peace of God, as with a garrison, from discouraging, distracting, and uncomfortable thoughts. For what is this peace else but a beam from the object of our faith, proceeding from the love of God to us ward, and the fruit of faith, as we feel it wrought in us by God?

This peace of God is two-fold, or one and the same in different degrees.

The first is an actual entering into, and mutual embracing of peace between God and man.

The second is the manifestation and expression of this peace.

The first is when God and man are made friends; which is, when God is pacified towards man, and when man is reconciled unto God, so that now God stands well affected towards men, and man has put off enmities against God; which mutual atonement and friendship, Christ Jesus, the only mediator between God and man, 1 Tim. ii. 5, has, by his satisfaction and intercession wrought for man, and by his Spirit applies unto, and works in man. For until this atonement be applied, God, in his just judgment and holy displeasure, is an enemy unto man for sin, Psa. v. 5; Rom. v. 10; and man in his evil mind, and unjust hatred is an enemy unto God, Col. i. 21, and unto all goodness, through sin.

The first peace, is peace of God with man, inherent in God, working the like disposition of peace in man towards God; and is the fountain from which the second floweth.

The second kind, or rather further degree, of peace of God, is the operation and manifestation of the former peace, which is a peace of God in man wrought by the Spirit of God, through the apprehension that God is at peace with him.

This peace is partly and most sensibly in the conscience, which is called peace of conscience, and may also be called peace of justification, according to that, being justified by faith, we have peace with God, &c., Rom. v. 1. And it is partly in the whole reasonable man, whereby the will and affections of the soul agree within themselves, and are subject to the enlightened mind, conspiring all of them against the common adversaries of God and the soul, i. e., the flesh and the devil: this may be called peace of sanctification: according to that of the apostle, Rom. vi. 22; being made free from sin, and become servants of God, you have your fruit unto holiness. This is the agreement of all the members, to become servants to righteousness unto holiness, Rom. vi. 19. Not but there will be warring always in our members, but it is not the warring so much of one member against another, as the warring of the flesh in every member against the Spirit, which Spirit also wars against the flesh. This conflict between the flesh and the Spirit, beginning in man, as soon as the Spirit has wrought the peace of holiness, in setting the soul in order.

Moreover, this peace of sanctification consists in this, that although a Christian must never be, nor ever is at peace with sin, so that it does not assault and molest him, or that he should subject himself to it, or have it absolutely subject to him in this life, yet he has a peace and quiet, in comparison, from sin, in as much as he is freed from the dominion and power of sin, Rom. vi. 14, 22, to condemn him, or to reduce him to his former bondage unto sin. Now so far as a man gets a conquest over his lusts, that they are kept under, and forbear to assault and molest him, so far he may be said to have this peace of sanctification.

The conscience, when it is awakened in the act of accusing and condemning man for sin, does withal prick, (Acts ii. 37; Prov. xviii. 14;) sting, and wound the heart with unutterable and inconceivable griefs, fears, and terrors, through the apprehension of God's infinite, eternal, and just wrath for sin.

Now, when God, by his Spirit, Rom. v. 1—5, gives any true hope and assurance unto a man that his justice is satisfied concerning him, through Christ: and that now all enmity and wrath is done away on God's part; and that he loves him in Christ, with a free, full, and everlasting love, Rom. viii. 16; hereby he speaks peace to the conscience, having done away all the guilt of sin, which before molested it through sense of God's anger and fear of punishment. Hence arises peace and comfort in the conscience, which therefore is called peace of conscience. Thus the mind ceases to be perplexed, and, by faith in Christ's death, through the Spirit, becomes quiet with an heavenly tranquillity, resting on the word of promise, and according to the measure of clear apprehension of God's love in Christ, in the same measure is at sweet agreement within itself, without fear or trouble, John xiv. 27; and in the same measure he has peace of conscience, flowing from the assurance of justification.

As soon also as a man begins actually to be at peace with God, his lusts do begin to be at war with him, rebelling against the law of his mind, which yet by little and little shall be subdued and conquered; which conquest, though it be imperfect in this life; yet by virtue of the peace now made with God, if he will improve it by seeking help of God, and taking to him the complete armour, fighting manfully under Christ's banner, Eph. vi. 10, he may so prevail against them, that they do not so often, nor so strongly, assault him as in former times. Now, so far as the powers and faculties of man agree in their fight against sin, and subdue it, that it does not assault and molest him, he may be said to have the peace of sanctification.

The first peace whereby God is pacified, and is

become propitious and gracious to man, is absolutely necessary to the being of a Christian.

The second, which rests from the manifestation of this peace unto man, and the sensible feeling of the operation of this peace in man, is not necessary to the being of a Christian, at least in a sensible degree of it, but to the well-being of a Christian it is necessary. For a man may be in the favour of God, and yet be without the sense of this peace in himself: because this peace of conscience does not flow necessarily from the being in God's favour, but from knowledge and assurance of being in his favour.

Now, a man, in many cases, may lose for a time his sense of God's favour, his faith being over clouded with fears and unbelief, as it was with David, after his adultery, &c. Psa. li. 11, 12, who yet was upheld secretly by his right hand, as the Psalmist was in another case, Psa. lxxiii. 23, by virtue of that first peace of God, yet, until God gave him the sense and feeling of his loving countenance, he could not enjoy the comfort of it; yea, though God by Nathan in the outward ministry of his word, had given him assurance of God's loving-kindness, saying, The Lord has put away thy sin, thou shalt not die, 2 Sam. xii. 13.

That first peace is absolute, and admits of no degree.

The second, which flows thence, both in respect of peace of conscience, and in respect of good agreement of the powers and faculties of man within themselves, and of freedom from assaults and molestations either of Satan from without, or from lusts within, is not absolute; but admits of several degrees. In the life to come this latter peace shall be perfect: for then all believers shall be perfectly freed from all trouble of conscience, and from all molestation by temptations; their victory shall be complete. But in this life their peace is but imperfect; it is true for substance, but is more or less, as the light they have received is more clear or dim; and as grace in them is more strong or more weak.

For although man's justification is absolute, and

admits not of degrees, yet the assurance of it, whereby a man has peace of conscience, is more or less according to the measure of his clear sight of Christ's love, and evidence of his faith. Hence it is that the dear children of God have interruptions and intermissions in their peace; have sometimes much peace, sometimes little or no peace; according as they have intermissions in their assurance of God's favour.

Thus it was with David and Asaph; sometimes his heart was quiet, and his spirit was glad, in assurance that his soul should rest in hope, Psa. xvi. 9; at other times, his soul was cast down and disquieted in him, Psa. xlii. 11, thinking that he was cast out of God's sight, Psa. xxxi. 22, fearing that God would show no more favour, Psa. lxxvii. 7. Yea, he was so perplexed, that he did almost faint, and his eyes failed with waiting for God, Psa. lxix. 3. For since the best assurance of believers is exercised with combating against doubting, their truest and best peace must needs be assaulted with disquiet. And as it is with a ship at anchor, so is the most stable peace of a Christian in this life, who has his hope as an anchor of his soul, sure and steadfast, Heb. vi. 19; who, though he cannot make utter shipwreck, yet he may be grievously tossed and affrighted with the waves and billows of manifold temptations and fears. Likewise, though peace of sanctification be true, yet it must needs be more or less, according as any man grows or decreases in holiness, and as God shall please to restrain his spiritual enemies, or give power to subdue them, more or less.

Now the peace of God, both in him to man, and from him manifested and wrought in man, does pass all understanding, and serves to keep the heart and mind of him that walks with God, and rests on him through Christ.

This peace it is which you must seek for, and embrace in believing, and if you would have true comfort and tranquillity in your mind, labour especially to get and keep the peace of a good conscience, which seems to be the peace that is chiefly, though not only, intended in this text.

SECT. 2 FURTHER EXCELLENCIES OF THE PEACE OF GOD.

That you may be induced with all diligence and earnestness to seek after this blessed peace, and may better perceive that this peace of God, for worth and use, passes all understanding, take these reasons in particular.

First, That must needs be an excellent peace which God will please to take into his holy title, calling himself the God of peace, Heb. xiii. 20, calling Christ the Prince of peace, Isa. ix. 6.

Secondly, That peace must needs be of infinite value, passing all understanding, for which Christ gave himself; paying the price of his own most precious blood for it, 1 Peter i. 18, 19.

Thirdly, This peace cannot but pass all understanding, because the cause from whence it comes, namely, Christ's love, Eph. iii. 18, 19, and the effect which it works, namely, joy in the Holy Ghost, 1 Peter i. 8, do, as the apostles affirm, pass knowledge, and are unspeakable.

Fourthly, This peace was that first congratulation, wherewith the holy angels saluted the church, at Christ's birth, giving her joy in her new-born husband and Saviour, Luke ii. 10, 11, 14. And it was that special legacy which Christ Jesus did bequeath to his church, leaving that as the best token of his love to it, a little before his death: saying, My peace I leave with you, John xiv. 27.

Fifthly, This peace is one of the principal parts of the kingdom of God, which consists, as the apostle says, of righteousness, peace, and joy in the Holy Ghost, Rom. xiv. 17.

Sixthly, By as much as the evils and mischiefs that come to a man by having God to be his enemy, which draws upon him God's wrath, justice, power, and all God's creatures to be against him: and by as much as the grievous and intolerable anguish of the wounded spirit passes understanding; by so much the peace of God, which frees him from all these, must of necessity, pass all understanding, Prov. xviii. 14.

Now that it is a fearful thing to have God to be an enemy, it is said, He is a consuming fire, Heb. xii. 29, and it is a fearful thing to fall into the hands of the living God, Heb. x. 21. It appears likewise by Christ's compassion and grief for Jerusalem, who neglected the time of making and accepting of peace with God; for he wept over it, and said, Luke xix. 42, "If thou hadst known, even thou, at least in this thy day, the things which belong to thy peace! but now they are hid from thine eyes." But what it is to have God to be an enemy, is seen most fully by Christ's trouble and grief in his passion and agony in the garden, and in the extremity of his conflict with God's wrath on the cross, when God showed himself to be an enemy, and did for man's sin pour on him the fierceness of his wrath. It made him, though he was God, being man, to sweat, for very anguish, as it were drops of blood, Luke xxii. 44, and to cry, If it be possible, let this cup pass, Matt. xxvi. 39, and, My God, my God, why hast thou forsaken me? Matt. xxvii. 46.

Moreover, if you do observe the complaints of such distressed souls that have had terror of conscience, if you have not had experience thereof in yourself, how that they were at their wits' end, pricked at heart, as it were with the point of a spear, or sting of a serpent, Acts ii. 37, pained like men whose bones are broken and out of joint, Psa. li. 8, making them to roar, and to consume their spirits for very heaviness, Psa. xxxii. 4, then you will say that peace of conscience does pass all understanding.

Seventhly, When God and a man's own conscience are for him, and God's grace in some good measure has subdued sin and Satan in him, this brings with it assurance that all other things, whose peace are worth having, are also at peace with him, Hosea ii. 18—20. For if God be for us, who can be against us? Rom. viii. 31, 32. This peace must of necessity bring with it all things which will make us happy, even all things which pertain to life, godliness, and glory, 2 Peter i. 3.

Lastly, Consider this, that as the worth and sense of peace with God is unutterable and inconceivable, so the time of it is indeterminable, it is everlasting, and has no end, Isa. ix. 7. Compare this with the former, and it cannot be denied, but that the peace of God does every way pass understanding.

CHAPTER XIV.

CONCERNING THE IMPEDIMENTS TO PEACE: FALSE HOPES, AND FALSE FEARS.

THE KINDS OF IMPEDIMENTS THAT HINDER PEACE.

First, If you would enjoy this happy peace, you must remove and avoid the impediments. Secondly, You must use all helps and furtherances which serve to procure and keep it.

I reduce the impediments unto two heads.

First, A false opinion and hope that all is well with a man, and that all shall be well with him in respect of his salvation, when yet indeed God is not reconciled to him. Hence will follow a quietness of heart, somewhat like to peace of conscience; which yet is but a false peace.

Secondly, Causeless doubting, and false fear, that a man's estate with respect to his salvation is not good; although God be indeed at peace with him; hence follows trouble and anguish of heart, somewhat like unto that of hellish despair, disturbing his true peace.

Either of these do hinder peace.

The first hinders the having, the second hinders the feeling and comfortable enjoying of peace.

It has been an old device of Satan, when he would keep any man from that which is true, to obtrude upon him that which shall seem to be true, but is false. Thus he did in the first calling of the Jews, Matt. xxiv. 5. When he saw they had an expectation of the true Christ, he, to divert and seduce them from the true Christ, sets up false Christs. Even so in the matter of peace: if he can so delude men that they shall content themselves with a false peace, he knows that they will never seek for that which is true. It is a common practice with the devil to endeavour to make all who are not in a state of grace, to presume that they are.

Also, such is his cunning and malice, that when any man is in the state of grace, he will labour by all means to distress and perplex the soul with unreasonable fears and suspicions, to make that estate doubtful and uncomfortable, to vex and to weary him, if he cannot drive him to despair, 2 Cor. ii. 7, 11. Now the heart of man, so far as it is unsanctified, being deceitful above all things, Jer. xvii. 9, is most apt to yield to Satan in both these cases. Whence it is, that there are very many who boast of much peace, and yet have none of it. And many fear they have no peace, who yet have much of it.

Wherefore the rule is, "Believe not either your deceitful heart, nor the devil, when they tell you either that you are in a state of salvation, or in a state of damnation: but believe the Scripture, what it saith in either."

You may know when these persuasions come from your deceitful heart, or from the devil, thus:

First, If the means to persuade you to either be from false grounds, or from misapplication of true grounds.

Secondly, If the conclusions, inferred from either persuasion, be to keep you in a sinful course, and to keep you, or to drive you from God, as if you need not be so strict in godliness, or that now it is in vain, or too late to turn and seek unto God; then it is from

Satan and from a deceitful heart, and you must not believe them. But if these persuasions be from a right application of true grounds, and do produce these good effects, to drive you to God, in praise or prayer, and unto a care to please God, they are from his gracious Spirit.

SECT. 2. THE CAUSES OF PRESUMPTION, OR FALSE PEACE.

The false peace and evil quiet of conscience arises from these three causes:

First, From gross ignorance of the danger wherein a man lives because of sin, Eph. iv. 18, 19, whence follows a blind conscience.

Secondly, From groundless security and presumption that all shall be well with him, notwithstanding that he knows he has sinned, Deut. xxix. 19, and knows that sin is damnable; whence he has a deluded conscience.

Thirdly, From obstinacy, through delight and custom in sin, Jer. xliv. 16, 17, whence comes hardness and insensibility of heart, which is a seared conscience.

Wheresoever any of these evils reign, although God has said there is no peace to the wicked, Isa. lvii. 21, that is, no true peace; yet such fear no evil; but promise to themselves peace and safety, 1 Thess. v. 3, like those of whom the prophet spake, who had made a covenant with death, and with hell were at an agreement, Isa. xxviii. 15. Yea, though they hear all the curses against sinners, which are in God's book denounced against them; yet will they bless themselves in their heart, and say, they shall have peace, though they walk in the stubbornness of their hearts, Deut. xxix. 19. But whosoever is thus quiet in himself through a false peace, it is a sign that the strong man keeps the house, Luke xi. 21, and that he, continuing in this fool's paradise, is not far from sudden and fearful destruction from the Almighty, 1 Thess. v. 3. Deut. xxix. 20.

Whosoever therefore would have true peace of God, must know and be thoroughly convinced that by nature, by reason of Adam's first transgression, which is justly imputed to him, Rom. v. 12, and because of his own inherent wickedness of heart and life, Rom. vii. 18; Psa. li. 3, 5, of omission and commission, in thought, word, and deed, he is in a state of sin and condemnation, having God for his enemy, yea, is an heir of wrath, Eph. ii. 3, and of eternal vengeance of hell-fire: according to that of the apostle, All have sinned, and are become guilty before God, and have come short of the glory of God, Rom. iii. 19, 23. Ignorance of danger may give quiet to the mind for a time, but it can give no safety. Is not he foolishly secure that rests quietly in a ruinous house, not knowing his danger, until it fall upon him? Whereas, if he had known it, he would have had more fear and disquiet; but less danger.

SECT. 3. GROUNDS OF FALSE HOPE DISCOVERED AND REMOVED.

LET no man presume upon weak and false grounds, that he shall escape the vengeance of hell, or attain to the happiness of heaven. How weak and vain are the foundations on which many build their hopes of salvation! and from thence their peace will appear by that which follows.

1. Some think that because God made them, surely he will not damn them. True, if they should have continued good, as he made them. God made the devil good, yea an excellent creature, yet, who knows not, that he shall be damned? Matt. xxv. 41. If God spared not his holy angels, Jude 6, after that they became sinful; shall man think that he will spare him? A sinful man shall be judged at the last day, not according to what he was by God's first making; but as he shall be found defiled and corrupted by the devil, and by his own lusts. When Judah became a people of no understanding, it is said, He who made them

will show them no mercy, and he that formed them will show them no favour, Isa. xxvii. 11. Thus it is spoken to every sinner remaining in his sin, notwithstanding that God made him.

2. Some say their afflictions have been so many, so great, and so lasting, that they hope they have had their hell in this life; whence it is that their hearts are quiet in respect of any fear of wrath and judgment at the last day.

I would ask such, whether they, being thus afflicted, have returned to God that smote them, Isa. ix. 13; and whether their afflictions have made them better; or whether, like Solomon's fool brayed in a mortar, Prov. xxvii. 22, their sin and folly is not departed from them? If so, they must know, the more they have been, and now are afflicted, if they be not reformed by it, this does presage that there is the more and worse behind; as it was in the case of Judah, Isa. i. 5. v. 12—14; Amos iv. 6—13. Many have been often and extremely corrected by their parents, &c. yet, remaining incorrigible, have at last suffered public execution.

3. Some, though their ways be never so evil, yet because to them God's judgments are far above, out of their sight, and because they have no changes, Psa. lv. 19; Eccles. viii. 11, God forbearing to execute his judgments upon them speedily, they persuade themselves that God sees not, or that he is not angry with them, or that he regards not, Psa. x. 6, 11, 13, and that he will neither do good nor bad, thinking that God has forgotten, or that he is like them, Psa. l. 21, well enough pleased with them; hereby they lay their consciences asleep, promising unto themselves immunity from punishment, and that they shall never be moved, Psa. x. 6.

Know ye, that God's forbearance of his wrath is not because he sees not, or because he has forgotten, or regarded not your wickedness; but because he would give you time and means of repentance; it is because he would not have you perish, but come to repentance, that you may be saved, 2 Peter iii. 9;

which if you do not, this his bounty and long-suffering makes way for his justice, and serves to leave you without excuse; and to heap up wrath for you against the day of judgment, the day of the revelation of the just judgment of God, Rom. ii. 4—6, who shall render to every man according to his works. For God knows how to reserve the wicked to the day of judgment, to be punished, 2 Peter ii. 9. He will take his time to hear and afflict you, Psa. lv. 19, when he shall set all the sins of you that forget him, in order before you, Psa. l. 21, 22; then, if your speedy repentance do not now prevent it, he will tear you in pieces when there shall be none to deliver. The longer he was in fetching his blow, the more deadly will his stroke be when it comes. Many malefactors are not so much as called at a petty sessions, when less offenders are both called and punished; yet have they no cause to promise safety to themselves, for they are reserved for a more solemn trial, and execution, at the grand assizes. So wicked men that are not afflicted here, are reserved for the last judgment, at the great and terrible day of the Lord.

4. There are some who hope that God does love them, and that he does intend to save them; for they prosper in every thing, and are not in trouble and distress as other men; hereupon their consciences are quiet, and without fear.

Let me tell you who thus think, that this is a poor foundation to build your hope upon. What are you the better for your prosperity? Are you more thankful and more obedient? Do you the more good, by as much as you prosper more? If so, well; if not, know, as Solomon, by the spirit of truth, tells you that no man can know God's love or hatred by all that is before him; be it prosperity or adversity, Eccles. ix. 1, 2. In these things there may be one and the same event to the righteous and to the wicked. Know, moreover, that the wicked, for the most part, thrive most in this world; God giving them their portion in this life, Psa. xvii. 14, wherewith they nourish themselves against the day of slaughter, making their own table their snare, Psa. lxix. 22, and their prosperity their ruin, Prov. i. 32.

5. There are many who compare themselves with themselves, passing by their own manifold sins, looking only upon their own hypocritical and civil good purposes and deeds; comparing also their sins with the notorious sins of God's people committed before their conversion, Luke xviii. 11, and with the gross sins of Noah, Abraham, Lot, Peter, and other godly men, after conversion. They hence conclude, that since such are saved, they must entertain a good opinion of themselves, and hope they shall be saved; they think that all is well with them, being such of whom our Saviour speaks that need no repentance, Luke xv. 7.

I would have these to know, that they who thus compare themselves with themselves are not wise, 2 Cor. x. 12; and they that think well of themselves, and commend themselves, are not approved, 2 Cor. x. 18; but those only whom the Lord commends. Moreover, the slips and falls of the people of God, both before and after conversion, did serve for their own humbling, and for a warning to all that should hear thereof. God knows how to reprove and chasten his own that offend, giving them repentance to life and salvation; and yet justly will condemn all those that shall presumptuously stumble at their falls, and wilfully lie in their sins, being fallen. It is not safe following the best men in all their actions, for in many things they sin all, James iii. 2, not only before, but after conversion. And as the cloud that guided the Israelites, Exod. xiv. 20, had two sides, the one bright and shining, the other black and dark, such is the cloud of examples of godly men. Those who will be directed by the light side thereof, shall, with the children of Israel, pass safely towards the heavenly Canaan; but those that will follow the dark side shall all perish with the Egyptians in the Red Sea of destruction. Whatsoever any were before conversion, or whatsoever gross sin they fall into after conversion, if they are humble and truly penitent, none of them are laid to their charge, because they are done away by Christ Jesus. These are in better state than those who for matter never committed so great sins, if pharisee-like they repent not of their lesser sins, as they esteem them, and are proud of their supposed goodness and well doing. For God, in justifying the humble publican rather than the proud pharisee, Luke xviii. 10, 11, shows that proud innocency is always worse than humble guiltiness.

6. There are likewise some others who are guilty to themselves of damnable sins, yet hope to be saved by the goodness of other men, by pardons from the Pope, by absolutions of priests, and by certain penitential external acts of their own, and by good works, such as alms, &c. These, if they may hope of the Pope's indulgences, and a priest's absolution, if they fulfil their penance enjoined, if they are devout in certain superstitions, in their will-worship and voluntary religion, Col. ii. 18, their conscience is quiet for a time, notwithstanding their foul and black sins, even their abominable idolatries.

I make known to these, that all this is but a blindfolding, smothering, and stupefying the conscience for a time, laying a double, and a far greater guilt upon it, and is far from being any means truly to pacify it. For how can a man have true peace from any, or from all such actions as are in themselves an actual denying of the true head of the church, Jesus Christ, Col. ii. 19, and are a cleaving to a false head, which is antichrist? And how can any man merit for himself, when our Saviour saith, Luke xvii. 10, He who has done all that is commanded, is an unprofitable servant, and has done but his duty, which thing he must say and acknowledge? All these before-mentioned build their hopes upon false grounds. Those that follow build their presumptuous and false hopes upon a misapplication of true grounds.

7. Many acknowledge that they have sinned and do deserve eternal damnation; but they say God is merciful, therefore their heart is quiet, without all fear of condemnation.

It is true, that God is most merciful; but how? Know, he is not necessarily merciful, as if he could not choose but show it to all men. He is voluntarily merciful, showing mercy only to those unto whom he will show mercy, Rom. ix. 18. God could and did hate, and in his justice condemned Esau, Rom. ix. 13, notwithstanding his love and mercy to Jacob. God is all justice, as well as all mercy; but he has his several objects of justice and mercy, and has his several vessels of wrath and mercy, Rom. ix. 22, 23, into which respectively he does pour his wrath or mercy. When God speaks of obstinate sinners, he says, That he will not be merciful to their iniquities, Isa. xxvii. 11; and again, He that made them will not have mercy on them. And David prays with a prophetical spirit, saying to God, Be not merciful to wicked transgressors, Psa. lix. 5; and who are these, but such as hate to be reformed, Psa. l. 17, 22, who are presumptuous, and turn the grace of God into wantonness, Jude 4? Nay, concerning them that always err in their heart, he has in effect sworn that he will show them no mercy; for he has sworn that they shall not enter into his rest, Heb. iii. 10, 11.

8. Some others go further; they acknowledge that God's justice must be satisfied, and they think it is satisfied for them, dreaming of universal redemption, by Christ, who indeed is said to die to take away the sins of the world, John i. 29. This causes their conscience to be quiet, notwithstanding that they live in sin.

It must be granted, that Christ gave himself a ransom for all, 1 Tim. ii. 6. This ransom may be called general, and for all, in some sense: but how? namely, in respect of the common nature of man, which he took, and of the common cause of mankind, which he undertook; and in itself it was of sufficient price to redeem all men; and because applicable to all, without exception, by the preaching and ministry of the gospel. And it was so intended by Christ, that the plaster should be as large as the sore, and that there should be no defect in the remedy, that is, in the price, or sacrifice of himself offered upon the cross, by which man should be saved, but that all men, and each particular man, might in that respect become salvable by Christ.

Yet does not the salvation of all men necessarily follow hereupon; nor must any part of the price which Christ paid, be held to be superfluous, though many be not saved by it.

But know, that the application of the remedy, and the actual fruit of this all-sufficient ransom, redounds to those who are saved only by that way and means which God was pleased to appoint, which in the case of adults, is faith, John iii. 16, i. 12, by which Christ is actually applied. Which condition, many, to whom the gospel does come, make impossible to themselves, through a wilful refusal of the gospel, and salvation itself by Christ, upon those terms which God does offer it.

Upon this sufficiency of Christ's ransom, and intention of God and Christ, that it should be sufficient to save all, is founded that general offer of Christ to all and to each particular person, to whom the Lord shall be pleased to reveal the gospel, Matt. xvi. 15, xxviii. 19; likewise that universal precept of the gospel, commanding every man to repent, and believe in Christ Jesus, Matt. iii. 2, 7, 8; Mark i. 15; Acts xvii. 30; as also the universal promise of salvation, made to every one that shall believe in Christ Jesus, John iii. 16.

Although, in one sense, it is true, Christ may be said to have died for all, yet let no one think to enjoy the benefits of his precious death and sacrifice, without serious diligence to make their calling and election sure. For God did intend this all-sufficient price for all, otherwise to his elect in Christ, than to those whom he passed by and did not elect; for he intended this not only out of a general and common love to mankind, but out of a peculiar love to his elect. He gave not Christ equally and alike to save all; and Christ did not so lay down his life for the reprobate as for the elect. Christ so died for all, that his death might be applicable to all. He so died for the elect, that his death might be actually applied unto them. He so died for all, that they might have an object of faith, and that if they should believe in Christ, they might be saved. But he so died for the elect that they might actually believe, and be saved. Hence it is that Christ's death becomes effectual to them, and not to the other, though sufficient for all. Now that many believe not, they having the means of faith, the fault is in themselves, Matt. xiii. 14, 15; Acts xxviii. 26, 27; Isa. vi. 9; through their wilfulness or negligence; but that any believe to salvation, is of God's grace, Matt. xiii. 11, attending his election, Acts xiii. 48, and Christ's dying out of his especial love for them; and not of the power of man's free-will: God sending his gospel, and giving the grace of faith and new obedience to those whom of his free grace he has ordained to eternal life, both where he pleases and when he pleases, John iii. 8.

Furthermore, it must be considered that notwithstanding the all-sufficiency of Christ's death, whereby the new covenant of grace is ratified and confirmed, the covenant is not absolute, but conditional. Now what God proposes conditionally, no man must take absolutely. For God has not said that all men without exception shall be saved by Christ's death: although he saith, Christ died for all; but salvation is promised to those only who repent and believe, Mark i. 15, xvi. 6.

Wherefore, notwithstanding Christ's infinite merit, whereby he satisfied for mankind; and notwithstanding the universality of the offer of salvation to all to whom the gospel is preached; both Scripture and experience show, that not all, nor yet the most, shall be saved, and that because the number of them who repent, and unfeignedly believe, whereby they make particular and actual application of Christ and his merits to themselves, are fewest. For of those many that are called, few are chosen, Matt. xx. 16. Wherefore let none ignorantly dream of an absolute, universal redemption, as many simple people do. For though Christ be said to suffer to take away the sins of the whole world, John i. 29; 1 John ii. 2; yet the Scripture saith, that the whole world of unbelievers and of ungodly men shall perish eternally, 2 Peter ii. 5; Jude 14, 15.

9. Many will yield that they must have faith and repentance, and that they must be ingrafted into Christ and become new creatures else they cannot hope to be saved; but they think they are all this already; whence follows quiet of conscience. Whereas when it comes to the trial, their faith and repentance are found not to be sound. As will thus appear:

They think they have faith, (1.) Because they believe the whole Scripture to be the good word of God. (2.) They believe not only that there is a God, but that Jesus Christ is the Son of God, and Saviour of the world, yea, according to the letter, they believe all the articles of the Christian faith. (3.) They think they are believers, because they have been baptized, and have given their names unto Christ. They profess the only true religion, they have the very true form of godliness in all the external exercises of religion, Luke xiii. 26. Whereas if they believe no more, nor better, they may know that their faith is only an historical and general faith, or only a temporary faith at the best, necessary indeed to salvation, but not sufficient to save. The devils believe as much as the first, James ii. 29, and very hypocrites may, and do profess and do as much as the second and third. The apostle Paul, having to do with hypocritical Jews, who because of their form of knowledge, and profession, though without practice, did nourish in themselves a vain persuasion that they should be saved; removed this false ground of their hope, thus, saying, He is not a Jew who is one outwardly, but he is a Jew which is one inwardly; neither is that circumcision which is outward in the flesh, but that which is of the heart, in the spirit and not in the letter, whose praise is not of men but of God, Rom. ii. 28, 29. In like manner, Peter assures all Christians, that the baptism, which is only a putting away of the filth of the flesh, does not save, 1 Peter iii. 21; but that baptism which gives proof that the heart is sprinkled from an evil conscience, as well as the body washed with pure water, Heb. x. 22, showing itself by the answer which a good conscience makes in believing the truth, 1 Peter iii. 21, consenting unto, and embracing the new covenant, whereof baptism is a seal, of which anciently men of years made profession when they were baptized. Neither is it any thing worth, to have the form of godliness in profession, when the power thereof is denied by an evil conversation, 2 Tim. iii. 5. For however such as these are most apt to claim an interest in Christ, Luke xiii. 26, yet so long as their faith is not a particular faith, drawing with it affiance, and sole reliance on Christ for salvation, declaring its truth and life by endeavouring to perform the new covenant on their part, by new obedience, in all manner of good works; our Saviour professes that he knows them not, but bids them depart from him, because they were workers of iniquity, Luke xiii. 27.

But many of these presume further, that their faith is a lively and saving faith, because, as they think, they have repented, and are become new creatures. And all because they had such enlightening as by nature man cannot attain unto; nay, the word has affected them much, and somewhat altered them from what they were, namely, (1.) When they were hearing a sermon, or when God's rod was over them, they have mourned, wept, and showed some kind of humiliation. (2.) At the hearing of God's precious promises in the gospel, in the glad tidings of salvation, they have felt a taste of the heavenly gift, and of the good word of God, and of the powers of the world to come. And (3.) They find that they do not commit many of those sins which they were used to commit; and that they do many good duties towards God and man, which they were used not to do.

But what of all this? These men, as near as they come, yet going no further, are far from salvation. For the common gifts of God's Spirit, given unto men in the ministry of the gospel, may elevate a man higher, and carry him further towards heaven, than nature, art. or mere human industry can do; and yet if the saving graces of the same Spirit be not added, he will be left far short of heaven. Mere oratory in some pathetical preachers, when they speak of matters doleful and terrible, will move the affections, and draw tears from some hearers. Likewise a plain, powerful conviction of the certainty of God's wrath denounced, and sense of some just judgment of God, may bring forth some tears, some humiliation, yea some kind of reformation. Did not Felix tremble, when Paul reasoned of righteousness, temperance, and judgment to come? Acts xxiv. 25. Did not Ahab humble himself, when the prophet denounced God's judgments against him and against his house? 1 Kings xxi. 21, 27, 29. Did not the Israelties oft, when they were in distress, and when God did not only warn them with his word, but smote them with his rod, return and seek early after God? Psa. lxxviii. 34.

And whereas they say, they have tasted of the heavenly gift, and of the good word of God, and of the powers of the world to come; they may know, that such is the sweetness of God's promises, and such is the evidence and goodness of God's truth in the glad tidings of salvation, that, the common gift of the Spirit going with it, all the fore-mentioned feelings may be wrought in men altogether destitute of saving grace. For did not the seed sown in stony and thorny ground go thus far? Matt. xiii. 20—22. Did not those mentioned in the Hebrews, who notwithstanding all this might fall away irrecoverably, attain to thus much? Heb. vi. 4—6.

Now if men not in a state of grace may go so far, as has been proved, then it must not be marvelled that even such, with Herod, may also reform many things, Mark vi. 20.

Besides, they mistake, when they say, they are changed and reformed, if still they retain any bosom and beloved sin, as Herod did. To change sins, one sin into another, is no change of the man, for he changes the prodigality of his youth into covetousness in old age, remaining a notorious sinner before God as well now, as then; judge the like of all other; likewise to forbear the act of any sin, because they have not the like power, occasions, temptations, or means, to commit sin as in former time, this is no change: sin in these respects has left them, not they it.

For true conversion and repentance does consist of a true and thorough change of the whole man, whereby not only some actions are changed, but first and chiefly the whole frame and disposition of the heart is changed and set aright towards God, from evil to good, as well as from darkness to light, Eph. iv. 22—24; Rom. xii. 2. And whereas man is naturally earthly-minded, and makes himself his utmost end; so that either he only minds earthly things, or if he mind heavenly things, it is in an earthly manner, and to an earthly end, as did Jehu, 2 Kings x, if this man have truly repented, and be indeed converted, he becomes heavenly-minded, Col. iii. 1, 2, he makes God and his glory his chief and highest end; insomuch that when he has cause to mind earthly things, his will and desire is to mind them in an heavenly manner, and to an heavenly end. If you would judge more fully and clearly of this true change, see at large the description, and signs of uprightness, before delivered, chap. xi. page 153, et seq.

Last of all, there are many who presume, that although as yet they have no saving faith in Christ, nor sound repentance, God will give them space and grace to repent and believe before they die. Whence it is they have peace for the present.

These must give me leave to tell them, that they put themselves upon a desperate hazard and adventure.

1. Who can promise unto himself one minute of time more than the present, since every man's breath is in his nostrils, ready to expire every moment? Besides, the Spirit saith, God does bring wicked men to desolation as in a moment, Psa. lxxiii. 19. And again, He that being often warned, hardens his neck, shall suddenly be destroyed without remedy, Prov. xxix. 1.

2. Suppose they may have time, yet whether they

shall have grace to believe and repent, is much to be doubted.

For the longer repentance is delayed, the heart is more hardened, and indisposed to repentance, through the deceitfulness of sin, Heb. iii. 12, 13, 15, 19. And it is a judgment of God upon such, as are not led to repentance by the riches of God's goodness, forbearance, and long-suffering, that he should leave them to their impenitent hearts, that cannot repent; so treasuring up unto themselves wrath against the day of wrath, Rom. ii. 6. Custom in sin does so root and habituate it in man, that it will be as hard for him by his own will and power to repent hereafter, he neglecting God's present call and offer of grace, as it is for the Ethiopian to change his skin, or the leopard his spots, Jer. xiii. 23.

It cannot be denied, but that God is free, and if he please, may open a door of hope and gate of mercy unto the most obstinate sinner, who has deferred his repentance to his old age, Hosea ii, 15; wherefore, if such an one find his heart to be broken with remorse for his former sins, and is troubled in conscience for this his sin of not accepting of God's grace when it was offered; I wish him to humble himself before God, and entertain hope. For God has promised pardon to the penitent, whensoever they repent, Ezek. xviii. 21, 22. And though no man can repent when he will, yet such an one may hope that God is now giving him repentance, in that he has touched his heart, and made it to be burdened with sin.

Yet for all this hope which I give to such a man, know, that it is very seldom to be found, that those who continued to despise grace until old age, did ever repent: but God left them justly to perish in their impenitency, because they despised the means of grace, and the season in which he did call them to repentance, and offered to them his grace, whereby they might repent. God deals with all sinners usually, as he said he would do, and as he did to Judah: Because I would have purged thee, said he, that is, I took the only course to purge thee, and bring thee to repentance.

and thou wast not purged, therefore thou shalt not be purged from thy filthiness any more, till I have caused my fury to rest on thee, Ezek. xxiv. 13.

Thus I have endeavoured to discover and remove the false grounds, and misapplication of true grounds, whereby the conscience is deluded, and brought into a dangerous and false peace.

To conclude, he that would not be deceived with a false peace instead of a true, must beware of obstinacy, delight in, and senselessness of sin. For this sears the conscience as with a hot iron, 1 Tim. iv. 2. Now a seared conscience is quiet with a false peace; not because there is no danger, but because it does not feel it. Great care must be taken therefore, lest the conscience be seared, being made senseless and hard; for then it does altogether, or for the most part, forbear to check or accuse for sin, be it never so heinous.

This searedness is caused by a wilful customary living in any sin; but especially by living in any gross sin, or in the allowance of, and delight in any known sin; also by allowed hypocrisy, and dissimulation in any thing, 1 Tim. iv. 2, and by doing any thing contrary to the clear light of nature, planted in a man's own head or heart, Rom. i. 27; Jude 10; Eph. iv. 18, 19; or contrary to the clear light of grace, shining in the motions of the Spirit, in the checks of conscience, and in the instructions of the word, Heb. x. 26.

Keep therefore the conscience tender by all means; (1.) By hearkening readily to the voice of the word; (2.) By a careful survey of your ways daily. (3.) By keeping the conscience soft with godly sorrow for sin. (4.) By hearkening to the voice of conscience admonishing and checking for sin.

Either of these three kinds of conscience, viz. the blind, presumptuous, and seared conscience, will admit of a kind of peace, or truce rather, for a while, while it sleeps; but what God said of Cain's sin, must be conceived of all sin: If thou dost not well, sin lies at the door, Gen. iv. 7. And upon what terms soever it lies still, and troubles not the conscience for a time, yet it will awake in its time, and then by as much as it did admit of some peace and quiet, it will grow more turbulent, mad, and furious; and, if God give not repentance, this false peace ends for the most part either in a reprobate mind, Rom. i. 21—29, or a desperate end, Matt. xxvii. 5, even in this life, besides the hellish horrors in that which is to come.

Now to the end that no man should quiet his heart in this false and dangerous peace, whether it proceed from the aforementioned causes, or any other; I would advise him to try his peace, whether it be not false, by these infallible marks:

1. Is any man at peace with God's enemies, allowing himself in the love of those things or persons which hate God and which are hated of God, such as are the world and the things of the world, whereby he denies the power of godliness: delighting in any evil company, or living in any wilful or gross sin, as vain or false swearing, open profanation of the sabbath, malice, adultery, theft, lying, or in any of those mentioned, 2 Tim. iii. 2, 3, or in any known sin with allowance? The Holy Ghost saith of such, that the love of God is not in them, therefore the peace of God is not in them, 1 John ii. 15; and whosoever makes himself a friend to his lusts and to the world, makes himself an enemy of God, James iv. 4. If any man be at peace with the flesh, the world, and the devil, he is not at true peace with God, nor God with him. If any such expect peace, and should ask, Is it peace? answer may be made like to that which Jehu made, What have you to do with peace? What peace, so long as your notorious sins and rebellions, wherein you delight, are so many? 2 Kings ix. 19, 22. For he that cares not to keep a good conscience towards God and towards men, cannot have true peace of conscience, Heb. xiii. 18. For there is no true peace but in a good conscience.

2. Is any man not at peace, but at war rather, with God's friends, and with the things which God loves; being out of love with spiritual and devout prayer, hearing the word, the company of God's people, and the like? If any man despise the things that God

commands and loves, certainly, God and he are not reconciled, 2 Tim. iii. 5; and whatsoever his form of godliness be, God esteems him to be yet in a state of perdition. For whosoever saith he knows God, but yet loves not, and keeps not his commandments, he is a liar, 1 John ii. 4. And if any man love not his brother, whatsoever show of peace and friendship is between God and him, I am sure God saith, He that does not righteousness is not of God, neither he that loves not his brother, 1 John iii. 10; he is a child of the devil, and therefore has no true peace with God.

3. He whose quiet of heart and conscience is from false peace, is willing to take it for granted, that his peace is sound and good; and cannot abide to look into, or to inquire into his peace, to try whether it be true, or whether it be false or no; being, as it seems, afraid lest stirring the mud and filth that lies in the bottom of his heart, he should disquiet it. And for this cause it is that such an one cannot endure a searching ministry, 2 Chron. xxxvi. 16; Acts vii. 54, nor will like that minister who will dive into the conscience, by laying the heart and conscience open to the light and purity of God's word.

Thus I have showed you what is a first and chief impediment to be removed, viz. presumption and false hope, if you would have true peace; for false hopes breed only false peace.

CHAPTER XV.

CONCERNING FALSE FEARS.

The second head to which I reduced impediments to true peace, is false fear; for if you doubt, fear, or despair of your estate without cause, it will much disturb and hinder your peace.

SECT. 1. OF NEEDFUL HOLY FEAR.

There is an holy fear and despair wrought in man, when God first convinces his heart and conscience of sin; whereupon, through sense of God's wrath and heavy displeasure, together with a sense of his own disability in himself to satisfy and appease God's wrath, he is in great perplexity; being out of all hopes to obtain God's favour, or to escape the vengeance of hell by any thing which he of himself can do or procure. This is wrought more or less in every man of years before conversion, as in those which were pricked to the heart at Peter's sermon, Acts ii. 37, and in Paul himself, Acts ix. 9, and in the jailor, Acts xvi. 29. This is a good necessary fear, serving to prepare a man to his conversion. For in God's order of working, he first sends the spirit of bondage to fear, before he sends the spirit of adoption to enable a man to cry, Abba, Father, Rom. viii. 15. This fear, and trouble of conscience arising from it, is good; and makes way to true peace.

Moreover, after that a man is converted, though he have no cause to fear damnation, yet he has much matter of fear, for as much as he is yet subject unto many evils both of sin and pain; as, lest he offend God, and cause his angry countenance, and his judgments; also, lest he should fall back from some degrees of grace received, and lest he fall into some dangerous sin, and so lose his evidence of heaven, and comforts of the Spirit. Wherefore we are commanded to work out our salvation with fear and trembling, Phil. ii. 12, and to pass the whole time of our sojourning here in fear, 1 Peter i. 17.

This fear, while it keeps due measure, causes a man to be circumspect and watchful lest he fall; it excites him to repent, and quickens him to ask pardon and grace to recover, when he is fallen; yea, is an excellent means to prevent trouble, and to procure peace of conscience. But the fear of which I am to speak, and which, because it disturbs true peace, is to be removed, is a groundless and causeless fear that a man is not in a state of grace, although he has yielded himself to Christ, by true faith and conversion; and has not only given good hope to others, but, if he would see it, has cause to conceive good hope that he is indeed in the state of grace.

SECT. 2. OF CAUSELESS FEAR, AND THE SPRINGS THEREOF.

THIS fear may arise either from natural distempers, Satan joining with them; or from spiritual temptations, arising from causeless doubts.

(1.) Of fears which arise from natural distempers.

By natural distempers, I mean a disposition to frenzy or melancholy, in which states of body the spirits are corrupted through superabundance of choler and melancholy, whereby first the brain, where all notions of things are framed, is distempered, and the power of imagination corrupted, whence arise strange fancies, doubts, and fearful thoughts. Then, secondly, by reason of the intercourse of the spirits between the head and the heart, the heart is distempered and filled with grief, despair, and horror, through manifold fears of danger, yea, of damnation, especially when Satan concurs with those humours, which as he easily can, so he readily will do, if God permit.

Where there is trouble of this sort, it usually brings forth strange and violent effects, both in body and mind, and that in him who is regenerate, as well as in him that is unregenerate. Yea, so far, that, which is fearful to think, even those who, when they were fully themselves, did truly fear God, have, in the fits of their distemper, through impotency of their use of reason, and through the devil's forcible instigation, had thoughts, and attempts of laying violent hands upon themselves and others, and when they have not well known what they have done or said, have been heard to break out into oaths, cursing, and other evil speeches, who were never heard to do the like before.

These troubles may be known from true trouble of conscience, by the strangeness, unreasonableness, and senselessness of their conceits in other things; as to think that they have no heart, and to say they cannot do that which indeed they do, and a thousand other odd conceits, which standers-by see to be most false. Whereby any man may see that the root of this disturbance is in the fancy, and not in the heart.

Although both the regenerate and unregenerate, according as they are in a like degree distempered, are in most things alike; yet in this they differ; some beams of holiness will glance forth now and then in the regenerate, which do not in the unregenerate. especially in the intermission of their fits. Their desires will be found to be different, and if they both recover, the one returns to his usual course of holiness with increase: the other, except God work with the affliction to conversion, continues in his accustomed wickedness. It pleases God, that for the most part his own children who are thus distempered, have the strength of their melancholy worn out and subdued before they die, at which time they have some sense of God's favour to their comfort; but if their disease continue, it is possible they may die lunatics, and, if you judge by their speeches, despairing, which is not to be imputed unto them, but to their disease, or unto Satan, working by the disease; if they have given good testimony of holiness in former times.

When these troubles are merely from bodily distempers, though they be not troubles of conscience, yet they make a man incapable of the sense of peace of conscience. Therefore, whosoever would enjoy the benefit of the peace of his conscience, must do what in him lies, to prevent or remove these distempers. And because they grow for the most part from natural causes, therefore natural as well as spiritual remedies must be used.

1. Take heed of all such things as feed those humours of choler and melancholy, which must be learned of experienced men, and of skilful physicians, and, when need is, take physic.

2 Avoid all unnecessary solitude, and, as much as may be, keep company with such as truly fear God, especially with those who are wise, full of cheerfulness and joy in the Lord.

3. Forbear all such things as stir up these humours; as, over much study, and musing too much upon any thing, likewise all sudden and violent passions of anger, immoderate grief, &c.

4. Shun idleness, and, according to strength and means, be fully employed in some lawful business.

5. Out of the fit, the party thus affected must not oppress his heart with fear of falling into it again, any otherwise than to quicken him to prayer, and to cause him to cast himself upon God.

6. Out of the fits, and in them also, if the party distempered be capable, spiritual counsel is to be given out of the word, wisely, according as the party is fit for it, whether to humble him, if he has not been sufficiently humbled, or to build him up and comfort him, if he be already humbled.

7. Lastly, Remember always that when the troubled person is himself, he be moved to prayer, and that others then pray much with him, and at all times pray much for him.

When these troubles are mixed, coming partly from natural distemper, and partly from spiritual temptation, then the remedy must be mixed of helps natural and spiritual. What the natural helps are, has been shown, also what the spiritual in general, and shall be shown more particularly, in removing false fears arising from spiritual temptations.

The fears which rise for the most part from distemper of body, may be known from those which for the most part, or only, rise from the spiritual temptation, thus: When the first sort are clearly resolved of their doubts, and brought unto some good degree of cheerfulness and comfort, they will yet, it may be, within a day or two, sometimes within an hour or two, upon every slight occasion and discouragement, return to their old complaints, and will need the same means to recover them again. But those whose trouble is merely out of spiritual temptation and trouble of conscience, although for the time it be very grievous, and hardly removed, and sometimes long before they receive a satisfying answer to their doubts; yet when oncc they receive satisfaction and comfort, it does hold and last until there fall out some new temptation, and new matter of fear. This is because their fancies and memories are not disturbed in such a manner as the others' are.

The seeming grounds of fears that a man is not in a state of grace, when yet he is, are for variety almost infinite. I have reduced them into this order, and unto these heads.

First, They who are taken with false fears, think their sins to be greater than can be pardoned.

Secondly, When they are driven from that, they say they fear God will not pardon. When they are driven from this, by causing them to take notice of the signs of God's actual love to them, which give proof that he will save them, then,

Thirdly, They will question the truth of God's love and favour. But being put upon the trial whether God has not already justified them, and given them faith in Christ, which are sufficient proofs of his love; then,

Fourthly, They will seem to have grounds to doubt whether they have faith, from which they are driven, by putting them to the trial of their sanctification; then,

Fifthly, They doubt, and will object strongly that they are not sanctified, which being undeniably proved: then,

Sixthly and lastly, They fear they shall fall away, and not persevere to the end. Which fear being taken away also, and all is come to this good issue, they shall have no cause of disquiet or fear.

This is the easiest, most familiar, and the most natural method, so far as I can judge, both in proposing, and in removing false fears.

(2.) Of fears which arise from thoughts of the greatness of punishment and sin.

First, Some in their fits of despair, speak almost in Cain's words, saying, that their punishment, which they partly feel, and which they most of all fear, is greater than they can bear, or than can be forgiven, Gen. iv. 14.

I answer such: If sense and fear of wrath and punishment be your trouble, I would have you not to busy your thoughts about the punishment; but fix them upon your sins, which are the only cause of punishment; for get deliverance from the guilt and power of sin, and in one and the same work you free yourself from the punishment. Labour therefore that your heart may bleed with godly sorrow for sin, cry out, as David did against his sin, Psa. li. 4, 5, so do you against yours, confess them to God, strike at the root of sin, at the sin of your nature, wherein you were conceived, aggravate your actual sins, hide none, spare none, find out, arraign, accuse, condemn your sins, and yourself for them, grow first into an utter detestation of your sins, which have brought present punishment, and a sense and fear of the eternal vengeance of hell-fire; then likewise grow into a dislike with yourself for sin, loath yourself in your own sight for your iniquities, and for your abominations, Ezek. xxxvi. 31. Now when you are as a prisoner at the bar, who has received sentence of condemnation, when you are in your own apprehension a damned wretch, fearing every day to be executed; O, then it concerns you, and it is your part and duty to turn to God, the king of kings, whose name and nature is to forgive iniquity, transgression, and sins; and, that you may be accepted, go to him by Jesus Christ, whose office is to take away your sins, and to present you without sin to his Father; whose office is also to procure and sue out your pardon. Wherefore in Christ's name pray, and ask pardon of God, for his Son Jesus Christ's sake, and withal be as earnest in asking grace and power against your sin, that you may serve him in all well pleasing. Do this, as for your life, with all truth and earnestness; then you may, nay, ought to believe that God for Christ's sake has pardoned your sin, and has done away the punishment thereof. For this is according to the word of truth, even as true as God is, who has commanded you to do thus, and to believe in him.

But some will reply, this putting me into a consideration of my sins, breeds all my distress and fear, for I find them greater and more than can be pardoned.

O! say not so; for you can hardly commit a greater sin than indeed to think and to say so. It is blasphemy against God; yet this sin, if you will follow God's counsel, and all others, may, and shall be pardoned. I intend not to extenuate and lessen you sin: but you must give me leave to magnify God's truth and mercy, and to extol Christ's love and merit. However, it is true, that because sin is a transgression of a law of infinite holiness and equity; and, in respect of the evil disposition of the heart, is of infinite intention, and would perpetuate itself infinitely, if it had time and means; and because God, the person against whom sin is committed, is infinite; therefore sin must needs contract an infinite guilt, and deserve infinite punishment.

Secondly, Consider that the price to satisfy God's justice, namely, the death of Christ, the only begotten Son of God, does exceed all sin in infiniteness of satisfaction of God's justice and wrath due for sin. For if Christ's death be a sufficient ransom for the sins of all God's elect in general; then much more of thine in particular, whosoever thou be, and how great, and how many sins soever thou hast committed.

Thirdly, Know that the mercy of God, the forgiver of sin, is absolutely and every way infinite. For mercy in God is not a quality, but is his very nature, as is clear by the description of his name, proclaimed, Exod. xxxiv. 6, which rightly understood and believed, removes all the objections which a fearful heart can make against itself, from the consideration of his sins.

1. He is merciful, that is, he is compassionate, and to speak after the manner of man, is one that has bowels of pity, which yearn within him at the behold-

20

ing of thy miseries, not willing to punish and put thee to pain, but ready to succour and do thee good.

But I am so vile and so ill-deserving, that there is nothing in me to move him to pity me, and do me good!

2. He is gracious; whom he loves, he loves freely, of his own gracious disposition, Hosea xiv. 4; I, even I, am he that blots out thy transgressions for mine own sake, and will not remember thy sins, Isa. xliii. 25. And when God saith he would sprinkle clean water upon sinners, and that he would give them a new heart, &c. not for your sakes do I this, saith the Lord God, Ezek. xxxvi. 25, 26, 32. That you should be sensible of your own misery, and then, in the sense thereof, that God may be inquired after, and sought unto for mercy, is all which he expects from you to move him to pity and mercy, Ezek. xxxvi. 37; and such is his graciousness, that he will work this sense and this desire in you, that he may have mercy.

But I have a long time provoked him!

3. He is long-suffering towards you, not willing that you should perish, but that you should come to repentance, 2 Peter iii. 9, 15; he waits still for your repentance and reformation, that you may be saved.

Yea, but I am destitute of all goodness and grace to turn unto him, or do any thing that may please him!

4. He is abundant in goodness and kindness; he that has been abundant towards others heretofore in giving them grace, and making them good, his store is not diminished, but he has all grace and goodness to communicate to you also, and to make you good.

Yea, but I fear, though God can, yet God will not forgive me, and give me grace!

5. He is abundant in truth; not only the goodness of his gracious disposition makes him willing, but the abundance of his truth binds him to be willing, and does give sufficient proof unto you that he is willing. He has made sure promises to take away your sin, and to forgive it; and not yours only, but reserves mercy for thousands. Believe therefore that God both can and will forgive you.

Yea, but my sins are such and such; innumerable heinous, and most abominable. I am guilty of sins of all sorts!

6. He forgives iniquity, transgression, and sin. He is the God that will subdue all your iniquities, and cast all your sins into the bottom of the sea, Micah vii. 19.

Yea, but I renew my sins daily!

7. I answer out of the Psalm—His mercy is an everlasting mercy, his mercy endures for ever, Psa. cxviii. 1; he bids you to ask forgiveness of sin daily, Matt. vi. 11; therefore he can and will forgive sin daily; yea, if you sin seventy times seven in a day, Luke xvii. 4; Matt. xviii. 22; and shall confess it to God with a penitent heart, he will forgive; for he that bids you be so merciful to your brother, will himself forgive much more, when you seek unto him.

But I have not only committed open and gross sins, both before and since I had knowledge of God; but I have been a very hypocrite, making profession of God, and yet daily committing grievous sins against him!

8. What then? Will you say your sins are unpardonable? God forbid. But say, I will follow the counsel which God gave to such abominable hypocrites. Wash ye, make you clean, Isa. i. 16. I will, by God's grace, wash my heart from iniquity, and my hands from wickedness, Jer. iv. 14, by washing myself in the laver of regeneration, bathing myself in Christ's blood, and in the pure water of the word of truth, applying myself to them, and them to me by faith. Say in this case I will hear what God will speak, Psa. lxxxv. 8. And know, that if you will follow his counsel, if you will hearken to his reasoning, and embrace his gracious offer made to you in Christ Jesus, the issue will be this, though your sins have been most gross, double dyed, even as crimson and scarlet; they shall be as wool, even white as snow, Isa. i. 18. God will then speak peace unto you, as unto others of his saints; only he will forbid you to return to folly.

For not only those who committed gross sins through ignorance before their conversion, as did Abraham in idolatry, and Paul in persecuting: nor yet only those who committed gross sins through infirmity after their conversion, as did Noah by drunkenness, Gen. ix. 21, and Lot by incest also, Gen. xix. 33, and Peter by denying and forswearing his master Jesus Christ, Matt. xxvi. 47, obtained mercy, because they sinned ignorantly and of infirmity; but also those that sinned against knowledge and conscience, both before and after conversion; sinning with a high hand as Manasses before, 2 Chron. xxxiii. 6, 10, 12, 15, and in the matter of Uriah, 1 Kings xv. 5, David after conversion, they obtained like mercy, and had all their sins forgiven. Why are these examples recorded in Scripture, but for patterns to sinners, yea to most notorious sinners of all sorts, who should in after-times believe in Christ Jesus unto eternal life? 1 Tim. i. 15, 16.

Be willing therefore to be beholden to God for forgiveness, and believe in Christ for forgiveness, which when you do, you may be assured that you never yet committed any sin which is not, and which shall not be forgiven.

For was it not the end, why Christ came into the world, that he might save sinners, yea, the chief of sinners, as well as others? 1 Tim. i. 15, 16. Was he not wounded for transgressions, viz., of all sorts? Isa. liii. 5. Is not the end of his coming in his gospel to call sinners to repentance? Luke v. 32. What sinners does he mean there, but such as you are, who are laden and burdened with your sin? Does he not say, If any man sin, observe, if any man, we have an advocate with the Father, Jesus Christ the righteous? 1 John ii. 1. Who by being made a curse for you, has redeemed you from the curse of the whole law, Gal. iii. 13; therefore from the curse due unto you for your greatest sin.

However, it is impossible for a notorious sinner, yea, for any sinner, by his own power or worth, to enter into the kingdom of heaven; yet know, what is impossible with man, is possible with God, Matt. xix. 26. Is any thing too hard for the Lord, Gen. xviii. 14. He can alter and renew you, and give you faith and repentance; he can make these things possible to you that believe; yea, all things are possible to him that believes, Mark ix. 23.

Yes, you will say, If I did believe. Why, what if you do not believe? It is not hard with him, if you come to his means of faith, if you hearken to the precepts and promises of the word, and consider that the God of truth speaks in them; I say, it is not hard for him, in the use of these means, to cause you to believe.

Wherefore neither greatness of sin, nor multitude of sins should, because of their greatness and multitude make you utterly despair of salvation, or fear damnation; when once you can believe, or but will and desire to obey and believe, Isa. i. 19, the great cause of fear is past.

I know if you never had sinned, you would not fear damnation. Now to a man whose sins are remitted, his sins, though sin dwell in him, Rom. vii. 20, are as if they were not, or never had been. For they are blotted out of God's remembrance. I, even I am he, saith God, that blots out thy transgressions, for my name's sake, and will not remember thy sins, Isa. xliii. 25. And who is like thee, saith the prophet, that pardons iniquities, &c.; he will have compassion upon us, he will subdue our iniquities, and will cast all our sins into the bottom of the sea, Micah vii. 18, 19. A debt when it is paid by the surety, puts the principal out of debt, though he paid never a penny of it himself. The Holy Ghost speaks most comfortably, saying, that God finds no sin in them whose sins are pardoned, Rev. xiv. 5. In those days, and at that time, saith the Lord, the iniquity of Israel shall be sought for, and there shall be none, and the sins of Judah, and they shall not be found; but how may this be? He gives the reason, for I will pardon them whom I reserve, Jer. l. 20.

If you believe that God can pardon any sin, even the least, you have like reason to believe that God can pardon all, yea, the greatest; for if God can do any thing, he can do every thing, because he is infinite. He can as easily say, Thy sins are forgiven thee, all thy sins are forgiven thee, as to say, Rise and walk, Matt. ix. 5. He can as well save one that has been long dead, rotten, and stinking in his sin, as one newly fallen into sin. For he can as easily say, Lazarus, come forth, John xi. 43, as, Damsel, I say to thee, arise, Mark v. 41.

Lastly, To make an end of removing this fear, I ask thee, who art troubled with the greatness of thy sins past, and with fear that they can never be pardoned, How stand you affected to present sins? Do you hate and loath them? Do you use what means you can to be free from them? Are you out of love with yourself, and humbled because you have indulged them to God's dishonour, and your own hurt? And do you resolve through faith in Christ Jesus, to return from your evil ways, Ezek. xviii. 21, 22, and to enter upon a holy course of life, if God shall please to enable you; and is it your hearty desire to have this grace to be able? And are you afraid, and have you now a care lest you fall knowingly into sin; then, let Satan, and a fearful heart object what they can, you may say, though my sins have been great and heinous, for which I loath myself and am ashamed, yet now I see that they were not only pardonable, but are already, through the rich mercy of God, pardoned, Ezek. xxxvi. 25—33. For these are signs of a new heart and a new mind. Now to whomsoever God gives the least measure of saving grace, to them has he first given pardon of sin, and will yet abundantly pardon. For he saith, Let the wicked forsake his way, and the unrighteous man his thoughts: and let him return to the Lord, and he will have mercy upon him, and to our God, for he will abundantly pardon, Isa. lv. 7.

SECT. 3. FEARS CONCERNING NOT BEING ELECTED, REMOVED.

There are others who make no doubt of God's power. They believe he can forgive them; but they fear, yea, strongly conclude, that he will not pardon them, and that because they are reprobates, as they say, for they see no signs of election, but much to the contrary.

I answer these thus. When your consciences are first wounded with a sense of God's wrath for sin, it is very like, that before you have believed and repented, you cannot discern any signs of God's favour, but of his anger; for as yet you are not actually in a state of grace, and in his favour. And oftentimes after the Christian does believe, though there be always matter enough to give proof of his election, yet he cannot always see it. If you be in either of these states, suppose the worst, yet you have no reason to conclude that you are reprobates.

It is true, that God, before the foundation of the world, fully determined with himself, whom to choose to salvation by grace, to which also he ordained them; and whom to pass by, and leave in their sins, for which he determined in his just wrath to condemn them. But who these be, is a secret, which even the elect themselves cannot know, until they be effectually called, nay, nor being called, until by some experience and proofs of their faith and holiness, they do understand the witness of the Spirit, which testifies to their spirits, that they are the children of God; and do make their calling and election, which was always sure in God, sure to themselves, 2 Peter i. 5, 10. But in point of reprobation, namely, that God has passed them by, to perish everlastingly in their wickedness, no man living can know it, except he know that he has sinned the sin against the Holy Ghost, that unpardonable sin.

For God calls men at all ages and times, some in

their youth, some in their middle age, some in their old age; yea, some have been called at their last hour, Luke xxiii, 42, 43. Now let it be granted, that you cannot, by searching into yourselves, find the signs of effectual calling, which yet may be in you, though your dim eyes cannot perceive them; nay, suppose that you are not yet effectually called, here is no cause for you utterly to despair, and say, you are reprobates. How know you that God will not call you before you die?

It were a far wiser and better course for you, who will be thus hasty in judging yourselves to be reprobates, to busy yourselves first with other things. Acquaint yourselves with God's revealed will in his word. Learn to know what God has commanded you to do, and do that; also what he has threatened, and fear that; and what he has promised, and believe and rest on that. After you have done this, you may look into yourselves, and there you shall read your election written in golden and great letters.

For God never intended that the first lesson which a Christian should learn, should be the hardest, and highest that can be learned, taken out of the book of his eternal counsel and decree; and so to descend to the A. B. C. of Christianity; which were a course most perplexed and preposterous. But his will is, that his scholars and children should learn out of his written word here on earth, first, that God made all things. Gen. i. 31, and that he made man good, and that men, hearkening to Satan, found out evil devices, Eccles. vii. 29, and so fell from grace, and from God, and so both they, and the whole world that came of their loins, became liable to eternal damnation. Next, God would have you to learn, that he, in his infinite wisdom, goodness, and mercy, thought of, and concluded a new covenant of grace, Gen. iii. 15; xvii. 1, 2, 11; Rom. iv. 11; Jer. xxxi. 31, 32; for the effecting whereof, he found out and appointed a way and means to pacify his wrath, by satisfying his justice, punishing sin in man's nature, by which he opened a way unto his mercy, to show it to whom he would; namely, He gave his only Son, very God, to become very man, Phil. ii. 6—11, and being made a common person and surety in man's stead, died, and endured the punishment due to the sin of man, and rose again, and was exalted to sit at God's right hand to reign, having all authority committed unto him. Thus he made the new covenant of grace, established in his Son Jesus Christ; the tenor and condition whereof required on man's part is, that man accept of, and enter into this covenant, believing in Christ, in whom it is established; then, whosoever believeth in him, shall not die, but have everlasting life, John iii. 16. This God did in his wisdom, justice, mercy, and love to man, that he himself might be just, and yet a justifier of him that is of the faith of Jesus, Rom. iii. 26. And he has therefore given his word and sacraments, and has called, and has given gifts to his ministers, Eph. iv. 8, thereby to beget, and increase faith in men, by publishing this good news, and by commanding them, as in Christ's stead, in God's name, to believe, and to be reconciled to God, 2 Cor. v. 20, and to live no longer according to the will of their old masters, the devil, the world, and the flesh, under whom they were in cursed bondage; but according to the will of him that redeemed them, in holiness and righteousness, whose service is a perfect and blessed freedom.

Now when you have learned these lessons first, and by looking into yourselves can find faith and new obedience, 2 Peter i. 5, 10, 11, then by this your effectual calling, you may safely ascend to that high point of your predestination, which will give you comfort, through assurance that you shall never fall away.

When you observe this order in learning your election to life, it will not minister unto you matter of curious and dangerous dispute, either with God or man; but of high admiration, thanksgiving, and unspeakable comfort, causing you to cry out with the apostle, O the depth of the riches both of the wisdom and knowledge of God, &c. Rom. xi. 33. And Blessed be the God and Father of our Lord Jesus

Christ, who has chosen us in him before the foundation of the world, that we should be holy and without blame before him in love, having predestinated us unto the adoption of children, by Jesus Christ to himself, according to the good pleasure of his will, to the praise of the glory of his grace, wherein he has made us accepted in his well beloved, &c. Eph. i. 3, 5, 6.

SECT. 4. OF FEARS CONCERNING THE SIN AGAINST THE HOLY GHOST.

There are yet some, who having heard that there is a sin against the Holy Ghost, and that it is unpardonable, are full of fears that they have committed that sin, thence concluding that they are reprobates, for they say, that they have sinned wilfully against knowledge and conscience, since they received the knowledge of the truth, and tasted of the heavenly gift, and of the good word of God.

If you who thus object, have sinned against knowledge and conscience, you have much cause for humbling yourself before God; confessing it to him, asking pardon of him, and grace to believe and repent, both which you must endeavour by all means. Yet I see no cause why you should conclude so desperately, that you have sinned against the Holy Ghost, and are a reprobate. For as few in comparison, though too many, commit this sin, so few know what it is.

All sin against knowledge and conscience is not this sin, 1 Kings xv. 5; 2 Sam. xi. 4, 6, 10, 15, 25. Nor yet all wilful sinning. It is not any one sin against the law, nor yet the direct breach of the whole law, nor every malicious opposing of the gospel, Heb. x. 28, if it be of ignorance; neither is it every blasphemy, 1 Tim. i. 13, or persecution of the gospel, and of those that profess the truth, if these be done out of ignorance or passion; nor yet is it every apostasy, 2 Chron. xvi. 10; 1 Kings xi. 4—6; Heb. x. 28, 29, and falling into gross sins of divers sorts, though done against knowledge and conscience; yet this sin against the Holy Ghost, contains all these, and more. It is a sin against the gospel, and free offer and dispensation of grace and salvation by Christ, through the Spirit. Yet, it is not any particular sin against the gospel, nor yet a rejecting of the whole gospel, if in ignorance, Luke xxiii. 34; nor yet every denying of Christ, or sudden revolting from the outward profession of the gospel, when it is of infirmity, through fear, and such like temptation, Matt. xxvi. 69, 70, 74; neither is it called the sin against the Holy Ghost, and is unpardonable, because it is committed against the essence, or person of the Holy Ghost, for the essence of the three persons in the Trinity is all one; and the person of the Holy Ghost is not more excellent than the person of the Father and the Son; but it is called the sin against the Holy Ghost, and becomes unpardonable, because it is against the office of the Holy Ghost, and against the gracious operations of the Holy Ghost, and therein against the whole blessed Trinity, all whose works, out of themselves, are consummate, and perfected in the work of the Holy Ghost. Moreover, know that it is unpardonable, not in respect of God's power, but in respect of his will; he having, in his holy wisdom, determined never to pardon it. And good reason why he should will not to pardon it, in respect of the kind of the sin, if you will observe it; it being a wilful and malicious refusing of pardon upon such terms as the gospel does offer it, scorning to be beholden unto God for it. You may perceive what it is, by this description:

The sin against the Holy Ghost is an utter, wilful, and spiteful rejection of the gospel of salvation by Christ, together with an advised and absolute falling away from the profession of it, so far, that, against former knowledge and conscience, Heb. vi. 4—6, a man does maliciously oppose and blaspheme the Spirit of Christ, in the word and ordinances of the gospel, and motions of the Spirit in them; having resisted, rejected, and utterly quenched all those common and more inward gifts and motions wrought upon their hearts and affections, which sometimes were entertained by them; insomuch, that out of hatred of the Spirit of life in Christ, they crucify to themselves afresh the Son of God, and do put him, both in his ordinances of religion, and in his members, to open shame, treading under foot the Son of God, counting the blood of the covenant, wherewith he was sanctified, an unholy thing; doing despite to the Spirit of grace, Heb. x. 26—29. If you carefully look into those places of the Scripture, which speak of this sin, and also observe the opposition which the apostle makes between sinning against the law, and sinning against the gospel, you will clearly find out the nature of this sin. Matt. xii. 24, 31, 32; Mark iii. 28—30; Luke xii. 10; Heb. vi. 4—6; x. 26—29.

But to resolve you out of this doubt, if you be not overcome with melancholy, for then you will answer you know not what, which is to be pitied rather than regarded, I would ask you, who think you have committed the sin against the Holy Ghost, these questions: Does it grieve you, that you have committed it? Could you wish that you had not committed it? If it were to be committed, would you not forbear it, if you could choose? Should you esteem yourself beholden to God, if he would make you partaker of the blood and Spirit of his Son, thereby to pardon and purge your sin, and to give you grace to repent? Nay, are you troubled that you cannot bring your heart unto a sense of desire of pardon and grace? If you can say, Yea; then, although the sin or sins which trouble you, may be some fearful sin, of which you must be exhorted speedily to repent, yet certainly it is not the sin against the Holy Ghost; it is not that unpardonable sin, that sin unto death. For he who commits this sin cannot relent, neither will he be beholden to God for pardon and grace, by Christ's blood and Spirit; he cannot desire to repent: but he is given over, in God's just judgment, unto such a reprobacy of mind, deadness of conscience, and rebellion of will, and to such an height of hatred and malice, that he is so blasphemously and despitefully bent against the Spirit of holiness, Heb. x. 29, that it much pleases him, rather than any way troubles him, that he has so maliciously and blasphemously rejected, or fallen from, persecuted, and spoken blasphemously against the good way of salvation by Christ, and against the gracious operations of the Spirit, and against the members of Christ; although he was once convinced clearly, that this is the only way of salvation, and that those graces and gifts were from God, and that they were the dear children of God, whom he now despises.

SECT. 5. OF FEARS ARISING FROM AN ACCUSING CONSCIENCE.

Others, if not the same persons, object thus: God will certainly condemn, because John has said, If their hearts condemn them, God is greater than their hearts, 1 John iii. 20. Hence they infer, God will condemn them much more. For, they say, their hearts do condemn them.

There is a double judgment by the heart and conscience. It judges a man's state or person, whether he be in a state of grace, or no. Also, it judges a man's own particular actions, whether they be good or no. I take it, that this place of John is not to be understood of judging or condemning the person; for God in his final judgment does not judge according to what a man's weak and erroneous conscience judges, making it the rule of his judgment to condemn or absolve any. For many a man, in his presumption, justifies himself in his life, when yet God will condemn him in the world to come, Hosea xii. 8; Luke xviii. 11; and many a distressed soul, like the prodigal, Luke xv. 18, 19, and humble publican, Luke xviii. 13, 14, condemns himself, when yet God will absolve him. For a man may have peace with God, yet God, for reasons best known to his wisdom, does not presently speak peace to his conscience, as it was with David; in which case, man judges of his estate otherwise than God does.

This place, 1 John iii. 20, is to be understood of judging of particular actions, namely, whether a man love his brother, not in word and tongue only, but in deed and truth, according to the exhortation, 1 John iii. 18—22; which, if his conscience could testify for him, then it might assure his heart before God, and give it boldness to pray unto him, in confidence to receive whatsoever he did ask according to his will. But if his own conscience could condemn him of not loving his brother in deed and in truth; then God, who is greater than his heart, knowing all things, must needs condemn him therein much more. This is the full scope of the place. Yet this I must needs say, that the Holy Ghost has instanced in such an act, namely, of hearty loving the brethren, which is an infallible sign of being in a state of grace: whereby, except in case of extreme melancholy, or violent temptation, a man may judge, whether at present he be translated from death to life.

If any shall think the place to be understood of judging the person, he must distinguish between that judgment which the heart gives rightly, and that which it gives erroneously. But suppose that, you trying yourselves by this, your hearts do condemn you of not loving the brethren, can you conclude hence, that you shall be finally damned? God forbid. All that you can infer, is this; you cannot have boldness to pray unto him until you love them; nor can you assure yourselves that you will have your petitions granted. And the worst you can conclude is, that now for the present, you are not in a state of grace, or at least you want proof of being in a state of grace. You must then use all God's means of being ingrafted into Christ, and must love the children of God, that you may have proof thereof. Did Paul love the brethren, when he breathed out threatening, Acts viii. 3; and was, as he himself saith, Acts xxvi. 10, 11, mad against them? Was he at that time a reprobate? Acts ix. 15. Did he not afterwards, being converted, so love God's people, that he could be content to spend, and be spent himself, for them? 2 Cor. xii. 15. So, many thousands, whose consciences for the present may justly condemn them of not loving those that are indeed God's children, may yet love them hereafter, as dearly as their own souls.

Some will yet say, Certainly we are reprobates: for we have, according to the command of the apostle, tried whether we be in the faith, or no; and whether Christ be in us; but we find neither: the apostle saith, We know these to be in us, else we are reprobates, 2 Cor. xiii. 5.

By reprobate, in this place, is not meant one that is not elect; for none of the elect can before their conversion know, by any search, that they are in the faith, or that Christ is in them; for that cannot be known which yet is not. Many are not converted until they be thirty, forty, or fifty years old. Will you say, these in their younger years were reprobates? You may say, they then were in a state of condemnation, and children of wrath, but no reprobates. Besides, a man must not be said not to be in the faith, and not to have Christ in him, because he does not know so much. For many have faith and are in Christ, yet do not always know it.

The word reprobate, because it is ordinarily understood, by our common people, for a man ordained to condemnation, is too harsh.

The words now rendered "except ye be reprobates," may, as I judge, rather be translated thus: "except you be unapproved, or except you be without proof," namely, of your being in the faith, and of Christ's being in you, whereof you outwardly make profession. As if the apostle had said, If upon trial you cannot find that you are in the faith, &c., you are unapproved Christians. Either you have yet only a mere form of Christianity, and, like false coin, or reprobate silver, are but hypocrites and counterfeits; or if you be Christians in truth, yet you are unexperienced Christians, and without proof of it to yourselves.

Some may reply, If I find upon trial that I am a counterfeit, may I not then judge myself to be a reprobate?

No. For, first, you may err in judging of yourself. Secondly, if you do not err, you can judge only this, that you are not yet in a state of grace: but in the use of the means, you may be. God can as well convert an hypocrite, as a pagan. For though now you be dross and refuse, you may ere long be pure gold. For God, in making vessels of honour, does more than all earthly kings, and all their goldsmiths can do: for they, by their prerogative and skill, can make current coin, and rich vessels, if they have pure metal to work upon: but they cannot make good metal of base stuff, nor make gold of brass. But such is the power of God's word and Spirit, that whereas they find you base and drossy stuff, they, by imprinting the character and stamp of God's image upon your hearts, do transform you into the same image, from glory to glory, even as by the Spirit of the Lord, 2 Cor. iii. 18. As soon as you are truly anointed with this Spirit, you shall become good gold and silver vessels of honour, fitted for the Lord's use whereunto you were appointed.

SECT. 6. FEARS ARISING FROM LATE REPENTANCE, ANSWERED.

There are yet others, who object fearfully, saying that they are cast-aways, and that God will not have mercy on them, because now it is too late; they have passed the time and date of their conversion, they therefore will not use, or at least have no heart in using God's means to convert them, such as prayer, reading, hearing the word, &c.; nor yet willingly will suffer others to pray either with them, or for them; and all because they think it is now too late, and in vain; mistaking this, and such other Scriptures: Because I have called, saith God, and you have refused—they shall call on me, and I will not answer, Prov. i. 24, 28; and because they think they sin when they pray, and hear the word, and that the more means is used to save them, their condemnation shall be the more increased. Thus Satan and a fearful heart delude many.

It must be acknowledged, that God would have all men walk and work while they have light, John xii 35, because the night will come on, when no man can work, John ix. 4. And whilst it is called to-day, Heb. iii. 15, he would have every one return, and accept of grace offered, and not to harden their hearts against it. And our Saviour bewails Jerusalem, because they despised the day of their visitation, Luke xix. 42—44. All which shows, that God has his set period of time, between his first and last offer of grace, which being passed, he will offer it no more; and that justly, because they took not his offer when they might. And this time is kept so secret with God, that if he offer grace to-day, who can tell whether he will offer it to-morrow; or whether he will offer it again? Who knows whether God will take him from the means of salvation, or will take the means of salvation from him? All this our holy and wise God has revealed in his word, to make men wise to take the opportunity and time of grace while it is offered. Wherefore, whosoever have neglected their first times and offers of grace, have sinned and played the fool egregiously; for which they have cause to be much humbled. But for you to conclude hence that the date and time of your conversion is out, this has no sufficient ground. For it is not possible for you to know, that your time of conversion is past all recovery. But you should rather for the present time believe, and hope that it is not passed. Indeed, presumptuously to put off receiving grace until to-morrow, is foolish and dangerous; but if God give you time till to-morrow, that you live, and it can be said to-day; so long as you yet live, and the means of salvation are not from you, either in their exercise, or out of your remembrance; but you do yet live to hear what God has commanded you to do, and to hear what good things he yet offers unto you with Christ; or if the means be taken from you, or you are detained from them by sickness, &c., so long as you yet live to call to remembrance what God has commanded you to believe and do, you cannot say the time is too late, if you do yet condemn yourselves for refusing grace heretofore, and are now willing and desirous to accept of it. Moreover, would you now, with all your heart, use the means of salvation, and endeavour to believe and repent, if you thought it were not too late? And does it grieve you that you have neglected the opportunity? And would you gain and redeem that lost time, if you knew how? Then, I dare in the name of God assure you, that the date of your conversion is not expired. It is not too late for you to turn unto the Lord. While it is to-day, I may boldly say, harden not your heart: which, if you do not, you must know that now is an acceptable time, now is the day and time of your salvation, Heb. iii. 15. At what time soever God does send his ministers unto you, by whom God does beseech you, they entreating you, as now I do, in Christ's stead, that you would be reconciled to God, 2 Cor. v. 20, this is the accepted day, if you will be intreated by them, 2 Cor. vi. 2; the day wherein God will accept of you is not passed. Moreover, at what time soever, and by what means soever, any man shall humble himself for sin, and seek the grace of God in Christ Jesus, the date of God's acceptance of him is not expired. Learn this in the example of Manasses, and many others, who refused grace in their younger time, yet were converted in their age, 2 Chron. xxxiii. 10, 12, 13. You have God's express word for it, who saith, From the days of your fathers, that is, for a long time, ye are gone away from mine ordinances, and have not kept them; return unto me, and I will return unto, saith the Lord of Hosts, Mal. iii. 7.

But may not a man pray too late, and seek repentance, in vain, as Esau did, who found no place of repentance, though he sought it carefully with tears? Heb. xii. 17. Did not the foolish virgins seek to enter into the bride chamber, but were not admitted? Matt. xxv. 11, 12. And did not our Saviour say, Many shall seek to enter in, and shall not be able? Luke xiii. 24.

No man can ask grace and forgiveness of sins too

late, if he ask for grace and power against sin heartily: but a man may ask a temporal blessing, or the removal of a temporal evil, when it may be too late.

As for Esau's careful seeking of repentance, you must understand it not of his own repentance from his profaneness, and from other dead works, but of his father Isaac's repentance: he would have had his father to change his mind, and to have given him the birth-right, which was already bestowed upon Jacob, Gen. xxvii. 34, 38.

Whereas the foolish virgins did seek to enter into the bride chamber when the door was shut; know, that this is a parable, and must not be urged beyond its general scope, which is to show, that insincere professors of Christianity, such as have only a form of godliness, without the power of it, although they will not live the life of the righteous, yet they wish their end might be like theirs, Numb. xxiii. 10; and because of their outward profession of Christ's name in this life, they securely expect eternal life; but forasmuch before their death, they did not provide the oil of truth and holiness, therefore at the day of judgment, they shall be disappointed of entering into heaven, which in the time of their life, they did so much presume on.

The same answer may be given unto that place, Luke xiii. 24. Yet you mistake, when you say, that Christ saith, Many shall strive to enter, and shall not be able. He saith strive to enter in at the strait gate, for many, I say to you, shall *seek* to enter, and shall not be able; he does not say, Many shall *strive* to enter.

There is a great difference in the signification of the words striving and seeking; seeking imports only a bare professing of Christ, hearing the word, and receiving the sacraments. For thus did the men spoken of by our Saviour, who are said not to be able to enter. But to strive to enter, is to do all these and more; it is to strive in seeking for him, so that they take up their cross and follow him; they give their hearts to him, as well as their names; they are hearty and sincere in praying, hearing, receiving; they strive to subdue their lusts, which offend Christ, and strive to be obedient to his will, as well as to believe his promises, and to hope for happiness; this is to strive, Hosea vi. 3. Now never one did thus strive in seeking to enter, though it were the last day in his life, that was rejected, and not received. Wherefore say not, It is too late, but say, The more time I have lost, the more cause is there now that I should seek my salvation in earnest, and not lose time in questioning, whether I may be accepted or not.

SECT. 7. FEARS OF MISUSING THE MEANS OF GRACE, REMOVED.

And whereas you said, you are afraid to use the means of salvation, for fear of increasing your guilt and condemnation thereby; hereby you may see, that this is but the malice and subtilty of the devil, by keeping you from the means, to keep you from salvation, Psa. lxxiii. 13. For it is most false to say, that to pray, hear the word, &c., is to increase your sins, because you cannot perform them as you should, and as you would. I am sure, it is a greater sin in you to forbear these necessary duties out of despair that they shall not profit you, or that you shall not be accepted of God. You should think thus: If I do not use the means of salvation, I shall certainly perish everlastingly; but if I do pray, hear, &c., I may be saved; therefore, in obedience to God, I will do as well as I can. But little does a man know how well he may do through the strength of Christ, if he would endeavour; neither can a man conceive how acceptable a little endeavour shall be, if he do but desire to be true in his endeavour. For as God's power is seen in a man's weakness, so is God's grace seen in a man's insufficiency, 2 Chron. xxx. 18—20; 2 Cor. xii. 9, 10. When we are weak, then God in us can be strong. And when we in humility like our services worst, then, through Christ, God may be best pleased with them. But, whatever you do, neglect not, nor absent yourselves from exercises of religion; for the weakest observances, where is truth, are far more acceptable than entire omissions. Wherefore, if as you say, you would not increase your sin, and thereby your damnation, be willing to use, and to join with others in the use of all good means of salvation; then, if you be not saved, yet you shall have the less punishment. But you may be assured, that if in obedience to God's commandments you shall pray, hear the word, receive the sacrament, and have communion and conversation with those that fear God, you shall be saved in the end, believing in Christ Jesus.

If you do not yet feel benefit and comfort, when you use these means of salvation, according to your desire, yet you must wait the good hour both of grace and comfort, even as the impotent people did, who lay waiting for the angel's coming to move the waters, that they might be healed of their diseases, at the pool of Bethesda, John v. 3. For if, when God hides his face, you will wait and look for him, Isa. viii. 17, then God will wait his time to be gracious, and blessed shall you be that wait for him, Isa. xxx. 18. It may be, it comes justly upon you, that God should make you wait his leisure, and cause you to buy wisdom with dear experience, because you did once account it an easy matter to believe and repent, and therefore you did not take the first offers, but made God wait If it were thus, yet despair not of grace; only be humbled. For God does not deal with us after our sins, nor reward us after our iniquities, Psa. ciii. 10, but according to his rich mercy and promise, made to us in Christ Jesus.

SECT. 8. FEARS ARISING FROM DOUBTS OF GOD'S LOVE REMOVED.

There are many, who have true proofs that they are the chosen of God, and have reason to think that God not only can, but will do them good; yet because they will deny that to be bestowed upon them, and to be in them, which indeed is, therefore they fear, and are causelessly disquieted. I would have such to consider, first, whether they have not in them already evident proofs and signs of God's effectual love towards them in Christ? These will acknowledge, that it is most true, that if they were sure God did love them, they should not fear; but this is all their doubt, that God does not love them.

(1.) Doubts of God's love because of afflictions, removed.

Some give this reason of their doubt; God has and still does severely afflict them; yea, ever since they have professed the name of Christ, they are in something or other chastened daily; insomuch, that they seem to be in the condition of those whom God threatened to curse in every thing, they put their hands unto, Deut. xxviii. 20. Therefore, say they, God does not love us.

Such weak and inconsiderate reasonings are incident to those whom God truly loves. Did not the holy men of God reason, and conclude thus? But when God's children do thus, it is in their haste; before they are well advised what they think or say, Psa. xxxi. 22; cxvi. 11. And whence is it? Is it not from their ignorance and weakness, being carried away by sense? So foolish was I, and ignorant, saith the prophet, &c. Psa. lxxiii. 13, 14, 22. But when they come to themselves, and learn by God's word and Spirit, that it is not outward prosperity will make wicked men happy, neither is it outward affliction that can make a good man miserable; then they will neither applaud nor envy the prosperity of the wicked, nor yet misconstrue, nor repine at their own afflictions. For they learn, that no man can know God's love or hatred by any outward thing, that does befall the sons of men in this life, Eccles. ix. 1, &c.

They learn, that God does often smile on his enemies, and that he does often frown upon, is angry with, and does correct those whom he dearly loves, even as a father does his children, Prov. iii. 12; Rev. iii. 19.

They learn by the word likewise, that God has

excellent ends in all this, even in respect of them, and for their good; namely, for trial of their graces, for prevention of sin, and to remove sin, by bringing them to repentance, Rom. vii. 12, that they might be made partakers of his holiness, Heb. xii. 10. Besides, herein he does much glorify himself, showing that he is wonderful in counsel, excellent in working, Isa. xxviii. 29; causing the affliction to work for his glory, in his people's good; yea, you may learn by your own experience, that the child of God in his infirmity and passion, when he is under the rod, may let go his hold of God; yet, that God, in his love and compassion towards his people, will hold him fast by his right hand, and will not leave him; but will guide him with his counsel, until he receive him into glory, Psa. lxxiii. 23, 24. This is God's method with his children; wherefore none from hence has cause to question God's love, but rather to conclude it.

(2.) Fears of the want of grace, on account of worldly prosperity.

There are others, and it may be the same, when the tide of affection is turned, who, because they prosper, and are not in trouble as other men, conceive that God does not love them. For it is said, As many as he loves, he rebukes and chastens, Rev. iii. 19, and he chastens every son whom he receives, Heb. xii. 6.

See, a fearful and doubtful heart will draw matter to feed its fears and doubts out of any thing. But know, God is a wise and good Father; he knows when to strike, and when to hold his hands.

In such cases as the following, God does not usually afflict his children with his heavy rod.

First, When they are infants, babes in Christ, or, if they be grown to years, when they are spiritually weak or sick, and cannot bear correction, then, though they be froward, and deserve strokes, God does forbear, and is inclined rather to pity.

Secondly, When they are good children; that is, when they show that they would please him, by endeavouring to do what they are able, though it be with much imperfection; then God will not strike, but spares them, as a father spares his only son, that serves him, Mal. iii. 17.

Thirdly, When forbearance of punishment, and when fruits and tokens of kindness will reclaim his children from evil, and prove sufficient incentives unto good; God in this case also, like a wise and loving Father, had rather draw them by the cords of love, Hosea xi. 4, than drive them with the lashes of his displeasure. Thus you see God may love his children, and not be always afflicting them.

Well, do you prosper? Then take notice of God's goodness towards you with thanksgiving; study and endeavour therefore to be the more obedient. If you cannot, yet grieve because you cannot be more thankful and more obedient. Then, because prosperity has made you to be better, or at least to desire to be better, hence you may assure yourselves, that your prosperity is not given you in wrath, but in love. But take heed; quarrel not with God, because he forbears to afflict you; either make this use, that you be good, and amend without blows; or else be sure the more is behind.

(3.) Doubts of God's love, from inward horrors, and distresses of mind, removed.

As the fore-mentioned persons did question God's love, from considerations taken from their outward conditions; so there are very many, who, besides what they conclude from outward crosses, conclude also from their inward horrors and distresses of conscience, and from their intolerable perplexities of soul, that God does not love them; they think that their distress is other, or greater than the affliction of any of God's children; therefore they want peace, fearing that God does not love them.

Those to whom God does bear special love, may be so far perplexed with inward and strange terrors and discomforts, that they may think themselves to be forsaken of God. Thus David complains: Will the Lord cast off for ever? and will he be favourable no more? Psa. lxxvii. 7, 8, 9. Yea, not only David, but Christ Jesus himself, and his church, did in their sense and feeling, take themselves to be forsaken of God, Matt. xxvii. 46; Cant. v. 6; Isa. xlix. 14; yet none that are wise will say, that these were destitute of God's love, or were ever quite forsaken, though never so much perplexed and cast down; yet, in their own feeling and sense, in the agony of their spirits, they did thus think or speak, 2 Cor. iv. 8, 9.

God has most holy and blessed ends, in many times leading and leaving his children in such straits, that they are altogether without any sense of his love.

First, It may be a just correction of them, for their not showing love to God, and because they do in part forsake him by their sins. This is therefore to humble them, and to make them know themselves, and to bring them to repentance. God may be pacified towards them in the main, yet for a time show them no countenance: as David, though his anger was appeased towards Absalom, yet for a time he would not let him see his love, for he would not let him come into his sight; that Absalom might be more humbled, and might the more detest his sin, 2 Sam. xiv. 24.

Secondly, God exercises his beloved ones with many fears, horrors, and doubts, to prevent that spiritual pride which else would be in them, and that self-sufficiency which else they would conceive to be in themselves. If they should always have a sense of inward and spiritual comforts, and should not sometimes have pricks in the flesh, and buffetings of Satan, they would be exalted above measure, and would be something in themselves, in their own opinion, 2 Cor. xii. 7. But when there is such difficulty in getting and keeping of grace and comfort, and when they find what need they have of both, and how neither can be had but from God, in and by Christ, it will make them empty themselves of all things in themselves, that they may be something in Christ. And then, when they have grace and comfort, they will acknowledge themselves to be beholden to God for the same.

Thirdly, God withholds from his children, the sense of his favour to try the sincerity and truth of their sole dependence on him; trying, whether because God seems to forsake them, they will forsake him; whether, like king Joram, they will say, Why shall they wait upon God any longer? 2 Kings vi. 33; and whether they will, with Saul, betake them to unlawful means of help, 1 Sam. xxviii. 6, 7. Or whether, on the other side, they will say, with Job and David, Though God kill us, or forget us, yet we will trust in him, hope in him, and praise him, Job xiii. 15; Psa. xlii. 9, 11, who, they are persuaded, is, and will show himself to be, the health of their countenance, and their God. God uses to leave his children, as, in another case, he left Hezekiah, to try them, and to know what is in their hearts, 2 Chron. xxxii. 31.

Fourthly, God withdraws himself for a time, that they may learn to esteem more highly of his favour, and to desire it more, when by the want of it, they find by experience, what a hell it is to be without it; and that they may be more thankful for it, and be more careful, by studying to please God, to keep it when they have it. This holy use, David and the church made of God's forsaking them, as they thought, for a time, Psa. lxxx. 18, 19; Cant. iii. 2, 5, 6; ii. 7; viii. 4. It made them seek more diligently after God, promising that if he would turn to them, they would not go back from him; resolving, by his grace, to cleave more closely unto him.

But know this to your comfort, when God most withdraws himself and forsakes you, it is but in part, in appearance only, and but for a time. He may, for the cause before mentioned, turn away his face, and forbear to show his loving countenance; but he will not take his loving-kindness utterly from you, nor suffer his faithfulness to fail, Psa. lxxxix. 32—34. What God said to his afflicted church, that he saith to every afflicted member thereof: For a small moment have I forsaken thee: but with great mercies will I gather thee. In a little wrath have I hid my face from thee for a moment; but with everlasting kindness will I have mercy on thee, saith the Lord thy Redeemer, Isa. liv. 7, 8. Hence it is that in your greatest extremities, your faith and hope shall secretly, though you feel not their work, preserve you from utter despair. As it was with David, and with our blessed Saviour, Psa. xxii. 1, who, although these words of theirs to God, Why hast thou forsaken me, argue fear, and want of sense of God's love; yet these words, My God, my God, do argue a secret alliance and hope, Matt. xxvii. 46.

(4.) Doubts of God's love on account of extraordinary afflictions, removed.

And whereas you say, that no man's grief or troubles are like yours, partly by reason of outward afflictions, and partly by inward temptations and distresses, give me leave to deal plainly with you, it is a foolish and a false speech. Talk with a thousand thus troubled, they will also say thus: No man's case was ever as mine is, nor as bad. Will any that have but common sense, think this to be true? Most of these must needs be deceived. You feel your own distresses, but you cannot fully know what another feels.

If you would rightly look into the distresses of others, who were better than yourselves, as they are recorded in Scripture, you would not think thus. As for outward afflictions, upon whom did God ever lay his hand more heavy than on his servant Job? Job i. Had not Paul also his trouble without, of all sorts, and terrors within, &c. 2 Cor. xi. 23—33. And, if you consider sorrows, fears, and distresses of all sorts, were yours such as David's were, or more than his? I pray, what mean these, and many more such speeches? My bones are vexed; my soul is vexed. But thou, O Lord, how long?—I am weary with my groaning.—Mine eye is consumed with grief, it waxes old, Psa. vi. 2, 3, 6, 7. Why standest thou afar off, O Lord? Why hidest thou thyself in time of trouble? Psa. x. 1. How long wilt thou forget me, O Lord; for ever? How long wilt thou hide thy face from me? Psa. xiii. 1. I am poured out like water, and all my bones are out of joint. My heart is like wax, it is melted in the midst of my bowels. My strength is dried up like a potsherd; my tongue cleaves to my jaws, and thou hast brought me to the dust of death, Psa. xxii. 14, 15. My bones waxed old, through my roaring all the day. For day and night thy hand was heavy upon me, Psa. xxxii. 3, 4. There is no soundness in my flesh because of thine anger, neither is there any rest in my bones, because of my sin. Mine iniquities, that is, the punishment of mine iniquities, are gone over my head, they are too heavy for me, Psa. xxxviii. 3, 4. Thus, and much more, does he complain: I am weary of my crying, my throat is dry. Mine eyes fail while I wait for my God, Psa. lxix. 3. So Asaph, My sore ran, and ceased not, my soul refused to be comforted, Psa. lxxvii. 2.

What think you now? Were not Job, Paul, and David, in God's love and favour, notwithstanding all this? It may be, you will reply, However the matter of their trouble might be greater than yours, yet they could remember God, they could pray to him, they had faith and confidence in God in their distresses, all which you want; therefore herein your case is worse than theirs.

Consider yourselves well, I speak only to you that are truly humbled for sin, and it is to be hoped that in some measure you shall find the like grace, faith, and confidence in you, as was in them; if you see it not, be grieved for the want thereof; endeavour to do as you see they did in their distresses, only be not discouraged, and all shall be well. But take notice, I pray you, that sometimes David neither did, nor could pray, as he conceived of his own prayer, any otherwise than in roaring and complaining, at which time, he saith he kept silence, Psa. xxxii. 3. But when he could confess his sins, and pray, then he had some apprehension that God had forgiven him his sin, Psa. xxxii. 5; and for all Asaph's remembering of God, yet even then he was troubled, and his spirit was overwhelmed, and he saith, his soul refused comfort, Psa. lxxvii. 2, 3; and David saith unto God, When wilt thou comfort me? Psa. cxix. 82. I grant, it was his fault, yet it was such a fault as was incident to one beloved of God. Moreover, I deny not that Job and David had faith and hope in God; but these graces in them were oftentimes overclouded with unbelief and distrust; as appears in their various passionate exclamations; at which times their faith appeared to others in their good speeches and actions, rather than to themselves. And the Psalmist confesses, that those his faithless complaints were in his haste, and from his infirmities, Psa. xxxi. 22; lxxvii. 10.

How say you now? Is it not thus with you? Are you not like others of God's children, off and on, up and down? You would pray and cannot; you would believe, but, as you think, cannot; you would have comfort, but cannot feel it. Only you feel a secret support now and then: and now and then, you see and feel a glimpse of God's light and comfort; for which you must be thankful, which you must cherish by all means, and with which you must rest contented, waiting until God give you more.

You should know and consider, that this is an old device of Satan, to make you believe that your case is worse, or at least much different from the case of any others, because he knows that while he fixes this upon your mind, no common remedy, which did cure and comfort others, can cure and comfort you. For you will still ask, Was ever any as I am? And if God's ministers and people cannot say, yea; and that such an instruction, and such a promise in the word did help him; then you conclude that you are incurable.

But, last of all, let it be supposed that your case is worse than any body's else, is there not a sovereign balm in God's word, Jer. viii. 22, a catholicon, or universal remedy, that will heal all spiritual diseases? God's word is like himself, to a believer, an omnipotent word, Mark ix. 23. Is any thing too hard for the Lord? Gen. xviii. 14. Neither is there any spiritual disease too hard for his word and Spirit to cure. When Christ healed the people with his word, did it not heal even such as were never known to be cured before?

They made no question, whether he cured the same before. Indeed, Martha failed in this; for she said of her brother Lazarus, being dead, Lord, he stinketh, for he has been dead four days, John xi. 39, 40; she conceived her brother's case to be desperate, and that none in his state could be restored to life. But Christ blamed her for want of faith; and by his word he as easily raised Lazarus from being dead so long, as he cured Peter's wife's mother, when only sick of a fever, Mark i. 31.

It is not the greatness of any man's distress whatever, that can hinder from help and comfort; but only, as then in curing men's bodies, so now in curing and comforting men's souls, nothing hinders the cure, but the greatness of unbelief in the party to be cured, Mark vi. 5, 6; for all things are possible to him that believes, Mark ix. 23.

You will yet reply, Indeed, here lies the difficulty, in unbelief.

Well, be it so. If unbelief be your disease and trouble, do you think that God cannot cure you of unbelief, as well as of any other sin? But know, that if, with the poor man in the gospel, you feel your unbelief, and complain of it, and confess it unto God, saying, Lord, I have cause to believe; Lord, I do, I would believe, help thou my unbelief, Mark ix. 24; if you also will wait until God give you power to believe, and to enjoy comfort in believing, for faith makes no haste; this is both to believe in truth, and is a certain means to increase in believing.

Wherefore let not Satan, nor yet a fearful heart, make you to judge your case to be desperate and remediless, either in respect of God's power or will, though you are yet in distress, and feel in you much fear and unbelief. Seek to God, and with patience wait the good time of deliverance and comfort; and in due time, you shall have help and comfort, as well as others.

(5.) Doubts of God's love, because prayers are not answered, removed.

There are yet some, that fear God does not love

them, because they have prayed often and much: but God has rejected their prayers, and not answered them.

There are many just causes, why God may reject, or at least not grant your prayers; and yet may love your persons.

For, first, It may be you ask amiss, either asking things unlawful, or asking things inconvenient for the present, or in asking to have good things, temporal, or spiritual, in that quantity and degree, which God does not see fit for you as yet; or you ask good things to an ill end, as to satisfy some lust, as pride, voluptuousness, covetousness, &c., James iv. 3. Or, lastly though you failed in neither of the former, yet you failed in this, you were doubtful, you did not ask in faith, you did not believe you should have the things so asked; whosoever thus fails in asking, let them not think to receive any thing in favour from the Lord, James i. 6, 7. And it is a fruit of God's love, when he does not answer prayers so made; for it will cause you to seek him and to pray to him in a better manner, that you may be heard.

Secondly, God does many times in love and mercy hear his children's prayers, when they think he does not. God hears prayers many ways; you must observe this, else you will judge that he does not hear your prayers, when yet indeed he does. Sometimes, yea, always when it is good for you, he gives the very thing which you pray for. Sometimes he gives not that thing which you ask; but something much better. As, when you ask earthly and temporal good things, he grants them not, but instead thereof gives you things spiritual and eternal; likewise, when you ask grace in some special degree, such as joy, or comfort in God, or the like, it may please him not to let it appear that he gives the same unto you; but instead thereof, he does enlarge your desires, and he gives humility and patience, to wait his leisure, which will do you more good than that which you prayed for So, likewise, when you pray that God would free you from such a temptation; God does not always rid and ease you of it; but he, instead thereof, gives you strength to withstand it, and keeps you, that you are not overcome by it. Thus Christ was heard in that he feared, Heb. v. 7; so he said to the apostle, My grace is sufficient for thee, 2 Cor. xii. 9; which is better than to have your particular request. For now God's power is seen in your weakness, and God has the glory of it; and you hereby have experience of God's power, which experience is of excellent use.

Likewise you may desire to have such or such a cross or affliction removed; yet God may suffer the cross to remain for a time, but he gives you strength and patience to bear it; wisdom and grace to be less earthly, and more heavenly-minded by reason of it. There was never any, that, with an humble and holy heart, made lawful requests, according to the will of Christ, believing he should be heard, but, though he were a man of many failings in himself, and did discover many weaknesses in his prayer, was heard in that he prayed, either in what he asked of God or in what he should rather have asked; either in the very thing, or in a better.

I would have you therefore leave objecting, and questioning whether God loves you; consider this: has he not loved you, who has given his only begotten Son for you and to you, John iii. 16; who has washed you in his blood, Rev. i. 5; having given him to die for your sins, and to rise again for your justification, Rom. iv. 25; and has hereby translated you into the kingdom of his dear Son, Col. i. 13; having also given unto you to believe in his name, Phil. i. 29; hereby making you his children, inheritors with the saints in light? Col. i. 12. What greater sign can there be of the love of God towards you; and what better evidence can you have of God's love in justifying you, than the evidence of your faith, Heb. xi. 1, whereby you are justified? Rom. iii. 28.

(6.) A removal of false fears, from the deficiency or weakness of faith.

All men will grant, that if they were sure they had faith, they should not doubt of their justification, nor of God's love to them in Christ. But many doubt that they have no faith, or if they have any, it is so little, that it cannot be sufficient to carry them through all oppositions to the end, unto salvation.

If you have any faith, though no more than as a grain of mustard-seed, you should not fear your final estate, nor yet doubt of God's love; for it is not the great quantity and measure of faith that saves; but the excellent property and use of faith, though never so small; Luke xvii. 6. For a man is not saved by the worth of his faith, by which he believes; but by the worth of Christ, the person on whom he believes. Now the least true faith does apprehend Christ entirely, to all the purposes of salvation; even as a little hand may hold a jewel of infinite worth, as well, though not so strongly, as a larger. The least infant is as truly a man, as soon as ever it is endued with a reasonable soul, as afterward, when it is able to show forth the operations of it, though not so strong a man; even so it is in the state of regeneration. Now you should consider, that God has babes in Christ, as well as old men, 1 John ii. 12; feeble-minded as well as strong; sick children as well as healthy, in his family, 1 Thess. v. 14; Rom. xiv. 1, xv. 1. And those that have least strength, and are weakest, of whom the Holy Ghost saith, they have a little strength, in comparison, yet they have so much as, through God, will enable them in the time of greatest trials, to keep God's word, and that they shall not deny Christ's name, Rev. iii. 8. Also know, God, like a tender father, does not cast off such as are little, feeble, and weak, but has given special charge concerning the cherishing, supporting, and comforting of these more than others, 1 Thess. v. 14. And Christ Jesus will confirm and increase, and not quench, the least spark of faith, Matt. xii. 20.

This which I have said in commendation of little faith, is only to keep him that has no more from despair. Let none hereby please or content himself with his little faith, not striving to grow, and to be strong in faith. If he do, it is to be feared that he has none at all; or if he have, yet he must know that he will have much to do to live, when he has no more than can keep life and soul together, and his life will be very unprofitable and uncomfortable, in comparison of him that has a strong faith.

SECT. 9. REASONS WHY CHRISTIANS THINK THEY HAVE NO FAITH, CONSIDERED.

But you will say, (1.) you are so full of fears and doubtings; (2.) you are so fearful to die, and to hear of coming to judgment; and (3.) you cannot feel that you have faith, you cannot feel joy and comfort in believing, therefore you fear you have no faith.

First, If you, having so sure a word and promise, do yet doubt and fear so much as you say, it is your great sin, and I must blame you now, in our Saviour's name, as he did his disciples then, saying, Why are ye fearful, why are ye doubtful, O ye of little faith? Matt. viii. 26. But, to your reformation and comfort, observe it, he does not argue them to be of no faith, but only of little faith, saying, O ye of little faith, Matt. xiv. 31. Thus you see that some fears and doubtings do not argue no faith.

Secondly, Concerning fear of death and judgment, some fear does not exclude all faith. Many from their natural constitution are more fearful of death than others. Yea, pure nature will startle and shrink to think of the separation of two so near, so ancient, and such dear friends as the soul and body have been. Good men, such as David and Hezekiah, have showed their unwillingness to die. And many, upon a mistake, conceiving the pangs and pains of death in the parting of the soul and body to be most torturing and unsufferable, are afraid to die. Whereas unto many, the nearer they are to their end, the less is their extremity of pain; and very many go away in a quiet swoon, without pain.

And as for being moved with some fear, at the thought of the day of judgment; who can think of that great appearance, before so glorious a Majesty, such as Christ shall appear in, Matt. xvi. 27, to answer for all the things he has done in his body, without trembling? 2 Cor. v. 10, 11. The apostle calls the thoughts thereof "the terrors of the Lord." Indeed, to be perplexed with the thoughts of the one or the other, argues imperfection of faith and hope, but not an utter absence of either.

You have other and better things to do in this case, than to make such dangerous conclusions, viz. that you have no faith, &c., upon such weak grounds. You should rather, when you feel this over-fearfulness to die and come to judgment, labour to find out the ground of your error, and study to endeavour to reform it.

Unwillingness to die, may proceed from these causes:

First, From too high an estimation of, and too great a love to earthly things of some kind or other; which makes you afraid, and unwilling to part with them.

Secondly, You may be unwilling to die, because of ignorance of the superabundant and inconceivable excellencies of the happiness of saints departed, which if you knew, you would be willing.

Thirdly, Fear of death and coming to judgment, does, for the most part, rise from a conscience fearful of the sentence of condemnation, being without assurance, that when they die they shall go to heaven.

Wherefore if you would be free from troublesome fear of death and judgment, learn,

(1.) To think meanly and basely of the world, in comparison of those better things, provided for them that love God; and use all the things of the world accordingly, without setting your heart upon them, Psa. lxii. 10, as if you used them not, 1 Cor. vii. 29—31. (2.) While you live here on earth, take yourselves aside often in your thoughts from the cares and business of the world, and enter into heaven, and contemplate deeply the joys thereof. (3.) Give all diligence to make your calling and election, and right unto heaven, sure to yourselves, 2 Peter i. 10, 11:

but let me give you this needful item, that you be willing and ready to judge it to be sure, when it is sure, and when you have cause so to judge. Let your care be only, through faith in Christ Jesus, to live well, joining unto faith, virtue, &c., and you cannot but die well, 2 Peter i. 5. Death at first appearance, like a serpent, seems terrible; but by faith, you may see this serpent's sting taken out, which when you consider, you may, for your refreshment, receive it into your bosom. The sting of death is sin, the strength of sin is the law, 1 Cor. xv. 55, 56; but the law of the spirit of life in Christ has freed you from the law of sin and death, Rom. viii. 2. I confess, that when you see this pale horse, death, approaching, it may cause nature to shrink, but when you consider that his errand is to carry you with speed to your desired home, unto a state of glory, how can you but desire he should remove you out of this vale of misery, that mortality might be swallowed up of life? 2 Cor. v. 4.

If you would do this in earnest, you would be so far from fearing death, that you would, if it were put to your choice, with the apostle, choose to be dissolved, and to be with Christ, which is the best of all, Phil. i. 23; and so far from fearing the day of judgment, you would love and long for Christ's appearing, 2 Tim. iv. 8; waiting with patience and cheerfulness, when your change shall be, Job xiv. 14. Endeavour to follow these directions; then, if you cannot prevent those fears, and conquer them as you would, yet be not discouraged, for fears and doubts in this kind do flow many times from strength of temptations, rather than from weakness of faith.

Moreover, what if you cannot attain to so high a pitch in your faith as Paul had, are you so ambitious that no other degrees of faith shall satisfy you? Or are you so foolish, as thence to conclude that you have no faith?

Thirdly, Whereas you say, you are without feeling, therefore you fear you have no faith; I acknowledge that want of a feeling sense of God's favour, is that which does more trouble God's tender-hearted children, and make them more doubt of God's love, and of their justification, than any thing else; whereas I know nothing that gives them less cause.

(1.) In what true faith consists.

For first, What do you mean by feeling? If you mean the enjoyment of the things promised, and hoped for, by inward sense; this is to overthrow the nature, and put an end to the use of faith and hope. For faith is the substance of things hoped for, and the evidence of things not seen, Heb. xi. 1. And the apostle saith, Hope that is seen, is not hope, Rom. viii. 24. Indeed, faith gives a present being of the thing promised to the believer, but it is a being, not in sense, but in hope and assured expectation of the thing promised: wherefore the apostle, speaking of our spiritual conversation on earth, saith, We walk by faith, not by sight, 2 Cor. v. 7. These two, faith and feeling, are opposite one to the other in this sense; for when we shall live by sight and feeling, then we shall cease to live by faith.

(2.) The difference between faith and assurance.

Secondly, If by feeling you mean a joyous and comfortable assurance that you are in God's favour, and that you shall be saved, and therefore, because you want this joyous assurance, you think you have no faith, you must know this is a false conclusion.

For faith, whereby you are saved and brought into a state of grace, and this comfortable assurance that you are in a state of grace and shall be saved, differ much from each other. It is true, assurance is an effect of faith. Yet it is not inseparable from the very being of faith, at all times. For you may have saving faith, yet at sometimes be without the comfortable assurance of salvation.

To believe in Christ to salvation is one thing, and to know assuredly that you shall be saved is another. For faith is a direct act of the reasonable soul, receiving Christ, and salvation offered by God with him. Assurance rises from a reflex act of the soul, namely, when the soul by self-inquiry, and the help of God's Spirit, can witness that it has the afore-mentioned grace of faith, whereby it can say, I know that I believe in Christ Jesus; and I know that the promises of the gospel belong unto me. The holy Scriptures are written for both these ends, that first faith, and then assurance of faith and hope should be wrought in men. These things are written, saith John in his gospel, that ye might believe, that Jesus is the Christ, the Son of God, and that believing ye might have life through his name, John xx. 31. Again, these things have I written, saith the same apostle in his Epistles, to ye who believe on the name of the Son of God, that ye may know that ye have eternal life, and that ye may believe, that is, continue to believe, and increase in believing, on the name of the Son of God, 1 John v. 13.

A man is saved by faith, but has comfort in hope of salvation by assurance; so that the being of spiritual life, in respect of us, does subsist in faith, not in assurance and feeling. And that is the strongest and most approved faith, which cleaves to Christ, and to his promises, and rests upon his truth and faithfulness, without the help of feeling. For, although assurance gives unto us a more evident certainty of our good estate, yet faith, even without this, will certainly preserve us in this good estate, whether we be assured or not.

Hence it is, that although reason, as it is now corrupt, will still be objecting, and will be satisfied with nothing but what it may know by sense, John iii. 4; yet faith, even above and against sense, and all natural reasoning, Rom. iv. 19, 20, from a reverence to God's command, who bids to believe and trust in him, and a persuasion of the truth and goodness of the promises will give credit unto, and rest upon the bare, naked divine witness of the word of God, for his sake that does speak it, Heb. xi. 8, 11.

Secondly, There is a certainty of evidence; namely, when the thing believed is not only said to be true and good, but a man does find it so to be by sense and experience, and is so evident to man's reason, convincing it by force of argument, taken from the causes, effects, properties, signs, and the like, that it has nothing to object against the thing proposed to be believed. The certainty of adherence is the certainty of faith. The certainty of evidence is the certainty of assurance.

This certainty of assurance and evidence is of excellent use, for it makes the Christian fruitful in good works, and does fill him full of joy and comfort: therefore it must by all means be sought after, yet it is not of itself so strong, nor so constant, nor so infallible as the certainty of faith and adherence is, 2 Peter i. 8, 10. For sense and reason since the fall even in the regenerate, are weak, variable, and their conclusions are not so certain, as those of pure faith; because faith builds only upon divine testimony, concluding without reasoning or disputing, yea, many times against reasoning, Rom. iv. 18, 19; Heb. xi. 11.

So that notwithstanding the excellent and needful use of assurance, it is faith and adherence to Christ and his promises, which, even in fears and doubts, must be the cable we must hold by, lest we make shipwreck of all, when we are assaulted with our greatest temptations; for then many times our assurance leaves us to the mercy of the winds and seas, as mariners speak. If you have faith, though you have little or no feeling, your salvation is yet sure in truth, though not in your own apprehension. When both can be had, it is best, for then you gain most strength and most comfort, giving you cheerfulness in all your troubles; but the power and grace of the Lord Jesus Christ, and faith in his naked word and promise, is that to which you must trust.

See this in the examples of most faithful men; for when they have been put to it, it was this that upheld them, and in this was their faith commended. Abraham against all present sense and reason, even against hope, believed in hope, both in the matter of receiving a son, and in going about to offer him again unto God in sacrifice. He denied sense and reason, he considered not the unlikelihoods, and seeming impossibilities in the judgment of reason, that ever he should have a seed, he being old, and Sarah being old and barren;

or having a seed, that he should be saved by that seed, since he was to kill him in sacrifice, Heb. xi. 17—19. He only considered the almighty power, faithfulness, and sovereignty of him that had promised, he knew it was his duty to obey and wait, and so let all the matter concerning it rest on God's promise, Rom. iv. 18—21. For this his faith is commended, and he is said to be strong in faith, Rom. iv. 20.

Job and David, or Asaph, showed most strength of faith, when they had little or no feeling of God's favour, but rather the contrary. Job had little feeling of God's favour, when for pain of body he said, Wherefore do I take my flesh in my teeth, and in anguish of soul he said, Wherefore hidest thou thy face, and takest me for thine enemy, Job xiii. 14, 24. Yet then this adherence of faith caused him to cleave unto God, and say in the same chapter, Though he slay me, yet will I trust in him, verse 15. When David said to God, Why hast thou forgotten me? Psa. xlii. 9, his assurance was weak; yet even then his faith discovered itself, when he saith to his soul, Why art thou disquieted within me? Hope thou in God, who is the health of my countenance, and my God, Psa. xlii. 11. You see then that the excellency of faith lies not in your feeling, but, as the Psalmist speaks by experience, in cleaving close unto the promise, and relying on God for it, upon his bare word. For he saith, It is good for me to draw near to God. I have put my trust in the Lord God, Psa. lxxiii. 28. This was that which secretly upheld him, and kept him in possession, when his evidences and assurance were to seek.

Wherefore, Believe God's promises made to you in Christ, and rest on him; even when you want joy, and feeling comfort. For having faith, you are sure of heaven, though you be not so fully assured of it as you desire. It will be your greatest commendation, when you will be dutiful servants and children at God's commandment, though you have not present wages, when you will take God's word for that. Those are bad servants and children, who cannot go on cheerfully in doing their master's or father's will, except they may receive the promised wages, in good part aforehand, or every day; or except they may have a good part of the promised inheritance presently, and in hand. Feeling of comfort is part of a Christian's wages and inheritance, to be received at the good pleasure of God, that freely gives it, rather than a Christian duty. To comfort and stay ourselves on God in distress, is a duty, but this joyful sense and feeling of God's favour, is a gracious favour of God towards us, not a duty of ours towards God. It argues too much distrust in God, and too much self-respect, when we have no heart to go about his work, except we be full of feeling of his favour. He is the best child, or servant, that will obey out of love, duty, and conscience; and will trust in God and wait on him for his wages and recompence.

Thirdly, When you say, you cannot feel that you have faith or hope, you mean, as indeed many good souls do, you cannot find and perceive, that these graces are in you in truth, which, if you did, you would not doubt of your salvation. My answer is, if faith and hope be in you, then if you would judiciously inquire into yourselves, and feel for them, you may find and feel them, and know that you have them; for, as certainly as he that sees bodily, may know that he sees; so he that has the spiritual sight of faith, may know that he has faith. Wherefore try and feel for your faith, and you shall find whether it be in you, yea or no.

(3.) The nature and properties of saving faith.

For this cause, (1.) Try whether you ever had the necessary preparatives, which ordinarily make way for the seed of faith to take root in the soul. (2.) Consider the nature of saving faith, and whether it has wrought in you accordingly. (3.) Consider some consequents and certain effects thereof.

First, Concerning the preparatives to faith. Has the law shut you up, in your apprehension, under the curse, so that you have been afraid of hell? And has the Spirit also convinced you of sin by the gospel, to the wounding of your conscience, and to the working of true humiliation, causing the heart to relent, and to desire to know how to be saved? And if after this you have denied yourself, as to your own wisdom and will, power and goodness, and received and rested on Christ alone for salvation, according to the nature of true faith, as follows, then you have faith.

If you doubt you were never sufficiently humbled, then read Section X., of this Chapter.

Secondly, Consider rightly the nature and proper acts of faith, lest you conceive that to be faith which is not, and that to be no faith which is.

You may know wherein true saving faith consists, by this which follows: whereas, man being fallen into a state of condemnation by reason of sin, thereby broke the covenant of works, it pleased God to ordain a new covenant, the covenant of grace, establishing it in his only Son, Christ Jesus, expressing the full tenor of this his covenant in the gospel, wherein he makes a gracious and free offer of the Lord Jesus Christ, in whom this covenant is established, and with him the covenant itself, with all its unspeakable blessings, unto man. Now when a man burdened with his sin, understanding this offer, gives credit, and assents thereunto, because it is true, and approves it, and consents to it, both because it is good for him to embrace it, and because it is the will and commandment of God, that he should consent for his part, and trust to it; when therefore a man receives Christ Jesus thus offered, together with the whole covenant, in all its duties and privileges, so far as he understands it; resolving to rest on that part of the covenant made and promised on God's part, and to stand to every branch of the covenant, to be performed on his part; thus to embrace the covenant of grace, and to receive Christ, in whom it is confirmed, is to believe.

This offer of Christ, and the receiving him by faith, may clearly be expressed by an offer of peace and favour, made by a king unto a woman, that is a rebellious subject; by making offer of a marriage between her and his only son, the heir apparent to the crown, who, to make way to this match, undertakes by his father's appointment, to make full satisfaction to his father's justice in her behalf, and to make her every way fit to be a daughter to a king. And for effecting this match between them, the son, with the consent and appointment of his father, sends his chief servants a wooing to this unworthy woman; making offer of marriage in their master's behalf, with the clearest proofs of their master's good-will to her, and with the greatest earnestness and entreaties, to obtain her good-will, that may be. This woman at first, being a bond-woman unto this king's mortal enemy, and being in love with base slaves like herself, companions in her rebellion, she aptly sets light by this offer; or, if she consider well of it, she may doubt of the truth of this offer, the match being so unequal and so unlikely on her part; knowing herself to be so base and unworthy, she may think the motion to be too good to be true; yet, if upon more advised thoughts, she does take notice of the danger she is in while she stands out against so powerful a king in her rebellion, and does also see and believe, that the king's son is in earnest in his offer to reconcile her to his father, and that he would indeed match with her; thereupon she considers also that it will be good for her to forsake all others, and take him; and that especially because his person is so lovely, and every way worthy of her esteem. Now when she can bring herself to believe this, and resolve thus, though she comes to it with some difficulty, yet if she give a true and hearty consent to have him, and to forsake all other, and to take him as he is, to obey him as her lord, and to take part with him in all conditions better or worse; though she come to this resolution with much ado, then the match is as good as made between them; for hereupon follow the mutual embracing of, and interest in each other.

The application is easy throughout: I will only apply so much as is for my purpose, to show the nature of justifying faith.

God offers his only begotten Son, Jesus Christ, yea, Christ Jesus by his ministers offers himself in the gospel, unto rebellious man, to match with him, 1 Cor. ii. 2; only on this condition, that forsaking his kindred and father's house, Psa. xlv. 10, forsaking all that he is in himself, he will receive him as his head, husband, Lord, and Saviour, Rom. vii. 4. Now, when any man understands this motion, so far as to yield assent and consent to it, and to receive Christ, and cleave to him, John i. 12, then he believes to salvation; then the match is made between Christ and that man; then they are betrothed, nay married, and are no longer two, but are become one spirit, 1 Cor. vi. 17.

By all this you may see, that in saving faith there are these two acts:

1. An assent to the truth of the gospel, not only believing in general, that there is a Christ, believing also what manner of person he is, and upon what condition he offered himself to man as a Saviour; but also believing that this Christ graciously offers his love and himself to the Christian's self in particular.

2. An hearty approbation of this offer of Christ, with consenting, and hearty embracing of it, as our own peculiar duty and privilege; resolving to take him wholly, and fully as he is; accepting of him according to the full tenor of the marriage covenant, not only as a man's Saviour, to defend him from evil, and to save him and bring him to glory; but as his head to be ruled by him, as his Lord and King, to worship and obey him, Psa. xlv. 11; believing in him, not only as his priest to satisfy, and to make intercession for him, but also as his prophet to teach, and as his king to govern him; cleaving to him in all estates, taking part with him in all the evils that accompany the profession of Christ's name, as well as in the good, Luke ix. 23.

The first act is not enough to save any; the second act cannot be without the former: where both these are, there is a right receiving of the gospel, there is true faith.

The principal matter lies in the consent and determination of the will in receiving of Christ; which, that may be without exception, know,

1. It must be with an advised and considerate will; it must not be rash, and on a sudden, in your ignorance, before you well know what you do. You must be well advised, and consider well of the person to whom you give your consent, that you know him, and that you know the nature of this spiritual union, and what you are bound unto by virtue of it, and what it will cost you, if you give yourself to Christ, Luke xiv. 28, 31.

2. Your consent must be with a determinate and complete will; with a present receiving him, even with all the heart, Acts viii. 37. It must not be a faint consent, in an indifferency whether you consent or no; it must not be in a purpose, that you will receive him hereafter; but you must give your hand and heart to him for the present, else, it is no match.

3. Your consent must be with a free and ready will; it must not be with a forced and constrained yielding, against the will; but, howsoever, it may be with much opposition and conflict, yet, you must so beat down the opposition, that when you give consent, you bring your will to do it readily and freely, with thankful acknowledging yourselves unspeakably obliged to the Lord Jesus Christ all the days of your life, that he will vouchsafe to make you such an offer.

When consent is rash, faint, and forced, this will not hold good any long time; but when your consent is advised, full, and free, out of true love to Christ, as well as for your own benefit; the knot of marriage between Christ and you is knit so fast, that all the lusts of the flesh, all the allurements of the world, and all the powers of hell, shall not be able to break it.

By this which has been said concerning the nature of faith, many, who thought they had faith, may see that yet they have none.

For they only believe in general that there is a Christ and a Saviour, who offers grace and salvation to mankind, and hereupon they presume. This general faith is needful, but that is not enough; it must be a persuasion of God's offer of Christ to a man in particular, that the will in particular may be induced to consent. There must likewise be that particular consent of will, and accepting of Christ, upon such terms as he is offered. They that receive Christ aright enter into the marriage covenant, resolving to forsake all others, and obey him, and to take up his cross, and to endure all hardships with him, and for him, as shame, disgrace, poverty, hatred in the world, and all manner of reproach; this they consent to, and resolve upon for the present, and from this time forward, for the whole time of their life; which things many neither did, nor intended to do, when they gave their names to Christ; they only received him as their Jesus, one by whom they hoped to be saved and honoured, expecting that he should endow them with a fair jointure of heaven, but they did not receive him as their Lord. In doing thus, they erred in the essentials of marriage. For they erred in the person, taking an idol Christ, for the true Christ. They erred in the form of marriage; they took him not for the present, nor absolutely, for better for worse, as we say, in sickness and health, in good report and ill report, in persecution and in peace, forsaking all other, never to part, no not at death. Wherefore Christ does not own those foolish virgins, when they would enter the bride-chamber, but saith, I know you not, Matt. xxv. 12, because there was no true consent on their part, they had no faith; and their contract or marriage with Christ was only in speech, but was never legal, or consummated.

By this which has been said, others who have faith indeed, may know they have it, namely, if they so believe the covenant of grace established in Christ, that with all their hearts they accept of him and it, so that they sincerely desire and purpose to stand to it on their parts, as they are able, and rest on it so far as it concerns Christ to fulfil it. For this is faith.

Unto this, some fearful souls will reply; If we have no faith, except to an assent unto the truth, we do also receive Christ offered, with a deliberate, entire, and free consent, to rest on him, to be ruled by him, and to take part with him in all conditions; then we doubt that we have no faith, because we so hardly brought ourselves to consent, and find ourselves so weak in our consent, and have been so unfaithful in keeping promise with Christ.

Truth, fulness and firmness of consent of the will to receive Christ may stand with many doubtings, and with much weakness and sense of difficulty, in bringing the heart to consent. For so long as there is a law in your members warring against the law of your mind, you can never do as you would, Rom. vii. 23. If you can bring your hearts to will to consent and obey, in spite of all oppositions, this argues hearty and full consent, and a true faith, Isa. i. 19. Nay, if you can bring the heart but to desire to receive Christ, and to enter into covenant with God, made mutually between God and you in Christ, and that it may stand according to the offer which he makes unto you in his word, even this argues a true and firm consent, and makes up the match between Christ and you. Even as when Jacob related the particulars of an earthly covenant, into which he would have Laban enter with him, Laban's saying, I would it might be according to thy word, gave proof of his consent, and did ratify the covenant between them, Gen. xxx. 34. If you can therefore, when God offers unto you the covenant of grace, commanding you to receive Christ, in whom it is established, and to enter into this covenant; if, I say, you can with all your heart, say to God, I would it might be according to thy word; the covenant is mutually entered into, and the match is made between Christ and you.

And whereas it troubles you, that you cannot be so faithful to Christ, as your covenant binds you, it is well you are troubled, if you did not also make it an argument that you have no faith; for in that it heartily grieves you, that you cannot believe, nor perform all faithfulness to Christ, it is an evident sign that you have faith. You must not think that after you are truly married to Christ, you shall be free from evil solicitations by your old lovers; nay, sometimes a kind of violence may be offered, by spiritual wickedness, unto you, so that you are forced to many evils against your will, Rom. vii. 19; as it may befall a faithful wife, to be forced by one stronger than she; yet if you give not full consent unto them, and suffer not your heart to follow them, your husband Christ will not impute these forced evils unto you.

Yet, let none by this take liberty to offend Christ in the least thing, for though Christ love you more tenderly and more mercifully than any husband can love his wife, yet know, he does not dote on you; he can see the smallest faults, and will sharply, though kindly, rebuke and correct you for them, if you do them presumptuously. But he esteems none to break spiritual wedlock, so as to dissolve marriage, but those whose hearts are wholly departed from him, and are set upon, and given to something else, Heb. iii. 12. If you thus look into the nature of faith, I speak to a soul troubled for sin, you may know and feel that you have it.

(4.) True faith may be discerned by its effects.

You may know a lively faith likewise, by most certain consequences and effects, I mean not comfort and joy, which are sometimes felt, and sometimes not; but by such effects, which are most constant, and more certain, and may be no less felt than joy and comfort, if you would search for them; amongst others, I reckon these:

1. You may know you have faith by your grieving for, and opposing of the contrary; if you feel a fight and conflict between believing and doubting, fear and distrust; and in that combat you take part with believing, hope, and confidence, or at least desire heartily that these should prevail, and are grieved at heart, when the other gets the better; if you feel this, do not say, you have no feeling. Do not say, you have no faith.

This conflict, and desire to have faith, gave proof, that the man in the gospel, who came to Christ to cure his child, had faith. I believe, Lord, saith he, Lord, help mine unbelief, Mark ix. 24. Do not say, as I have heard many, This man could say, I believe; but we cannot say so. I tell you, if you can heartily say, Lord, help my unbelief, I am sure, any of you may say, I believe. For, whence is this sense of unbelief, and desire to believe, but from faith?

2. You may know you have faith, I speak still to an afflicted soul, which dare not sin wilfully, inasmuch as you will not part with that faith which you have upon any terms. I will ask you, who have given hope to others, that you do believe, and that doubt you have not truth of faith and hope in God, only these questions, and as your heart can answer them, so you may judge. Will you part with that faith and hope which you call none, for any price? Would you change present states with those who presume they have a strong faith, whose consciences do not trouble them, but are at quiet, though they live in all manner of wickedness? or at best are merely civilly honest? Nay, would you, if it were possible, forego all that faith, and hope, and other graces of the Spirit, which you call none at all, and return to that former state, wherein you were in the days of your vanity, before you endeavoured to leave sin, and to seek the mercy of God in Christ Jesus in good earnest? Would you lay any other foundation to build upon, than what you have already laid? Or is there any other person or thing, whereon you desire to rest for salvation and direction, besides Christ Jesus? If you can answer, No; but can say, with Peter, To whom should we go? Christ only has the words of eternal life, John vi. 68; you know no other foundation to lay, than what you have laid, and have willed, and desired to lay it right; you resolve never to pull down what you have built, though it be but a little; and it is your grief that you build no faster upon it. By this answer you may see, that your conscience, before you are aware, witnesses for you, and will make you confess that you have some true faith and hope in God, or at least hope that you have. For, let men say what they will to the contrary, "they always think they have those things, which by no means they can be brought to part with."

3. If you would have sensible proof of your faith and justification; look for it in the most certain effect, which is in your sanctification. Do you feel your-

selves loaded and burthened with sin; and your hearts distressed with sorrow for it? And do you also perceive yourselves to be altered from what you were? Do you not bear good will to God's word and ordinances? And do you desire the pure word of God, that you may grow in grace by it? 1 Peter ii. 2. Do you love and consort with God's people, because you think they fear God? 1 John iii. 14. Is it your desire to approve yourselves to God, in holy obedience? And is it your trouble, that you cannot do it? Then certainly you have faith, you have an effectual faith. For what are all these but the very pulse, breath, and motions of faith? James ii. 22, 26. If you feel grace to be in you, it is a better feeling, than feeling of comfort; for grace, in men of understanding, is never separated from effectual faith, but comfort many times is; for that may rise from presumption and false faith. Grace, only from the Spirit of God, and from true faith.

SECT. 10. FEARS CONCERNING THE TRUTH OF SANCTIFICATION, REMOVED.

It is granted by all, that if they are truly sanctified, then they know that they have faith, and are justified; but many fear they are not sanctified, and that for these seeming reasons:

(1.) Fears of not being sanctified for want of deep humiliation, answered.

First, Some fear they are not sanctified, because they do not remember, that ever they felt those wounds and terrors of conscience, which are first wrought in men to make way to conversion; as it was in them who were pricked to the heart at Peter's sermon, Acts ii. 37; and in Paul, Acts ix. 6; and in the jailor, Acts xvi. 29. Or if they felt any terrors, they fear they were but certain flashes, and forerunners of hellish torments; like those of Cain, Gen. iv. 13, and Judas, Matt. xxvii. 3, 4.

As it is in the natural birth, with the mother, so it

is in the spiritual birth with the child. There is no birth without some travail and pain, but not all alike. Thus it is in the new birth with all that are come to years of discretion. Some have so much grief, fear, and horror, that it is intolerable, and leaves so deep an impression, that it can never be forgotten; others have some true sense of grief and fear, but nothing to the former in comparison, which may easily be forgotten.

There are causes, why some feel more grief and fear in their first conversion than others:

1. Some have committed more gross and heinous sins than others; therefore, they have more cause and need to have more terror and humiliation than others.

2. God sets some apart for greater employments than others, such as will require a man of great trust and experience; wherefore God, to prepare them, exercises such with the greatest trials, for their deep humiliation, and for their more speedy and full reformation, that all necessary graces might be more deeply and firmly rooted in them.

3. Some have been religiously brought up from their infancy, whereby, as they were kept from gross sins; so their sins were subdued by little and little, without any sensible impression of horror; grace and comfort being instilled into them almost insensibly.

4. Some by natural constitution and temper of body, are more fearful, and more sensible of anguish than others, which may cause that although they may be alike wounded in conscience for sin, yet they may not feel it all alike.

5. There may be the like fear and terror wrought in the conscience, of sin, in one as well as another; yet it may not leave the like lasting sense and impression in the memory of the one, as in the other; because God may show himself gracious in discovering a remedy, and giving comfort to one, sooner than the other. As two men may be in peril of their lives by enemies; the one, as soon as he sees his danger sees an impregnable castle to step into, or an army of friends to rescue him; this man's fear is quickly over

and forgotten; the other does not only see great danger, but is surprised by his enemies, is taken and carried captive, and is a long time in cruel bondage and fear of his life, till at length he is redeemed out of their hand. Such a fear as this can never be forgotten.

You may evidently know, whether you had sufficient grief and fear in your first conversion, by these signs. Had you ever such, and so much grief for sin, that it made you dislike sin, and to dislike yourself for it, and to be weary and heavy laden with it; so as to make you heartily confess your sins unto God, and to ask of him mercy and forgiveness? Has it made you to look better to your ways, and more careful to please God? Then be sure, it was a competent and sufficient grief; because it was a godly sorrow to repentance, never to be repented of, 2 Cor. vii. 10.

Again, are you now grieved and troubled, when you fall into particular sins? Then you may be certain, that there was a time when you were sufficiently humbled in your conversion; for this latter grief is but putting that grief into further act; whereof you received an habit in your first conversion.

If you can for the present find any proof of conversion, it should not trouble you, though you know not when, or by whom, or how you were converted; any more than thus, that you know God has wrought it by his word and Spirit. When any field brings forth a crop of good corn, this proves that it was sufficiently ploughed: for God does never sow, until the fallow ground of men's hearts is sufficiently broken up.

Now as for those who remember that they have had terrors of conscience, and, it may be, ever and anon feel them still, who fear that these were not beginnings of conversion, but rather beginnings of desperations and hellish torments; you should know, that there is a great difference between these and those.

1. Those fears and horrors, which are only flashes and beginnings of hellish torments, are wrought only by the law and spirit of bondage, giving not so much as a secret hope of salvation. But those fears, which make way unto, and which are the beginnings of conver

sion, are indeed first wrought by the law also, yet not only, for the gospel has, at least, some share with them; partly to melt the heart, broken by the law, partly to support the heart, causing it by some little glimpse of light, to entertain a possibility of mercy. Compare the terrors of Cain and Judas, with those of the men pricked at Peter's sermon, with Paul's and the jailor's, and you shall see both this, and the following differences.

2. The former terrors and troubles are caused, either only for fear of hell, and the fierce wrath of God, but not for sin; or, if at all for sin, it is only in respect of the punishment. These tending to conversion, are also caused through fear of hell, but not only; the heart of one thus troubled is grieved because of his sin, and that not only because it deserves hell, but because by it he has offended and dishonoured God.

3. Those, who are troubled in the first sort continue headstrong and obstinate, retaining their usual hatred against God, and against such as fear God, as also their love to wickedness: only, it may be, they may conceal and smother their rancour, through the spirit of restraint, that for the time it does not appear; but in the other will appear some alteration towards goodness; as, whatsoever their opinions and speeches were of God's people before, now they begin to think better of them, and of their ways. So did they in the Acts; before they were pricked at heart, they did scoff at the apostles, and derided God's gifts in them, Acts ii. 13; but afterwards said. Men and brethren; they thought reverently of them, and spoke reverently to them, Acts ii. 37. See the same in Paul, in his readiness to do whatsoever Christ should enjoin him, Acts ix. 6. The jailor, also in this case, quickly became well affected to Paul and Silas, Acts xvi. 24, 30, 33.

4. The former sort, when they are troubled with horror of conscience, fly from God, and seek no remedy, but such as is worldly and carnal, as company-keeping, music, and other earthly delights, as in building, and in their lands and livings, according as

their own corrupt hearts, and their vain companions advise them, whereby sometimes they stupefy and deaden their conscience, and lay it asleep for a time. Thus Cain and Saul allayed their distempered spirits, Gen. iv. 17, &c., 1 Sam. xvi. 17. And if they had some godly friends, who shall bring them to God's ministers, or do themselves minister to them the instructions of the word, this is tedious and irksome to them; they cannot relish these means, nor take any satisfaction in them. But the other are willing to seek to God, by seeking to his ministers, Acts ii. 37, to whom God has given the tongue of the learned, to minister a word in season, to the soul that is weary; and though they cannot presently receive comfort, will not utterly reject them, except in case of melancholy, which must not be imputed to them, but to their disease, Isa. l. 4.

And in application of the remedy, as there were two parts of the grief, so they must find remedies for both, or they cannot be fully satisfied. First, They were filled with grief for fear of hell; for the removing of which, the blood of Christ is applied, together with God's promise of forgiveness to him that believes, and a commandment to believe; all this is applied to take away the guilt and punishment of sin. Secondly, They were troubled for sin, whereby they dishonoured and displeased God; now unless also they feel in some measure, the grace of Christ's Spirit healing the wound of sin, and subduing the power of it, and enabling them at least to will and strive to please God, they cannot be satisfied. As it was with David, though God had said by the prophet, The Lord has put away thy sin, that is, forgiven it, 2 Sam. xii. 13; yet he had no comfort until God had created in him a clean heart, and renewed a right spirit within him, Psa. li. 10. Whereas if fear of hell be removed, it is all that the former sort care for.

5. As for the first sort, it may be, while they were afraid to be damned, they had some restraint of sin, and, it may be, made some essays towards reformation; but when their terrors are over and forgotten, then like the dog, they return to their vomit, and like

the sow that was washed, to their wallowing in the mire of their wonted ungodliness, 2 Peter ii. 22. But as for them, whose terrors were preparations to conversion, when they obtain peace of conscience, they are exceedingly thankful for it, and are made by it more fearful to offend. And although they may, and often do fall into some particular sin or sins, for which they renew their grief and repentance; yet, they do not fall into an allowed course of sin any more. Thus much in answer to the first doubt of sanctification.

(2.) Fears of not being sanctified from the intrusion of many evil thoughts.

Secondly, There are many who doubt they are not sanctified, because of those swarms of evil thoughts which are in them; some whereof, which is fearful for them to think or speak, are blasphemous, unnatural, and inhuman; calling God's being, truth, power, and providence into question; doubting whether the Scripture be the word of God, and others of this nature, having also thoughts of laying violent hands upon themselves and others, with many more of that and other kinds of evil and blasphemous thoughts, such as they never felt at all, or not so much, in their known state of unregeneracy, before they made a more strict profession of godliness; and such as, they think, none that are truly sanctified are troubled with.

To resolve this doubt, know that evil thoughts are either put into men from without, as when Satan does suggest, or wicked men do solicit to evil, 1 Chron. xxi. 1; thus Job's wife, Curse God and die, Job ii. 9; or they rise from within, out of the evil concupiscence of man's own heart, Matt. xv. 19; and sometimes they are mixed, coming both from within and without.

Those which come only from Satan, may usually be known from them that arise out of man's heart, by their suddenness and incessantness; namely, when they are repelled they will sometimes return again a hundred times in a day. Also they are unreasonable and unnatural; strange and violent in their motions; receiving no check, but by violent resistance. Whereas, those which altogether, or in great part, are from man's own corrupt heart, they usually arise by occasion of some external object, or from some natural cause, and are not so sudden and incessant, nor so unnatural and violent.

Now all those evil thoughts, or thoughts of evil rather, which are from Satan; if you consent not unto them, but abhor and resist them with detestation, they are not your sins, but Satan's, and theirs that put them into you. They are your crosses, because they are matter of trouble to you, but they are not your sins, because they leave no guilt upon you. They are no more your sins than these thoughts, Cast thyself down headlong, and fall down and worship me, viz. the devil, were Christ's sins, if you consent not, but resist them, as Christ did, Matt. iv. 5, 9.

You should carefully observe this. For if the devil was so malicious and presumptuous, as to assault our blessed Saviour with such devilish temptations, injecting into him such vile and blasphemous notions and thoughts; should you think it strange that he does perplex you with the like? And for all this, you have no cause to doubt, whether Christ were the Son of God or no, though the devil made an *if* of it, and it was the thing the devil aimed at; why then should it be doubted that any of Christ's members may be thus assaulted? And yet, surely, they have no cause for this to question, whether they be sanctified, or in a state of grace. For these vain thoughts in them are so far from being abominable evils, that, being not consented to, they are, as I said, not their sins.

It is a piece of the devil's cunning, first to fill a man full of abominable thoughts, and then to be the first that shall put in this accusation and doubt, viz. Is it possible for any child of God, that is sanctified with God's Holy Spirit, to have such thoughts? But consider well, that an innocent Benjamin may have Joseph's cup put into his sack's mouth, without his knowledge or consent, by him, who for his own ends intended thereby to accuse Benjamin of theft and ingratitude, Gen. xliv. 2, 4, 15. Was Benjamin any thing the more dishonest or ungrateful for this? No! Satan does not want malice or cunning in this kind to play his feats. Where he cannot corrupt men, yet there he will vex and perplex them.

But let it be granted, that these blasphemous and abominable thoughts, which trouble you are indeed your sins, either because they arise from your own evil heart, or because you did consent to them. If so, then you have much cause to grieve and repent, but not to despair, or to say you are not God's child; for it is possible for a sanctified man to be made guilty, either by outward act, or by consent and approbation, or by some means or other, of any one sin, except that against the Holy Ghost; and yet if he confess and bewail his sin, and repent, believe, and ask mercy, it shall be forgiven him; for he has our Saviour's word for it. Matt. xii. 31, 32.

And whereas you say you were not troubled with such abominable thoughts before you made profession of a holy life; I answer, this is not to be wondered at. For before that time, the devil and you were friends; then he thought it enough to suffer you to be proud of your civil honesty, or, it may be, to content yourself with a mere form of godliness, because that you were free from notorious crimes, as adultery, lying, swearing, &c. For when he could by these more plausible ways lead you captive at his will, he saw you were his sure enough already, what need was there then, that he should solicit you any further, or disturb your quiet? But now, that you have renounced him in earnest, and that he and you are opposites; you may be sure, that he will attempt by all means to reduce you into your old state: or if he fail of that, yet as long as you live, so far as God shall permit, he will do what he can to disturb your peace, by vexing and molesting you.

Moreover, God does permit this, for divers holy purposes:

1. To discover the devil's malice.

2. To chasten his children, and to humble them, because they were too well conceited of the goodness

of their nature in their unregeneracy, or might he too uncharitable and censorious of others; and too presumptuous of their own strength, since they were regenerate.

3. God likewise permits these buffettings and winnowings of Satan, both to prevent pride, and other sins, and to exercise and try the graces of his children; to give them experience of their own weakness, and of his grace towards them, and strength in them, even in their weakness; preserving them from being vanquished, although they fight with principalities and powers, and spiritual wickedness. For God's strength is made perfect in man's weakness, 2 Cor. xii. 9.

Remedies against evil and blasphemous thoughts.

That Christians who are troubled with blasphemous, and other abominable thoughts, may be less troubled, or at least not hurt by them, follow these directions.

(1.) Proofs of the being of God.

First, Arm yourself with evident proofs that there is a God, that there is a divine, spiritual, absolute, and independent Being, from whom, and to whom are all things, and by whom all things consist.—Next, Confirm yourself in a sure persuasion that the Bible and holy Scriptures are the pure word of this only true God. Then labour with your heart, that it so reverence and love God and his will, as to be always ready to rise against every motion to sin, especially these of the worse kind, with loathing and detestation.

First, To be assured that there is a God, consider first the creation, preservation, and order of the creatures. How could it be possible that such a world could be made and upheld, or that there should be such an order, or subordination among creatures, if there were not a God? The heavens give their influence into the air, water, and earth; these, by virtue hereof, afford means of comfort and support to all living creatures, Psa. xix. 1; civ. The creatures without sense serve for the use of the sensitive; and all serve for the use of man; who, although he be an excellent creature, yet of himself he is so impotent, that he cannot add one cubit to his stature, Luke xii.

25; nay, he cannot make one hair white or black, therefore could not be the maker of these things, Matt. v. 36.

Moreover, if the creatures were not limited and ordered by a superior Being, they would devour one another, in such a manner as to bring all to confusion. For the savage beasts would eat up and destroy all the tame and gentle, the strong would consume the weak, Job xxxviii. 10, 11; the sea, if it had not bounds set to its proud waves, would stand above the mountains, Psa. civ. 6; and the devil, who hates mankind, would not suffer a man to live at any quiet, if there were not a God, one stronger than the strongest creatures, to restrain Satan, and to confine every thing to its place and order. How could there be a continual vicissitude of things? How could we have rain and fruitful seasons, and our souls be filled with food and gladness, if there were no God? Acts xiv. 15—17. Thus, by the creation, the invisible things of God, that is, his eternal power and Godhead, are clearly seen, Rom. i. 20; for by these things, which are thus made, and thus preserved, he has not left himself without witness, that God is, and that he made all things for himself, even for his own glory, Prov. xvi. 4.

Secondly, If all things came by nature, and not from a God of nature, how then have miracles, which are many times against nature, and do always transcend and exceed the order and power of nature, been wrought? For nature in itself does always work even in its greatest works, in one and the same manner and order. For nature is nothing else but the power of God in the creatures, to support them, and to produce their effects in due order. Wherefore if any thing be from nature, or from miracle, it is from God; the one from his power in things ordinary, the other from his power in things extraordinary; wherefore, whether you look on things natural, or above nature, you may see there is a God.

Thirdly, Look into the admirable workmanship of but one of the creatures, namely, your own soul, and particularly into your conscience: whence are your fears that you shall be damned? What need it; nay, how could it trouble you, for your blasphemous thoughts and other sins, if it were not privy to itself, that there is a God, who will bring every thought into judgment? Eccles. xii. 14.

Fourthly, Make use of the eye of faith, whereby you may see God, who is invisible, and that more distinctly, more certainly, and more fully, Heb. xi. 27. Remember that it is the first principle of all religion, which is first to be learned, namely, That God is, that all things are made by him, and that he is a rewarder of all those, who so believe this, that they diligently seek him, Heb. xi. 3, 6.

2d. Proofs of the divinity of the Scriptures.

1. That you may assure yourselves, that the Scriptures are the word of God; consider, first, how infallibly true they relate things past, according as they were many hundred years before; also in foretelling things to come many hundred of years after, which you may see to have come to pass, and daily do come to pass accordingly; which they would not do if they were not God's word.

2. They lay open the particular and most secret thoughts and affections of man's heart, which they could not do, if they were not the word of him, that knows all things; in whose sight all things are naked and open, Heb. iv. 12, 13.

3. They command all duties of piety, sobriety, and equity, and do prohibit all vice, in such a manner as all the writings and laws of all men laid together, neither do, nor can do, Psa. xix. 7.

4. As the Scriptures discover a state of eternal damnation unto man, and condemn him to it for sin, Gal. iii. 22; so they reveal a sure way of salvation, Rom. i. 17; which is such a way as could never enter into the imagination and heart of any man, or of all men together, without the word and revelation of the Spirit of God, who in his wisdom found out, and ordained this way, 1 Cor. ii. 9.

5. The Scriptures are a word of power, almighty beyond the power of any creature; pulling down

strong holds; casting down imaginations, and every high thing that exalts itself against the knowledge of God, and brings into captivity every thought to the obedience of Christ, 2 Cor. x. 4—6.

6. Lastly, The Scriptures have an universal consent with themselves, though penned by divers men; which proves that they are not of any private interpretation; but that these holy men of God spake as they were moved by the Holy Ghost, 2 Peter i. 20, 21. Much more might be said to this point, but this may suffice.

Helps against unnatural and violent suggestions.

Against temptations to lay violent hands upon yourself and others, you must have these or the like Scriptures in readiness: Thou shalt not kill, Exod. xx. 13; and, See thou do thyself no harm, Acts xvi. 28; and such like. And that you may be prepared against all other vile temptations, possess your heart beforehand with this, that these are great wickednesses against God, against your God. When Joseph could say, Shall I commit this great wickedness and sin against God? no temptations could prevail against him, Gen. xxxix. 9. Thus much for forearming yourselves against blasphemous and vile thoughts and temptations.

In the second place; when you are thus armed, whensoever these blasphemous and fearful thoughts rise in you, or are forced upon you, take heed of two extremes:

First, Do not contemn them, so as to set light by them; for this gives strength to sin, and advantage to Satan.

Secondly, Be not discouraged, nor yet faint through despair of being free from them, in due time; or of withstanding them in the meantime. For then Satan has his end, and his will of you.

But carry yourself in a middle course; pore not too much on them, dispute not too much with them presume not of your own strength; but by lifting up of your hearts in prayer, call in God's aid to resist and withstand them; present some suitable Scripture to your mind, such as is directed against them, whereby you may with a holy detestation resist them, according to Christ's example, with, It is written, Matt. iv. 6, 7. Now when you have done this, then, if it be possible, think on them no more.

Thirdly, Endeavour at all times to make conscience in the whole course of your life of your thoughts, even of the least thoughts of evil, yea of all thoughts, and this will be a good means to keep out all evil thoughts, 2 Cor. x. 5. If it cannot prevail thus far; yet you shall have this benefit by it, when your heart can testify for you, that you would in every thing please God, and that you make conscience of less sinful thoughts than those vile ones with which you are troubled; then you may be sure that you may be, and are God's children, and are sanctified, notwithstanding those blasphemous thoughts and devilish temptations.

(3.) Doubts of sanctification from the prevalence of some gross sin.

Again, Some doubt they are not sanctified, because they have fallen into some gross sin; it may be, into worse than those which they committed in their state of unregeneracy.

I answer such; you are in a very ill case, if you do not belie yourselves; and if so, you are in an ill case, because you do belie yourselves. I advise you that have thus sinned in either, to repent speedily, and to ask forgiveness. God by his Spirit does as well call you to it, as he did Israel, saying, Return to the Lord—thou hast fallen by thine iniquity,—take with you words, and turn unto the Lord, and say unto him, Take away all our iniquity, and receive us graciously; then will God answer, I will heal your backsliding, I will love you freely, Hosea xiv. 1—4. You say, that you are backslidden; suppose it were so, he saith, I will heal your backslidings, &c., read Jer. iii. 12, 13. Micah vii. 18, 19.

You must not doubt, but that gross sins committed after a man is effectually called, are pardonable. It is the devil's policy to cast these doubts into your heads, so wholly to drive you to despair, by shutting out all hope of grace and mercy, that you might have no thought of returning and seeking unto God again; but believe him not: he is a liar, John viii. 44. For it may befall one that is in a state of grace, to commit the same gross sins after conversion, which he did before, if not greater than the same. Did not David, by his adultery and murder, exceed all the sins that ever he committed before his conversion? 2 Sam. xi; 1 Kings xv. 5. Did not Solomon worse in his old age than ever in his younger days? 1 Kings xi. 4, 5. Did Peter commit any sin like that of denying and forswearing his master, before his conversion? Matt. xxvi. 74. Why were the falls of these worthies written, but for examples to us, on whom the ends of the earth are come? 1 Cor. x. 11, 12.

First, That every one who stands should take heed lest he fall.

Secondly, That if any are fallen into any sin by any occasion, he might rise again as they did, and not despair of mercy.

No man, though converted, has any assurance, except he is specially watchful, and except he have special assistance of God's grace, to be preserved from any sin, except that against the Holy Ghost; but if he be watchful over his ways, and do improve the grace of God in him after conversion, seeking unto God for increase of grace, then he, as well as the apostle Paul, may be kept from such gross sins as are of the foulest nature, otherwise not, 1 Cor. iv. 4.

Indeed, they that are born of God, have received the sanctifying influences of God's Spirit, that seed of grace, which ever remains in them. Whence it is that they sin otherwise in a state of regeneracy than they did before; insomuch that the Scripture of truth, notwithstanding the after sin, saith, that whosoever is born of God sins not, 1 John iii. 9; not that they are free from the act and guilt of sin, for in many things we sin all, saith James, James iii. 2; but because they sin not with full consent, Rom. vii. 15, &c. They are not servants to sin; they do not make a trade of sin, as they did in their unregeneracy, John viii. 34.

Rom. vi. 16, 18, 19, 20, 22. Neither do they sin the sin unto death, 1 John v. 17, 18, which all unregenerate men may, and some do. Yet for all this, it may, and often does come to pass, that, partly from Satan's malice and power, partly from the remains of corrupt nature, and partly from God's just judgments on many, because of their negligence and presumption, their conceit of their own strength, or their censoriousness and unmercifulness to them that had fallen, that true Christians may fall into some particular gross sin or sins, for matter, greater than ever before conversion.

(4.) Doubts of sanctification from the want of affectionate sorrow for sin; and the defects of repentance.

Others yet complain and say, they fear they have not repented, they feel that they cannot repent; for they cannot grieve as they ought. They can pour out floods of tears, more than enough for crosses, but many times they cannot shed one tear for sin. They do nothing as they ought to do. They live in their sins still. How then can they be said to have repented, and to be sanctified?

If by doing as you ought, you mean perfectly fulfilling every point and circumstance of the law, never any mere man did thus; if you could do as you ought, what need have you of Christ Jesus as a Saviour and an Advocate?

But if by doing as you ought, you mean a doing according as God, now, qualifying the rigor of the law by the graciousness of the gospel, does require of you, and in Christ will accept of you; namely, to will and endeavour in truth to do the whole will of God; then, if you will, desire, and endeavour to mourn for sin, to repent, and obey as you should, you may truly be said to do as you ought, Isa. i. 19. And in this case, look by faith to the perfect obedience of the Lord Jesus Christ, your surety and redeemer, Rom. viii. 4.

And as for weeping at crosses, sooner or more than for sins, this does not always argue more grief for one than for the other: for weeping is an effect of the body, following much the temper thereof; also sense apprehends a natural object, or matter of bodily grief, in such a manner, that the body is wrought upon more sensibly, than when a spiritual object of grief is only apprehended by faith. Wherefore bodily tears flow easily from sense of crosses, and more hardly from thoughts of sin: for spiritual objects do not ordinarily work passions in the body so soon, nor so much, as bodily and sensible objects do. Grief for a cross is more outward and passionate; thence tears; but spiritual grief is more inward and deep, in which cases, tears lie so far off, and the organs of tears are so much contracted, and shut up, that they cannot be fetched or wrung out, but with much labour. When you are bidden in Scripture to mourn and weep for your sins, nothing else is meant, but to grieve much, and to grieve heartily, as they do, who weep much at outward calamities. Besides it is not unknown that even in natural grief, dry grief is many times greater than that which is moistened, and overflows with tears. And some soft effeminate spirits can weep at any thing, when some harder spirits can weep at nothing, As the greatest spiritual joy is not expressed in laughter, so neither is the greatest spiritual grief expressed in tears. God regards the inward sighing of a contrite heart, more than the outward tears of the eyes, Psa. li. 17. An hypocritical Saul, being overcome with kindness, 1 Sam. xxiv. 17, 18, and a false-hearted Ahab, being upon the rack of fear, 1 Kings xxi. 27, 29, may in their qualms and passions weep, and externally humble themselves, and that in part for sin; when a dear child of God may not be able to command one tear. The time when God's children have most plenty of tears, is when the extremity and anguish of grief is well over, namely, when their hearts begin to melt through hope of mercy, Zech. xii. 10.

And as for leaving sin altogether; Who ever did it in this life? Who ever shall? Since there is no man that lives, and sins not, 2 Chron. vi. 36. But mistake not, you may through God's grace have left sin, when yet sin has not left you. For whosoever hates sin, and resolves against it, and in the law of his mind would not commit it; but is drawn to it by Satan, and by the law of his members: and, after it is done, does not allow it, but disclaims it with grief; this man has left sin, Rom. vii. 23. And if this be your case, it may be said of you, as the apostle said of himself: it is not you that do evil; but it is sin, that dwells in you, Rom. vii. 20.

(5.) Doubts of sanctification on account of dulness in spiritual duties.

Many yet complain, they cannot pray, read, hear, meditate, nor get any good by the best companies, or best conferences which they can meet with. They are so dull, so forgetful, so full of distraction, and so unfruitful, when they go about, or have been about any thing that is good, that they fear they have no grace at all in them; yea, it makes them sometimes to forbear these duties; and for the most part to go about them without heart.

It is not strange that it should be so with you; so long as there is a Satan to hinder you, and so long as you carry about the old man and body of sin in you. Moreover, do you not many times go about these holy duties remissly, negligently, only customarily, without preparation thereunto, not looking to your feet, and putting off your shoes before you approach unto God's holy things, and holy presence? Do you not many times set upon those holy duties in the power of your own might, and not in the power of God's might; or have you not been proud, or too well conceited of yourselves, when you have felt that you have performed good duties with some life; or, are you sure, that you should not be spiritually proud, if you had your desire in doing all these? Further, do you not miscall things; calling that, no prayer, no hearing, &c., or no fruit, because you do them not so well, nor bring forth so much, as in your spiritually covetous desires you long to do, and have? If it be thus with you, then first mend all these faults, confess them to God, and ask mercy. Next be thankful for your desires, to pray, read, hear, &c., and for your longing to do all these as you should; prosecute those desires, but always in the sense of your own insufficiency, and in the power of God's might; then all the fore-mentioned duties will be performed with less difficulty and more fruit and comfort.

Yet because in all these duties you travel heavenward up the hill, and your passage is against wind and tide, and with a strong opposition of enemies in the way; you must never look to perform them without sense of much difficulty and little progress in comparison of what you aim at in your desires. It concerns you therefore to ply your oars, and to apply yourselves by all means, to work out your salvation with fear and trembling; I mean, with fear to offend in any of the afore-mentioned duties, not in fear that you have no grace, because you cannot perform them as well as you should, and would, Phil. ii. 12. For since you feel and bewail your dulness, deadness, and unprofitableness in holy services, it argues that you have life, because no man feels corruption, and dislikes it, by corruption, but by grace. I am sure that such as have no true grace, can, and do daily, fail in all these duties, but either they find not their failings, or if they do, yet they complain not of them with grief and dislike. If you heartily grieve, because you do no better, your desires to do as you should do, are a true sign of grace in you, Neh. i. 11. For this duty is always well done, in God's account, where there is truth of endeavour to do well, and true grief that it is done no better.

And whereas you say, that by reason of want of spiritual life in holy duties, you have been made to neglect them altogether, I pray, what have you got thereby, but much grief, and uneasiness? But tell me, how is it with you? are you pleased with yourself in your neglect; or is it so that you can have no peace in your heart until you set yourselves diligently to do those duties again, as well as you can? If so, it is a sign that you are not quite destitute of saving grace.

(6.) Doubts of sanctification from sudden dulness after duties.

Others, when they have been at holy exercises, and

in good company have felt joy, and sweet comfort therein; but afterward, oftentimes much dulness has suddenly seized upon them; which makes them fear they have no root in themselves, and that their joys and comforts were not sound.

This dulness after fresh comforts may, and often does befall those, in whom is truth of grace, but commonly through their own fault. And to speak freely to you; it may be you were not thankful to God for your joys and comforts when you had them; but did ascribe too much to yourselves, or unto the outward means by which you had them. Or it may be, you did too soon let go your hold of these spiritual comforts, betaking yourself to worldly business, or to other thoughts, before you had sufficiently digested these, and before you had committed them under safe custody, insomuch that the devil finding your comforts lie loose, and unguarded, steals them from you; or else haply the Lord knows that you are not able to bear the continuance of your joys and comforts, but your hearts will be over-light and overjoyed, and exalted above measure, 2 Cor. xii. 7; therefore in his just chastisements, or in his loving wisdom, God may suffer deadness in this sort to seize you.

(7.) Doubts of sanctification on account of being outdone by others.

There are also some, who, when they perceive that some new converts to religion, who have not had half of the time or means to be good as they have had, yet outstrip them in knowledge, faith, mortification, and willingness to die; doubt of the truth of their own graces.

It is more than you can certainly know, whether they have more saving grace than you; for when with a charitable eye you look upon the outside of another's behaviour, and shall look with a severe and searching eye into the corruptions of your own heart, you may easily, through modesty and charity, think others better than yourselves, and it is good for you so to do; an error in that case, if you do commit it, is tolerable. Many also can utter what they have, it may be, better than you, and can make a small matter seem much, and a little to go far, when many times you, in modesty, may not set forth yourself, or, if you would, could not.

But let it be granted, that many of short standing in the school of Christianity, have got the start of you in grace: if it was through God's grace accompanying their diligence, and from his just hand upon you, following your negligence, then they are to be commended, and you are to be humbled, and to be provoked unto an holy emulation by them to quicken your pace, and to double your diligence. But take heed that it be not your pride and self love, which causes you not to bear it, that others should be better than yourselves.

It may be that it is not your fault, but it is from God's abundant grace unto others, above that which you have received: for the Scriptures make it evident, that God gives unto several men differently, according to his good pleasure, Eph. iv. 7; Rom. xii. 3; 1 Cor. xii. 11. Hence it was, that David became wiser than his teachers, and ancients, Psa. cxix. 99, 100, and the apostle Paul attained more grace than those that were in Christ before him. God gives unto some five talents, when he gives unto others but two; he that has most given him, gains in the same space of time, twice as much as the other, yet he that gained but two talents had his commendation, and his proportionable reward of well-doing. For the Lord saith unto him also, Well done, faithful servant, enter into thy Master's joy, Matt. xxv. 21—23. For he improved his talents according to the measure of grace received, though he gained not so much as the other.

Take heed that your eye be not evil, because God is good, Matt. xx. 15. May not he give as much unto the last as unto the first, and more, if he please? We should rather be thankful for the increase of grace in others, than either to repine at them, or, without ground, to conclude against the truth of our own. For we are much the better, if we would see it, for other's graces; God's kingdom is enlarged and strengthened thereby; the common good of Christ's body, which is the church, gains by it. Now the more excellent any member of the body is, according to his gifts and place, the rest of the members should therein the more rejoice, 1 Cor. xii. 26.

(8.) Doubts of sanctification from a sense of the hardness of the heart.

Lastly, Many yet will say, that their hearts remain hard and stony, yea, they say, that they grow harder and harder; wherefore they think that the stony heart was never taken out of them, and that they remain unsanctified.

Know, that there are two sorts of hard hearts.

One total and not felt, which will not be broken, nor brought unto remorse either by God's threats, commandments, promises, judgments, or mercies; Zech. vii. 11; but obstinately stands out in a course of sin, being past feeling, Eph. iv. 19.

The second is, a hardness mixed with some softness, which is felt and bewailed; this is incident to God's children: of this the church complains, saying unto God, Why hast thou hardened our hearts against thy fear? Isa. lxiii. 17. Now when the heart feels its hardness, and complains of it, is grieved, and dislikes it, and would that it were tender like Josiah's, 2 Chron. xxxiv. 27, so that it could melt at the hearing of the word; this is a sure proof that the heart is regenerate and not altogether hard, but has some measure of true softness; for it is by softness that hardness of heart is felt, witness your own experience; for before the hammer and fire of the word were applied to your hearts, you had no sense of it, and never complained thereof.

You must not call a heavy heart, a hard heart; you must not call a heart wherein is a sense of indisposition to good, a hard heart; except only in comparison of that softness, which is in it sometimes, and which it shall attain unto, when it shall be perfectly sanctified; in which respect it may be called hard. Whosoever has his will so wrought upon by the word, that it is bent to obey God's will, if he knew how, and if he had power; this man, whatsoever hardness he feels, his heart is soft, not hard. The apostle had a heart held in, and clogged with the flesh, and the law of his members, that it made him to think himself wretched, because he could not be fully delivered from it; yet we know his heart was a sound heart, Rom. vii. 24.

Among those that are sanctified, there remains more hardness in the heart of some than in others; and what with the committing of gross sins, and cursory and slight doing of good duties, and through neglect of means to soften it, the same men's hearts are harder at one time than at another, of which they have cause to complain, and for which they have cause to be humbled, and to use all means to soften it; but it is false and dangerous, hence to conclude that such are not in a state of grace because of such hardness in the heart; for as God's most perfect children on earth know but in part, and believe but in part; so their hearts are softened but in part, 1 Cor. xiii. 9.

SECT. 11. FEARS OF APOSTASY, REMOVED.

There yet remain many, who though they cannot reply to the answers given to take away their false fears and doubts; but are forced to yield, that they find they now are, or at least have been in a state of grace; yet, this they fear, that they are already fallen, or shall not persevere, but shall fall away before they die.

(1.) What kind of Christians may apostatize.

Concerning falling away from grace, first know, that of those that give their names to Christ in outward profession, there are two sorts:

The first sort are such, who have received only the common gifts of the Spirit; as first, illumination of the mind to know the mystery of salvation by Christ, and truly to assent unto it, Heb. vi. 4, 5.

Secondly, Together with this knowledge, is wrought in them by the same Spirit a lighter impression upon the affections, which the Scripture calls a taste of the heavenly gift, and of the good word of God, and of the powers of the world to come, Heb. vi. 4, 5. By these gifts of the Spirit, the souls of these men are raised to an ability to do more than nature, and mere education can help them unto; carrying them further than nature or art can do, by working in them a kind of spiritual change in their affections, and a kind of reformation of their lives.

But yet all this while they are not ingrafted into Christ; neither are deeply rooted, as the corn in good ground, nor yet are thoroughly changed and renewed in the inward man, Matt. xiii. 21; they have at best only a form of godliness, but have not the power thereof, 2 Tim. iii. 5.

Now these men may, and often do fall away, not into some particular gross sins, of which they were sometime after a sort washed; but into a course of sinning; falling from the very form of godliness, and may so utterly lose those gifts received, that they may in the end become very apostates; yet this is not properly a falling from grace. It is only a falling away from the common graces or gifts of the Spirit, and from those graces which they did seem to have, and which the church out of her charity did judge them to have; but they fall not from true saving grace, for they never had any, Luke viii. 18. For if ever they had been indeed incorporated into Christ Jesus, and had been sound members of his body, and in this sense had ever been of us, as the apostle John speaks, then they would never have departed from us, but should no doubt have continued with us, 1 John ii. 19.

(2.) Of such Christians as shall persevere.

The second sort of those that have given their names to Christ, are such as are endued with true justifying faith, and saving knowledge, and are renewed in the spirit of their mind; whereby, through the gracious and powerful working of the sanctifying Spirit, the word makes a deeper impression upon the will and the affections, causing them not only to taste, but, which is much more, to feed and to drink deep of the heavenly gift, and of the good word of God, and of the powers of the world to come; so as to digest them unto the very changing and transforming them, by the renewing of their minds, Rom. xii. 2, and unto the sanctifying of them throughout in their whole man, both in spirit, soul, and body, 1 Thess. v. 23; so that Christ is indeed formed in them, and they are become new creatures, 2 Cor. v. 17; being made partakers of the divine nature, 2 Peter i. 4.

Now concerning these: it is not possible that any of them should fall away, either wholly or for ever.

(3.) How far a Christian may decline in grace, and the causes thereof.

Yet it must be granted, that they may decline and fall back so far, as to grieve the good Spirit of God, and to offend and provoke God very much against them, and to make themselves deserving of eternal death. They may fall so far as to interrupt the exercise of their faith, Psa. xxxii. 3, wound their conscience, Psa. li. 8—11, and may lose for a time the sense of God's favour, and may cause him, like a wise and good father, in his just anger, to chide, correct, and threaten them, so that they may have cause to think that he will utterly reject them, and never receive them into his heavenly kingdom; until, by renewing their faith and repentance, they return into the right way, and do recover God's loving kindness toward them again.

That you may understand and believe this the better, consider what grace God gives unto his elect, and how, and from what they may fall: also you must observe well the difference there is between the sinning of the regenerate and unregenerate, together with the different condition wherein they stand, while they are in their sins.

In the first act of conversion, I speak of men of years and discretion, God by his word through his Holy Spirit does infuse a habit of holiness, namely, a habit of faith, and all other saving graces; thus, every child of God receives that holy anointing of the

Spirit, 1 John ii. 20; that which the Scripture calls the seed remaining in him, 1 John iii. 9.

Secondly, God by his gracious means and ordinances of the gospel does increase this habit and these graces.

Now, because every man that is truly regenerate, carries about with him the body of sin and corruption, and lies open daily unto the temptations of the world and the devil; a truly regenerate man may be drawn, not only into sins of ignorance and common frailty, but into gross sins; whereby the light and warmth of God's Spirit may be so chilled and darkened, that he may break out into presumptuous sins. Yea, upon his negligent use, or omission of the means of the spiritual life and strength, God may justly give him over to a fearful declension in grace and backsliding; yet the truly regenerate fall only from some degrees of holiness, and from certain acts of holiness; but not from the infused habit of holiness; that blessed seed ever remains in him, 1 John iii. 9. His falling is either only into particular sins, and into much failing in particular good duties; or if it be towards a more general defection, yet it is never universal from the general purpose of well-doing, into a general course of evil.

For the regenerate man does never so sin, as the unregenerate man does, although for matter their sins may be alike, yea, sometimes those of the regenerate, greater. There is great difference in their sins, and manner of sinning.

1. Regenerate men may sin through ignorance, but they are not willingly and wilfully ignorant, as are the unregenerate in some things or other, 2 Peter iii. 5.

2. Regenerate men may commit, not only the common sins of infirmity; into which, by reason of the remains of the lusts of the flesh, they fall often; such as rash anger, discontent, doubts, fears, dulness, and deadness of heart in spiritual exercises, and inward evil thoughts and motions of all sorts; but they may also commit gross sins, such as an open and direct breach of God's commandments; yet those are done

against their general purpose, as David did, for he had said, he would look to his ways, Psa. xxxix. 1; and he had determined to keep God's righteous judgments, Psa. cxix. 106. Yea, many times they are done against their particular purposes, as Peter's denial of of his Master, Matt. xxvi. 35. They are not usually contrived, or thought on before, but fallen into by occasion, or are forced thereunto by the violent corruption of the affections, or sensual appetites, 2 Sam. xi. 2; 2 Gal. vi. 1. Moreover, they do not make a trade and custom of sin; these kinds of sins do not pass them any long time unobserved; but are seen, bewailed, confessed to God, and prayed against; and are burdensome and grievous to them, making them to think worse of themselves, and to become base in their own eyes because of them. But it is usually directly otherwise with the unregenerate in all these particulars, Gen. xxvii. 41; Micah ii. 1.

3. The regenerate may not only commit sins gross for matter, but presumptuous for manner, namely, they may commit them not only against knowledge and consent, but with a premeditated deliberation, and determination of will, as David did in the murder of Uriah, 2 Sam. xi. 8—25. But it is seldom that a child of God does commit presumptuous sins; his general determination and prayer is against them, Psa. xix. 13. It is with much strife and reluctance of will, and with little delight and content, in comparison. He never sins presumptuously, but when he is drawn thereunto, or forced thereupon by some overstrong corruption and violent temptation for the time, as David was, being over eagerly bent to hide his sin, and to save his credit: for if he could by any means have gotten Uriah home to his wife, he would never have caused him to be slain, 2 Sam. xi. 8—13. And although presumptuous sins cast him into a deadness and numbness of heart and spirit, in which he may lie for a time speechless and prayerless, as it was with David; yet he feels that all is not well with him, until he have again made his peace with God, Psa. xxxii. 3, 4. And when he has the ministry of God's power-

ful word, to make him plainly see his sin, then he will humble himself and reform it, 2 Sam. xii. 18; Psa. li. 12, 13. The unregenerate are not so.

4. Lastly. A regenerate man may fall one degree further, namely, he may so lose his first love, that he may, though not fall into utter apostasy, yet decline from good very far, even to a coldness and remissness in good duties, even in the exercises of religion, if not to an utter omission of them for a time. The life and vigour of his graces may suffer sensible eclipses and decay. Asa, though a good king, went apace this way, 2 Chron. xv. 17, as appears by his imprisoning the good prophet, and in oppressing the people in his latter days, and in trusting to the physicians, and not seeking to God to be cured of his disease, 2 Chron. xvi. 10, 12. And Solomon, the truly beloved of God in his youth, went further back, 2 Sam. xii. 24; Neh. xiii. 26; giving himself to all manner of vanities, Eccles. ii. and in his old age did so doat upon his many wives, that he fell to idolatry, or at least became accessary, by building them idol temples, and accompanying them to idolatrous services, insomuch that it is said, they turned away his heart after other gods, and his heart was not perfect with the Lord his God, as was the heart of David his father, 1 Kings xi. 3—20. Yet there is a wide difference between these backslidings, and the apostasies of men unregenerate. For these do not approve nor applaud themselves in those evil courses, into which they are backslidden, when, out of the heat of temptation, they do think of them; neither have the regenerate full content in them, but find vanity and vexation in them, as Solomon did even in the days of his vanity. They do not in this their declined estate, hate the good generally, which once they loved, but look back upon it with approbation; and their heart secretly inclines unto a liking of it, and of them who are, as they once were; so that in the midst of their bad estate, they have a mind to return, but that they are yet so hampered, and entangled with the snares of sin, that they cannot get out. Lastly, They in God's good time, by his

grace, do break forth out of this eclipse of grace, by the light whereof they see their wretchedness and folly, and are ashamed of their backsliding and revolting; and they again do their first works; and with much ado, recover their former joys and comforts, though it may be never with that life, lustre, and beauty, as in former times; and this as a just correction of their sin, that they may be kept humble, and be made to look better to their standing all the days of their life by it. It is not so with the hypocritical professors, who were never truly regenerate; but quite contrary, as you may observe in the apostasies of Saul, 1 Sam. xxviii. 3, 6, 7, &c. and of king Joash, 2 Chron. xxiv. 11, 18—23, and Simon Magus, and others.

(4.) The differences between the falls of the sincere and insincere.

These differences rise hence, because that the common graces of the unregenerate are but as flashes of lightning, or as the fading light of meteors, which blaze but for a while; and are like the waters of land-floods, which because they have no spring to feed them, run not long, and in time may be quite dried up.

But the saving graces of the regenerate receive their light, warmth, and life from the Sun of righteousness, therefore can never be totally or finally eclipsed. And they rise from that well and spring of living water which cannot be drawn dry, or so dammed up, or stopped, but that it will run more or less, unto eternal life, John iv. 14,

As the regenerate man does not sin in such a manner as the unregenerate, with all his heart, so neither is he, when he has sinned, in the same state and condition, which the unregenerate is in. He is in the condition of a son, who notwithstanding his failings, abides in the house for ever. But not so the other; who, being no son, but a servant, is for his misdemeanour turned out, and abides not in the house for ever, 1 John viii. 35.

Although the regenerate as well as the unregenerate draw upon themselves, by their sins, the simple guilt of eternal death, yet this guilt is not accounted, neither does it redound to the person of the truly regenerate, as it does to the others; because Christ Jesus has so satisfied, and does make intercession for his own, that his death is made effectual for them, but not for the others, John xvii, 9, 15, 20. Their justification and adoption by Christ remain unaltered; although many benefits flowing from thence are, for a while, justly suspended; they remain children still, though under their Father's anger; as Absalom remained a son uncast off, not disinherited by David, when yet his father would not let him come into his presence, 2 Sam. xiv. 24. This spiritual leprosy of sin, into which God's children fall, may cause them to be suspended from the use and comfortable possession of the kingdom of God, and from the enjoyment of the privileges thereof, until they be cleansed of their sin by renewed faith and repentance. Yet, as the leper in the law, had still right to his house and goods, although he was shut out of the city for his leprosy; so the truly regenerate never lose their right to the kingdom of heaven by their sins, Lev. xiii. 46; 2 Chron. xxvi. 21. For every true member of Christ is knit unto Christ by such everlasting bonds, whether we respect the relative union of Christ with his members by faith to justification, which, after it is once made by the Spirit of adoption, admits of no breach or alteration by any means, Rom. viii. 15—17, 35; or whether we respect the real union of the Spirit, whence flows sanctification, which though it may suffer decay, and admits of some alteration of degrees, being not so strong at one time as at another, yet can never quite be broken off, as has been proved, 1 John ii. 27; iii. 9; these bonds, I say, are so strong and lasting, that all the powers of sin, Satan, and hell itself, cannot separate the weakest true member from Christ, or from his love, or from God's love towards him in Christ, Rom. viii. 33 to the end.

This strength of grace, that keeps men from falling totally or finally from Christ, does not depend upon the strength or will of him that stands, but on the election and determination of him that calls, Rom. ix. 11.

(5.) Why the faithful shall not finally apostatize.

And whereas it may be demanded why a man, who being at his highest degree of holiness, did yet fall back more than half way, may not as well, or rather fall quite away?

I answer, it is not in respect of the nature of inherent holiness in him; for Adam had holiness in perfection, yet fell quite from it, Gen. i. 27; iii. 6. There is nothing in the nature of this grace and holiness, excepting only in the root whence it springs, but that a man may now also fall wholly from it. But it is because grace is now settled in man on better terms. For the little strength we receive in regeneration, is, in point of perseverance, stronger than the great strength which the first Adam received in his creation. Adam was perfectly, but changeably holy; God's children in regeneration are made imperfectly, but unchangeably holy, Jer. xxxii. 40. This stability of grace now consists in this, in that all who, by faith and by the Holy Spirit, are ingrafted and incorporated into Christ, the second Adam, have the spring and root of their grace founded in him, and not in themselves, as the first Adam had, 2 Cor. i. 21, 22. They are established in Christ, Eph. i. 4. Wherefore, all that are actual members of Christ cannot fall from grace altogether; for as Christ died to sin once, and being raised from the dead dies no more, Rom. vi. 5—12, so every true member of Christ, having part with him in the first resurrection, dies no more, but lives for ever with Christ. For all that are once begotten again unto a lively faith and hope, by the resurrection of Jesus Christ from the dead, to an inheritance incorruptible, are kept, not by their own power, unto salvation, but by the power of God through faith in Christ Jesus, 1 Peter i. 3—5.

Now, that a man effectually called, can never fall wholly or for ever from a state of grace, I, in a few words, reason thus. If God's counsel, on which man's salvation is founded, be sure and unchangeable, 2 Tim.

ii. 19, and if his calling be without repentance, Rom. xi. 29:

If God's love be unchangeable and alters not, but whom God once loves actually, him he loves to the end, John xiii. 1:

If Christ's office of prophet, priest, and king, in his teaching, satisfying, and making intercession for, and in his governing his people, be after the order of Melchisedec, unchangeable and everlasting, he ever living to make intercession for them, Heb. vii. 21, 24, 25; and if his undertaking, in all these respects, with his Father, not to lose any whom he gives him, cannot be frustrated, John vi. 39; Luke xxii. 32; John xvii. 15:

If the seal and earnest of the Spirit be a constant seal, which cannot be razed; but seals all in whom it dwells unto the day of redemption, Eph. i. 13, 14:

If the word of truth wherewith the regenerate are begotten, be an immortal seed, which when once it has taken root, lives for ever, 1 Peter i. 23, 25:

If God be constant and faithful in his promise, and omnipotent in his power, to make good this his word and promise, saying, I will make an everlasting covenant with them, that I will not turn away from my people and children to do them good, but I will put my fear in their hearts, that they shall not depart from me, Jer. xxxii. 40:

Then from all, and from each of these propositions, I conclude, that a man once indeed a member of Christ, and indeed in a state of grace, shall never totally or finally fall away.

The patrons of the doctrine of falling from grace, when they cannot answer the invincible arguments which are brought to prove the certainty of a man's standing in a state of salvation; make a loud cry in certain popular objections, such as are very apt to take with simple and unstable people.

They first come with suppositions, and ask this and like questions: If David and Peter had died in the act of their gross sins, whether should they have been saved or no?

I answer, we have an English proverb, What if the sky fall? Propositions are but weakly grounded on mere suppositions. Should they ask, What if they had died in the act of their sins? Well, say they had died in the act of their sin, they could not die in their impenitence; they in an instant might return to God, and rely on Christ: or at least, if sudden death had surprised them; their general repentance and faith in Christ which they had before their fall, would have been sufficient for them. For their justification and adoption were not impaired, though their sanctification was diminished. But we must believe God's promise, and the issue will be this, though we cannot always tell how, that God will so guide his children with his counsel, that afterwards he will receive them to glory, Psa. lxxiii. 24.

Secondly, They object violently, that this doctrine of not falling wholly from God, and of certainty of salvation, after a man is once in a state of grace, is a doctrine of licentiousness and carnal liberty, causing men to be negligent in the use of means of grace, and careless in their Christian course; for when they once know they shall not be damned, they will live as they list; say they.

First, I appeal to ancient and daily experience, both in ministers and people. For those who have been most assured of God's favour, and of their salvation, have been and are more frequent in preaching, more diligent in hearing, and in the use of all good means of salvation, than those of the other opinion, and have been most holy and more strict in their lives. But the doctrine of these, that teach falling totally and finally from grace, they being the patrons of free-will, on which all the fabric of their building hangs, is rather a doctrine opening a door to licentiousness. For thinking that they may repent if they will, they judge themselves not so unwise but that they will and shall repent before they die, therefore they take liberty to live as they list in the mean time.

Secondly, The Scriptures, the nature of saving faith, and all sound judgment, do reason quite contrary; for the certainty of the end does not hinder, but excite and encourage men in the use of all good means which conduce unto that end. Christ knew certainly that he should attain his end of mediatorship, viz. the salvation of men's souls; but this was no cause why he might be negligent in the means. Was there ever any more earnest in prayer, or more longing to finish his work, than our blessed Saviour, although he was infallibly certain that he should save and glorify man, and that God would glorify him? John xvii. 1. When Daniel knew certainly the time of deliverance out of captivity, he was not hereby carnally secure, and careless in the use of all good means to hasten it; but betook himself to fasting and prayers, that God's people might be delivered, Dan. ix. x. Because God assured David that he would build him a house, therefore, saith he, thy servant has found in his heart to pray, viz. that thou wouldest establish it, 2 Sam. vii. 27. What child is there, that has an ingenuous disposition, or any real goodness in him, will slight, and neglect to please his father, because he has assured him of a large inheritance, or because his inheritance is entailed upon him?

None but those who are indeed destitute of grace will ever wrest and pervert the doctrines of grace, making them to be unto them licences, and occasions of wantonness and sin, Jude 4; Rom. v. 21; so as to say, If where sin abounded, grace abounded much more; then, let us sin that grace may abound; and if we are not under the law but under grace, then let us sin, because we are not under the law but under grace, Rom. vi. 1, 15.

But as any man has truth of grace, the more he knows it, the more he reasons otherwise. Ezra having not only a hope, but the possession of that which God has promised; he does not say, Now we may live as we list, but saith, Should we again break thy commandments? Ezra ix. 13, 14. An honest heart makes the same inference from spiritual deliverances. The Scripture, from the abundance of God's grace, and from the certainty of it, does reason for grace and for obedience. How shall we, that are dead to sin, live yet therein? Rom. vi. 2. And in another place the apostle John saith, We know that we are the children of God, &c., but what is the inference? Is it, we may now sin, and live as we list, because we know that when Christ shall appear, we shall be like him? No, the holy apostle infers this, He that has this hope, purifieth himself as he is pure, 1 John iii. 1—3.

SECT. 12. SUNDRY DOUBTS REMOVED; IN PARTICULAR, ABOUT FALLING FROM GRACE.

NOTWITHSTANDING all that has been said, concerning the certainty of perseverance in grace, after the Christian has been truly converted unto God; yet many will doubt they shall fall away;

(1.) Because they fear that all their religion has been but in hypocrisy, and in form only, but not in power; now such may fall away, as has been said.

If it were true, that all which you have done were in hypocrisy, then until you repent of your hypocrisy, and be upright, you may justly fear as much; yet you must not desperately conclude, that you shall fall away from your profession; but should rather be quickened and stirred up by this fear to abandon hypocrisy and to serve the Lord in sincerity; and hereby make your calling and election sure, that you may not fall; and then you have God's word for it, that you shall never finally perish, Psa. xv. 1, 2, 5.

Many think that they are hypocrites, who yet are sincere; wherefore try whether you be an hypocrite or upright, by the signs of uprightness before delivered, Chap. XI., Sect. I.

Only, for the present, note this; when was it known, that an hypocrite did so see his hypocrisy, as to have it a burden to him, and to be weary of it, and to confess it, and bewail it, and to ask forgiveness heartily of God; and above all things to labour to be upright? If you find yourselves thus disposed against hypocrisy and for uprightness, although I would have you humbled for the remainder of hypocrisy which you discern to be in you; yet chiefly I would have you to be thankful to God, and to take comfort in this, that you feel it, and dislike it: thank God therefore for your uprightness, comfort yourselves in it, and cherish and nourish it in you, and fear not.

(2.) Fears because of the decay of grace and comfort removed.

Others object, that they are already fallen far backward in religion; they do not feel so much zeal and fervency of affection to goodness, nor against wickedness; nor do they now enjoy those comforts and clear apprehensions of God's favour towards them, as they did in their first conversion.

It may be that you are declined in the ways of godliness, and have lost your first love, from whence all those inconveniences have arisen; but may it not befall any child of God to have lost his first love, as well as a whole church, the church of Ephesus? Rev. ii. 4. You could not from thence conclude that Ephesus was no church, neither can you hence conclude, that you are none of God's children, or that you shall not hold out unto the end. But if it be so, be willing to see your sin, and to be humbled, and repent heartily of it; following the counsel of Christ, remember whence you are fallen, repent and do your first works, Rev. ii. 5; and certainly God's child shall have grace to repent, Psa. lxxiii. 24, lxxxix. 30, 32; then you, enduring to the end, shall not be hurt of the second death, notwithstanding that sin of yours in losing your first love, Rev. ii. 11.

But it may, and often does happen, that a true child of God does in his own feeling think he has less grace now than at first, when it is not so; the reasons of his mistake may be these:

At the first a truly regenerate man does not see so much as afterwards he does. At first you had, indeed, the light of the Sun, but as at the first dawning of the day, whereby you saw your greater enormities, and reformed many things, yea, as you thought, all; but now since the Sun being risen higher towards the perfect day, shines more clearly, it comes to pass, that in these beams of the sun, as when it shines into a house, you may see more motes, and very many things amiss in your heart and life, which were not discovered nor discerned before; you must not say you had less sin then, because you saw it not, or more sin now, because you see more. For as the eye of your mind sees every day more clearly, and as your hearts grow every day more holy; so will sin appear unto you every day more and more, for your constant humiliation and daily reformation. For a Christian, if he go not backward, sees in his advanced lifetime more clearly, what is yet before him to be done, and with what an high degree of affection he ought to serve God, and to what an height of perfection he ought to raise his thoughts in his holy aim, which in the infancy of his Christianity he could not see; hence his error. Even as it is usual for a novice in the University, when he has read over a few systems of the arts, &c., to conceit better of himself for scholarship, than when he has more profound knowledge in those arts afterwards, for then he sees his difficulties, which his weak knowledge not being able to pry into, passed over with presumption of his knowing all.

Secondly, Good desires, and enjoyments of comforts are sudden, new, and strange at first, which suddenness, strangeness, and newness of change, out of a state of corruption and death, into the state of grace and life, is more sensible, and leaves behind a deeper impression, than can possibly be made, after such time that a man is accustomed to it: or than can be added by the increase of the same grace. A man that comes out of a close, dark, and stinking dungeon, is more sensible of the benefit of a sweet air, of light and liberty the first week, than he is seven years after he has enjoyed these to the full. Let a mean man be raised suddenly and undeservedly to the state and glory of a king, he will be more sensible of the change, and will be more ravished with the glory of his estate for the first week or month, than at ten years' end when he is accustomed to the heart and state of a king, yea more, than if at ten years' end, double power and glory should be conferred on him.

Thirdly, God for special causes is peculiarly tender of his scholars, when they first enter into Christ's school: in like manner does he deal with his babes in Christ, before they can go alone, Isa. lxvi. 12.

Do not wise schoolmasters, the better to encourage their young and fearful scholars, show more outward expressions of affection and kindness towards them the first week that they come to school, yea, it may be, show more countenance and familiarity towards them their first week than ever after, until the time that they send them to the University? And has not a young child more attendance, and fewer falls in his or her infancy, while carried in the arms, or led in the hands of his father or mother, than when it goes alone? But when it goes alone, it receives many a fall, and many a knock; yet this does not argue less love in the parents, or less strength in the child now, than when it was but one or two years old.

Fourthly, Although God's trees, planted in his courts, Psa. xcii. 14, always should, and usually do in their advanced years, bear more and better fruit, than they did or could do in their youth; yet these, through a false apprehension of things, may judge themselves to be more barren in their age, than they were in their youth. It may be, you feel not in you that vigour, heat, and ability to perform good duties now in age, as you did in your younger days; but may not this arise from natural defects, as from want of memory, quickness of thought, or of natural heat and vigour of your spirits, all which are excellent handmaids to grace? You may observe this in older Christians, who have long walked with God, that in their age, they have these natural defects recompensed with better and more lasting fruit; as with more fixedness and soundness of judgment, more humility, more patience and experience, wherewith their grey hairs are crowned in the way of righteousness, 1 John ii. 12, 13. Look for these, and labour to improve yourselves in them in your age, and they will prove more beneficial to you, than your fresh feelings, and your sensibly felt zeal in your younger times, Prov. xvi. 31.

(3.) Fears of backsliding and apostasy, from the examples of others, removed.

There are yet others, it may be the same, when they observe that many who are of longer standing than themselves, who have had much more knowledge, and have made a further progress in the practice of godliness than they, are yet fallen fearfully into some gross sin or sins; yea, some of them are departed from the faith, and have embraced with Demas this present world, either in the lust of the flesh, the lust of the eye, or pride of life, 2 Tim. iv. 10. They are some of them fallen to popery, or some other false religion; wherefore they fear that they shall fall away also, and that their hearts will deceive them in the end.

That the falls of others should make all that stand to take heed lest they fall, is the express will of God, 1 Cor. x. 12. It is a high point of wisdom for you to observe and do it. Likewise to fear so much as to quicken you to watchfulness and prayer, is an holy and commendable fear; but to fear your total or final falling away, only because some that have made profession of the same religion are fallen, is without ground.

For it may be, those whom you see to be fallen away, never had any other than a form of godliness, and never had more than the common graces and gifts of the Spirit. For if they be quite fallen from the faith, it is because they were never soundly of the faith, 1 John ii. 19. Moreover, grant some of them who are fallen, had saving grace; may they not, with David, Psa. li, and Solomon, Eccles., recover their falls? This you should hope and pray for, rather than by occasion of their falls, to trouble yourself with false and fruitless fear.

(4.) Fears of apostasy in times of persecution.

Lastly, Some yet fear, that if persecution should come because of the word and religion which they profess, they should never hold out, but shall fall away.

Do you thus fear? Then buckle close unto you the complete armour with the girdle of sincerity, exercise yourselves beforehand at your spiritual weapons: with all watchfulness preserve your peace with God, under whom, at such times, you must shelter yourselves, and by whose power it is that you must stand in that evil day, Ephes. vi. 11, 13, 14. But know that a child of God need not fear persecution with such discouraging and distrustful fear, neither should you; for this will but give advantage to your enemies of all sorts, and will make your hands feeble, and your hearts faint. Raise up your spirits, and chase away your fears thus: Consider the goodness of your cause. Consider the wisdom, valour, and power of him that has already redeemed you with his blood, who has already led captivity captive, who is your champion, and has engaged himself for you, until he has brought you to glory; I mean Christ Jesus, who is Lord of Hosts, under whose banner you fight in the whole Christian warfare. Consider likewise the faithfulness of God's promise, made to all his children, concerning his presence and help in time of persecution; commanding them not to take thought concerning it, having promised to give them a mouth and wisdom, which all their adversaries shall not be able to resist, Luke xxi. 14, 15. Consider, last of all, the blessed experience which the holy martyrs have had of God's love and help, according to his promise, in their greatest persecutions and fiery trials. Observe the wisdom and courage of those who in their own nature were but simple and fearful. Read the Book of Martyrs next after the Scriptures for this purpose, and through God's grace, though you were naturally as fearful as hares, when you shall be called to it, you shall be as courageous as lions.

It is not hard for you to know now, whether you shall be able in time of persecution to stand fast and not fall away. If you now, in the peace of the gospel, can deny yourselves in your lusts, through love to God, and for conscience sake towards him, and can rather part with them, than with the sincere adherence to

Christ, then you shall be able, and you will deny yourselves in the matter of your life, if you be put to it in time of persecution, rather than deny Christ. For this first is as difficult as the latter; and the same love to God, and conscience of duty, which does now uphold you, and bear you through the one, will then rather uphold and bear you through the other. For in times of trial and suffering for his name, you may look for his more special assistance.

Wherefore I wish all who are trouhled with false fears, to rest satisfied in these answers to their doubts: and I would have them give over calling their election, God's love, their justification, or their final perseverance into question: but rather fill yourselves with hope and assurance of God's favour, I speak still to burthened consciences, comforting yourselves therein; abounding in thanksgiving to God for what you have, rather than repining in yourselves, for what you want.

(5.) Fears arising from the deceitfulness of the heart, removed.

Yet I know there are some, as if they were made all of doubting, will object, my heart is deceitful I doubt all is not, I doubt all will not be, well with me.

If your heart be deceitful, why then do you believe it, when it casts in these doubts? and why do you trust to it more than unto the evidence of the word of God, and the judgment of his faithful ministers; who, by the word, give most satisfying resolutions to your doubts; which also administer unto you matter of assured hope and comfort?

(6.) Doubts from present weakness and fears answered.

Another will say, I do even faint in my troubles, and in my fears, and I am ready to give all over. What shall I do? What would you have me to do?

Your case is not singular, many others have been, and are in this case; it is no otherwise with you than it was with the Psalmist and Jonah; do as they in that case did: First, Give not over, but remember God, call upon him, give him no rest. Secondly Trust on him, and wait until you have comfort, Psa.

xxvii. 23, 24. That holy man of God said, My flesh and my heart fails, but God is the strength of my heart, and my portion for ever, Psa. lxxiii. 26. Likewise Jonah; I said, I am cast out of thy sight, yet I will look again towards thine holy temple, Jonah ii. 4, 7. And again, When my soul fainted within me, I remembered the Lord, and my prayer came up unto thee, into thine holy temple; that is, as if he had said unto God, I prayed unto thee in the name of Christ, and thou didst hear me. When you walk in the darkness of affliction and inward discontent, He, to whom God gave the tongue of the learned, to speak a word in due season to him that is weary, gives you counsel, saying, Who is among you that fears the Lord, and obeys the voice of his servant, that walks in darkness and has no light? Let him trust in the name of the Lord, and stay upon his God, Isa. l. 4, 10. Psa. xxvii. 23, 24. Observe it, he that fears the Lord, and obeys his voice, yet may be in darkness and have no light; what darkness is this, but that spoken of ver. 4. viz. an afflicted weary soul, without light or comfort? And men, thus distressed, must trust in the Lord, and stay upon their God.

(7.) Fears of not enjoying the promises, for not sufficiently performing the conditions.

Yet these poor souls, who, whether they should be sharply reproved, or pitied more, is hard to say; I am sure they deserve both, will yet object strongly, It is true, they that fear God and obey him, may trust in the Lord, and stay upon God. And he has made most rich promises to them that know him, and do fear and obey him. See, here is a promise with condition, saith one, I must fear the Lord, I must obey him, I know God will do his part, if I could do mine, but these I do not; what warrant then have I to look for comfort, or any thing at God's hand, for his promises belong not to me?

I know well, that with this doubt the devil does much perplex the afflicted souls of many of God's dearest children, and by it keeps off all the remedies which God's word can afford, so that they fasten not upon them to do them good. For the propositions of the word are easily assented unto; but all the matter lies in the application to the wound. It is still put off with, This is true which you say, but it belongs not to me, for I do not fulfil the condition required on my part.

Wherefore that I may, by God's help, fully satisfy this doubt, and quite remove this scruple of scruples, it must be carefully observed, that God makes some promises with condition; and that he makes some absolute promises, without any condition on man's part. Would you know what promises only are made with condition to be fulfilled on man's part, and what promises are absolute?

Know that many promises in the word concern the end of man's faith, which is salvation itself, and the recompence and reward of well-doing, whether corporal or spiritual, whether it be temporal or eternal. These are made with condition; namely, to those, and only to those who believe in the name of God, and that love, fear, and obey him. For it does not consist with the wisdom and holiness of God, to bestow heaven and his good blessings upon any, until they be thus qualified and made meet to receive them.

Know secondly, and observe it diligently, that there are many promises in the word which concern God's free giving the said grace of fear and obedience, required as means to obtain the former promises of good things, even an ability to perform the condition in the fore-mentioned promises; I mean not such a power as that they may fulfil the condition if they will, or if they will not they may choose. But God has made absolute promises to give men powei actually to will and to do the things required in the conditional promises, in such a manner that he will accept both will and deed, and in some cases the will for the deed, so as to fulfil those his conditional promises of salvation, &c. Heb. viii. 10; Phil. ii. 12.

That you may understand me fully, I will instance in some of the chief promises in this kind, made to every member of Christ, without exception. This is the covenant that I will make with the house of Israel, that is, with the whole church of God, Heb. viii. 10, a new covenant—and I will put my law into their inward parts, and write it in their hearts, and I will be their God, and they shall be my people, Jer. xxxi. 31, 33. He does not say, he will be their God if they will be his people, but saith absolutely, they shall be my people. Which that they might be, both there and elsewhere, he has said absolutely, without condition, they shall be all taught of God, Isa. liv. 13; John vi. 45. He promises likewise, saying, I will sprinkle clean water upon you, and you shall be clean; from all your filthiness, and from all your idols, I will cleanse you. A new heart also will I give you, and a new spirit will I put into you, and I will take away the stony heart out of your flesh, and I will give you an heart of flesh. And I will put my Spirit within you, and cause you to walk in my statutes, and ye shall keep my judgments and do them, &c. Ezek. xxxvi. 25—27. And not for your sakes do I this, saith he, be it known to you, be ashamed and confounded for your own ways, O house of Israel, ver. 32. And again he saith, I will make an everlasting covenant with them, that I will not turn from them to do them good; but I will put my fear in their hearts, that they shall not depart from me, Jer. xxxii. 40. Note this also, in very many places, God promises his blessing to them that fear him and keep his commandments; there he promises with condition; here he absolutely promises those on whom he intends to bestow these blessings, that he will put his fear in their heart, that they may be capable of them: and, which is more to the end that men might repent, believe, and live godly, which is the condition to which the promise of forgivness and salvation is made, God declares that he has raised Christ, and exalted him to be a Prince and Saviour, to give this faith and repentance, that their sins may be forgiven, and their souls saved by him, Acts v. 30, 31. I pray consider well whether all these promises of this sort be not made absolutely on God's part, and without any condition on man's part.

Wherefore, whereas God has made many excellent promises of free and great rewards; as, to hear the prayers, and fulfil the desire of them that fear him, and to give life and glory to them that believe and obey him, and that hold fast the confidence, and the rejoicing of the hope to the end; you see that here are promises of the first sort made with a kind of condition; but that God will give his people both to will and to do these things required in the condition, he has absolutely promised; as has been clearly proved.

If you yet reply and say, Are not these latter promises made under condition of our well using the outward means thereof, such as hearing of the word, prayer? &c.

God, indeed, commanded these means to be used; and, if we perform them aright, God will not fail to bless the good use of these means; but this well using them is not in our own power, neither is it a condition for which God is necessarily bound to give faith, and to plant his fear in our hearts, any otherwise than by his promise; but it is a condition by which he has ordained usually to give these graces to all who in the use of them shall wait upon him for them. For both the giving of his word, and the giving us minds to hear the word, and the opening of the heart to attend, and the convincing and alluring of the heart to obey; depend all upon those absolute promises, They shall be taught of God, and the rest before-mentioned, Isa. liv. 13.

Wherefore, let none of years think that without hearing, praying, and the right using of God's ordinances, that ever they shall have faith, and the fear of God wrought in them, or shall ever come to heaven. For we are commanded to pray, hear, &c. and that in faith, or else we can never look to receive any thing of the Lord, Heb. iv. 2; James i. 7. And doing what lies in man's power, in the right using the means of salvation is of great consequence, although it be not a sufficient cause to move God necessarily to give grace; for I am persuaded that the best should have more grace, if they would do what in them lay continually to make good use of the outward means of grace; and the worst should be guilty of less sin, if they would do what in them lay to profit by the good use of the said means. And the neglect, or the abusing of the means, is a sufficient cause why God should not only withhold grace, but condemn men for refusing it, Psa. lxxxi. 11, 12; Matt. xxi. 43.

(8.) Fears of salvation, for want of such graces as God has promised, removed.

But some will yet say, Let all that has been said be granted, yet I find that God has not fulfilled these his absolute promises to me; for I do not yet fear God and obey him. How can I hope? How can I but fear my estate to be bad?

Let this for the time be granted, that God has not planted his fear in your heart, &c. as yet; may he not do it hereafter? Since he has made such excellent and absolute promises of grace, will you not attend to the appointed means of grace, and hope for the blessing of God in his own time? and will you not wait, and be glad if they may be fulfilled at any time? Times and seasons of God's communicating his graces are reserved to be at his own disposing, not at ours. It should be your care diligently to attend upon God's ordinances, and when you read or hear the word or will of God, to endeavour to believe and obey it; as when he saith, Thou shalt love the Lord thy God with all thy heart. Thou shalt believe in the name of the Lord thy God, and trust in his name. Thou shalt obey the voice of the Lord thy God, and serve him, and such like. Attend to the word carefully, and because his word is infallibly true, and excellently good, labour to believe and to approve it; and say within yourselves, These are true, these are good, this I ought to do, this I would believe and do, Lord help me, and I will do it; O that my ways were directed to keep thy statutes, Psa. cxix. 5. In such exercises of the reasonable soul, it pleases God to give his grace, both to will and to do his commandments.

But, secondly, do not say you have not faith, nor the fear of God, and love to him, when in truth you have them. For what kind of duties are these, think you? Are they legal, which require perfect, exact, and full degrees of faith, fear, and love? Or are they not evangelical, such as require truth and sincerity in all these, and not full and absolute perfection? If you have true desire to fear him, which is the one measure of the fear of God's people, Neh. i. 11; so if you desire to believe, Matt. ix. 24, and have a will to obey, Isa. i. 19, in the inmost longing of your soul, according to the measure and strength of grace in you; this, according to the tenor of the blessed gospel of our Lord Jesus Christ, is true and acceptable through Christ, for whose sake God does accept the will for the deed, in all such cases wherein there is truth of will and endeavour, but not power to do, 2 Cor. viii. 12.

Furthermore, if you think that it is your well-doing which must make you acceptable to God, you are in a proud and dangerous error. Indeed God will not accept of you, if you do not endeavour to do his will; but you must propose to yourself another end, than to be accepted for your well-doing; you must do your duty to show your obedience to God, and to show your thankfulness, that God has pleased, and does please to accept you in his Son Christ; and that it is your desire to be accepted through him.

But I would have you, who are pressed with the load of your sins, to look judiciously and impartially into yourself; it may be, you have more faith, fear of God, and obedience, than you are aware of. Can you grieve, and does it trouble you that you have so little faith, so little fear of God, and that you show so little obedience? And is it your desire and endeavour to have more, and to do as well as you can; though you cannot do so well as you should? Then you have much faith, fear, and obedience. For to grieve for little faith, fear, and obedience, is an evident sign of much faith, fear, and obedience. For whence is this trouble and grief, but from God's saving grace? And to grieve for little, shows that you long for and would have much.

Let this suffice for a full answer to the principal

doubts, wherewith fearful hearts distress themselves continually. Never yield to your fears, wait on God still for resolution of your doubts in his best time; for it is not man that can, but it is God that both can and will speak peace to his people, not only outward, but inward peace, Psa. lxxxv. 8.

In the mean time, though you can have no feeling comfort in any of God's promises, yet consider God is the Lord, and that Christ is Lord of all, and you are his creature, owing to him all obedience, faith, and love: wherefore, you will, as much as you can, keep yourself from iniquity, and diligently strive to do his will, let him do with you as he pleases; yea, though he kill you, or though he give you no comfort till death, you will trust in him, and will obey him, and it is your desire to rest and hope in him as in your Redeemer; then, whether you know that God is yours or no, I am sure he knows you to be his; this is an argument of strong faith, and you are upon sure ground: The foundation of God remains sure: The Lord knows his; and who are they? Even all who professing his name, depart from iniquity, 2 Tim. ii. 19. And whosoever in his heart would, he in truth does depart from iniquity.

(9.) Fears arising from manifold temptations, removed.

Something remains yet to be answered. Many say that, do what they can, they are assaulted still so thick with temptations, that they cannot have an hour's quiet.

What of that? Does it hinder your peace with God, that the devil, the world, and your lusts, God's sworn enemies, are not at peace with you? So long as you have peace of sanctification in this degree, that the faculties of soul and body do not mutiny against God's holy will, but hold a good correspondence in joining together against the fleshy lusts, which fight against the soul, you are in good case; I mean, when the understanding, conscience, and affections, are all willing to do their part against sin, their common enemy: not but that you will find a sensible warring and opposition in all these, while you live here, even when you have most peace in this kind, but how? The unsanctified part of the understanding is against the sanctified part of the understanding; and the unsanctified will against the sanctified will, and so in all other faculties of the soul; the flesh in every part lusts against the Spirit, and the Spirit in every part lusts against the flesh, Gal. v. 17. Now if your faculties and powers be ruled all by one spirit, you have a good agreement and peace within you, notwithstanding that the flesh does so violently war against the Spirit; for this warring of sin in your members against the Spirit, and the warring of the Spirit against sin, proves clearly that you have peace with God, and this war continued, will in time beget perfect peace.

But let no man ever look to have peace of sanctification perfect in this life; for the best are sanctified but in part, 1 Cor. xiii. 9; wherefore let no man, professing Christ, think that he shall be freed from temptations and assaults arising from within, or coming from without, so long as he lives in this world. Are not Christians called to be soldiers? Wherefore we must arm ourselves, that we may stand by the power of God's might, and quit ourselves like men against the assaults of our spiritual enemies, 1 Cor. xvi. 13.

Is it any other than the common case of all God's children? 1 Cor. x. 13. Was not Christ himself tempted, that he might succour those that are tempted? Heb. ii. 18. Have you not a promise not to be tempted above that you are able? 1 Cor. x. 13. It is but resisting and enduring a while, yea a little while, 1 Peter v. 10; Heb. x. 37. Is there any temptation out of which God will not give a good issue? Has not Christ prayed that your faith fail not? Luke xxii. 32; John xvii. 15, 20.

Let us therefore keep peace in ourselves, that the whole man may be at agreement, and let us keep peace one with another, fighting against the common enemy, and the God of peace shall tread Satan and all enemies under foot shortly, Rom. xvi. 20; and then, through Christ, ye shall be more than conquerors,

Rom. viii. 37. You shall not only hold what you have obtained, but shall possess all that Christ has won for you. And the more battles you have fought, and in them, through Christ, have overcome, the greater triumph you shall have in glory.

SECT. 13. THE CHRISTIAN'S GROUND OF HOPE AND CONFIDENCE IN GOD, AGAINST ALL KINDS OF FEAR.

Now as a surplusage to all that has been said against groundless fears, which deprive poor souls of heavenly comfort; if any yet cannot be satisfied, but still fear that God is not at peace with them, I will propose a few questions, to which, if any soul can answer affirmatively, he may be assured of God's peace and love, and of his own salvation, whatsoever his fears or feelings may for the present be.

1. How stand you affected to sin? Are you afraid to offend God thereby? Is it so that you dare not wilfully sin? Is it your grief and burden that you cannot abstain from sin, get the victory over it, or deliver yourself from it so soon as you would, when you are fallen into it?

2. How stand you affected towards holiness and goodness, and unto the power of godliness? Is it your hearty desire to know God's will, that you may do it? Do you desire to fear him, and please him in all things? And is it your grief and trouble when you fail in well doing? And is it any joy to you to do well in any true measure?

3. How stand you affected to the church and religion of God? Are you glad when things go well in the church, though it go ill with you in your own particular? And are you grieved when things go ill in the church, when it may happen to be with you, as it was with good Nehemiah, Neh. i. 4; or Ichabod's mother, that all things go very well, or at least tolerably well, as to your own personal concern? 1 Sam. iv. 20, 21.

4. How stand you affected to men? Is it so that

you cannot delight in wicked men, because of their wickedness, but dislike them? Psa. xv. 4. Whereas, otherwise their parts and conditions are such, that you could much desire their company, Psa. xvi. 4. Do you love those that fear the Lord, and that delight in goodness, because you think they are good and are beloved of God? 1 John iii. 14; Psa. xvi. 3.

5. Can you endure to have your soul ripped up, and your beloved sin smitten by a searching minister, approving that ministry, and liking that minister so much the more? And do you, with David, desire that the righteous should reprove you? Psa. cxli. 5. And would you have an obedient ear to a wise reprover? Prov. xxv. 12.

6. Lastly, Though you have not always that feeling sense of your good estate, which is the certainty of evidence; nay, say you have it but seldom, or it may be, you can scarcely tell whether you have it at all; do you yet resolve, or is it your desire, and will you, as you are able, resolve to cleave to God, and depend upon Christ, and upon God's merciful promises made to you in him, seeking salvation in Christ by faith, and by none other, nor by any other means?

If you can answer, Yea, to all, or any one of these, you may assure yourselves that you are in God's favour, and in a state of grace. What though you cannot feel in yourselves, that you have this so sure as you would, by a full certainty of evidence, but it is your fault that you have it not so; yet you have it sure by the best certainty, namely, by a true faith in Christ, and an upright cleaving unto God. For when you are resolved not to sin wilfully and allowedly against God, and not to depart from him, whatever becomes of you; and it is your longing desire to please him: when I say, you stand thus resolved, and thus affected, then certainly God and you are joined together by an inseparable bond. When you hate what God hates, and love what God loves, and will what God wills; are not God and you at peace? Are you not nearly and firmly united one to another? What though this bond be somewhat secret and unseen to yourselves; yet it is certain; God knows you to be actually his, and will own you, when you seem to doubt it; and will always hold you by your right hand, whether you feel it or no, Psa. lxxiii. 23. But why should you think that you are without evidence, when you cannot but feel that in truth you cleave thus to God, and stand thus affected to him? Hence, if you were not wanting to yourselves, you might gain a most peaceable and joyous assurance, that you are in God's favour, and shall be saved. Thus much of removing the impediments.

CHAPTER XVI.

SHOWING THE MEANS TO ATTAIN THIS PEACE OF GOD.

It yet remains, that I should show the helps and means to attain and keep this true peace of God, which passes all understanding.

SECT. 1. CAUSES OF ERROR IN MISJUDGING OF A PERSON'S STATE.

Men often err in judging of their own estates, and in like manner in concluding that they have true peace, or not. If you would judge rightly, you must know what is necessary to the very being of a Christian, what not; and this is to be learned only by the word of God. For many err herein, because they think that such and such things are necessary to the being in a state of grace, which are not; and such and such things are sufficient to the being of a Christian, which are not.

Now you shall find, that it is truth of faith and other saving graces, not the great degree and quantity of them, that makes a Christian. And that it is not the most forward profession and form of godliness, without the power and truth thereof, that will do it.

Nothing is more common than for persons to be in truth otherwise than they judge. For every man's own spirit, so far as it is sinful, is apt to give a false testimony of itself. David said, he was cut off from God, when he was not, Psa. xxxi. 22. The Laodiceans thought themselves in a good state, when Christ said they were wretched and miserable, Rev. iii. 17.

Now that you may not err in this great point, you must use all good means to have your judgment rightly informed: and then be willing to judge of yourself as you are, and of your peace with God as it is.

I told you, that the holy Scripture must be your guide, in judging what you should be, and what you are; I mean the Scripture rightly understood. Now to attain a right understanding of the Scripture, and ability to judge yourself by it, whether you be in a state of grace, from the knowledge whereof comes peace, look back to chap. viii. sect. iii. adding unto them the following directions.

SECT. 2. RULES FOR A RIGHT JUDGMENT OF OURSELVES.

1. OBSERVE a difference and distinction in true Christians, both in their different manner of calling, and estate after calling. Some are called in infancy, as Samuel, and John the Baptist: some are in middle and old age, as Abraham, and Zaccheus. Some called without sensible terrors of conscience, as those before mentioned. Some with violent heart-ache and anguish, as Paul and the jailor. In some, these terrors abide longer, in some a shorter time. And after conversion, all are not of like growth and strength. Some are babes, weak in judgment and affections; some strong men, strong in grace generally; but strong also in corruption in some particular. Some old men, so well grounded in knowledge, and confirmed in grace, that no lust gets head to prevail in them: also one and the same man may be sometimes in spiritual health

and strong, sometimes under a temptation, weak and feeble; sometimes can pray, &c., and enjoy comfort, sometimes not. Now, none must conclude he is no Christian, because he is not in every thing like others, nor at all times like himself.

2. Trust not your own judgment or sense, in your own case: whosoever would understand, and be wise according to the Scripture, must deny himself, and not lean to his own sense or wisdom, Prov. iii. 5; but must be a fool that he may be wise, 1 Cor. iii. 18; you must bring your judgment to be ordered and framed by the Scriptures. You must not presume to put a sense of your own into the Scripture; but always take the sense and meaning out of it. It is presumption of a man's own opinion, and obstinacy in his own conceits, which spoils all in this case. And whence is this, but from his folly and pride? O, if you who are troubled in conscience could be every way nothing in yourselves; if you could be humbled, and not nourish this in you, you should soon know your state and comfort.

I know many of you will wonder that I should charge you with pride; you judging yourselves to be so base and vile as you do. Well, for all that, I will now prove to your faces, that it is humility you want; and that if you were not proud, you would judge of things otherwise than you do.

For you cannot believe in Christ, you say, because you cannot obey him, and be dutiful to him; if you could obey, then you could believe that he is yours, and you his; whereas, you must first believe in Christ, and take him for your Saviour and Lord, and believe he is yours, before you can obey him. Can a woman, or should a woman obey a man, and carry herself towards him as to her husband, before she believes that he is her husband? If you could obey as you should; O, then you think Christ would love you. It were well if you could love Christ, and obey him, as it is your duty. But to think he will not save you, because you have no goodness or worth in you to cause him to love you; is not this because you would be something in yourself, for which Christ should bestow his love upon you? Christ marries you, not because you are good, but that he may make you good, and that you may know him, &c. Hosea ii. 19, 20.

But you do not see his work of grace in you, that he has made you good, therefore you doubt.

I answer, Though it may be in you, yet Christ hides it from you, because you would not renounce your own righteousness, and believe his mercy, power, and faithfulness. Bring your heart to this, and you have reason for it, for the Father gives him, and he gives himself to you in the word and sacraments; then you will love him, and obey him abundantly. Is not she a proud and foolish woman, who may have a king's son, upon condition that she strip herself of all her own goods, and let him endow her at his pleasure, yet will be whining and discontented with herself, because she has nothing of her own to bring to him, for which he should love her?

But you will still say, Christ has not endowed you with so much grace, as to be able to do as you would.

Content yourselves; if you could but see that he has married you to himself, you then would use the means which he has appointed, whereby he gives his graces; you would be thankful for what you have, you would pray and wait his pleasure for more, relying on his wisdom for how much, and when. If you do not thus, then you show your pride in preferring your own wisdom before his.

Let it be supposed that you are not proud, nor standing upon terms of having any goodness in you, for which Christ should love you; but you would with all your hearts be all that you are in him, and would be beholden to him for taking you, poor and base, as you are. Is there no other pride, think you, but when you judge well of yourselves, or would be thought well of for your goodness? Yes, there is another kind of pride, still as dangerous in this case of causeless doubting; and that is, to be well conceited of, and wedded unto your own knowledge, and to your own opinion in judging yourselves. For instance, the holy Scriptures give you to understand, I speak still to such only as with all their souls would please God, yet can feel no comfort, that your state, in point of salvation, is good. And God's experienced children, yea, his faithful ministers, who dare not lie for God, much less to ease you, assure you according to the Scriptures, that your state is not as you say it is; but you think otherwise, and, having no sensible comfort, in your own judgment it is otherwise than either the Scripture or the ministers speak. Now when you will prefer your own opinion and sense, such as it is, before the judgment of God's word of truth, and before the judgment of God's ministers, judging according to his word, are you not highly conceited of your own opinion? And are you not strongly proud? Though, it may be, you thought otherwise.

Wherefore, if you understand things aright, you must have a mean conceit of your own understanding, of your own opinion, and of your own sense. For as you must deny your goodness, and be poor in respect of conceit of any goodness in you, if you would ever expect to have any goodness from Christ; so you must deny your own opinion, knowledge, sense, and wisdom, if you would know spiritual things aright, and become wise through Christ.

And that it may appear that you are not too well conceited of your own opinion concerning your spiritual condition, make use in this case of experienced Christians, but especially of judicious and godly ministers. Let no fear either of troubliug them, nor yet of shaming yourself, hinder you. But do it according to these directions.

SECT. 3. DIRECTIONS FOR TROUBLED CONSCIENCES, THEIR APPLICATION TO MINISTERS, OR OTHERS.

First, Acquaint such an one with your case betimes; keep it not to yourself too long. For then like a bone long out of joint, and a festered wound, it will not be so well, nor so easily cured; beside the vexation in the meantime.

Secondly, Deal plainly, truly, and fully, in showing the cause of your trouble; not doing as many, telling one part of your grief, and not another, which has been the cause that they have gone away without comfort. Either tell all or none in this case. If you think him not faithful, reveal nothing therefore to him: if you judge him a fit man, then show, as you would do your bodily maladies and diseases to a surgeon, or physician, if you would have them cured.

Thirdly, Believe them rather than yourselves in this case; hearken to them, and make use of their judgment and experience, and be not presumptuous of your own understanding and feeling. In times of your fears and doubts, be not rash and sudden in judging yourselves. The devil is a juggler, and your eyes are dazzled, and of all men you are the most unfit and incompetent to judge of yourselves in this case, for when groundless suspicion, and causeless fears have, like a headstrong colt, caught the bit in their teeth, they will, like to other passions, carry you headlong whither they list, contrary to all right reason and understanding. In such suspicion and fear of your estate, you are like a woman in the fit of her jealousy, who will pick matter out of every thing her husband does to increase her suspicion of him; if he be somewhat strange and austere, then, she says, he loves her not, but others better. If he be kind to her, then she thinks that this is but to dazzle and blind her eyes, that he may without suspicion give himself to others. Deal now ingenuously, and answer whether it is not, or whether it has not been so with you. I pray observe your absurd and contrary reasonings. When you prosper, thence you infer, Sure, God does not love me, for whom he loves, he corrects. When God corrects you, and lays upon you grievous afflictions, thence you conclude, Sure, God is wroth with me, and does not love me. If you be troubled in conscience, O, then God writes bitter things against you, you can have no peace. And when he gives you quiet of mind, O, then you fear all arises from presumption, your case is naught, aad it was better with you when you had trouble of mind. Is it not thus? Are you not ashamed that you have been thus senseless, and absurd in your own reasonings; and yet, this understanding, reason, and sense of yours must be hearkened unto, before the truth of God's word, and before the judgment of all men, though never so judicious. Will any body that is wise trust such a judgment? If an excellent physician for others, is seldom found to be the best physician for himself in a dangerous sickness, but will make use of one, it may be, inferior in judgment in physic to himself; for his own direction is not so well to be trusted in his own case; then, methinks, it should be your wisdom to make use of the judgment of others, and not follow your own sense.

But you will say, Shall I think otherwise of myself than I feel?

I answer; Aye, in some cases, or else you will be counted a wilful fool. As in the case of an ague, you taste your drink to be of an odd savour; before you had your ague, you knew it was well relished, and those who bring it, tell you it is the same; standers-by taste it for you, and say it is the same, and that it is excellently well relished. I hope you are wiser in such a case as this, than to conclude according to your feeling and taste; every one sees that the fault was in your palate, not in the drink. Even so is it with you, when the understanding is distempered with a shaking fit of groundless and faithless fear: wherefore, in this state, deny your own sense, and trust not your own judgment; but hearken unto the judgment of other men. And the rather, because God does therefore comfort men, and give them experience of his consolations, that they may comfort others in like cases, 2 Cor. i. 4. Also, he has given commandment to his more understanding and confirmed children, that they should comfort you, 1 Thess. v. 14; giving them to understand how it is with you in the matter of your soul, better than you can know of yourselves. Nay, God has given to his ministers the tongue of the learned, to speak a word in due season to the soul that is weary, Isa. l. 4. Should not the judgments of these be regarded? But, which is most of all, God has not only given to ministers skill, to discern your state better than yourselves, but it is the duty of their office to declare to you, being penitent, the remission of your sins, John xx. 23; and to assure you, that, if it be with you, according as you thus relate your state to be, you are in God's favour, and in a state of grace.

I mean not that you should rest your faith upon any man's judgment; but when judicious men, being in better case to judge of you, than you are to judge of yourselves, shall by the word of God, and by authority from him, give you hope and comfort, you ought to comfort yourselves by these means.

Thus much I have said, that your judgment might be fitted to understand aright in what state you stand; which, if you will observe, it will be an excellent means towards the obtaining of peace.

Now I will show you by what means, you may have just cause and matter of your judgment to work upon, whence it may give you peace and comfort.

SECT. 4. MEANS TO GET AND PRESERVE TRUE PEACE.

If you would have peace and comfort in your souls, then first and chiefly you must get and cherish the Spirit of God in you, that it may speak peace to you, and may give you matter for your spirit to work upon; whereby you may conclude, you are in God's favour. For, though I grant, that you can have no sure evidences of your adoption, say whatever can be said, until your spirits can witness that you are God's children; yet your spirits are not to be trusted in their witnessing, but only so far as the Spirit of God does witness to your spirits that it is so; that you are indeed his children, Whatsoever comfortable apprehension a man may have in himself of his good estate in grace, he can have no true joy and comfort, but by the Holy Ghost, whose proper work it is to comfort, and who is therefore called the Comforter, John xiv 16. For by him only a man can know, and by him a man may know, the things which are given him of God, 1 Cor. ii. 12.

But it will be said, The Spirit blows where it lists, how is it possible for any man by any means to get it?

In respect of man's own ability, it is as impossible for him to obtain the divine Spirit to dwell and work in his heart, as it was for those impotent folk, who lay waiting at the pool of Bethesda for the angel's coming to move the waters, to cause the said moving of the waters; yet they waited, the waters were moved, and they that continued patiently waiting at the pool were benefitted, John v. 3, 4. Thus if men will wait in the use of the means wherein and whereby God does give and continue his Holy Spirit to men, they may hope to enjoy this unspeakable blessing.

The first means to get the Spirit, is humility. To be sensible of the loss of that which once you had in Adam, you must mourn, and hunger and thirst after the Spirit, Matt. v. 3—6. If you will do thus, you may hope to receive the Spirit. For God saith, that he will pour water upon him that is thirsty, &c. I will pour my Spirit upon thy seed, saith he to the church, Isa. xliv. 3.

Secondly, That your heart may be stirred up to long for the Spirit, you must know that there is a Holy Ghost, and not only so, but you must know him to be God, and you must believe him to be the Comforter; and give him this honour and glory, as to believe in him, and conceive of him as the proper author of sanctification and comfort; this is the way to have the Spirit, and to be sure of it that you have it. Our Saviour saith, that the not knowing or believing hereof, is the cause why the world receive not the Spirit, John xiv. 17.

Thirdly, Be constant and diligent in waiting for the having, and for the increase of the gifts of the Spirit, in the holy exercises of religion, as, reading and meditating on the word of God, especially on the blessed truths and promises of the gospel, &c. You must wait for it in the motions and stirring of God's word in you by God's means; then, as Cornelius and his company received it at Peter's sermon, Acts x. 44, and as the Galatians, at the hearing of faith, Gal. iii. 5, so may you. For the gospel is called the ministry of the Spirit, 2 Cor. iii. 6, 8.

Fourthly, Pray for the Spirit; and though you cannot pray well without the Spirit, yet since it is God's will that you should pray for it, set about prayer for it as well as you can; then God will enable you to pray for the Spirit, and you shall have it. For Christ saith, If ye that are evil know how to give good gifts to your children, how much more shall your heavenly Father give the Holy Spirit to them that ask him? Luke xi. 13. As these are means to get the Spirit, so they are means to continue, nourish, and increase the graces of the Spirit.

Fifthly, If you would keep and nourish this Spirit, you must take part with it, in its conflicts with the flesh and sin: you must not resist, but willingly receive the comforts and motions of the Spirit, and must do your best to bring forth the fruits of the Spirit. You must take heed that you neither grieve nor quench the Spirit. It is grieved, when it is resisted, crossed, or opposed any way, Eph. iv. 30, 31; 1 Thess. v. 19, 20. It is quenched as fire is, First, by throwing on water; all sinful actions, as they be greater or smaller, are as water. They do accordingly more or less quench and abate the Spirit's operations. Secondly, Fire may be quenched and put out by withdrawing of wood and fuel; all neglect, or negligent using of the word, sacraments, prayer, meditation, and holy conference, and communion of saints, do much offend and quench the Spirit; whereas the daily and diligent use of all these, through his concurring grace, does much increase and strengthen the life of God in the soul; whence must needs follow much peace and comfort.

Now when you have gotten this Holy Spirit, and have any proofs of the Holy Spirit's being in you, then you ought to rest satisfied in the Spirit's witness to your spirit; your spirit should doubt no more. For even in this that God has given you his Spirit, the very being of it in you is a real proof, and the greatest confirmation that can be of your being in a state of grace. For when you have this Spirit, you are anointed, 1 John ii. 27; what greater confirmation would you have of being made kings and priests to God? Rev. i. 6. You are also by this Spirit sealed to the day of redemption, Eph. iv. 30. What greater confirmation can there be of God's covenant, and of his will and testament towards you? It is likewise the earnest of your inheritance, 2 Cor. i. 22; Eph. i. 14, which gives present being, and the beginning to the enjoyment of the blessings, and is the sure evidence of the full possession in due time. You are so surely God's, when he has given you his Spirit, that unless you can think he will lose his Spirit, the earnest of which he gave you, you can have no cause to think that he will lose, or not fulfil the promise of salvation made unto you, whereof his Spirit is the earnest, and part of the covenant.

This Spirit does witness to a man, that he is the child of God, two ways:

First, By immediate witness and suggestion. Secondly, By necessary inferences, by signs from the infallible fruits of the said Spirit. By which latter witness you may know the former to be a true testimony from God's Spirit, the Spirit of adoption, and not from a spirit of error and presumption. For this Spirit of adoption is a spirit of grace and supplication, Zech. xii. 10; Rom. viii. 26; it is a spirit of holy fear, Isa. lxi. 3; and it is a spirit of holy joy, Acts viii. 8. Where it does testify that you are God's children, there it will give you new hearts, causing you to desire and endeavour to live like God's children, in reverend fear and love; leading you in the right way, checking you and calling you back from the way of sin; stirring you up to prayer, with sighs, desires, and inward groans; at least making you to confess your sins, and to ask and hope for pardon in the name of Christ, Gal. v. 22; Acts xxiv. 16; Isa. xxx. 21. And it will still be quickening and strengthening you in the ways of godliness, and giving you no rest if you walk not therein. Thus much of the first and principal means of getting true peace and comfort.

Secondly, If you would have the invaluable jewel of peace, then abstain as much as possible from all gross and presumptuous sins; and from the allowance of any sin: for sin will produce fear, even as the shadow follows the body. And the more sin, the more guilt; and the less sin, the lest guilt: now, the less guilt lies upon the conscience, the more peace, Psa. li. 14.

Thirdly, When you fall into sin, for who lives and sins not? then with all speed affect your heart with godly sorrow for it, cause it to be a burden, and a load and weariness to the conscience; but withal, comfort your heart with hope of mercy, forgiveness, and grace through Christ. Then with all humble submission you must seek unto God, the God of peace, but come to him by Christ Jesus, the Prince of peace, Isa. ix. 6, upon whom lay the chastisement of your peace, Isa. liii. 5. Ask repentance, grace, and new obedience. Believe in Christ. If you do all this, then you come unto Christ, and unto God by Christ, according to his commandment, and you have his sure promise, that you shall have rest to your souls, Matt. xi. 29. This do, for in Christ only can you have peace, John xvi. 33. This true application of Christ's blood and satisfaction, will so sprinkle the conscience from the guilt of sin, Heb. ix. 14, x. 12, that there shall remain no more conscience for sin, Heb. x. 2, that is, no more guilt which shall draw upon you any punishment for sin; whence must needs follow peace of conscience; because the conscience has nothing to accuse you of, guiltiness being washed away by Christ's blood, Heb. ix. 14. As soon as David, after his foul sins, could come thus to God, his heart had ease, Psa. xxxii. 1—5.

But when you have thus gotten a good and clear conscience, take heed of defiling it again, or giving it any manner of uneasiness; be as tender in keeping your conscience unspotted and unwounded, as you are of the apple of your eye. Sin not against knowledge and conscience, and in any case smother not the good checks and motions of your conscience. For if being washed, you do again defile it, this will cause new trouble of heart, and you must again apply yourselves to this last prescribed remedy.

Fourthly, Christ having taken upon him the burden of your sins, which was intolerable, you must take upon you, and submit unto the yoke of Christ's service, which is light and easy, Matt. xi. 29. You must endeavour to do whatsoever he has commanded in his word and gospel, following his steps in all his imitable actions; in all humility and meekness, in all spiritual and heavenly mindedness. When you can thus subject yourselves to Christ in holiness, you shall have peace. For the Holy Ghost saith, The work of righteousness is peace, Isa. xxxii. 17; and again, he saith, To be spiritually-minded is peace, that is, brings with it peace, Rom. viii. 6. I comprehend Christ's yoke of the gospel in these three things, faith, hope, and love. As these three be in you, and abound, in the same degrees shall peace be in you, and shall abound.

Having faith in Christ, saith the apostle, we have peace with God, Rom. v. 1. It is God that justifies who shall lay any thing to your charge? Rom. viii. 33. For justifying faith is the ground and spring, from which only sound and true comfort does flow.

Hope will make you wait, and expect with patience, the accomplishment of God's sure promises, whereby it will hold you as steady, and as sure from wreck of soul, as any anchor can hold a ship. God does therefore give hope, that it may be as an anchor, sure and steadfast, Rom. viii. 25. Though while you are in the sea of this world it does not keep you so quiet, but that you may be in some measure tossed and disquieted with the waves and billows of fear and doubt, to try the goodness of your vessel, and strength of your anchor, &c., yet you shall be sure not to make shipwreck of faith and a good conscience, if you shall lay hold upon this hope set before you, Heb. vi. 18, 19.

And as for love, they that love the Lord shall have peace: you must therefore love God; love his ordinances and his people; love God with all your heart; love your neighbours as yourselves; love God's commandments. For great peace shall they have, saith the prophet, that love God's law, and nothing shall offend them, Psa. cxix. 165.

Whoever shall thus take up Christ's yoke, and follow him, shall find rest to their souls, Matt. xi. 29; and peace shall be upon them, as upon the Israel of God, Gal. vi. 16.

Fifthly, If you would have peace, use all good means whereby you may be often put in remembrance of the exhortations and consolations of God. They in the Hebrews were therefore disquieted, and ready to faint in their minds, because they forgot the exhortation, which said, My son, despise not the chastening of the Lord, &c., and because they forgot the consolation, which saith, Whom the Lord loves, he chastens, Heb. xii. 5, 6.

The principal means of being put in mind of God's consolations, are these following:

1. You must be much conversant in the Scriptures, by reading, hearing, and meditating thereon. For they were all written to that end, that through patience and comfort of the Scriptures, you might have hope, Rom. xv. 4.

The Scriptures of God, they are the very wells and breasts of consolation and salvation, Isa. xii. 3. lxvi. 11. The law discovers sin, and by its threats against you, and by relating judgments executed upon others, does drive you to Christ, Gal. iii. 24. The promises of the gospel made to you, and the accomplishment thereof to others, do settle and confirm you in Christ, whereby your heart is filled with joy and consolations The gospel is called the gospel of peace, and the ministers of the gospel are said to bring glad tidings of this peace, Rom. x. 15. It is the bright shining light in the gospel, which will guide your feet in the way of peace, Luke i. 79.

2. Be much in good company, especially in theirs,

who are full of joy and peace in believing, whose example and counsel will mind you of joy and comfort, and will be of excellent use unto you, to establish you in peace.

Sixthly, and lastly, Acquaint yourself with God, concerning the course he uses to take with his children in bringing them to glory; acquaint yourself with God also in praying much for peace, unto him who is the God of peace, the Father of mercies, and the God of all consolation; then you shall have peace, and much good shall be unto you, Job xxii. 21. For it is God that speaks peace to his people, Psa. lxxxv. 8; wherefore assuredly his answer to him that asks peace, will be an answer of peace, even this peace which passes all understanding. God shall give you peace, and with it glory, even a glorious peace.

Thus, I have shown you the excellency of peace, together with the impediments, furtherances, and means of peace. Shun the impediments, improve the furtherances; and, I dare assure you, that although in this life you may still feel a conflict between faith and doubting, between hope and fear, between peace and trouble of mind; yet in the end you shall have perfect peace, Psa. xxxvii. 47. In the mean time, though I cannot promise you to have always that peace, which will afford you sense of joy; yet God has promised, that you shall have that which shall keep your hearts and minds in Christ Jesus; and what would you have more?

I thank God, I have reaped much benefit to myself in studying, and penning these directions. I pray God that you may reap much good in reading of them. Now the God of hope fill you with all joy and peace in believing, Rom. xv. 13. "And the God of peace that brought again from the dead our Lord Jesus, that great Shepherd of the sheep, through the blood of the everlasting covenant, make you perfect in every good work, to do his will, working in you that which is well pleasing in his sight, through Jesus Christ; to whom be glory for ever and ever. Amen." Heb. xiii. 21.

THE END.